31 July 2010

For Dad

from Jenny

for your 83rd birthday
(late as usual!)

x x x

The Making of the Geological Society of London

Geological Society books refereeing procedures

The Society makes every effort to ensure that the scientific and production quality of its books matches that of its journals. Since 1997, all book proposals have been refereed by specialist reviewers as well as by the Society's Books Editorial Committee. If the referees identify weaknesses in the proposal, these must be addressed before the proposal is accepted.

Once the book is accepted, the Society Book Editors ensure that the volume editors follow strict guidelines on refereeing and quality control. We insist that individual papers can only be accepted after satisfactory review by two independent referees. The questions on the review forms are similar to those for *Journal of the Geological Society*. The referees' forms and comments must be available to the Society's Book Editors on request.

Although many of the books result from meetings, the editors are expected to commission papers that were not presented at the meeting to ensure that the book provides a balanced coverage of the subject. Being accepted for presentation at the meeting does not guarantee inclusion in the book.

More information about submitting a proposal and producing a book for the Society can be found on its web site: www.geolsoc.org.uk.

It is recommended that reference to all or part of this book should be made in one of the following ways:

LEWIS, C. L. E. & KNELL, S. J. (eds) 2009. *The Making of the Geological Society of London*. Geological Society, London, Special Publications, **317**.

KHAIN, V. E. & MALAKHOVA, I. G. 2009. Scientific institutions and the beginnings of geology in Russia. *In*: LEWIS, C. L. E. & KNELL, S. J. (eds) *The Making of the Geological Society of London*. Geological Society, London, Special Publications, **317**, 203–211.

GEOLOGICAL SOCIETY SPECIAL PUBLICATION NO. 317

The Making of the Geological Society of London

EDITED BY

C. L. E. LEWIS
University of Bristol, UK
and
S. J. KNELL
University of Leicester, UK

2009
Published by
The Geological Society
London

THE GEOLOGICAL SOCIETY

The Geological Society of London (GSL) was founded in 1807. It is the oldest national geological society in the world and the largest in Europe. It was incorporated under Royal Charter in 1825 and is Registered Charity 210161.

The Society is the UK's national learned and professional society for geology with a worldwide Fellowship (FGS) of over 9000. The Society has the power to confer Chartered status on suitably qualified Fellows, and about 2000 of the Fellowship carry the title (CGeol). Chartered Geologists may also obtain the equivalent European title European Geologist (EurGeol). One fifth of the Society's fellowship resides outside the UK. To find out more about the Society, log on to www.geolsoc.org.uk.

The Geological Society Publishing House (Bath, UK) produces the Society's international journals and books, and acts as European distributor for selected publications of the American Association of Petroleum Geologists (AAPG), the Indonesian Petroleum Association (IPA), the Geological Society of America (GSA), the Society for Sedimentary Geology (SEPM) and the Geologists' Association (GA). Joint marketing agreements ensure that GSL Fellows may purchase these societies' publications at a discount. The Society's online bookshop (accessible from www.geolsoc.org.uk) offers secure book purchasing with your credit or debit card.

To find out about joining the Society and benefiting from substantial discounts on publications of GSL and other societies worldwide, consult www.geolsoc.org.uk, or contact the Fellowship Department at: The Geological Society, Burlington House, Piccadilly, London W1J 0BG: Tel. +44 (0)20 7434 9944; Fax +44 (0)20 7439 8975; E-mail: enquiries@geolsoc.org.uk.

For information about the Society's meetings, consult *Events* on www.geolsoc.org.uk. To find out more about the Society's Corporate Affiliates Scheme, write to enquiries@geolsoc.org.uk.

Published by The Geological Society from:
The Geological Society Publishing House, Unit 7, Brassmill Enterprise Centre, Brassmill Lane, Bath BA1 3JN, UK

(*Orders*: Tel. +44 (0)1225 445046, Fax +44 (0)1225 442836)
Online bookshop: www.geolsoc.org.uk/bookshop

British Library Cataloguing in Publication Data

A catalogue record for this book is available from the British Library.
ISBN 978-1-86239-277-9

Typeset by Techset Composition Ltd., Salisbury, UK
Printed by MPG Books Ltd, Bodmin, UK

Distributors

North America
For trade and institutional orders:
The Geological Society, c/o AIDC, 82 Winter Sport Lane, Williston, VT 05495, USA
Orders: Tel. +1 800-972-9892
Fax +1 802-864-7626
E-mail: gsl.orders@aidcvt.com

For individual and corporate orders:
AAPG Bookstore, PO Box 979, Tulsa, OK 74101-0979, USA
Orders: Tel. +1 918-584-2555
Fax +1 918-560-2652
E-mail: bookstore@aapg.org
Website: http://bookstore.aapg.org

India
Affiliated East-West Press Private Ltd, Marketing Division, G-1/16 Ansari Road, Darya Ganj, New Delhi 110 002, India
Orders: Tel. +91 11 2327-9113/2326-4180
Fax +91 11 2326-0538
E-mail: affiliat@vsnl.com

Contents

The Bicentenary

Appendices

Preface: in the Footsteps of the Founding Fathers

Davy asserted at our first dinner at the Freemasons Tavern, that he never knew anything prosper that was begun upon a Friday [the 13th], and our little band, besides, was of the bad omened number 13, at it's start; well may we laugh at the prejudices of mankind.

James Laird, 1834
The Geological Society's first Secretary

It was inevitable that in the Geological Society's Bicentenary year, the History of Geology Group (fondly known as HOGG) would hold an event to mark this historic occasion. HOGG is affiliated to the Geological Society and meets, free of charge, in its rooms in Burlington House. Ties to the Society are therefore strong, although many HOGG members are not Society Fellows. As the Society itself had chosen to look forward to the future during its Bicentenary year, it was fitting that HOGG should look back and celebrate the events, 200 years ago, that led to the Society being where it is today. Accordingly, HOGG planned a five-day celebratory event (9–13 November 2007) entitled 'In the Footsteps of the Founding Fathers'.

The occasion commenced with a field trip to the Isle of Wight – Walk with the Founding Fathers – that followed in the footsteps of Thomas Webster, the Society's first paid officer, as he tried to unravel the island's geological history in 1814. Martin Rudwick's text and the beautiful illustrations from the trip's field notes are reproduced in the Bicentenary section of this volume, in case others would like to follow in our footsteps. On returning to London from the Isle of Wight, I convened a two-day international conference in Burlington House – Talk with the Founding Fathers – that explored the status of geology around 1807. The first day looked at the geological context into which the Geological Society was born – geology in Europe around 1800, and that was followed by talks on the contributions made to the Society by each of the Founding Fathers. The second day covered some of the people and significant geological events of the following decades. On the evening of 12 November, a fabulous dinner in period costume – Dine with the Founding Fathers – was held in the New Connaught Rooms, the site of the Freemasons' Tavern where the Society was founded. During the evening a plaque was unveiled by the Society's President, Richard Fortey, to commemorate the day 200 years ago when it all began. The year's events culminated on 13 November 2007 – the Bicentenary date – with another dinner for all Society Fellows, underneath the dinosaurs in the Natural History Museum.

With no Society archivist in place – there hasn't been one for some years, a rather lamentable situation in its Bicentenary year – I was concerned that a record of these events would be lost. This book therefore attempts to document them as William Watts did for the Society's Centenary, although his 1909 account did a much more thorough job for that event than we have space for here. So while the majority of these pages are taken up with papers presented at the conference – although substantially rewritten for publication – and others that were subsequently commissioned, the 'Bicentenary' section provides an account of HOGG's field trip and dinner, as well as a summary of all the Society's Bicentenary events held during 2007. The volume also includes, in the Appendix, the text of the Society's first two publications, both of which are now rare documents that are hard to come by.

Within weeks of setting up the Society, the founders sent out a small booklet entitled *Geological Inquiries* (Appendix I) to the 42 Honorary Members around the country that it had appointed at its first meeting, with the intention of gathering information for the geological maps of Britain they were to make. This fascinating document provides a valuable insight into the way the founders were thinking about geology at that time. Appendix II is a translation of Comte de Bournon's *Discours préliminaire*, prefixed to his treatise on calcite and aragonite. As it refers specifically to events preceding the foundation of the Geological Society, and since publication of de Bournon's treatise is considered by many to be the event that triggered the founding of the Society, we felt it was particularly relevant to include it here. This is the first time it has been translated into English and I am extremely grateful to Margaret Morgan, Royal Cornwall Museum, who generously responded to my plea to HOGG members – at a very late stage in this book's progress – for someone to do the translation. The section covering de Bournon's ideas on crystallography and geology has been omitted owing to lack of space and time, but I hope it will become available in due course. I am indebted to the Geological Society of London for allowing us to reproduce these important documents here.

In the same way that many of the Founding Fathers enabled the publication of Comte de Bournon's treatise on mineralogy by subscribing to it beforehand, all who attended the Founding Father's conference were offered the opportunity of subscribing to this volume at a subsidized rate. By the end of the conference we had enough subscribers to proceed and HOGG made up the financial difference. I am therefore very grateful to the HOGG committee for making this possible. In true nineteenth-century style, the names of those who subscribed are printed on the pages following this Preface.

Simon Knell and I were particularly keen that this volume would look in more depth at the Founding Fathers and the contributions each made to the Society, as well as the context of the Society's birth and the various developments that caused it to come into being. We hoped to draw our readers away from the conventional narrative about how the Society was inaugurated and to challenge some of the myths that have grown up over the past 200 years. The story so far has relied heavily on the account written by the first President, George Bellas Greenough, at least 25 years after the event, but I felt certain there were other voices to be heard.

Greenough's recollections were first interpreted by the Society's historian, Horace B. Woodward, in his history of the Society (1908), written for the Centenary celebrations in 1907, but Woodward had not discovered the fantastic Greenough archive, now at University College London, which Martin Rudwick first used in 1963 when writing his version of the Society's origins. These two accounts, along with the less well-known but equally important papers by Paul Weindling (1979 and 1983) on the prehistory of the Society, have become essential reading for anyone seriously interested in piecing together the Society's early history, and many in this volume refer to them. All of these works, and now this volume, will become, or continue to be, the starting point for anyone looking back to these events during the next 100 years. As a consequence, Simon and I could not help feeling, at times, somewhat conscious of the responsibility imposed on us from the future.

In writing our own papers, we found such contradictory accounts of various events in those works mentioned above that we were enticed into the archives to take a look for ourselves at what the records really said. What we found was revealing. A quick cross-reference of Greenough's history of the Society with the Society's minutes showed that his memory could not always be relied on; dates and events had become confused, committee members' names misremembered, places forgotten. So if his memory couldn't be depended on for the things we could check, what reliance could we place on those we couldn't check? Even at the time Greenough wrote his recollections, James Laird, the Society's first Secretary and Greenough's right-hand man for the first four years of his Presidency, disagreed with Greenough's version of how the Society had formed. Laird apparently wrote his own account, now sadly lost – what a gem that will be for future historians to find. But none of this was very surprising – who does remember things perfectly the next day, let alone 25 years later? And when do two people ever remember the same event exactly the same way? Furthermore, facts become confounded in the retelling. So although there is no such thing as a 'right' or 'wrong' account of what happened, we do need to be aware of what Greenough's motives might have been in writing his recollections, so carefully left in the Society's archives for posterity to find.

Like the other historians before us, Simon and I have put our own interpretation on what we found in the Greenough archives. While we both read some of the same material, we also read different letters, according to what we were seeking for own papers, and the writers of those had their own things to say about what happened, which inevitably influenced our separate interpretations. Furthermore, there was material we didn't have time to look at, including two large cardboard boxes of letters, bills, lists, etc., each torn into four neat pieces, held together with a paper clip. So did Greenough tear up material not of 'historical' interest, but then never throw it away, or did someone else do that after his death? And who clipped all the pieces together? There's still plenty of scope for future research.

And so history is written. By individuals with their own agendas, in their own times, interpreting what little material is available according to their own belief systems and prejudices. And we are not the only ones to cover the 'founding' in our essays. In compiling a book of this nature where many authors must refer to the same event, there is bound to be some duplication, particularly as these papers need to stand alone as individual pieces of work outside the context of this book. However, I feel these different accounts of how the Society formed only add to the book's richness, allowing future readers a selection of interpretations. Who knows, perhaps like us, it will even entice a few to go back into the archives and take a look for themselves.

When I first asked Simon, about a year ago, if he would co-edit this volume with me, I don't think either of us had the first idea about the monster we were taking on. We had successfully worked together on a previous volume, *The Age of the Earth: From 4004 BC to AD 2002* (Lewis & Knell 2001), seven years ago, which, through the mists of time, had seemed a relatively straightforward task. Not that this volume was particularly complicated – just immensely time-consuming. Or perhaps it was now harder to fit it in around our daily responsibilities, which undoubtedly had grown in those seven years. So it is just as well that it rained all year and I wasn't tempted into the garden.

Despite, or perhaps because of, only meeting face-to-face twice during the year, we finished on time and I think it would have been hard to find anyone other than Simon so in tune with my hopes for this book. As he said, we shared 'a capacity for irony and laughing in the face of, well, just about everything, were efficient and yet self-indulgent, obsessive, ridiculously ambitious and broke more rules than all the other authors put together' – just looking at the length of our papers says how long they took us. Throughout it all, Simon was immensely supportive and I thank him profoundly for everything he has put into this project. I also thank Sarah Gibbs, our very patient and understanding editor, as well as all the authors who helped us realize our ambitions for this volume and for their immensely interesting papers. I am sure you will enjoy them too.

DR CHERRY LEWIS
HOGG Chairman, 2004–2007
8 September 2008

References

LEWIS, C. L. E. & KNELL, S. (eds) 2001. *The Age of the Earth: from 4004 BC to AD 2002*. Geological Society, London, Special Publications, **190**.

RUDWICK, M. 1963. The foundation of the Geological Society of London: its scheme for co-operative research and its struggle for independence. *British Journal for the History of Science*, **1**, 325–355.

WATTS, W. W. 1909. *The Centenary of the Geological Society of London*. Longmans, Green and Co., London.

WEINDLING, P. J. 1979. Geological controversy and its historiography: the prehistory of the Geological Society of London. *In*: JORDANOVA, L. J. & PORTER, R. S. (eds) *Images of the Earth: Essays in the History of the Environmental Sciences*. British Society for the History of Science, Monographs, **1**, 248–271.

WEINDLING, P. J. 1983. The British Mineralogical Society: a case in science and social improvement. *In*: INKSTER, I. & MORRELL, J. (eds) *Metropolis and Province: Science in British Culture, 1780–1850*. Hutchinson, London.

WOODWARD, H. B. 1908. *The History of the Geological Society of London*. Longmans, Green and Co., London.

List of Subscribers

Stuart Baldwin	Baldwin's Scientific Books
Richard Bateman	The Geological Society of London
Jenny Bennett	University of Exeter
Nic Bilham	The Geological Society of London
Jay Bosanquet	Alnwick, Northumberland
Alan Bowden	National Museums Liverpool
Patrick Boylen	City University, London
David Branagan	The University of Sydney, Australia
Anthony Brook	Worthing, West Sussex
Susan Brown	Geologists' Association
Cynthia Burek	University of Chester
Wendy Cawthorne	The Geological Society of London
Barrie Chacksfield	British Geological Survey, Keyworth
Renee Clary	Mississippi State University, USA
Chris Cleal	National Museum of Wales
Beris Cox	Stanton-on-the-Wolds, Keyworth
Peter Crowther	National Museums Northern Ireland
Alan Cutler	Black Country Geological Society
Margaret Dobson	Ascot, Berkshire
Peter Dolan	Twickenham, Middlesex
Philip Doughty	Belfast, Northern Ireland
Douglas Fleming	Forfar, Scotland
Gerald Friedman	The Science Foundation, USA
Sue Friedman	The Science Foundation, USA
Bill George	Barking, Essex
Martin Guntau	Literaturhaus Rostock, Germany
C. John Henry	19th Century Geological Maps, London
Noah Heringman	University of Missouri, USA
Thomas Hose	Buckinghamshire New University
Richard Howarth	University College London
Frank James	The Royal Institution, London
Gilbert Kelling	Keele University
Anthony King	Bromley, Kent
Simon Knell	University of Leicester
David Knight	Durham University
Martina Kölbl-Ebert	Jura-Museum Eichstätt, Germany
Alan Lane	Haywards Heath, West Sussex
Janet Lane	Haywards Heath, West Sussex
Cherry Lewis	University of Bristol
Irena Malakova	Russian Academy of Sciences, Russia
John Mather	Royal Holloway, University of London
Sandra McDonald	St Andrews, Bristol
Richard Moody	Kingston University
Margaret Morgan	Royal Cornwall Museum
Julie Newell	Southern Polytechnic State University, USA
Anne O'Connor	Durham University
Ralph O'Connor	University of Aberdeen, Scotland
Michael Richardson	University of Bristol Library Special Collections
David Ridgeway	St Albans, Hertfordshire
Edward Rose	Royal Holloway, University of London
Martin Rudwick	University of Cambridge
Tom Sharpe	National Museum of Wales
Jessica Shepherd	Plymouth City Museum
John Smallwood	Hess Ltd, London
Diana Smith	Great Missenden, Buckinghamshire
Anthony Spencer	Statoil, Norway
Mike Sumbler	Stanton-on-the-Wolds, Keyworth
Bob Symes	Sidmouth, Devon
Richard Symonds	ENI UK Ltd
Peter Tandy	The Natural History Museum, London

Philippe Taquet	Muséum National d'Histoire Naturelle, Paris
Ray Thomasson	Thomasson Partner Associates Inc, USA
Hugh Torrens	Madeley, Staffordshire, UK
Gian Battista Vai	Università di Bologna, Italy
A. Bowdoin Van Riper	Southern Polytechnic State University, USA
Leucha Veneer	University of Leeds
John White	Hawkes Bay, New Zealand
Julian Wilson	Maggs Bros. Ltd, London

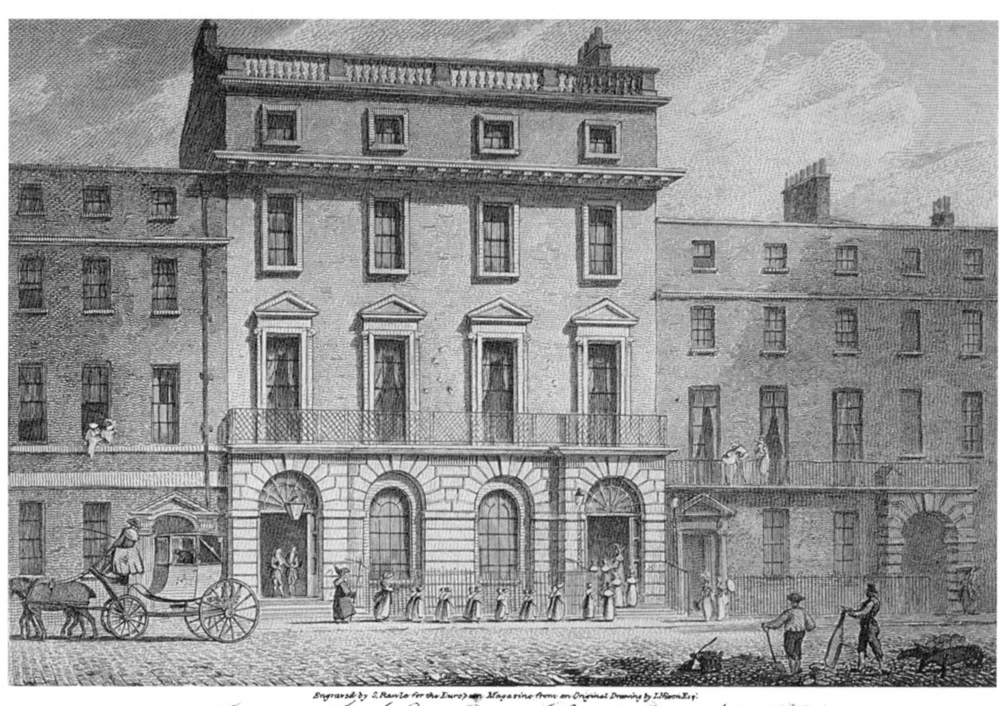

Freemasons Tavern Great Queen Street Lincolns Inn Fields.

Freemason's Tavern as it looked on 13 November 1807, when the Geological Society was founded there. *Source*: Guildhall Library.

The road to Smith: how the Geological Society came to possess English geology

SIMON J. KNELL

Department of Museum Studies, University of Leicester, Leicester LE1 7LG, UK
(e-mail: sjk8@le.ac.uk)

Abstract: The modern image of the Geological Society owes much to William Smith whom the Society used, in 1831, to claim ascendency over European rivals. At its birth, however, the Society pursued a science adopted from the Continent, which privileged field data and saw mineralogy and chemistry as the sciences of the Earth. The Society's birth mobilized the nation; its co-operative, mobile, investigative, subtly theoretical and didactic vigour materialized in the production of Greenough's geological map of England. Yet Smith's geology spread virus-like, converting the membership in various ways, some acknowledging Smith, others denying him. In possession of Smith's geology, and impressed by his publications, the Society men emerged from a philosophical wilderness, only to break out in a competitive fever to write an *Elements of Geology*. The Society's great supporter, John Farey, broke free, disillusioned and determined to destroy Greenough. Nevertheless, Greenough pushed forward with his map, competing directly with Smith and intent on surpassing him. However, following the development of a powerbase for his geology in Yorkshire, Smith rode into London to be crowned the father of a peculiarly English science. Smith's map now became the national icon of English geology, less than a decade after the Society had rendered it obsolete. Next to it, Greenough's map – the Society's 'glory' – symbolized the Society's co-operative spirit and political acumen, attributes no less important to the science's advance.

> Is it possible, in the final analysis, for one human being to achieve a perfect understanding of another? We can invest enormous time and energy in serious efforts to know another person, but in the end, how close can we come to that person's essence? We convince ourselves that we know the other person well, but do we really know anything important about anyone?
>
> (Haruki Murakami 2003, *The Wind-up Bird Chronicle*, p. 24)[1]

Learned societies are both concrete and chameleon. They are made material in their buildings, reports, collections and other objects, but the meanings of these things are at the command of an ever-changing cast of actors. As historians we add to a sense of tangibility by converting this intangible element into published sketches and portraits; we turn the ambivalent and ephemeral into a series of plausible and persistent narratives. By such means a learned society acquires far greater material form than it has any right to claim. Indeed, the actuality of a society comes to us as if the subject of an exhibition of material objects. In the case of the Geological Society of London, these objects might include some of the science's greatest heroes: William Smith (1769–1839), Roderick Murchison (1792–1871), Adam Sedgwick (1785–1873), Charles Lyell (1797–1875) and William Buckland (1784–1856). In an arrangement of individual portraits of this kind lies much of the image of the Society we know today. Of course, the political manipulation of images such as these is an important role for a society such as this and long has it been so. But behind its carefully curated image lies a Society which is and was culturally fluid, adapting to the changing social, political and intellectual worlds of which it is a part.

Science and its learned societies are socialized worlds. In order to operate, they deploy scientific *and* social values and processes. The breathtaking intellectual transformation of geology in the early nineteenth century was paralleled, in ways now largely invisible to us, by equally radical shifts in how the science was socialized. This paper considers the relationship between these two aspects of the science and does so by considering a very familiar story – that of Smith's geology. It is a story I intend to make more central to understanding the development of geology in Britain in the early nineteenth century by revealing its hidden nuances. I shall do so in four 'Acts'. The first considers the Society's modern use of William Smith as an iconic image in British science, tracing the origins of this use to 1831 and the presentation of the first Wollaston Medal. The second considers the period before the Society fell in love with Smith, when its own theoretical ideas and

[1] Here Toru Okada considers how well he knows the wife who, within a few pages, will leave him.

From: LEWIS, C. L. E. & KNELL, S. J. (eds) *The Making of the Geological Society of London*. The Geological Society, London, Special Publications, **317**, 1–47.
DOI: 10.1144/SP317.1 0305-8719/09/$15.00

extraordinary dynamism proved a new and effective way to do geology in Britain. The third considers how Smith entered the Society's thinking, suggesting that he did so at a time when the Society itself was entering a period of social turmoil. Perhaps he was the catalyst for this, for he demonstrated that the geological wilderness, which many Society men felt encircled them, was not really there. The final part demonstrates that in order to possess Smith and his 'English geology', history itself had to be rewritten. At that moment, however, there was a developing sense that a Society could not possess and control geology anyway.

The opening quote, taken from Murakami's acclaimed novel, offers a central theme for this essay for, above all else, this is an essay about not knowing but nevertheless believing. In the political turmoil of the Society's early years, this indefinite aspect of life was hugely influential in shaping relationships and spawning a culture of social interaction that aided the science's rapid advance. It has also shaped the histories that have been written, including those deployed by the Society to construct its present-day image. It is with these points I shall begin.

ACT 1. On the uses of history

Histories of societies

In using William Smith as an avenue of approach to understand the making of the Society, I have to take into account the considerable attention he has received from historians. Indeed, according to the present incumbent, Hugh Torrens (2001, p. 62), every generation has had its Smith biographer: William Stephen Mitchell (1840–1892), Arthur George Davis (1892–1957), Leslie Reginald Cox (1897–1965) and Joan Mary Eyles (1907–1986), to name but a few. Each has felt it necessary to fight for Smith against detractors and romantics. In doing so, these historians differ little from many of Smith's contemporary advocates who, in their different ways, fought battles for his recognition (most notably, John Farey (1766–1826), William Henry Fitton (1780–1861), Joseph Townsend (1739–1816), James Parkinson (1755–1824) and William Phillips (1773–1828)). Both historian and advocate have used a rich empiricism to defend their man. Some have also sought that Smithian Holy Grail, a document that proves Smith's influence on those Continental geologists who seemed to possess an almost identical idea. This has not, however, simply been an argument about priority; it has also been variously modelled as a battle on behalf of the little man, and the practical and utilitarian, and an argument against the power of social and intellectual elites.

Few historical figures have attracted such compassion and support. Smith, we might believe, is the patron saint and guardian of the true religion. 'Any attempt to set him in his context is too readily misinterpreted as an iconoclastic attempt to topple him from his pedestal', Rudwick (2005, p. 443) complained. If, then, an anti-Christ can be imagined, and it can be a society of individuals, then that society is the subject of this essay for, as Torrens (2001, p. 61) has put it, in the baldest of terms:

> the founding fathers of the Geological Society were unconvinced of the reality or utility of Smith's discoveries. Its leaders at first did not believe he had uncovered anything of significance and then simply stole much of it.

From his own point of view, and that of his supporters, Smith was undoubtedly visible to the Society's members, his proven ideas stolen by them for their own purposes, as Torrens has shown so well. I have also argued for the pervasiveness of Smith's influence in nineteenth-century England (Knell 2000). However, the challenge in this present essay is to see things from the other side: to see Smith through the window of the Geological Society. Might events look a little different from this side of the glass? The Society was not born a den of thieves and it did not pursue its goals at any cost; social mores, so important to the interactions that characterize a new society, would, we might imagine, have prevented this. Indeed, might it be true, at least from the Society's perspective, that Smith was a relatively invisible figure or of no great significance, and the victim of no crime? And might our answer change depending on the period of which we ask the question? The Society, locked in a reflexive relationship with the world around it, was bound to change. In doing so, the meanings of people, ideas and objects also seemed to change. It was these changing perceptions, I shall demonstrate, which changed the way the science was socialized and in turn affected the way the science progressed.[2] By these means the Society came to perceive Smith as the embodiment of English geology in a way it had not done previously. By these means, too, once innocent actions then appeared felonious.

Historians sometimes forget that questions are often asked in worlds far removed from those in

[2] By 'socialized' I am referring to the social aspects of the science and not to political socialism.

which they are answered. These are worlds divided not so much by geographical distance as by time and culture. Change meant that people saw differently and found different things to see. In the emergent geological world discussed in this essay, no individual or society controlled the development of the field; thus, it is wrong to believe that someone like Smith really needed the Geological Society to be considered, or legitimized as, a 'geologist'. Equally, there is no reason to believe that the early Geological Society saw Smith as particularly relevant to its cause; after all, it wished to be an extraordinarily exclusive, private, gentlemen's club, and it imagined geology as something different from that narrow and practical field which it believed occupied Smith.

In 1807, both parties possessed preconceived and incompatible ideas of what geology was or was to be. In this unfettered world, our Society geologists adopted goals, but no sooner had they begun along a path to achieve them, than the path itself was affected by new information, theoretical ideas and changing social configurations. The science of English geology did not emerge from a simple narrative of discovery, but through a succession of unscripted interactions that gave it unpredictability. Smith appears in a number of these interactions, sometimes in person but at other times only in essence. In the end, the Society arrived at Smith, but in doing so its leading commentators were more than a little surprised. When they had looked into the future, in the chill of that November evening when the Society was born, they had never believed that the path they were to follow would lead to this humble surveyor.

If the world I have described seems rather more indefinite and chaotic than our historical imagination permits and a subject hard to contain within an easy narrative, let me add that our attempts to unravel it are further confounded by the inadequacy of our actors' understandings of each other. A geological society is only vaguely a unified entity; it is rather more an assemblage of disparate and variably political individuals. In the correspondence of George Greenough (1778–1855), the Society's first President, for example, it is clear that he never had a complete knowledge of his correspondents or they of him. By all accounts, he frequently missed or avoided meetings; like us all, he doubtless wished to know certain individuals in certain ways. Where he did have close relationships, although still largely conducted at a distance, such as with William Conybeare (1787–1857), the haze of formality, respect and jest concealed from each man the kinds of thoughts that turned them from the closest of friends into something akin to petulant lovers. The politics of social relationships, so easily reconfigured by rumour and gossip, particularly as a result of competing ambitions, enhances this sense that each individual operated within a world affected by myth and belief.

Like most early geologists, Greenough was emotionally affected by disputes and misunderstandings of this kind. Few of these new geologists avoided them and none fully understood them; indeed, there was no single understanding to be had. For instance, in the eyes of one of the Society's curators, Thomas Webster (1772–1844), Greenough came to embody all that was wrong in the science. In contrast, he believed that the Secretary at the time, Fitton, was honest and trustworthy. Yet, Fitton was with Greenough in believing Webster failing.[3] However, we cannot know that Fitton was always straight with Greenough or Greenough with Fitton, for these two also fell out.

Of course, both Greenough and Webster had their admirers and supporters but these too probably said and thought quite different things to and about the individuals they supported. Some, like Henry Warburton (*c.* 1784–1858), who stood as mediator between Greenough and Conybeare, could barely comprehend the cause and depth of the disagreement that separated his friends. This was, however, but a moment in their lives. One day Greenough despised Conybeare while wishing to like him; the next he was more disposed to understand him. My aim here is not simply to reveal a dispute, which I shall briefly return to later, but to suggest that relationships and personalities are considerably more complex than we generally credit.

Consider, for example, our images of Smith and Greenough shaped by events in the late 1810s; one impoverished and heading for debtors' prison, and the other an empowered and wealthy gentleman. Yet, just a few years later, Greenough was embroiled in a succession of jealous disputes and surrounded by critics. The hurt was so strong that he temporarily turned his back on geology. Meanwhile, Smith was in Kirkby Lonsdale in Yorkshire: 'In this sweet retirement Mr. Smith long and

[3] 'It is our object to employ Mr. Webster in the manner you propose – (i.e. we shall tell him to do certain things) – but my great fear is that he has acquired such habits that it will be difficult to bring him really to work in his vocation. The experiment, however, it does appear to me, ought to be fairly made before we sever a connexion that has existed so long.' UCL: Fitton to Greenough, 15 September 1828, Add. 7918, 611.

willingly lingered, feeding all the best qualities of his mind by calm meditations, not unmixed with the poetic impressions which seem perpetually to haunt the romantic banks of the Lune' (Phillips 1844, p. 104). Most people suffer tragedy but few live tragic lives, and if anyone was fitted not to do so, it was William Smith. Although Greenough possessed none of Smith's financial worries, he also appears to have lacked Smith's happy resilience.

Relationship-building was critical to the birth of the new science, but in the early years of the Society it was often between individuals who knew each other insufficiently. This was dealt with using diplomacy – that social concern with illusions, even if for very noble reasons. By the late 1810s, however, these illusions had evolved and resulted from concealment through economical disclosure. Individuals may have known each other better, but they knew rather less about each other's projects and ambitions.

The Geological Society we must imagine, then, if we are to make sense of these disputes and the place of Smith, was composed of indefinite individuals, constantly reconfiguring their relationships, while components in their world – people, objects, science – were in constant flux. My point in referring to this complexity is to note that the historical interpretations we make of this world are political (Macintyre & Clark 2003). How we cast our historical actors is largely up to us. The Society itself is, in this respect, also a historian. It publishes its own history in the very manner of its being. Let us look at that history.

The Society exhibited

If we display a solitary object in an architectural space, we provide an opportunity for the construction of meanings. If we juxtapose several objects, a narrative might result. This is the basis upon which museums have operated for centuries. What, then, should we make of the modern-day entrance hall to the Geological Society's grand Burlington House apartments? Here, it seems, a museum has been reinstated in an institution which, in 1911, decided not to have one (Moore *et al.* 1991, p. 51). Indeed, many of the artefacts on display here once found a place in that long-lost museum.

Here, the bust of founder and President, George Bellas Greenough, stands next to his fine, but contentious, geological map of November 1819. Those who penetrate the apartments a little further will find John MacCulloch's (1773–1835) equally fine geological map of Scotland published in 1836, but really a product of the years before 1832. Not far away is his bust. MacCulloch was an early member of the Society, but, unlike Greenough's, his map was not a product of the Society's efforts (Bowden 2009). Like Greenough's map, however, MacCulloch's also conceals a contemporary dispute, this time about the uses of public money (Cumming 1985).

Today, however, these are mere backdrops to the great map of William Smith, published in August 1815. Indeed, one might at first suffer the illusion that one has entered a *Smithian* museum, for here is his map, his portrait, his bust, and even his nephew and heir, John Phillips (1800–1874). Amongst the few books on sale at the Society's reception desk, that are not produced by its publishing house, is Simon Winchester's *The Map that Changed the World* (Winchester 2001). An accessible, populist and romantic fusion of fact and fiction,[4] it uses for a plotline that corruption of the biblical narrative that has fuelled a hundred heroic Hollywood sporting biopics.[5] In it, the humble Smith is initially engaged in relating psalms to people in all walks of life, the profundity of his revelations fundamentally altering the thinking of those who heard him. However, at that very point when he should have been raised aloft he was crucified in a debtor's prison, only then to have his 40 days in the wilderness. It was while in that wilderness that he found himself and others found him, and he rode into London, not many years before his death, to be resurrected and pronounced 'the father of English geology'!

This is the Smith of popular belief and as a result of Winchester's bestseller we *can* say there is a *popular* belief, whatever the book's historiographic merits. Winchester has done the Geological Society the service of placing it on the London tourist trail and a new public is now admitted to these private rooms to gaze up at Smith's extraordinary achievement, doubtless casting an eye at the evil Greenough who looks on. They then pull back the curtain on Greenough's map to see the perfect image of a theft, perhaps not realizing that the rocks themselves constrain the art. But if it was not a theft, was it not an injustice for Greenough to publish his improved map so soon after Smith had published his? Smith lost out on the possibilities of long-term sales, not least from successive editions – or so some claimed at the time. Greenough's map, which

[4] Hugely promoted and the subject of considerable positive press, Winchester's book has nevertheless attracted criticism from historians. See, for example, Oldroyd (2001) and McBirney (2005).

[5] Part of an important publishing phenomenon, it is discussed by Miller (2002) who coins it 'The Sobel Effect'.

apparently cost £3000 for the first two editions,[6] was the result of the co-operative efforts of the Society's members and supporters, and published using private funds. Of all the maps which today hang in Burlington House, only Greenough's speaks of the Society and its aspirations in 1807. Yet, for nearly 200 years, its achievement has been overshadowed by rumours doubting its originality and the integrity of its maker (see, for example, Laudan 1977).

Whatever goes on in the minds of those who, today, look at these maps, the performance here is not just for them; it is also for the Society, for the Society has needed Smith since 1831. Here, in the entrance hall, then, history is presented to evoke a sense of institutional memory, and to suggest an impeccable geological pedigree stretching back over generations to the very founders of the science. History is here performing for the sake of the Society's membership and its science, constructing an institution which best serves them both. The Society must, besides maintaining the business of income and services, construct a political identity for a nation's science. This must be an identity that Fellows find attractive – for they will make it part of their own identities and which has impact beyond the Society's walls. Fellows sign up to this identity but they also exploit it. Who amongst the Fellows, for example, pays their annual membership fee at least partly for the sake of what the Fellowship represents rather than the use that will be made of the Society's publications and services? How many work in occupations where this Fellowship casts a small spell over their employers? Indeed, in some cases employers reimburse the fee, aware, too, that the Fellowship might also cast a small spell over others.

These things are far removed from what we might normally consider 'geology', but they are nevertheless part of the science. Only when we grasp this can we assess the place of Smith or the value of a year of bicentennial celebrations, or indeed, the tenor of the Society's new biography, *Whatever is Under the Earth*, written for the membership, rather than for historians,[7] by historian of science, Gordon Herries Davies (2007).

If, then, we distinguish the purposes of a society from more narrow conceptions of geological science, what should we make of Smith's map, hung on the Society's wall? Here, Smith's great masterpiece works with the architecture of the space, the interior décor, the prestigious London location, the modern lecture theatre and a fabulous old library (with its very modern Lyell Collection), to construct an image of geology befitting a national institution. Everything, down to the light fittings and wall coverings, contributes to this image.[8] Only by these means can this private Society seek to stand alongside London's great public institutions and command an equal level of authority and respect.

'Geology', as it is embodied within the very being of the Society, is thus, in part, dependent upon the image the past can be made to present. By these means, Fellows might imagine that geology is one of those things that made Britain 'Great'. Of course, we might perceive this without ever inquiring what precisely a society does or whether it really does contribute in any significant way to the science as more narrowly understood. Or, indeed, without considering where the Society stops and the individual begins.[9] Smith is, then, situated in the entrance hall so that Fellows and visitors alike can imagine and believe. And, if we want, we can imagine that he *is* the father. It is a use of the past that requires no appeal to rational thought or rigorous historiography (as ironic as that might seem, given the empirical nature of the science of geology which these historical beliefs seek to support).

Actuality, however, is unimportant here, for what we are discussing are identities – the science's identity, the Society's identity and Fellows' identities – which are not the simple product of objective logic but result from myth, belief and perception. They are in these senses aesthetic. It is here, then, that Winchester's book finds a role, for, in the construction of myths of identity, rigorous histories are rather less important than convincing stories.

There are many national societies, some with far larger memberships, who lack this history or London premises of this quality. For those societies, members cannot walk into long-established spaces and sense that their discipline permeates its every pore. They cannot look up at artefacts like Smith's map, which with every passing year approaches the status of a religious icon. In possessing these historical things, then, the modern Society should consider itself blessed.

[6] UCL: Greenough, main collection, mss. 5. However, an earlier account of the first edition (UCL: Warburton to Greenough, April 1823, Add. 7918, 1696) suggests that this was not an unprofitable venture and that Greenough's £3000 may include the subscriptions of members.

[7] Herries Davies (pers. comm. 13 November 2007).

[8] This kind of analysis owes much to work in the interdisciplinary field of museum studies, which itself has drawn on work in sociology, anthropology, literary criticism and so on. See, for example, Pearce (1990, 1995), Preziosi (1989, 1998) and Knell (2007*c*).

[9] Consider, for example, Branagan's (2009) geologists.

But the Society does not possess these things because Smith was a prominent member; he was not a member at all. It possesses these things because in 1831 the then President, Adam Sedgwick, saw Smith's political potential. Unmatched in his eloquence,[10] Sedgwick took Smith and remade him, turning him from provincial folk hero into a major icon of British science. In doing so, he implicitly recognized a major transformation in the Society's conceptualization of geology. In order to understand this we must first be clear about what Sedgwick actually did.

Claiming a geology for England

On 18 February 1831 William Smith travelled to London to be given what might be deemed, with hindsight, to be the greatest honour of his life: the Society's first Wollaston Medal. It assured Smith of his immortality and John Phillips of his intellectual inheritance. The decision had been made barely a month earlier, at a Council Meeting on 11 January 1831:

> That the first Wollaston Medal be given to Mr. William Smith, in consideration of his being a great original discoverer in English Geology; and especially for his having been the first, in this country, to discover and to teach the identification of strata, and to determine their succession by means of their embedded fossils.

We should consider what 'English geology' means in this minute, for Adam Sedgwick tells us that his intention was to give Smith the medal and the 'purse of twenty guineas' with little oratory. The Council minute tells us that its members made no claim beyond the nation – Smith was merely a contributor to a history of geology in England. But Sedgwick had a few hours before the presentation to probe Smith about his discoveries (information about these had been sent by Phillips in advance but had only made it as far as Cambridge (Morgan 2007, p. 14)). Fortunately, Smith had arrived carrying a number of key historical documents that enabled Sedgwick to see an opportunity to place Smith not merely as a labourer in an English field of study but in 'the correct history of European geology' (Sedgwick 1831, p. 271). Sedgwick was to turn Smith himself into a Smithian fossil and place him low down in the succession of European discoverers. In doing so, he was to redefine the meaning of 'English geology' in nationalistic terms.

In his attempts to locate the earliest possible date for Smith's discoveries, it was natural for Sedgwick to soften that moment of initial discovery by suggesting that Smith was a child of the rocks of Oxfordshire, whose innate qualities act 'powerfully on peculiar minds, so as to influence the whole tenour of after-life' (p. 271). From the age of 18, in 1787, Smith was travelling, observing, collecting and drawing sections. 'I think these facts of great importance', Sedgwick remarked, 'as they contain the germ of all Mr. Smith's future discoveries' (p. 274). In 1795, Smith had created the first stratigraphic collection of fossils, which Sedgwick thought a defining action unmatched by any Continental geologist. By 1800, according to Sedgwick, Smith had produced that critical correlation which linked the strata of the SW of England with those of the NE. Thus, in a period of prehistory, so to speak, long before the invention of the Geological Society, Sedgwick was willing to assert:

> For eight or nine years he [Smith] had been steadily and resolutely advancing, but without aid, and almost without sympathy; for he was so far before the rest of our geologists, if indeed they deserved the name, that they could not even comprehend the importance of what he had done.
>
> (p. 275)

This view was corroborated by a letter from Rev. Benjamin Richardson, who had been Smith's friend and supporter since 1799. Indeed, Richardson admitted to disseminating Smith's table of strata of that year to correspondents across Europe in 1801.[11] Now that table was laid out before the assembled audience as a physical proof of this remarkable claim. Beside it was Smith's 1801 prospectus for a map and description of strata. Sedgwick emphasized that all this was before Cuvier and Brongniart's 'magnificent' fossil-based correlations in the Paris Basin (Taquet 2009). Sedgwick's use of 'magnificent' here had the dual purpose of softening any offence given to two esteemed foreign colleagues whilst elevating an achievement for which he now stole priority. He was also able to show that fossil-based geology had entered the Society's publications under the influence of Smith in an era before the establishment of palaeontology. All this belonged to an age so distant that it even predated septuagenarian Joseph Townsend's odd and much derided book, *The Character of Moses Established for Veracity as an Historian* (1813). Now that book, with its glowing appreciation of Smith, was an important historical document that had said nearly 20 years earlier what Sedgwick was saying now. How blind they had all been! Or, rather, how blind they wished to be seen to be!

While we cannot, from the published account, know precisely what Sedgwick said, in these

[10] For which, see Smith's record of the event published by Morgan (2007).

[11] Benjamin Richardson, Farley [Farleigh Hungerford] to Sedgwick, 10 February 1831 (Sedgwick 1831, pp. 275–276).

statements are echoed claims made for Smith by Society members in the mid-1810s. Yet, Sedgwick reworks them and uses them to make a claim for England. In conferring on Smith the title of 'the Father of English Geology', one senses, then, not a modest claim that the Scots, French and Germans had their own fathers and now the English had theirs, but rather a national assertion that the fossil-based study of strata that had come to the fore in recent years was to be understood as 'English geology'; some had thought it French. This gives an entirely different meaning to the event, the appellation, and to the notion of English geology, and makes understandable John Phillips' claim, in 1843, that Wales was the 'only remaining Crown that English Geology can offer' (Knell 2000, p. 272). Smith was the father not of geology *in* England (a common interpretation that has forced recent historians to rail against the use of the term) but of an internationally important genre of geology which the national Geological Society could now be said to possess. It was this geology that would enable the likes of Sedgwick and Murchison to claim much of the stratigraphic column.

It was proposed by Smith's admirer, Fitton, and seconded by Smith's supposed nemesis, Greenough, that Sedgwick's address be published. The Society had long established a belief that publication alone provided the means to make and defend claims; there can be no doubt that Sedgwick's speech contained a significant claim, although not everyone might have signed up to it. Greenough's role here may well have been politically symbolic: perhaps it was hoped that this noble act would put an end to all the speculation and accusation. It was followed by Fitton writing a history of geology in England (and Europe) that gave Smith the status Sedgwick desired. I shall come to this history in due course, for, in order to write this history, it was necessary for Fitton to recast events. First, however, we must understand something of the Society itself, its conceptualization of geology without Smith, and its early efforts to establish that science in Britain.

ACT 2. The Geological Society

A geological powerhouse

The Geological Society began with tremendous energy and co-operation; its birth was a high impact event that seemed to mobilize the nation.[12] At its birth it was simply 'The Geological Society'; it had no need of its 'of London' until forced by its imitators.

The birth of the Society has been described as the logical culmination of a social and intellectual trajectory which sees interests and overlapping memberships flowing through the British Mineralogical Society, the Askesian Society and the Geological Society (Weindling 1979, 1983; Torrens 2009; Veneer 2009). The ephemeral nature of these earlier incarnations of a 'geological' society is important to understanding the growing possibilities for socialized participation in this emerging field, but we cannot say that one begat the other. The linkages here are purely of our own making – a distillation of considerable social interchange in London at the turn of the century. An alternative trajectory, for example, might be drawn through Guy's Hospital and the Medical and Chirurgical Society, or, indeed, any number of London societies, for these men were engaged in many (Lewis 2009*b*). It was in such locations that professional interests in chemistry and mineralogy were already established.

A third trajectory, and the now standard narrative of the Society's birth, came from Greenough himself and has been the basis for all subsequent histories (Woodward 1907; Rudwick 1963; Herries Davies 2007), although it was doubted by Weindling (1979, p. 249) and had been contested by the Society's first Secretary, James Laird (1779–1841), when Greenough first gave voice to it in his Presidential Address of 1834 (Greenough 1834, p. 42).[13] It was on that occasion that Greenough spoke of the contribution of founder member, Dr William Babington (1756–1833), who had recently died. In doing so, he described the Geological Society as originating in weekly teatime meetings, held at Babington's house, which pushed forward the publication of the Comte de Bournon's (1751–1825) *Traité complet de la chaux carbonatée auquel on a joint une Introduction à la Minéralogie [...] une Théorie de la Cristallisation* (1808) (Herries Davies 2009; Taquet 2009). The group consisted of Babington, three Quakers – William Allen (1770–1843), Richard Phillips (1778–1851) (who had been trained by Allen) and his brother William (Torrens 2009) – and about 12 others. Finding their meetings enjoyable they continued, but as an informal breakfast club that met as early as 7 am; their

[12] Laudan (1977) mistakenly, in my view, interpreted the birth of the Society as having little or no impact on the development of the science.

[13] GSL: Greenough, 'Papers connected with the Geological Society', a history of the first three years with copies of letters and lists (*c.* 1850), LDGSL 960 (hereafter simply, Greenough).

discussion, observation and chemical analysis centred on Babington's mineral collection. From these discussions came the Geological Society (discussed more fully by Lewis 2009*b*).

Here, then, are three possible trajectories leading to the birth of the Society that should force us to dispense with any simple creation narrative. Clearly a society, as a confluence of individuals, is also a confluence of diverse interests and social relations – enhanced by personal networks and rife nepotism (Lewis 2009*b*). There were multiple paths that led to the formation of this dining club which then became a learned society. Indeed, to the moment of its birth, we can no more claim the influence of the Askesian, Bournon's book or Guy's Hospital than we can the chill of the November mornings or the quality of British beef; all had their part to play.

Of more fundamental importance was population; both its densities and demographics. As Fitton recognized in 1812, it affected one's profession, one's intellectual development and one's leisure. He was then a prisoner of his medical practice in Northampton, where he found 'no food for mineralogists'. Although no geological novice, the county offered a kind of geology – that of strata – of which he was wholly ignorant. He had yet to discover what lay beneath his feet or how to begin to discover it, yet he was already a well-established contributor to the Society.[14] The situation was made worse by his scientific isolation. As he told Greenough, 'There is no scientific society in or near this place [...] I feel very severely the difference between London & the country, in the privation of that liberal & delightful intercourse with the members of my own profession, which I fear is only to be found in great cities'.[15] Fitton's trips 'to Town' (London) would take in the meetings of the Geological Society and the Medical and Chirurgical (i.e. surgical) Society, which was just two years older, and Sir Joseph Banks' Sunday meetings.[16] London was an intellectual oasis where he could freshen up his not unrelated medical and geological knowledge.

The Society was, then, an elite and very expensive private club which feared losing its exclusivity. Leonard Horner (1785–1864), an early Secretary, told Greenough this less than 18 months after its establishment:

> Allen has hinted to me, that we are thought to be going on too rapidly in the admission of members, and are not sufficiently discriminate. Now that we form so respectable a body, we certainly ought to be more nice,[17] & render the admission more difficult perhaps than it is at present – it will to a certain extent add to its respectability.[18]

To progress, the Society certainly had to conform to contemporary expectations; if it was to have influence then it certainly needed respectability and this was most easily obtained by social discrimination. That does not mean, however, that members necessarily possessed an overly developed sense of their superiority; it was simply a matter of playing by rules that are not so very different from those which distinguish the Society today. Certainly, Greenough, for example, communicated with modesty and on equal terms with many of his supposedly social inferiors. The Society also had to fight for its own status, but its early tussle with the Royal Society (Knight 2009; Lewis 2009*b*) seemed only to make the geologists more united: 'I am quite satisfied, that if we chuse, this business will go a great way to establish our Society on a firmer footing'.[19]

The Society rapidly acquired a sense of those factors and signifiers that would aid its advance. Horner, for example, soon laid out a detailed plan for the Society to enter into publishing, perhaps along the lines of the *Journal de Mines* of Paris. In 1811, the Society published the first volume of its *Transactions*. It was a remarkable achievement for a society so young. Its pages perfectly embodied the energy and spirit, mobility and philosophical position of its highly co-operative membership. Fitton captured honestly the naïve enthusiasm of 'plain men [...] who felt the importance of a subject about which they knew very little in detail; and, guided only by a sincere desire to learn' ([Fitton] 1817, p. 70). Greenough imagined that there was hardly 'any one in the Kingdom beyond the pale of our little coterie who aspired to be thought a Geologist',[20] but in believing this he was perhaps reflecting the novelty of the

[14] UCL: Fitton to Greenough, 12 November 1812, Add. 7918, 592.
[15] UCL: Fitton to Greenough, 12 November 1812, Add. 7918, 592.
[16] UCL: Fitton to Greenough, 12 November 1812, Add. 7918, 592; 3 February 1813, Add. 7918, 595.
[17] Delicate in judgement.
[18] UCL: Horner to Greenough, 4 April 1809, Add. 7918, 823.
[19] UCL: Horner to Greenough, 4 April 1809, Add. 7918, 823.
[20] GSL: Greenough, p. 3.

name rather than the rarity of the practice (Rudwick 2009). Certainly throughout the Society's first 15 years, the term 'mineralogical' was used quite commonly, but not without confusion, to mean or embrace the 'geological'. As Fitton told Greenough, in 1812, 'mineralogy (I mean the term to include also geology)'.[21] This mineralogical perspective owed everything to the inescapable influence of Abraham Gottlob Werner (1749–1817) whose writings were at the height of their powers (Rudwick 2005, p. 422) and to whom I shall return a little later.

The Society established itself as a network, its central body initially made up of, and restricted to, Ordinary Members who were by definition situated in London, with tentacles formed of a far greater number of Honorary Members distributed across the country. This network personified geology as it was imaged in 1807. It was composed of men of wealth and land who had interests in those aspects of property that expose the underlying geology (mines, quarries, engineering projects, agriculture) and who had workforces of labourers and engineers fluent in the language of such places. To these were added men of those sciences thought to be in command of the material study of the Earth (principally chemists and mineralogists, the two being generally inseparable) (Knight 2009). Medical men were an important source of scientific expertise, primarily for their knowledge of, and interest in, chemistry and mineralogy; an interest in anatomy or palaeontology was then a rather minor concern (Lewis 2009*b*). Above all, the Society aimed to be useful and it only gave honours in order to receive some payment; Honorary Members were so honoured in order to encourage their contributions. These contributions were to fuel debate in the City.

However, there were others, like Fitton, Buckland, Conybeare, Farey, Thomas Meade of Chatley Lodge, near Bath, and Gideon Mantell, medical man and protégé of James Parkinson, who appear as extraordinarily active correspondents long before being recorded as members, if, indeed, they ever did join. Indeed, it is doubtful that anything meaningful can be drawn from the dates on which individuals became members; the Society was bigger than its membership. Nor can we suppose to know much of the business or activity of the Society from the record of its meetings or its committee minute books. There was, it is true, a core in London, made up of the likes of Laird, Horner and other officers, who contributed significantly. Like Greenough, many of these were inveterate travellers. Some also had bases distant from London. However, this dynamic aspect only becomes apparent in what survives of Greenough's correspondence (as first President he was the recipient of much of the Society's correspondence). Here we see an investigative and mobile organization in pursuit of a grand project to define the discipline and resources of an English geology. For many in the provinces, Greenough's identity, and doubtless those of other Society men, was closely associated with the Society. Greenough, in particular, was seen as a dynamo, driving, and making possible, this new field of investigation: 'We soon became fired with our subject, & pushed on vigorously', he later recalled.[22]

In December 1807, Aikin and Greenough proposed to write to every member with 'a series of questions relating to the most essential points in Geology'.[23] These *Geological Inquiries* written by Aikin and modified by Greenough were printed and circulated to members two months later.[24] Greenough believed this indicated 'how early a period I had contemplated a work which readily passed on afterwards to completion' (his map).[25] These *Inquiries*, however, support no such claim. They clearly articulate the chemical and mineralogical visionary apparatus of the two authors, and Greenough's latent geographical interests.

The Society soon adopted a systematic approach, establishing committees in focused areas of study. A committee was established in May 1808, for example, to resolve the problems of nomenclature in geology and mineralogy. Members were charged with gathering up provincial terms so as to eradicate synonymy and gain a unified understanding of the nation's geology. By June 1810, the Society possessed eight committees including those concerned with the selection of papers to be read at meetings; to form and arrange the collection of minerals; to begin the production of maps; to undertake chemical analyses; and to investigate 'extraneous fossils'.[26] There seems to have been an order to their establishment, as, following the committee for chemical analysis, Greenough recalled:

> Soon after this two other committees were appointed which were considered useful not so much in regard

[21] UCL: Fitton to Greenough, 17 November 1812, Add. 7918, 593.

[22] GSL: Greenough, p. 5.

[23] GSL: Ordinary Minute Book 1, 1 January 1808.

[24] These are reproduced in this Special Publication.

[25] GSL: Greenough p. 2. Pencil note added later and reflecting the purposeful nature of this record; also p. 39.

[26] GSL: Greenough, p. 36, pencil note.

to the subject matter referred to them, but as inducing our members to seek each other's acquaintance & to induce them to take an interest in what was going on, & to stimulate their desire to be what their proposers had declared them likely to be to the Society viz useful and valuable members.

The first of these was a Committee of Maps and Sections. Bearing in mind that Greenough was making a record for posterity, he later recalled:

> The result of this Committee was the execution by Mr Atkinson of an etched outline map of England extracted from the larger map of Cary which it was intended to circulate freely over the country among the honorary members in order that they might lay down upon it the boundary lines of the strata in their several neighbourhoods. The outline map did not prove so useful as was anticipated but I refer to it as shewing the zeal by which we were all animated at a period when geological mapping was a novelty. The etching was performed gratuitously by Mr Atkinson & the President of the day [Greenough] who had from the very first aspired to be the Author of a geological map of England most willingly paid for the paper & printing.

The proposal to establish this committee was made at a meeting on 3 March 1809, with a request for volunteers.[27] On the 7 April it was constituted: 'That the construction of Mineralogical Maps and collections of Drawings, Models, Sections of Mines & c, are well worthy [of] the attention of the Geological Society, and cannot fail materially to promote the objects which they had in view in associating'. It is noteworthy that Greenough was not, initially at least, a member of this committee.

On the same day, a Committee of Extraneous Fossils was established. Greenough recalled:

> It consisted of Capt Apsley who afterwards by will bequeathed to the Society a collection of fossil shells, the Count de Bournon [...], Mr Carlisle who as a physiologist was likely to treat such a subject in the way most wanted. Mr Lambert in natural history an infectious enthusiast, Dr McCulloch a man of extensive acquaintance with physics & better acquainted with geology than most of his colleagues. Dr Marcet an excellent chemist. Mr Parkinson not merely the best but almost the only fossilist of his day at least in England. Dr Roget even then distinguished as a comparative physiologist. Dr Saunders a physician investigative practise. Mr Sowerby already the greatest authority in conchology (with the exception perhaps of Mr Parkinson) & well versed in other branches of natural history. Dr Warren who in his own profession was considered by a colleague who wished to satirise it as the best guesser in the college, & Mr Wilkinson who lived in the neighbourhood of Bath & who starting under the direction of Mr Meade, Mr Richardson & Mr Townsend was known to be the possessor of one of the most valuable collections of fossils in the only county almost in which such collections were frequent.[28]

While Greenough later used this to indicate that the Society had not been ignorant of the value of fossils, at the time in question, this committee had no immediate effect on the Society's collecting or palaeontological interests. It was only with the absorption of Smith's ideas that this changed. This is, however, for the moment unimportant. First we must re-imagine the Society as dynamic and connected, actively pursuing its broadly conceived and mineralogically tinged geology.

Data, empiricism and theory

> It has been remarked by critics, that the want of education is sometimes of advantage to a man of genius, who is thus left free to the suggestions of invention, and is neither biassed in favour of erroneous maxims, nor deterred from the trial of his own powers by names of high authority. On this principle, it is evident that the members of the Geological Society have derived great benefit from their want of systematical instruction.
>
> ([Fitton] 1817, p. 70)

The greatest provincial concentration of support for Greenough's project was at Oxford University. A bond of mutual respect and affection became rapidly established, as is apparent from young William Conybeare's teasing of the slightly older Greenough, as here in June 1811, when all seemed to be peace and conviviality in this nascent geological world:

> Far from using my influence and solicitations with Kidd to visit you in London I regard the request as perfectly shameless – the whole nomenclature of mineralogy affords not a term sufficiently harsh to stigmatize it by [...] Are you not conscious that it is positively incumbent on you to visit him and your other friends here [...] In our kindness however we allow you one chance of redeeming your character by spending the week of the Commemoration which is nearly approaching among us.[29]

This reprimand was accompanied by a circular invitation recording the names of the 'Members of the Independent Rag Formation of Oxford' (Fig. 1). However, Greenough could not attend, which called upon Conybeare to respond playfully, 'I regret extremely that the Gnome of the mines was of so stubborn a nature as placed him beyond the

[27] GSL: Ordinary Minute Book 1.

[28] GSL: Greenough, p. 41. Lewis (2009*b*) discusses the medical make-up of this committee.

[29] UCL: Conybeare to Greenough, 12 June 1811, Add. 7918, 431.

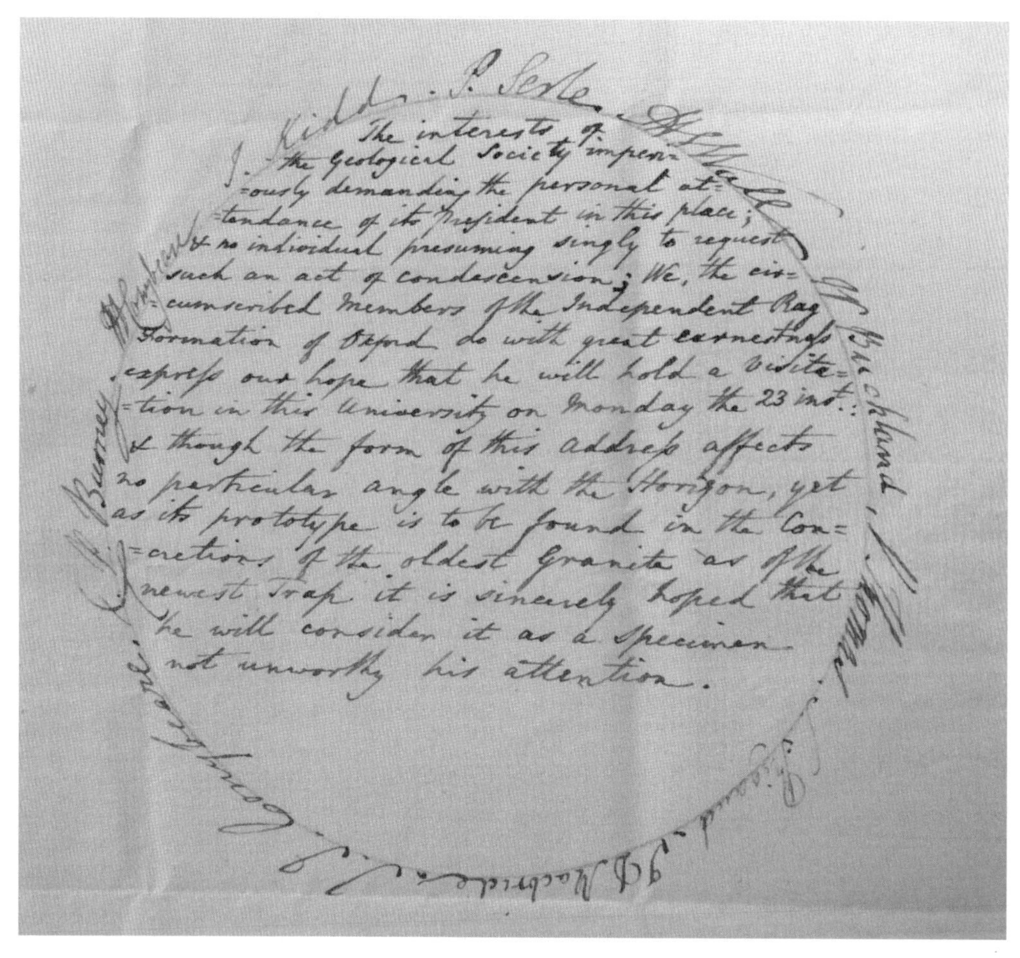

The interests of
the Geological Society imperi-
=ously demanding the personal at-
=tendance of its President in this place;
& no individual presuming singly to request
such an act of condescension; We, the cir-
=cumscribed members of the Independent Rag
Formation of Oxford do with great earnestness
express our hope that he will hold a Visita=
=tion in their University on Monday the 23 inst.:
& though the form of this address affects
no particular angle with the Horizon, yet
as its prototype is to be found in the Con=
=cretions of the oldest Granite as of the
newest Trap it is sincerely hoped that
he will consider it as a Specimen
not unworthy his attention.

Fig. 1. William Conybeare's invitation to Greenough, June 1811, reads: 'The interests of the Geological Society imperiously demanding the personal attendance of its President in this place; & no individual presuming singly to request such an act of condescension; we, the circumscribed members of the Independent Rag Formation of Oxford do with great earnestness express our hope that he will hold a visitation in this University on Monday the 23 inst. & though the form of this address affects no particular angle with the Horizon, yet as its prototype is to be found in the Concretions of the oldest Granites as of the newest Trap it is sincerely hoped that he will consider it as a specimen not unworthy his attention'. (Courtesy of UCL Special Collections).[31]

influence of our magic circle to which every sprite of more gentle frame and occupation must necessarily have been obedient'.[30]

Their correspondence records the extraordinary respect afforded Greenough and the degree to which the Oxford geologists were on-board the Society's project, even if 'Independent'. The intellectually sharp Conybeare regarded Greenough as gifted in enterprise. Indeed, all the members of the Rag Formation were correspondents with Greenough, but most diligent in this relationship were: Conybeare, who exceeded all others in the supply of data; John Kidd (1775–1851), who became Greenough's great correspondent and friend, and who took a leading role in nurturing the new science at Oxford; and Buckland, who was soon to succeed Kidd and who became one of Greenough's closest friends.

Kidd was, interestingly enough, yet another product of Guy's Hospital (see Lewis 2009*b*), where he had trained in medicine from 1797. Lecturer in Chemistry at Oxford from 1801, he was

[30] UCL: Conybeare to Greenough, 21 June 1811, Add. 7918, 433.

[31] UCL: Conybeare to Greenough, 21 June 1811, Add. 7918, 433.

soon a leading light in the emergence of geology there, and wrote to Greenough admiringly, congratulating him on the stature he had given the science through his efforts as the Society's President. Greenough was seen not simply as a hub gathering information on British (and European) geology, and working out the principles of geology itself; he was also credited with developing a culture of geology across the land. 'A Geological Committee met at Buckland's Rooms this afternoon at 4 o/clock; &, having dined with that worthy, proceeded to Sir G. Bowyer's Coal-pit at Foxcomb. As you can well guess we met with nothing there but Blue Clay.' This was the failed coal trial at Bagley Wood, discussed by Torrens (2001, pp. 74 and 76), which led to Bowyer's ultimate ruin.[32] The pit fell naturally into Kidd's local network, the possession of a gentleman of that social stratum that populated the university. Kidd continued:

> We flatter ourselves that you will be much pleased with the improved state of the Mineralogical Room, & Cabinets, of the Ashmole: the collection of which has within these two days been enriched by the addition of the upper jaw of a fossil Rhinoceros, which had been placed amongst & catalogued as the jaw of the recent animal. The subject seems to flourish here – I began my lectures on Monday, & have a very good Class.[33]

Both Greenough and Fitton were early associates of this Oxford circle, and Fitton in particular was considerably impressed by the contribution it was making. The subject was so new, however, that respect and information flowed both ways; Oxford was finding its geological feet too. In May 1813, for example, Kidd and Buckland sat up till the early hours discussing two lectures written by Greenough.[34] The nature of these lectures is probably indicated by Greenough's draft lecture notes on strata, which still survive. These reveal Greenough's awareness of Werner, Saussure and other earlier workers, and consider quite fundamental questions about the structure and field relations of rock masses.[35] Like those in Oxford with whom he corresponded, Greenough aspired to make a fundamental breakthrough concerning the structure of the Earth.

Kidd, who in June 1814 was planning to visit Greenough's collection, was preparing to publish his own lectures. These he refrained from calling 'a System of Geology' but rather 'An Essay' or 'Essays, on Geology' (Kidd 1815). In this essay he wished to make two main points:

> The two points which I kept constantly in view during my lectures were these – 1st that there is such a degree of uniformity & regularity in the phenomena of Geology as to afford upon the whole a practical certainty in the operations of mining, & c – and, 2dly that the character of the phenomena is, generally speaking, such as to render abortive any attempt at explaining them by the analogy of any effects produced by the operation of existing causes.[36]

There was nothing odd in Kidd's position. His view of stratification placed granite as the lowermost stratum and he told Greenough, 'I shall certainly attend to your hint of inserting such notices as may be in my power respecting the Chemical Analysis of the Strata'. Like Greenough, Kidd was a 'Neptunist' who believed in the aqueous origin of granite (Kölbl-Ebert 2009); by contrast, Fitton and Conybeare were clearly 'vulcanists' or 'plutonists' by the mid 1810s, if not before ([Fitton] 1817, p. 80). This reflected an old debate still raging across Europe but it had little bearing on how the interior of the Earth might be investigated, and both perspectives suggested the validity of chemical analysis. But these were not the only ideas shaping the outlook of these nascent geologists; the subject had a history that Fitton preferred to see as composed of a 'chaos of philosophies' ([Fitton] 1818, p. 313): Wernerian geognosy, Neptunism, Volcanism, uniformitarianism, catastrophism, mineralogy, chemistry, de Luc's Christian philosophy and countless theories of the Earth (Rudwick 2005, 2009; O'Connor 2009). It was partly from these building blocks that the new 'geology' was to be constructed. It gave each individual his own outlook, which resulted in the presence of multiple and overlapping 'geologies'. And while these resulted in what appear much like modern-day disputes over fact and theory, they were more fundamental as they concerned different conceptions of what geology was and how it should be investigated, including its natural disciplinary allegiances.

However, the Society men were determined not to let their theoretical favourites and biases affect their social interactions as they had 'north of the

[32] Torrens (pers. comm.) suggests this trial may have contributed to the Oxonians' change of mind concerning the reliability of Smith's methods for at some point Smith appears to have visited the trial (which closed around 1815). It is, however, also apparent that Wernerians, Kidd and Greenough, also seemed to doubt the presence of coal.

[33] UCL: Kidd to Greenough 8 May 1812, Add. 7918, 1067.

[34] UCL: Kidd to Greenough, 25 May 1813, Add. 7918, 1068.

[35] GSL: Greenough lectures on strata, *c*. 1813, LDGSL 955.

[36] UCL: Kidd to Greenough, 22 June 1814, Add. 7918, 1069.

border'. Onetime Society supporter, John Farey, was not alone in believing the Scots had fallen foul of this theorizing: 'Edinburghians, who hitherto have so readily and warmly entered into disputes on Geological *Theories* [...] the task of defending, each their own set of whimsical Dogmas, against the facts of Nature, and the published Observations of several Writers' ([Farey] 1817). Farey thought the whole thing laughable. Evidently, the leadership of the Society thought so too for their response was to adopt a strongly anti-theoretical inductivism, which they maintained throughout the first two decades of the Society's existence. It was essentially an argument against *a priori* reasoning, a position to which Werner himself subscribed. But like Werner, the Society men could do little to remove the theoretically calibrated spectacles through which they made sense of their data. While they were suspicious of the publications of Edinburgh Wernerian, Robert Jameson (1774–1854), including his widely consulted three-volume *A System of Mineralogy*, published in successive editions between 1804 and 1816, there is no doubt that the Society's geology was also, for the first decade, naturally situated in a Wernerian world.

Werner espoused a view of the Earth that saw rocks as precipitates. Rudwick (2005, p. 175) tells us that Werner did not claim this idea as original, his most celebrated essay being 'carefully entitled "A Brief Classification and Description...", *not* "Theory of the Earth"' (see also Albrecht & Ladwig 2002). Werner's descriptions of strata were widely influential, and shaped thinking in Germany, Russia and elsewhere (Guntau 2009; Khain & Malakhova 2009), and certainly had a great impact on the British before 1820. Werner's *New Theory of Mineral Veins* (1791) had become more accessible with its translation into French in 1802 and English seven years later. His mineralogical *Treatise on the External Characters of Fossils*, which did not discuss 'fossils' in the modern sense, had first appeared in English in 1805. Still alive in the first decade of the Society's existence, Werner offered a thoroughly worked and aesthetically appealing system of universal application. Rudwick, again:

> Werner thought that the two dozen 'species' [of rock] described in his *Brief Classification* probably included all those reported so far, or likely to be reported in the future, from anywhere in the world. In that sense they were 'universal', and he expected that his classification – with due refinement – would prove to be valid everywhere. This was a conclusion no more unrealistic, or conceited, than the confidence of mineralogists that they had described the broad outlines of mineral diversity on a global scale. *Gebirge* [mountains or the rock mass of which mountain's are made] of granite and gneiss, coal and gypsum, were 'universal' in the same sense as minerals such as quartz and feldspar, hornblende and mica.
>
> (Rudwick 2005, pp. 94–95)

This refined descriptive and classificatory work was least contested by the Society men as it seemed to have validity beyond the 'Neptunist' theory upon which it rested. Yet, as François Ellenberger (1999, p. 278) noted, when reviewing a contemporary statement of the Wernerian position, even its most theoretical aspects seemed unproblematic: 'This theoretical genetic system, summarized in this manner in 1802, strikes us today by its perfect logic'. Werner was, however, a topic for theological debate rather than for religious adoption in the Geological Society. He affected the Society's outlook and made theory a natural aspiration. Farey was amongst those supporters who understood this, but also understood that this needed to emerge through scientific rigour and objectivity, as he told his correspondent, Greenough, in August 1810:

> I am anxious Sir, to see the scientific caution now so happily established in Chemistry, of always repeating & verifying new Experiments, by different persons, before they are admitted as data, on which to ground any reasoning, or extension in the Theory, introduced & rigidly enforced in Geology.[37]

Farey valued the ways of chemists so fully (Farey 1811) that he, cheekily perhaps, expected Greenough to verify his own work in Derbyshire.

The empiricism to be found in the fieldwork of the Society's correspondents tended to result in discrete pieces of data. The facts collected were rather bald and disconnected. Correlation proved difficult or impossible. In this respect they used a kind of naïve empiricism different from that applied by Smith. Smith built his knowledge out from an initial base, extending and testing it as he moved further afield. Like the Society geologists he, too, was cautious but Smith did not rely upon a network and therefore gathered his facts from an observed geological continuum. Most critically, he developed his knowledge in an internalized mental database without the need to deploy language or illustration to communicate meaning. He observed, absorbed, correlated and verbally communicated his conclusions in the field and in the museum or collection where real things underlined the truth of his words. And, as an ultimate proof, he could use his knowledge to predict and thus prove. The Society men were, by comparison, considerably

[37] UCL: Farey to Greenough, 16 August 1810, Add. 7918, 549.

disadvantaged and not only by theoretical spectacles and attempts at naïve induction; they also had to deploy language in a medium where terms were ambiguous and uncontrolled. In 1812 Farey and Greenough were in agreement that a basic terminology (stratum, bed, seam, etc.) was required but that it was not yet present because of Wernerian influences from Scotland and the local terminologies of 'Quarry-men, Colliers, Miners, Wellsinkers & c.'[38] William Phillips, in 1816, wrote, 'The Geognosy of Jameson is altogether a scientific work, not well adapted to the learner; in as much as a preponderating anxiety for the support of a favourite theory, has caused the introduction of many terms not hitherto adopted by English mineralogists' (Phillips 1816, p. vi).

Data arrived at the Society's headquarters in disjointed form, sent from individuals uneven in their scientific capabilities and perhaps attached to different nomenclatures. Although this kind of fact-collecting network was used repeatedly in the early nineteenth century, such as in the elucidation of the geology of Yorkshire in the 1820s, in the 'Great Devonian Controversy' in the 1830s and in the geological survey of Wales in the 1840s, each of which drew upon John Phillips to achieve the fossil correlations, all in the end required Phillips to enter the field himself (Knell 2000). In doing so, he would deploy the mode of investigation that Smith had used that assured that facts were seen in context and, as well as being collected physically, were also accessioned into a mental database (Knell 2000).

Greenough may not have theorized these implicit weaknesses in the fact-collecting network but he would certainly have experienced them. He could only counter them by being, like Smith, an inveterate traveller. Indeed, his wealth and mobility enabled him to see far more of British, Irish and Continental geology than Smith ever would, although never, of course, in the same depth as a man who lives in the field. Thus, while Conybeare would later refer to Greenough as a mere compiler, this he was not. Greenough studied the rocks first hand. As Fitton told him:

> Your account of your travels (if that word be not too humble) is quite intriguing. By this time you must certainly be so far acquainted with the general structure of England & Ireland that you can tell to any persons disposed to minute investigation the place where most is to be learned. A kind of information excellently adapted to a president of your Society – past & future.[39]

Like Smith, Greenough also possessed a mental image of British geology as only by its possession could he make sense of the data of his correspondents. However, Greenough's database was not constructed on Smithian principles nor did it develop from a secure local foundation. Fitton might have considered Greenough's tourism a little superficial in this regard, for he pushed him to find his own locality or topic and publish in detail and by these means make a major contribution 'which would do much for Theory'.

Fitton knew, however, that there was also a manly aestheticism to the science that brought another kind of reward that working one's own patch might not achieve: 'Geology has this great advantage, of which not even Botany partakes more largely, that it leads continually to healthful and active exertion, amidst the grandest and most animating scenery of Nature' ([Fitton] 1817 p. 74) (Fig. 2). It was a science that called for tourism. This aesthetic aspect permeated the experience of geology: its objects and landscape effused emotions of time and distance, which were remodelled in minds trained in poetics and the picturesque (Heringman 2009; O'Connor 2009). Fitton's hopes that the Society's project would lead to 'a rational theory of the earth' were similarly permeated with a manly philosophical aesthetic ([Fitton] 1817 p. 73). This was simply a different way of seeing and we should not presume that Greenough acquired an inferior sense of the nation's geology as a result.

Making the Society's map

Greenough and his co-workers recognized early on that 'mineralogical maps' were the most perfect means to capture the nation's geology empirically and without the contamination of theory. The identification and location of the 'sienite' at Mountsorrel in Leicestershire, for example, were by these means kept separate from the debate over what these facts actually meant. Sir James Hall believed these rocks represented the peaks of a unified mass projecting through the red clay; Farey saw them as discrete masses floating in the clay.[40] The treatment of these still-contentious igneous (as we know them) masses, however, posed challenges for representation on a map: 'Some geologists are of opinion that basaltic rocks are homogeneous with Neptunian beds with which they are associated; others consider them wholly distinct and parasitical. Not wishing to make myself a party to either opinion, I have

[38] UCL: Farey to Greenough 7 January 1812 Add. 7918, 554.

[39] UCL: Fitton to Greenough, 28 November 1813, Add. 7918, 598.

[40] UCL: Farey to Greenough, 8 March 1808, Add. 7918, 548.

Fig. 2. Thomas Webster, 'Western Lines, Isle of Wight', sepia watercolour, *c.* 6 March 1812 (LDGSL 400, 20), later engraved for Webster (1814).

kept the basaltic substances by themselves, but so distinguished, that each may be referred, without difficulty, to the group in which it occurs' (Greenough 1820, p. 3). The production of the map, then, became central to the Society's mission to record before it sought to theoretically understand.

As President and with *Geological Inquiries* acting as a catalyst, Greenough soon found himself under an avalanche of mail. From this he needed to extract data for his map. He had various paper systems for doing this, sometimes apparently cutting up correspondence, at other times transcribing it and drawing a vertical line through those parts so copied to prevent him accidentally repeating the exercise. As there are examples of letters retained out of historical interest, it seems possible that he also destroyed a good deal that he had 'used'. While some of this information came in extensive measured sections and drawings, the bulk arrived in narrative prose and often in a difficult hand.

This was still, for most, a period of raw discovery. Much was beyond reliable articulation. Visual methods – museum, map and section – were by far the easiest means of communicating the current state of knowledge but few were versed in their construction. Take, for example, Farey's 'Memorandums at Hunstanton Cliff, 24 June 1804', which he later communicated to Greenough. Although even then a disciple of Smith, it seems this knowledge helped him not at all. Those who know this famous section of the Norfolk coast will recognize the elements in Farey's description, but perhaps be overwhelmed by the seemingly disorganized complexity of his detail. How could this description be of any use to Greenough, unless Greenough had other sources, expert knowledge or saw the section first hand?

> beginning S at the Road which comes down to the beach. – first, some alluvial Clay with pebbles in it; in about a furlong proceeding N along the Beach the Geodetic Sandstone comes up above the Beach, (the upper beds having retired some poles[41] in Land) & is covered for $1\frac{1}{2}$ furlongs with the alluvial Clay, 8 or 10 feet thick; when about 15 feet of it is up, it stands in columns, & has Arches & Caves formed thro it, by the waves; alluvial clay still covering it: the Caves are formed by the washing away of the nodules & the sand-stone above, which begins at 2 feet & increases

[41] A pole is a measure of length, identical to a perch, and equivalent to 5.03 m.

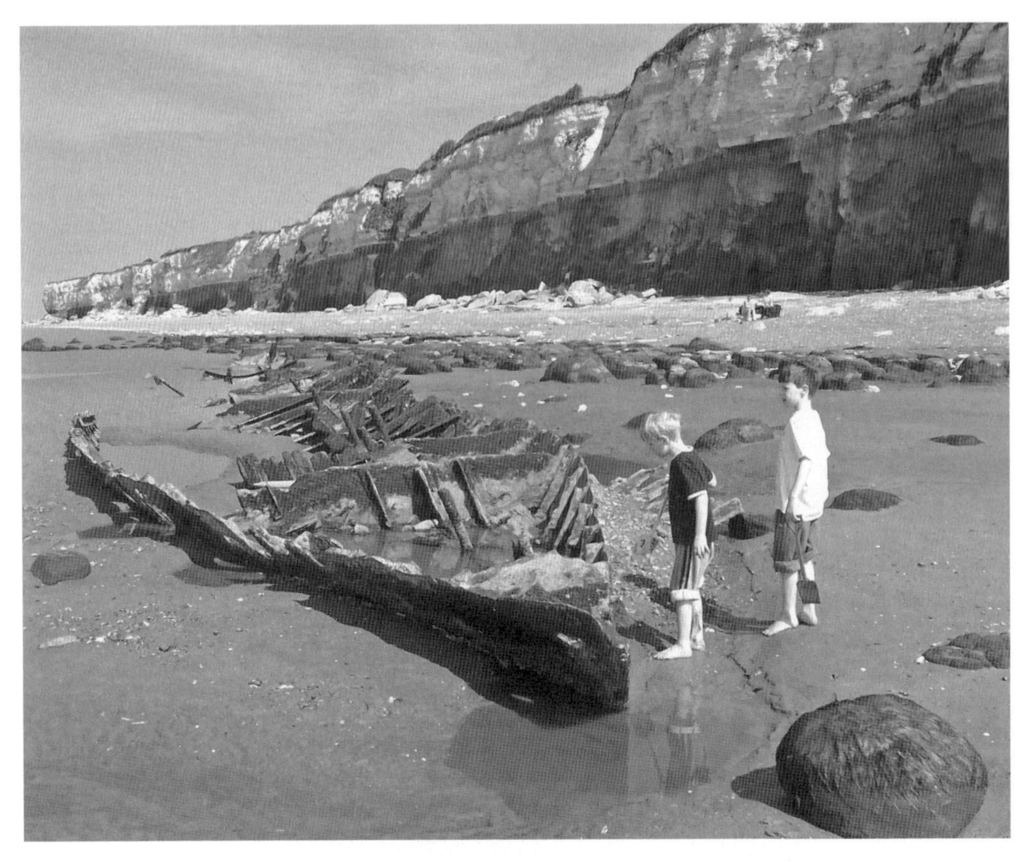

Fig. 3. The striking cliffs at Old Hunstanton reveal with clarity the interior of the landscape. The coast provided critical information for the construction of the geological maps of England (author's photograph).

to 25 feet thick, cracking down in columns. This stratum has many pea-stones in it; the upper 10 feet is darker, & has the blue or purple joints, then in 3 furlongs, comes on a yellowish v. soft sandstone (which all washes away before the waves, & falls down) the last increases, till the whole Cliff is 4 feet. – Then at $3\frac{1}{2}$ furlongs, the Red Rock comes on, covered for 50 yds with a red rubble, the chalk rubble begins to succeed it; and at $\frac{3}{4}$ furlongs, the hard bed of white stone (containing fibrous shells, broken) begins, where the Geodetic nodules goes down; the thick cake is decomposed, & has laid the harder stone (full of pea-stones) bare, like ladus helmontia. – Above the Red is an 18 inch hard stone, containing branches of softer matter than the stone, like Coraloids, then is the best view of the hard stone, containing fibrous shells, then one $2\frac{1}{2}$ feet thick, then another $2\frac{1}{4}$ less perfect, then a 2 feet bed nearly of the same kind, & then a Yard more, disposed to rubble: above this are several beds of perfect stone, & most probably the Tottenhoe-stone; very little if any chalk appears; – between the red and the hard bed above it, are the very deep red lumps, & some kind of fossils, & very large branching Coraloids, branching in all directions, the Tottenhoe-stone, lays about 20 feet above the Red, the upper sand is a little nodulous, but none of the pea-stones appears in it; the red strata are about 5 feet thick all together; near a foot of the bottom is clay & sand, without any fossils, but a few pea-stones; about 2 feet high many [. . .].[42]

To modern, informed, eyes, the striking cliffs at Hunstanton do not pose any descriptive difficulty (Fig. 3). Farey is, however, attempting an objective interpretation in which is woven the detail necessary to corroborate the identification of each stratum. There is here a sophisticated awareness of geological method based on a belief in regular succession demonstrated to him by Smith but in which many in the Society did not believe. In possessing such beliefs, Farey was ahead of the game, and by 1806 – a year before the Society's birth – he was beginning to draw extraordinarily fine geological

[42] UCL: Farey, Memorandums at Hunstanton Cliff, 24 June 1804, Add. 7918, 547.

Fig. 4. Strata seen through Smithian eyes: coloured geological sections drawn by John Farey revealing how his early adoption of Smith's belief in a natural and repeated geological succession enabled him to achieve correlation considerably more rapidly that his Geological Society peers. Note he, like Smith, used numbers to distinguish strata before such strata were named. (**a**) 'Section from Brampton & Chesterfield to Bolsover & Langwith'.

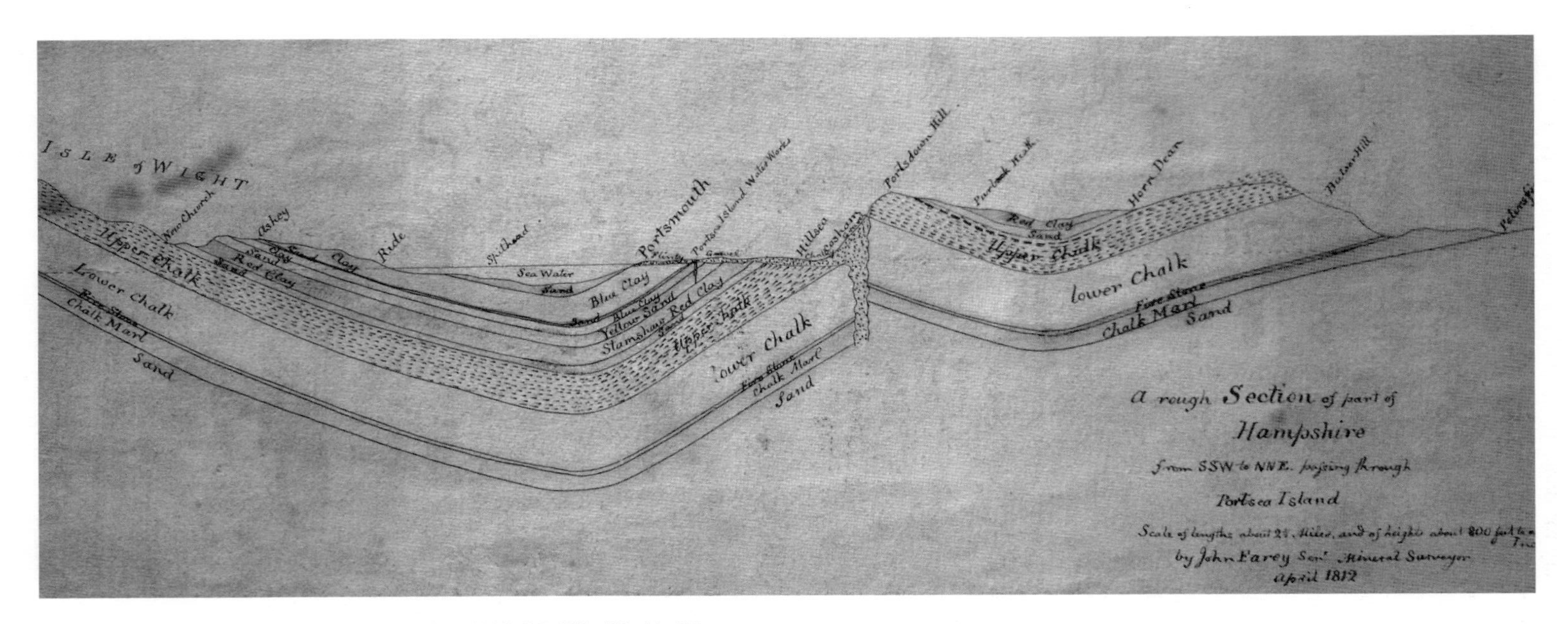

Fig. 4. (**b**) 'A rough section of part of Hampshire', 1812 (LDGSL 450, 43, 39).

sections across the country for Banks (Ford 1967). These named the rocks he knew, but he also gave working names to other strata, numbering similar strata in order to distinguish them (Fig. 4).

Such cross-country correlation, however, was slow to appear in the research of other core supporters of the Society. Conybeare, for example, sent Greenough 12 dense pages of text describing a geological expedition on foot to the western extremity of South Wales, which he had taken with Kidd during the summer of 1811.[43] It was an exceptional contribution – as Greenough must have told Conybeare – but Conybeare admitted that he made no attempt at an understanding of the relationships between the isolated facts he had recorded. Kidd, who published the results of this trip in 1814, could locate no order in the rocks and saw no fossils; his visionary apparatus was entirely mineralogical (Kidd 1814). It seems that the response from those in London was to wonder why Conybeare had not adopted a Wernerian classification. Conybeare explained:

> I must object to the manner in which the Wernerian classification of rocks is applied to the geological appearances of countries foreign to that in which it was originally formed [...] we cannot be supposed from the limited induction which has as yet been made of geological phenomena to be in possession of any thing like a complete catalogue of rock formations [...] I am therefore unwilling to apply too hastily names by which [Werner] designates individual beds such as old red sandstone [...] to any of our strata & wish to proceed cautiously & institute a more complete comparison of the geological observations which having been made within the limits of our own island may most easily be verified by ourselves before we proceed so far as to frame any thing which shall pretend to be a permanent nomenclature.[44]

Greenough bracketed these comments in pencil probably for reading at the next Society meeting.

In his long follow-up letter, Conybeare demonstrated the value of his comparative thinking. However, it is also apparent that Jameson, as Conybeare's theoretical nemesis, rather than Smith, is shaping his fieldwork. At this time, Conybeare seems not to be aware of Smith's methods, and sees fossils as useful only as something akin to those other mineralogical components (colour, lithology and so on) that indicate specific rocks. Thus, if he writes, 'I think most certainly lias – it contains Ammonites, Pectines, & Gryphitae the latter in greatest abundance',[45] he is merely observing a characteristic of those particular rocks long recognized by fossil collectors even before Smith; he is not practising Smithian geology. This was typical; if members were not engaging with the field occurrence of cabinet minerals, then they were describing rocks lithologically. Their descriptions of strata considered the mineral components and disposition of the rocks, and paid no attention to fossils (Fig. 5). Nevertheless, we should not fool ourselves into believing these investigations naïve or unsophisticated. The plates to be found in the first five volumes of *Transactions*, published in the decade beginning 1811, are sufficient to demonstrate an analytical richness and precision that suggested a science rapidly advancing without Smith.

Nor should we expect fossils to dominate descriptions by Smith's supporters. Smith's method concerned three determinants: lithology, position and, lastly, fossils, and relied on a belief in a natural succession (Knell 2000, pp. 20–23). Fossils were not always needed to resolve matters, nor was it necessary to scientifically describe them; one simply needed to recognize and value them. Once something of the natural order of rocks had been learned, it was possible to follow strata in the lay of the land, or in field relationships. Greenough could follow these strata in the landscape just as well as Smith and his supporters. Farey, this time firmly Smithian, told Greenough in the same year:

> This section will give some faint idea of the thickness & importance of the several Red Marle strata in Britain, for these are not the same strata which regularly [emerge] from under the Lias Clay (producing Gypsum, Salt, & c) & over-lay Coals near Bath; or the same as occur under the Basaltic & Lime Coal Series of Newcastle on Tyne, & produce Gypsum, red sandstone & c; and the task will prove a very arduous one I fear, to distinguish these three or more parts of the Red Marle from each other, in many instances, owing to their anomalous beds, & want of Reliquia [fossils].[46]

He accompanied this statement with a measured section listing 58 strata ranging in thickness from 800 yards to four inches, copied by Farey's son from data possessed by David Mushett, coal mine proprietor at Coleford, in the Forest of Dean. Again, it is typical of the relationship between practical mining and geological knowledge; Farey was here interested only in geological intelligence, not in improving mining.

[43] UCL: Conybeare to Greenough, 15 November 1811, Add. 7918, 435.
[44] UCL: Conybeare to Greenough, 4 December 1811, Add. 7918, 436.
[45] UCL: Conybeare to Greenough, 4 December 1811, Add. 7918, 436.
[46] UCL: Farey to Greenough, 5 May 1811, Add. 7918, 553. See also Torrens (2001, p. 68) for Smith's later comments on this problem.

Fig. 5. Strata seen through Society eyes: (**a**) John Josiah Conybeare's (1779–1824) watercolour, 'Curvature in Killas on the Coast immediately below St. Agnes' (1813) (top); (**b**) MacCulloch's watercolour, 'a part of the rock upon which Stirling Castle is built' (1812) with its strata of 'two substances' juxtaposed with Trap (bottom).

Fig. 5. (**c**) MacCulloch's pencil sketch, 'Curvature of the coal strata near Sandyfoot' (1809). The Society men saw structure, mineralogy and lithology, and aspired to grand explanation (LDGSL 400, 84, 37 and 18, respectively).

The Society's map-making was made more effective by the distribution of fragments of the map to be coloured locally. In 1813, for example, Fitton asked for the 'Skeleton map engraved for the Geol. Soc [. . .] such a portion of one as includes Northamptonshire & some of the adjoining country, on which I may, hereafter, insert what I shall observe'.[47] In his two-way exchanges, Greenough also sent his own colourings of the map – actually maps – out for corroboration. Conybeare, for example, in response to such a request stated on viewing Greenough's initial attempt:

> In reviewing your maps I find that I differ much from you as to Somerset & Devon. I have enclosed sketches of those counties made out after my own fashion – on one side you will find part of Gloucester & the Nth of Somerset on the other the South of Somerset & East of Devon. My colours being packed up I have endeavoured to make them out as clearly as I could by different sorts of lines.[48]

In another letter referring to these efforts, he writes:

> I am engaged at present in colouring with Buckland's assistance the contiguous squares of parts of Gloucester Somerset with Dorset & East Devon = from the Outline Map = in proceeding with this work I find that I was myself mistaken in attributing error to Martin's account of the limestone in the South of the Nailsen Coal Field by Blackwell Barrow & Broadfield downs – which really are mountain lime – & not lias as I have been induced to suppose from their appearing to form a continuation of the ridge on wh. Dundry beacon stands. When my map is finished it will be much at yr. service – that district seemed the most imperfect part of yours.[49]

By means of a friendly and open network of this kind, centred on the Society's enterprising President, the map progressed as a representation of the Society's social aspirations. But it did so for only a short time. Smith was beginning to enter the Society's thinking. He had first appeared implicitly written into Farey's communications, some of which came through Banks. Now Smith's influence became more pervasive and in doing so opened up that controversy from which the Society has never fully escaped.

ACT 3. The Smith controversy

Smith's virus

In the early years of the Society, Smith was situated in a blind spot in the visual apparatus of most of the membership. He was known to them, but

[47] UCL: Fitton to Greenough, 19 February [1813], Add. 7918, 596.
[48] UCL: Conybeare to Greenough, [undated], Add. 7918, 439.
[49] UCL: Conybeare to Greenough, 18 February 1813, Add. 7918, 455.

nevertheless occupied another place in geology – indeed, a different kind of geology – which the Society ultimately recognized as true and important. Before 1813, Smith was considered to possess no novelties. In March 1808, for example, Farey mentions Smith to Greenough in passing, knowing Greenough had been part of a small delegation from the Society that had visited Smith on that day (Torrens 1998*b*, p. 111). Farey thought Greenough was assembling a complete collection of British fossils, and thus assumed that he knew the nature of Smith's discoveries and how fundamentally they had shaped the organisation of geological collections.[50] On another occasion, in 1811, Thomas Meade, a correspondent of Smith and another of Greenough's informants, and no less enthusiastic in his support for Greenough than Farey or Conybeare, sought to extract intelligence from Benjamin Richardson's collections, which had been curated on Smithian principles. He wrote:

> I called on Mr Richardson in pursuance of my intention to examine with him & to make a Catalogue of his Fossils of the respective Strata. But he assured me that he had promised Mr Smith not to join or take a part in such a Catalogue; but he will allow me access to his Drawers; which will in some degree answer any purposes altho' it may increase the trouble, for he might assist me much.[51]

The closing phrase of this quotation suggests that Meade wished to interrogate these collections unobserved. This might be seen as spying whether intended by Meade himself or suggested by Greenough, or motivated by Meade's somewhat futile attempts to ingratiate himself with Greenough. It is possible that Greenough saw it as morally legitimate, that – as the President of a national society engaged in the free movement of data – he had a right to data. Perhaps he believed Smith's restrictions irrational or felt Smith's data unlikely to enter the public domain; he certainly doubted that Smith possessed any views so novel as to warrant protectionism.

Smith's newly acquired protectiveness may have come about as a result of his earlier contact with the Society or been pricked by his early relationship with John Farey. Farey had, from first meeting Smith in 1801, become his disciple and, at times, his publicist. At the time of their meeting, Farey was a land steward at Woburn (Torrens 1994) and no threat to Smith. But in 1807 he was publishing Smith's discoveries in Rees's *New Cyclopaedia* and had accepted a commission from the Board of Agriculture to undertake a Smithian survey of Derbyshire. Perhaps unthinkingly, he asked Smith for the lines of strata he had recorded in that county (Ford & Torrens 1989). Smith now saw Farey as a 'scientific pilferer' (Cox 1942, p. 40) and as a result also realized the consequences of the open communication of his ideas. Farey would publish his first report – on the strata of Derbyshire – in 1811. By, then, he had effectively become a Society man, although not a member.

Smith was not a self-publicist. His omission from much of the surviving correspondence of Greenough is not simply a result of censorship in public discourse. There are continuous series of private letters from individuals who one would have expected to have referred to Smith. Yet, Smith is hardly there. Even Farey's letters make little reference to him. Given the long and copious communications received from others, it is quite easy to understand why Smith should have received such scant attention, particularly if he was not believed. From the outset, the Society's most active members became incredibly busy with its inquiries, with travelling, with meetings of various kinds, with reading, compilation and so on. This was in addition to busy social and business lives. Theirs was an open and enquiring field of exchange amongst social and Society networks – amongst friends and strangers. It was a visible, particular and part-time network affected by all manner of other cultural influences (Heringman 2009).

Meanwhile, Smith's geology was spreading, virus-like, as a whisper and as reportage. All those to whom he spoke became infected, and he knew it.[52] Yet, Smith's ideas make no real appearance in the correspondence of the Society's key 'inquirers' before the early 1810s. The change comes, however, not long after the publication of Townsend's book in 1813. Here, it was not the adulation Townsend heaped on Smith that first caught the attention of those who read it, but rather the correctness of many of Townsend's facts where he had used Smith's methods. The effect was doubtless helped by a paper on Derbyshire strata by Farey presented to the Society in that year; his report having already been in print for two years. Conybeare, for example, is suddenly transformed in 1813. It came at a time when a visit to Oxford by Greenough had done much to excite the 'Members of the Rag Formation'. As Conybeare told him:

> Since your visit the Geological ardour produced by it has blazed forth [?] with such vehemence. Scarcely a day has passed in which I have not undertaken either

[50] UCL: Farey to Greenough, 8 March 1808, Add. 7918, 548. See also Smith (1818) in Sheppard (1917, p. 216).
[51] UCL: Meade to Greenough, 7 November 1811, Add. 7918, 1211.
[52] See Smith's 1818 account in Sheppard (1917, pp. 214–220).

> on foot or horseback some expedition to explore the mysteries of hidden junctions & compel reluctant mountains to reveal their whole History. Shotover was the first to yield to my prowess [...] the key to all the neighbouring country.[53]

But Shotover's yielding owed little to Greenough's science and everything to Smith, whose name goes unmentioned: a 'rubbly limestone [...] beautifully characterized by all its distinguishing fossils. These are also contained in the concretions of the Sand beneath which may therefore be considered as the lowest member of this formation'. Drawing upon Townsend (and perhaps intelligence from the Bagley Wood coal trial (Torrens pers. comm.)), in this single letter Conybeare makes frequent mention of fossil characterization and in doing so reveals what was for him an entirely new way of thinking.

Conybeare's work had a direct impact on Fitton who, since arriving in Northamptonshire, was sure that the county could only be understood as a continuation of strata seen elsewhere. From 1814, he had joined the Oxford club in the field where the possibilities of this form of geologizing were plainly obvious: 'The beautiful regularity of the long continued line, as it were, of coast, which is seen from Shotover hill, at the termination of the chalk strata to the east of Oxford, has no doubt powerfully assisted the zealous geologists of that University, in making converts to their favourite pursuites' ([Fitton] 1817, p. 73). But it seems likely that Fitton had also been influenced by Parkinson or by Farey, as he told Greenough in 1812, before the publication of Townsend's book and without need of explanation, 'Of petrefactions (or rather fossil organized remains) there are great abundance & variety in this country – Some characteristic of the strata, & others interesting in themselves'.[54]

Townsend's advocacy of Smith had been preempted by founder member, James Parkinson, who in the first volume of *Transactions*, in 1811, wrote of Smith: 'he was the first who noticed, that certain fossils are peculiar to, and are only found in, particular strata; and first ascertained the constancy in the order of superposition, and the continuity of strata of our island'. In doing so, he signalled that Cuvier and Brongniart were, by comparison, latecomers (Parkinson 1811, p. 325; on Parkinson, see Lewis 2009*a*, *b*). Parkinson's assertion of the importance of fossils correlates well with a decision on the part of the Society to establish a committee to investigate them. Greenough, in his reflections late in life, would forget that Smith's ideas were involved, preferring to see the committee as asserting the Society's independent interest in fossils. Webster's fieldwork on the Isle of Wight was also furthering the cause of fossil-based geology, and Webster thought it influential within the Society (Webster 1814).

What is certain is that Smith's ideas entered the Society's walls at this time, not with a new membership of so-called 'young turks' (Morrell 2005, p. 83) but by the conversion of long-time supporters, including two founder members. Full recognition of Smith's achievement came with the publication of his map in 1815: the Society was utterly changed by it. All who saw it were amazed by the achievement (Fig. 6a).

William Phillips, the Society's publisher – and publisher of numerous individual works by members – had himself compiled a number of books on geology. In 1816, in his general *Outline of Mineralogy and Geology*, he wrote, without reference to Smith or Cuvier:

> Every part of the globe distinctly bears the impress of these great and terrible events. The appearances of change and ruin are stamped on every feature. Change and ruin *by which not a particle of the creation has been lost*, but which have been *repeated*, and are *distinctly marked by the genera and species of the organic remains they enclose*.
>
> Thus, those fossils and petrifactions which heretofore were carefully collected as curiosities, now possess a value greater than as *mere* curiosities. They are to the globe what coins are to the history of its inhabitants; they denote the period of revolution; they ascertain at least comparative dates.
>
> (Phillips 1816, p. 189)

To this, the second edition of this work, he appended with the author's permission a miniature version of Smith's map (Fig. 6b). Its lines, however, were made to both simplify and extend those on Smith's map using information published in the Society's *Transactions* and elsewhere; this was no mere replica but a reflection of a continuing attempt to move beyond the limitations of Smith's lines. Smith was also improving his map at this time, each version he had coloured seemingly an improvement on the last (Eyles & Eyles 1938) (Fig. 6a). Like Fitton, Phillips was in awe of Smith's achievement. If it was to inspire Fitton to become a historian so it inspired Phillips to produce the first edition of his *A Selection of Facts from the Best Authorities Arranged so as to Form an Outline of the Geology of England and Wales* (1818) rather than a third edition of his more general mineralogical work. And while Phillips'

---

[53] UCL: Conybeare to Greenough, [nd], Add. 7918, 446

[54] UCL: Fitton to Greenough, 17 November 1812, Add. 7918, 593.

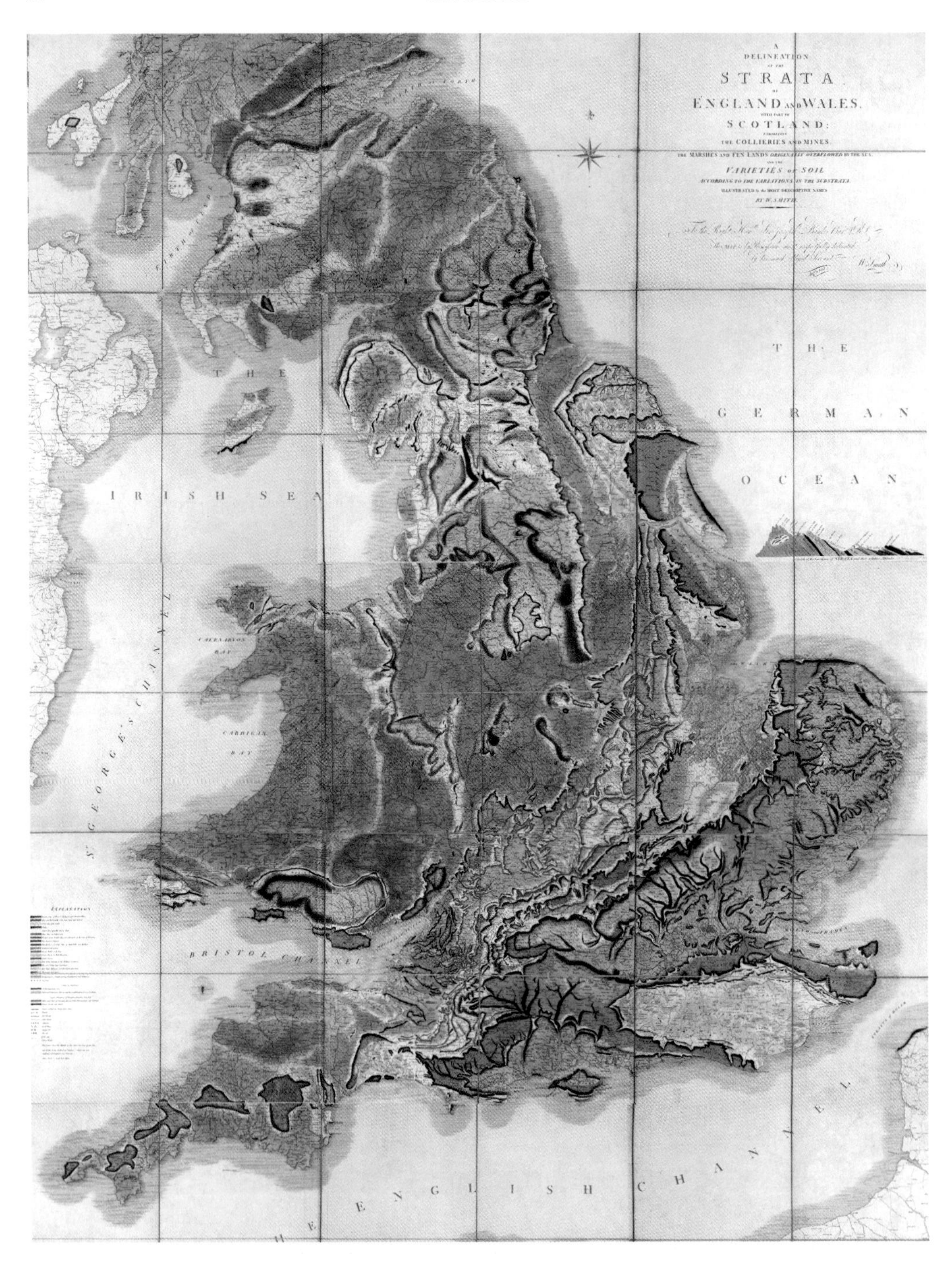

Fig. 6. The evolving geological map of England and Wales. After Smith's pioneering attempt, all subsequent versions become genetically blurred. (**a**) William Smith's (1815) large-scale map. (Courtesy of the National Museums of Wales.) Smith, while in competition with Greenough's forthcoming map, repeatedly enhanced his own map.

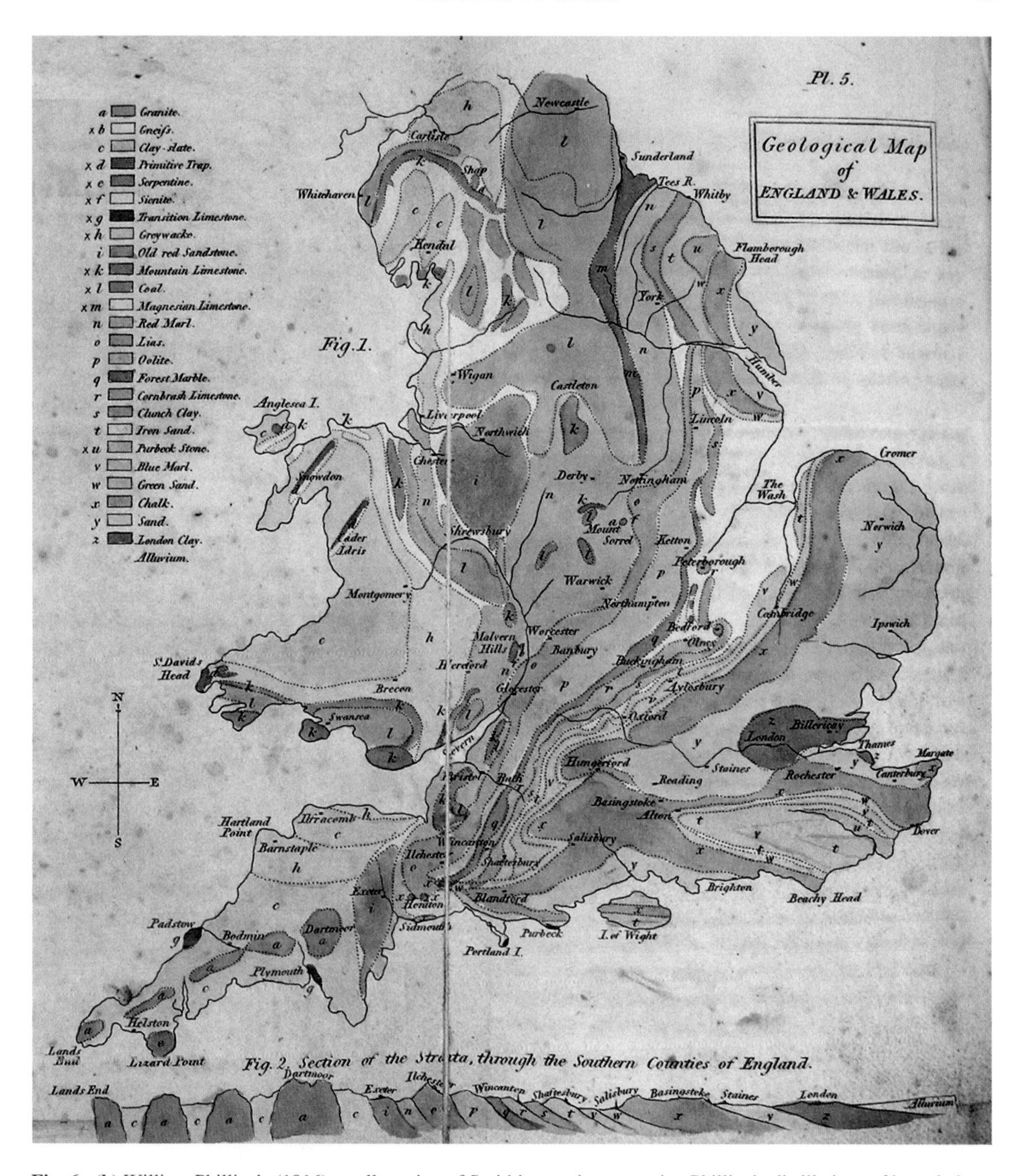

Fig. 6. (**b**) William Phillips's (1816) small version of Smith's map incorporating Phillips's distillations of knowledge taken in part from the Society's publications.

Outline of Mineralogy and Geology had been based upon the ideas of Jameson, Cuvier, Aikin and papers in the Society's *Transactions*, his later *Selection of Facts*, or *Outline of the Geology of England and Wales*, as it would become known, followed the non-theoretical philosophy of Smith who Phillips claimed as an original observer. In this volume, too, Phillips added a version of Smith's map (Fig. 6c), this time extended still further, and in his description of the strata he took care to list their fossil contents. Conybeare responded to this book by flooding Phillips with new intelligence, and by this means acquired the position of senior 'Editor' for an increasingly necessary revision, published under the shortened title and in larger format as *Outlines of the Geology of England and Wales* in 1822. This book gave extended voice to William Phillips' admiration for Smith, but in it Smith's map (in Phillips's improved form) was replaced by the map to which Conybeare had made such a significant contribution: Greenough's (Figs 6d and 6e).

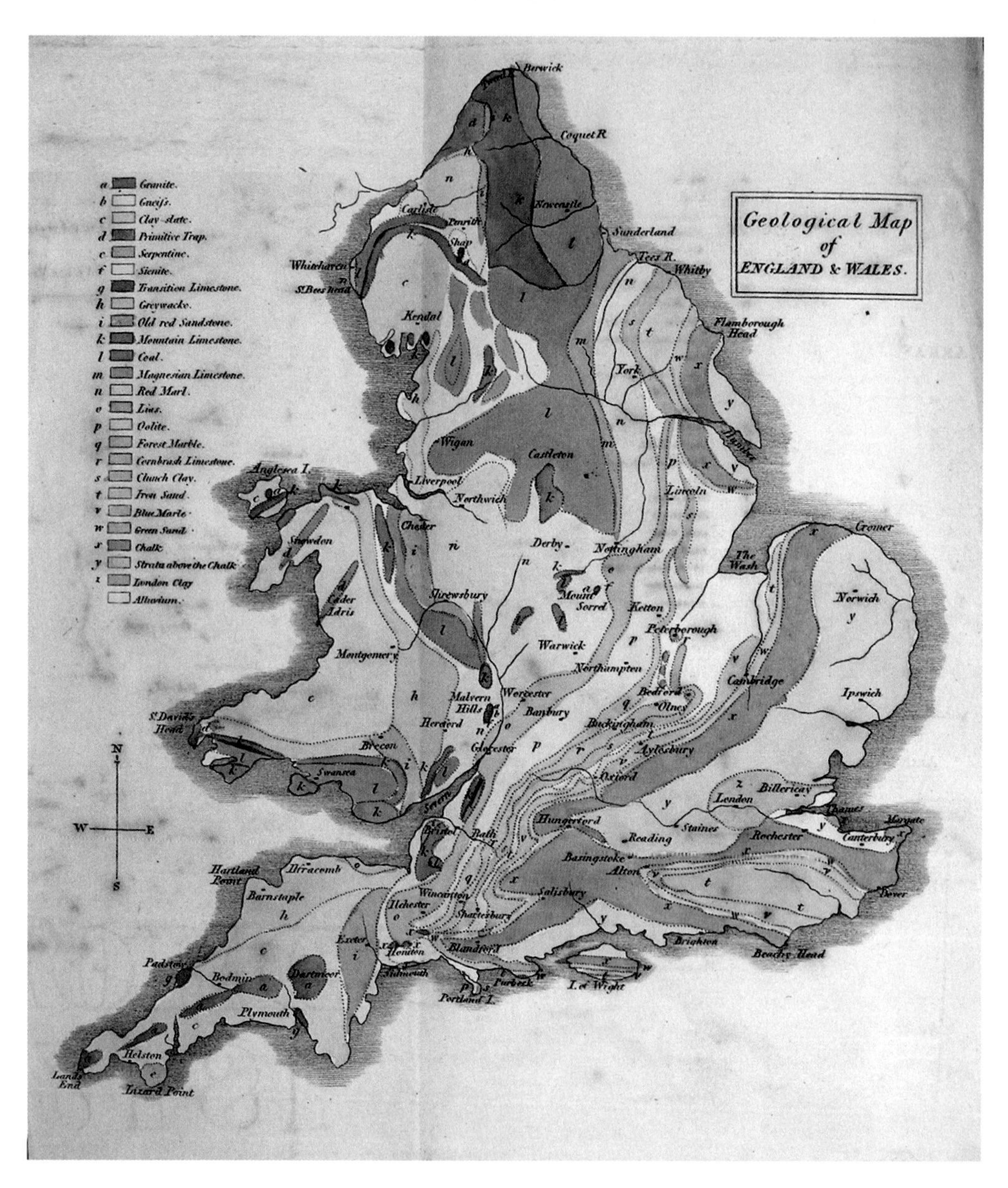

Fig. 6. (**c**) William Phillips's (1818) similarly extended version of Smith's map as used in an entirely different and very Smithian volume.

Even by the beginning of 1814, Greenough's map had been well advanced, at least according to its author. As Fitton comments, 'I have great pleasure in learning that your Geological map of England, which will doubtless be a very valuable acquisition to mineralogists, has made such progress, and I shall be very happy if it were in my own power to contribute to its completion'.[55] Greenough's role was in part that of compiler, just as William Phillips had 'compiled' his own geological books from the works of others without full reference. But Phillips' admitted that his books were for beginners – for popular digestion – not claims to any kind of scientific merit. In the early 1810s, however, the boundary between the plagiarized and original was not clearly drawn, and many understood that the

[55] UCL: Fitton to Greenough, 28 February 1814, Add. 7918, 599.

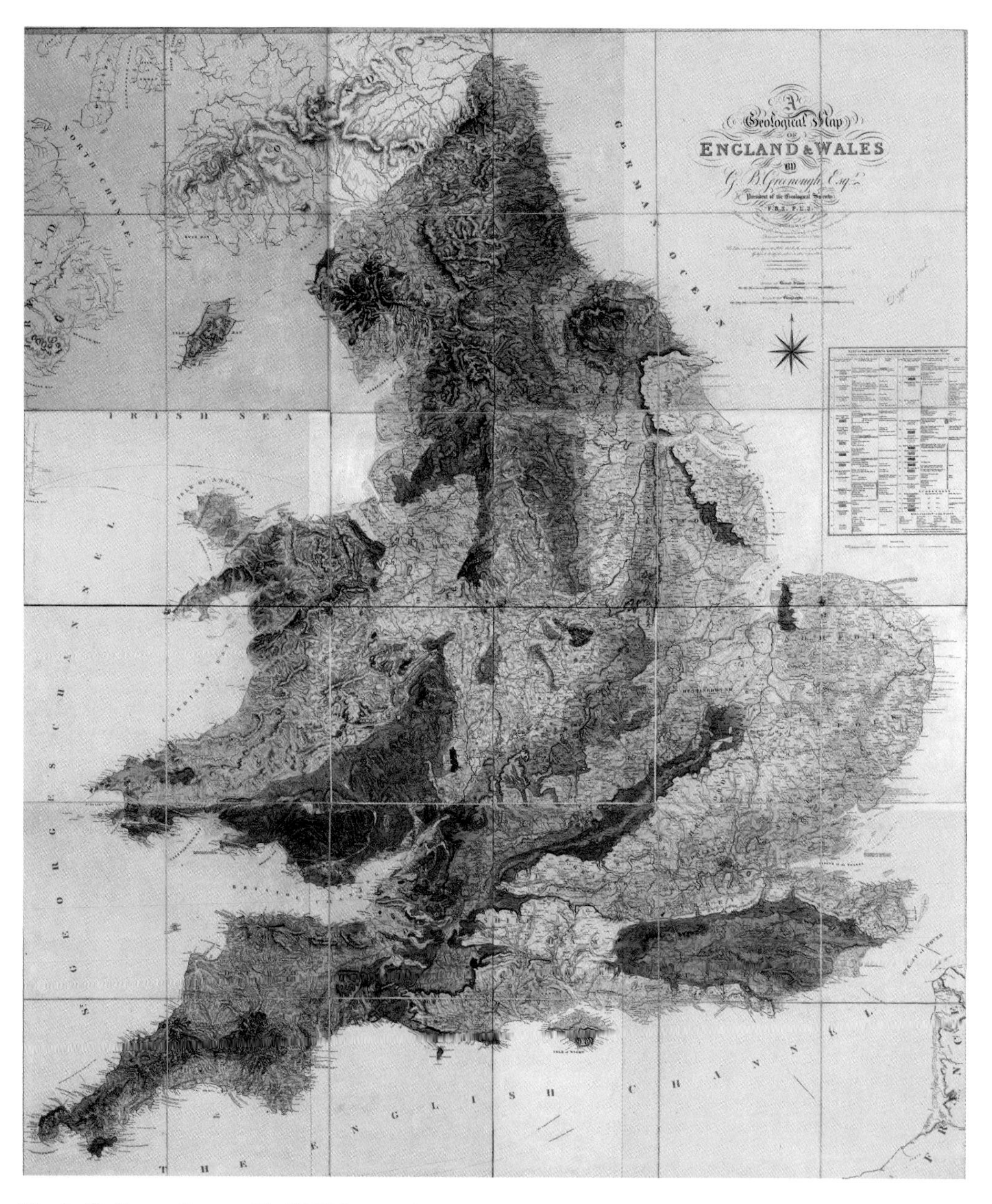

Fig. 6. (**d**) George Greenough's (1819) large-scale map, which aimed to be superior to Smith's. (Courtesy of the National Museum of Wales.)

science would progress in steps involving successive revision and improvement. However, such 'understanding' stretched social relations to the point of snapping and the Society's previously highly effective co-operative spirit evaporated. I shall come to this shortly, but first it is necessary to examine its most extreme manifestation: Farey's war with the Society. Were he and Smith catalysts in the Society's social undoing?

Farey's war

In his relationship with the Society, Farey had always been self-assured and very much his own man. He was in tune with the Society's new project and its search for system and theory. He was no agent or minion of Smith or Greenough. His relationship with the latter blossomed and was going splendidly until he came to submit a paper on Derbyshire

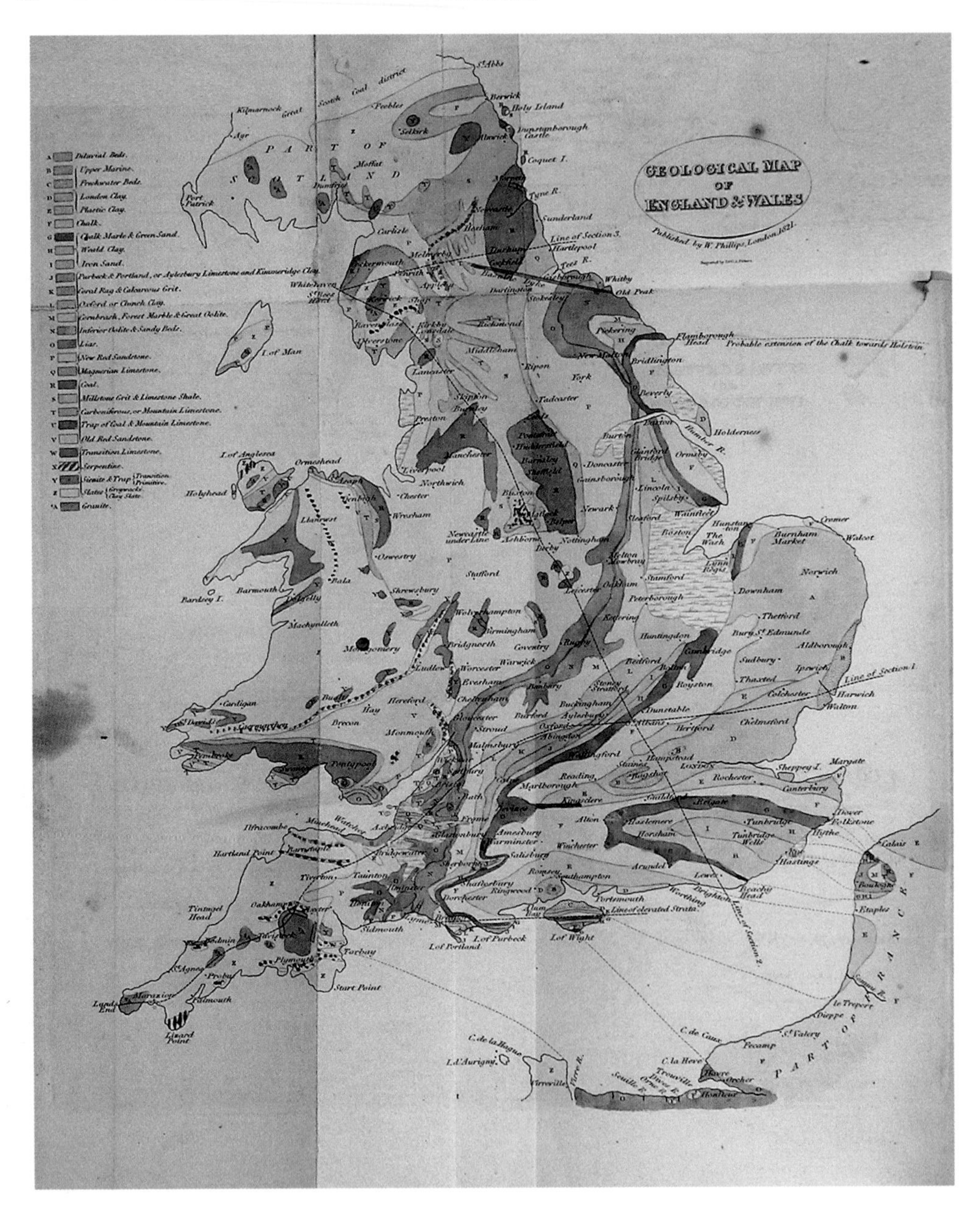

Fig. 6. (**e**) Conybeare & Phillips's (1822) small version of Greenough's map in which some of Greenough's more speculative interpretations have been edited out by Conybeare (Conybeare & Phillips 1822, unpaginated preliminary notice).

to the Society's *Transactions*. Greenough advised Farey on the first draft, suggesting some judicious cutting and editing but this did little to save Farey from the disaster which befell his paper on its submission.[56] Council wished to 'shorten it'; the copy returned containing 'an abridgment of the whole & several transpositions & omissions'.[57] At least one of these abridgements was of material

[56] UCL: Farey to Greenough, 23 January 1813, Add. 7918, 562.

[57] UCL: Farey to Greenough, 8 April 1813. Add. 7918, 567.

Greenough had asked Farey to research; now Farey doubted the messages he was receiving.[58] More serious was the manner in which it was proposed to revise his paper as it completely disenfranchised the author:

> Much as I see reason to disapprove the work adopted by your Geological Council (if I rightly understand it), of referring my paper to be new cast throughout, & reduced as 4 to 1 in length, by strangers to me & my subject, without their being directed, to confer first with me, on the best & least objectionable mode of abridging it; or, what on all accounts was most proper, to have requested you to so confer with me.[59]

He begged to have his paper serialized instead.[60] Despite implicating Greenough in the conspiracy, Farey remained on good terms with him. Greenough loaned Farey books to improve his work and Farey offered some of his own in return. Farey visited Greenough's house and Greenough Farey's, all the while engrossed in discussing the stratification of England. Although never possessing a relationship of the kind Greenough reserved for his social equals, like his Oxonian friends, it is clear that the two men held common interests and mutually benefitted from their relationship. But beyond this relationship there was much ill feeling. Henry Warburton, who was behind the revision, felt its heat most of all. He told Greenough:

> He seems to be angry beyond measure. Perhaps if I were to meet him at your house at breakfast on Thursday morning, something might be done. I do not wish to bring him, as he lives on the fruits of his labour, out of the way to my part of the town, as this may form a new subject of accusation. Otherwise I would beg you to meet him.[61]

Warburton was at a loss: 'I can make nothing of the man; at one time pacified, then flying off in a tangent'.[62] In early June 1813, with Greenough's mediation, Farey started to believe that the dispute might find satisfactory resolution but this turned out to be a false hope.[63] Farey withdrew his paper (although Warburton felt the paper, having been read, was now the property of the Society).[64] Warburton asked Greenough to explain matters to Farey's patron, Banks, who had been behind the Derbyshire commission but on reflection thought Farey angry with Banks too. Farey had, however, been as reasonable – certainly with Greenough – as he could allow, never permitting this unpleasant business to come between them. But in the end he understood that the Society's desire to take absolute control of his paper jeopardized his career as a mineral surveyor and geological writer.

In the middle of this debacle, while still on good terms, Greenough lent Farey a copy of Townsend's book, in which Greenough had made notes in the margin. Farey unthinkingly picked up where Greenough had left off. While complimentary of Townsend's knowledge of the geology of Wiltshire, Somerset and Gloucestershire, Farey found the rest of the book contaminated with errors, poor illustration and ludicrous theories:

> Poor Moses, of whom so ridiculous a parade is made in his title & c, seems after all most unfairly dealt by the primitive waters, under which the Strata were formed, & dislocated, & c, are confounded with the Deluge which Moses describes, as happening long after this, & after Men & all our present races of Beings had long existed on the earth, tho' such confounding is in violation of every sentence that is handed to us as Moses' description of his Deluge, which I maintain to be an object of faith or belief only, and incapable of the least philosophical proof [...] and that this work cannot fail, I think, laying Moses open, to more marked neglect or contempt by practical Geologists & Philosophers, than any which has preceeded it: the facts of the Strata are utterly inconsistent with any material convulsions or violent operations upon them since Man and his contemporary Beings were created, and it is only false theories, this & De Luc's, that pretend so.[65]

Farey was not unusual in openly ridiculing Jean André de Luc's (1727–1817) Christian interpretations of Earth history (Rudwick 2005, pp. 150–158; Torrens 1998*c*). Indeed, Greenough bracketed Farey's most entertaining remarks possibly for reading at a Society meeting; the Society men were not above jovial acts of ridicule. It is more telling that Farey appended a 'p.s.' to his letter: 'If you think it right to communicate the whole or any extracts from this letter, to Mr Townsend, I

[58] UCL: Farey to Greenough, 29 May 1813. Add. 7918, 570.
[59] UCL: Farey to Greenough, 30 May 1813, Add. 7918, 571.
[60] UCL: Farey to Greenough, 8 April 1813. Add. 7918, 567.
[61] UCL: Warburton to Greenough, [c. June 1813], Add. 7918, 1684.
[62] UCL: Warburton to Greenough [June 1813], Add. 7918, 1685.
[63] UCL: Greenough to Farey (author's copy), [31 May 1813], Add. 7918, 572; Farey to Greenough, 2 June 1813, Add. 7918, 573.
[64] UCL: Farey to Greenough, 4 June 1813, Add. 7918, 574; Farey to Greenough, 15 June 1813, Add. 7918, 575. For further background and interpretation, see Torrens (1994, p. 67).
[65] UCL: Farey to Greenough, 12 March 1813, Add. 7918, 564.

have not the least objection'.[66] Of course, he may have written this aware that Greenough's notes confirmed his view. However, according to one obituary, Farey was 'of very retiring habits; rarely mixing in society' (Ford & Torrens 1989). He certainly had an unconstrained pugilistic tendency, which, when combined with his honest openness and lack of tact, tended to undermine the clear-sightedness and intellectual sophistication of his message. As a critic he lacked that guarded professionalism possessed by Greenough and his gentlemen friends. Greenough, for example, received a similar but rather more understanding review of Townsend's book from Conybeare, who, after cataloguing a profusion of errors, dealt with his subject more delicately:

> Notwithstanding this mixture of Error & nonsense I still think that much may be gathered from the book. Many of the observations & facts stated in it appear perfectly accurate & by no means devoid of interest or importance. Geology in [its] present state must not be fastidious or capricious but thankfully receive any contribution of authentic facts in however slovenly a form they may be presented. To indicate the more favorable part of my opinion of the Pewsey Geologist I would particularly cite the chapter on the thickness of the strata and that on springs. This last strikes me as containing a new and ingenious application of Geological investigation.[67]

From 1813, with his paper sunk, Farey's relationship with the Society took a most remarkable turn. The Society, and Greenough in particular, now became the focus of this most sharp-tongued of critics.[68] He now reflected on his past relationship, reinterpreting the intelligence he had freely given to the Society as spying, and turning Smith into an offensive weapon. He had, of course, seen no difficulty in borrowing from Smith himself or of ridiculing the efforts of others, but now he turned his eloquence in vitriol on the Society:

> instead of the least patronage or countenance being given to Mr. Smith, every means, direct and indirect, were soon resorted to, by a leading Individual [Greenough] therein, in particular, *to obtain his materials and delineate them on a new Map*, pretended, at first, to be for the private use of the Society; but after it had twice or thrice been copied, to correct its first egregious errors, as new materials were quickly collected, with inconceivably less pains or cost than Mr. Smith's materials were originally obtained, and I was repeatedly applied to for contribution to this new Map, I began to suspect, that all was not right, and determined on putting the question plainly, whether the design was not really entertained, *of publishing this rival Map?*, and this not being longer denied, then, whether it was intended in such publication, to make *the acknowledgements* so justly due to Mr. Smith, for his long priority in the research, and his materials, obtained as above mentioned?, when I was unblushingly told, that theirs being a Map begun and altogether *made on Wernerian principles!*, no such acknowledges as I asked, would be made!!
>
> (Farey 1815*a*, p. 337)

Farey claimed in public that he had been assured repeatedly in 1811, when supplying unpublished materials belonging to Smith, that the Society's map would not be published (Farey 1820, p. 380) and thus not compete with Smith's. How Farey came to show Smith's unpublished materials to Greenough in 1811 is unknown, as we know that Smith was highly protective of his data in that year. Farey pushed forward his advantage, hoping to pre-form opinion against the successor map: 'so much confidence that the fear of the exposure I could make, and the consequent shame and disgrace that must attach to the actors herein, would restrain them, that I had determined to delay submitting the present statements to the public' (Farey 1815*a*, p. 338). Greenough remained silent and did not answer the charge. Fitton, however, was moved to act, piqued by Farey's savaging of Kidd's *Geological Essay* (1815), which carried a dedication to Greenough:

> Which, I would ask, is most to be deplored, *the ignorance still prevailing* in the chief *Seats of Learning* among us, as to *the most obvious Geological facts around them?*, or *the pitiful design* manifested in Dr. K's Book, in the revival of these excusable mistakes of former great Men, of depriving a deserving, although non-academical Individual [Smith], of the merited rewards of his labours?
>
> (Farey 1815*a*, p. 340)

It was not to defend Greenough, however, that Fitton acted, but rather to protect his Oxford friend who had earlier supported him in his attempts to obtain a lucrative medical appointment in Northampton.[69] Determined to ascertain whether the claims made for Smith were true, Fitton turned up at Smith's house in Buckingham Street, London, to probe the surveyor about his past doings, taking in the process 'a few desultory notes' (Farey 1818, p. 184n; [Farey 1818*a*]). Neither Smith nor Farey, who was there, knew anything of this mysterious stranger, but Farey aided his desire to separate the

[66] UCL: Farey to Greenough, 12 March 1813, Add. 7918, 564.

[67] UCL: Conybeare to Greenough, 18 February 1813, Add. 7918, 445.

[68] His articles in *Philosophical Magazine* critical of the Society fall into the years, 1814–1822, a period strangely missing in the Society's own collection of this publication.

[69] UCL: Fitton to Greenough, 14 March 1814, Add. 7918, 600.

discoveries of Werner and Smith by uncovering details of the latter's past. Farey did not know then that it was Fitton's intention to use this information against him in a defence of Kidd and he could hardly believe it when he found out. Fitton, however, chose not to pursue his quarry and the review he wrote surprised Farey to such an extent that he was willing to overlook Fitton's criticism of him as an advocate for Smith: 'But the patronage of this gentleman is really a little too vehement, and of such a sort, that if we wished to ensure the failure of a valuable performance, we should begin by recommending it to his protection' ([Fitton] 1818, p. 312). Indeed, Fitton had earlier claimed Smith's great map, *A Delineation of the Strata of England and Wales* (1815), as 'a work which it would be unjust to mention, without adding, that it is of great and original value; indeed, regarding it as the production of an unassisted individual, of most extraordinary merit' ([Fitton] 1817, p. 71). Fitton had, to his apparent surprise, become a supporter of Smith, impressed by his map and his other recent publications. He recommended that the Society take up Smith's project, with its difficult and barbaric names, and refine it ([Fitton] 1818, p. 336), believing that Smith's nomenclature could be improved using the 'proper scientific names of the substances composing the strata' rather than a terminology intelligible only to locals. Fitton also thought Smith had chosen an impractical scale: 'The scale of the map, five miles to an inch (the whole occupying a space of $5\frac{1}{2}$ feet by $7\frac{1}{2}$), is, we think, considerably too large. It would have been much better if the general map had been more portable, and the minuter details left for separate maps of counties: and, if a second edition be published, we recommend the adoption of this plan' ([Fitton] 1818, p. 336; see Sheppard 1917, pp. 161–163, for discussion of a smaller version of Smith's map published by Cary in 1820; see also Eyles & Eyles 1938, p. 208).

The third volume of Farey's *General View of the Agriculture of Derbyshire* arrived in print in 1817, before Fitton's review. A report concerned with sheep, roads and canals, Farey used its preface to launch a scathing attack on 'a Geognostic Society' and its thefts and deceptions:

> the principle part of his [Smith's] hard-earned materials, having been surreptitiously obtained, and the progress of a rival Map of our Strata, unblushingly announced, by the same Party, *while his was engraving and colouring!* – and since its completion, an extensive sale of it has been prevented, such as was necessary *to repay the outlay* of Engraving and Colouring [. . .] by the sinister arts employed to under-value the same, and dissuade, even the few who are disposed *to purchase a Map of the Strata*, to wait for the improved Map of a *ci-devant* Member of Parliament! [Greenough] Who, it may be proper to observe, entered on this pursuit years after Mr. Smith had, not only commenced, but brought his Map almost to its present state of approach to perfection, had freely shewn it to hundreds, and was soliciting Subscribers' *names*, for its publication.
>
> (Farey 1817, p. ix)

It is true that Smith's map was well advanced and known to literati like Banks in 1802 (Torrens 1994, p. 59) but now Farey's rage against the Society embraced every aspect of its being. In Farey's eyes, the Society could do no good. He complained of the Society's '*refusing to publish*' and detention of his paper, and of the Society's treatment of his friend, 'the Fossilist and Petrifaction-Worker of Castleton', Elias Hall's model of strata that was so colourful 'as to raise *a laugh*, by the far-fetched and contemptible joke, that "a tray of Guts and Garbage in a Fishmonger's or Poulterer's Shop," rather than any thing else, was called to mind!' (Farey 1815*b*, p. x, 1817, p. viii). The Society's corrupting influence seemed to pervade the whole science:

> The conduct of the gentleman [a 'London professor', who Torrens has identified as the chemist, William Thomas Brande (1788–1866)] now alluded to, has even been still worse towards my friend *Mr. Smith*, whose *Geological Map* has been in the [Royal] Institution Library since its publication, and during the delivery of several of the Lectures alluded to, was actually *hung up at the back of the Lecturer*, and made the diagram of his local descriptions of English Mineralogy and Geology; and yet, in two volumes which this 'learned Professor' has since put forth, in 1816 and 1817, detailing the Geological facts of England, *not the least mention or allusion is made to Mr. Smith or his labours*, through the last 28 years!!
>
> (Farey 1818, p. 186)

Perhaps at Smith's request, Farey took a rather different tack in November 1817, after he attended Sir Joseph Banks' Conversazione with Smith. A group of those present met at Farey's house to draft a chronology of Smith's discoveries. Fitton was invited to join them but declined (Farey 1818, p. 184n); according to Farey, he just disappeared ([Farey] 1818*a*, p. 174) – but Fitton had no time for Farey's abusive crusade. Nevertheless, it had been Fitton's initial enquiries at Buckingham Street that had stimulated Farey to gather together a potted history in the first place ([Farey] 1818*a*, p. 173). This was published anonymously and as objective facts early the next year in the *Philosophical Magazine* and *Annals of Philosophy* ([Farey] 1818*a*, *b*). A duplicated effort, it revealed that even the publishers, who probably liked Farey's rather scandalous contributions, were tiring of Farey's relentless and well-known campaign. It was no longer news; their whole readership, if not the whole of England, knew of Smith's plight.

Greenough finally broke his silence in 1819, answering Farey's accusations in a polite and understated manner, which said something of how practiced he was in that slippery kind of argument and diplomacy prevalent even then in Westminster. He presented himself as a modest man, who was willing to admit to his mistakes and to put in the public domain his appreciation of Smith. That appreciation, however, was for the man and not his science:

> Your correspondent considers me, in common with many other persons, actuated by feelings of hostility towards Mr. Smith. Now my feelings towards that gentleman are directly the reverse. I respect him for the important services he has rendered to geology, and I esteem him for the example of dignity, meekness, modesty, and candour, which he continually, though ineffectually, exhibits to his self-appointed champion.

In Greenough's view, Farey was merely a blind advocate of Smith's method, unable to prove the truth of what Smith proposed:

> [Farey] has examined Derbyshire with very laudable industry; will he take the trouble to mention, what the fossils are, by which he is enabled to distinguish the different limestones in that county, or the different sandstones, or the different shales? There will be time to discuss the originality of the doctrine when its truth is established. If its truth cannot be established, I beg very respectfully to ask Mr. Farey, whether he can hope to exalt the character of his teacher by proving him the first discoverer of that which does not exist?
> (Greenough 1819*b*)

It was a masterful response. Given all that Farey had written, it was a telling criticism to suggest that Farey's ground-breaking 532 page geological report (Farey 1811) did not *illustrate* Smithian principles. Greenough knew this work well and knew that Farey would not be able to easily muster a defence from it, even if it was commonly believed that its completion relied upon Smith's teachings (Torrens 1994, p. 62). Indeed, Greenough knew that, but for a handful of mentions of Smith, it espoused a geology that seemed little different from that practised by his other informants. It may have been the most detailed report on British strata published to date, but it was not an exposition on the importance of fossils. But, as I have explained, the Smithian method did not rely on palaeontological description, merely on indication, memory and recollection. However, while Greenough's accusations were in press, Farey published a list of fossils and their stratigraphic positions in the magazine in which Greenough's article was to appear. If that did not pull the carpet from beneath Greenough's feet, then the publication of William Phillips's *Selection of Facts* the previous year should have done so, for it also illustrated the specific occurrences of fossils (Phillips 1818; Farey 1819).

Greenough may have looked at the situation and thought he, rather than Smith, was the victim. Farey's attacks were aimed at Greenough's philosophy, friends, Society, morality, honesty, class and even his ambition. This was not simply a defence of Smith; Farey was out for revenge. In the manner of his arguments, Farey captured nothing of the man he defended: Smith's affable good humour had won him friends and admirers in every quarter; Farey was, by comparison, isolated by his anger. Greenough could look at the Society's huge efforts since 1807 to record and map the nation's geology, and legitimately see the world differently. The Society was in receipt of many maps, huge correspondence and numerous reports from its members and supporters. Its officers and correspondents had travelled widely. Whatever Greenough had acquired of Smith's may well have had no striking place in this assemblage of data, and no-one else in that earlier period seemed to be making individual claims on the basis of their contribution to Greenough's plan. It was true, too, that his efforts embodied a belief that Wernerian principles would enable rocks to be understood, correlated and mapped. Yet, many of his correspondents had without acknowledgement converted themselves into Smithian geologists – using fossils and believing in regular succession.

Soon Smith's map was in the public domain, and who could tell what the contemporary public might make of Greenough's very similar map when it finally appeared? How can one defend against charges of plagiarism in the production of maps? Greenough was in a difficult position and Farey knew it. Greenough's only defence was to do better. Indeed, if he could compile an even better map using his more abundant resources, was he not morally obliged to do so for the sake of science? Or was he morally obligated not to compete with a map produced by a single man, as Farey claimed? On every front, science advanced by the first principle. Reputations were made by refuting the claims of others, and by mere improvement. Nearly everyone in this new geological world knew it, participated in it, fell victim to it and despised it. Nevertheless, it gave the field huge positive energy.

Whatever Greenough felt, he kept it to himself, at least until the publication of his map, at which point he could hardly avoid answering the charges against him. He did so in his *Memoir of a Geological Map of England* (1820). In doing so, he used that same political ability to disguise what had really happened. He would not answer every charge, only those where others might corroborate his view or about which only he would know. We cannot read into this a belief on Greenough's part that he had acted unethically but we can infer that, in the changed world in which he finally published his map, history needed careful curatorship if it was not to generate a false impression. Only in this memoir did he give his map's chronology, admitting to sketching a number of geological maps from 1808, each with a single stratum marked upon it. When these were presented to the Society in 1812 (the year after Farey said he had received assurances that the Society possessed no plans to publish a map) he was 'requested [...] to construct a new map of England upon a larger scale'. This, he admits, was the point of origin of the published map. Greenough admitted that he knew of Smith's projected map from 1804 but exclaimed:

> but I appeal to all the friends of Mr. Smith, with whom I have conversed upon the subject, and especially to the individual who complains of my conduct, whether he, and they did not, for a long time afterwards, in consequence of a variety of circumstances which it is unnecessary to detail, consider its completion, and still more its publication, hopeless. In the belief that the work had been virtually abandoned by Mr. Smith, it was undertaken by me [...] it had been more than a twelvemonth in the hands of the engraver when Mr. Smith published his [...] but I do not think it [Smith's map] of a nature to render the publication of mine superfluous. I do not admit that any consideration of justice or delicacy required me either to abstain from constructing my map, in the first instance, or, when partly engraved, to request permission of the Geological Society to withdraw it. Mr. Smith's map was not seen by me till after its publication, and the use I have since made of it has been very limited. The two maps agree in many respects, not because the one has been copied from the other, but because both are correct; and they differ in many, not from an unworthy apprehension on my part of being deemed a plagiarist, but because it is impossible that the views, the opportunities, the reasonings of two persons engaged on the same subject should be invariably the same.
>
> (Greenough 1820, p. 4)

Now it may appear from the convenience of the dates and Greenough's wording that he was hiding behind a Society façade; that he was conducting work that had been requested of him, when as President we presume him all powerful. But, yet, the map committee and Council did indeed 'request' Greenough to undertake tasks, even such menial tasks as mounting maps on canvas.[70] We do not, however, possess any record of the Council asking Greenough to prepare his map. But in 1812, the map committee did decide to meet every two weeks.[71] It made its Secretary Thomas Webster, who later that year was appointed curator and draughtsman; it was Webster who drew Greenough's map. The meetings appear to have been unminuted, but there is no reason to doubt the timing of Greenough's decision to proceed, regardless of the strength of any independent request that he should do so. While Farey's indignation concerning Greenough's change of plans was justified, Greenough claimed that he desired to be the author of such a map from the outset of the Society. It was a *personal* ambition and for very human reasons he doubtless saw Smith as a difficulty to be overcome; Farey had earlier shown his own disregard for the feelings of Smith. That ambition seems to have risen to a point of open competition by 5 December 1814, when Greenough's map was sufficiently complete to employ the services of an engraver.[72] The stubs of money paid and received for 1815 reveal that Webster was then very active in drawing, and Greenough active in promoting, the map, in the year in which Smith published. Smith's first complete map was produced in February, with full production beginning in August (Eyles & Eyles 1938, p. 210). Those early maps went directly to his many subscribers, but future sales must certainly have been hindered by the news that an improved map was close to completion. Farey's claims concerning this competition, then, were also justified and it may have been this that motivated Smith to continue to improve his own map with each colouring. But what could Greenough do? The work had certainly been done. If he had been faced with the competition of another gentleman, society or company, there would be no question of Greenough quitting. But the difficulty here was that the arguments were not about science but morality. It was a David and Goliath battle in which the little man always has the moral advantage (who ever cared about Goliath whatever his moral claims?). Greenough never asked to be cast in this oppositional role, his very being reconfigured as the antithesis of the exemplary character and achievement of the man with whom he found himself in competition. Yet, Greenough no longer lived in that convivial world that was once the preserve of geology. Doing geology was now a cut-throat business; the Society had been consumed by a fever of competition.

If Greenough's map alone was not sufficient to finish off Smith's when it was published in

[70] GSL: Council Minute book 1, 11 March 1812.

[71] GSL: Council Minute book 1, 11 March 1812.

[72] GSL: Geological map of England and Wales: Agreements, accounts and drafts 1814–1820, four folders, LDGSL 974.

1819, William Conybeare, in his contribution to his and William Phillips's *Outlines of the Geology of England and Wales* (Conybeare & Phillips 1822) seemed determined to see it superseded. Although Conybeare continued to dispute some of Greenough's interpretations, as he had done during the map's preparation, he nevertheless included Greenough's map in the book (suitably modified). In doing so, he hid behind a façade of objectivity and disinterestedness, and unashamedly promoted it:

> Subsequently to the publication of Mr. Smith's map in 1819, another on nearly the same scale was published by Mr. Greenough; the execution of this is more minute and delicate, and the details more exactly laboured; the general configuration of the surface of the country, its hills and vallies, are represented with far more precision than had previously been attempted in any general map of the island, – points which did not enter into the construction of Mr. Smith's map; and many of the imperfections of the former are removed; to this therefore we have referred as a general standard throughout the work, and have therefore studiously noted every remaining incorrectness which a careful collation of it in the course of our enquiries with materials derived from subsequent observation has enabled us to detect; from this also we have copied the slight outline map prefixed to this volume, with a trifling change in the system of colouring which a different view of the division of a part of the carboniferous series of rocks has obliged us to introduce.
>
> (Conybeare in Conybeare & Phillips 1822, p. xlvii)

Conybeare had, despite only a few pages earlier singing the praises of Smith, now undermined him in the sale of his map. He had not only made Greenough's map the natural accompaniment to this, the new cornerstone of British descriptive geology, but also asserted that Smith's map simply would not do. From 1822, only a fool would buy it. Greenough had his supporters in influential places; after all, the Society had begun as a network of the distinguished. Conybeare told Greenough in 1825, 'We have now a very flourishing institution at Bristol of wh by the way I took care that you shd be early made an honorary member hoping thereby to attract a copy of your map as one of our members has given us Smith's'.[73]

We cannot doubt that geology had become politicized. Data once shared was now to be protected. Consider, then, how things changed for Smith and his supporters. Smith was then a survivor of an older generation, who had developed his ideas in a thinly populated rural world where aristocratic landowner and middling man came into contact and collaboration. In that world, his geological knowledge alone was of limited use to others because it supported a practical occupation; knowing the principles was not enough to supplant this *field-experienced* surveyor. Equally, the interests of leisured gentlemen naturalists were hardly likely to interfere with Smith's business ambitions. However, this tussle over maps took place after geology had been transformed into a fashionable enterprise capable of generating personal reputation and income in new ways. In no small measure did Smith's own publications permit the Society men to see this potential in the science for they saw these as indicative and preliminary rather than definitive. They could see progress beyond them and, thus, locate a laudable goal. There were at the time numerous mineralogical books in circulation that differed little in content, although each author doubtless thought he had advanced the subject and its dissemination in some way. Why not, then, multiple geological maps? If we lived in this nepotistic world of favour, where reputation was everything and poverty everywhere, where all endlessly pursued elevation and the competitive thrust of life was widely felt, how differently might we view the ambiguous boundaries of plagiarism, compilation and incremental advance – particularly as the rules of the scientific game were far from agreed?

It was Eyles & Eyles (1938, p. 212) who first wondered why the Society had not simply aided Smith in the production of his map. The answer should be apparent from what has already been written: Smith was at first doubted, disbelieved and considered unlikely to complete; he was later simply a competitor in an already competitive world. Gentlemen were not inclined to wholly fund such projects. The established mode of using subscription to progress an expensive project could have proven successful in the case of Smith's map – he had an ample list of subscribers – had not a series of unfortunate circumstances delayed its completion and later interfered with its sale. However, balancing demand and the high production costs of illustrated works proved difficult and most of Smith's publications failed to reach completion. As gentlemen funded their own social and scientific advancement – and many struggled to do so – there was a presumption against posts and support that fostered an individual's advance at the expense of others. There were exceptions, of course, and even Greenough found himself defending paid curatorship against the jealous attacks of self-funded individuals. It was a complex competitive and cut-throat world that legitimized selfish individualism.

[73] UCL: Conybeare to Greenough, 4 January 1825, Add. 7918, 463.

The Society's fever

From the mid-1810s, the Society had become increasingly embroiled in territorial politics affecting fame, reputation and income. This need not be solely inferred from actions; the actors in this drama said so overtly and almost to a man. By 1815, it was becoming increasingly possible to imagine the production of a highly profitable geological text. William Phillips's compilations were one manifestation of this; Arthur Aikin's *A Manual of Mineralogy* (1814) was yet another; and Kidd's *Geological Essay* (1815) still another. Indeed, the competition was becoming so fierce it was difficult to keep ahead of the game, and Kidd, for example, found himself pre-empted on several points by Cuvier and de Luc, and could only feebly claim to have drawn his own conclusions before reading theirs (Kidd 1815, p. v).

Kidd's *Essay* was the culmination of years of summer fieldwork, which he had undertaken to support his 'mineralogical' (geological) lectures. But he long desired giving up those lectures and in November 1814, having completed his last fieldtrip, he did so. The publication of his book was his final act: 'In offering this Essay to the public, I take a final leave of the pursuit of Mineralogy' (Kidd 1815, p. viii). His quiet exit, however, was far from painless, and the pain came not only from Farey's review. He had asked Conybeare and Buckland if either would replace him. Conybeare declined but Buckland accepted. Kidd, however, was unprepared for the political consequences of his decision. Having abdicated his power to Buckland, Buckland now reconfigured the world that had once been Kidd's. He now progressed his own lecture course without ever discussing it with Kidd; Kidd had openly discussed his ideas. Buckland also closed the museum door that Kidd had always left open. When in the summer vacation Kidd requested a key, Buckland was hesitant; Kidd had to remind him that it was a public collection. Buckland, however, was determined to protect his materials knowing that Kidd had attended his lectures. Protecting knowledge in a world increasingly populated with able and aspiring geologists was a new problem, which became apparent around 1815. The emergent science still had everything to say, and all in the geological community vied to say it. The world was shifting from open collaboration and sharing of information to closed and competitive possession. Yet, the culture of compilation, which fed upon that earlier co-operative world, could not be so easily displaced and Buckland's fear of plagiarism was a fair response. Kidd was clearly distressed by his disempowerment and opened his heart to Greenough, telling him to burn the letter once he had read it.[74] Still only a young man, Kidd's premature departure from geology enabled his return to a full-time interest in anatomy where he continued a highly successful and profitable career.

At the time of his departure, then, Kidd felt the competition of a friend who feared plagiarism while his *Essay* did the rounds chased by Farey's assertion that he had knowingly plagiarized Smith. It was not that Kidd was a usurper of other people's material, but that the rules had changed. This change in the way the science was socialized caught others in its trap, including Smith, Greenough and Farey. Actions, values and contexts, which once seemed so right, now lacked compatibility. The fever now spread throughout the geological community. A pencil draft of a letter by Greenough shows that he and Fitton had come head-to-head over a project to write 'a popular work illustrating the first principles of geology', or, as Fitton preferred, an 'Elements of Geology'. Fitton had reluctantly taken up the project in 1816 in order to relieve his financial pressures. A trip home to Dublin that year relieved that pressure but not before the damage had been done.[75] Greenough, too, thought the project a money-spinner; it seems one could never have enough money. Intrigue had then begun between them, with Buckland implicated, and Fitton in the awkward position of having borrowed a large number of Greenough's books in order, apparently, to supplant him in the project.[76] Greenough accused Fitton of competing, but then giving up and urging Greenough's close friend Buckland to take up the plan, in full knowledge that it would interfere with Greenough's ambitions. Fitton defended himself by telling Greenough that his conversation with Buckland arose accidentally, and referred to the publication of Buckland's lectures. However, in 1817, Fitton ([Fitton] 1817, p. 93) had in print, implicitly and anonymously, suggested that such a work would not arise from the Society but from lectureships in mineralogy at Oxford, Cambridge, and the Dublin Society, or professorships in natural history at Dublin and Edinburgh. It would, he said, answer the questions: '*First*, Is there any certain order of succession in the rock formations? – and, *Secondly*, What is the series?'. In 1817, he still desired that mineralogical logic, believing the book would come from the kind

[74] UCL: Kidd to Greenough, 7 November 1814, Add. 7918, 1070.

[75] UCL: Fitton to Greenough, 15 November 1816, Add. 7918, 601.

[76] UCL: Greenough pencil draft of letter to Fitton, [26 December 1817], Add. 7918, 603.

of 'analyses' that had produced such great advances in chemistry and mineralogy. His ideal model for this book was 'Dr Thomson's admirable System of Chemistry'.[77] Fitton, and apparently everyone in the Society, thought Greenough was merely writing a memoir to his geological map, which would only include an introduction discussing the principles.[78] Greenough was, however, working on his sceptical *Critical Examination* (1819*a*) and drawing upon the counsel of his long-time friend, Henry Warburton, to negotiate his way through that 'chaos of philosophies'. Warburton, however, doubted Greenough's capacity to do this negotiating and said of an early draft:

> Had I written a similar work myself I do not know that I should have condescended to notice with so much [compliance?] many absurd and inconclusive reasonings which have sprung from the hand of Geological zealots; and I should perhaps have endeavoured to distinguish in more pointed manner those arguments which are valid and conclusive, by whatever [sort?] produced, from those which rest upon weak analysis; I should have treated Strata with somewhat more respect; less sceptically & more practically.[79]

Thus, while they fought for the right to address this great opportunity, desiring money and reputation, neither Fitton nor Greenough was really up to the task. And no one seemed to be openly conversing on what they were doing until a spreading rumour or conflict of interests forced them to do so.

The feud between Greenough and Conybeare, which erupted in January 1823, was nearly identical, with Buckland again implicated. By then, these disputes could be imaged as territorial. For Conybeare and Greenough that territory was European (i.e. non-British) geology. Conybeare claimed this as a right, having read up on the subject in 1814 when the country was still at war, and before any savants, including Greenough, had entered the Continent, although Greenough did so in that year. From this Conybeare constructed a geological section of Europe that was still on display in Buckland's museum. Although it contained errors, he had communicated his map to Greenough in 1814. Now, however, Conybeare had heard via Buckland that Greenough planned to publish a work on the same subject. Conybeare claimed he was still working on his book, nearly nine years after his initial interest. It was commonplace in these disputes to believe that once one had produced a work of some kind, an individual automatically obtained a God-given right to possess that territory. Now they realized that no such rights existed.

The Greenough–Conybeare dispute lasted two years. At one point it was agreed that Greenough would produce a map and Conybeare a memoir. Greenough imagined he could repeat the same trick as he had for his map of England, although Warburton suggested the involvement of Continental geologists. But then Conybeare published a small version of a map he had been working on with De la Beche, claiming to suppress a large-scale version in deference to Greenough. Matters were made worse by Conybeare's use of French maps belonging to Greenough and were not helped by describing Greenough as a mere compiler![80] The dispute, which was only resolved when he gave up the project, almost caused Greenough to quit geology entirely.[81]

In this new competitive environment, individuals would refuse to see drafts of competing works or engage in anything that might later bring about a charge of theft or plagiarism. Early on there were attempts at chivalrous resolution, but soon the battle became too intense and was increasingly between strangers. There are countless examples of the colourful language and spiteful commentary that seem to define the culture of parts of the Society as it entered the 1820s. John Phillips, one of the most level-headed and diplomatic geologists of his day, commented as a newcomer to London in 1831: 'The jealousy among the men of Science here is wonderful and you feel to walk on a cavity, and to be grasped by a hand of friendship no firmer than a ghost's shadow' (Knell 2000, pp. 30–33).

It was into this world that Fitton pondered injecting a proposal, in March 1822, to have Smith elected an Honorary Member. We should note, however, that the Society had begun life with huge numbers of such members who were distinguished by living outside London and not having to suffer the high costs of membership and participation. The purpose had been to get free labour and support, and it had worked very well. Fitton's proposal, then, was not the act of ennoblement it might at

[77] Jameson (1816) reviews the literature and notes the importance of Thomson's work, which devoted one volume to mineralogy. Fitton doubtless used Jameson as an important resource.

[78] UCL: Fitton to Greenough, 11 January 1818, Add. 7918, 604.

[79] UCL: Warburton to Greenough, 30 May 1817, Add. 7918, 1691.

[80] UCL: Conybeare to Greenough, 24 January 1823, Add. 7918, 459; Copy letter, Greenough to Conybeare, 26 January 1823, Add. 7918, 460. Warburton to Greenough, 14 December 1824, Add. 7918, 1697 enclosing a fragment from a letter from Conybeare, 11 December 1824. Draft of Greenough's response, Add. 7918, 1699.

[81] UCL: Copy of letter, Greenough to Conybeare [December 1824/January 1825?], Add. 7918, 463a.

first seem. Famously, Webster commented: 'What do you think of [Fitton's] declared intention of proposing Smith an Honorary Member of the Geological Society on account of the services he has rendered to geology, and that Greenough [then the President] shall be the first to sign his certificate? Do you think Greenough will find this pill as difficult to swallow as any of his uncle's?!' (Torrens 1990, p. 8). One can well imagine Greenough's difficulty and the loss of face that would result: he had quite legitimately led the disbelievers, deniers and competitors. But now the legitimacy of those actions began to dissolve as a result of cultural change, the incoming of youthful minds and the rapid loss of the other founders.

Geology beyond London

If the Society's increasing tendency towards competitive dialogue seems to begin around the time of the publication of Smith's map in 1815, it must also begin at the close of a far more influential event: the Napoleonic Wars. Cessation of hostilities resulted in a mass exchange of savants, each crossing the Channel to engage in intelligence gathering and even geological trade (Knell 2000, pp. 28–30). This certainly had a hugely competitive aspect, which may have combined with the dissolution of a 'wartime spirit' to alter the perspective of Society members. In the years after the war, then, geology was changing socially and intellectually. Greenough's *Critical Examination* in 1819 was almost the last gasp of the philosophical approach prevalent at the Society's founding (Greenough 1819*a*); it now appeared as a relic from a former world. As the science entered the 1820s, it was on a new footing. There was an increasingly unified sense of what geology was and how it should be prosecuted, and indeed what were its major questions, even if the membership itself had become rather less unified and more individualistic. The science had continued to become increasingly fashionable. These factors now contributed to a development that effectively removed the ability of the Society to be the single geological hub of the nation, for a geological society, of sorts, began to spring up in almost every town in England. It was in these new geological centres, rather than in London, that Smith presented himself as the harbinger of the future of the science.

Escaping his creditors, the taxman and the politics of geology in the metropolis, in the late 1810s Smith had headed for Yorkshire. His timing could not have been more perfect, for post-war northern England was becoming the heartland for a reinvigorated movement to establish literary and philosophical societies interested, as their name suggests, in libraries, science and urban cohesion. Smith was welcomed amongst these elite associations, for he was known, by name at least, as the great originator in English geology. Here he began a new life as a peripatetic lecturer soon accompanied on stage by John Phillips, who proved to be one of the most eloquent and captivating geological lecturers of his age. Smith may have been, by comparison, meandering and anecdotal, but he was nevertheless appreciated as the embodiment of invention and rich experience. There was also a humble wit and charm to the man that had never failed to attract the friendship and admiration of others. Phillips penned an appreciative portrait of his uncle on one of those few occasions when Smith's health failed him, which captures something of Smith's lectures:

> It was a singular spectacle, to witness the delivery of lectures which required continual reference to large maps and numerous diagrams, by a man who could not stand, but was forced to read his address from a chair [...] but Mr. Smith would have thought it unworthy of his resolved mind and firm trust in Providence, to have abated one jot of his accustomed cheerfulness, shortened one of the innumerable playful stories which were always springing to his lips from the rich treasure-house of his memory or turned his meditations from his favourite subjects.
>
> (Phillips 1844, p. 111)

Many of these societies were concerned above all else with geology: the society in York admitted to being a geological society; the societies in Whitby and Scarborough were also strongly geological. The museum was established as a central scientific resource, and it was this geologically attuned movement that gave the nation its provincial museum culture. Between lectures, in the field and in society collections, Phillips, sometimes with Smith, performed a stratigraphic magic that was entirely new to these audiences (Knell 2000). Local naturalists and geologists were astounded; this was revelatory science. It was revelatory not just because of the order it instilled, but also for its philosophical certainty; there were heads in this part of the world still dreaming up theories of the Earth. In 1824, in lectures in York, Scarborough and Hull, audiences were converted to Smithian geology. Leeds had earlier fallen, and other towns across the county would follow. These truly were conversions, for, on the day after the lecture, audiences woke to see the world differently. Many began to collect fossils for the first time or with a new purpose. Phillips referred to these events in evangelical terms: 'I may flatter myself with a good many converts out of an audience of several hundreds' (Edmonds 1975, p. 389). This was the final phase in Smith's conversion of the nation. Fitton may later suggest that Smith should have published in order to make his claim, but, true to his character and resources, Smith had converted the nation in person. The Geological Society had used its networks to connect to land and mine owners; Smith had visited them himself, often proving himself in commissioned work. Here, in Yorkshire, Smith found patronage from the leaders of the philosophical movement in York and Scarborough. The county could boast some of the most politically influential landowners and religious figures in the country; both groups were connected to these new societies. They now took possession of Smith as a national treasure; they had no desire to use him politically. Similarly, John Phillips, then in his early 20s, was adopted by the Yorkshire Philosophical Society as its curator. In that role he became the living embodiment of science as a youthful moral and cultural force in provincial England. It was Phillips who now achieved the intellectual goals of the York society, revealing with wonderful eloquence and sophistication the power of Smith's geology in his *Illustrations of the Geology of Yorkshire* (Phillips 1829; Morrell 2005, p. 66). This achievement was recognized by Geological Society stalwarts, Horner, Fitton, Murchison, Buckland, Conybeare and De la Beche.

Chief among Smith's supporters in Yorkshire were William Vernon (later Vernon Harcourt) (1789–1871), first President of the Yorkshire Philosophical Society, and his brother-in-law, Sir John Johnstone (1799–1869), patron of Smith and the Scarborough society. Vernon was an Oxford graduate and friend of Buckland and Conybeare. Indeed, Conybeare was creating his own institution at Bristol at that time, and here Smith was given honorary membership with Conybeare's backing (Torrens 1990, p. 10). It was acquired, however, after Smith's patronage by Conybeare's Yorkshire friends. In 1826, Murchison made a pilgrimage to Yorkshire to be educated by Smith and Phillips, and he too was converted. In 1829, the recently formed Literary and Philosophical Society in Scarborough opened the Scarborough Museum. Its circular form was an architectural manifestation of Smith's notion of a geological museum. It was a monument to Smith's geology:

> The collection is very valuable for a provincial town; and is particularly rich in Geological specimens. Their arrangement is strictly natural, and when adopted was one of the first of the kind in Great Britain. The fossils are placed on sloping shelves, in vertical succession, so as to represent their position in the different stratifications from which they have been removed.
>
> (Hume & Evans 1853, p. 154)[82]

At its opening, Smith was for the first time acclaimed the 'father of English geology'. In the following year, at the opening of the grander, but less architecturally innovative, Yorkshire Museum, in York, Smith was again addressed in these terms (Morrell 2005, pp. 73–74).

The Geological Society's recognition of Smith, then, did not emanate from an extraordinarily long period of digestion spanning the period between Fitton's first published expression of appreciation for him in 1818 and Sedgwick's address. Work had been done elsewhere and Smith had already gained the recognition that the Society would later give. However, the Yorkshire geologists saw the Geological Society as pre-eminent in the science and thus capable of awarding Smith the highest recognition. They did not know that their fellow Yorkshireman, Adam Sedgwick, would give still greater recognition to their man.

ACT 4. The difficult legacy

Rewriting history

Sedgwick's presentation of the Wollaston Medal to Smith in 1831 (the medal was actually put in his possession in June 1832 (Phillips 1844, p. 117)) reflected a fundamental transformation in the Society's values and perceptions, but clearly this was not something a Society would wish to make public knowledge. Change had made some of its members' past actions incomprehensible, even reprehensible. A little historical concealment would be necessary if Smith's claim was to be substantiated without embarrassment to the Society. There are hints of this in the changes Fitton made to his 1818 review of English geology, which he republished in 1832–1833. Both reviews were written with the aim of understanding Smith's achievements; the second was, however, to secure Sedgwick's claim. Historians have seen these articles as important acts of advocacy for Smith, but they are just as much acts of concealment.

Fitton's first review greeted, after decades of near silence in print (as a result of failed attempts at publication), a spate of remarkable publications by Smith. Following the publication of his map in 1815, Smith also published his *Memoir of the Map and Delineation* (1815), his *Strata Identified by Organized Fossils* (1816), his extraordinary

[82] This *natural* arrangement was not in itself an invention of Smith. Whitehurst (1786, pp. 204–205), had suggested that a geological museum could be arranged along these lines in order to display the nature and productions of strata as they are found in the field. He makes no particular mention of fossils or their utility, however.

Geological Section from London to Snowdon (1817), his series of county maps (1817), which were now surpassing his national map in utility, and his *Stratigraphical System of Organized Fossils* (1817), which described his collection, recently sold under financial duress to the British Museum (Sheppard 1917, p. 118ff; Eyles 1967). In this remarkable turnaround in his publications, he had been aided by the arrival in London of the precociously talented 14-year-old John Phillips, in 1815.

Fitton was unreserved in recommending these publications to his readers, but he felt his review needed to do more that this. It was necessary once and for all to understand if these works rested upon principles Smith himself had invented. In order to conduct his review, therefore, Fitton had no other course open to him but to construct a 'chronology of science' based on 'the relative order of publication'. But here he faced a problem, for Smith had not published his ideas until recently, despite possessing and utilizing them for many years. In 1818 Fitton was willing to overlook this problem, but in 1832, with Smith now formally recognized by the Society, he felt compelled to use this failing to cover up the Society's long denial of Smith. He did this by gently blaming Smith for not communicating his ideas through the Royal Society and thus giving them a 'most dignified and authentic form'. While it is true that, from 1801, Smith did have the ear of the Royal's president, Sir Joseph Banks (1743–1820) (Eyles 1985; Torrens 1994), Smith's ambitions would never have been directed towards that particular society. Smith did, however, harbour rather grand ambitions to give comprehensive expression to his ideas in monograph and map (Cox 1942). While he might be considered neither a writer nor an orator, he was proven in print, as shown by his 1801 *Prospectus* and 1806 book *Observations on the Utility, Form and Management of Water Meadows* (Sheppard 1917, pp. 110–117). But he realized, on completing this latter book, the full cost of publication:

> The Subject of Strata [...] now appears to be a most gigantic Undertaking, the accomplishing of which would almost alarm me to despair [...] I had no Idea of the depth to which I am imersed. The time and attention necessary to the completion of such a small work as that on Irrigation has fully convinced me of the much greater difficulties attending the publication of a voluminous Work on the Strata.
>
> (Smith, 14 December 1806, quoted in Torrens 2001, p. 66)

His preference was for the immediacy of verbal communication and the power of empirically produced visual aids. These were suited to the business matters that occupied his life and made no call upon those literary or academic talents which his education had failed to give him (Phillips 1844, pp. 127–130). By these means he converted so many to his cause (Torrens 2001). Costs of publication were, anyway, prohibitive, especially for a man with pictorial ambitions. His geological map of England took more than a decade to reach production both because of these difficulties and the scale of the project; it is not an unreasonable timescale for so major a solo project even today. Only in a competitive world of claim and counterclaim would Smith have sacrificed his grand and practical ambitions merely to stake his own claim, although even then he would not have been the man to do so through the Royal Society. This competitive geological world did not exist in 1801 or 1807, and by all accounts it never occurred to him that his failure to publish might have implications for him in later life. Despite his silence, however, others did put Smith's ideas into print in this early period (Sheppard 1917, p. 215; Eyles 1967, 1985; Ford & Torrens 1989; Knell 2000, pp. 14–15).

In a few unreasonable words, then, Fitton had used the Society's preoccupation with publication, then established as critical to priority claims in a highly competitive science, to legitimize its own supposed ignorance. He now needed to admit that some members of the Society did, in fact, adopt Smith's ideas without crediting him. He did this by recognizing that Smith's verbal dissemination through his Bath contacts had been successful: his 'knowledge so diffused has had a most important, though unobserved, effect upon the labours of all succeeding inquirers; who were, perhaps unconsciously, but not less really, indebted to the author for very essential assistance in their progress' ([Fitton] 1818, p. 313; Fitton 1832–1833, p. 149). This excused the failure to acknowledge Smith, although it could only be believed if the reader had never come across the writings of Farey or any of the early adopters of Smith, including some in the Society itself. The real reason why Smith remained uncredited was not that his name was lost to the Society men, but that most believed he did not invent the principles he preached, while a number of others, including Greenough and Kidd, thought them unproven and useless. If either charge were true then Smith possessed nothing that could be stolen or plagiarized and he deserved no mention. Of course, this scepticism was quite legitimate in the mid-1810s, when the discipline itself was still in a process of formation; but Fitton could not admit to it in 1832, as it would make the Society seem at best incompetent and at worst thieving. Instead, Fitton admitted that there was a sense that several earlier authors had made a 'near approach' to the same ideas; in other words, that this scepticism was legitimate.

Most prominent amongst these earlier authors was the 'ingenious' John Michell (1724–1793), who in Fitton's 1818 mind appeared to have discovered almost everything that was then claimed for Smith. Michell clearly had an extraordinary knowledge of stratification far beyond the borders of Britain. And in his discussion of the tracing of strata he seemed to make his most profound of statements: 'and this often makes it difficult to trace the appearances I have been relating; which, *without a general knowledge of the fossil bodies of a large tract of country, it is hardly possible to do*' (emphasis, [Fitton] 1818, p. 318). In 1818, Fitton found this an insurmountable obstacle to proving Smith's originality.

On reading this, Farey immediately appealed to the University of Cambridge, where Michell had been appointed Woodwardian Professor in 1762, for access to any surviving records. These, however, were not forthcoming, so he decided, instead, to republish Michell's essay in full in one of the popular magazines of the day, accompanying it with a critique which tenaciously defended Smith (Farey 1818, p. 184). By permitting readers to see Michell's comments in context, it should have been clear to many that Michell could not have been suggesting a Smithian methodology. As Farey had not come across Michell's essay – but, nevertheless, knew of Michell's ordering of strata (Farey 1811, p. 109n.) – until reading Fitton's piece, he must have been relieved by what he found:

> Mr. Smith's detractors would fain make it out, that Mr Michell here meant *organized remains*; if this were apparent, I would not hesitate an instant, in giving him the praise due to so important a suggestion: plainly however, Mr. M. intimates, that the requisite knowledge of the 'fossil bodies', of whatever kind, *was not then possessed by him*: – my highly-injured Friend Mr. Smith, did possess the useful knowledge, and was liberal *in communicating it*, of the unorganized and organized bodies 'of a large tract of country', years before Geognosts, or any other their hand-specimen mineralogical Theory of the Earth, was to be heard or read of in this country.
>
> (Farey 1818, p. 260)

Fitton's indecision seemed unwarranted, perhaps even a deception. For the rewrite, however, Sedgwick reassured him that there was no evidence in Michell's collections at Cambridge that he signified 'organized remains' (animal or plant fossils) by his use of the term 'fossil bodies' or that he organized collections stratigraphically. But Fitton chose to also hide behind a historiographic nicety, remarking:

> it is only candid to allow, that the passages which bear upon these points might possibly have slept much longer in the volumes which contain them, if the attention excited by Mr. Smith's publications had not led to their detection; and that the light in which they now appear to us is very different from what it would have been without such assistance.
>
> (Fitton 1832–1933, p. 151)

This was tantamount to condemning Michell to the same fate that had befallen Smith: condemning his invention for the manner of its dissemination. It seemed to suggest that Fitton still did not believe in Smith's originality, but it was more probably his way of excusing himself for the effects of his earlier erroneous interpretation. Now, rather than contradict himself, he suggested that Michell could simply be ignored.

Fitton's 1818 essay had begun by drawing upon the 'near truth' of a dismissive speech from the *Vicar of Wakefield*, to suggest that investigations of the Earth, before the modern age, had merely produced chaos ([Fitton] 1818, p. 313). In 1832 he was, by comparison, conducting a 'scientific' argument of some import, which he knew would be scrutinized by his friends. Those authors, whom he had earlier thought absurd, now had eminence and merit. Only now could the project of the geological map be traced back to Martin Lister's (*c.* 1639–1712) 'An ingenious proposal for a new sort of maps of countries' published in the *Philosophical Transactions* in 1684. With some judicious editing of quotes, Fitton could argue that although Lister had not produced the geological map he proposed, his interpretation of the 'soiles' of Yorkshire mirrored the spatial divisions of modern day maps and accounts, such as those produced by Smith and John Phillips in the 1820s. Fitton's quotation of John Woodward (1665–1728) was similarly expanded to reveal that the notion of strata was key to his assertion that rocks extended beyond the bounds of Britain and into Europe. In 1818, the Society's possessed rather less interest in strata, believing that the big answer would arise from mineral geography. In 1818, then, Fitton noted the value of the 'collection of minerals and fossils' Woodward had made and left to the University of Cambridge. However, in 1832, in this same passage, fossils were all and minerals made no showing. In this simple piece of editing is recorded a fundamental reconceptualization of the central resources of the science. Rocks and minerals were still collected, but their significance was by then much reduced; in 1807 minerals had had the prominence he now gave to fossils.

In 1832 Fitton's Smithian spectacles were firmly attached, and he sought out those workers whom he believed were Smith's intellectual ancestors by detecting the general arrangement of strata in Britain and in some degree their order. These were William Stukeley (1687–1765) and John Strachey (1681–1743), (Fitton 1832–1833, p. 156). The insertion of Strachey permitted Fitton to strike a blow against German mineralogist, Johann Gottlob Lehmann (1719–1767), who suffered the illusion that he was 'the first to observe and describe correctly the structure of stratified countries' ([Fitton] 1818, p. 317; Fitton 1832–1833, p. 160).

Even in 1818 Fitton recognized that Smith offered a workable future, but Fitton had difficulty in extracting himself from the kind of geology he had always known and which so shaped the early outlook of the Society. Then, and in 1832, he had obligations and friendships within the Society that called for his careful curation of the past. His essays cleverly attempted to negotiate a present in which Smith, the Society and earlier sceptics could all be brought together in an atmosphere of peace and reconciliation. One cannot doubt his honest appreciation of Smith, but he, like Farey, was never an uncomplicated advocate. Indeed, other factors were far more influential in the Society's recognition of Smith: Farey's early death in 1826 meant that the Society could make a U-turn without losing face; Smith's near silence in these disputes meant he could be visualized as a political neutral; Smith's 'inestimable' character made him an uncontroversial subject for adoption; the rise of Smithian geology amongst the membership had proceeded steadily, and by the early 1820s most understood that they were indebted to Smith and wished to recognize him; John Phillips's recent proof of Smith's method; and Smith's adoption by the Yorkshire elite, which made him an unstoppable force ripe for recognition by the best-known and most intellectually talented Yorkshire geologist of them all, Sedgwick.

Dissipation and reinvention

From the moment of its birth, the Geological Society had become the intellectual hub and powerhouse of the new science. It embodied a national desire, and tapped into the social, economic and intellectual networks for which the project had immediate and obvious worth. It was not alone for long. The Edinburgh-based Wernerian Natural History Society formed in January 1808, at that moment when the Geological was transforming itself into a proper learned society. It was no imitator; it was established to take forward the teachings of Werner. It became one hub of a Scottish debate that shaped the anti-theoretical resolve of the London men. If the Society believed it possessed the nation's geology, it knew that nation was England. Yet, it believed it followed the only proper philosophical course, and in doing so doubtless also believed that its ideas and methods would eventually displace those of the Scots. In 1811 Greenough and Buckland planned trips into Scotland and Ireland. Referring to the famous old English ballad, *Chevy Chase*, in which the King of Scotland mistakes, for an invasion, an English hunt in the Cheviot Hills, Conybeare joked:

> May nothing inauspicious befall your expedition. May no jealousy inspire the Wernerian Society of Edinburgh to array full twenty hundred Scottish hammers under the banners of Jameson to resent the Intrusion of English interlopers on the only quarries which Cheviot now furnishes. Should any fatal accident betide our President I fear we could not console ourselves so easily as King Harry [:]
>
> we trust we have within our realm
> five hundred good as he.[83]

The presence of the Wernerian should make us doubt that a geological society alone could really define and control a nation's science. Geological debate and the control of geological territories were not restricted to societies, and especially not to geological societies. There were other society-like associations such as the regular readership and correspondents of a journal like *Philosophical Magazine*. The average member experienced the Geological Society much as they would a magazine subscription. This is not to devalue the Geological Society's status or achievements, but to place them in a richer context. The Geological Society may have believed from the outset that it was the nation's geological society, and it certainly was for those who participated in its activities, but there was more to geology than this. If the likes of Smith were outsiders, they were in good company.

Six years after the Wernerian, the Royal Geological Society of Cornwall was established, in February 1814. Again, while this signals a peculiar concentration of practical geological interest in that county, this Cornish innovation was rather less extraordinary than it now seems. In towns and cities across the nation, there were attempts to formalize associations into philosophical and literary clubs and societies. Most frequently these failed, formally, but existed in networks that may well have been geological. Smith's contacts around Bath suggest such a concentration. Farey's work for the Board of Agriculture in Derbyshire further weakens the firmness of the boundaries that we might put in place to separate societies from individuals and government bodies. Here, too, we might add the almost formalized social network that surrounded men like Banks, to whom so many were indebted for patronage and various opportunities for socialized science beyond the more rarefied activities of the Royal Society. The primary

[83] UCL: Conybeare to Greenough, 28 June 1811, Add. 7918, 434.

building blocks for these different social configurations were, of course, the individuals who wished to belong and join. In this ambition they were aided by such works as Hume & Evans' (1853) directory of learned societies. Interestingly, appended to the entry on the Geological Society was a long note on the Geological Survey. Plainly the Survey was not a Society, but the compilers of the book thought the differences rather less definite than we might presume. It had, for example, offered casual employment to local gentlemen.

In the 1820s Britain became populated with philosophical societies, a large number of which had strong – sometimes dominant – interests in geology. The curatorial class they now sponsored was primarily made up of geologists. In 1831 the British Association for the Advancement of Science was established as a roving body outside the control of a single clique. Involving Jameson, but first meeting in the neutral territory of York, where it was controlled by Vernon Harcourt, its opening impromptu lecture was on geology performed by that embodiment of youthful science, John Phillips, who also became a controlling influence in the Association. Thus, at the moment when Sedgwick defined and elevated 'English geology' and took possession of it for the Society, another national scientific 'society' was formed that reconfigured the disciplinary landscape. It may have represented an association of interests, and was thus unable to usurp the Geological Society's pre-eminence, but it became a public forum for debate simply because it lay outside the Society's grasp (Rudwick 1985).

In many ways, all these developments endorsed the achievements of the Geological Society. They were a product of that individualistic, investigative and aspiring culture of geology propagated by the Society in its early years. But at the same time, if success brings imitation and diversification, then for the parent Society it must also produce a degree of dissipation. With the founding of the philosophical societies, for example, it had to defer, in part, to the authority of local networks and collections, and it began to doubt the necessity of its own museum. The biggest challenge, however, came with the formation of the Geological Survey of Great Britain. This, when reconfigured at the beginning of the 1840s, effectively removed geology from the competitive culture that the Geological Society had engendered. It was an intentional political move – a cultural revolution – that changed the science fundamentally and the role of all societies within it (Knell 2007*a*, *b*). The Survey, skilfully managed by De la Beche, now enclosed and internalized those disputed aspects of the science that had given it forward momentum. The Society's role must, by this move, have been weakened still further, but by then the governance of the country was changing and taxation could increasingly fund geological enterprise. By then also, geology had been normalized as a subject and was thus naturally brought into the work of numerous organizations. Buckland, in 1840, noted the spread of geology into all areas of public life (Buckland 1840). Geology had also established itself in the literature, and courted an expanded public who wished to speculate and imagine (O'Connor 2007, 2009).

Following the 50th anniversary celebrations of the Society, yet another geological society was established in London. This was the Geologists' Association, which was formed to perform a rather humbler kind of geology for the everyman (Sweeting 1958). However, it drew considerable support from the Society's Fellows who liked its active field-based programme. The Association was no talking shop: it was proud of Britain's rich geological heritage and made engagement participatory. Across the nation field clubs were established and, once again, the socialization of geology changed and with it the place of the Society.

Smith, too, would be subject to the same dissipating influences. Buckland, in 1840, with the second edition of Greenough's map newly mounted in the meeting room, inadvertently began the assault on Smith. Then President, Buckland, who had been active in seeing through the revision of the map, referred to it as 'the glory of this Society' (Buckland 1840, p. 222). Ironically, but perhaps intentionally, he did so on the occasion of recording the death of William Smith. At that moment, it seems, his task was to care for the living and particularly for his great friend, Greenough, for by then one could not look at Greenough's map or think of William Smith without believing that the Society, and Greenough in particular, had orchestrated an injustice. Farey's claims against the Society remained unanswered and in festering away doubtless produced rumours particularly amongst those who never knew the Society in the 'old days'.

But Buckland did not stop there, for he also withdrew a little of the status the Society had conferred on Smith less than a decade before, suggesting implicitly that Smith's parentage of English geology needed to be taken with a pinch of salt:

> But it must not be forgotten, that both in this country and on the continent, other investigators, many of them no doubt unknown to him, were simultaneously collecting similar evidence in support of this great physical generalization. [...] In our just admiration of our countryman, therefore, we must not lose sight of the merits of his contemporary labourers on the continent; and whilst we honour him as the father of English Geology, let us also pay just homage to those

who had started before him in the same course, wherein it was his undisputed merit to have arrived first at the goal.

(Buckland 1840, pp. 251–252)

Buckland's tribute to Smith, then, for all its warmth, is also the beginning of a counter-narrative, by which a succession of historians and commentators have sought to question or reconfigure Smith's status. This has not been a co-ordinated effort, just the product of revisionist ambitions or simple forgetting, but it has been met by a seemingly co-ordinated response from generations of historians who have seen the defence of Smith as an act of almost religious devotion. Perhaps it is a defence against the inevitable effects of the Second Law of Thermodynamics: the past requires constant curation if it is not to fall into disorder. A reviewer of the Society's *Transactions* in 1830, for example, wrote: 'Mr Smith's views were taken up by the Geological Society of London, and prosecuted with a degree of zeal, liberality, and success, which does them infinite honour' (Anon. 1830). Smith had his advocates and admirers in a way that Greenough did not. One could well romanticize the John-Bullish Smith who seemed to epitomize the English folk hero.

In Greenough's life, those who had contributed to his map – men such as Buckland and Conybeare – also helped promote it. By these efforts, Greenough's map became the Society's map. However, when the third edition of the map was produced in 1865 its title seemed to finally admit that Greenough had plagiarized Smith (Challinor 1964–1965, II, p. 164). This, too, is an erroneous interpretation.

In July 1857 a committee was formed to revise the recently deceased Greenough's map, which, in its second edition of 1840, was still actively used by contemporary scientists. It is noteworthy that this committee was made up of a clique of Smithians: Murchison, John Phillips, Robert Alfred Cloyne Godwin-Austin (1808–1884), John Morris (1810–1886), Joseph Prestwich (1812–1896) and Colonel James. Of these, Phillips, who was soon to become President, was to superintend the project. It was to be funded in part from a legacy of £500 left by Greenough. Council gave the committee permission to give the map a new title, and it was resolved on 7 December 1864, that this be 'A Geological Map of England and Wales by G. B. Greenough Esq. F.R.S. (on the Basis of the original Map of Mr Smith 1815) revised and improved with the results of the Geological Survey 1836–63, on MS additions by Sir Roderick Murchison, Professor Phillips, Joseph Prestwich Esq. and R. Godwin Austen Esq. under the Supervision of a Committee of the Geological Society'.[84] What had happened was that the committee had credited itself, with contributors like Murchison, who was not alone in appreciating the memorializing nature of publication, considering the legacy of his own earlier work. Indeed, given that the basis for this new map was the Geological Survey's maps, in what sense was this even Greenough's map? Those Survey maps had, of course, been developed from earlier maps – earlier sources were plagiarized because one could never know precisely who had first drawn and coloured which line or done so correctly. Geological maps were cumulative ventures. This is accepted today, but was used to doubt Greenough in 1819. In hardly crediting his contributors, Greenough had set a standard for map-making or rather adopted one already long established. Those he did mention in his first *Memoir* did not appear in his second edition, yet clearly their contributions remained on the map itself. Smith finds his way onto the third edition of the map primarily because those living wished to see themselves credited for their earlier work and therefore they could not fail to mention the pioneering work of Smith himself. In doing so they did not mean to intimate that Greenough had stolen Smith's data. It was an act of self-acknowledgement that caused much debate and argument following the map's production, many others apparently believing that they too deserved credit.

It is ironic indeed that Smith, a 'happy farmer' or 'a plain farmer-looking man',[85] who was by all accounts a simple, modest and generous man, should become such a significant political object in the development and history of geology. It is perhaps because of all the things the science discovered in that period, none was more fundamental than Smith himself. If one could claim a discovery and a reputation by possessing a fossil, then by possessing Smith, as Sedgwick did, one might also believe, and lead others to imagine, that one possessed the science itself. Today, there is no better material representation of his very English achievement than his map. It rightly stands for scientific aspiration, but in a rather humble and self-effacing English manner. However, if we wish to understand the role of the Society in this science then we must instead look to Greenough's map, for it both embodies the tremendous co-operative spirit that saw the Society

[84] GSL: Geological Map Revision Committee details the process of completing the 1865 map.

[85] For 'happy farmer' as described by John Phillips in 1828 (Knell 2000, p. 167); 'plain farmer-looking man', as described by John Eddowes Bowman, 1836 (Edmonds & Beardmore 1955, p. 104).

come into being and the competitive politics of socialized scientific engagement which the Society nurtured so ably and which drove the science to achieve so much in its first 50 years. It was this, too, that enabled the Society men to capture Smith and his map for their own purposes.

I would like to thank my referees, Martin Rudwick and Pietro Corsi, for their helpful comments and suggestions on what became an early and rather different version of this paper. Mike Taylor kindly agreed to read and comment on two later versions of this paper, and Hugh Torrens offered a number of useful corrections to the manuscript in the final hours of its preparation. In thanking them, I must stress that the views expressed here are my own and not necessarily theirs. The expertise of GSL librarian, Wendy Cawthorne, has been vital to this project as she has been able to locate many of the primary sources upon which this paper is based. The staff at the Special Collections at UCL gave considerable help in locating most of the Greenough-related materials used in this paper and I thank them for permission to quote from and illustrate them. Michael Richardson and the University of Bristol Library Special Collections aided Cherry Lewis in obtaining photographs of the Phillips maps illustrated here. Tom Sharpe, at the National Museum of Wales, kindly supplied me with the images of Smith's and Greenough's maps. I am grateful to the History of Geology Group for generously supporting the printing of the coloured illustrations, and to the University of Leicester for further financial support. Without the help of these individuals and organisations this paper would not have been possible. Finally, and most importantly, I must thank Cherry for her constant encouragement during the seemingly never ending writing of this paper, which took much of 2008.

Archives

GSL: Geological Society of London.

UCL: University College London, Special Collections, Greenough papers.

References

AIKIN, A. 1814. *A Manual of Mineralogy*. Longman, London.

ALBRECHT, H. & LADWIG, R. (eds) 2002. *Abraham Gottlob Werner und die Begründung der Geowissenschaften*. Technische Universität Bergakademie, Freiberg.

ANON. 1830. Transactions of the Geological Society of London, Second Series. *Edinburgh Review*, **52**, 43–72.

BOWDEN, A. J. 2009. Geology at the crossroads: aspects of the geological career of Dr John MacCulloch. *In*: LEWIS, C. L. E. & KNELL, S. J. (eds) *The Making of the Geological Society of London*. Geological Society, London, Special Publications, **317**, 255–278.

BRANAGAN, D. F. 2009. The Geological Society on the other side of the world. *In*: LEWIS, C. L. E. & KNELL, S. J. (eds) *The Making of the Geological Society of London*. Geological Society, London, Special Publications, **317**, 341–371.

BUCKLAND, W. 1840. Address to the Geological Society, delivered at the Anniversary, on the 21st of February. *Proceedings of the Geological Society of London*, **3**, 210–267.

CHALLINOR, J. 1964–1965. Some correspondence of Thomas Webster, geologist I–VI (1773–1844). *Annals of Science*, **17**, 175–195; **18**, 147–175; **19**, 49–79, 285–297; **20**, 59–80, 143–163.

CONYBEARE, W. D. & PHILLIPS, W. 1822. *Outlines of Geology of England and Wales*. Phillips, London.

COX, L. R. 1942. New light on William Smith and his work. *Proceedings of the Yorkshire Geological Society*, **25**, 1–99.

CUMMING, D. A. 1985. John MacCulloch, blackguard, thief and high priest, reassessed. *In*: WHEELER, A. & PRICE, J. H. (eds) *From Linnaeus to Darwin: Commentaries on the History of Biology and Geology*. Society for the History of Natural History, London, 77–88.

EDMONDS, J. M. 1975. The geological lecture-courses given in Yorkshire by William Smith and John Phillips, 1824–1825. *Proceedings of the Yorkshire Geological Society*, **40**(3), 373–412.

EDMONDS, J. M. & BEARDMORE, P. A. 1955. John Phillips and the early meetings of the British Association. *The Advancement of Science*, **12**, 97–104.

ELLENBERGER, F. 1999. *History of Geology*, Volume 2. Balkema, Rotterdam.

EYLES, J. M. 1967. William Smith: The sale of his collection to the British Museum. *Annals of Science*, **23**, 177–212.

EYLES, J. M. 1985. William Smith, Sir Joseph Banks and the French Geologists. *In*: WHEELER, A. & PRICE, J. H. (eds) *From Linnaeus to Darwin: Commentaries on the History of Biology and Geology*. Society for the History of Natural History, London, 37–50.

EYLES, V. A. & EYLES, J. M. 1938. On the different issues of the first geological map of England and Wales. *Annals of Science*, **3**, 190–212.

FAREY, J. 1811. *General View of the Agriculture and Minerals of Derbyshire*, Volume I. Sherwood, Neely and Jones, London.

FAREY, J. 1815*a*. Observations on the priority of Mr. Smith's investigations of the strata of England; on the very unhandsome conduct of certain persons in detracting from his merit therein; and the endeavours of others to supplant him in the sale of his maps. *Philosophical Magazine*, **45**, 333–444.

FAREY, J. 1815*b*. *General View of the Agriculture of Derbyshire*, Volume II. Sherwood, Neely and Jones, London.

FAREY, J. 1817. *General View of the Agriculture of Derbyshire*, Volume III. Sherwood, Neely and Jones, London.

[FAREY, J.] (A correspondent). 1817. Geological queries to Mr Westgarth Forster, Mr Winch, Mr Fryer &c regarding the basaltic and other strata of Durham, Northumberland, & c & c. *Philosophical Magazine*, **50**, 45–50.

[FAREY, J.] 1818*a*. Mr Smith's geological claims stated. *Philosophical Magazine*, **51**, 173–180.

[FAREY, J.] 1818*b*. Mr W Smith's discoveries in geology. *Annals of Philosophy*, **11**, 359–364.

FAREY, J. 1818. On the very correct notions concerning the structure of the Earth, entertained by the Rev. John Michell as early as the year 1760; and the great neglect which his publication of the same has received from later writers on geology, and regarding the treatment of Mr. Smith by certain persons. *Philosophical Magazine*, **52**, 183–195, 254–270, 323–341.

FAREY, J. 1819. On the importance of knowing and accurately discriminating fossil-shells, as the means of identifying particular beds of strata, etc. *Philosophical Magazine*, **53**, 112–132.

FAREY, J. 1820. Free remarks on Mr. Greenough's geological map, lately published under the direction of the Geological Society of London. *Philosophical Magazine*, **55**, 379–383.

[FITTON, W. H.] 1817. Transactions of the Geological Society, established November 1807. *Edinburgh Review*, **18**, 70–94.

[FITTON, W. H.] 1818. Smith's geological map of England. *Edinburgh Review*, **29**, 310–337.

FITTON, W. H. 1832–1833. Notes on the history of geology. *Philosophical Magazine and Journal*, **1**, 147–160, 268–275, 442–450; **2**, 37–57.

FORD, T. D. 1967. The first detailed geological sections across England, by John Farey, 1806–8. *Mercian Geologist*, **2**, 41–49.

FORD, T. D. & TORRENS, H. S. 1989. John Farey (1766–1826): An unrecognized polymath. *In*: FAREY, J. *General View of the Agriculture and Minerals of Derbyshire* (1811). Peak District Historical Society reprint.

GREENOUGH, G. B. 1819*a*. *A Critical Examination of the First Principles of Geology*. Longman, Hurst, Rees, Orme and Brown, London.

GREENOUGH, G. B. 1819*b*. Observations on certain free remarks by Mr. Farey published in the last number of the Philosophical Magazine. *Philosophical Magazine*, **54**, 205–206.

GREENOUGH, G. B. 1820. *Memoir of a Geological Map of England*. Longman, Hurst, Rees, Orme and Brown, London.

GREENOUGH, G. B. 1834. Presidential address. *Proceedings of the Geological Society*, **2**(35), 42.

GUNTAU, M. 2009. The rise of geology as a science in Germany around 1800. *In*: LEWIS, C. L. E. & KNELL, S. J. (eds) *The Making of the Geological Society of London*. Geological Society, London, Special Publications, **317**, 163–177.

HERINGMAN, N. 2009. Picturesque ruin and geological antiquity: Thomas Webster and Sir Henry Englefield on the Isle of Wight. *In*: LEWIS, C. L. E. & KNELL, S. J. (eds) *The Making of the Geological Society of London*. Geological Society, London, Special Publications, **317**, 299–317.

HERRIES DAVIES, G. 2007. *Whatever is Under the Earth*. Geological Society, London.

HERRIES DAVIES, G. 2009. Jacques-Louis, Comte de Bournon. *In*: LEWIS, C. L. E. & KNELL, S. J. (eds) *The Making of the Geological Society of London*. Geological Society, London, Special Publications, **317**, 105–113.

HUME, A. & EVANS, A. I. 1853. *The Learned Societies and Printing Clubs of the United Kingdom*. G. Willis, London.

JAMESON, R. 1816. *A System of Mineralogy* (3 volumes), 2nd edn. Archibald Constable, Edinburgh.

KHAIN, V. E. & MALAKHOVA, I. G. 2009. Scientific institutions and the beginnings of geology in Russia. *In*: LEWIS, C. L. E. & KNELL, S. J. (eds) *The Making of the Geological Society of London*. Geological Society, London, Special Publications, **317**, 203–211.

KIDD, J. 1814. Notes on the mineralogy of the neighbourhood of St. David's, Pembrokeshire. *Transaction of the Geological Society*, **2**, 79–93.

KIDD, J. 1815. *A Geological Essay on the Imperfect Evidence in support of a Theory of the Earth, deducible either from its General Structure or from the Changes Produced on its Surface by the Operation of Existing Causes*. Oxford University, Oxford.

KNELL, S. J. 2000. *The Culture of English Geology, 1815–1851: A Science Revealed Through Its Collecting*. Ashgate, Aldershot.

KNELL, S. J. 2007*a*. Museums, fossils and the cultural revolution of science: Mapping change in the politics of knowledge in early nineteenth-century Britain. *In*: KNELL, S. J., MACLEOD, S. & WATSON, S. E. R. (eds) *Museum Revolutions: How Museums Change And Are Changed*. Routledge, London, 28–47.

KNELL, S. J. 2007*b*. The sustainability of geological map-making: The case of the Geological Survey of Great Britain. *Earth Science History*, **26**(1), 13–29.

KNELL, S. J. 2007*c*. Museums, reality and the material world. *In*: KNELL, S. J. (ed.) *Museums in the Material World*. Routledge, London, 1–28.

KNIGHT, D. 2009. Chemists get down to Earth. *In*: LEWIS, C. L. E. & KNELL, S. J. (eds) *The Making of the Geological Society of London*. Geological Society, London, Special Publications, **317**, 93–103.

KÖLBL-EBERT, M. 2009. George Bellas Greenough's 'Theory of the Earth' and its impact on the early Geological Society. *In*: LEWIS, C. L. E. & KNELL, S. J. (eds) *The Making of the Geological Society of London*. Geological Society, London, Special Publications, **317**, 115–128.

LAUDAN, R. 1977. Ideas and organizations in British geology: A case study of institutional history. *Isis*, **66**, 527–538.

LAUDAN, R. 1987. *From Mineralogy to Geology: The Foundations of a Science, 1650–1830*. University of Chicago, Chicago, IL.

LEWIS, C. L. E. 2009*a*. 'Our favourite science': Lord Bute and James Parkinson searching for a Theory of the Earth. *In*: KÖLBL-EBERT, M. (ed.) *Geology and Religion: A History of Harmony and Hostility*. Geological Society, London, Special Publications, **310**, 111–126.

LEWIS, C. L. E. 2009*b*. Doctoring geology: The medical origins of the Geological Society. *In*: LEWIS, C. L. E. & KNELL, S. J. (eds) *The Making of the Geological Society of London*. Geological Society, London, Special Publications, **317**, 49–92.

MACINTYRE, S. & CLARK, A. 2003. *The History Wars*. Melbourne University, Melbourne.

MCBIRNEY, A. 2005. Essay Review: Simon Winchester, *The Map That Changed the World*. *Earth Sciences History*, **24**, 127–130.

MILLER, D. P. 2002. The 'Sobel Effect'. *Metascience*, **2**, 185–200.

MOORE, D. T., THACKRAY, J. C. & MORGAN, D. L. 1991. A short history of the Museum of the Geological Society of London 1807–1911, with a catalogue of the British and Irish accessions, and notes on surviving collections. *Bulletin of the British Museum (Natural History) Historical Series*, **19**(1), 51–160.
MORGAN, N. 2007. True confessions. *Geoscientist*, **17**, (5), 14–18.
MORRELL, J. 2005. *John Phillips and the Business of Victorian Science*. Ashgate, Aldershot.
MURAKAMI, H. 2003. *The Wind-up Bird Chronicle*. Vintage, London.
NEWELL, J. R. 2009. A story of things yet-to-be: the status of geology in the United States in 1807. *In*: LEWIS, C. L. E. & KNELL, S. J. (eds) *The Making of the Geological Society of London*. Geological Society, London, Special Publications, **317**, 213–217.
O'CONNOR, R. 2007. *The Earth on Show*. Chicago University Press, Chicago, IL.
O'CONNOR, R. 2009. Facts and fancies: the Geological Society of London and the wider public, 1807–1837. *In*: LEWIS, C. L. E. & KNELL, S. J. (eds) *The Making of the Geological Society of London*. Geological Society, London, Special Publications, **317**, 331–340.
OLDROYD, D. 2001. The story of Strata-Smith. *Science*, **293**, 1439–1440.
PARKINSON, J. 1811. Observations on some of the strata in the neighbourhood of London, and on the fossil remains contained in them. *Transactions of the Geological Society*, **1**, 324–354.
PEARCE, S. M. 1990. Objects as meaning; or narrating the past. *In*: PEARCE, S. M. (ed.) *Objects of Knowledge*. Athlone, London, 125–140.
PEARCE, S. M. 1995. *On Collecting*. Routledge, London.
PREZIOSI, D. 1989. *Rethinking Art History*. Yale University Press, New Haven, CT.
PREZIOSI, D. (ed.). 1998. *The Art of Art History: A Critical Anthology*. Oxford University Press, Oxford.
PHILLIPS, J. 1829. *Illustrations of the Geology of Yorkshire or, a Description of the Strata and Organic Remains of the Yorkshire Coast accompanied by a Geological Map, Sections, and Plates of the Fossil Plants and Animals*. Wilson, York.
PHILLIPS, J. 1844. *Memoirs of William Smith*. Murray, London.
PHILLIPS, W. 1816. *Outlines of Mineralogy and Geology, intended for the use of those who may desire to become acquainted with the Elements of those Sciences; especially Young Persons*, 2nd edn. W. Phillips, London.
PHILLIPS, W. 1818. *A Selection of Facts from the best authorities, arranged so as to form an outline of the Geology of England and Wales*. W. Phillips, London.
RUDWICK, M. J. S. 1963. The foundation of the Geological Society of London: Its scheme for co-operative research and its struggle for independence. *British Journal for the History of Science*, **1**, 325–355.
RUDWICK, M. J. S. 1985. *The Great Devonian Controversy*. University of Chicago, Chicago, IL.
RUDWICK, M. J. S. 1998. Lyell and the *Principles of Geology*. *In*: BLUNDELL, D. J. & SCOTT, A. C. (eds) *Lyell: The Past is the Key to the Present*. Geological Society, London, Special Publications, **143**, 3–15.
RUDWICK, M. J. S. 2005. *Bursting the Limits of Time: The Reconstruction of Geohistory in the Age of Revolution*. Chicago University, Chicago, IL.
RUDWICK, M. J. S. 2009. The early Geological Society in its international context. *In*: LEWIS, C. L. E. & KNELL, S. J. (eds) *The Making of the Geological Society of London*. Geological Society, London, Special Publications, **317**, 145–153.
SEDGWICK, A. 1831. [Address on the award of the Wollaton Prize]. *Proceedings of the Geological Society of London*, **1**, 270–280.
SHEPPARD, T. 1917. William Smith: His maps and memoirs. *Proceedings of the Yorkshire Geological Society*, **19**, 75–253.
SWEETING, G. S. 1958. *The Geologists' Association 1858–1958*. Benham, Colchester.
TAQUET, P. 2009. Geology beyond the Channel: The beginnings of geohistory in early nineteenth-century France. *In*: LEWIS, C. L. E. & KNELL, S. J. (eds) *The Making of the Geological Society of London*. Geological Society, London, Special Publications, **317**, 155–162.
TORRENS, H. S. 1990. The scientific ancestry and historiography of *The Silurian System*. *Journal of the Geological Society*, **147**, 1–17.
TORRENS, H. S. 1994. Patronage and problems: Banks and the Earth sciences. *In*: BANKS, R. E. R., ELLIOTT, B., HAWKES, J. G., KING-HELE, D. & LUCAS, G. L. (eds) *Sir Joseph Banks: A Global Perspective*. Royal Botanic Gardens, Kew, 49–75.
TORRENS, H. S. 1998*a*. Geology in peace time: An English visit to study German mineralogy and geology (and visit Goethe, Werner and Raumer) in 1816. *In*: FRITSCHER, B. & HENDERSON, F. (eds) *Towards a History of Mineralogy, Petrology, and Geochemistry*. Instut für Geschichte der Naturwissenshaften, Munich, 147–175.
TORRENS, H. S. 1998*b*. Le 'nouvel art de prospection minire' de William Smith et le project de houillère de Brewham: Un essai malencontreux de recherche de charbon dans le sud-ouest de L'Angleterre entre 1803 et 1810. *In*: GAUDANT, J. (ed.) *De la Géologie à son Histoire: Livre Jubilaire pour Franois Ellenberger*. CTHS Presse, Paris, 101–118.
TORRENS, H. S. 1998*c*. Geology and the natural sciences: Some contributions to archaeology in Britain 1780–1850. *In*: BRAND, V. (ed.) *The Study of the Past in the Victorian Age*. Oxbow Monograph, **73**, 35–59.
TORRENS, H. S. 2001. Timeless order: William Smith (1769–1839) and the search for raw materials, 1800–1820. *In*: LEWIS, C. L. E. & KNELL, S. J. (eds) *The Age of the Earth: From 4004 BC to AD 2002*. Geological Society, London, Special Publications, **190**, 61–83.
TORRENS, H. S. 2009. Dissenting science: the Quakers among the founding fathers. *In*: LEWIS, C. L. E. & KNELL, S. J. (eds) *The Making of the Geological Society of London*. Geological Society, London, Special Publications, **317**, 129–144.
TOWNSEND, J. 1813. *The Character of Moses Established for Veracity as an Historian, Recording Events from the Creation of the Deluge*. Gye, Bath and Longman, London.

VENEER, L. 2009. Practical geology and the early Geological Society. *In*: LEWIS, C. L. E. & KNELL, S. J. (eds) *The Making of the Geological Society of London*. Geological Society, London, Special Publications, **317**, 243–253.

WEBSTER, T. 1814. On the freshwater formations of the Isle of Wight, with some observations on the strata over the Chalk in the south-east part of England. *Transactions of the Geological Society*, **2**, 161–254.

WEINDLING, P. J. 1979. Geological controversy and its historiography: The prehistory of the Geological Society of London. *In*: JORDANOVA, L. J. & PORTER, R. S. (eds) *Images of the Earth*. British Society for the History of Science, 248–271.

WEINDLING, P. J. 1983. The British Mineralogical Society: a case study in science and social improvement. *In*: INKSTER, I. & MORRELL, J. B. (eds) *Metropolis and Province: Science in British Culture 1780–1850*. Hutchinson, London, 120–150.

WERNER, A. G. 1791. *Neue Theorie von der Entstehung der Gänge mit Anwendung auf dem Bergbau besonders den Freibergischen*. Freiberg.

WERNER, A. G. 1805. *A Treatise on the External Characters of Fossils*. (Trans. Weaver, T.). Mahon, London.

WHITEHURST, J. 1786. *An Inquiry into the Original State and Formation of the Earth*. W. Bent, London.

WINCHESTER, S. 2001. *The Map That Changed the World*. Viking, London.

WOODWARD, H. B. 1907. *The History of the Geological Society of London*. Geological Society, London.

Doctoring geology: the medical origins of the Geological Society

CHERRY L. E. LEWIS

Senate House, University of Bristol, Tyndall Avenue, Bristol BS8 1TH, UK (e-mail: Cherry.lewis@bristol.ac.uk)

Abstract: Four of the Geological Society's 13 founders were medical men: William Babington, James Parkinson, James Franck and James Laird, the Society's first Secretary. All were physicians and mineralogists except Parkinson, an apothecary surgeon and fossilist. At least 20 percent of the Society's early members were also medical practitioners whose prime interest was mineralogy. The subject was taught as part of medical training, required as it was in the fabrication of medicines, thus medical men were drawn into mineralogy and on into geology. In 1805 a number of medical practitioners broke away from the constraints of their parent body, the Medical Society of London, to form the Medical and Chirurgical Society, which became a role model for the young Geological Society when challenged by its parent body, the Royal Society. Driven by wealthy mineral collectors and patrons of science like Charles Greville, one reason – perhaps *the* reason – for founding the Society was to map the mineralogical history of Britain. Towards this endeavour, Babington's expertise in mineralogy brought people together, Laird organized them and Parkinson was invited because he was *not* a mineralogist. Franck was unable to participate significantly, being away at war for much of the time. The contribution made to the founding of the Geological Society by each of the medical founders is examined, and a biographical sketch of each man reveals the close relationship between medicine and the emergence of this new science of geology.

> There is no walk of life in which the struggle with difficulties is so sure to be hard, or the array of circumstances to be disheartening, as in the profession of medicine. To succeed in this most arduous of all professions, requires qualities of a high order [...].
>
> (Jackson 1845, p. xvii)

King George III (1738–1820) came to the throne in 1760 in the middle of the Seven Years War (1756–1763), which was soon followed by the American War of Independence (1775–1783), the French Revolutionary Wars (1793–1802) and the Napoleonic Wars (1804–1815). Each war cost almost twice the previous one, hugely increasing the national debt (Morgan 1796). In order to help repay this debt, everything was heavily taxed – from hair powder to candles – and, it seemed, the greater share of this burden fell upon the working man: 'More than two thirds of every shilling we earn, is torn from us by Taxes laid on the articles most necessary to the support of life' ([Parkinson] Old Hubert *c*. 1795) complained the apothecary surgeon, James Parkinson (1755–1824). Not only was the population heavily taxed to meet the cost of war, but in the 1790s wages fell and the price of food rose sharply:

> While the wages of the poor manufacturer fell, in some instances, a full third, every article of consumption rose in double proportion. Oats rose from 1790 to 1795, 75 per cent, and hay and every article of pulse kept pace with this increase.
>
> (Thelwall 1837, p. 319)

Throughout the eighteenth century, the population of London rose by 400 000, standing at just under one million in 1800 (Porter 2000, p. 157). As the City became more prosperous, residential areas were taken over for business purposes and houses were demolished to make room for factories, warehouses and offices, which forced the displaced residents to find homes beyond the City walls. Rents soared. Poor families found they could no longer afford a home of their own and were obliged to share with others, resulting in gross overcrowding (Porter 2000; Ackroyd 2001; Picard 2003). With overcrowding and hunger came diseases such as small pox, typhoid and malaria (the ague). These ran unchecked through the population at frequent intervals and for decades, the mortality of children under the age of two was 50 percent. It was, of course, the medical practitioners who dealt with this misery on a daily basis (Porter 1991, 1997).

Medical practice in the eighteenth century

There were three types of medical practitioners in the eighteenth century – physicians, surgeons and apothecaries – and each was overseen by its parent body: the Royal College of Physicians; the Company of Surgeons (which in 1800 became the Royal College of Surgeons of London) and the Worshipful Society of Apothecaries. There was a clearly demarcated hierarchy between the three

From: LEWIS, C. L. E. & KNELL, S. J. (eds) *The Making of the Geological Society of London*. The Geological Society, London, Special Publications, **317**, 49–92.
DOI: 10.1144/SP317.2 0305-8719/09/$15.00

Table 1. *Birth and death dates for the 13 founder members of the Geological Society, showing their age on 13 November 1807, their age at death and the year they became FRS (shaded grey). Their lives spanned more than a century*

Founders in chronological order	Born	Died	1807	Age at death
J.L. de Bournon	21 Jan 1751	24 Aug 1825	56	74
James Parkinson	11 Apr 1755	21 Dec 1824	52	69
William Babington	21 May 1756	29 Apr 1833	51	76
James Franck	29 Jun 1768	27 Jan 1843	39	75
Richard Knight	26 May 1768	21 Feb 1844	39	75
William Allen	29 Aug 1770	30 Dec 1843	37	73
William Phillips	9 May 1773	2 Apr 1828	34	54
Arthur Aikin	19 May 1773	15 Apr 1854	34	80
William H. Pepys	23 Mar 1775	17 Aug 1856	32	81
George B. Greenough	18 Jan 1778	2 Apr 1855	29	77
Richard Phillips	21 Nov 1778	11 May 1851	28	73
Humphry Davy	17 Dec 1778	29 May 1829	28	51
James Laird	17 Dec 1779	3 Jan 1841	27	62

Timeline axis: 1750, 1755, 1760, 1765, 1770, 1775, 1780, 1785, 1790, 1795, 1800, 1807, 1810, 1815, 1820, 1825, 1830, 1835, 1840, 1845, 1850, 1855

professions, with apothecaries at the bottom and physicians at the top. A Fellow of the Royal College of Physicians had to have a medical degree from either Oxford or Cambridge, and of the four medical founders of the Geological Society, only James Franck (1768–1843) fell into that category. This elite group of men supplied health care to the rich and were consulted in difficult cases (Porter 2006, p. 110) by the likes of the apothecary surgeon, James Parkinson (1755–1824). They were forbidden to practice a 'trade' such as apothecary, surgeon or midwife. Below them came physicians whose religious beliefs, or financial abilities, barred them from Oxford and Cambridge, and who therefore had to obtain their medical degrees from universities such as Edinburgh, Aberdeen or Leyden. Medical men with these degrees, such as William Babington (1756–1833) and James Laird (1779–1841) were not Fellows but 'Licentiates' of the Royal College of Physicians. Their degree licensed them to practice as physicians and to 'dabble in trade', but they could not vote on College matters. Apothecaries tended to pursue formal apprenticeships, but as a group they remained largely unregulated until the Apothecaries Act of 1815. On completion of their apprenticeship, apothecaries could take an oral exam at the College of Surgeons that qualified them to practice as apothecary surgeons. Many did not – although still practised surgery.

Despite the common medical – and geological – threads running throughout the careers of these four men, their lives, ages (Table 1) and personalities varied greatly. All took very different routes that led them to the Freemasons' Tavern on the 13 November 1807, for the founding of the Geological Society. Each man's path will now be followed to see how he reached that point; en route, the bi-ways linking medicine and geology will be explored.

William Babington

William Babington was born at Port Glenone, 'on the banks of the river Ban, near Coleraine, in Ireland' (Smyth 1843, p. 93). He was the son of a clergyman, Humphrey Babington (*c.* 1722–*c.* 1770) and his wife, Anne Buttle (born *c.* 1734), who 'having a large number of sons, distributed them amongst the various professions, and made the choice of that of medicine for William, who was apprenticed at an early age to a practitioner in Londonderry' (Smyth 1843, p. 93). In March 1777 Babington entered Guy's Hospital (Fig. 1) in London to complete his medical education, where he was assigned as a 'dresser' to the surgeon James Franck senior (died 1783), uncle to the James Franck involved in the founding of the Geological Society. Apprentices like Babington frequently came to the London hospitals for clinical experience and would either 'walk' the wards,

Fig. 1. North Front of Guy's Hospital, 1815. Fry Collection, University of Bristol Library Special Collections.

William Saunders, M.D.
First President of the Medico-Chirurgical Society, 1805.

Fig. 2. William Saunders. Wellcome Library, London.

benefiting only by observing the surgeons or physicians in action, or, for a higher fee – about £50 a year or £30 for six months – they became 'dressers'. This involved assisting the surgeon to dress wounds, reduce fractures and perform some of the lesser operations of surgery under the surgeon's direction (Lawrence 1996, p. 127). On his arrival, Babington was befriended by William Saunders (1743–1817), the senior physician at Guy's who was to become hugely influential not only in Babington's life, but also in James Franck junior's and, indirectly, in James Laird's – three of the four medical founders of the Geological Society.

Saunders (Fig. 2) came from Scotland and was educated at the University of Edinburgh, where he was a pupil and friend of William Cullen (1710–1790). He graduated MD in 1765 with a thesis on antimony, then came to London and began teaching chemistry and pharmacy in the private medical schools (Moore 2004). In 1770 he was appointed as a physician to Guy's Hospital on the recommendation of Sir George Baker (*c.* 1723–1809) who considered him 'an ingenious gentleman who teaches chemistry' (Coley 1988). Saunders had assisted Baker with his chemical experiments which showed that the high incidence of 'Devonshire colic' among cider drinkers was, in fact, a form of poisoning, caused by lead used in the cider press. In 1790, Saunders was made a full Fellow of the College of Physicians, despite only having a degree from a Scottish university. It was a rare mark of distinction. He became a Fellow of the Royal Society in 1793 and was appointed Physician Extraordinaire to the Prince Regent in 1807.

In 1768 Guy's and St Thomas's hospitals, which were located across the road from one other and known as the United Hospitals of the Borough, formalized their arrangements for teaching students and began to share a medical school. In 1770 Saunders began to deliver a broadly based syllabus of medical lectures and 'under his direction the reputation of the Hospital, as a medical school, first commenced, and gained its high character; he stood himself at the head of the city practice in a few years; and his countenance and support was courted in every liberal and scientific undertaking that went on' (Anon. 1817, p. 389 ff).

Until the mid-nineteenth century there were three types of student: apprentices and dressers were assigned to a particular physician, surgeon or apothecary, but 'pupils' were not attached to anyone in particular. As with the doctors, there was a strict hierarchy: 'The only persons more priveledged than Dressers are the Apprentices, who in general are such fine Gentlemen, they do not chuse to dirt their fingers, or skreen their Panteloons with a stuff apron' (Anon. 1800, p. 11). Pupil numbers were unrestricted and they paid about half the fees of dressers, but tended to be on the periphery of surgical procedures. University students or 'schollars', as they were known in the hospitals, became physician's pupils, thereby maintaining the hierarchy they were to follow once qualified.

Apprentices, dressers and pupils attended courses of lectures for which they paid a fee of between £3 and £4 for each course (Lawrence 1996, p. 169). St Thomas's delivered the anatomical and surgical lectures, which were those most in demand and for which all pupils were prepared to pay fees, while Guy's established courses in *materia medica* (pharmacy), chemistry, botany, physiology, practice of medicine and natural philosophy (Fig. 3). In 1769 the wards were officially opened to students by a Governors' resolution, which also proposed that the surgeons of the hospitals should occasionally give practical lessons on surgery.[1] Students had absolute autonomy to tailor their attendance at lectures, according to their own

[1] KCL: GB 0100 G/Students. Student records of Guy's Hospital Medical School. www.kcl.ac.uk/depsta/iss/archives/collect/10gu7125.html.

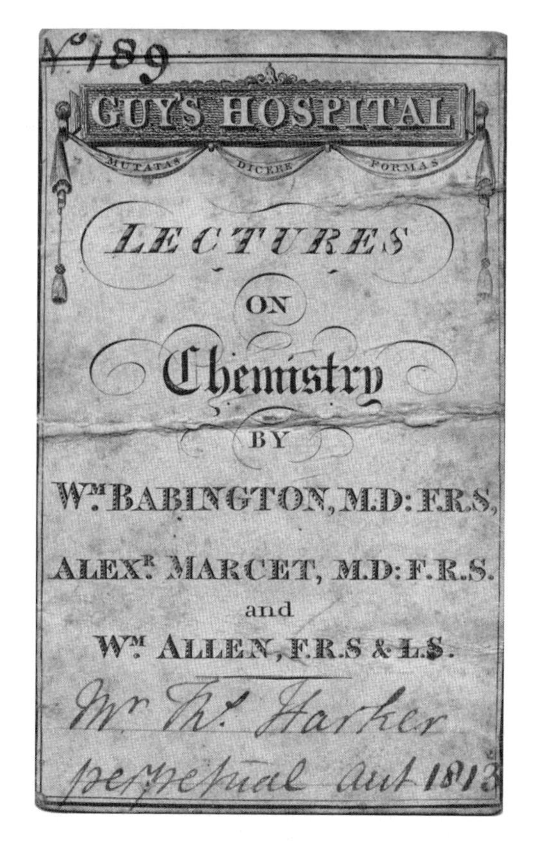

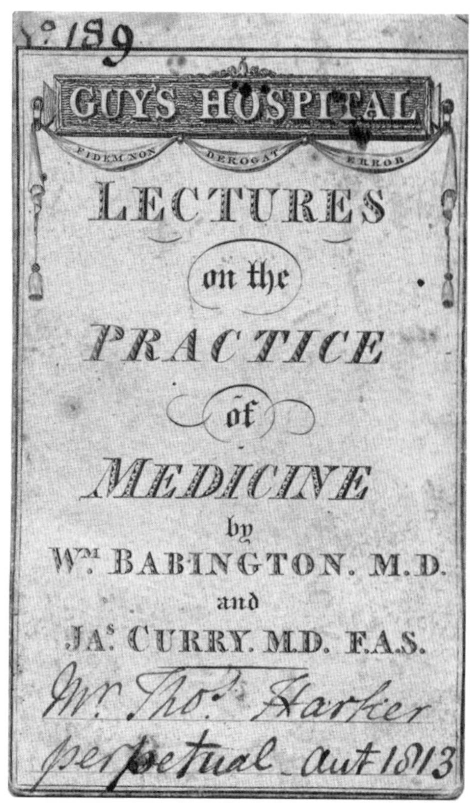

Fig. 3. Admission cards for Thomas Harker to Guy's Hospital lectures in 1813. Kings College London Archives. G/PP3/3 1813.

interests, ambitions and their income. Lecturers too had complete freedom over where and when they gave courses, and many built highly successful businesses lecturing both within and outside the hospitals. Saunders, for example, is said to have earned £1000 a year from his medical courses (Wilks & Bettany 1892). Students attended operations at both hospitals, although the operating theatres were completely inadequate for the numbers wishing to attend:

> The general arrangement of all the theatres was the same, a semicircular floor and rows of semicircular standings, rising one above the other to the large skylight which lighted the theatre. On the floor the surgeon operating, with his dressers, the surgeons and apprentices of both hospitals, and the visitors, stood about the table, upon which the patient lay, and so placed that the best possible view of what was going on was given to all present. The floor was separated by a partition from the rising stand-places, the first two rows of which were occupied by the other dressers, and behind a second partition stood the pupils, packed like herrings in a basket, but not so quiet, [...]. There was also a continual calling out of 'Heads, heads' to those about the table, whose heads interfered with the sight-seers, [...]. The confusion and crushing was indeed at all times very great, especially when any operation of importance was to be performed, and I have often known even the floor so crowded that the surgeon could not operate till it had been partially cleared.
>
> (South 1970, p. 127)[2]

It was also not uncommon for young dressers, new to the whole ghastly procedure of surgery without anaesthetics, to faint during an operation.

Babington remained as dresser to James Franck for six months, towards the end of which 'through the influence of kind friends' (Munk 1878, vol. 2, p. 452) he acquired a post as assistant surgeon at the Haslar Naval Hospital in Gosport. Saunders was famous for his patronage of 'the young and unbefriended of the profession', and had connections at the highest levels in the Army and Navy

[2] John Flint South was a student at St Thomas's and the operating theatre at that time was in the roof space of St Thomas's Church. Today the theatre is preserved as a museum, much as it was in South's time, see: The Old Operating Theatre, Museum and Herb Garret at www.thegarret.org.uk.

due to the many students who had passed through his hands and were grateful for the patronage he had shown them: 'At the breaking out of the late war [with France in 1793], so long and disastrous, several of his pupils possessed high departments in the Army and Navy Boards, and their respect to their former preceptor made them pay every attention to his recommendations. In one year four army physicians owed their preferment solely to his interest' (Anon. 1817, p. 390). It therefore seems likely that Saunders was the 'kind friend' who recommended Babington for the position at the Haslar, considered an excellent training ground for many young physicians: 'This gentleman [Babington] [...] owes his rise in life entirely to his [Saunders] introduction' (Anon. 1817, p. 391). Babington remained at the Haslar for four years.

In 1781, aged 25, Babington returned to Guy's as its resident apothecary. Each of the seven London hospitals had only one such position and Babington was the third person to hold the post at Guy's, the first having been appointed in 1725. He was chosen by a ballot of the 21 governors who served on the Court of Committees that managed the hospital (Horne 1976). To fulfil the requirements of the post, Babington had to be single, to live in the hospital and to forgo private practice. If he wanted to leave the hospital for any reason, he had to ask permission of the Treasurer. For this he received a salary (around £35 a year), living quarters, board and coals, and essentially had the status of an upper servant, similar to that of the matron (Lawrence 1996, p. 57). Although responsible for patient care when staff physicians and surgeons were not in the hospital, the apothecary took orders from them about treatment.

Babington spent much of his day overseeing the compounding of medicines, although he probably had several assistants and an apprentice or two to help him. The apothecary's domain consisted of a suite of rooms, comprising a laboratory where the medicines were made, a shop that was open throughout the day for the dispensing of medicines, with store-rooms underneath, and a little room at the back where the apothecary lived (Ford 1987, p. 19). On taking-in day[3] the apothecary would be particularly busy, as all new patients would need fresh medicines. Guy's had a tradition of not buying in anything that could be made on the premises (including its own beer) and a contemporary account of the hospital (Howard 1789) reports how 'Medicines are principally prepared under the care of an ingenious gentleman [Babington], who had been furnished with a laboratory and mills, that they may be certain of having their drugs free from adulteration'. Two built-in marble mortars with eight-foot-long pestles remained there until 1935 (Horne 1976).

Babington remained as apothecary to Guy's for 13 years (Table 2), during which time he helped Saunders with his chemical experiments (Fig. 4) and lectures at the medical school. At various times he taught *materia medica*, the practice of medicine, mineralogy, experimental philosophy and chemistry (Ford 1987, p. 25). One of his early chemistry lectures, in the 1790s, opened as follows: 'In searching for a satisfactory definition of the subject we are about to investigate there appears, amongst the most celebrated Authors who have treated of it, this remarkable difference, that by some it is regarded as little more than an Art, whilst others consider it as a most important branch of Science'.[4] To Babington, however, there is no doubt that chemistry was a science; 'a new science expressed in a new language' (Babington & Allen 1802, p. iv). Remarkably, this new science progressed so fast that only 14 years later, 'chemistry is now so intimately connected with various departments of Science, and with most of the Arts and Manufactures, whether useful or ornamental [...] it has become in some degree necessary in the general system of education' (Babington *et al.* 1816).

Babington published his first *Syllabus of a course of lectures, read at Guy's Hospital* in 1789 and continued to do so periodically, and with various colleagues, until 1816. These were not simply an outline of the course, but a discourse on the topic with alternate pages left blank for the student to make notes on during the lecture. As Babington explained:

> In every Science taught by Lectures, a Syllabus of the Course has been found of advantage to the Student. [...] if, for instance, we take a retrospective view of the state of Chemical Science eight or ten years ago, it will appear that a number of discoveries have been made since that period, which have opened new fields of investigation, and have in some instances pointed out the imperfection of our former systems.
>
> On these grounds it has been thought right to draw up a Syllabus of the Lectures delivered at this hospital, and to renew it from time to time, as the progress of the Science may appear to require.
>
> (Babington *et al.* 1816)

[3] Hospitals admitted non-urgent patients one day a week; Wednesday into Guy's and Thursday into St Thomas's. Friday was usually the day operations were performed (Anon. 1800, p. 14).

[4] KCL: G/PP2/3. Babington's Lectures. Undated manuscript volume of lectures given by [William] Babington on chemistry [*c.* 1790s].

Table 2. *The election, promotion and resignation of physicians at Guy's Hospital between 1770 and 1824, illustrating how new appointments and moving up the ranks was entirely dependent on someone above you resigning or dying (highlighted grey)*

Year	Position held
1770	William **Saunders** elected full physician (to replace Nicholas Munckley)
1779	James **Hervey** elected full physician (to replace Henry Hinckley)
1781	William **Babington** elected resident apothecary
1789	John **Relph** (1750–1804) elected full physician (to replace Thomas Skeete)
1794	**Babington** resigns to do medical degree
1794	Richard **Stocker** replaces **Babington** as apothecary
1795	**Babington** returns with MD to become Guy's first assistant physician (new post)
1802	**Saunders** resigns as full physician
1802	**Babington** replaces **Saunders** as full physician
1802	James **Curry** elected to replace **Babington** as assistant physician
1802	**Hervey** resigns as full physician
1802	**Curry** replaces **Hervey** as full physician
1802	Alexander **Marcet** elected to replace **Curry** as assistant physician
1804	**Relph** dies
1804	Alexander **Marcet** replaces **Relph** as full physician
1804	James **Cholmeley** replaces **Marcet** as assistant physician*
1811	**Babington** resigns to concentrate on practice
1811	**Cholmeley** replaces Babington as full physician
1811	**Babington** persuaded to come back as assistant physician
1813	**Babington** resigns again
1813	James **Laird** elected to replace **Babington** as assistant physician
1819	**Marcet** and **Curry** resign as full physicians
1819	**Laird** moves up to full physician (only one physician is replaced)
1819	Richard **Bright** elected to replace **Laird** as assistant physician
1824	**Laird** resigns through ill health
1824	**Bright** replaces **Laird** as full physician

*John Yelloly, a close friend of Marcet's, contested this post and lost, indicating it was not always possible to have your favoured candidate replace you.

In 1802 William Allen (1770–1843), another founder of the Geological Society (Torrens 2009), was invited by Babington to become a partner in his now highly profitable chemistry lectures. Already a successful manufacturing chemist concerned with the preparation of fine chemicals and drugs on a large scale (Coley 1988), Allen had enrolled as a physician's pupil at St Thomas's in 1795 (Allen 1847, vol. 1, p. 20). Whether this was with a view to Allen becoming a physician, or just to improve his knowledge to support his business is unclear, but in 1800 Allen was still taking lectures in anatomy, surgery, physiology and chronic diseases (Allen 1847, vol. 1, p. 34), and complaining frequently that the 'press of business' kept him from attending them. Allen and Babington became firm friends, sharing a mutual love of chemistry which, after Allen's appointment as lecturer, assumed an even greater significance in the medical curriculum at Guy's. In the syllabus published by Babington and Allen in 1802 the new chemical nomenclature of Antoine Lavoisier (1743–1794) and Guyton de Morveau (1737–1816) was used for the first time, with tables relating the new names to the traditional ones.

Required as he was to make all the hospital's medicines and ensure they were free of impurities, Babington would have needed an excellent knowledge of both chemistry and pharmacy, two subjects that have been inextricably linked since medieval times. Pharmacopoeias and other apothecary books of the eighteenth century included recipes for hundreds of chemical medicines. Conversely, almost all eighteenth-century chemical textbooks presented numerous recipes for the fabrication of medicines, describing their properties and medical

Fig. 4. The chemistry laboratory at Guy's Hospital. Frontispiece to Babington *et al.* (1811). Wellcome Library, London.

virtues (Klein 2004). In chemistry textbooks, descriptions of the properties of minerals, metals, salts and earths would be followed by 'organic chemistry', which included inquiries into the natural history and physiology of plants and animals. Consequently, Joseph Franz Jacquin's 1799 *Elements of Chemistry* had three main headings: The Mineral Kingdom, The Plant Kingdom and The Animal Kingdom. Mineralogy, also a crucial component of the physician's and apothecary's repertoire, was taught as part of the *materia medica* and chemistry courses offered in the medical schools. It is therefore no coincidence that many early geologists had a medical background, having been introduced to mineralogy during their medical courses. Hampton Weekes, for example, one of Babington's students in 1802, was so inspired by Babington's lectures on mineralogy, given as part of the medical course at Guy's, that Weekes started his own collection (Ford 1987, pp. 103 and 186). Undoubtedly, many other medical men discovered mineralogy through a similar route. In the first four years of the Geological Society's history, *c.* 20 percent of its membership (including Honorary Members) had MDs (Woodward 1908, p. 268 ff), and there were other medical men like James Parkinson, without that formal qualification, who are harder to identify.[5]

On 21 July 1787, William Babington married Martha Elizabeth Hough (*c.* 1766–1849) and by 1794 they had four children.[6] He was now nearly 40 and still having to spend most of his time in the hospital, which could not have been conducive to family life. On the advice of William Saunders (Hale-White 1934, p. 260), Babington resigned his post in order to take a medical degree, which he duly obtained from Aberdeen the following year. On returning to Guy's with his MD, Babington was offered the post of Assistant Physician, a position newly created for him. A year later, he became a Licentiate of the College of Physicians. This status, and the payment of a fee of £31 10s for the privilege of practising in 'Town',[7] allowed Babington to start his own medical practice, which

[5] See also the many doctor-geologists discussed in Wills (1935) and Sakula (1990).

[6] William and Martha had 12 children in total, seven sons and five daughters between 1788 and 1809, eight of whom survived into adulthood. His fourth son, Benjamin Guy Babington (1794–1866), became a famous physician, following in his father's footsteps as assistant physician to Guy's Hospital in 1837 and as full physician in 1840 (Payne 2004*b*). One of Babington's daughters, Martha Lydon (1802–1823) married Richard Bright (1789–1858) in 1822, who had come to Guy's in 1810 as physician's clerk to her father. Tragically, she died only a year after the wedding, but Bright went on to an illustrious career and to identify Bright's disease, which afflicts the kidneys. The three other Babington daughters married three Peile brothers of Tottenham Green; the eldest two couples marrying on the same day.

[7] i.e. within a seven-mile radius of the City (of London) (Ford 1987, p. 162).

he did in Freeman's Court, Cornhill.[8] The family lived in Cornhill until about 1796, in Basinghall Street until 1801[9] and then in a large house in Aldermanbury until 1830 (Payne 2004*a*), after which Babington retired to his house in Devonshire Street, Portland Place, although his patients followed him there 'and calls upon his time continued as heavy as ever' (Smyth 1843, p. 94). The Babingtons also had a country house in Tottenham Green.

In all the seven 'great' London hospitals at this time, there were only 22 posts for full physicians and 23 for surgeons (Lawrence 1996, p. 58), so for young and ambitious men like Babington, opportunities to move up the hierarchy were few. They had to wait until someone above them either resigned or died. William Blizard (bap. 1744–1835), for example, surgeon at the London Hospital, did not retire until he was 93, having been in the post for 54 years (Lawrence 1996, p. 58). While this may have been an extreme case, Babington still had to wait seven years before becoming full physician when William Saunders retired from Guy's in 1802 (Table 2). On the other hand, James Curry, who stepped in to replace Babington as assistant physician, only had to wait a few months before James Hervey (*c*. 1750–1824) resigned in 1802 and Curry was appointed full physician to replace him – age and relevant experience for the post seemed immaterial.

Assistant and full physicians gave freely of their time to the hospital, earning only a token fee that was no more than the apothecary's salary. They supplemented this meagre income by taking dressing pupils and apprentices, by giving lectures and by developing their own private practice. Nevertheless, these hospital posts were highly sought after, and could be contested as fiercely and as expensively as those in a parliamentary election. An article from *The Times* explains why, and although this was commenting on the situation some decades later, the circumstances it relates had not changed since the beginning of the century:

> Connexion with a great hospital is an object of primary ambition to a London physician or surgeon. It gives professional status; it brings fees for tuition which are serviceable during the time of waiting for fame; it often leads to a large and lucrative practice; and, indeed, without it there is scarcely a possibility of a very high position being attained.[10]

Physicians and surgeons needed to build their reputations in the hospitals as this attracted the paying public to their practices.

By 1811, Babington's practice had become so successful that he announced he wished to retire from Guy's because his patients were too demanding of his time. His position at Guy's had clearly served its purpose. On receiving his letter of resignation, the hospital governors at once resolved that they should try to 'avail themselves of the services of Dr. Babington if they can be afforded in any situation consistent with his professional engagements' (Lawrence 1996, p. 66). The very next week, Babington (Fig. 5) agreed to come back in the lower post of assistant physician, presumably expecting the demands on his time to be less than as full physician. The tables had been turned and now it was the hospital that benefited from the reputation of the physician, and they did not want to let him go – although note the governors' concern about being able to afford him. A few years later, when Saunders retired from private practice in 1814, Babington apparently took over his enormous and highly lucrative practice[11] (Anon. 1817, p. 591)

Fig. 5. William Babington. Stipple engraving by W. T. Fry after J. Tannock. Wellcome Library, London.

[8] It was not Freemason's Court, as stated by Hale-White (1934, p. 259).

[9] Herries Davies (2007, p. 3) says 1807, but this does not concur so well with other facts.

[10] *The Times*. London hospitals and dispensaries. 30 January 1869.

[11] It is likely that Babington would have paid Saunders a substantial sum for this privilege.

and succeeded Saunders as the acknowledged head of his profession in the City. Perhaps anticipation of this event explains why Babington finally resigned as physician from Guy's in 1813, although he continued to lecture there for many years afterwards.

The success of Babington's practice stemmed not only from his considerable experience as a physician, but from his sympathetic and warm personality, remarked upon by almost everyone who knew him:

> History does not supply us with a physician more loved or more respected than was Dr Babington. Dr Gooch, writing in Dr Babington's lifetime (and many still alive echo the sentiment) describes him as 'a man who, to the cultivation of modern sciences, adds the simplicity of ancient manners; whose eminent reputation and rare benevolence of heart have long shed a graceful lustre over a profession which looks up to him with a mingled feeling of respect, confidence, and regard'.
>
> (Munk 1878, p. 452)

His virtues were similarly extolled by his one-time son-in-law, Richard Bright (1789–1858), in an obituary of Babington in the *London Medical Gazette*: 'The character of Dr Babington was probably as nearly without fault as is consistent with human nature' (Bright 1833). And John Flint South (1797–1882), who started his illustrious medical career as a pupil at St Thomas's in 1814 and used to attend Babington's lectures, gives a fine portrait of his teacher:

> Dr Babington, a good tempered, kindly Irishman, [...] had no pretensions to oratory, but he was a very excellent practical teacher, who was listened to with great pleasure and advantage and his lectures were full of experience and practical good sense which the very simplest minds could receive and understand. He used to wear black, with silk stockings, was a very untidy dresser, and rejoiced in dirty hands, but he was gentle and pleasant with every one, and always ready with some funny anecdote, and always in a hurry, for which his large and well earned practice was a just excuse.
>
> (South 1970, p. 58)[12]

James Franck

James Franck senior married Frances Clifton on 29 September 1772, but they appear not to have had any children. In his will, Franck senior left his estate to be shared equally by his wife and his nephew, James Franck junior (Table 1), who was the son of his brother, William Franck, a sea captain from Surrey.[13] As the will was written in 1780 and, as the will states, William Franck was deceased, James Franck junior, born 28 June 1768, must have been younger than 12 when his father died. The equal division of Franck senior's estate between his wife and nephew suggests that the fatherless James Franck junior was cared for by the childless James Franck senior. Young Franck did have a sister Sarah, but she is not mentioned in Franck senior's will, suggesting the children may have been split up on the death of their father.

Franck senior had already been a surgeon for two years at Guy's Hospital when William Saunders joined as physician in 1770 (Lawrence 1996, Appendix 1A), thus the two had worked with each other for at least 15 years by the time Franck senior died in 1783. It seems they became close associates – Saunders was a witness to Franck's will, for example. The younger Franck had been educated at the Merchant Taylors' School in London, from where he went to Pembroke College, Cambridge, in 1788 and obtained his bachelor of medicine in 1792. With his influence in the appointment of medical staff to the military, it seems likely that Saunders was involved in Franck's appointment, on 10 June 1794, to the post of Physician in the British Army (Rose 2009).

The war with France had commenced the year before Franck joined the army, but little is known of his early career, other than that he served in Corsica for a time. In 1800 he was promoted to the rank of Inspector of Hospitals (McGrigor 1861, p. 112) and became involved in the Egyptian Campaign, a major encounter between French and English troops in Egypt. Napoleon had decided to attack Egypt, hoping to impact on Britain's trade with India, and early in 1801 Franck found himself based at Giza on the west bank of the Nile, from where he initially supervised the British medical arrangements. He was subsequently joined by Thomas Young, the Director-General of Hospitals, who took over as medical officer in charge as he had greater war-time experience, although Franck stayed on to assist him. After a major battle at Aboukir Bay near Alexandria where the French attempted, but failed, to oust the British troops, over 1000 men were evacuated to stations in the Mediterranean and others returned to England, but more than 2500 sick or wounded remained in Egypt for Young and Franck and their assistants to look after. Equipment, bedding and staff were in desperately short supply as it had been estimated that only 10 percent of the force

[12] The fact that Babington 'rejoiced in dirty hands' is not necessarily a criticism, but acknowledgement that he did not mind getting his hands dirty like some 'fine gentlemen' (see above).

[13] Cambridge University Alumni records, 1261–1900.

would be hospitalized at any one time. In fact, the figure was closer to 25 percent, and before the end of the campaign in 1802 another 3000 wounded would have to be hospitalized.

Gunshot wounds almost certainly led to amputation, without anaesthetics, although field operations had a much lower death rate (10 percent) than those carried out in the hospitals (more than 50 percent), which led to a recommendation that amputations should be performed within 24 hours of the injury (Ayliffe & English 2003, p. 69). Nevertheless, thousands of men died from hospital gangrene as their wounds became infected and, although the dangers of overcrowding were recognized, infection was endemic as a result of poor hospital hygiene, which involved such common practices as using a single sponge to clean up everything and everyone. Dysentery and the plague, both frequently fatal, added to the burden of the medical staff. From the description given in reports of 'suppurating buboes' preceded by a fever, the 'Egyptian plague', as it was known, was undoubtedly the bubonic plague and, for those who contracted it, the death rate was about 45 percent. Its cause was still unknown and doctors were perplexed as to why some who came in contact with sufferers contracted it, while others did not. This caused medical opinion to be divided about whether it was contagious or not. James Curry, a senior physician at Guy's Hospital (Table 2) who probably did not see that many cases, considered its sudden appearance was due to a 'peculiar state of atmosphere, which he thinks greatly to depend upon electric influence' (Select Committee 1820, p. 119). Franck, on the other hand, who had seen hundreds of cases in Egypt, was convinced of its contagious nature, in the sense that it could spread from person to person, but he was not aware of the mechanism by which this happened:[14]

> He [Franck] could distinctly trace its progress, by contagion, from one person to the sick and [the] hospital attendants, who afterwards became infected; whilst none of the wounded, who were in a separate hospital, about half a mile from the former, had the disease. He has seen several instances of contact with impunity, and has himself handled plague patients. His opinion is, that the plague of London in 1665 was imported, and that circumstances might arise in this country favourable to its existing again.
>
> (Select Committee 1820, p. 122)

Franck was correct in thinking that the 1665 plague of London had been 'imported' and right to be concerned that another outbreak could occur any time. However, it was 1894 before Alexandre Yersin (1863–1943), among others, recognized the true nature of the disease and identified the bubonic plague bacillus, now called *Yersinia pestis*.

In 1802, during the short period of peace, Franck was back in England and able to obtain his MD at Cambridge. A year later, he was admitted as a full Fellow to the Royal College of Physicians where he gave the prestigious Goulstonian lecture in 1804[15] and became Censor[16] in 1805. In November 1807 Franck became a founder member of the Geological Society just as Napoleon's army was marching through Spain and into Portugal, occupying Lisbon on 1 December. Thus, early in 1808 Franck was back on foreign soil, serving in the Peninsular War, an alliance between Spain, the United Kingdom and Portugal against France. He was now Principal Medical Officer (Ackroyd *et al.* 2006, p. 176) under Sir Arthur Wellesley (1769–1852), who later became the Duke of Wellington. The instructions given to Wellesley had been to liberate Portugal and Spain, and secure the expulsion of French forces from the Peninsula – an almost impossible task (Gash 2004).

By January 1809 things had gone badly and the British decided to withdraw from Spain on learning that Napoleon's far larger army was only 20 miles away. The retreat over the Cantabrian Mountains in northern Spain, towards the port of Corunna, was a shambles, carried out in terrible conditions amidst slushy snow and ice. The roads quickly turned into quagmires beneath the tramping of thousands of feet and the troops suffered dreadful hardships in the bitterly cold winter weather. During the 18-day retreat several thousand men, their horses and bullocks, as well as women and children that accompanied the army, were frozen to death or captured by the pursuing French army. When the troops finally arrived at Corunna on 11 January, it was to find their ships had been delayed owing to bad weather and they were forced to wait three days, during which time the enemy arrived. The ensuing battles was extremely fierce and, although the English eventually fought off the French, almost 4000 troops had died either in battle, by the roadside or in the hospital, and more than 2000 were captured

[14] The bubonic plague (or Black Death) was caused by being bitten by an infected flea, thus it was not 'contagious' in the sense that you could 'catch' it like a cold. It is now considered that the bubonic plague may have originated in Egypt (Panagiotakopulu 2004).

[15] The Goulstonian lecture was an annual lecture read over three days by one of the four youngest Fellows of the College between Michaelmas and Easter. Originally it would have been 'read on some dead body, which was to be dissected for the diseases treated of' (pers. comm. Archivist, Heritage Centre, Royal College of Physicians).

[16] An officer of the College appointed to help the President with examination and disciplinary matters.

by the French. By the time the 28 000 survivors finally disembarked at Portsmouth, typhus fever had spread widely among them because the sick and healthy were indiscriminately mixed together. This severely overloaded the local hospitals and the Haslar Naval Hospital at Gosport, normally able to accommodate 4000 sick and wounded seamen, was entirely given over to the army. The situation was so desperate that an appeal was made for medical assistance from London (Kaufman 2001, p. 62).

It seems that Franck was part of this retreat back to England, where he must have remained for several months, for on 7 April 1809 he is recorded as being present at the Geological Society.[17] However, this respite was not for long. In May 1809 Wellesley decided to attack French forces in the coastal city of Oporto in Portugal, taking them completely by surprise. Although British losses were relatively slight, the general hospital was 100 miles away in Coimbra, too far to transport the wounded. The 2600 patients from previous battles therefore had to be left in the Coimbra hospital, in the care of assistant surgeons, while the senior medical staff went to establish another general hospital at Orporto. Both hospitals were put under the direction of Franck (Kaufman 2001, p. 62), who had by then returned to the Peninsular.

Wellesley then turned his attention to Spain to where the French had fled after the surprise attack at Oporto and, after negotiations with the Spanish commander, a joint advance was planned against Madrid. Wellesley and his troops moved south, and on 28 July 1809, having joined forces with the Spanish army, he confronted the French troops in the battle of Talavera, a town about 60 miles SW of Madrid. Casualties were extremely high – 1500 British dead with 4000 wounded. They overwhelmed the medical staff, who had set up a field hospital in a tent some 700 yards beyond the range of the cannons and musket fire. It took two days to remove the wounded from the battlefield, but Franck was unable to supervise this dreadful task, having become extremely ill with dysentery (Kaufman 2001 p. 63). Despite the human cost, the victory at Talavera earned Wellesley a peerage and a pension as Viscount Wellington.

Two years later Wellington wrote from the allied cantonment in Frenada, Portugal, to Lieutenant-Colonel Torrens, Military Secretary to the Commander in Chief, on 30 October 1811:

> My Dear Torrens
>
> I am sorry to tell you that Dr Frank [sic], the Inspector of Hospitals, is so unwell as to be obliged to go home; and the department under him is so important, that if, as I fear, he should not be able to come out again, it will be necessary that we should have the most active and intelligent person that can be found to fill his station.
>
> (Kaufman 2001, p. 42 n. 96)

Wellington's fears proved correct; Franck did not return to active service and in October 1811 he was retired from the army due to ill health on half pay, although he continued to be paid at the full rate until he died in 1843 (Ackroyd *et al.* 2006, p. 176), despite some controversy over his capabilities as Wellington's Principal Medical Officer. His successor was James McGrigor (1771–1858) who, over the next two years, brought relative order and efficiency to a demoralized and poorly run service, successfully reducing the number of troops reporting unfit for duty, which, under Franck, had been spiralling out of control (Ackroyd *et al.* 2006, p. 37). Wellington, however, appears to have been supportive of Franck (Rose 2009), so it is difficult to assess whether these problems were due to Franck's ineptitude or to his frequent illnesses, coupled with impossible conditions. Unlike McGrigor, however, he was not knighted for his services to the war.

James Laird

James Laird (Table 1) was born on 17 December 1779, in the parish of St Elizabeth, Jamaica, where his father, Henry Laird (died 1801), owned a plantation of 102 acres – undoubtedly a sugar plantation – called Prospect Estate.[18] James was the last of at least five children[19] and became the youngest founder of the Geological Society; he and Humphry Davy (1778–1829) sharing the same birthday, but a year apart (Table 1).

It is not known exactly when Laird came to England, but his father's will states that at the time of writing (1792) James was already in Britain, where he must have been at school. The Laird family appears to have had connections with Bristol, a port strongly involved in the trade in sugar and slaves, where James's aunt Letitia, wife

[17] GSL: Ordinary Meeting Minutes 1807–1818.

[18] *Jamaica Almanac* for 1811. In 1817, the *Slave Registers of Former British Colonial Dependencies* lists over 100 slaves registered to Laird's elder brother Henry, who then owned Prospect Estate.

[19] On 8 January 1765 Henry Laird married Mary in the parish of St Elizabeth. Their four other children were: Mary (died 1818), Emily (dates unknown), Letitia Isabella (born 27 January 1775) and Henry (born 21 November 1776, died *c.*1817).

of his father's brother James, was buried in 1824.[20] Laird was also a close family friend of the Bright family, who lived in Bristol, where he seems to have been mentor to the son, Richard Bright (Berry & MacKenzie 1992, p. 66), who was to become Babington's son-in-law. Bright went to school at the dissenting academy in Warrington, Cheshire, and, as he seems to have followed Laird in many other aspects of his career, it is just possible that Laird had been there before him. Evidence in support of this is that the only donations of rocks made by Laird to the Geological Society's 'cabinet' were from Cheshire and Bristol (Geological Society 1811, pp. 409 and 412).

James Laird began his matriculation at Edinburgh University in 1799[21] and obtained his MD in 1803 with a thesis entitled *De stomacho, ejusque morbis.* That same year he was one of the four presidents elected each year to Edinburgh's prestigious Royal Medical Society. A contemporary of Laird's at Edinburgh was Thomas Bateman (1778–1821) who, a year older and therefore a year ahead in his studies, had also been President of the Royal Medical Society. Bateman and Laird were to become close friends and work together for more than 20 years (Rumsey 1826, p. 124). Laird's medical education in Edinburgh also overlapped with Alexander Marcet (1770–1822) and John Yelloly (1774–1842), both of whom were to feature in the Geological Society's early history.

It was standard practice for Edinburgh's medical students to seek clinical experience in a hospital and more than 50 percent of them came to London (Lawrence 1996, p. 108) to get this expertise, either before they qualified, like Laird, or afterwards, like Marcet. In 1801, when Laird arrived at Guy's as a physician's pupil, the two physicians were James Hervey and William Saunders and the assistant physician was Babington. It is probably during this period that Laird became Babington's 'protégée' (Buckland 1841) and they established their close friendship.

On leaving Edinburgh in 1801, Thomas Bateman became assistant physician to Dr Robert Willan (1757–1812) at the Public Dispensary in Carey Street (Asherson 2004), and was elected full physician there in 1804 (even though he did not become a licentiate of the Royal College until 1805). A publication by Laird in 1807 gives Laird's affiliation as 'physician to the Carey Street Dispensary', thus it seems likely that he stepped into Bateman's position as assistant physician there in 1804, becoming full physician in 1806 when he, too, became a licentiate of the Royal College of Physicians. That same year, 1806, the Public Dispensary moved from Carey Street, Lincoln's Inn Fields, to Bishop's Court, Chancery Lane, where it remained until 1850 when it moved back to Carey Street (Master of the Rolls 1854). Most dispensaries started in rented accommodation and moved to larger premises, often purpose built, as patient numbers grew (Loudon 1981). As Bishop's Court was just around the corner from Carey Street, the move in 1806 was probably to larger premises which required an extra physician, hence Laird's promotion to full physician. In his obituary of Laird, William Buckland (1784–1856) reports that he 'was the intimate friend of Dr. Babington, with whom he passed much of his time, assisting him in his correspondence, and occasionally aiding him in his profession' (Buckland 1841, p. 525). Hale-White (1834, p. 264) further claims that for a time Laird lived with the Babingtons in Basinghall Street. He may have lodged with them and certainly, between at least 1809 and 1812, Laird's address was 15 Basinghall Street, but by that time the Babingtons had moved a few streets away to Aldermanbury.

Dispensaries (Fig. 6) were voluntary institutions funded by the 'Nobility, Gentry and others' (Anon. 1795). In the mid-eighteenth century they sprung up

Fig. 6. St Marylebone dispensary, Welbeck Street, London. Sepia photo-lithograph after A. B. Pite, 1892. Wellcome Library, London.

[20] Her grave can still be seen at the front of St Paul's church, Portland Square, Bristol.

[21] University of Edinburgh student record cards.

all over the City, championed by those who argued that providing inexpensive advice and medicines to out-patients had decided benefits over in-patient care in hospitals. Their distinctive feature was that as well as offering a local centre where minor ailments could be treated and medicines dispensed, they provided medical care in the home. From then on all patients within a certain radius and unable to come to a dispensary could be visited by qualified physicians like Laird and Bateman. As in the hospitals, physicians and surgeons gave freely of their time to the dispensaries for a nominal retainer fee and thus had to earn a living elsewhere, but working for a dispensary was seen as a stepping-stone to a position in a major hospital, so posts in dispensaries were keenly sought after.

In 1813, when Babington retired from Guy's, it was undoubtedly due to his patronage that Laird stepped in to replace him as assistant physician (Table 2). In his turn, Laird became patron to the next generation of Guy's men. Richard Bright, following Laird's advice (Berry 2004), also worked in the Public Dispensary while waiting for a post in a London hospital, and it was on Laird's recommendation that Bright obtained a position at the London Fever Hospital, where Laird also worked. In 1819 when Alexander Marcet retired from Guy's, Laird became full physician and Bright replaced him as assistant physician (Table 2) with particular responsibility for Laird's patients. Bright, having married one of Babington's daughters in the meantime, duly replaced Laird as full physician when he retired from Guy's in 1824. In 1819 both Laird and Bright were practising in Bloomsbury Square.[22] It was a small, tight, nepotistic world.

Apparently, Laird was an 'excellent mineralogist' (Buckland 1841), possibly because when Laird was attending Edinburgh's Medical School it was one of the best places in Britain to learn mineralogy (Eddy 2002, p. 412). Furthermore, Laird's time in that city immediately followed James Hutton's death (1797) when John Playfair (1748–1819) was vigorously defending and amplifying his friend's work, so it is quite possible Laird attended Playfair's lectures, 'so chaste in style, so clear in demonstration', as his young friend Richard Bright did a few years later (Berry & MacKenzie 1992, p. 35).

However, Laird's letters to George Bellas Greenough (1778–1855), the Society's first President, reveal a man whose real forte lay in organizing the Society rather than participating in its geological activities. As Greenough commented, Laird was one of the 'most active of our body [...] & eminently qualified for the post assigned to him that of secretary'.[23] Leonard Horner (1785–1864), on the other hand, who assisted Laird as Junior Secretary from 1810 filled his letters to Greenough with excited comments on geology (Fig. 7):

> I have seen a rock with the most indubitable proofs in its structure that four worlds must have existed before the formation of the present. This interesting geological fact (so decisive of the truth of the fundamental position of the Huttonian Theory) I have happily procured in a specimen about four pounds weight [he describes and sketches the specimen]. I leave you to make out the four worlds – they are obvious to every one.[24]

Laird's only publication, apart from his thesis, was a short medical paper entitled 'Case of small-pox occurring in the foetus' (Laird 1807), where he reported on a woman who contracted small pox while she was pregnant. She subsequently gave premature birth to a dead baby covered in small pox pustules. Laird's recommended treatment for the mother while she had the pox consisted of keeping the bowels open, acidulated drinks and opium at night. The paper is typical of its time – the author observes, categorizes and then affirms his observations with details of the remedy and its outcome, but he offers no pathological analysis.

James Parkinson

James Parkinson, born on 11 April 1755 (Table 1), was the eldest of three children: James and his sister Mary had a middle brother, William, who died aged 21. At the age of 16 James was apprenticed to his father, John Parkinson (1726–1784), then the apothecary surgeon in Hoxton, in the parish of Shoreditch, to learn the 'art and mystery' of being an apothecary. He later recalled what a waste of time he felt his seven years as apprentice had been:

> The first four or five years are almost entirely appropriated to the compounding of medicines; the art of which [...] might just as well be obtained in as many months. The remaining years of his apprenticeship bring with them the acquisition of the art of bleeding, of dressing a blister, and, for the completion of the climax – of exhibiting an enema.
>
> (Parkinson 1800*a*, p. 32)

---

[22] Bright was at No. 14, but it has not been possible to establish which number Laird was at, or whether they shared a practice.

[23] GSL: LDGSL 960. Greenough's 'Papers connected with the Geological Society', a history of the first three years with copies of letters and lists. Hereafter: Greenough Papers.

[24] UCL: AD7981 824. Letter from Horner to Greenough, 26 May 1809.

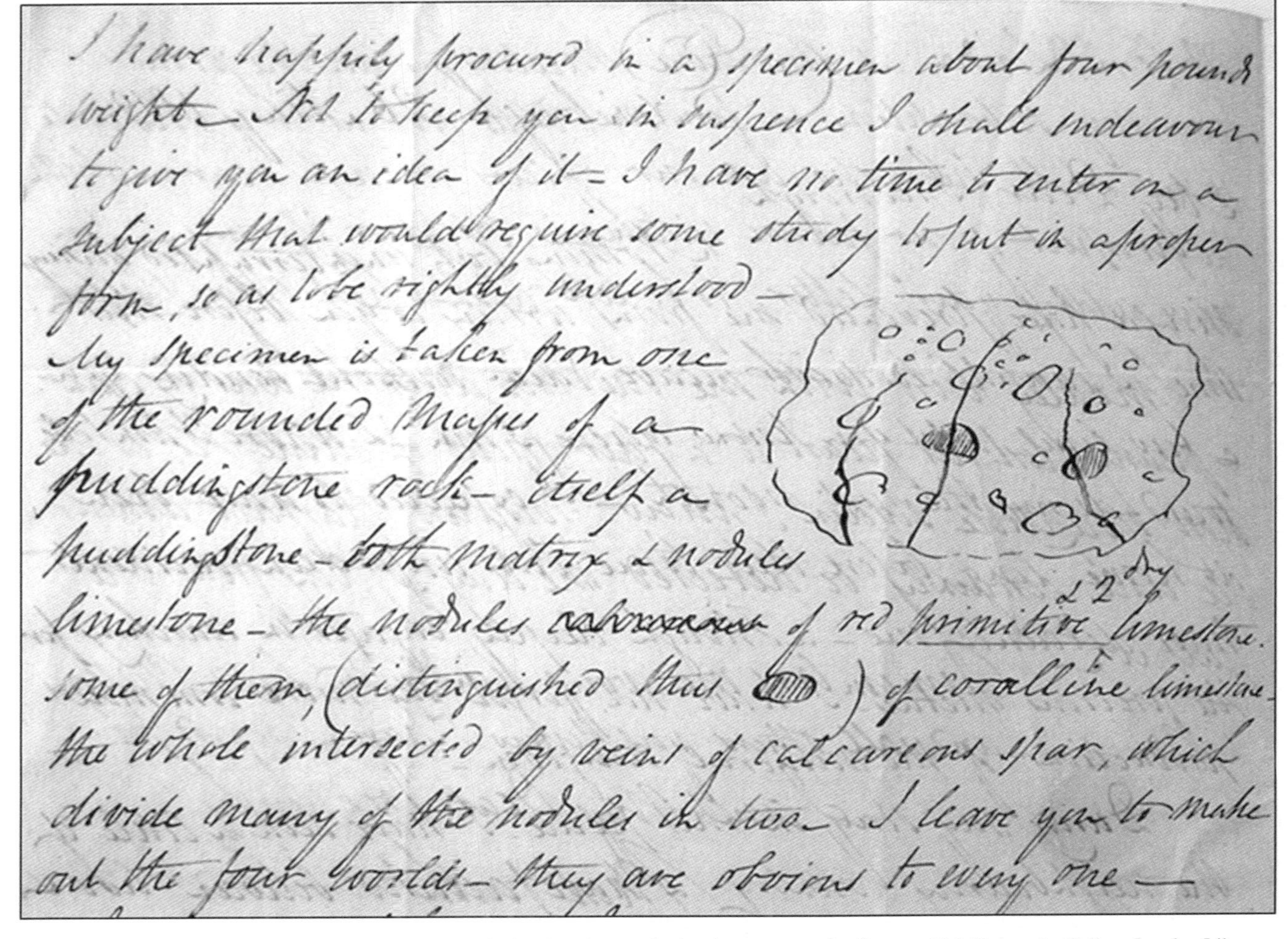

I have happily procured for you a specimen about four pounds weight – Not to keep you in suspence I shall endeavour to give you an idea of it – I have no time to enter on a subject that would require some study to put in a proper form, so as to be rightly understood –

My specimen is taken from one of the rounded Masses of a puddingstone rock – itself a puddingstone – both matrix & nodules limestone – the nodules of red primitive & 2dry limestone. some of them (distinguished thus) of coralline limestone – the whole intersected by veins of calcareous spar, which divide many of the nodules in two – I leave you to make out the four worlds – they are obvious to every one –

Fig. 7. Part of a letter from Horner to Greenough, dated 26 May 1809, showing the sample that demonstrates the 'four worlds'. University College London Library.

Fig. 8. The London Hospital, Whitechapel, seen from the northern side of the Whitechapel Road. Wood engraving, after an engraving of *c.* 1753. Wellcome Library, London.

Towards the end of his apprenticeship, Parkinson spent six months at the London Hospital on Whitechapel Road, in London's East End (Fig. 8), as dresser to the surgeon Richard Grindall. He would have supplemented what he learnt there with lectures and anatomy classes, taken outside the Infirmary.[25] Completion of his apprenticeship in 1778 enabled Parkinson to join his father as a partner in the Hoxton practice and to fully share in its duties, but when John Parkinson died, only six years later, James was left to manage the practice single-handedly. He immediately set about improving his medical knowledge and status, knowing full well that if he did not provide adequate medical care for his patients that they would soon take their allegiance elsewhere (Ford 1987, p. 8).

Parkinson was highly critical of the traditional methods of teaching medicine and considered his medical education to have been 'mis-managed' (Parkinson 1800*a*, p. 36). Even after completing his six months as a dresser at the London Hospital, he still felt 'miserably ignorant'. In particular, he believed that 'a knowledge of anatomy ought to precede even the study of surgery, and [. . .] that the principles of surgery, as a science, ought to be indelibly impressed on the mind before any attempt should be made to employ it as an art' (Parkinson 1800*a*, p. 36). Many aspired to perfect this 'art', inspired by the fame and fortune won by such surgeons as John Hunter (1728–1793), but Parkinson warned against such folly:

> [. . .] our student anxiously endeavours to inform himself of every step of an operation, which appears to him so easy to imitate, and which, he thinks, would, nevertheless, secure him that fame for which he pants. But, my young friend, I must apprize you, that celebrity is not thus cheaply purchased. Much is to be learnt, before the skill you have witnessed, can be imitated.
>
> (Parkinson 1800*a*, p. 83)

[25] It was not until 1785 – nearly a decade after Parkinson had completed his education – that the London Hospital Medical College (the first hospital-based medical school in England) was founded by William Blizard and James Maddocks.

Three months after his father's death Parkinson took the oral exam at Surgeon's Hall, from which he obtained a diploma that qualified him to become a member of the College of Surgeons and to practice as a surgeon. The following year, he enrolled on a course of lectures on the 'Principles of Surgery' given by John Hunter. Each one-hour long, these were delivered three evenings each week in Hunter's operating theatre in Castle Street, Leicester Square, between October 1785 and the following April (Roberts 1997). In all, this course consisted of 68 lectures which Parkinson took down in shorthand, thereby providing us with a verbatim account of Hunter's teaching methods. Given that much of Hunter's writings were subsequently lost in a fire, these notes have become a valuable resource. Parkinson transcribed them either when there was a 'hiatus' during lectures, or when he got home. The great benefit, he argued, of taking shorthand notes was the 'unremitting attention which they necessarily excite; acting as a constant monitor and preserving the mind from straying, so that nothing material can escape unrecorded' (Parkinson 1800*a*, p. 75). After James's death, his son John Parkinson (1785–1838), also an apothecary surgeon, edited the transcriptions and had them published under the title *Hunterian Reminiscences* (Parkinson 1833).

The radical Mr Parkinson

During the 1790s Parkinson became heavily involved in radical politics, joining first the Society for Constitutional Information and then, shortly after it was initiated in 1792, the London Corresponding Society (LCS) that was founded to campaign for the reform of parliament (Davis 2002). Despite William Pitt the Younger (1759–1806) having been elected Prime Minister in 1783 on a promise of social reform, a decade later the situation had not changed and only male owners of freeholds who paid more than 40 shillings land tax a year were eligible to vote. This small number represented about two percent of the population, which left the vast majority disenfranchized (Thale 1983; Watson 1985).

James had married Mary Dale (1757–1838) on 5 December 1781, when he was 26. By the 1790s, and now in his late 30s, his family responsibilities were increasing.[26] Furthermore, on 21 January 1792, only a few days before the LCS was founded (on 25 January), his three-year-old daughter died, which affected him deeply (Parkinson 1800*c*). In his practice, Parkinson was constantly reminded of the close relationship between sickness and poverty. A philanthropist characteristic of the Enlightenment, he genuinely cared about improving both the welfare and the morals of the poor. In 1799 he was made a trustee for the poor of the liberty of Hoxton (Bevan 2004); he also became Secretary to the Sunday school at St Leonard's, believing that 'a mere literary education without the inculcation of moral and religious principles would prove highly injurious' (Morris 1989, p. 17), and for 25 years was medical attendant to the private mad-house at Holly House, Hoxton. In addition, Parkinson was quick to take up new medical developments. He was a disciple of Edward Jenner's (1749–1823) method of smallpox vaccination, becoming one of London's first inoculators in 1802 and campaigning vigorously to have the treatment taken up across the City. He was the first to introduce fever wards into a workhouse, thereby separating contagious patients from others. He also campaigned for more humane methods of treating lunatics (Parkinson 1811*a*), and to change the laws on child labour (Morris 1989, Appendix I, p. 83).

With his high moral code and a genuine desire to improve the lot of the working man, Parkinson was an obvious candidate for the new reform movement. Under the pseudonym of 'Old Hubert', he wrote many political pamphlets that advocated representation of the people in the House of Commons, the institution of annual parliaments and universal suffrage. He also ridiculed and harangued a corrupt and incompetent Government for the soaring cost of food and high taxes, incurred during the war with France. In June and July 1795, the shortage and cost of food led to riots across Britain:

> The hour of Calamity is arrived: Famine is already at our doors. [...] In their endeavours to enslave Frenchmen, they have prodigally wasted the Bread which should now feed us and our children; by their ineffectual attempts to Starve those they could not Conquer,
>
> [...] the evidence is all at hand, and incontestably proves, that our calamities are the result of an Abominable System of Corruption and profusion, conducted for the sole benefit of its administrators: – a set of men, whose enormous Salaries [are] voted to each other from the purse of a distressed Nation.
>
> ([Parkinson] Old Hubert *c*. 1795)

William Pitt's Government, convinced that Britain was about to follow the French Revolution, became increasingly concerned by the activities of radical societies. In May 1794, believing they

[26] Parkinson's first child, James John, was born in February 1783, but died shortly after birth. In July 1785, John William Keys was born, named after James's close friend, John Keys, who was by then his brother-in-law. James then had two daughters – Emma, born in 1788, and Jane the following year (it was Jane who died in 1792). Two further children, Henry and Mary, were born in 1791 and 1794, respectively (Gardner-Thorpe 1987).

were planning to dispose of the King and establish a republic (Emsley 2000), the Government arrested the 12 leaders of the LCS and the Society for Constitutional Information, and charged them with high treason, a charge which could see them hung, drawn and quartered. A few days later Parliament suspended the *Habeas Corpus* Act, enabling the Government to detain its prisoners without trial. With the leaders in prison, Parkinson became Secretary to a secret committee set up in their absence, its prime objective being to raise money for the wives and children of those imprisoned. It was during one of these committee meetings that a government spy who had infiltrated the LCS became aggrieved at the way he was being treated – he was accused of deliberately setting fire to his house in order to obtain the insurance money. In retaliation, the spy reported to government officials a trumped-up account of a plot to kill the King. The 'Pop Gun Plot', as it became known, alleged that these radicals conspired to shoot a poisoned dart at the King through a blow-pipe, while he was at the theatre. On learning of the arrests, Parkinson immediately offered to testify before the Privy Council on behalf of his friends, hoping to demonstrate the ludicrous nature of the charges:

> A poisoned arrow was to destroy the King, and to what purpose? would not another King succeed; or if instant death had not followed the unerring dart of wonder-working poison, to what advantage to any conspirators could have been the illness of a King? Has he not been for several months unable to hold the reins of government yet the nation was not, in consequence of his madness, in a state of anarchy and confusion? What could the plotters gain by this attempt?
>
> (Parkinson in Smith 1795, p. 69)

On Tuesday 7 October 1794, 'Mr Parkinson, that worthy citizen who did not scruple to come forward in the midst of danger, and assert, before the Privy Council, the innocence of the supposed conspirators' (Smith 1795, p. 41), was examined under oath by the Prime Minister William Pitt, the Attorney General and the Lords of the Privy Council. In the prevailing political climate, Parkinson must have realized that he ran a serious risk of being implicated in the plot and that he could be putting his life in danger. Earlier that year, five men in Scotland had been condemned to transportation to Australia for up to 14 years, having been found guilty of the lesser charge of sedition (Cockburn 1888).

A verbatim account of Parkinson's spirited defence of his friends, his confession that he was on the secret committee of the LCS and the author of several radical papers is covered in Smith (1795), along with Smith's account of his own arrest and time in prison. Fortunately, owing to the good sense of the juries, both trials eventually collapsed and everyone was acquitted. But this only incensed the Government further and led to further controls on the corresponding societies. By 1799, the Seditious Societies Act required any place such as a coffee house or tavern 'in which any lecture or discourse shall be publicly delivered, or any public debate shall be had on any subject whatever for the purpose of raising or collecting money' to be licensed (Weindling 1980). Being unable to raise funds, this effectively put a stop to the corresponding societies' activities, driving radicals to increasingly dispersed forms of organization to evade the Acts. Parkinson, perhaps alarmed by his brush with the law, turned to science.

The prolific author

Trained as an apothecary, Parkinson had an excellent knowledge of chemistry and recognized the importance of it to the medical practitioner, recommending the medical student take at least two courses during his studies (Parkinson 1800*a*, p. 52):

> But I fear I have too much reason to say, that pupils are too frequently satisfied with, a much less portion of instruction in CHEMISTRY and NATURAL PHILOSOPHY, than is really necessary for the successful prosecution of their studies [. . . and] that the want of sufficient intelligence, in this respect, may injure them, when established in their profession.

To these reluctant students Parkinson offered his *Chemical Pocket-Book* (1800*b*), 'arranged in a Compendium of Chemistry, according to the latest discoveries', hoping it would 'point out the indispensable connection between Chemistry and most other sciences' (Parkinson 1800*b*, p. iii). A definition in the Preface describes chemistry as 'the Science which discovers the constituent principles of bodies, the results of their various combinations, and the laws by which those combinations are effected'. However, the speed at which discoveries in chemistry were being made, alluded to earlier by Babington, made it difficult to keep up. For example, while Parkinson followed Lavoisier in describing heat, 'caloric', as an elementary substance, he also gave the alternative and most up-to-date view, that it was a phenomenon caused by the motion of minute particles of matter as advocated by Humphry Davy and Thomas Beddoes (1760–1808). The nomenclature too was changing, thus the current spelling of 'gaz' and 'oxyd' can be found in the first half of the book, followed by the modern spelling, 'gas' and 'oxide', in the second half. As with many of his published works and those of his contemporaries, he read extensively and then distilled his facts into a format readily accessible to students and a wider public.

The year before the *Chemical Pocket-Book* appeared, Parkinson published a substantial medical work in two volumes: *Medical Admonitions to*

Families, on Domestic Medicine, the Preservation of Health, and the Treatment of the Sick, which contained a 'Table of Symptoms, pointing out such as distinguish one disease from another, and the degree of danger they manifest' (Parkinson 1799). This work was aimed at helping people assess in times of sickness, whether or not to call in the apothecary or physician. It sold extremely well, going into five editions over the next 10 years, its key points even being distilled into a single sheet to hang by the fireside for emergencies. Similar works, addressed to parents 'on the treatment of children, sobriety, industry, etc' (Parkinson 1800*c*), to children warning them against the 'wanton, careless, or mischievous Exposure to those Situations from which alarming Injuries so often Proceed' (Parkinson 1800*d*) and to medical students pointing out the perils of idleness (Parkinson 1800*a*), also appeared in 1800, aimed at informing and 'improving' the general public.

With these same principles in mind, in 1804 Parkinson published the first of his three-volume work on fossils: *Organic Remains of a Former World*: *an examination of the mineralised remains of the vegetables and animals of the antediluvian world; generally termed Extraneous Fossils* (1804, 1808, 1811). Having become interested in fossils,[27] Parkinson had tried to find publications in English that would help him understand their significance and what they represented (Lewis 2009). As little was available at the time, and assuming that others must be having similar difficulties, he decided that he would write the definitive book on fossils for 'readers in general' (Parkinson 1804, p. vii). *Organic Remains* was written in epistolary form, a popular format that many of his readers would have been familiar with (O'Connor 2007, p. 52). But despite being aimed at a popular readership, the work reveals a man fully conversant with contemporary geological ideas being propounded on the Continent which were, according to Rudwick (2005; Taquet 2009), in advance of those in Britain. But even here, however, he continues to 'improve' his readers by chastizing those who waste their leisure hours, and by demonstrating how everything that has happened to the Earth since the Deluge has been for the benefit of mankind (Lewis 2009).

The Medical and Chirurgical Society

Despite the very disparate backgrounds and careers of these four men – Babington, Franck, Parkinson and Laird – they were linked by common interests in medicine, chemistry and geology, which ultimately led to them participating in the founding of the Geological Society. But prior to that, they were also all involved in another new society which was to play an important role in their lives and that of the young Geological Society.

Societies flourished in response to the Enlightenment. Intellectuals, natural historians, philosophers, mathematicians, astronomers, physicians and surgeons all questioned, searched and experimented, founding clubs and societies to present and discuss their findings. The movement grew up alongside the tradition of debating societies, so fashionable and prevalent during the last quarter of the eighteenth century (Thale 1989; Andrew 1994), but which, by the end of the century, were virtually shut down by the Government's attempt to stamp out radicalism. But what could not be suppressed was the urge for compatible gentlemen to get together in convivial surroundings to discuss their favourite science.[28] Presiding over them all was the Royal Society, setting an example of what should be aspired to, while keeping her brood tucked close under her wing. Medical science in particular was fostered by the Royal Society – some 20 percent of papers up to 1850 were on medical matters (Hunting 2002) – and its primacy in this field was rivalled only by the Medical Society of Edinburgh. While many of the leading medical figures of the time had gained their medical education in Edinburgh and been active members of its medical society, it was London that dominated the nation's professional life (Hays 1983), attracting these same medical men to its opportunities. However, on coming to practice in London before their reputations were established, not all were eligible for the Royal Society and so they joined the Medical Society of London.

In the early years of the nineteenth century, things were not well with the Medical Society of London. Dr James Sims who had joined only two months after it was founded in 1773 had, for the past 22 years, been its autocratic President. Now there was a desire to put the society on a more democratic footing, but Sims blocked all proposals for reform and nothing changed. After a stormy meeting on 29 April 1805 many of the reformers resigned, among whom were Alexander Marcet and his mentor William Saunders from Guy's Hosptial, John Yelloly, physician to the General Dispensary, and the recently knighted surgeon, Sir William Blizard from the London Hospital. Three weeks later, on the 22 May 1805, a large number of

[27] His great friend, John Keys, who was also a naturalist, lived in Sewardstone, Essex, near the gravel pits in the Lea Valley where they might well have hunted for fossils.

[28] Edward Jenner, for example, founded a society entitled the Medico-Convivial Society (Hunting 2002, p. 62).

Table 3. *Officers and council members of the Medical and Chirurgical Society*

Name	Status	Position	Member of Geological Society
William Saunders	MD FRS	President	1808
John Abernethy	FRS	Vice-President	
William Babington*	MD FRS	Vice-President	Founder
Sir William Blizard	FRS	Vice-President	
John Cooke	MD FRS	Vice-President	
Charles Rochemont Aikin		Secretary	
John Yelloly	MD	Secretary	1811
Alexander Marcet	MD	Foreign Secretary	1808
Astley Cooper	FRS	Treasurer	
Matthew Baillie	MD FRS	Council Member	
Thomas Bateman	MD	Council Member	
Gilbert Blane	MD FRS	Council Member	
James Curry	MD	Council Member	1810
Sir Walter Farquhar	MD	Council Member	
Thompson Forster		Council Member	
Algernon Frampton	MD	Council Member	
John Heaviside	FRS	Council Member	
David Pitcairne	MD FRS	Council Member	
Henry Revell Reynolds	MD FRS	Council Member	
Leigh Thomas		Council Member	
James Wilson	FRS	Council Member	

*In fact, Babington did not become FRS until eight days after the MCS was founded.

physicians, surgeons and apothecaries met at the Freemasons' Tavern in Great Queen Street where it was unanimously resolved 'that a Society comprehending the several branches of the medical profession be established in London, for the purpose of conversation on professional subjects, for the reception of communications, and for the formation of a library' (Hunting 2002, p. 8).

From this meeting emerged the Medical and Chirurgical[29] Society, from which today's Royal Society of Medicine is descended. Its first President was William Saunders, with William Babington elected one of the Vice-Presidents (Table 3). Five of the founders of this society were to become members of the Geological Society, and more than half were Fellows of the Royal Society. As the *Medical and Physical Journal* noted, the 'names of the officers and council will justify the highest expectations of the advantages to science which are likely to result from this institution'(Anon. 1805, p. 191). Indeed, eight days later, on 30 May 1805, William Babington 'a Gentleman eminently learned in Chemistry and Mineralogy'[30] was also elected a Fellow of the Royal Society; an approbation of his contribution to the London scientific scene. New members were sought by personal nomination and their suitability balloted upon. However, to put the young Medical and Chirurgical Society on its feet, prominent medical figures, including James Parkinson and Everard Home (1756–1832), were simply invited to join (Hunting 2002, p. 8). James Laird and James Franck became members shortly thereafter.

The Medical and Chirurgical Society made a point of being open to all branches of the medical profession and it further decided that its Honorary Members would not just include medical men who lived out of town, but 'gentlemen who have eminently distinguished themselves in sciences connected with medicine, but who are *not* of the medical profession' (my italics). Humphry Davy, William Wollaston (1766–1828), Charles Hatchett (1765–1847)[31] and Smithson Tennant (1761–1815),[32] all notable chemists and mineralogists, fell into this category. Wollaston, who had an MD from Cambridge, saw the invitation to join as recognition 'of my endeavours to promote sciences connected with medicine' (Hunting 2002, p. 9).

[29] Chirurgical is derived from the Latin *chirurgia*, meaning surgery.

[30] RS: Babington, William. Certificate of Election and Candidature.

[31] In 1801 Hatchett discovered a new metal in a mineral from North America which he called columbium, later identified as niobium.

[32] In 1797 Tennant demonstrated that diamond has the same chemical composition as charcoal.

All these men were to become members of the Geological Society, and Dr Wollaston went on to endow it with its most prestigious medal. Originally made of palladium, a metal discovered by Wollaston, the Wollaston medal is today the highest honour granted by the Geological Society of London. It was first awarded in 1831 to William Smith (1769–1839) (Knell 2009).

While these medical men had found a new home for their discussions, it was but one forum where they might meet. London in the early nineteenth century teemed with societies and aspiring gentlemen patronized any number of them (Table 4). Bound together by the common threads that link medicine, chemistry, mineralogy and geology, these men would have met regularly and known each other well. So although the role played by societies such as the Askesian and the British Mineralogical to the founding of the Geological Society has long been recognized (Weindling 1979, 1983; Torrens 2009), the contribution of the medical societies has not previously been acknowledged. Guy's Hospital Physical Society, in particular, was a place where members such as Richard Bright, Astley Cooper, Everard Home, John Hunter, Edward Jenner, Alexander Marcet, John Yelloly, William Saunders and many others shared an interest in the emerging science of geology with those founders-to-be, Allen, Babington, Davy, Laird and Parkinson (Table 4).

Even less well known is the important contribution made by wealthy and influential mineral collectors of the upper classes who, in general, did not participate in such societies, but reserved their attendance for the exclusivity of the Royal Society. So when Babington became a Fellow of the Royal Society in 1805, his long-standing connections with such men became a vital ingredient in the Geological Society's early success. He recognized that it was important for new societies to gain the support and patronage of such men, and without it the young Society might have remained just a dining club for a few like-minded friends.

A physician among the mineral collectors

A passion for mineralogy was flourishing in both Britain and France at the end of the eighteenth century. Abbé Rene Just Haüy (1743–1822) and Jean Rome de l'Islé (1736–1790) had developed the fundamental principles of modern crystallography, while Lavoisier and his colleagues laid out the new chemical concepts that enabled analysts to begin defining mineralogy as a branch of chemistry, rather than of mining science or pharmacy (Wilson 1994). Jacques-Louis, Comte de Bournon (1751–1825) (another of the Geological Society's founders and discussed in this volume by Herries Davies) had studied mineralogy in Paris under Romé de l'Isle and had amassed his own collection 'of which the number of pieces rose to nearly 14 thousand' (Bournon 1808). He was also familiar with 'a large number of the French cabinets, several of the

Table 4. *Societies and clubs, noting members who were founders of the Geological Society. This is by no means an exhaustive list*

Name of Society	Date founded	Geological Society founders known to be members
Royal Society	1665	Before 13 November 1807: de Bournon, Davy, Babington, Greenough; after 13 November 1807: Allen, Pepys, Franck, Phillips brothers
Guy's Hospital Physical Society	1771	Allen, Babington, Davy, Knight, Laird, Parkinson
Medical Society of London	1773	Babington, Laird, Franck, Parkinson
Athletae	*c.* 1780	Babington
Lyceum Medicum Londinense	1785	Babington (probably others)
Linnean Society	1788	Allen
Askesian Society	*c.* 1796	Aikin, Allen, Babington, Knight, Pepys, Phillips brothers
British Mineralogical Society	1799	Aikin, Allen, Babington, Knight, Pepys
Royal Jennerian Society	1803	Allen, Babington, Parkinson, Phillips brothers
Medical and Chirurgical Society	1805	Babington, Davy, Franck, Laird, Parkinson
MCS Club	*c.* 1805	Babington, Davy, Franck, Laird, Parkinson
Chemical Club	*c.* 1807	Babington
Animal Chemistry Club	1808	Babington, Davy
Hunterian Society	1819	Babington, Parkinson
Athenaeum Club	1823	Davy, Franck

German ones and all those in Paris worthy of being cited' (Bournon 1808). Through his search for minerals, de Bournon had become an accomplished geologist, although, like others in the new Geological Society, he used the term 'mineralogy' to embrace many of our modern notions of 'geology' (Knell 2009; Rudwick 2009):

> Observing these same extinct volcanoes [of Auvergne] would, I think, also have a great effect on the opinion of those mineralogists [geologists] who, having adopted the Neptunist system in all its exclusive rigour, would make up their minds to travel through them and study them before giving an opinion.
>
> (Bournon 1808)

Thus, among the collectors of minerals on the Continent, de Bournon ranked as one of the most famous and his name and collection undoubtedly would have been known to those interested in the subject across the Channel. Unfortunately, as a wealthy count and supporter of Louis XVI, de Bournon was forced into exile during the French Revolution, becoming 'without country, without fortune, without status'. After 'wandering for a long time in Germany with my family', he finally landed in England, 'the shore of true liberty' (Bournon 1808).

The arrival in London in 1794 of this French nobleman and refugee stirred much interest in scientific circles and soon afterwards de Bournon appears to have met Babington who introduced him to the London mineralogists. Within a short time 'nearly all the cabinets in London were successfully placed in my charge to put them in order and arrange them, with the exception of that at the British Museum: the sole direction of the three largest was given me by their owners, Mr Greville, Sir John St Aubyn and Sir Abraham Hume' (Bournon 1813, p. iii; English translation provided pers. comm. by Jessica Shepherd, Plymouth City Museum).

Sir Charles Greville (1749–1809), who was a patron of science almost on a par with Joseph Banks, became de Bournon's principal supporter and by 1803 Greville had one of the finest collections of minerals in the world, much of its excellence due to de Bournon's influence (Cooper 2004). Sir Abraham Hume, MP (1749–1838), owned an outstanding assortment of diamonds and precious stones that de Bournon also arranged. Hume was to become one of the Geological Society's 'most strenuous friends and most liberal supporters', standing by its side throughout the difficulties of its early years, such that many at the time considered him to be one of the Society's founders (Whewell 1842 p. 65). Sir John St Aubyn (1758–1839) who, with his family background in the heartland of the Cornish mining industry, had developed a great interest in mineralogy, employed de Bournon on a part-time basis to curate and catalogue his collection.

Greville was to become the new Geological Society's first patron and Hume, St Aubyn and Babington its first Vice-Presidents, but the fourth Vice-President was the now little-known MP, Robert Ferguson (1767–1840). Ferguson was born into a wealthy Scottish family and when a boy in Edinburgh had been tutored by John Playfair, from whom he probably acquired his interest in geology (Lloyd 2000). As a young man, Ferguson spent 10 years travelling in Europe meeting other mineralogists and building up a collection that was 'surpassed by few private collections in the kingdom' (Society for the Benefit of the Sons and Daughters of the Clergy 1845, p. 153). In 1803 when Ferguson returned from his travels, he soon became a significant figure on London's mineralogical scene,[33] although de Bournon does not appear to have worked on Ferguson's collection, perhaps because it was housed in Scotland.

In 1795 Babington published *A Systematic Arrangement of Minerals*. He had developed his system while 'employed in arranging a Cabinet selected from the very extensive Collection of Minerals which he had an opportunity of purchasing a few years ago' (Babington 1795). This collection of more than 2000 British and foreign minerals had originally belonged to John Stuart, the third Earl of Bute (1713–1792). In his lifetime, Lord Bute had owned what was probably the largest collection of minerals in the world (Wilson 1994). After his death in 1792, it was sold at auction over 14 days in 1793 and 1794 (Turner 1967, p. 213; Lewis 2009). Babington does not state whether he acquired his collection at the auction or from Lord Bute himself, but given the time it took him to do the work[34] and that in 1795 he says he bought the collection 'several years ago', it seems most likely that he purchased the minerals from Bute before he died. Bute had been the patron of the Irish chemist and mineralogist Peter Woulfe (*c.* 1727–1803) who carried out chemical experiments for him in Bute's laboratory in London. Woulfe's rooms were so cluttered with apparatus and specimens that when Babington visited he complained that on

[33] Around 1810, when his father died and Ferguson had to return to Scotland to manage the estate, the collection disappeared from public knowledge until being rediscovered and sold by his descendants in 1997 (Lloyd 2000).

[34] The second edition, 1799, says it was a 'considerable' length of time.

having put his hat down he was unable to find it again (Campbell 2004). As Babington evidently knew Woulfe, he probably knew Bute as well – the circle of mineralogists in Britain was small.

Babington based his system for cataloguing the minerals 'on the joint consideration of their chemical, physical and external characters', following that devised by Baron Ignace von Born (1742–1791), 'the Transylvanian mineralogist and Fellow of the Royal Society' (Townson 1797, p. 410[35]), whose enlightened use of Lavoisierian principles had been developed when arranging the collection of minerals (termed 'fossils' in the work) of Eléonoré de Raab:

> The Classes, Orders, Genera and Species being founded on chemical distinctions, and the Varieties on external character. On the subject of Crystallisation particular pains have been taken to associate the external figure of substances with their internal structure, so as to reduce them into more connected series than has hitherto been done in any attempt of the kind.
> (Babington 1795)

In the Advertisement to this work, Babington explained how he had not originally intended it to be published, but had compiled it for his own use:

> But as that arrangement occupied a length of time, and a degree of attention, much beyond what was at first conceived to be sufficient for such an undertaking, the Author [...] thinks, that in publishing it, he may not only considerably abridge the labour of those who shall hereafter engage in a similar task, but also render an acceptable service to many who wish to acquire a comprehensive knowledge of Mineralogy, but who have neither leisure nor inclination to turn over the numerous works on the subject, in which the necessary information lies scattered.
> (Babington 1795)

In April 1797, Babington sold his 'Butean selected cabinet of minerals' at auction[36] and it was purchased by St Aubyn for £3000.[37] This large sum reflects how mineral prices had risen as collecting them became more and more fashionable in the last decade of the eighteenth century and as the Napoleonic Wars created a shortage of supply (Cleevely 2004). The whole of Bute's collection, some 1600 lots, had only fetched £1225 when sold at auction just a few years earlier (Turner 1967). Two years after St Aubyn's purchase, in 1799, Babington published a revised edition of his 1795 catalogue now entitled *A New System of Mineralogy*.[38] Babington was assisted in this work by Richard Kirwan (1733–1812) who performed the chemical analyses necessary to distinguish the 'different species', by Richard Stocker who became his successor at Guy's (Table 2) and by the young Arthur Aikin (1773–1854) who, like Babington, was destined to become one the 13 founder members of the Geological Society (Knight 2009).

St Aubyn had allowed Babington to consult the Butean collection for this publication and in gratitude Babington dedicated the new work to him. The major difference between the catalogues of 1795 and 1799 is that in the later one Babington follows Rome de l'Islé as his 'guide on the subject of crystallography'. Babington had been taking lessons in mineralogy from de Bournon (Woodward 1908, p. 11) who, as a one-time student of Rome de l'Islé, clearly influenced him in this respect. In fact, it seems the collection remained at Babington's house for many years after St Aubyn had purchased it and that de Bournon used it as his teaching material while cataloguing it for St Aubyn. Greenough recollects examining the collection shortly before the Geological Society was formed: 'the opportunity which he [Babington] afforded us of examining the most interesting specimens which it contained, furnished to some a powerful motive and to others a fair excuse to resort once a week or oftener to his house as early as 7 in the morning'.[39] It was undoubtedly de Bournon who recommended St Aubyn should purchase the Butean collection – which de Bournon then went on to catalogue. Unlike Babington's classification of the Butean cabinet, de Bournon's arrangement is not structured on orders, genera and species. Instead, there is a general subdivision, which is in the form of different codes and symbols on the specimen labels.[40]

---

[35] This work includes a biographical sketch of Baron Born.

[36] Advert in the *Glasgow Courier*, Vol. VI, No. 849, Tuesday 31 January 1797, says the auction will be held in April.

[37] In a copy of Babington's *New System of Mineralogy*, held at the Plymouth City Museum and Art Gallery, Isaiah Deck (1792–1853), who helped to auction St Aubyn's collection when he died, wrote beneath Babington's dedication to St Aubyn: 'Sir John St Aubyn afterwards purchased the whole of Dr. Babington's collection of which this catalogue is the basis, and for which he gave £3000 over'. See: http://www.plymouth.gov.uk/a_new_system_of_mineralogy.pdf.

[38] A contemporary review of Babington's *New System* concluded 'we have no doubt that Dr. Babington's New System of Mineralogy will prove of great utility, and [is] well calculated to promote the advancement of mineralogical knowledge' (Anon. 1801).

[39] GSL. LDGSL 960. Greenough Papers.

[40] Some of the collection, with the original labels, can still be seen at the Plymouth City Museum and Art Gallery.

Thus, for well over a decade before the founding of the Geological Society, Babington, 'one of ye. best mineralogysts in this Metropolis' (Ford 1987, p. 186) and de Bournon 'the first man in this country who understands it [crystallography]'[41] appear to have been at the centre of mineralogy circles in London, cataloguing the collections of the wealthy and influential, and advising them on future purchases. Babington's house, with its laboratory close by, used primarily for the preparation of medicines for his patients,[42] but which could also facilitate the chemical analysis of minerals, became a focal point for gatherings of mineralogists. De Bournon gave mineralogy lessons there, meetings of the merged British Mineralogical and Askesian societies continued there (Weindling 1979; Torrens 2009), and the equipment and minerals belonging to each group were taken to his laboratory (Weindling 1983, p. 130).

Greville, Hume and St Aubyn paid de Bournon handsomely for his services – in 1804, for example, St Aubyn was paying him 75 guineas every six months, over a period of 18 months, to complete his work on a cabinet.[43] De Bournon also had 'a valuable Traffick'[44] with dealers from which he no doubt made a profit that helped sustain him and his family. Nevertheless, by 1807 de Bournon seems to have been finding it difficult to support his family. St Aubyn had taken his collection back to Cornwall, possibly because it was being 'injured by the sulfureous air of London'[45] and had 'withdrawn some of his bounty'.[46] And perhaps de Bournon's work on the other collections was also nearing completion since he had been working on them for some 12 years. Early in 1807 he finished writing two volumes of a three-volume treatise on carbonate of lime, crystallography and mineralogy (Bournon 1808), but his topic was highly specialized and had limited appeal:

> [de Bournon] is so deliberate and particular in his descriptions, that he fatigues his readers who are but novices in that particular branch of the science, before they are got half way through his Paper or Lecture. He is an excellent Scholar after Rome de Lisle but very few of the moderns have attended to that kind of Arithmatick. It is almost necessary for a Mineralogist, & as that study is so much the fashion, it ought to be more cultivated.[47]

De Bournon had tried to get the work published because 'my position and that of my family, to the support of which my work has been devoted, was making this an obligation and indeed a duty', but he was having little success. On the point of giving up and turning to another occupation for employment, he was one day relating his difficulties to Babington and William Allen:

> and talking with them about having found it impossible to have my work printed in any profitable way, and thus about the necessity with which I would probably have to give it up, these two friends, joined at the same instant by Mr W. and Mr. R Phillips, offered, in a manner that was as noble as it was generous, to put themselves at the head of a subscription list which they proposed to limit to sixteen persons, the purpose of which would be to underwrite the expenses of my work, while assigning me the whole of the profit.

William Phillips (1773–1828), who was a printer (Torrens 2009), agreed to print the work if his expenses were underwritten. De Bournon goes on:

> To this generous proposal they added some extremely flattering remarks about their certainty that they would see this subscription list completed very swiftly by the mineralogists in London, and that I should consider myself completely detached from the matter, since they alone would take care of it.
>
> (Bournon 1808)

Indeed, that is exactly what happened. A committee of management was formed to admi nister its publication,[48] a subscription fee of £50 (Rudwick 1963) was settled on and, as de Bournon says, the group determined that 16 subscribers were required. Notable among those in England who quickly committed funds are the four wealthy mineral collectors discussed earlier, but word soon spread abroad. In 1802 Alexander Crichton (1763–1856) had been one of the Fellows of the Royal Society who nominated de Bournon to that august society. Crichton had been a physician at the Westminster Hospital, but like Babington he was also a renowned chemist and mineralogist. In August 1804 Crichton moved with his family

[41] UoBL: Eyles Collection DM1186. Phillip Rashleigh (1729–1811) to James Sowerby (1757–1822), 17 January 1804.
[42] Some of Babington's medicinal 'recipes' can be found in the Wellcome Archives: MS.7832.
[43] Pers. comm. Jessica Shepherd, Plymouth City Museum and Art Gallery, 12 September 2008.
[44] UoBL: Eyles Collection DM1186. Phillip Rashleigh to James Sowerby, 15 April 1806.
[45] Cornwall Records Office, Truro: Rashleigh Letters DDR/5757/1/92. John Hawkins to Phillip Rashleigh, 9 January 1799.
[46] UoBL: Eyles Collection DM1186. Phillip Rashleigh to James Sowerby, 15 April 1806.
[47] UoBL: Eyles Collection DM1186. Phillip Rashleigh to James Sowerby, 15 April 1806.
[48] GSL. LDGSL 960. Greenough Papers.

to St Petersburg to take up the post of physician-in-ordinary to the Emperor of Russia and the Dowager Empress (Appleby 1999). It is unclear whether Crichton himself subscribed to de Bournon's treatise, although he was sent a copy,[49] but on learning 'that in London there was a list of subscriptions for the printing of my work' (Bournon 1808), Crichton informed his Russian colleagues. Subscriptions were soon forthcoming from three eminent individuals: Nicolas Novossiltzoff (1770–1838), President of the St Petersburg Academy, and Count Paul de Strogonoff, the Academy's Director of the Fine Arts, as well as from the Polish Prince Udan Czartorinsky, a Minister and close friend of the Emperor. Furthermore, these gentlemen offered to find more funds if what they sent was not enough to complete the project. In gratitude, de Bournon dedicated the work to 'His Imperial Majesty, Alexandre I', while hoping, no doubt, that such an endorsement would help sales. The other subscribers 'to whom today I am able publicly to pay homage' were:[50]

William Allen, Esq. F.R.S.
Sir J. St Aubyn, Bart. M.P. F.R.S.
Wm. Babington, M.D. F.R.S.
Robert Ferguson, Esq.
Rt. Hon. Charles Greville, F.R.S.
Geo. B. Greenough, M.P. F.R.S.
Sir Abm. Hume, Bart. M.P. F.R.S.
Chas. Hatchett, Esq. F.R.S.
Luke Howard, Esq.
Richard Knight, Esq.
Richard [James] Laird, M.D.
William Phillips, Esq.
Richard Phillips, Esq.
John Williams, Junr, Esq.

As Allen was to contribute to the fund in kind by printing the work, the remaining 13 men listed above, plus the three Russians, made up the 16 subscribers being sought, and publication of de Bournon's treatise soon commenced.[51]

Founding the Geological Society

In Greenough's later years, probably around the time of Babington's death in 1833, he wrote an account of the founding of the Society which is now in the Society's archives.[52] However, on comparing this account with events recorded in the Society's minute books, it is evident that, 25 years later, Greenough's memory of events was not as flawless as he thought it was:

> It has been pointed out to me as one of my peculiarities that I exhibit great fidelity in relating a story – friends have noted my love of truth by calling upon me to relate the same adventure or anecdote at distant periods and time and find that tho' I may change the words or phrases, the facts and incidents invariably are the same.[53]

Nevertheless, Greenough's is the only contemporary account we have and, as such, is of great value. However, the success of the Geological Society in its early years meant that during their lifetimes the founders would have been aware of the significance of what they had achieved and Greenough in particular, ever the politician, probably pondered how posterity would view his 'interpretation' of events. James Laird, for one, disagreed with the version Greenough gave in his 1834 Presidential Address in which he recounts the founding of the Society in an obituary of Babington, and which corresponds closely with his later account:

> Bognor
> July 3 1834
>
> Dear Greenough
>
> I am tempted by an opportunity of writing to Town to trouble you at this time with a few lines, differing as we do, in no material respect, however, with respect to the origin of the Geological Society, which, in your interesting Anniversary Address, you state to have resulted from a continuance of the meetings which were connected with the Bournon fund. In my little MSt [manuscript], which I now doubly regret that Bright did not lay before you,[54] I have given my impressions upon this subject; and in recently adverting to this discrepancy in our recollections, or rather view of the

[49] Wellcome: MS.7887/1-2. Letter from William Babington to Richard Phillips, May 22 1809. This letter requests that copies of the work be sent to 'Dr Crichton, Prince Udan Czertovinsky, Mr Novossillzoff, and count Paul de Strogonoff, free of expence [sic]'. It also asks Richard Phillips to attend a meeting to inform the subscribers how often the work had been advertised, suggesting that sales were not going very well.

[50] Those in bold went on to become founders of the Geological Society. James Laird was erroneously listed as Richard.

[51] Since having the English translation of the Preface to Bournon (1808), it is evident that there were in fact at least 16 subscribers and not 14, as all previous accounts have related. Parts of this translation are now available in Appendix II to this volume (Lewis & Knell 2009).

[52] GSL: LDGSL 960. Greenough Papers.

[53] UCL: 24/1. Greenough Autobiographical Notes.

[54] When Babington died in 1833, Richard Bright proposed writing a memoir of Babington and asked Laird to write an account of the origin of the Society for him. This Laird did, asking Bright to send it on to Greenough knowing,

> matter, in a letter to Dr Benjn Babington,[55] I have noticed the important circumstance that Davy did not join in the Bournon fund (and with him I couple the dinner at the Freemasons Tavern in which arrangement of Geological communings arose, if I mistake not, our Society.[56]

In support of Laird's argument, only half those who subscribed to the 'Bournon Fund' as he calls it, went on to become founder members of the Geological Society: Allen, Babington, the Phillips brothers, Greenough, Richard Knight (1768–1844) and Laird, plus, of course, de Bournon himself.[57] But as Laird later admits, whether the Society was formed as a consequence of these meetings 'or had, according to my version, another but nearly contemporaneous formative nisus, is comparatively a trifle'.[58]

Greenough's version of how the Society formed states that once the matter of the subscription to de Bournon's treatise had been settled, the meetings they had established 'were found too agreeable to be discontinued. The effect of them was to connect together in bonds of permanent esteem parties who had hitherto been pursuing the same studies in solitude'. Of the subscribers, Greenough notes, 'the most constant attendants upon these occasions were Mr Allen, Mr Knight, Dr Laird & the two Phillips' but the numbers were soon augmented 'by the attendance of some non subscribers, among whom the most eminent was Sir Humphrey [sic] Davy'. Aikin, Franck and de Bournon also attended regularly.[59] Many of these meetings were held at 7 o'clock in the morning over breakfast, partly to accommodate Babington's hectic schedule, although he is not mentioned as being a regular attendee, and partly, as alluded to earlier, because they were able to study St Aubyn's Butean mineral collection, still kept at Babington's house,[60] and perform chemical analysis in Babington's laboratory:

> Our studies were carried on with so much spirit and enthusiasm that at an early[61] meeting (Nov 13th 1807) at which I happened not[62] to be present that which had hitherto been only as a club was voted a society – and I was appointed without the slightest notice its President.

Thus, it appears that *prior* to the dinner meeting at the Freemasons' Tavern on the evening of 13 November, there was a breakfast meeting that morning where those present decided they should form a geological *society*, rather than a dining club as appears to have been discussed previously. It seems Humphry Davy was also not present at this breakfast meeting. An undated note from him to Greenough, recorded in Greenough's recollections, excuses Davy for not being present at the breakfast meeting 'tomorrow' and suggests that instead of holding breakfast meetings they should meet for dinner at a tavern (Knight 2009).[63] Although we cannot be certain that it was the breakfast meeting on the 13th that Davy was excusing himself from, in a note to William Hasledine Pepys (1775–1856), written during that day Davy says, 'We are forming a little talking Geological Dinner Club, of which I hope you will be a member. I shall propose you today'. Clearly, Davy still thought they were forming a dining club right up until they sat down to dine, and the fact that they did not remained a bone of contention for this precocious 28-year-old Secretary of the Royal Society, right up until the day he resigned from the Geological Society. But would Laird have gone to the dinner,

presumably, that he too would be writing an obituary of Babington for his Anniversary Address. But apparently Bright did not do this (see letters from Laird to Greenough, UCL: AD7981 1099 and 1100). In the Bright Collection at the Norfolk Records Office (NRO) are 500 letters, amongst which might be this valuable document. Unfortunately, the collection is in such a poor state that it cannot be handled. It awaits conservation, endless dedicated time and, of course, considerable funds which the NRO does not currently have.

[55] Benjamin Guy Babington, William Babington's son.

[56] UCL: AD7981 1102. Laird to Greenough, July 3 1834.

[57] Greville, however, became its first patron, and Ferguson and Hume were on the first committee of trustees. John Williams (1777–1849) became an Honorary Member in 1808 and, of the British subscribers, only Luke Howard (1772–1864) never joined the Society.

[58] UCL: AD7981 1102. Laird to Greenough, July 3 1834.

[59] GSL: LDGSL 960. Greenough Papers.

[60] Although Greenough mistakenly remembers the collection as still belonging to Babington.

[61] Much hinges on what Greenough meant by an 'early' meeting. Others have thought he meant a meeting early in the Society's history, i.e. the meeting on 4 December where Greenough was elected President. However, the Society minutes clearly record that Greenough *was* present at that meeting, so this is not likely to be the case. I consider he meant a *breakfast* meeting held, as he states, early on the morning of 13 November.

[62] The word 'not' has been inserted afterwards, but in Greenough's handwriting.

[63] GSL: LDGSL 960. Greenough Papers. A copy of this note from Davy is on p. 4 opposite where Greenough talks about the founding, of the Society. Greenough used the left-hand pages of his recollections to 'illustrate' what he was saying on the right-hand page.

pen and ink in hand and ready to take minutes, if he wasn't already expecting to be Secretary of the new society? Would that first resolution as to the Society's intentions, so clearly stated in the minutes taken that night (Woodward 1908, pp. 15–16), have been so succinct had some members not already given it considerable thought? Furthermore, if the Society had arisen as Greenough described, why was someone like Parkinson invited to become a founder, when he was not a member of any of the precursor mineralogical gatherings and did not subscribe to de Bournon's treatise?

While it called itself a tavern – which today we associate with 'pub' – the Freemasons' Tavern was, in fact, the front half of the Grand Lodge of the Freemasons Hall in which Masonic meetings were held (Moody 2009). The Freemasons' Tavern was much more like a gentleman's club than a pub. Dinner there cost 15 shillings – more than a week's wages for many – the high price of which both Davy[64] and Greville (Rudwick 1963) complained. Of the founders, only Parkinson was actually a Freemason,[65] although St Aubyn was a Provincial Grandmaster of the Freemasons for 54 years (Wilson 2008).

In the event, 11 men sat down to dine at 5 o'clock at the Freemasons' Tavern on the 13 November 1807: Arthur Aikin, William Allen, William Babington, Jacques Louis Comte de Bournon, Humphry Davy, James Franck, George Bellas Greenough, Richard Knight, James Laird, James Parkinson and Richard Phillips. William Hasledine Pepys, who was not expected, and William Phillips, who did not attend, were absent in body if not spirit, bringing the total number of founders to 13. Years later Laird recalled how:

> Davy asserted at our first dinner at the Freemasons Tavern, that he never knew anything prosper that was begun upon a Friday [the 13th], and our little band, besides, was of the bad omened number 13, at it's start; well may we laugh at the prejudices of mankind [...].[66]

By the end of the evening, a formal decision had been taken to inaugurate a Society, as Allen recorded in his diary: '13th. – Dined at the Freemasons' Tavern, about five o'clock, with Davy, Dr. Babington, &c., &c., about eleven in all. – Instituted a Geological Society' (Allen 1847, p. 66). The minutes from that evening record that for some reason Greenough was elected Treasurer at this meeting, not President, with Laird as Secretary. It was not until the meeting on December 4 that Greenough was elected President, with Pepys replacing him as Treasurer. It is noteworthy that two of the youngest members of the group took on the two most demanding roles – neither Greenough nor Laird were yet 30 years old (Table 1).

Torrens (2009) considers it a mystery why this group, consisting predominantly of mineralogists, should form a geological society, and, indeed, Greenough claims that he and Davy overruled a proposal that evening that they form a mineralogical society.[67] Whether that actually happened we shall never know, but as the minutes of the Askesian Society noted: 'A London Institution having been formed and most of the Members of the Two Societys [Askesian and British Mineralogical] having become Members, It was deemed not requisite to continue the Societys meetings, till further notice be given' (Weindling 1983, p. 131). The opening of the London Institution in 1806, with its extensive lecture programme, splendid reference library, reading rooms, laboratory and scheme for public mineral analysis, had rendered the British Mineralogical Society redundant and so it had merged with the Askesian. Furthermore, the Royal Institution had a museum of more than 3000 mineral and fossil specimens, including a special collection of minerals presented by Davy, as well as the patronage of Greville, Hume and St Aubyn (Woodward 1908, p. 9). So what would be the point of forming yet another mineralogical society, when one had failed so recently due to lack of support?[68] While some at the dinner on the 13th may have wanted to *call* it a mineralogical society, because the term was more familiar and many still used mineralogy to describe geology, what they wanted to *do* was geology.

Indeed, at least 10 of the 13 founders had already demonstrated a considerable interest in geology. As early as 1796 Aikin had undertaken a geological

---

[64] GSL: LDGSL 960. Greenough Papers.

[65] Parkinson was initiated on 15 June 1791 as a member of the Lodge of Freedom and Ease, No. 176, which met at the Three Jolly Butchers, Old Street Road, Hoxton, from 1790, before relocating to the New London Tavern, Cheapside, by 1803. The New London Tavern was later known as the Moira Lodge, No. 92, but there is no record of Parkinson having been a member of this Lodge, thus it is possible he only remained a Freemason while the Lodge met near to his home. Pers. comm. Susan Snell, Archivist and Records Manager, Library and Museum of Freemasonry.

[66] UCL: AD7981 1100. Laird to Greenough, April 4 1834. Missing and inappropriate apostrophes as original.

[67] GSL: LDGSL 960. Greenough Papers.

[68] There is now some evidence that the British Mineralogical Society was subsequently resurrected and that informal gatherings of the merged Askesian and BMS continued at Babington's for several more years (see Torrens 2009).

tour of Wales and Shropshire (Herries Davies 2007) and in 1805 Davy had given a series of 10 lectures on geology at the Royal Institution (Siegfried & Dott 1976). These covered the appearance and composition of primary and secondary rocks, mineral veins and mining, and a discussion on volcanoes that lasted over two lectures. He also considered the Neptunist theory of Werner and the Plutonist theory of Hutton, criticizing them both, although Hutton fared somewhat better than Werner. That summer Greenough toured Scotland 'Determined to make myself master of the Huttonian theory', and while there he had 'associated with Playfair'.[69] He too concluded that neither the Neptunist nor the Plutonist theory was entirely satisfactory as an explanation of the observable facts (Rudwick 1962). Laird had also been in Edinburgh, between 1799 and 1803, where he could hardly have avoided the controversy. He later donated a copy of Playfair's *A Comparative View of the Huttonian and Neptunian System of Geology* (1802) to the Society.[70] In 1806, Greenough and Davy travelled together through Ireland 'studying nature on the great scale',[71] while at the same time Allen toured Wales. Knight, in the first paper ever read to the Society, demonstrated an interest in the formation of igneous rocks (Knight 2009) and James Parkinson had already published the first volume of *Organic Remains of a Former World* (1804), which was to put palaeontology on the scientific map of Britain. De Bournon, having studied the Alps 'the mountains of Auvergne, together with those of Forez, of Velay, of Vivarais and of the Cévennes' and developed an 'important interest in the coal mines' (Bournon 1808), was also of the opinion 'that Geology would be a more attractive & engaging study than mineralogy'.[72] Even Babington is known to have visited quarries and collected rocks (rather than minerals) for his cabinets (Weindling 1983). Furthermore, Greenough recalled how 'Davy & Franck & Dr Babington were all expert & zealous anglers & I cannot but be amused at recollecting with what eagerness they contended for the close affinity that subsisted between the various descriptions of rock & the various kinds of trout'.[73] Thus it is evident that most, if not all, the founders already understood the importance of geology, and in turning down mineralogy that evening, they embraced geology with zeal.

At the Geological Society's first meeting, on 4 December 1807, it was resolved that Greville, now Vice-President of the Royal Society (1800–1809), be invited to become the new Society's patron, a role he accepted – on condition he did not have to attend meetings – and he was accordingly elected a member at the next meeting on 1 January 1808. Greville had first proposed the idea of making a mineralogical map of Britain as far back as 1790, and by 1804 he had formulated a proposal for a National Collection and Office of Assay to be based at the Royal Institution (Weindling 1979). The purpose was to create a centre for applied geology from which the nation's mineral wealth would be surveyed. The proposal was backed by St Aubyn and Hume, but by the middle of 1806, following 'the total failure of a subscription for an extensive collection of minerals and for an additional laboratory of assay',[74] Greville's plans were put on hold. They would eventually be implemented through the Geological Society, as Horner confirms in a motion he drafted for a Society meeting in January 1809: 'the undersigned [Geological Society committee members] together with several gentlemen [...] have associated themselves for the purpose of promoting the Science of Geology and Mineralogy, *particularly as relating to the Mineral History of the British Isles* [my italics]'.[75] Furthermore, at the very first meeting in December 1807, no fewer than 42 gentlemen, spread far and wide around the country, were designated Honorary Members (without being asked). A few weeks later they were each sent a booklet entitled *Geological Inquiries*, which, if answers to the questions it contained were forthcoming, would achieve the following important objectives: 'that Mineralogical maps of districts, which are now so much wanting, may be supplied; that the nomenclature of the science may be gradually amended [...]; that theoretical opinions may be compared with the appearances of Nature, and above all, a fund of practical information obtained applicable to purposes of public improvement and utility'.[76]

[69] UCL: 24/1. Greenough Autobiographical Notes.

[70] GSL: Ordinary Meeting Minutes 1807–1818.

[71] GSL: LDGSL 960. Greenough Papers.

[72] GSL: LDGSL 960. Greenough Papers.

[73] GSL: LDGSL 960. Greenough Papers.

[74] Royal Institution: 6/3/6. Letter to the Managers of the Royal Institution, signed by Hume, Greville and St Aubyn, *c.* June 1806.

[75] UCL: AD7981 821. Horner to Greenough, Monday, January 1809.

[76] GSL: LDGSL 352. Geological Inquiries. A copy of this is now available as Appendix I in Lewis & Knell (2009).

Within a few months, a Committee for Nomenclature had been formed because 'the present state of the Nomenclature of Geology & Mineralogy renders it highly necessary that this Society should adopt some measures which shall tend to remove the confusion which now prevails'.[77] A Committee of Chemical Analysis followed and, in April 1809, a Committee of Maps was established.[78] The objective of this committee was to make a mineralogical map of Britain and Greenough '*who had from the very first aspired to be the Author of a geological map of England* most willingly paid for the paper & printing' (my italics).[79]

According to Greenough's autobiographical notes, he became 'intimate with Greville'[80] in 1805 just after Greenough returned from his geological tour of Scotland, and it was around this time that Greville's plan to establish a centre at the Royal Institution, from which the nation's mineral wealth would be surveyed, had failed. It therefore seems highly probable that Greville discussed his plans with Greenough, from which a clear objective for the young society emerged. Steered by Greville's desire to have a mineralogical map of Britain and driven by Greenough's ambition to be the author of such a map, the founders immediately set out to gather 'practical information' that would facilitate a greater understanding of where the mineral wealth of Britain lay buried. Thus the Society's birth was hardly 'the effect of accident rather than design', as romantically stated in Greenough's opening words to his recollections.[81] Quite the contrary.

Medical men in the Society

Apart from subscribing to de Bournon's treatise, James Franck's chief contribution to the Geological Society (and science in general) seems to have been that he saved the life of Humphry Davy. In November 1807, just a week or so after the founding of the Society, Davy fell 'seriously & even alarmingly ill'[82] following a visit to Newgate prison (Knight 2009). By the middle of December, William Allen reported in his diary that Davy was 'dangerously ill'. He remained unwell for nine weeks but eventually, according to Greenough, 'Dr Franck cured him'.[83] Soon afterwards, Franck was called abroad and, aside from donating some rocks from Sweden, seems to have had little active involvement in the Society, other than serving on Council for two years from 1812 after his retirement from the Peninsular War. In April 1809 he is listed as being a member of the Committee of Maps although, as has been shown, he was called to supervise medical arrangements in the Peninsular War in Portugal just a few weeks later.

Among the founders, James Parkinson is something of an anomaly, not having a particular interest in mineralogy. Without the advantage of a medical degree, he may not have been exposed to mineralogy to the same extent as the other medical founders, nevertheless, mineralogy was an important part of pharmacy and his expertise in that area is undoubted. He certainly knew most, if not all, the other founders well before 1807, thus it seems likely that he was invited to join them precisely because the mineralogists recognized that they needed someone with his specific *geological* expertise. As Greenough said, 'Mr Parkinson [was] not merely the best but almost the only fossilist of his day at least in England'.[84] Indeed, Parkinson's understanding of William Smith's (1769–1839) notions of strata would have proved invaluable when he was elected onto the Committee of Maps. But not only did he understand 'that immense beds, composed of the spoils of these animals [ammonites] extend for many miles under ground' (Parkinson 1804, p. 8), in the first volume of *Organic Remains* he also drew attention to the work of the French comparative anatomist, Georges Cuvier (1769–1832). Cuvier argued that comparative anatomy was an essential tool for establishing a theory of the Earth: 'in a word, it is only with the help of anatomy that *geology* can establish in a sure manner several of the facts that serve as its foundations' (Rudwick 1997, p. 21). By deciphering the language of fossil bones found

[77] GSL: Ordinary Meeting Minutes 1807–1818. Members of the Committee for Nomenclature were: Aikin, Davy, Greenough, Sir James Hall (1761–1832), Horner, Laird, Wilson Lowry (bap. 1760–1824), William MacMichael (1783–1839), John McCulloch (1773–1835) and William Phillips.

[78] GSL: Ordinary Meeting Minutes 1807–1818. Members of Committee of Maps were: William Atkinson [1774/5–1839], Rev Edward John Burrow [1785–1861], [Gilbert] Giddy Davies (1767–1839), Franck, Horner, Lowry, MacCulloch, MacMichael, Pepys, W. Phillips, Parkinson and James Shuter.

[79] GSL: LDGSL 960. Greenough Papers.

[80] UCL: 24/1. Greenough Autobiographical Notes.

[81] GSL: LDGSL 960. Greenough Papers.

[82] GSL: LDGSL 960. Greenough Papers.

[83] GSL: LDGSL 960. Greenough Papers.

[84] GSL: LDGSL 960. Greenough Papers.

in the Paris Basin, Cuvier was making it possible to reconstruct a detailed history of the Earth and the life that had inhabited it. As both a fossilist and a surgeon, Parkinson fully appreciated the significance of Cuvier's work on comparative anatomy and foresaw the impact it was to have on geology: 'From this work so much information is to be expected, that, I doubt not, its publication will prove an important epoch in the history of this science' (Parkinson 1804, p. 28). Indeed it did.

Parkinson was also appointed to the Committee of Extraneous Fossils, of which he became Secretary, after finishing the third volume of *Organic Remains* (1811). On 12 March 1812 the Secretary reported that the committee had met on the 9th and appointed Mr Parkinson as its new Secretary. At the same meeting, they had agreed to 'arrange all the organic remains in the possession of the Society, according to the System adopted in Mr Parkinson's work, and to meet every Monday until their objects should be accomplished'.[85] It is notable that seven of the 12 members of this committee were medical men and several had a particular interest in anatomy, suggesting that others besides Cuvier recognized that comparative anatomy was a tool for understanding geology. These medical men were:[86]

(Sir) Anthony Carlisle (1768–1840) FRS, a physiologist and Professor of Anatomy at the Royal Academy.

Dr John McCulloch (Bowden 2009), a surgeon who Greenough described as 'better acquainted with geology than most of his colleagues'.[87]

Dr Peter Roget (1779–1869), today renowned for his famous thesaurus, but then distinguished for his lectures on comparative physiology at the Royal Institution (Jock Murray 2004) – he borrowed fossils from the Society with which to illustrate them.[88] He was also an early member of the Medical and Chirurgical Society.

Drs William Saunders and Alexander Marcet, both physicians at Guy's Hospital and Medical and Chirurgical Society founders.

Dr Pelham Warren (1778–1835), a physician at St George's Hospital who, according to his contemporaries, was an accurate and careful observer of disease (Webb 2004), which probably inspired Greenough's anecdote about him: 'in his own profession [he] was considered by a colleague who wished to satyrise it[,] as the best guesser in the college'.[89]

But Parkinson's involvement with the Society was not extensive. He only read two papers there. The first, in June 1811 was his 'Observations on some of the Strata in the Neighbourhood of London, and the Fossil Remains contained in thems', 'which was published in the first volume of the *Transactions* (Parkinson 1811*b*). In this seminal work he made deductions about the formations of the London and Paris basins, concluding that there was a 'continuity of the stratification' between England and France, well in advance of anyone else in Britain at that time (Rudwick 2005, p. 512; Lewis 2009). In this paper he also extolled the work of William Smith, and insisted that a study of fossils and the strata in which they were found could provide information about former worlds in a way that nothing else could, implying that even by 1811 geologists in general still did not really understand the importance of fossils or the order of strata. His second paper (Parkinson 1821) – remarks on fossils found by William Phillips – was read over two dates early in 1818. Parkinson did serve on Council between 1813 and 1815, once he had completed *Organic Remains* a work that had kept him busy during the Society's early years.

William Babington, ever busy with his practice, took on the undemanding role of a Vice-President on the Society's first committee, a position he maintained until 1813, only becoming President in 1822 when he was 66 years old. He did not read any papers to the Society and his name does not appear on any committees; he was more a figurehead than an active participant. But in October 1808, just a year after the founding of the Geological Society, he and Davy became founder members of yet another society – the Society for the Improvement of Animal Chemistry (Coley 1967) (Table 4) – that was to have a profound impact on the young Geological Society. The other founders included the mineralogist Charles Hatchett, the chemist William Brande (1788–1866), and the surgeons Everard Home and Benjamin Brodie (1783–1862). Only Brande and Brodie, who were still in their early 20s, were not Fellows of the Royal Society. The founders considered that the branch of chemistry that included analysis and examination of animal substances was not sufficiently developed

[85] GSL: Council Minutes for 1812.

[86] The five other committee members included a Captain Apsley; de Bournon; the botanist Aylmer Bourke Lambert (1761–1842); the artist and naturalist James Sowerby (1757–1822); and a Charles Wilkinson.

[87] GSL: LDGSL 960. Greenough Papers.

[88] GSL: Council Minutes for 1812.

[89] GSL: LDGSL 960. Greenough Papers.

Fig. 9. Sir Joseph Banks. Line engraving by N. Schiavonetti, 1812, after T. Phillips. Wellcome Library, London.

and that important discoveries awaited those prepared to investigate it further.

Sir Joseph Banks (1743–1820) was then President of the Royal Society (Fig. 9) and his opposition to the rise of independent societies had been well known since the Linnean Society had formed some 20 years earlier in 1788. Initially Banks had given the Linnean his wholehearted support, allowing its members full use of his extensive personal collection of specimens, but once he realized that the Linnean would rival the Royal Society in the field of Natural History, Banks became wary of allowing other specialist groups to flourish. At the same time, with Royal Society members sleeping through papers that did not interest them and no time being allowed for their critical examination (Coley 1967), it was becoming increasingly apparent that the Royal Society was not providing the opportunities needed for all the specialized branches of science that were now emerging. Banks, however, maintained that any new societies that stemmed from the parent body – in other words, that had a significant number of Royal Society Fellows in its membership – should become what he called an 'Assistant Association' and remain wholly subordinate to the Royal Society. Under this arrangement, the Royal Society was to have first refusal on all papers. Those papers deemed worthy of publication in Royal's *Philosophical Transactions* could only be published by the new society's journal a year later. Those rejected, however, could be published immediately by the new society. There would also be two grades of membership – 'actual' members who were already Royal Society Fellows and 'probationary' members who were not Fellows (although the term implied that they might be one day).

The Animal Chemistry Society seems to have been the only society ever to have accepted this arrangement, but in doing so it set a precedent that Banks expected the Geological Society to follow, particularly since Babington and Davy were founders of both societies. Banks had become a member of the Geological Society in January 1808, and throughout that year the membership grew rapidly. But when it was resolved at a meeting on 2 December 1808 'That it is expedient that rooms be engaged by the Society for the reception of its cabinet and for the meetings of the committees',[90] Banks expressed his displeasure. Nevertheless, a month later rooms were found at 4 Garden Court, Temple, for an annual rent of £61 6s 2d, and at a meeting on the 6 January 1809 the Committee of Trustees was empowered to take the rooms.[91] A struggle for ownership of the Geological Society then ensued (Woodward 1908; Rudwick 1963; Herries Davies 2007; Knight 2009).

As a founder of both the Geological Society and the Animal Chemistry Society, Babington was in a difficult position when it came to deciding where his loyalties lay and which route the Geological Society should take. On Monday 13 February 1809, Laird wrote to Greenough in alarm:

> Urgent business which I cannot set aside prevents my taking the chance of meeting with you in Parliament Square and the subject I wish to discuss with you is one which I am afraid will more affect the interests of our Society than any which has yet come before us.[92]

The subject he wished to discuss was Babington's declared intention to resign from the Geological Society. But although Greenough 'strenuously urged Laird [...] to prevent if possible so lamentable a catastrophe', on 16 February Laird wrote again, having been unable to meet with Babington who seemed to be avoiding him: 'I fear that I plainly perceive there is no chance of his opinion being [dis]abused and the discussion of the point is one that will be unpleasant to him'.[93]

[90] GSL: Ordinary Meeting Minutes 1807–1818.
[91] GSL: Ordinary Meeting Minutes 1807–1818.
[92] GSL: LDGSL 960. Greenough Papers.
[93] GSL: LDGSL 960. Greenough Papers.

A few weeks later, on 10 March 1809, a Special General Meeting of the Geological Society was held at the Freemasons' Tavern. Twenty members – including Sir Abraham Hume – were present to discuss a resolution proposed by Greville for 'consolidating the Geological Society with the Royal Society as an Assistant Society', but Babington was not among them.[94] Greville's plan called for two classes of subscribing members: Fellows of the Royal Society would be in the First Class and the others would form an Assistant Class. The Royal Society was also to have first pick of any papers written by members for its *Philosophical Transactions*. Furthermore, the arrangement was to be forever: 'the ratification of these resolutions by the Councils of the Royal Society and the Assistant Geological Society shall be permanently binding in honour to both Societies, and shall not be subject to repeal as all other resolutions, or byelaws are declared to be'.[95] When it came to the vote, Greville's proposal was unanimously rejected, furthermore an amendment was added to the resolution: 'That this Society does not consider itself dependent upon nor subservient to any other'. A copy of the resolutions taken at that meeting was transmitted to the President of the Royal Society and to each of the Ordinary Members of the Geological Society. They were burning their bridges and there would be no going back.

Banks and Davy, followed by Greville and Hatchett (the latter having joined only two months previously), resigned, but Babington did not.[96] When he finally had to make a choice – and had seen which way the wind was blowing – he sided with the geologists. Four years earlier he had been one of the founders of the Medical and Chirurgical Society when it broke away from the Medical Society of London, a body almost equivalent in stature to the Royal Society. In those four years that new society had gone from strength to strength, and Babington saw no reason to doubt that the Geological Society would be similarly successful. The two societies shared many members, quite a number of whom had been at the meeting that rejected intervention by the Royal Society. Indeed, by April 24 that year Laird was able to inform Greenough 'We have 96 Honorary and 52 Ordinary Members'.[97] In the same letter he notes that: 'Mr Greville breathed his last breath yesterday', but expresses no regret. Greville died intestate and consequently, as Horner anticipated, 'his collection must be sold I fear – had he left it to the Geological Society he would have been immortalized – that I fear he has not much chance of now'.[98] Parliament asked Babington, Hatchett, Ferguson and Wollaston to place a value on Greville's fabulous mineral collection, which was then purchased by the British Museum for £13 727 (Cooper 2007). Greville's death coincided with, and perhaps contributed to, a turning point in the history of the Geological Society. Within a few weeks of the Special General Meeting, the Society had moved into their rooms in Garden Court, Temple, but a year later it was necessary to move again as they were already running out of space for their cabinets.

The Medical and Chirurgical Society had rooms at 2 Verlum Buildings, Gray's Inn Road, but in January 1810 they were given notice to quit as the landlord wanted to sell the building. Alexander Marcet (Fig. 10), an active member of the Geological Society, approached Laird and the committee with the idea of the two societies sharing premises, since they also shared many members. Yelloly had identified a suitable property at 3 Holborn-Row, which was being offered on a 14-year lease and an annual rent of £115 10 s. Located on the north side of Lincoln's Inn Fields, it was an eminently suitable and prestigious location, a short distance from the Freemasons' Tavern.

Greenough lent both parties the necessary capital for refurbishments, to be repaid with interest,[99] and a joint committee of trustees was formed to administer the arrangements.[100] Under the agreement drawn up, the first room on the ground floor was for the library of the Medical and Chirurgical Society, the whole of the second floor was allocated

[94] GSL: Ordinary Meeting Minutes 1807–1818. Those present included eight of the original founders and 12 others: G. B. Greenough Esq. M. P. President in the chair, Mr Aikin, Count de Bournon, Rev. C. J. Burrows, Dr Franck, Mr Garnier, Mr Horner, Dr Hue, Abraham Hume Bart, Mr Lambert, Mr Lowry, Mr Parkinson, Mr Rich[d] Phillips, Mr W[m] Phillips, Mr Shuter, Mr Tilloch, Mr Warburton, Mr Williams, Mr Woods and Dr Laird, Secretary.

[95] GSL: Ordinary Meeting Minutes 1807–1818.

[96] The only other resignation was John Walker MD (1759–1830), a notorious troublemaker who had already caused the demise of the Royal Jennerian Society – he was not even an FRS.

[97] UCL: AD7981 1083. Laird to Greenough, 24 April 1809.

[98] UCL: AD7981 824. Horner to Greenough, 26 May 1809.

[99] The Medical and Chirurgical Society did not repay their share of £165 until 1819 (Hunting 2002, p. 59).

[100] The trustees were Babington, Ferguson, Greenough, Francis Horner, MP (1778–1817), Hume, David Ricardo (1772–1823) and Samuel Woods, representing the Geological Society. Acting on behalf of the Medical Chirurgical Society were Drs James Curry, Alexander Marcet, John Yelloly, Peter Roget and Thomas Young (1773–1829).

Fig. 10. Alexander John Gaspard Marcet. Stipple engraving by H. Meyer after Sir H. Raeburn. Wellcome Library, London.

to the Geological Society for their cabinets and collections, while all the other rooms, including the first floor meeting room, were shared. The Geological Society held its first meeting there on 1 June 1810, but it wasn't long before their collections spilled over into the meeting room where they were displayed in two large cabinets, either side of the fireplace (Hunting 2002, p. 59).

Through those early years at Lincoln's Inn Fields, Laird and Horner, who had been elected Junior Secretary shortly before they moved in, were largely responsible for turning the Society into a professional body and for conceiving what they called *Young Hopeful*, the first volume of the Society's *Transactions*. It was a long gestation, but at last, on 7 August 1811, Horner was able to write to Greenough with good news:

> My Dear Greenough
>
> At length may the hour when Young Hopeful will be delivered into the world be declared. Violent throes indicate his approach and in another week I hope my labour pains will cease [...] for they have been violent and incessant. I trust no accident will happen to it as it issues from the womb, for the whole responsibility now rests with me, my brother Doctor [Laird] having gone to visit his Mother at Weymouth.[101]

Despite a committee having been formed for overseeing the publication of *Young Hopeful* it seems Horner and Laird did most of the work. Horner's letter continues:

> On Friday last I summoned all the members of the Committee of Papers in town to attend in order to settle the price of the volume. I particularly requested their attendance but no one came but Laird. So we had all to do. [...] the volume cannot be sold to the Public for less than £2. 0. 0. – we have fixed the price at that and if the 750 copies are sold by the end of four years, there will be a balance in the Society's favour of £117 [...].

Heedless of this advice, it seems that in the end the volume was sold for 30 shillings, putting this venture on a loss-making path from the start.

Laird was responsible for choosing the Society's motto printed on *Young Hopeful's* title page, which was taken from the Preface to Francis Bacon's (1561–1626) major philosophical work, *Novum organum*. A summarized translation from the Latin invites 'any human being who earnestly desires to push on to new discoveries, instead of just retaining and using the old [...] to join our ranks as a true son of Science'. A few months later, having seen the Society through its crucial first four years and the birth of *Young Hopeful*, Laird wrote formally to Greenough, still the Society's first President, with his resignation:

> 15 Basinghall Street
>
> Jany 15 1812
>
> Dear Sir
>
> As the day of our annual election will shortly arrive, I deem it incumbent on me previously to communicate to you my desire of being allowed to withdraw from the office to which it has been the pleasure of the society so repeatedly to call me, and I request that you will accordingly have the kindness respectfully to tender my resignation. The more important claims of my professional pursuits and my sense of what is due to the interests of the society will not permit me any longer to hesitate respecting the adoption of this measure.[102]

Arthur Aikin joined Horner as the new Secretary and the minutes of the AGM on 7 February 1812 record: 'That the thanks of this Society be given to Dr Laird for the zeal, diligence and ability with which he has discharged the important duty of Secretary from the first establishment of the Society to the present period'. Laird remained on Council for a further three years, but by the AGM of 1817, the last recorded in the Society's first minute book and 10 years since its founding, only

[101] UCL: AD7981 831. Horner to Greenough, 7 August 1811.

[102] UCL: AD7981 1098. Laird to Greenough, 15 January 1812.

Greenough, Pepys and Aikin of the 13 original founders were still on Council.

For five years the two societies successfully lived side-by-side in Lincoln's Inn Fields, but in July 1815 the Geological Society gave notice to the Medical and Chirurgical Society that it was leaving because once again it needed more space. Shortly afterwards the geologists parted company with the doctors for good, symbolic perhaps, that geology was now a science in its own right, able to stand on its own two feet. Similarly, the Medical and Chirurgical Society went on to greater things. In 1834 it received a Royal Charter, and in 1907 it incorporated 17 others medical societies to form today's Royal Society of Medicine.

Epitaph for the founding doctors

James Parkinson

At the end of the Napoleonic Wars (1815), England was again experiencing social, economic and political upheaval. 'New-fangled' machinery greatly reduced employment, threw thousands out of work and left many without a legitimate means of sustaining a living. When a bad harvest exacerbated the situation, distress reached its climax and the Luddites started smashing industrial machines developed for use in the textile industries of the north of England. Such incidents were invariably followed by execution of the culprits, enraging the work force even further. In 1817, again fearing revolution, the Government passed the two infamous 'Gagging Acts'; on 3 March *Habeas Corpus* was suspended, and on 27 March Lord Sidmouth ordered the Lords Lieutenant to apprehend all printers, writers and demagogues responsible for seditious and blasphemous material (Hone 1982). The Seditious Meetings Act again required societies to become licensed, such that on 15 July 1817 even the, by then, highly respectable Medical and Chirurgical Society felt compelled to apply for one to allow its members to meet:

> Dr John Yelloly of Finsbury Square, Dr Alexander Marcet of Russell Square, and Peter Mark Roget of Bernard Street, Fellows of the Royal Society, and officers of a Society called the Medical and Chirurgical Society, which was instituted in the year 1805, and holds its meetings at No. 3 Lincoln's Inn Fields, apply to this court for a License for the said Society. The present President of the Society is Dr Babington; the Trustees are Dr Baillie, Sir Walter Farquhar, Drs Marcet and Yelloly, Henry Cline Esq, and John Abernethy Esq. A list of the Members, and a copy of the Laws are in court, ready for Inspection, and one of the Volumes (the 7th) of the Society's Transactions is also here.
>
> (Weindling 1980, p. 144)

The parallels with events of 1794 must have been very evident to James Parkinson. His passion for justice, undimmed by the 25 years since he had first joined the London Corresponding Society, meant he was present at a meeting of Hoxton parishioners on 13 February 1817, where it was resolved to take the necessary action:

> for obtaining a redress of the numerous grievances under which the country groans and a full real and true representation in Parliament of such portions of the English people as are either misrepresented or not represented.

Parkinson, now aged 62, was elected to a committee of gentlemen 'authorised to carry into effect the above resolutions in such a manner as may appear to them most expedient' (Morris 1989, p. 77). In the event, Parkinson did not live to see the introduction of even the first Reform Bill in 1832. Universal suffrage on equal terms for men and women finally arrived in 1928.

Later that year (1817) Parkinson was elected President of the Association of Apothecaries and, while it was a mark of his standing in the medical community, there were also those who hoped that his radical politics would spill over into the field of medical reform (Loudon 1987, p. 131). Despite the Apothecaries Act having been introduced in 1815, general practice, as it was becoming known, was still in need of considerable regulation. However, not everyone appreciated Parkinson's political radicalism. During his long tenure as President of the Royal Society, Joseph Banks greatly strengthened the links between the Royal Society and the political establishment. The price for this was a reluctance on his part to allow the election of those suspected of political radicalism (Gascoigne 2008). So while there is no evidence that Parkinson ever wanted to become a Fellow – like others at the time, he may well have considered it just an arm of the establishment – it seems possible that he was never elected because of his radical politics, despite his scientific achievements being far greater than the likes of Franck and some others.

Today Parkinson is best known for the disease that eponymously bears his name. His *Essay on the Shaking Palsy* (1817) has deservedly become a medical classic and, of all his writings, it is this short work that has endured, both for the simplicity of its style and for the accuracy with which the clinical history of the disease was described. However, its real significance lies in the fact that Parkinson was able to differentiate the symptoms that defined *paralysis agitans* from other forms of tremor with which it had hitherto been confused, thereby enabling it to be classified for the first time as a distinct disease. Following its publication, the *Essay* was favourably reviewed in the medical press,[103] although it was not until 1877, 60 years after publication, that the shaking palsy was first called *maladie de Parkinson* by Jean-Martin Charcot, then a famous French neurologist (Charcot 1877, p. 129).

In 1823 the Royal College of Surgeons awarded Parkinson their first gold medal. The College had established the award in 1802, but more than 20 years had passed without anyone being considered of sufficient calibre to be its recipient. The presentation was made at a ceremony held at the College on 11th April 1823 – Parkinson's 68th birthday. In his address, the President, Sir William Blizard, referred to the College's wish to acknowledge achievement in any field of natural science, since all scientific progress was beneficial to 'the cultivation, the improvement and the honour of surgery', thus Parkinson was being rewarded for his work in palaeontology and chemistry, as well as his contribution to medical science. In his reply, Parkinson acknowledged his debt to present and past members, including the great surgeon John Hunter whose collection of fossils was housed in the College's museum. He agreed that the various branches of natural science were able to contribute to each other's progress, thus it was possible that by solving some of the problems in oryctology,[104] a better understanding would be gained of human anatomy and physiology (Roberts 1997, p. 111). This was a clever twist on Cuvier's argument that an understanding of anatomy was an indispensable tool for the geologist.

Parkinson continued writing until the end. In 1822 he published a small work dedicated to 'Admirers of Fossils' who wished to delve a little deeper into the subject and study 'the numerous organized beings with which the earth was peopled before the creation of man'. In addition, he intended to 'point out, from the strata in which they exist, the order in which they were probably formed'. *Outlines of Oryctology* (Parkinson 1822), designed to accompany Conybeare & Phillips's (1822) *Outlines of the Geology of England and Wales*, became another best seller. His last publication, just a few months before his death, was a short note on typhus fever (Parkinson 1824).

James Parkinson was the first of the Geological Society founders to die – on 21 December 1824, aged 69 years, after a severe stroke from which he never recovered. In his will he bequeathed houses he owned in Stratford Langthorne – then in the county of Essex, now just south of the Jubilee tube station – to his wife and sons; the apothecary's shop attached to the house in Hoxton Square to his son John; and his collection of organic remains to his wife Mary, the majority of which was sold at auction in 1827 – a catalogue of the sale has never been found. He was buried in St Leonard's Church, Shoreditch, where a memorial plaque was erected in 1955 by the nursing staff of St Leonard's hospital, to commemorate the bicentenary of his birth. A plaque has also been erected on his house, No. 1 Hoxton Square, which today is a trendy bar-restaurant called 'Bluu'. The plaque calls him a physician; it is a common mistake.

In 1850, almost 25 years after his death, Parkinson's young friend Gideon Mantell (1790–1852) reprinted the plates from *Organic Remains* in his *Pictorial Atlas of Fossil Remains* (Fig. 11):

> Although nearly forty years have since elapsed [since publication of *Organic Remains*], and hundreds of geological works of all kinds and degrees of merit, have subsequently been issued, Mr. Parkinson's plates, owing to their fidelity and beauty, are still in such request as to induce the proprietor, Mr. Bohn, now that the work is out of print, to publish them, with the descriptions and modern names of the fossils represented.

And it is to Mantell that we owe the only known description of James Parkinson:

> Mr Parkinson was rather below the middle stature, with an energetic, intelligent, and pleasing expression of countenance, and of mild and courteous manners; readily imparting information, either on his favourite science, or on professional subjects [...].

William Babington

When William Babington finished his term as the Geological Society's President in 1824 he took lessons in geology from Thomas Webster (1772–1844), then the Society's part-time curator, librarian and draughtsman (Heringman 2009). Then, in the winter of 1832–1833, at the age of 75, he enrolled at University College London as a student and attended a seven-month course in chemistry (Hale-White 1934), following which he overhauled the collection of fossils in the Society's museum. As Greenough said in his obituary of Babington (Greenough 1838), he had an 'enlightened passion for knowledge', a passion which never left him, despite the demands of a busy practice and a large family. In 1819 he became a governor of Guy's hospital and in 1827, like his mentor William Saunders before him, he was made a Fellow of the College of Physicians by special grace; a rare honour. On 26 March 1833 he presided over a dinner at the Freemasons' Tavern attended by 120 people, held to

[103] See, for example, the review in the *The Medico-Chirurgical Journal and Review* (1817).

[104] The study of fossils or palaeontology.

Fig. 11. Fossil drawing from the title page of *Organic Remains of a Former World*, Volume I. Eyles Collection, University of Bristol Library Special Collections.

commemorate the centenary of the birth of the chemist Joseph Priestley (1733–1804). Less than a month later, Babington was dead. An epidemic of influenza had begun in the City earlier that month and an attack had left Babington with a bad cough. On Wednesday 24 April he returned home at 7 o'clock in the evening after visiting patients, including some 'whose circumstances prevented him from accepting any remuneration' (Munk 1878). Munk relates his final hours:

> He was then much oppressed and extremely weak, but a committee for preparing the new Pharmacopoeia having been appointed to sit at his house that evening, he insisted on joining it, and up to eleven o'clock that night was occupied at what proved the last of his professional labours. He then went to bed exhausted, became delirious, and was next morning in a hopeless state; the chest affection rapidly assuming the character of peripneumonia notha;[105] and the lungs becoming oppressed with mucus, which he was unable to expectorate, he died (at his residence in Devonshire-street, Portland-place) on the 29th April 1833, in the seventy-seventh year of his age.
>
> (Munk 1978, vol. 2, p. 454)

The eulogies to this much-beloved man were numerous and his loss so keenly felt that a public subscription was collected to erect a monument in his memory. The fashionable portrait sculptor William Behnes (*c*. 1794–1864) was commissioned to create a full-length, larger-than-life statue of Babington wearing the academic gown of his degree (Fig. 12). It was completed in 1837 and erected in St Paul's cathedral, where it still stands in the southwest transept. George Lewis Smyth, in his book on *Biographical illustrations of St Paul's Cathedral* (1843, p. 93), described the statue: 'Old

[105] Otherwise known at the time as Catarrhus Suffocativus or Bastard Peripneumony; it was probably what we call double pneumonia.

Fig. 12. Statue of William Babington in St Paul's Cathedral. © Dan Salter.
The memorial plaque beneath it reads:
WILLIAM BABINGTON, M.D.F.R.S.
Fellow of the Royal College of Physicians,
Born May 21st 1756, died April 29th 1833.
Eminently distinguished for science;
Beloved for the simplicity of his manners, and the benevolence of his heart,
Respected for his inflexible integrity, and his pure and unaffected piety,
In all the relations of his professional life
He was sagacious, candid, diligent, and humane,
Firm in purpose, gentle in execution;
Justly confident in his own judgment
Yet generously open to the opinion of others;
Liberal and indulgent to his brethren
But ever mindful of his duty to the public.
To record their admiration of so rare a union of intellectual excellence and moral worth,
And to extend to future generations
The salutary influence which his living example can no longer diffuse,
This monument has been erected
By the public subscription of his contemporaries,
A.D. 1837

age and acute suffering are painfully expressed in the countenance. On the whole, the monument is a well-finished but not a pleasing work of art'. The long inscription was written by Dr John Ayrton Paris (*c*. 1785–1856), founder of the Royal Geological Society of Cornwall and another physician with a passion for geology.

Babington, who outlived most of his children, left all his estates and effects to 'my dear wife Martha Elizabeth Babington'[106] without itemizing any of them, so it is difficult to assess his wealth at death. However, when Franck died in 1843 he left £1000 to Martha and various sums to three of her remaining children, although they were now adult, suggesting Babington's wealth may not have been very substantial. Babington was buried in the family vault at St Mary-the-Virgin, Aldermanbury, which was destroyed by a bomb in 1940 during World War II.

James Laird

When James Laird's father died in 1801, James inherited a third of his father's estate in Jamaica, which he shared equally with his mother Mary and his elder brother Henry. His brother managed the estate but when he died, around 1817, another Henry Laird, presumably his son, became the registered owner. As joint owner, it must be assumed that James collected a private income from the estate that helped finance his education and ventures such as his £50 subscription towards Comte de Bournon's treatise on mineralogy. Although subscriptions to the treatise might today be seen as a stepping-stone towards the founding of the Society, in a letter to Greenough Laird recalls how the many unsold copies were eventually divided up between the original subscribers. 'I unadvisedly had all my copies completed with plates, and put into bounds costing me £20 additional[;] some of my copies I gave away and when I was packing up my library in leaving Bloomsbury Square, I sent to Mr. Caddell 20 copies for a trade sale, at which they brought 12/6 per copy [...]'.[107] It seems a considerable loss was made on the venture. A similar fate befell *Young Hopeful*, the first volume of the Society's *Transactions*, but this time Laird was determined not to lose out financially. In 1834, some 23 years after publication, he was still refusing to allow the 113 unsold copies to be divided up among the original subscribers – half of whom were already dead – or to be sold to existing members: 'I cannot see how our recent members should expect to get at a guinea what in common

[106] Prerogative Court of Canterbury will.
[107] UCL: AD7981 1100. Laird to Greenough, 4 April 1834. The original volumes had cost three guineas.

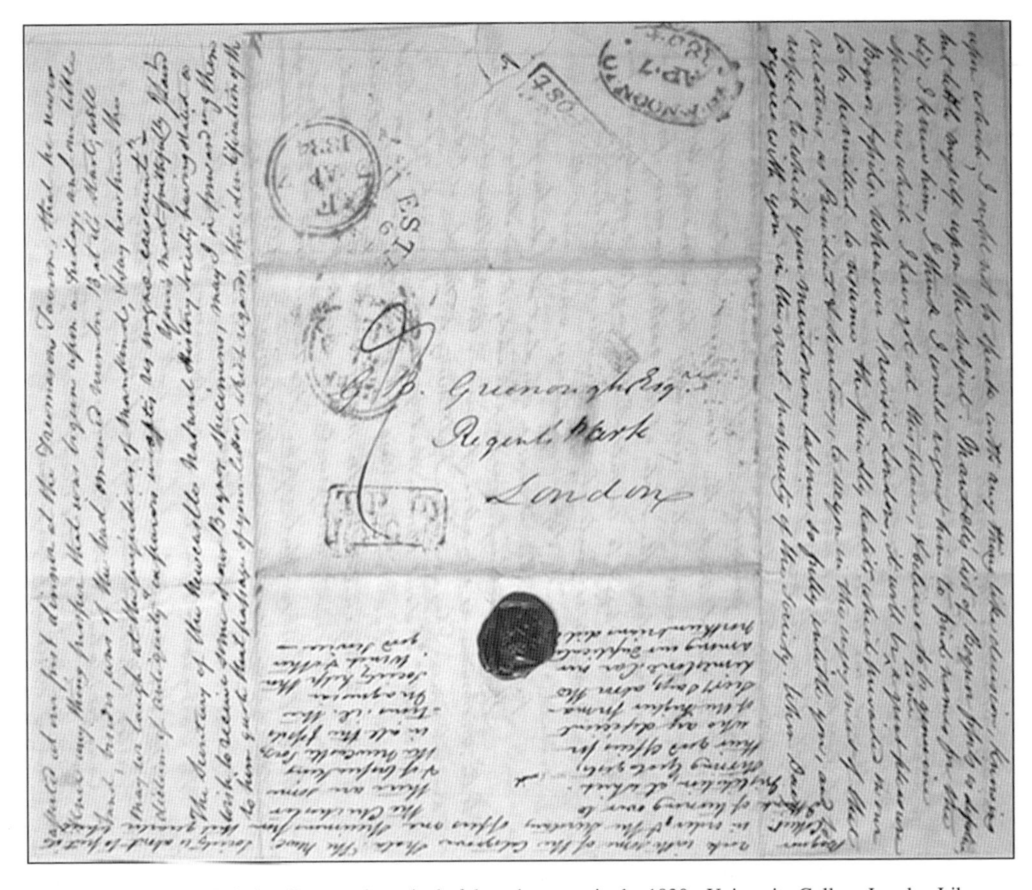

G. B. Greenough Esq.
Regent's Park
London

Fig. 13. A letter from Laird to Greenough, typical of those he wrote in the 1830s. University College London Library.

with all our members I paid 30/- for'.[108] Perhaps he felt it now had historical value: 'Our first volume is interesting as shewing the very imperfect state of Geological Science at the commencement of our Society's proceedings'.[109]

It seems that late in 1823 Laird suffered a 'paralytic stroke' which forced him to pass 'the remainder of his days in close retirement' (Buckland 1841), and on 14 January 1824 he resigned from his post as physician at Guy's on the grounds of ill health; he was only 45 years old. He moved to Bognor where he spent the rest of his life. Laird himself described his illness as an 'attack of Palsy', which seems to have partly paralysed him, although by 1829 he was much improved and able to write letters (Fig. 13), as well as keeping himself busy investigating the local geology; although in June 1832 he suffered a second stroke. His inability to work must have imposed considerable constraints on his finances and he mentions that his bankers 'suspended their payments' in 1826, which caused him to close his account. This suggests funds ceased arriving from Jamaica.

James Laird died on 7 January 1841 from 'paralysis', aged 62.[110] The person who registered his death was an H. J. Laird – possibly his nephew, Henry, who seems to have left Jamaica in 1840.[111] Following emancipation of the slaves in 1834, many estates were abandoned and went into bankruptcy, and the sugar industry fell into decay. James Laird appears not to have left a will and it is not known where he is buried.

[108] UCL: AD7981 1100. Laird to Greenough, 4 April 1834.

[109] UCL: AD7981 1099. Laird to Greenough, 18 March 1834.

[110] Death certificate.

[111] The *Jamaica Almanac* for 1840 is the last in which a Henry Laird is listed as owning property there.

James Franck

Franck eventually became an Inspector-General of Army Hospitals, although little is known of his later life once he returned from the Peninsular War. His address given in various medical society lists is Paper Buildings, Temple, where he probably had a private medical practice. In 1815, at the end of the Napoleonic Wars, he became a founder of the United Service Club for officers returning from the war:

> This club took its rise [...] when so many of the officers of the army and navy were thrown out of commission. Their habits, from old mess-room associations, being gregarious, and their reduced incomes no longer affording the luxuries of the camp or barrack-room on full pay, the late Lord Lynedoch, on their position being represented to him, was led to propose some such institution as a mess-room, in peace [time], for the benefit of his old companions in arms. A few other officers of influence in both branches of the service concurred, and the United Service Club was the result.
>
> (Walford 1878).

The United Service Club, considered to be one of the most commodious of all the London Club Houses, was reputed to be the Duke of Wellington's favourite club. No doubt Franck spent much of his time there for there is no evidence that he ever married.

Franck was evidently a patron of Richard Bright's, who dedicated Volume 2 of his renowned work *Reports of Medical Cases* to him and who named one of his children after him: James Franck Bright (1832–1920). Franck was elected to the Royal Society in 1821, being described in his Certificate of Election as 'a gentleman well versed in various branches of Natural knowledge'. His research field is given as geology, but it is notable that among his proposers are the eminent mineralogists William Babington, William Wollaston and Charles Hatchett, suggesting that, in line with the other medical founders, his interest had originally lain with mineralogy. The use of the word 'geology' to describe his activities probably reflects its wider use by the 1820s.

Franck died of 'diseases of the heart and bladder'[112] at his home in Hertford Street, Mayfair, then in the county of Middlesex, on 27 January 1843, aged 76. He left much of his substantial fortune (around £45 000, plus his property) to be shared in a complicated arrangement between his sister's son and his two cousins.[113] However, his two cooks, coachman, butler, housekeeper and housemaid, with various other individuals, also benefited. He was buried in the vault of Battersea Church in the county of Surrey.

Conclusion

As William Whewell (1794–1866) noted in his Presidential Address of 1842 (p. 66), mineralogy was the nursemaid to geology, destined to be:

> left in comparative obscurity by the growth and progress of her vigorous nursling [...] our warm acknowledgments must on all due occasions be paid to those who zealously cultivated mineralogy, when geology, as we now understand the term, hardly existed.

But what Whewell does not acknowledge is how many of these mineralogists also had a medical background; thus, it could well be argued that medicine has the higher claim as nursemaid to geology, since she trained the mineralogists.

It is tempting to view the founding of the Geological Society as a 'Eureka' moment in the history of science, rather than, as it was, simply part of a continuum of societies and clubs that were being founded on a regular basis by small groups of like-minded men throughout the early nineteenth century. These men had a desire to create sociable communities in which they could talk about shared interests – and so they created societies. Most lasted a few years and faded away as members became too busy or their interests changed; some lasted several decades; a few are still active today. And because geology was bound up with mineralogy, pharmacy, chemistry and anatomy, there was considerable overlap of individuals within these societies, enabling what had been learnt from the success, or failure, of one to be applied to the next. However, it was not enough for men like Babington, Franck, Laird and Parkinson just to be interested in mineralogy and geology, in order for the science to thrive they also had to harness the important connections with wealthy and socially prominent patrons like Greville and Hume with whom they shared this intellectual interest – and it was the formal social setting of a society which made that possible. Babington is an excellent example of someone who brought himself up by hard work and intense activity in societies that provided both intellectual and social advantages. Shared interests in 'improving' topics such as the sciences were a means of democratizing social differences.

The fact that the founders of the Geological Society were medical men, chemists or Quakers is of little relevance. What the achievements of the

[112] Death certificate.

[113] Prerogative Court of Canterbury will.

new Society really depended on were the personalities involved. The ambition of Greville, Hume and St Aubyn, along with Greenough, to map the mineral wealth of Britain provided a clear focus for the founders in the Society's early years. Facilitators such as Babington and Greenough used their networks to bring people together and set the Society in motion, while organizers such as Laird and Horner took it forward from there, getting the business done. Without the combined efforts of these individuals in particular, it is unlikely the Society would be where it is today.

> I am much pleased to find how greatly the Society has prospered, and I hope that it's useful and honourable labours will be continued with undiminished zeal and success.[114]
>
> (Laird to Greenough, 24 May 1834)

Many thanks to my reviewers, David Knight and Susan Lawrence, who made valuable suggestions. Huge thanks also to my editor, Simon Knell, who improved this paper immeasurably. In working with Simon on this book I have not only learnt a great deal about writing history, I have also had a lot of fun. Our email correspondence with each other, and with various authors of this book, has been deposited in the University of Bristol Special Collections. It contains background material we were not able to use, and much else, and will become available for public view in due course. Without the advice, enthusiasm and help of Sandra McDonald, whom I met through the Mormon Family History Centre in Bristol, I might never have found James Laird's family in Jamaica, and tracked down his birth and death dates. Now at last we have dates for all the founders, so very special thanks to Sandra. The archivists and librarians in all institutions were consistently helpful, and I thank them all for their dedication and permission to quote from works in their care, but three stand head and shoulders above the rest. Wendy Cawthorne who, in the absence of a Society archivist, has voluntarily taken on this role; Michael Richardson, at the University of Bristol Library Special Collections who unfailingly met my requests; and Jess Shepherd at the Plymouth City Museum who enthusiastically enlightened me on Babington and de Bournon. But for someone struggling to find a few moments for research after a hard day at the office, nothing proved more useful than Google Books. Like everything indispensable in life, you wonder how on Earth you ever managed without it. So to all those nameless people whose fingers I occasionally see on scans, I raise my glass.

Archives

GSL: Geological Society of London Archive.

KCL: King's College London Archives: Guy's Hospital Records.

RS: Royal Society online biographies, Sackler Archive Resource.

UoBL: University of Bristol Library Special Collections.

UCL: University College London Library, Greenough Papers, including uncatalogued material from Cambridge University.

Wellcome: Wellcome Library Archives.

References

ACKROYD, M., BROCKLISS, L., MOSS, M., RETFORD, K. & STEVENSON, J. 2006. *Advancing with the Army: Medicine, the Professions, and Social Mobility in the British Isles, 1790–1850.* Oxford University Press, Oxford.

ACKROYD, P. 2001. *London. The Biography.* Vintage, London.

ALLEN, W. 1847. *Life of William Allen: With Selections from His Correspondence* (2 vols). H. Longstreth, Philadelphia, PA.

ANDREW, D. T. 1994. *London Debating Societies 1776–1799.* London Record Society, London.

ANON, 1795. *Plan of the St Mary-le-Bone General Dispensary.* London.

ANON, 1800. *The Hospital Pupil's Guide Through London in a Series of Letters.* Printed by A. Seale, London.

ANON, 1801. [Review of Dr. Babington's New System of Mineralogy.] *The Monthly Review or Literary Journal,* **36**, 54–58.

ANON, 1805. [Founding of the Medical and Chirurgical Society.] *Medical and Physical Journal,* **XIV**, 192.

ANON, 1817. Account of Dr. William Saunders. *The London Medical and Physical Journal,* **XXXVIII**, (July–Dec), 388–392.

APPLEBY, J. H. 1999. Sir Alexander Crichton, F. R. S. (1763–1856), Imperial Russian Physician at Large. *Notes and Records of the Royal Society of London,* **53**(2), 219–230.

ASHERSON, G. L. 2004. Bateman, Thomas (1778–1821). *Oxford Dictionary of National Biography online.* Oxford University Press.

AYLIFFE, G. A. J. & ENGLISH, M. P. 2003. *Hospital Infection: From Miasmas to MRSA.* Cambridge University Press, Cambridge.

BABINGTON, W. 1789. *Syllabus of a course of lectures, read at Guy's Hospital by Wm. Babington.* Printed by H. Reynell, London.

BABINGTON, W. 1795. *A systematic arrangement of minerals: Founded on the joint consideration of their chemical, physical, and external characters; reduced to the form of tables, and exhibiting the analysis of such species as have hitherto been made the subject of experiment.* London.

BABINGTON, W. 1799. *A new system of mineralogy in the form of [a] Catalogue, after the manner of Baron Born's systematic catalogue of the collection of fossils of Mlle Éléonore de Raab.* Printed by T. Bensley, London.

BABINGTON, W. & ALLEN, W. 1802. *A Syllabus of a Course of Chemical Lectures Read at Guy's Hospital.* William Phillips, London.

[114] UCL: AD7981 1101. Laird to Greenough, 24 May 1834.

BABINGTON, W., MARCET, A. & ALLEN, W. 1811. *A Syllabus of a Course of Chemical Lectures Read at Guy's Hospital*. Printed at the Royal Free School Press, by J. Lancaster, London.

BABINGTON, W., MARCET, A. & ALLEN, W. 1816. *A Syllabus of a Course of Chemical Lectures Read at Guy's Hospital*. Printed by William Phillips, London.

BERRY, D. 2004. Bright, Richard (1789–1858). *Oxford Dictionary of National Biography online*. Oxford University Press.

BERRY, D. & MACKENZIE, C. 1992. *Richard Bright 1789–1858. A Physician in an Age of Revolution and Reform*. Royal Society of Medicine Services, London.

BEVAN, M. 2004. Parkinson, James (1755–1824). *Oxford Dictionary of National Biography online*. Oxford University Press.

BOURNON, J. L. DE. 1808. *Traité complet de la chaux carbonatée et de l'arragonite* (3 vols). William Phillips, London. [English translation of the 'Discours préliminaire'. *In*: LEWIS, C. L. E. & KNELL, S. J. (eds) *The Making of the Geological Society of London*. Geological Society, London, Special Publications, **317**, Appendix II, 457–464.]

BOURNON, J. L. DE. 1813. *Catalogue de la collection minéralogique du Cte de Bournon ... faites lui meme, et dans lequel sont placés plusieurs observations et faits interessants ... ainsi qu'une reponse au mémoire de M. l'abbé Hauy concemant la simplicité des lois auxquelles est soumise la structure des cristaux*. London.

BOWDEN, A. J. 2009. Geology at the crossroads: aspects of the geological career of Dr John MacCulloch. *In*: LEWIS, C. L. E. & KNELL, S. J. (eds) *The Making of the Geological Society of London*. Geological Society, London, Special Publications, **317**, 255–278.

BRIGHT, R. 1833. Character of the late Dr. Babington. *London Medical Gazette*, **XII** (Vol. 2), 264–265.

BUCKLAND, W. 1841. Presidential Address. *Proceedings of the Geological Society*, **III**, 525–526.

CAMPBELL, A. 2004. Woulfe, Peter (1727?–1803). *Oxford Dictionary of National Biography online*. Oxford University Press.

CHARCOT, J. M. 1877. *Lectures on Diseases of the Nervous system*. Lecture V. The New Sydenham Society, London.

CLEEVELY, R. J. 2004. Rashleigh, Philip (1729–1811). *Oxford Dictionary of National Biography online*. Oxford University Press.

COCKBURN, H. 1888. *An Examination of the Trials for Sedition with have hitherto Occurred in Scotland* (2 vols). David Douglas, Edinburgh.

COLEY, N. G. 1967. The Animal Chemistry Club: Assistant Society to the Royal Society. *Notes and Records of the Royal Society of London*, **22**, 173–185.

COLEY, N. G. 1988. Medical chemistry at Guy's hospital (1770–1850). *Ambix*, **35**(3), 155–168.

CONYBEARE, W. D. & PHILLIPS, W. 1822. *Outlines of the geology of England and Wales: With an introductory compendium of the general principles of that science, and comparative views of the structure of foreign countries*. Published by W. Phillips, London.

COOPER, M. P. 2004. Greville, Charles Francis (1749–1809). *Oxford Dictionary of National Biography online*. Oxford University Press.

DAVIS, M. 2002. *London Corresponding Society 1792–1799* (6 vols). Pickering and Chatto, London.

EDDY, M. 2002. Scottish chemistry, classification and the early mineralogical career of the 'ingenious' Rev. Dr John Walker (1746 to 1779). *British Journal for the History of Science*, **35**(4), 411–438.

EMSLEY, C. 2000. *Britain and the French Revolution*. Longman, London.

FORD, J. M. T. 1987. *A Medical Student at St Thomas's Hospital, 1801–1802. The Weekes Family Letters*. Medical History, Supplement No. 7. Wellcome Institute for the History of Medicine, London.

GARDNER-THORPE, C. 1987. *James Parkinson 1755–1824*. Department of Neurology, Royal Devon and Exeter Hospital, Exeter.

GASCOIGNE, J. 2004. Banks, Sir Joseph, baronet (1743–1820). *Oxford Dictionary of National Biography online*. Oxford University Press.

GASH, N. 2004. Wellesley, Arthur, first duke of Wellington (1769–1852). *Oxford Dictionary of National Biography online*. Oxford University Press.

GREENOUGH, G. B. 1838. President's Address. *Proceedings of the Geological Society of London*, **II**, 42–44.

GEOLOGICAL SOCIETY. 1811. List of donations to the library, to the collection of maps, plans, sections, drawing and models; and to the cabinet of minerals, belonging to the Geological Society. *Transactions of the Geological Society*, **1**, 401–412.

HALE-WHITE, SIR WILLIAM. 1934. Three early nineteenth century Guy's physicians: Willam Babington, James Curry and James Laird. *Guy's Hospital Reports*, July 1934, 259–266.

HAYS, J. N. 1983. The London lecturing empire, 1800–50. *In*: INKSTER, I. & MORRELL, J. (eds) *Metropolis and Province: Science in British Culture, 1780–1850*. Hutchinson, London.

HERRIES DAVIES, G. L. 2007. *Whatever is Under the Earth. The Geological Society of London, 1807 to 2007*. Geological Society, London.

HERRIES DAVIES, G. L. 2009. Jacques-Louis, Comte de Bournon. *In*: LEWIS, C. L. E. & KNELL, S. J. (eds) *The Making of the Geological Society of London*. Geological Society, London, Special Publications, **317**, 105–113.

HERINGMAN, N. 2009. Picturesque ruin and geological antiquity: Thomas Webster and Sir Henry Englefield on the Isle of Wight. *In*: LEWIS, C. L. E. & KNELL, S. J. (eds) *The Making of the Geological Society of London*. Geological Society, London, Special Publications, **317**, 299–317.

HONE, A. 1982. *For the Cause of Truth: Radicalism in London, 1797–1821*. Clarendon Press, Oxford.

HORNE, R. W. 1976. Pharmacy at Guy's 1726–1976. *Pharmaceutical Journal*, **218**, 395–397.

HOWARD, J. 1789. *An Account of the Principal Lazarettos in Europe*. Printed by William Eyres, London.

HUNTING, P. 2002. *The History of the Royal Society of Medicine*. Royal Society of Medicine Press, London.

JACKSON, R. 1845. *A View of the Formation, Discipline and Economy of Armies*, 3rd edn, Parker, Furnivall and Parker, London.

JACQUIN, J. F. 1799. *Elements of Chemistry*. Printed by J. W. Myers, London.

JOCK MURRAY, T. 2004. Roget, Peter Mark (1779–1869). *Oxford Dictionary of National Biography online*. Oxford University Press.

KAUFMAN, M. H. 2001. *Surgeons at War: Medical Arrangements for the Treatment of the Sick and Wounded in the British Army During the late 18th and 19th Centuries*. Greenwood Press, Westport, CT.

KLEIN, U. 2004. Not a pure science: Chemistry in the 18th and 19th centuries. *Science*, **306**, 981–982.

KNELL, S. J. 2009. The road to Smith: how the Geological Society came to possess English geology. *In*: LEWIS, C. L. E. & KNELL, S. J. (eds) *The Making of the Geological Society of London*. Geological Society, London, Special Publications, **317**, 1–47.

KNIGHT, D. 2009. Chemists get down to Earth. *In*: LEWIS, C. L. E. & KNELL, S. J. (eds) *The Making of the Geological Society of London*. Geological Society, London, Special Publications, **317**, 93–103.

KÖLBL-EBERT, M. 2009. George Bellas Greenough's 'Theory of the Earth' and its impact on the early Geological Society. *In*: LEWIS, C. L. E. & KNELL, S. J. (eds) *The Making of the Geological Society of London*. Geological Society, London, Special Publications, **317**, 115–128.

LAIRD, J. 1803. *Dissertatio medica inauguralis de stomacho, ejusque morbis*. R. Allan, Edinburgh.

LAIRD, J. 1807. Case of small-pox occurring in the foetus. *Edinburgh Medical and Surgical Journal*, **3**, 155–157.

LAWRENCE, S. C. 1996. *Charitable Knowledge. Hopsital Pupils and Practitioners in Eighteenth-century London*. Cambridge University Press, Cambridge.

LEWIS, C. L. E. 2009. 'Our favourite science': Lord Bute and James Parkinson searching for a Theory of the Earth. *In*: KÖLBL-EBERT, M. (ed.) *Geology and Religion: Historical Views of an Intense Relationship Between Harmony and Hostility*. Geological Society, London, Special Publications, **310**, 111–126.

LEWIS, C. L. E. & KNELL, S. J. (eds). 2009. *The making of the Geological Society of London*. Geological Society, London, Special Publications, **317**.

LLOYD, B. 2000. The journals of Robert Ferguson (1767–1840). *Mineralogical Record*, **Sep/Oct**, 1–19.

LOUDON, I. 1987. *Medical Care and the General Practitioner, 1750–1850*. Oxford University Press, Oxford.

LOUDON, I. S. L. 1981. The origins and growth of the dispensary movement in England. *Bulletin of the History of Medicine*, **55**, 322–342.

MANTELL, G. A. 1850. *A Pictorial Atlas of Fossil Remains, Consisting of Coloured Illustrations Selected From Parkinson's "Organic Remains of a Former World", and Artis's "Antediluvian Phytology"*. H. G. Bohn, London.

MASTER OF THE ROLLS. 1854. King's College Hospital v. Whieldon. *The Legal Observer, and Solicitors' Journal*, **XLVII**, 263.

MCGRIGOR, J. 1861. *The Autobiography and Services of Sir James McGrigor, Bart: Late Director General of the Army Medical Department*. Longman, Green, Longman, and Roberts, London.

MOODY, R. T. J. 2009. Dining with the Founding Fathers: A personal view. *In*: LEWIS, C. L. E. & KNELL, S. J. (eds) *The Making of the Geological Society of London*. Geological Society, London, Special Publications, **317**, 439–448.

MOORE, N. 2004. Saunders, William (1743–1817). *Oxford Dictionary of National Biography online*. Oxford University Press.

MORRIS, A. D. 1989. *James Parkinson: His Life and Times*. Birkhauser, Boston, MD.

MORGAN, W. 1796. *Facts Addressed to the Serious Attention of the People of Great Britain Respecting the Expence of War and the State of the National Debt*, 2nd edn improved, J. Debrett, London.

MUNK, W. 1878. *The Roll of the Royal College of Physicians; Comprising Biographical Sketches* (3 vols), 2nd edn. Harrison & Sons, London.

O'CONNOR, R. 2007. *The Earth on Show: Fossils and the Poetics of Popular Science, 1802–1856*. University of Chicago Press, Chicago, IL.

PANAGIOTAKOPULU, E. 2004. Pharaonic Egypt and the origins of plague. *Journal of Biogeography*, **31**(2), 269–275.

PARKINSON, J. 1799. *Medical Admonitions to Families, on Domestic Medicine, the Preservation of Health, and the Treatment of the Sick. With a Table of Symptoms, pointing out such as distinguish one disease from another, and the degree of danger they manifest*. 5 edns, 1799–1809. London.

PARKINSON, J. 1800*a*. *The Hospital Pupil – An Essay Intended to Facilitate the Study of Medicine and Surgery*. H. D. Symonds, London.

PARKINSON, J. 1800*b*. *The Chemical Pocket-Book, or Memoranda Chemica*. H. D. [sic] Symonds, London.

PARKINSON, J. 1800*c*. *The Villager's Friend and Physician; or, A Familiar Address on the Preservation of Health, and the Removal of Disease on its First Appearance. Supposed to be Delivered by a Village Apothecary. With Cursory Observations on the Treatment of Children, on Sobriety, Industry, & intended for the Promotion of Domestic Happiness*. H. D. Symonds, London.

PARKINSON, J. 1800*d*. *Dangerous Sports. A Tale, Addressed to Children. Warning them Against the Wanton, Careless, or Mischievous Exposure to those Situations from which Alarming Injuries so often Proceed*. H. D. Symonds, London.

PARKINSON, J. 1804, 1808, 1811. *Organic Remains of a Former World. An Examination of the Mineralized Remains of the Vegetables and Animals of the Antediluvian World; Generally Termed Extraneous Fossils* (3 vols). C. Whittingham, London.

PARKINSON, J. 1811*a*. *Mad Houses: Observations on the Act for Regulating Madhouses*. Printed by Whittingham and Rowland, London.

PARKINSON, J. 1811*b*. Observations on some of the Strata in the Neighbourhood of London, and on the Fossil Remains contained in them. *Transactions of the Geological Society*, **1**, 324–354.

PARKINSON, J. 1817. *An Essay on the Shaking Palsy*. Sherwood, Neely and Jones, London.

PARKINSON, J. 1821. Remarks on the fossils collected by Mr Phillips near Dover and Folkestone.

Transactions of the Geological Society of London, **5**, 52–59.

PARKINSON, J. 1822. *Outlines of oryctology*. W. Phillips, London.

PARKINSON, J. 1824. On the treatment of Infections of typhoid fever. *Medical Repository*, **1**, 197.

[PARKINSON, J.] OLD HUBERT. *c.* 1795. *Ask and You Shall Have*. Printed by J. Burks, London.

PARKINSON, J. W. K. (ed.). 1833. *Hunterian reminiscences: Being the substance of a course of lectures on the principles and practice of surgery, delivered by the late Mr. John Hunter in the year 1785; taken in short-hand, and afterwards fairly transcribed, by Mr. James Parkinson*. Sherwood, Gilbert, and Piper, London.

PAYNE, J. F. 2004*a*. Babington, William (1756–1833). *Oxford Dictionary of National Biography online*. Oxford University Press.

PAYNE, J. F. 2004*b*. Babington, Benjamin Guy (1794–1866). *Oxford Dictionary of National Biography online*. Oxford University Press.

PICARD, L. 2003. *Dr Johnson's London*. Phoenix, London.

PLAYFAIR, J. 1802. *A Comparative View of the Huttonian and Neptunian Systems of Geology: In Answer to the Illustrations of the Huttonian Theory of the Earth*. Printed for Ross and Blackwood, and T. N. Longman and G. Rees, London.

PORTER, R. 1991. Cleaning up the Great Wen: public health in eighteenth century London. *In*: BYNUM, W. F. & PORTER, R. (eds) *Living and Dying in London*. Medical History, 11. (Supplement). Wellcome Institute for the History of Medicine, London.

PORTER, R. 1997. *The Greatest Benefit to Mankind. A Medical History of Humanity From Antiquity to the Present*. Fontana Press, London.

PORTER, R. 2000. *London. A Social History*. Penguin Books, London.

PORTER, R. (ed.). 2006. *The Cambridge History of Medicine*. Cambridge University Press, Cambridge.

REVIEW. 1817. An Essay on the Shaking Palsy. By James Parkinson, Member of the Royal College of Surgeons. *The Medico-Chirurgical Journal and Review*, **IV**, 401–408.

ROBERTS, S. 1997. *James Parkinson 1755–1824. From Apothecary to General Practitioner*. Royal Society of Medicine Press, London.

ROSE, E. P. F. 2009. Military men: Napoleonic warfare and early members of the Geological Society. *In*: LEWIS, C. L. E. & KNELL, S. J. (eds) *The Making of the Geological Society of London*. Geological Society, London, Special Publications, **317**, 219–241.

RUDWICK, M. 1962. Hutton and Werner compared: George Greenough's geological tour of Scotland in 1805. *British Journal for the History of Science*, **1**(2), 117–135.

RUDWICK, M. 1963. The foundation of the Geological society of London: Its scheme for co-operative research and its struggle for independence. *British Journal for the History of Science*, **1**(4), 325–355.

RUDWICK, M. 1997. *George Cuvier, Fossil Bones, and Geological Catastrophes*. University of Chicago Press, Chicago, IL.

RUDWICK, M. 2005. *Bursting the Limits of Time. The Reconstruction of History in the Age of Revolution*. University of Chicago Press, Chicago, IL.

RUDWICK, M. 2009. The early Geological Society in its international context. *In*: LEWIS, C. L. E. & KNELL, S. J. (eds) *The Making of the Geological Society of London*. Geological Society, London, Special Publications, **317**, 145–153.

RUMSEY, J. 1826. *The Life and Character of the Late Dr Thomas Bateman*. Longman, Orme, Rees, Brown and Green, London.

SAKULA, A. 1990. Gentlemen of the hammer: British medical geologists in the 19th century. *Journal of the Royal Society of Medicine*, **83**, 788–794.

SELECT COMMITTEE. 1820. Report on the Doctrine of the Contagion of the Plague. *Edinburgh Medical and Surgical Journal*, **LXII**, 109–124.

SIEGFRIED, R. & DOTT, J. H. JR. 1976. Humphry Davy as geologist, 1805–29. *The British Journal for the History of Science*, **9**, 219–227.

SMITH, J. 1795. *Assassination of the King!*, 2nd edn. Printed for John Smith, at the Pop-Gun, Lincoln's Inn Fields, London.

SMYTH, G. L. 1843. *Biographical Illustrations of St Paul's Cathedral*. Whittaker and Co., London.

SOCIETY FOR THE BENEFIT OF THE SONS AND DAUGHTERS OF THE CLERGY. 1845. *The New Statistical Account of Scotland*, Volume IX, William Blackwood and Sons, Edinburgh.

SOUTH, J. F. 1970. *Memorials*. Centaur Press, Sussex.

TAQUET, P. 2009. Geology beyond the Channel: the beginnings of geohistory in early nineteenth-century France. *In*: LEWIS, C. L. E. & KNELL, S. J. (eds) *The Making of the Geological Society of London*. Geological Society, London, Special Publications, **317**, 155–162.

THALE, M. (ed.). 1983. *London Corresponding Society: selections from the Papers of the London Corresponding Society, 1792–1799*. Cambridge University Press, Cambridge.

THALE, M. 1989. London debating societies in the 1790s. *Historical Journal*, **32**, 57–86.

THELWALL, MRS [CECIL BOYLE]. 1837. *Life of John Thelwall by his Widow*. London.

TORRENS, H. S. 2009. Dissenting science: the Quakers among the Founding Fathers. *In*: LEWIS, C. L. E. & KNELL, S. J. (eds) *The Making of the Geological Society of London*. Geological Society, London, Special Publications, **317**, 129–144.

TOWNSON, R. 1797. *Travels in Hungary with a Short Account of Vienna in the Year 1793*. Printed for G. G. and J. Robinson, London.

TURNER, G. L'E. 1967. The auction sales of the Earl of Bute's instruments, 1793. *Annals of Science*, **23**, 213–242.

WALFORD, E. 1878. Pall Mall—Clubland. *Old and New London*, **4**, 140–164. http://www.british-history.ac.uk/report.aspx?compid=45188.

WATSON, J. S. 1985. The Reign of George III, 1760–1815. *In*: CLARK, G. (ed.) *The Oxford History of England*. Clarendon Press, Oxford.

WEBB, W. W. 2004. Warren, Pelham (1778–1835). *Oxford Dictionary of National Biography online*. Oxford University Press.

WEINDLING, P. J. 1979. Geological controversy and its historiography: The prehistory of the Geological Society of London. *In*: JORDANOVA, L. J. & PORTER, R. S. (eds) *Images of the Earth: Essays in the History of the Environmental Sciences.* British Society for the History of Science, Monographs, **1**, 248–271.

WEINDLING, P. J. 1980. Science and sedition: How effective were the acts licensing lectures and meetings, 1795–1819? *British Journal for the History of Science*, **44**, 139–153.

WEINDLING, P. J. 1983. The British Mineralogical Society: A case in science and social improvement. *In*: INKSTER, I. & MORRELL, J. (eds) *Metropolis and Province: Science in British Culture, 1780–1850.* Hutchinson, London.

WHEWELL, REV. W. 1842. President's Anniversary Address. *Proceedings of the Geological Society*, **3**, 65–66.

WILKS, S. & BETTANY, G. T. 1892. *A Biographical History of Guy's Hospital.* Ward & Lock, London.

WILLS, R. G. 1935. President's Address: Contributions of British Medical Men to the Foundation of Geology. *Proceeding of the Liverpool Geological Society*, **XVI**, 199–219.

WILSON, W. E. 1994. *The History of Mineral Collecting: 1530–1799.* The Mineralogical Record, **25**, November.

WILSON, W. E. 2008. John St Aubyn (1758–1839). *Mineralogical Record Biographical Archive* at www.mineralogicalrecord.com.

WOODWARD, H. B. 1908. *The History of the Geological Society of London.* Longman, Greens and Co., London.

Chemists get down to Earth

DAVID KNIGHT

Durham University Philosophy Department, 50 Old Elvet, Durham DH1 3HN, UK
(e-mail: d.m.knight@durham.ac.uk)

Abstract: Four of the Founding Fathers of the Geological Society, Arthur Aikin, Richard Knight, William Hasledine Pepys and Humphry Davy, were chemists, coming to geology through mineralogy. The nature and status of chemistry in 1807 helps us to see why, and we note that chemists were down to earth and empirical minded in contrast to the speculative geologists of the eighteenth century. These four, a closely linked and coherent group, made various and important contributions to the scientific world in London generally, and to the Geological Society in particular. Nevertheless, Davy had very different expectations of the Society from the others: he wanted a dining club under the aegis of the Royal Society (of which he was Secretary), but they (successfully) sought a formal and separate body in which papers would be read and published. By the time of Davy's death in 1829, the Society was chartered and flourishing; and the rise of palaeontology had made chemistry much less central to geology.

O God! That one might read the book of fate,
And see the revolution of the times
Make mountains level, and the continent –
Weary of solid firmness, – melt itself
Into the sea! and, other times, to see
The beachy girdle of the ocean
Too wide for Neptune's hips. How chances mock,
And changes fill the cup of alteration
With divers liquors!

(Henry IV, part 2, III, 2, 45–53)

Thus spoke King Henry IV, or anyway Shakespeare's Henry IV; and if the Geological Society were, like the chemists, to look for a Royal patron, this Plantagenet usurper might qualify (Herries Davies 2007, p. 56). We tend to think of geology, after all, as chiefly concerned with Prospero's 'dark backward and abysm of time' (The Tempest I, 2, 50). But that was not how it struck some at least of the Founding Fathers 200 years ago. There had certainly been speculations in the seventeenth and eighteenth centuries, from Neptunists and Plutonists, about the Earth's history, in Germany, France, England and Scotland: but by 1800 there was a 'second scientific revolution' underway, and such 'systems' were frowned upon. Real science as seen in Paris, the world's centre of excellence, depended upon close observation, careful experiment and quantitative deductions. 'Hypotheses' were played down. And while mathematical physics was triumphant in the work of Pierre Simon Laplace (1749–1827) and his circle, it was a forbidding discipline. Chemistry was where both useful knowledge and accessible intellectual stimulus was to be found, despite the impact of revolutionary politics: 1794 saw the execution of Antoine Lavoisier (1743–1794), a powerful member of the hated body that collected taxes in France, and the departure of the radical dissenter Joseph Priestley (1733–1804) to America as a political refugee after his house had been sacked by a mob shouting 'Church and King'.

The status of chemistry

By 1800, it was clear that Abraham Gottlob Werner's (1749–1817) descriptions (Werner 1786) of minerals, based like Carl Linnaeus' (1707–1778) on external characteristics, were not enough to characterize them; chemists were moving into the territory of mineralogy, transforming it from a descriptive to an analytical study. Priestley had extended chemistry into the gaseous phase, after Joseph Black (1728–1799) had shown that solid limestone contained 'fixed air', while Lavoisier and his associates introduced a new theory of combustion and (in the wake of Linnaeus) a new nomenclature. Facts, theory and language were all in the melting pot: the novice could rapidly break into the science and expect to make discoveries. With new techniques of analysis, using smaller quantities and greater precision, new minerals were being identified, and new metals isolated from them, all over Europe. For the elitist Lavoisier, science was a matter for experts like him, with splendid and expensive apparatus; for the democratic Priestley, very important in our story, it was accessible to all with the enthusiasm for finding out about God's world, and he believed that the truth would make them free (Rivers & Wykes 2008). These chemists had moreover extended their science into physiology and botany, explaining (in principle, if not in detail) respiration and photosynthesis: chemistry

From: LEWIS, C. L. E. & KNELL, S. J. (eds) *The Making of the Geological Society of London*.
The Geological Society, London, Special Publications, **317**, 93–103.
DOI: 10.1144/SP317.3 0305-8719/09/$15.00 © The Geological Society Publishing House 2009.

was not just about analysis, but was a science that could provide dynamical accounts of processes. People, hoping to participate or just to wonder, flocked to chemistry lectures and demonstrations, joined societies, and bought books and periodicals like the *Philosophical Magazine* and *Nicholson's Journal*, whose editors sought to bind their readers into a kind of community (Averly 1986). More facts were being added every day, and moreover the science seemed exciting and fundamental. It is not surprising that among the founders of the Geological Society there should be eminent chemists; they were taking over the world.

The four Founding Fathers with whom we are concerned, Arthur Aikin, or Aiken (1773–1854), Richard Knight (1768–1844), William Hasledine Pepys (1775–1856), and Humphry Davy (1778–1829), were all chemists based in London, knew each other well and had connections with important institutions there at the heart of the scientific establishment. Chemistry was the 'materials science' of the day, and chemists particularly asked two questions of substances: what is it made of?, and what use is it and its components? (Klein & Lefèvre 2007, pp. 11–13 and 163–177). Chemists were, in their publications at least, careful to avoid being too much seduced by hypothesis or premature generalization, and William Nicholson (1753–1815) set an example of separating facts from theory in his standard textbook of the science (Nicholson 1792, 1796). He was critical of those who spent too much time arguing about the merits of phlogiston and oxygen in accounting for burning, rather than carrying out new experiments. Davy in his chemical papers read at the Royal Society always claimed that his main objective was to bring forward new facts, and that his subsequent interpretation of them was personal and provisional. The chemists' ideal of science was Baconian: their practice, notably Davy's, was sometimes very different, and like anyone else they made and tested hypotheses – but, if challenged, would admit that these were mere scaffolding, not really part of the structure of science. Their view would become that of the Founding Fathers, impatient with 'systems' and aware of the dangers of premature theorizing.

Chemists in Sweden, with its mining industry, had become expert in mineralogical analysis. Axel Fredrik Cronstedt (1722–1765) used an early 'portable laboratory', a case of apparatus that could be taken into the field for analyses (chiefly using a charcoal block and a blowpipe), thus blurring the line between laboratory and fieldwork. Torbern Bergman (1735–1784), and Davy's great rival Jons Jacob Berzelius (1779–1848), the leading chemist of the 1810s and 1820s, continued this tradition. New metals were isolated in triumphs of analytical method; most importantly, platinum, at first only a refractory grey powder, was made into a metallic form by hammering at high temperature (swaging) by William Hyde Wollaston (1766–1828) in 1800. His patent process enabled him to withdraw from medicine into full-time chemistry, make a fortune and become a benefactor of the Geological Society, bequeathing money for its now most prestigious medal. Platinum was not used for jewellery, where it might have been mistaken for silver, but in chemical research and manufacturing where it was almost magical. Vessels, notably crucibles, of platinum would resist heat and corrosion, and could be used industrially and even in the small blast-furnaces with which larger laboratories were equipped; platinum wires could be sealed through glass without cracking it for electrical experiments; and platinum spatulas enabled chemists to stir and spoon potent reactants. Wollaston became Britain's leading analyst, nicknamed 'the Pope' because he was infallible, and famous for working with ever-smaller quantities.

Priestley, only intermittently Baconian, had looked forward to a foreseeable future in which the glory of Sir Isaac Newton (1642–1727) would be eclipsed. Mechanics was involved with the surface of things: optics, electricity and chemistry would give us 'an inlet into their internal structure, on which all their sensible properties depend' (Priestley 1775, pp. 1 and xv). He was right. Wollaston's optical goniometer of 1809 was to bring precision into crystallography; and meanwhile, in 1800, Alessandro Volta (1745–1827) with his electric battery rang an alarm bell that got everyone investigating electrochemistry. In November 1806, Davy established that pure water is decomposed by an electric current into oxygen and hydrogen only – others had been perplexed by side reactions generating acid and alkali from dissolved nitrogen. He inferred that chemical affinity was electrical: and was awarded a prize by the Parisian Academy of Sciences – crucial international recognition, and in wartime too. Electricity then might be a powerful agent for analysis; and, indeed, on 19 November 1807, just a calendar month of hectic research and conversation after first performing the experiment (and dancing around the laboratory in ecstatic delight), Davy announced to the Royal Society the isolation of the extraordinary metals, potassium and sodium, about which he had written to a friend (Paris 1831, p. 179):

> I have decomposed, and recomposed the fixed alkalies, and discovered their bases to be two new inflammable substances very like metal; but one of them lighter than ether, and infinitely combustible. So there are two bodies decomposed, and two new elementary bodies found.

Light, when metals were supposed to be heavy, furiously reactive, they seemed like the alkahest of the alchemists (Knight 2007). To Davy, and to Berzelius, these discoveries established that chemical affinity was, indeed, electrical: chemistry was becoming the dynamical science envisaged by Priestley, based like Newtonian physics upon forces. But there was much analytical work to do; and minerals, durable, beautiful and interesting both chemically and industrially, were an obvious place to begin. And one might go on and ask how they had been come to be formed, and to be found in their present position.

Metropolitan science

We might wonder why Davy's researches were done in the autumn, and why the Founding Fathers met in dank and chilly November; and, indeed, why they sat down to dine at 5pm. It was because in November the London Season began with the Opening of Parliament, the Law Terms and entertainments. Landed gentry came up for sociability and fashion; fathers brought their marriageable daughters to London for balls; the scholarly and the eminent went to lectures, meetings of learned societies, clubs, and, perhaps, to fashionable bluestocking ladies' salons, dinner and supper parties. Midday dinner time, for this leisured class, had moved steadily later, and was to continue to do so through the century: at 5 pm they were not having a plebeian high tea, but the main meal for those who expected to stay up late. In the summer, about June, everyone in this upper crust dispersed to the country. That was the time for visiting, travel and field trips; and from September there was an interlude in which Davy did his research in romantic bursts of energy.

The Royal Society's AGM is on 30 November, St Andrew's Day; but their year might begin, as in 1807, one or two weeks earlier. Royal Society meetings were extremely formal, with Sir Joseph Banks (1743–1820) presiding in court dress from behind a mace. He was a genial and generous, but crusty and autocratic baronet who had been elected in 1778. He had made the extraordinary decision to devote his life to scientific administration when he might have been expected to go into parliament, or, perhaps, just become a squire and JP, playing an important role in the country. He built up collections and a library of international importance, and made them available to scholars at his London house in Soho Square (Carter 1988, pp. 334–335). He can be seen as presiding more widely over an English Enlightenment, and the growth of the British Empire (Banks *et al.* 1994; Gascoigne 1994, 1998). Wealthy and landed, in the 1790s as the French Revolution turned into the Terror, he was greatly concerned to ensure that science was not seen through Priestley's eyes as subversive of custom and authority, but as an ally of improvement and traditional good government. If men of science did not hang together, they would be sure to hang separately in those days of clamping down on French sympathizers (when the poet Samuel Taylor Coleridge (1772–1834), for example, was tailed by what we might call security forces). Fragmentation and overspecialization were thus to be avoided at all costs; nominations for Fellowship were carefully scrutinized to maintain the atmosphere of a congenial gentlemen's club.

The Royal Society lacked both a laboratory and a lecture room; and women could attend only on rare special occasions. Banks therefore welcomed and warmly supported in 1799 the founding of the Royal Institution in Albemarle Street in London's fashionable West End. Its leading figure was Benjamin Thompson, Count Rumford (1753–1814), an American Tory (or Loyalist) refugee, the Yankee at the Court of Bavaria who had modernized its economy and army, and founded the Englische Garten in Munich. Back in England he was promoting useful science (especially more efficient stoves and fireplaces, and improved diets) among both the leisured classes and artisans (James 2002). The Institution was open to both men and women. Its laboratory was intended to be for analyses of mineral waters, soils, minerals and chemical manufactures on behalf of members or outsiders paying a fee: but Davy turned it into a place for fundamental research, sometimes in public. The artisans were soon squeezed out.

Natural philosophers and natural historians in Britain, and elsewhere in Europe, were conscious that France (meaning Paris) was in a class by itself as far as science was concerned. There could be found the critical mass, the excellence right across the sciences visible in the Academy of Sciences (First Class of the Institute). Important men of science were to be found elsewhere, but they were relatively isolated figures who had, willy-nilly, to look to France. In Britain, undergoing the industrial revolution that would bring victory over Napoleon, the emphasis was utilitarian: Davy got the Copley Medal not for his electrical researches, but for his work on tanning. But he and his contemporaries saw no reason why science done in a free country should not be at least as good as that done in France, a failing state under the monarchy, then a centre of terror, and by 1807 a military dictatorship. They aspired to show the French up. They seem to have had no problem about acquiring French scientific publications: wartime communication with other European countries was more uncertain.

Adjoining the smart and leisured City of Westminster is the City of London, the square mile around St Paul's Cathedral, the world of work. London was, after all, an enormous and busy metropolis, a commercial and industrial centre where most people had a job to do and lived all the year round. There, in 1805, the London Institution was set up at Finsbury Circus, with similar objectives to the Royal Institution; another analogous body was the Surrey Institution in Blackfriars (Kurzer 2000, 2001). There was thus scope for would-be lecturers; and the London hospitals (notably Guy's, founded to deal with desperate diseases, the 'incurables' discharged from St Thomas's) also provided opportunities for chemical lecturers, even before attendance became compulsory for students in 1815 (following the Apothecaries Act): it was just becoming possible to support oneself by science, but very difficult (Knight 2006, pp. 29–43). There were also specialized groups, such as the Askesian Society founded in 1796 and the British Mineralogical Society of 1799, which coalesced and many of whose members joined the infant Geological Society (Torrens 2009). Less formal groups included the City Philosophical Society, a self-help body for apprentices that young Michael Faraday (1791–1867) frequented; and the informal mineralogical gatherings that took place at Dr William Babington's (1756–1833) house at 7 am, so that he could go on afterwards to his medical duties (Herries Davies 2007, p. 5; Lewis 2009).

Fig. 1. Arthur Aikin by William Brockendon, 1826, National Portrait Gallery ref. 2515(9). Reproduced by permission.

Chemists as Founding Fathers

Aikin: mineralogist, author and science administrator

Arthur Aikin (Fig. 1) was eminent enough to have entries both in the *Oxford Dictionary of National Biography* (Torrens 2004) and in the *Dictionary of Nineteenth Century British Scientists* (Tuttle 2004). He came from an eminent dissenting medical and literary family, and his father was an enthusiast for natural history. He was educated at Warrington, then under an uncle at Palgrave in Suffolk and then at the Dissenting Academy at Hackney, where Priestley, a refugee from the riots at Birmingham, was teaching. Entrance to Oxford and Cambridge (the only English universities) was restricted by their religious tests; Hackney and other Dissenting Academies, open to those excluded, gave a good and more up-to-date education including modern languages and natural philosophy. Priestley was a very prominent early Unitarian, who had begun by rejecting Calvinistic predestination and had gone on to deny the divinity of Christ along with other Platonic corruptions, as he saw it, that had crept into true Christianity. Unitarians, who saw themselves as 'rational dissenters', were prominent in scientific and philanthropic institutions around the country, and Arthur had been intended for the doctrinally undemanding Unitarian ministry. But although Unitarianism was famously described in the Darwin and Wedgwood families as a feather-bed to catch a falling Christian, Aikin lost his faith, abandoned the idea of becoming a minister, and moved into journalism, mineralogy and applied chemistry. In 1796 he undertook tours on foot in Shropshire, Wales and Scotland, investigating the 'masses of rock in their native beds' (Herries Davies 2007, p. 12). By 1802 he was toying with the idea of combining his observations with those of William Smith in a mineralogical map of Great Britain. From 1801 he was President of the British Mineralogical Society, until it merged with the Askesians in 1806. He lived precariously by writing, translating and lecturing, both to general audiences and at Guy's Hospital to the medical students. He never married. His journalism and editing were particularly important because he brought foreign scientific news to the interested public.

In 1805 Aikin became a Proprietor of the London Institution, founded by City of London men, citizens of credit and renown concerned with commerce and industry. In 1807 with his brother Charles (who had been adopted by his childless aunt, the poet Anna Laetitia Barbauld) he published a *Dictionary of Chemistry and Mineralogy*. He was one of the committee charged in December 1807 with making rules for the infant Geological Society. In 1808 with George Bellas Greenough (1778–1855) he prepared the Society's first publication, *Geological Inquiries*, a 20-page pamphlet of advice and queries (Herries Davies 2007, p. 36). From 1812 to 1816 he was one of the Secretaries of the Society, and in 1813–1814 he delivered lectures at the Society, published as his *Manual of Mineralogy* in 1814. In 1825 he was one of the five 'petitioners' named on the Society's Charter (Boylan 2009), and thus one of the first Fellows and a member of the first Council (Herries Davies 2007, p. 56). In 1817 he had secured the paid position of Secretary to the Society of Arts, concerned with the promotion of new techniques; and advised about patent law. An enthusiastic member of London's scientific community, he belonged to the Linnean Society, and was a founder member of the Chemical Society of London (now Royal Society of Chemistry) in 1841, becoming its Treasurer and later its second President (1843–1845). He published a number of papers on mineralogy. His *Prospectus for a Geological Survey, Section and Map, of Shropshire* had been published in 1810, but there were not enough subscribers; and the materials in due course passed to Roderick Impey Murchison (1792–1871) (recruited for geology by Davy, who persuaded him it was compatible with field sports).

Fig. 2. Richard Knight from a miniature painted by his daughter Elizabeth. Reproduced by permission from Peta Buchanan.

Knight: instrument maker and city gentleman

Richard Knight (Fig. 2) is much less familiar to historians of science, and does not feature in the standard dictionaries (Hunt & Buchanan 1984). In the course of a dissenting education, he was another protégé of Priestley, converted by him to Unitarianism and an enthusiasm for chemistry. He had grown up in Stoke Newington, then a village to the NE of London, afterwards moving to Spitalfields and then Hackney. A chemist and mineralogist, and friend of Aikin's, he was apprenticed to his grandfather and then his father in the family ironmongery business (William Knight and Son). After his father died in 1799, he entered into partnership with his brother George (as R. and G. Knight) and they transformed the business into suppliers of laboratory equipment, based in Foster Lane off Cheapside. They designed and sold apparatus, notably Pepys' and William Allen's for repeating Davy's isolation of potassium more conveniently. More mundanely, Davy recommended Knight's standard 'chemical chest' (or portable laboratory) in his agricultural lectures, and Banks used his apparatus for blowing up tree stumps on his estate at Revesby. Knight, a sociable man, belonged like other Founding Fathers to various less-formal precursors of the Geological Society: the Guy's Hospital Physical Society, the Askesians, the British Mineralogical Society and Babington's early-morning group of mineralogists (Herries Davies 2007, pp. 5–6; Lewis 2009).

In 1807 Knight read the first-ever paper to the Geological Society – on Rowley Rag, an igneous rock from Staffordshire. For comparison, he had (like Davy's friend Gregory Watt (1777–1804)) fused and slowly cooled a portion of the same mass of rock, bringing together laboratory and field study (Herries Davies 2007, p. 17). In 1810, with Pepys and others, he was unsurprisingly a member of the Society's committee for chemical analysis; but he was a busy man and seems to have had little time for research. By then a respected and well-known tradesman, Knight officially valued the Royal Institution's mineral collection (chiefly assembled by Davy) that year at £400; it is interesting to note that he reckoned its apparatus was worth only £150. This must have included platinum vessels and spatulas. In 1800 Knight had worked on platinum, but saw little commercial future for it and passed his supply over to Wollaston, who in due course was able to make good use of his 800 ounces of native platinum. Like Pepys, Knight was a prominent City figure; Master of the Skinners' Company in 1808. These city companies, originally craftsmen's guilds, had by the eighteenth century lost most of their connection with their original trades and become wealthy. They were essentially clubs with charitable pretensions, where city gentlemen could meet, and have grand dinners from time to time (Owen 1965, pp. 276–297). Their Masters or Wardens were important figures, who might be elected to office in the City's government. Knight made apparatus, bespoke and off the peg: a clinometer for the Swedish mineralogist William Hisinger (1766–1852), a frictional electrical machine for Erasmus Darwin junior (1804–1881) and chemical equipment for dye works, his wife's family business; also standard furnaces, eudiometers, equipment for etching glass with fluoric acid and electric batteries. He equipped the laboratories of both the London and Surrey institutions. He had five daughters and, hard working and successful, he died a wealthy man, bequeathing them a total of £35 000.

Pepys: city gentleman and natural philosopher

William Hasledine Pepys (Fig. 3) was a distant connection of Samuel, whose now-famous diary was undeciphered and unknown in the early nineteenth century. William Pepys has an entry in the *Oxford Dictionary of National Biography* (McConnell 2004), and there are papers and correspondence at the Royal Institution. His father was a surgical instrument maker and cutler, and William, a churchman, followed him into that prosperous business, and into the Cutlers' Company in the City of London, of which he was twice Master. He was from 1800 a Proprietor of the Royal Institution, and played a very important part in its administration, serving on the committee that supervised its chemical investigations, and being successively a Visitor and Manager: in 1830 he was awarded its Fuller Medal for chemistry. Pepys, stimulated at the Royal Institution by Davy, became a most expert chemical manipulator and deviser of apparatus, delighting in experimentation, and greatly respected as an analyst. On 13 November 1807 Davy wrote to him about recent researches Pepys had carried out with William Allen (1770–1843), the Quaker pharmacist, lecturer at Guy's Hospital and philanthropist (Torrens 2009):

> If you and Allen had been one person, the Council of the Royal Society would have voted you the Copleian Medal; but it is an indivisible thing, and cannot be given to two.
>
> (Paris 1831, p. 179)

Joint papers were still a rarity and research was supposed to be carried out, like Newton's, by voyagers in strange seas of thought, alone.

Pepys had been a member of the Society of Arts, the Askesians and the British Mineralogical Society; and was in 1808 elected a Fellow of the Royal Society. In May 1805 he had made a platinum fruit-knife and presented it to Banks. He was also a founder member of the London Institution, and its first Secretary. In 1819 he planned its laboratory, and Knight equipped it. It included an enormous battery, of 2000 plates, designed by Pepys. Davy and others came to work with it. Pepys researched, often with Allen, not only in mineralogy and inorganic chemistry, but also on respiration. In 1815 he married Lydia Walton, and they had seven children. When, in 1812, Davy was approached by an enthusiastic young bookbinder's apprentice who sent him a fair copy of lecture notes nicely bound up, it was to Pepys that he turned for advice (Knight 1998, p. 130):

> 'Pepys, what am I to *do*, here is a young man named Faraday; he has been attending my lectures and wants me to give him employment at the Royal Institution, *what can I do?*' '*Do*' replied Pepys, 'put him to wash bottles; if he's good for anything, he will do it directly; if he refuses he is good for nothing.' 'No, no' replied Davy, we must try him with something better than that.

Pepys was a useful patron, the upshot was, of course, extremely successful, and Faraday went (not without some difficulties) from lab boy and amanuensis through a kind of scientific apprenticeship and on to independent research. When in the 1820s Davy and others founded the Athenaeum

Fig. 3. William Hasledine Pepys by Walter, printed by Day and Haghe, 1836, Atheneum Portraits 34. Reproduced by permission from the Royal Institution.

Club, as part of a plan to transform Banks' club-like Royal Society into something more like an Academy of Sciences, Faraday became its first Secretary and Pepys became a member.

But the rest of Davy's 1807 letter is especially important for us here in relation to the Geological Society, for it went on:

> We are forming a little talking geological Dinner Club, of which I hope you will be a member. I shall propose you today. Some things have happened in the Chemical Club, which I think make it a less desirable meeting than usual, and I do not think you would find any gratification in being a member of it. [Charles] Hatchett never comes, and we sometimes meet only two or three. I hope to see you soon.
>
> (Paris 1831, p. 179)

Pepys couldn't make it on that memorable evening; we don't know why. But clearly, in the foundation of the Geological Society, Davy rang the bell that called the wits together. Pepys, a clubbable man, as well as a good experimentalist, went on to play a prominent role in the Society. He was with Aikin a member of the Society's committee set up in December to draft rules, and in 1824 they were both members of the Geological Society Club, an in-group or convivial dining club of up to 40 members. From December 1807 until February 1816 he was Treasurer of the Society; in 1810 he presented an engraving of the Gold Mines Valley in Ireland to the Society. When he died, in 1856, he was the last survivor of the Founding Fathers, Aikin having died the year before (Kurzer 2003).

Davy: essential but awkward

Humphry Davy (Fig. 4) was from Cornwall, his family impoverished but respectable and 'church' rather than dissenters. Contemporaries and biographers saw social mobility as the key to understanding his life in a society that was at once both snobbish and open to the talents. He was apprenticed to a local apothecary-surgeon. Then in 1797–1798 Gregory Watt came to board with his hard-up widowed mother because he was consumptive and Cornish winters were milder than those of Birmingham. This brought young Davy into contact with the Watts and Wedgwoods, from the Lunar Society which had included Priestley (Uglow 2002). When, in 1798, the radical chemist Thomas Beddoes (1760–1808) set up the Pneumatic Institution in Bristol, supported by Josiah Wedgwood (1769–1843) and James Watt (1736–1819), to administer Priestley's factitious airs to those with lung complaints, he appointed Davy as his assistant. In 1800 Davy published his study of the oxides of nitrogen, especially laughing gas, that attracted the attention of the ubiquitous Priestley – who saw from exile his mantle falling upon this young disciple – who wrote him a fine letter:

Fig. 4. Humphry Davy by Henry Howard, 1803, National Portrait Gallery ref. 4591. Reproduced by permission.

> I have read with admiration your excellent publications, and have received much instruction from them. It gives me peculiar satisfaction, that as I am now far advanced in life, and cannot expect to do much more, I shall leave so able a fellow-labourer of my own country in the great field of experimental philosophy [...] I rejoice that you are so young a man, and perceiving the ardour with which you begin your Career, I have no doubt of your success.
>
> (Schofield 1966, pp. 313–314)

In Bristol, Davy had become friends also with Coleridge, and then through him with William Wordsworth (1770–1850) and later Walter Scott (1771–1832): there were no 'two cultures' in Britain then. His work also came to the attention of Count Rumford: in 1801 he invited Davy to London.

The eloquent Davy turned out to be a brilliant lecturer who could compete with the entertainment industry, and who became a lion to invite to dinner parties and salons. Aikin's aunt Anna Laetitia Barbauld (1723–1825) wrote of him in her poem '1811' (Barbauld in Ashfield 1995, p. 25), imagining future tourists visiting the ruins of London and seeking to:

> Call up sages whose capacious mind
> Left in its course a track of light behind.

> Point to where crowds on Davy's lips reposed,
> And Nature's coyest secrets were disclosed.

Science as strip-tease was hard to resist. With the Continent closed to Britons (except for the brief half-time Peace of Amiens in 1802) the wealthy perforce had to get their culture during the London season, coming up in November and returning in June to their estates in the country. By 1807 the war was not going well. The threat of invasion had, indeed, been removed by Nelson's victory at Trafalgar in 1805, but Napoleon had turned his attention eastwards, defeating the armies of Austria, Prussia and Russia; and in 1807 the Peace of Tilsit had confirmed his domination of Europe. British trade was blockaded. At home, Wordsworth (the second edition of whose *Lyrical Ballads* Davy had punctuated and seen through the press) published his *Poems in Two Volumes*, including the famous 'Daffodils', and this new style of poetry with its simple language and Romantic sensibility was beginning to make its way. Davy had persuaded Wordsworth that truly dynamical science, like poetry, required the shaping spirit of imagination and was worthy of respect from poets. Davy himself had delighted in the sublime scenery of Cornwall, and then of Wales and the Lake District (where he visited the Wordsworths at Dove Cottage), and wrote much poetry, including a rhapsody (Knight 1998, p. 9):

> Oh, most magnificent and noble Nature!
> Have I not worshipped thee with such a love
> As never mortal man before displayed?
> Adored thee in thy majesty of visible creation,
> And searched into thy hidden and mysterious ways
> As Poet, as Philosopher, as Sage?

By 'philosopher' he meant, like his contemporaries, 'scientist'; a term not invented until after his death. He was, in fact, one of the first 'professional scientists' in Britain, living by his science in the Royal Institution, which he transformed into an institution for fundamental research largely financed by public lecturing.

Davy, Banks and the Royal Society

Davy was not satisfied with facts: as a natural philosopher he sought explanations, and a world-view. Unlike our other three, he was not content to stop at mineralogy. At the age of 28 he was the rising star of the Royal Society, its Copley Medallist in 1805, and by November 1807 its Secretary, editing the *Philosophical Transactions*. His salary, even though he was Professor at the Royal Institution, was small, but he received £100 a year from his Royal Society post, and another £100 from lectures annually delivered on agricultural chemistry. In both these things, he depended on the patronage of Sir Joseph Banks, who had been elected President of the Royal Society a few days before Davy was born. Banks was perturbed about the possible fragmentation of science, as the specialization that had made the French so successful began to be evident in Britain also; and both he and Davy, his protégé, were concerned about individual sciences declaring their independence of the Royal Society. They feared that geological papers would be siphoned off from the *Philosophical Transactions*, and that this might be the thin end of a wedge. The day of the expert was dawning: looking back over the nineteenth century on 1 January 1901 the sociologist Leonard Trelawney Hobhouse (1864–1929) wrote in *The Manchester Guardian:*

> To specialisation carried to this extreme we owe the efficiency and accuracy of modern science. To it we also owe a loss of freshness and interest, a weakening of the scientific imagination, and a great impairment of science as an instrument of education.

Banks' efforts to stem the process were vain, but not absurd. But having made the decision to devote his life to scientific administration and become, in effect, the spokesman for science, with close but not partisan links to government, Banks was busily boosting both the British Empire and his own Learned Empire, with the Royal Society (speaking for the whole scientific community) as the hub of both.

He, and Davy with him, therefore opposed developments outside the control, or at least benevolent and paternal oversight, of the Royal Society, which they had sworn to support (Herries Davies 2007, pp. 15 and 23–27). Thus (as in Davy's letter to Pepys about the first meeting), they envisaged any Geological Society as a subset of the Royal Society, a dining club for informal scientific conviviality, with any members who were not FRS being essentially associates or guests. This had been the pattern for the abortive Chemical Club, and was, indeed, the way the later Society (or Club) for Animal Chemistry developed, as a congenial, relaxed and informal discussion group, beginning to focus on what we would call biochemistry, but publishing no papers (Coley 1967; Averly 1986; Lewis 2009). We get some idea of its wide-ranging discussions from the dialogues in Davy's *Salmonia* and *Consolations in Travel* written at the end of his life, and in the 1855 *Psychological Inquiries* of Sir Benjamin Brodie (1783–1862), later President of the Royal Society. That was not what the geologists intended; they expected focused papers, like Knight's, and, although in due course they were to make their meetings much less stuffy than those of the Royal Society, they expected due formality while allowing discussion

(Thackray 2003). Such a display of independence outraged Banks: it was unfortunate, but probably inevitable, that this affable and clubbable man, generous, dedicated and cosmopolitan in his support of useful knowledge, should have emerged as the wicked uncle of the Geological Society, seeking to do it down. Neither he nor Davy could in the event support an independent Geological Society (Herries Davies 2007, pp. 15 and 27).

In 1807, however, Davy was a great catch who had been drawn into Babington's early morning group of mineralogists; but he was not at his best at that time of day and urged evening meetings for the Geological Dining Club he envisaged. But on November 23, after hectically investigating potassium, and perhaps from gaol fever picked up in Newgate following a visit to advise on ventilating it, he fell desperately ill. Babington was among the doctors called in, bulletins were posted, lectures cancelled and the Royal Institution's finances became even more tottery (Davy 1836, vol. 1, p. 386). It was nine weeks before he was well again, 'cured', according to Greenough, by James Franck (1768–1843), another of the Society's founders (Lewis 2009). When he resumed his investigations, they led to the isolation, in a race with Berzelius, of the alkaline earth metals. When he resumed attendance at the Geological Society, he was outraged to find that its character was very different from the convivial dining club he had envisaged. He behaved badly at meetings, grumbling like a spoiled brat, rudely interrupting the papers to make his point: and eventually in 1809, with Banks and others, resigned (Herries Davies 2007, pp. 21–23).

Davy did not abandon his geological interests. A keen mineralogist since his youthful days in Cornwall, he had included some geology in his agricultural lectures to landowners, and delivered what he claimed to be the first public course of geology lectures in London in 1805 (Siegfried & Dott 1976, 1980). He had made visits to Wales and Ireland, collecting minerals for the collection he was building up at the Royal Institution. In 1806 he had returned to Ireland, a country he loved, with Greenough. Now, intellectually restless, he moved beyond mineralogy into that study of processes that might be called mainstream geology. Bringing chemistry and geology together in a novel way, he suggested that volcanoes (which he believed to be always near the sea or a lake) had beneath them deposits of metallic potassium or sodium, which exploded on contact with water (Davy 1836, vol. 1, p. 397: this two-volume *Memoir* was written by his brother, John Davy).

In 1812, completing his Dick Whittington journey to prosperity, Davy married a wealthy widow, Jane Apreece, and was knighted: he could now give up the grind of lecturing and live the life of a gentleman of means, although never achieving social ease or much happiness in his childless marriage. The Davys toured the Continent (with a special 'passport' from Napoleon's government so that he could collect the prize awarded earlier for his electrochemical discoveries), visiting Paris and going on to Naples. On Vesuvius his chemical theory about volcanoes was duly falsified because, on analysis, the composition of the rocks was found to be incompatible with it. In 1820 Banks died, and Davy, as Sir Humphry, a baronet where Newton had only been a knight and the inventor in 1815 of the safety lamp that allowed coal mining to boom, was the unstoppable candidate to succeed Banks as President. Great, indeed impossible, things were hoped for from this new broom: he did speedily reverse Banks' policy of opposing specialized societies and in 1822 the Royal Society's Copley Medal went for the first time to a geologist, William Buckland, for his discoveries in Kirkdale Cave in North Yorkshire. He had taken Davy to see the cavern. Ironically, the thrust of Buckland's paper, and subsequent book *Reliquiae Diluvianae* (1823), was that his discovery of an extinct hyena's den was good evidence for the reality of Noah's Flood. But Davy (1839–1840, **7**, pp. 41–42) warned his hearers against brilliant speculations, hypotheses, dreams and visionary ideas. For the 'sound geologist':

> The more we study nature, the more we obtain proofs of divine power and beneficence; but the laws of nature and the principles of science were to be discovered by labour and industry, and have not been revealed to man; who with respect to philosophy, has been left to exert these god-like faculties, by which reason ultimately approaches, in its results, to inspiration.

However, Davy seemed haughty and bad-tempered, felt thwarted, and soon lost the confidence of the 'Gentlemen of Science' – by then prominent in the Geological and Royal Societies, and who would run the British Association for the Advancement of Science. Having rejoined in 1815, Davy resigned from the Geological Society again at the end of 1824. This was when the Society was applying for a charter and Davy would have felt that he could not serve two masters in the negotiations that required the goodwill of the Royal Society. He remained on excellent terms with Babington, dedicating *Salmonia* to him 'in remembrance of some delightful days passed in his society, and in gratitude for an uninterrupted friendship of a quarter of a century', and allegedly modelling a character in the dialogues upon him (Davy 1839–1840, **9**, pp. xv and 6). In his will he left Babington a legacy of £100, and asked for seals with a fish engraved on them to be given to Babington and Pepys in his memory (Paris 1831, pp. 545–546).

Davy had a stroke at the end of 1826, and his last two years were spent in Italy and in the Julian Alps (now Slovenia) and Carinthia in search of health. In Romantic vein, he loved mountains, the ascent to passes and the sources of rivers as much as the summits; and by now he was a geologist as much as a chemist. There, a dying sage prematurely aged in his 40s, he dictated essays (as a kind of testament) that were posthumously published as *Consolations in Travel* (1830) and sold rather well. Buckland acquired and annotated a copy. Later editions included an engraving after a drawing by Lady Murchison. Writing about the vast tracts of deep time, Davy declared his adherence to a *refined plutonic view*, referring to John Playfair (1748–1819) and Sir James Hall (1761–1832); but progressive, and with catastrophes and recreations separating the various epochs (Davy 1830, pp. 142–147). For this pretty mainstream view, Davy was castigated by Buckland's pupil Charles Lyell in his actualist *Principles of Geology* (Lyell 1830–1833, **1**, pp. 144–145). Being very eminent, safely dead, and unpopular among geologists and gentlemen of science, he was a good target: and the Geological Society had, indeed, become a centre for gentlemanly, but hard-hitting, debate. In Davy, especially, we can see chemists entering geology through mineralogy, but being drawn in deeper into the Earth's remote past. It was unfortunate that his relationship with the Society was stormy, but romantic geniuses are not necessarily easy to get on with. Like a catalyst, he had been essential in the processes bringing the Society into being, but once that was done he was no longer required. His death in 1829 was not officially noticed at the Society (Herries Davies 2007, p. 15).

Conclusion

What then had these chemists, or chemists more widely, brought to the infant Society? In the 1780s and early 1790s Priestley had probably been the best-known man of science in Britain, while in France Lavoisier was the dominant figure in the Academy of Sciences. Chemistry, with its ethos of careful experiment and cautious theorizing, was in many ways the leading and most popular science of the years around 1800, as its practitioners brought excitement and exactness into other branches of science. Their outlook was very important in the early Geological Society, which eschewed the speculation characteristic of eighteenth-century thinkers in favour of Baconian induction. They also provided mineralogy with a firm basis in chemical analysis, discovering new metals and minerals – even, like Knight, synthesizing minerals. By the 1820s things were different. Chemistry was firmly on medical syllabuses, but was duller: all those facts were hard to digest. In France, still the centre of things in science, the dominant figure in the Academy, and its Permanent Secretary, was Georges Cuvier. Comparative anatomy and the reconstruction of what James Parkinson called the 'organic remains of a former world' meant that palaeontology was becoming the scene of action. Meanwhile, the chemists among the Founding Fathers had contributed two members of the Society, both notable for almost half a century of commitment, another who had given its first paper and a maverick star-performer, whose role in geology was to be played out separately from the Society that he had been so prominent in founding.

References

AIKIN, A. & AIKIN, C. R. 1807. *A Dictionary of Chemistry and Mineralogy, with an Account of the Processes Employed in many of the most Important Chemical Manufactures*. John and Arthur Arch and William Phillips, London.

AIKIN, A. 1814. *A Manual of Mineralogy*. Longman, London.

ASHFIELD, A. (ed.). BARBAULD, A. L. 1995. '1811'. *Romantic Women Poets, 1770–1838: An Anthology*. Manchester University Press, Manchester.

AVERLY, G. 1986. The Social Chemists, *Ambix*, **33**, 99–128.

BANKS, R. E. R., ELLIOTT, B., HAWKES, J. G., KING-HELE, D. & LUCAS, G. L. (eds). 1994. *Sir Joseph Banks: A Global Perspective*. Royal Botanic Gardens, Kew, London.

BOYLAN, P. J. 2009. The Geological Society and its official recognition, 1824–1828. *In*: LEWIS, C. L. E. & KNELL, S. J. (eds) *The Making of the Geological Society of London*. Geological Society, London, Special Publications, **317**, 319–330.

BRODIE, B. C. 1855. *Psychological Inquiries: In a Series of Essays*, 2nd edn. Longman, London.

BUCKLAND, W. 1823. *Reliquiæ Diluvianiæ; or, Observations on the Organic Remains Contained in Caves, Fissures, and Diluvial Gravel, and on other Geological Phenomena*. Murray, London.

CARTER, H. B. 1988. *Sir Joseph Banks, 1743–1820*. British Museum (Natural History), London.

COLEY, N. 1967. The Animal Chemistry Club. *Notes and Records of the Royal Society*, **22**, 173–185.

DAVY, H. 1830, 1851. *Consolations in Travel: Or the Last Days of a Philosopher*. John Murray, London. (5th edn published in 1851.)

DAVY, H. 1839–1840. *Collected Works*, DAVY, J. (ed.). Smith, Elder, London.

DAVY, J. 1836. *Memoirs of the Life of Sir Humphry Davy* (2 vols) Longman, London.

GASCOIGNE, J. 1994. *Joseph Banks and the English Enlightenment: Useful Knowledge and Polite Culture*. Cambridge University Press, Cambridge.

GASCOIGNE, J. 1998. *Science in the Service of Empire: Joseph Banks, the British State and the Uses of Science in the Age of Revolutions*. Cambridge University Press, Cambridge.

HERRIES DAVIES, G. L. 2007. *Whatever is Under the Earth: The Geological Society of London 1807–2007*. Geological Society, London.

HUNT, L. B. & BUCHANAN, P. D. 1984. Richard Knight (1768–1844): A forgotten chemist and apparatus designer. *Ambix*, **31**, 57–67.

JAMES, F. A. J. L. (ed.). 2002. *The Common Purposes of Life: Science and Society at the Royal Institution, 1799–1844*. Ashgate, Aldershot.

KLEIN, U. & LEFÈVRE, W. 2007. *Materials in Eighteenth-century Science: A Historical Ontology*. M.I.T.Press, Cambridge, MA.

KNIGHT, D. M. 1998. *Humphry Davy: Science and Power*, 2nd edn. Cambridge University Press, Cambridge.

KNIGHT, D. M. 2006. *Public Understanding of Science: A History of Communicating Scientific Ideas*. Routledge, London.

KNIGHT, D. M. 2007. *Wheeler Lecture: Davy and the Placing of Potassium Among the Elements*. Royal Society of Chemistry, Historical Group Occasional Paper, **4**.

KURZER, F. 2000. History of the Surrey Institution. *Annals of Science*, **57**, 109–141.

KURZER, F. 2001. Chemistry and chemists at the London Institution, 1807–1912. *Annals of Science*, **58**, 163–201.

KURZER, F. 2003. William Hasledine Pepys FRS: A life in scientific research, learned societies and technical enterprise. *Annals of Science*, **60**, 137–183.

LEWIS, C. L. E. 2009. Doctoring geology: the medical origins of the Geological Society. *In*: LEWIS, C. L. E. & KNELL, S. J. (eds) *The Making of the Geological Society of London*. Geological Society, London, Special Publications, **317**, 49–92.

LYELL, C. 1830–1833. *Principles of Geology, Being an Attempt to Explian the Former Changes of the Earth's Surface, by Reference to Causes now in Operation*. John Murray, London.

MCCONNELL, A. 2004. Pepys, William Hasledine (1774–1856). *Oxford Dictionary of National Biography online*. Oxford University Press.

NICHOLSON, W. 1792, 1796. *The First Principles of Chemistry*. Robinson, London. (3rd edn published in 1796.)

OWEN, D. 1965. *English Philanthropy, 1660–1960*. Harvard University Press, Cambridge, MA.

PARIS, J. A. 1831. *The Life of Sir Humphry Davy*. Colburn and Bentley, London.

PRIESTLEY, J. 1966. *The History and Present State of Electricity*, [1775], 3rd edn, reprint. Johnson, New York.

RIVERS, I. & WYKES, D. L. (ed.). 2008. *Joseph Priestley Scientist, Philosopher, and Theologian*. Oxford University Press, Oxford.

SCHOFIELD, R. E. (ed.). 1966. *A Scientific Autobiography of Joseph Priestley*, M.I.T. Press, Cambridge, MA.

SIEGFRIED, R. & DOTT, R. H., JR. 1976. Humphry Davy as Geologist. *British Journal for the History of Science*, **9**, 219–227.

SIEGFRIED, R. & DOTT, R. H., JR. (ed.). 1980. *Humphry Davy on Geology: The 1805 Lectures*. Wisconsin University Press, Madison, WI.

THACKRAY, J. C. 2003. *To See the Fellows Fight: Eyewitness Accounts of Meetings of the Geological Society of London and its Club, 1822–1868*. British Society for the History of Science, Monograph, **12**.

TORRENS, H. S. 2004. Aikin, Arthur (1773–1854). *Oxford Dictionary of National Biography online*. Oxford University Press.

TORRENS, H. S. 2009. Dissenting science: the Quakers among the founding fathers. *In*: LEWIS, C. L. E. & KNELL, S. J. (eds) *The Making of the Geological Society of London*. Geological Society, London, Special Publications, **317**, 129–144.

TUTTLE, J. 2004. Aiken, Arthur (1773–1854). *In*: LIGHTMAN, B. (ed.) *Dictionary of Nineteenth-century British Scientists*. Thoemmes, Bristol.

UGLOW, J. 2002. *The Lunar Men*. Faber and Faber, London.

WERNER, A. G. 1786. *Short Classification and Description of the Various Rocks*. OSPOVAT, A. M. (ed. and translator). 1971. Hafner, New York.

Jacques-Louis, Comte de Bournon

GORDON L. HERRIES DAVIES

Trinity College, Dublin 2, Ireland

Abstract: Jacques-Louis, Comte de Bournon was a French soldier and mineralogist who, following the French Revolution, took refuge in England. There he was elected to the Royal Society and became a leading figure within the scientific circle of the metropolis. He conducted masterclasses in mineralogy and in 1807 he was one of the founders of the Geological Society. He remained a leading light of that Society until his return to France at the restoration of the Bourbon monarchy. His final years he passed as the Director-General of the Royal Mineral Cabinet.

Until it was taken by the Germans in 1870, the French frontier town of Metz enjoyed the sobriquet of 'La Pucelle' or 'The Maiden'. It was there that Jacques-Louis, Comte de Bournon (1751–1825), was born on 21 January 1751. His father was Jacques de Bournon, the seigneur de Gras et Retonfey and the proprietor of the Château de Fabert located in sight of the town's Vauban-inspired defences. His mother was Marie-Anne Martinet of Nibouville, and he was the eldest of four children. His sister Charlotte (1753–1830), achieved a reputation as the author of successful novels. At the Château de Fabert his father possessed a cabinet of minerals and this excited the interest of Jacques-Louis while he was but a child. The die had been cast. He began to move in the direction of science. As he later wrote:

> Dès les premières années du printemps de ma vie, l'étude de la nature a eu pour moi un charme infini.[1]
>
> (Bournon 1808, vol. I, p. i; see also Appendix II)

The young Jacques-Louis travelled extensively in eastern France, and he was to be found observing and collecting in the provinces of Auvergne, Bourgogne (Burgundy), Champagne, Dauphiné, Franche Compté, Lorraine and Lyonnais. At this stage it was botany and entomology that seem to have been his prime concerns, but as he journeyed he assembled for himself a mineral collection which, in both size and quality, soon surpassed that held by his father.

Increasingly fascinated by mineralogy, Jacques-Louis went off to Paris to explore the science further under the tutelage of Jean-Baptiste Romé de L'Isle (1736–1790) whose *Essai de Cristallographie* (1772) the young man had found to be inspirational. We are told that at Romé de L'Isle's feet the student from Metz soon became a favourite pupil. The first significant publication to come from the pen of Jacques-Louis was, nevertheless, not so much mineralogical in character as geological. He had spent some time studying the coal-bearing and other strata of the Loire coalfield, along the SE border of the Massif Central, and the result was his 1785 *Essai sur la lithologie des environs de Saint-Etienne-en-Forez et sur l'origine de ses charbons de pierre. Avec des observations sur les silex, pétro-silex, jaspes et granits*[2] (Bournon 1785). At this stage, rocks and minerals were for Jacques-Louis merely a favourite recreation. They offered a pursuit rather than a vocation. For his career, his aspirations were focused on the profession of arms. Perhaps he dreamed of military glory. He became an officer of artillery in the Régiment de Toul and for the moment his concern was with poudre à canon rather than with mineral streak, with canon de treize rather than with the blowpipe. His military career prospered. He was appointed to the rank of lieutenant de maréchaux de France and in that capacity the momentous year of 1789 found him serving as capitaine d'artillerie in the garrison of Grenoble, then the capital of Dauphiné.

For Jacques-Louis, as for so many others, *La Révolution* seemed to spell disaster. He was a fervent loyalist; a man of l'ancien régime. He saw himself as a servant of Louis XVI. He cared deeply about the Bourbon cause, even though l'ancien régime had thrown his sister into the Bastille for a time for alleged libels in her 1782

[1] From the first years of the springtime of my life, the study of nature has had infinite charm for me.

[2] *Essay on the lithology around St Etienne-en-Forez and on the origins of its coals. With observations on its cherts, felsites, jaspers and granites.* St Etienne was later to become the type locality of the Stephanian stage of the Carboniferous. There is some uncertainty about his use of the term *pétro-silex* in this context. 'Felsite' is the usual modern equivalent, although de Bournon may have simply been using the term to refer to any very fine-grained siliceous rock, such as occur frequently in the Coal Measures.

From: LEWIS, C. L. E. & KNELL, S. J. (eds) *The Making of the Geological Society of London.*
The Geological Society, London, Special Publications, **317**, 105–113.
DOI: 10.1144/SP317.4 0305-8719/09/$15.00

Le Fripon.[3] In his own words, published in 1808, he looked back to 'la révolution qui a couvert la France de sang, de cendres et de larmes'[4] (Bournon 1808, vol. I, p. vi; see also Appendix II). By 1791, the year of the royal family's flight and apprehension on the bridge at Varennes-en-Argonne (21 June), Jacques-Louis considered his life in France to have become intolerable. Moved by intense loyalty and deep emotion, he abandoned his career and his possessions and, with his family, he crossed the Rhine. His destination was Coblence, that city then being the headquarters for *l'émigration* and for counter-revolution. There he joined the army of Louis-Joseph, Prince of Condé (1736–1818), in its campaign against the forces of the new France. But before quitting France Jacques-Louis marked certain of the minerals from his cabinet – his choicest and favourite specimens – in such a manner that, at some future date, there might be some hope of his establishing a case to be their rightful owner.

He served with the army of Condé and alongside the Austrians, throughout 1792, but after that year's campaign he decided to renounce the profession of arms and to become a member of the émigré colony that had now taken root in London. There he will surely have received a warm welcome from his fellow nationals who were seeking impoverished refuge from the guillotine, which had removed the head of Louis XVI on 21 January 1793. But Jacques-Louis was more than just a fugitive nobleman holding military rank. He was also a scientist. As such, he was eagerly received into London's flourishing scientific circle. In particular, mineralogy was a science then much in vogue in Britain and here was a man who had studied the science at its Parisian epicentre. His expertise served as a key capable of opening stately English doors. With the Honourable Charles Francis Greville (1749–1809), the one-time 'protector' of Emma Hart, Lady Hamilton (1765–1815), he seems to have established a particularly warm relationship.

On 16 February 1796 he was elected a Fellow of the Linnean Society; on 25 February 1802 he was named as a Fellow of the Royal Society; on 13 November 1807 he was, aged 56, the oldest of the 11 gentlemen who dined at the Freemasons' Tavern and there founded the Geological Society (Woodward 1907; Herries Davies 2007; Lewis 2009); and during 1808 he was chosen as a Non-Resident Member of the new Wernerian Natural History Society of Edinburgh. At the time of his election to the Royal Society he was domiciled in Bryanston Street, off fashionable Portman Square, and six of his eight Royal Society sponsors were Richard Chenevix (1774–1830), (Sir) Alexander Crichton (1763–1856), the Hon. Charles. F. Greville, Charles Hatchett (1765–1847), Edward Charles Howard (1774–1816) and Sir John St Aubyn, Bart (1758–1839). In varying degree these six Fellows of the Royal Society had all discovered the fascination of minerals and, by 1811, all of them save for Howard, had joined de Bournon as Members of the Geological Society.

Amidst his new English friends, de Bournon soon found many a congenial task well suited to his talents. He became Curator of the mineral collection belonging to Greville, a collection said to have been one of the world's finest, and a collection whence, following its owner's death in 1809, some 14 800 minerals were sold to the nation for the princely sum of £13 727 (Fletcher 1904; Smith 1962–1969, 1977–1980). Greville's 5200 faceted gems were the subject of a separate sale. He arranged the cabinet of Sir John St Aubyn (Currey 1975) (Fig. 1), and he catalogued the fabulous diamond collection assembled by the art connoisseur Sir Abraham Hume (1749–1838), the catalogue being published in 1816 (Bournon 1815). Thirteen years earlier – in 1802 – he confessed that being accorded access to the riches of the Greville, St Aubyn and Hume cabinets 'diminishes my regret for the loss of my own' (Bournon 1802, p. 265 ff).

To the Royal Society he read papers on corundum from the East Indies, China and elsewhere (Bournon 1798, 1802), arseniates of copper and iron from Cornwall (Bournon 1801), carbonate of lime (Bournon 1803), and sulphuret of lead, antimony and copper from Cornwall (Bournon 1804). Arising from the latter paper, Robert Jameson (1774–1854) in 1805 gave to a new mineral, a sulphantimonite of lead, copper and antimony ($PbCuSbS_3$), the name 'Bournonite' in honour of its first describer (Jameson 1805, p. 579). This group of papers presented to the Royal Society were, for de Bournon, something rather more than just contributions to the advancement of science. They were a political tribute. He explained in 1801:

> [...] I offer the result of my observations to the Royal Society, as an acknowledgement of that gratitude which I and all Frenchmen, faithful to their king, ought to feel and profess to a country which has distinguished itself as the protector of honour and loyalty.
> (Bournon 1801, p. 169)

De Bournon also found himself to be in demand as a teacher of mineralogy. His expertise was sought after. He conducted masterclasses in the science.

[3] *The Rogue.*

[4] 'the revolution which has covered France with blood, with ashes, and with tears'.

Fig. 1. Micromount specimens, with detail, from the St Aubyn collection, Plymouth City Museum and Art Gallery ©. Within the St Aubyn collection is a subcollection of 400 beautifully turned wooden cups, which have small wax-mounted mineral and gemstone specimens set within them. The labels on the wooden cups were handwritten by de Bournon in French.

Among those who attended his prelections was the eminent London physician William Babington (1756–1833). In 1807 it was Babington who played a leading role in the foundation of the Geological Society. Indeed, in the words of the *Oxford Dictionary of National Biography* (Payne 2004), he 'has some claim to be regarded as the founder of the Geological Society', and as Babington's mentor de Bournon merits a second entry in the genealogy of the Geological Society.

Interestingly, de Bournon's attention was not confined to terrestrial mineralogy. In collaboration with E. C. Howard, he analysed material from several bodies that were said to have been observed falling to Earth from the heavens. Some of these specimens came from the cabinet of Sir Joseph Banks, Bart (1743–1820) and included fragments of the stone weighing 56 lbs (25 kg), which a Yorkshire ploughman claimed to have seen tumble from the sky at Wold Cottage Farm, Wold Newton, near to Filey, on 13 December 1795 (Howard 1802). During the Age of Reason many a scientist had chosen to doubt the reality of such events – they savoured the supernatural overmuch – and Banks hoped that mineralogical analysis might serve to settle the issue (Sears 1975; Marvin 1986). In the event, when they published their findings in 1802, Howard and de Bournon remained somewhat guarded, and it was the following year before the celestial origin of meteorites was conclusively demonstrated by a spectacular shower of material that descended upon L'Aigle, in Normandy, on 26 April 1803.

Shortly after his arrival in England, de Bournon played a part in the affair of a large natural history collection which stood awhile in the limbo of disputed ownership. The story is as follows (de Beer 1952). In August 1785 a French government expedition of exploration under the command of Jean-François de Galaup, Compte de la Pérouse (1741–1788), set sail for the Pacific. Three years later – early in 1788 – the two ships of the expedition disappeared somewhat mysteriously after leaving Australia, and during 1791 the French dispatched a new expedition in an effort – a fruitless effort as it transpired – to try to discover the fate of the missing navigators. The new expedition was under the command of Jospeh-Antoine Raymond Bruni d'Entrecasteaux (1739–1793) and there sailed with him, as a naturalist, Jacques-Julien Houtou de la Billardière (1755–1834) who in the South Seas soon brought a wealth of natural history specimens on board the two ships of the expedition. D'Entrecasteaux died during July 1793 and the expedition's officers faced a dilemna. Most of them – but not de la Billardière – were royalists, and France had now become a republic. It was therefore resolved to surrender the expedition to the Dutch in Java where all of de la Billardière's specimens eventually fell into the hands of the British as a prize of war.

The British shipped the collection off to London and there, for a while, it was left to languish in the dampness of a customs warehouse. But such neglect raised a storm of protest from several quarters, so the bulk of the collection was removed to Harcourt House and there placed under the administration of the émigré duc d'Harcourt, Ambassador of the French government in exile to the Court of St James's. At Harcourt House the collection was inspected by Banks and he reported it to consist of 'a vast Herbarium' of some 10 000 specimens, many dried birds, lizards and snakes, together with fish preserved in spirits. But the collection had now lost its insects. The insects had already been placed elsewhere. Upon the instruction of the duc d'Harcourt, the entire collection had earlier been resigned to the care of de Bournon and he had removed all the insects to his own residence. It would seem that his more recent affair with minerals had not entirely quenched his earlier passion for entomology.

As King Louis XVI had been guillotined and the young Louis XVII had died in prison on 8 June 1795, in British eyes the rightful owner of the collection was now deemed to be Louis XVIII who, during the 1790s, was leading the semi-nomadic existence of a monarch-in-exile on the Continent. Louis XVIII clearly had no immediate need of the collection, so he intimated his intention of presenting it in its entirety to the botanically inclined Queen Charlotte, the consort of George III. But the Queen felt little empathy for dried snakes or pickled fish, so, on 1 April 1796, she accepted merely the collection's botanical specimens and left de Bournon to decide the fate of the remainder of the collection. This rump he offered to Banks as a token of the esteem in which he was held by the French refugees who were present in London. But, in the event, all such planned disbursing of the collection came to naught. There were howls of remonstration from across the Channel, orchestrated by de la Billardière who had now returned to France after 18 months of imprisonment in Java. War or not – and with the blessing of Banks – it was to France that the entire collection, or what remained of it, was dispatched before the year 1796 was out.

Presumably the insects from de Bournon's residence went to France with the remainder of the de la Billardière collection, but it is interesting to note that a few mineral specimens consigned to de Bournon seem to have crossed La Manche in the opposite direction. It appears that his friends in France had, indeed, been able to identify a few of the minerals marked by their owner before he fled France in 1791, and these were now restored to him in London. There, by the time of the memorable Society-founding events of November 1807, he was a resident of Duke Street somewhere near its junction with Manchester Square, and presumably it was there that he maintained his phoenix of a mineral collection. An inveterate collector, by 1813 that new collection had attained a size sufficient to justify the publication of its own detailed catalogue (Bournon 1813). In the catalogue he paid tribute to his friend Sir Alexander Crichton by bestowing the name 'Crichtonite' upon a hitherto unnamed mineral. It was a mineral that de Bournon himself had discovered near to the town of Visan in the Alps of Duphiné.

If Babington has some claim to be *the* founder of the Geological Society, then it is de Bouron who has to be recognized as Babington's foil. At this point a scientific manuscript has to be brought to the centre of our stage. By 1807 de Bournon had completed the writing of a substantial work devoted to carbonate of lime, and to crystallography and mineralogy in general (see Appendix I). The completed manuscript must have been bulky. The study was comprehensive, and is now widely regarded as an important landmark in mineralogical research. It was expansive. It was recondite. And it was written in French. De Bournon had presumably acquired at least some mastery of spoken English, but with a pen in his hand he very naturally preferred to express himself in his native language. Now the rub. How was such a work – alien and abstruse – going to find an English publisher?

Babington and three of his Quaker mineralogical friends – William Allen (1770–1843), Richard Phillips (1778–1851) and Richard's elder brother, William Phillips (1773–1828) (Torrens 2009) – resolved to come to de Bournon's assistance. Early during 1807 these four convened a meeting whereat de Bournon's problem could be aired. That meeting was most probably held at Babington's recently purchased residence, 17 Aldermanbury, and the upshot of this and subsequent meetings was the adoption of three resolutions. First, that de Bournon's manuscript was scientifically important and deserving of publication. Second, that William Phillips, a publisher of George Yard, Lombard Street, would add the work to his list and oversee the actual task of publication (Torrens 2009). Third, that 14 of the gentlemen within the group would become the work's sponsors, each of them subscribing £50 towards the cost of publication.

The 14 gentlemen who now became the financial backers of de Bournon's text were: William Allen; Sir John St Aubyn; William Babington; Robert Ferguson (1767–1840); George Bellas Greenough (1778–1855); the Hon. C. F. Greville; Charles Hatchett; Luke Howard (1772–1864); Sir Abraham Hume; Richard Knight (1768–1844); James Laird (1779–1841); Richard Phillips; William Phillips; and John Williams (1778–1849) (Bournon 1808, vol. I, p. xxxvi, where James Laird is misnamed 'Richard'; see also Appendix II).[5] Of the 14 only Knight and Laird were never elected to the Royal Society, and only Howard never numbered among the Members of the Geological Society; Williams was elected an Honorary Member in 1808. De Bournon's treatise was shortly published in three substantial quarto volumes dated 1808 and dedicated to Alexander I, 'Empereur et Autocrate de Toutes les Russies'. The minutes of the Geological Society's Council Meeting of 3 March 1809 record de Bournon's presentation of the three volumes to the Society. At the same meeting a communication was received from Sir Joseph Banks. The youthful society was in turmoil. It is time to explore de Bournon's significance in the foundation of the Society and his activities during its early years.

The meetings held chez Babington to discuss the publication of de Bournon's manuscript proved to be singularly convivial occasions. Even after de Bournon's script was safely lodged with William Phillips over at George Yard, the group of friends continued to foregather for scientific discussion. In Greenough's words, the meetings 'were found too agreeable to be discontinued'. Indeed, new faces began to be seen at the gatherings, one of them being that of Humphry Davy Bart (1778–1829). Davy had just become one of the secretaries to the Royal Society (Knight 2009), and it was he who coined for the group the title 'the Geophilists'.[6] But in Davy's eyes the group had an unpleasant habit. It tended to meet in the early mornings over breakfast. Davy was a lie-a-bed; he detested early rising. In any case, he was convinced that evening meetings over dinner would be more conducive to scientific discussion. Roast beef and wine were more inspiring than tea and buttered buns (Knell 2009). Davy had his way, and so it was that de Bournon found himself at that memorable dinner table in the Freemasons' Tavern on 13 November 1807.

Among the Geophilists, de Bournon was evidently one of those who supported the notion that in character the new Society founded that evening should be geological rather than mineralogical.

[5] The total number of subscribers was in fact 16, see Lewis (2009).

[6] GSL: LDGSL 960. Greenough's 'Papers connected with the Geological Society', a history of the first three years with copies of letters and lists.

Perhaps he was remembering his youthful study at St Etienne. His conception of the task of geology is revealed in the following passage from his treatise of 1808:

> La géologie, telle qu'elle est aujourd'hui, et telle qu'elle doit être encore quelque temps, est le rassemblement d'un grand nombre de faits locaux et particuliers, qui doivent être un jour la base d'une théorie générale, mais sur lesquels il est encore impossible de l'établir.[7]
>
> (Bournon 1808, vol. I, p. xi; see also Appendix II)

At the second meeting of the newly constituted Society – a meeting held at the Freemasons' Tavern on 4 December 1807 – de Bournon presented and explained a display of feldspar minerals, and his was evidently the second communication offered to the Society. Whether he may have addressed the gathering in English or in French is not on record, but that he should thus have featured as one of the Society's earliest speakers needs evoke no surprise. Within the Society, he was a pivotal figure during those early years. He probably was the Society's most experienced geologist. Had he not been a foreign national, and had he not departed for France at the restoration of the Bourbons, then history must surely have accorded him a place of high distinction among the Society's Founding Fathers.

As an Original Member, de Bournon was witness to the serious difficulties that beset the Society during the earliest years of its existence. At the core of those difficulties there lay a single issue: what significance was to be attached to an oath of obligation? Some of those who joined the Geological Society at the outset thought that it was going to be essentially a geological dining club, but almost immediately the Society set itself upon a course to become a full-blown scientific society. For those members of the Society who were already Fellows of the Royal Society (by the close of 1808 there were 37 Members or Honorary Members – 27 percent of the roll – within this category) this development created a serious crisis of conscience. They had subscribed to the Royal Society's Oath of Obligation; they had sworn 'to promote the good of the Royal Society'. Such an oath was hardly to be reconciled with Membership of a rival geological society that enforced its own Oath of Obligation. Members of the new Society had to promise that they 'will endeavour to promote the honour and interest of the Geological Society'. There was dissension.

De Bournon was present at Greville's abode on Saturday 14 January 1809 for a dinner during which effort was made to resolve this conflict of interest. The effort failed. In consequence at the Geological Society's Council Meeting on 3 March 1809 – the meeting at which the Society received the three volumes of de Bournon's *Traité complet* – Sir Joseph Banks asked for his name to be withdrawn from the Society's roll. Three days later Charles Hatchett, one of de Bournon's eight sponsors at the Royal Society, resigned his Membership of the Geological Society because as loyal a FRS he could not support an upstart Geological Society of London.

What de Bournon's personal views upon this issue of conflicting loyalties may have been is not on record. Presumably he experienced the dilemma just as strongly as did any of his confrères. That he was a man of principle is amply demonstrated by his own life. At immense personal cost he had refused to abandon his loyalty to his monarch in favour of the allegiance demanded by a new France born of revolution. His presence in England was testament to the strength of his character. As we saw earlier, he believed himself now to be resident in a land distinguished 'as the protector of honour and loyalty'. But was the establishment of the Geological Society not perilously close to being a revolutionary act subversive of an existing scientific order wherein the status quo had been supported and sanctified by the Royal Society's Oath of Obligation? Were events at the Freemasons' Tavern on 13 November 1807 to be likened to the storming of the Bastille on 14 July 1789?

De Bournon may have preferred to write in his native French rather than in English, but his pen was nonetheless ever at the disposal of the youthful Geological Society. During the weeks around Christmastide 1807, for example, he prepared for Babington a summary of what he perceived to be salient questions to be placed on the agenda of the infant society.[8] His document fills 43 manuscript leaves in French, its opening words being as follows:

> Le project de la Société Géologique de Londres, la première qui ait eté établie sous ce titre, est de travailler à l'acquisition des connoissances propres à établier la Théorie de la terre.[9]

[7] 'Geology, such as it is today, and such that it should be for some time, is the assembling of a great number of local and specific facts, which should one day become the basis of a general theory, but which it is still today impossible to establish.'

[8] Archives of the Geological Society. LDGSL 1. Mémoire sur la Société Géologique Londres. 1808.

[9] 'The project of the Geological Society of London, the first that has been established under that name, has been established to work for the acquisition of appropriate knowledge for establishing a Theory of the Earth.'

De Bournon's document would seem to be closely related to the so-called *Geological Inquiries*, which was compiled by Arthur Aikin (1773–1854) and Greenough, laid before the Council on 5 February 1808, printed by William Phillips, and then circulated to all Members and Honorary Members during the March and April of 1808 (Geological Society of London 1808). The 20-page *Inquiries* was the Society's earliest publication (see Appendix I) .

When the Society resolved to embark upon the publication of something rather more substantial then the *Inquiries*, de Bournon's pen was again placed at the Society's disposal. The first volume of the Society's *Transactions*, published in 1811, contains two essays by him. One is on the subject of Laumonite (Bournon 1811*a*) and the other is devoted to sulphate of lime (Bournon 1811*b*). As was his custom, de Bournon wrote both essays in French, but it was in English that they featured within the *Transactions*. The Society may have chosen to translate de Bournon's French into English, but de Bournon himself must from time to time have found himself to be engaged in the reverse process because from 1810 to 1813 he served as the Society's Foreign Secretary. Throughout his exile he remained in close touch with the French scientific scene, publishing at least a dozen papers in the *Journal des Mines* between about 1796 and 1811, several of which were translations of Royal Society and other UK publications.

It is said that around 1810 de Bournon refused several offers of a highly prestigious and well-paid position in l'Institut (Wilson 1994) when Napoleon tried to entice him back to France. If this be true, then it is a clear sign of de Bournon's high standing within the world of science. But such blandishments as the Emperor may have uttered fell upon deaf ears. De Bournon remained resolute: his return to France must await a restoration of the monarchy. He, presumably, had some first-hand acquaintance with the man whom he now regarded as his rightful monarch. That man – Louis XVIII – was resident in England for some years, first at Gosfield near Halstead in Essex (1807–1809), and then at Hartwell near Aylesbury in Buckingham (1809–1814). At both sites de Bournon will surely have been a familiar figure within the court in exile. Then in 1814 there came that moment of which de Bournon must so often have dreamed. Napoleon was overthrown. The Emperor was packed off to Elba and, on 2 May 1814, Louis XVIII made his triumphal entry into Paris. The date of de Bournon's own return to France remains uncertain. The packing of his belongings, including the new mineral collection, took time and he was certainly still in London as late as July 1814.

Louis XVIII was now anxious to reward the devotion of those who had, over long years, remained loyal to the Bourbon cause. In the case of de Bournon the reward involved the implementation of a simple piece of legerdemain. It so happened that the King had developed a mild interest in mineralogy (whether this owed anything to the influence of de Bournon I just cannot say), and it was arranged that he would purchase de Bournon's London mineral collection 'à un prix vraiment royal'[10]. In return, the King appointed de Bournon to be the handsomely remunerated Director-General of the Royal Mineral Cabinet. As was the long-established French practice (Taquet 2009), the position also carried generous technical and financial support. He was even provided with the services of an Assistant Director! Officially, he assumed responsibility for what was now the royal collection during July 1814 and three years later published a catalogue for it. In essence, this was a second edition of the 1813 catalogue of what had once been de Bournon's own collection (Bournon 1817).

During the spring of 1815 de Bournon's world fell apart for the second time. Napoleon had promised to return to France with the violets of 1815. He was as good as his word. On 26 February 1815 he left Elba, landed in France and began his remarkable advance upon Paris. On 20 March he entered the capital. The 100 days had begun. De Bournon, like his King, saw prudence in flight. The mineral collection was still in its packing cases in Paris, but de Bournon felt it wise to rejoin his friends in London. There he remained until after Wellington had sealed Napoleon's fate upon the field of Waterloo. The King re-entered Paris on 8 July 1815 and de Bournon presumably returned to France around the same time.

The subsequent history of de Bournon and the royal mineral collection is a part of the history of French mineralogy rather than being an episode in the story of the Geological Society. Those who desire details of de Bournon's final years in France will find their satisfaction in the essays by Alfred Lacroix (1931, 1932). Here it suffices to note that de Bournon died in Versailles on 24 August 1825, leaving a daughter who married the Comte de Carbonnieres. The mineral collection that had been both his delight and his responsibility was divided between the Muséum d'Histoire Naturelle and the Collège de France, where the two parts survive more or less intact today (Wilson 1994). The archives of the Geological Society contain several of his manuscripts and his collection of

[10] 'at a really royal price'.

wooden crystal models is in London's Natural History Museum. Those, presumably, are the demonstration models that he used in the classes which he conducted for the benefit of the Geophilists assembled at Babington's residence during the months of the Geological Society's parturition.

Grateful thanks for assistance are due to Wendy Cawthorne of the Geological Society's Library, to Professor Boylan of City University, London, and to Gaston Godard of L'Institut de Physique du Globe de Paris.

Bibliography

Biographical details have been drawn from the following works:

Dictionary of Scientific Biography (1970)
Dictionnaire de Biographie Française (1954)

References

CURREY, D. A. 1975. The Plymouth City Museum mineral collection. *Newsletter of the Geological Curators Group*, **3**, 132–137.

DE BEER, G. R. 1952. The relations between Fellows of the Royal Society and French men of science when France and Britain were at war. *Notes and Records of the Royal Society of London*, **9**(2), 244–299.

DE BOURNON, J. L. 1785. *Essai sur la lithologie des environs de Saint-Etienne-en-Forez et sur l'origine de ses charbons de pierre. Avec des observations sur les silex, pétro-silex, jaspes et granits*. Lyon.

DE BOURNON, J. L. 1798. An analytical description of the crystalline forms of corundum, from the East Indies, and from China. *Philosophical Transactions of the Royal Society*, **II**, 428–448.

DE BOURNON, J. L. 1801. Description of the arseniates of copper, and of iron, from the County of Cornwall. *Philosophical Transactions of the Royal Society*, **I**, 169–192.

DE BOURNON, J. L. 1802. Description of the corundum stone, and its varieties, commonly known by the names of oriental ruby, sapphire & c with observations on some other mineral substances. *Philosophical Transactions of the Royal Society*, **II**, 233–326.

DE BOURNON, J. L. 1803. Observations on a new species of hard carbonate of lime; also on a new species of oxide of iron. *Philosophical Transactions of the Royal Society*, **II**, 325–338.

DE BOURNON, J. L. 1804. Description of a triple sulphuret of lead, antimony, and copper, from Cornwall; with some observations upon the various modes of attraction which influence the different kinds of sulphuret of copper. *Philosophical Transactions of the Royal Society*, **I**, 30–62.

DE BOURNON, J. L. 1808. *Traité complet de la chaux carbonatée et de l'arragonite* [sic], *auquel on a joint une introduction à la minéralogie en général, ainsi que celle du calcul, à la dètermination* [sic] *des formes cristallines de ces deux substances* (3 vols). William Phillips, London. [English translation of the 'Discours préliminaire'. *In*: Lewis, C. L. E. & Knell, S. J. (eds) *The Making of the Geological Society of London*. Geological Society, London, Special Publications, **317**, Appendix II, 457–464.]

DE BOURNON, J. L. 1811*a*. Memoir on the laumonite; in order to determine the differences that exist between it and bardiglione. *Transactions of the Geological Society of London*, **I**, 77–92.

DE BOURNON, J. L. 1811*b*. Memoir on bardiglione or sulphate of lime, containing a sketch of a theory of the true nature of plaster, as well as of its properties. *Transactions of the Geological Society of London*, **I**, 355–385.

DE BOURNON, J. L. 1813. *Catalogue de la collection minéralogique du Cte de Bournon ... faites* [sic] *par lui-même, et dans lequel sont placés plusieurs observations et faits interessants ... ainsi qu'une réponse au mémoire de M. l'abbé Haüy concemant la simplicité des lois auxquelles est soumise la structure des cristaux*. London.

DE BOURNON, J. L. 1815. *Catalogue raisonné des diamants dans le cabinet de Sir Abraham Hume, Bart....* John Murray, Londres [1816]. There is also a translation by Hume published in 1815 [1816].

DE BOURNON, J. L. 1817. *Catalogue de la collection minéralogique particulière du Roi*. A. Lance, Paris.

FLETCHER, L. 1904. *The Department of Minerals. The History of the Collections Contained in the Natural History Departments of the British Museum*, Volume **I**, 343–442.

GEOLOGICAL SOCIETY OF LONDON. 1808. *Geological Inquiries*, N.D. [1808]. William Phillips, London.

HERRIES DAVIES, G. L. 2007. *Whatever is Under the Earth: The Geological Society of London 1807–2007*. Geological Society, London.

HOWARD, E. C. 1802. Experiments and observations on certain stony and metalline substances, which at different times are said to have fallen on the Earth; also on various kinds of native iron. *Philosophical Transactions of the Royal Society*, **I**, 168–212.

JAMESON, R. 1805. *A System of Mineralogy, Comprehending Oryctognosy, Geognosy, Mineralogical Chemistry, Mineralogical Geography and Economical Mineralogy*, Volume **II**. Edinburgh, 579–582.

KNELL, S. J. 2009. The road to Smith: how the Geological Society came to possess English geology. *In*: LEWIS, C. L. E. & KNELL, S. J. (eds) *The Making of the Geological Society of London*. Geological Society, London, Special Publications, **317**, 1–47.

KNIGHT, D. 2009. Chemists get down to Earth. *In*: LEWIS, C. L. E. & KNELL, S. J. (eds) *The Making of the Geological Society of London*. Geological Society, London, Special Publications, **317**, 93–103.

LACROIX, A. 1931. Notice historique sur François-Sulpice Beudant et Alfred-Louis-Oliver Legrand des Cloizeaux. *Memoires de l'Academie des Sciences de l'Institut de France*, 2ème serie, **60**, 101. This essay contains a bibliography of de Bournon's writings (pp. lxxxiii–lxxxvi).

LACROIX, A. 1932. *Figures de Savants*, **I**, 169–176.

LEWIS, C. L. E. 2009. Doctoring geology: the medical origins of the Geological Society. *In*: LEWIS, C. L. E.

& KNELL, S. J. (eds) *The Making of the Geological Society of London*. Geological Society, London, Special Publications, **317**, 49–92.

MARVIN, U. B. 1986. Meteorites, the Moon and the history of geology. *Journal of Geological Education*. **34**(3), 140–165.

PAYNE, J. F. 2004. Babington, William 1756–1833. *Oxford Dictionary of National Biography online*. Oxford University Press.

ROMÉ DE L'ISLE, J. B. 1772. *Essai de cristallographie, ou description des figures géométriques, propes à différens corps du règne mineral, connus vulgairement sous le nom de cristaux*. Paris.

SEARS, D. W. 1975. Sketches in the history of meteoritics 1: The birth of the science. *Meteoritics*, **10**(3), 215–225.

SMITH, W. C. 1962–1969. A history of the first hundred years of the mineral collections in the British Museum. *Bulletin of the British Museum (Natural History) (Historical Series)*, **3**(8), 237–259.

SMITH, W. C. 1977–1980. Early mineralogy in Great Britain and Ireland. *Bulletin of the British Museum (Natural History) (Historical Series)*, **6**, 49–74.

TAQUET, P. 2009. Geology beyond the Channel: the beginnings of geohistory in early nineteenth-century France. *In*: LEWIS, C. L. E. & KNELL, S. J. (eds) *The Making of the Geological Society of London*. Geological Society, London, Special Publications, **317**, 155–162.

TORRENS, H. S. 2009. Dissenting science: the Quakers among the Founding Fathers. *In*: LEWIS, C. L. E. & KNELL, S. J. (eds) *The Making of the Geological Society of London*. Geological Society, London, Special Publications, **317**, 129–144.

WILSON, W. E. 1994. *The History of Mineral Collecting: 1530–1799. Jacques Louis de Bournon (1751–1825). The Mineralogical Record*, **25**, November.

WOODWARD, H. B. 1907. *The History of the Geological Society of London*. Geological Society, London.

George Bellas Greenough's 'Theory of the Earth' and its impact on the early Geological Society

MARTINA KÖLBL-EBERT

Jura-Museum Eichstätt, Willibaldsburg, D-85072 Eichstätt, Germany
(e-mail: Koelbl-Ebert@jura-museum.de)

Abstract: George Bellas Greenough, co-founder and first President of the Geological Society of London, became interested in geology when he went to study law in Göttingen. There he attended lectures given by Professor Johann Friedrich Blumenbach, an admirer of Jean-André de Luc, who greatly influenced Greenough's geological ideas. Greenough himself was not an original researcher, but saw his scientific task as a most diligent gatherer of information and as a critical and – as he felt – impartial reviewer of his fellows' research. In 1819 he published a book entitled *A Critical Examination of the First Principles of Geology in a Series of Essays*. In this book, as well as in many of his numerous private notes, he struggled with the question of how to develop a proper scientific method for the new science of geology, striving for firm principles and definitions as a basis for geological observations. Although he despised theorizing on general principles, especially when it concerned those whom he called the Huttonians or Plutonists, he himself was not free of a theoretical concept in which he judged the validity of data. This bias sometimes became a drag on scientific 'progress' because his preoccupation with his own Theory led him to dismiss important developments such as William Smith's biostratigraphy when they did not fall within his horizon of interest. Thus Greenough, and with him the new Society, was slow to recognize their importance and value.

The founding of the Geological Society (Rudwick 1963; Herries Davies 2007, pp. 1–53; Lewis 2009) was, as George Bellas Greenough (1778–1855) later recollected, 'the effect of accident rather than design' (Rudwick 1963, p. 326). It all started with informal meetings of wealthy mineral collectors for the purpose of furthering the publication of a mineralogical book by the Count de Bournon (1751–1825). After this goal was achieved, they continued to meet. More people joined in and, possibly because in November there is not much fun in getting up early and meeting for breakfast as they had done during the summer, they resorted to get together for dinner in the Freemasons' Tavern.

Right from the very first evening the diners decided to form a learned society, but – and this was unexpected – they were persuaded by Humphry Davy (1778–1829) and Greenough not to form a mineralogical society, but a geological one. Davy, it seems, had planned for a continuation of the informal breakfast meetings, albeit at dinner, but Greenough (Fig. 1) had something more ambitious in mind – and it was he who became first President of the new Geological Society until 1813. He was later re-elected from 1818 to 1820 and again from 1833 to 1835. His influence on the young society was immense, and his geological ideas and ideals – pursued with the ample time only a rich bachelor could afford – were responsible for much that was implemented as the Society's goals and methods during its formative years. And it is these years, up to about 1825, that will be the focus of this paper.

Greenough's geological education

> bright eyes – silver hair
> large mouth – ears – feet
> fondness for generaliz^{n}. for system and clearliness
> Sowerby's account of me: Ammonite
> great diligence – patience – zeal. – goodnature – but hasty – firmness of principle
> hand for gardening.[1]

When the Geological Society was founded, Greenough was a young man, nearly 30 years old, of independent means, who looked back on a rather bleak childhood. After a period of financial trouble his father, George Bellas, had died in 1784 when little George was only six years old. His mother, Sarah Bellas (1750–1784), née Greenough, followed her husband to the grave only a few months later. The orphan was handed down through several boarding schools and passed his

[1] UCL: Greenough Papers, Autobiographical Notes, 24/1. Greenough's characterization of himself.

From: LEWIS, C. L. E. & KNELL, S. J. (eds) *The Making of the Geological Society of London*.
The Geological Society, London, Special Publications, **317**, 115–128.
DOI: 10.1144/SP317.5 0305-8719/09/$15.00

Fig. 1. George Bellas Greenough in the 1830s (Rudwick 1985, fig. 4.2).

holidays under the roof of his maternal grandfather, the wealthy apothecary and chemist Thomas Greenough. Eventually, George went to Cambridge to study law for three years – but, as he was a dissenter, he never graduated (Wyatt 2004).

In 1795 Greenough's grandfather died and left him a considerable fortune, amassed in the field of popular patent medicines, which enabled him to pursue his many interests without needing to earn a living. In 1798 Greenough moved to Göttingen (Kurfürstentum Hannover) to continue his studies in law, but he soon discovered that his German was in need of improvement first. He took private lectures for German grammar and was advised to attend the lectures of Johann Friedrich Blumenbach (1752–1840), a professor of physiology and comparative anatomy, who was an engaging, inspiring and, owing to the natural history objects he presented to his students, an easy-to-follow lecturer. Blumenbach's geological ideas, although based on a German mineralogical tradition, were not identical to the Wernerian school, but had been greatly influenced by Jean-André De Luc (1727–1817), a persistent critic of James Hutton. De Luc provided Blumenbach with a series of letters concerning geology, which Blumenbach published in German translation (e.g. Blumenbach 1796; see also Rudwick 2002, 2005, pp. 326 ff.). Although Greenough retained a critical attitude towards much of what De Luc wrote, the general idea of an Earth's history closely tied to the Biblical Genesis and an Earth born from the primordial waters echoes back in Greenough's writings.

Blumenbach diverted Greenough completely from studying law and inspired in him a passion for mineralogy and geology. But, apart from these favourite passions, Greenough also remained interested in gardening, architecture, archaeology, ethnology and politics. Consequently, he even sat in the House of Commons from 1807 to 1822 – for the 'rotten' borough of Gatton. In addition, he served for 16 years in the Light Horse Volunteers of London and Westminster until he resigned for humanitarian reasons after the disaster of 'Peterloo' in 1819 (Rose 2000, 2009). From 1799 to 1806 Greenough travelled widely (Rudwick 1962) through Germany, the British Isles and Italy, meeting people like John Playfair (1748–1819) and Robert Jameson (1774–1854), and always with the intention of learning more about minerals and rocks. He also attended a few lectures by Humphry Davy and the mineralogist William H. Wollaston (1766–1828).[2]

Young Greenough received his geological, mineralogical and chemical education during a time of changing attitudes towards geological expertise. He knew the old ways of formulating 'Theories of the Earth' – all too often philosophical armchair exercises – as well as the more practical endeavours of surveyors out in the field, or chemists and mineralogists in the laboratory. While it seems that he himself was very interested in the more general, global questions that were tackled by the old-fashioned Theories of the Earth, at the same time he basically rejected all the available answers these theories gave as being much too speculative. He was especially critical of those he called the 'Plutonists' or, less frequently, 'Huttonians' (Kölbl-Ebert 2003). An anonymous obituary in *The Literary Gazette and Journal of Science and Art* (1855) characterized Greenough as 'an habitual doubter of theories', whose 'mind was of an essentially practical tendency, and he was extremely reluctant to believe anything that was not capable of being proved'. It even stated that 'for a long time [Greenough] was considered a sort of 'drag' on the progress of geological science'.[3]

However, these statements have to be viewed in the light of acknowledging that for Greenough

[2] For more biographical details see Hamilton (1856), Boulger (1908), Eyles & Eyles (1955), Golden (1981), Wyatt (1995, 2004) and Kölbl-Ebert (2003).

[3] UCL: Greenough Papers, 24/2.

'theory' was something very different than for us, and that he was also very much concerned with the question of what the correct scientific method should be. Instead of 'theory' (in the eighteenth-century sense of 'Theory of the Earth'), Greenough preferred to address the questions about the geological origin of the Earth empirically, and his industry in collecting information struck Roderick Murchison as the centrepiece of Greenough's achievements: '... his sagacity in detecting, and industry in collecting, all the scattered information that bore upon the physical geography, not of England alone, but of the globe, was in itself truly admirable' (Murchison 1855). Nevertheless, behind all his obsession with collecting little pieces of information, his philosophical project remained the (ultimate) answer to the question about the origin of the Earth. While in his 1819 book, his personal notes and the annual addresses he gave to the Geological Society, he loved to abuse colleagues who committed themselves to theorizing (Kölbl-Ebert 2003), he nevertheless cherished his very own Theory.

Greenough's 'Theory'

In 1819 Greenough published his only book: *A Critical Examination of the First Principles of Geology*. It contained eight essays dealing with the various geological features that were the focus of contemporary geological reasoning. For a working geologist of 1819 it must have been extremely frustrating to start reading Greenough's book, for in the very first essay, 'On Stratification', he basically deconstructed all contemporary knowledge about 'stratas', a term under which Greenough included any kind of lamination or parallel structures in rocks of any size (i.e. sedimentary layers as well as all other rocks with recognizable parallel structures, from schistosity to joint sets to all sorts of larger structures). Inevitably, the book was not very well received among Greenough's colleagues. William Buckland, for example, complained: 'I cannot move hence till my lectures are over at which I think I have a larger Class than ever in spite of all that Mr. Greenough has done in his Book to discourage Geology. His Map however which is just published makes ample compensation for the peccadillos of his Book; it is extremely beautiful'.[4]

In order to better understand Greenough's method, as well as his personal geological ideas and convictions, we must follow his main argument through the book (Greenough 1819): 'Stratum is a word so familiar to our ears that it requires some degree of manliness to acknowledge ourselves ignorant of its meaning [...]' (p. 1) and 'By way of illustration, I ask, whether granite is stratified' (p. 2). For several pages Greenough presents examples pro and contra the latter opinion, citing numerous authorities, coming to no definite conclusion: 'Whence this contrariety of opinion? Are our senses at variance, or our judgments? The cause I think is obvious. Every one uses the word stratum, no one enquires its meaning; the remedy is as obvious – definition' (p. 9). And – astonishingly from our point of view – he then proceeds to define strata, instead of defining granite, albeit without reaching a practical conclusion (p. 90). His reluctance to define 'granite', however, and his insistence that basically each and every rock is 'stratified' to some extent has method, as we shall see later.

That (some) time is needed to deposit all these rock strata is clear for Greenough: 'That an interval of time, greater or smaller, elapsed between the several depositions seem ascertained' (p. 26). And he presents several arguments to strengthen his opinion: there are layers of rounded pebbles (which obviously need some time to produce before being deposited) or surface structures and bioturbation; also there are changes between marine and freshwater deposits, and vice versa. Basically all rocks – in Greenough's opinion – are deposited in water because they are stratified, although he admits at least to some physical difficulties in this model: 'It is difficult to conceive the existence of so large a quantity of water as would be necessary, under present circumstances, to hold the entire globe in solution, that difficulty, if not removed, is at least much diminished, by supposing the different parts to have been in solution at different periods' (p. 27). In former times, he explains, the strata have been more continuous and less disrupted than today: 'The resemblances hitherto adduced as occurring on the opposite side of vallies [sic] have been brought forward only to prove that those intervals which we now find between mountains, have not been from the beginning; that vallies owe their origin to the removal of matter which once occupied them; that there was a time, when, to use a memorable expression of Sir J. Hall, vallies were not only submarine but subterranean' (p. 113).

Greenough felt himself drawn towards the general Neptunist idea of a primordial ocean, the standard geological model of the time (Rudwick 2005, p. 172 ff.), from which all strata eventually emerged. Nevertheless, he was not impressed when he met Abraham Gottlob Werner (1749–1817)

[4] NMW: Henry De La Beche Papers 153. Letter by William Buckland to Lady Mary Cole, 18 May 1820.

(Guntau 2009) for the first and only time in 1816 (Torrens 1998). In his book Greenough highlighted Werner's lack of cosmopolitan experience:

> Virgil's peasant fondly imagined his native village a miniature of imperial Rome. With a corresponding love of generalization, Werner imagined Saxony and Bohemia a miniature of the world. He set down, more or less correctly, the order in which different rocks had arranged themselves in the districts with which he was acquainted, and concluded, that such must be the order of their arrangement in every other district. His theory was useful as a standard of reference, an incitement to inquiry, a clue to observation; but, unfortunately for Werner, his pupils viewed it in a different light; it was represented by them not as an hypothesis to be tried, but as a system to be followed; a system mature at its first conception, and perfect in all its parts and proportions. To merit of so high an order Werner was not entitled.
>
> (Greenough 1819, pp. 234–235)

Nevertheless, Greenough also assumed the continuity of strata in former times for basalt in a truly Neptunistic manner: 'Strange, that some writers who admit the original continuity of all other rocks, however discontinuous we now find them, should so lose sight of analogy, when they speak of basaltic rocks, as to imagine the occurrence of these in insulated hummocks, a remarkable phenomenon, out of the common course of nature, and to be explained only by the capricious interference of their favourite idol Vulcan or Pluto!' (pp. 100–101). After some more quoting of authorities and discussing their opinions, Greenough concludes: 'From all these considerations, I think we are justified in concluding that vallies have in general been formed by the action of running water; and consequently, that mountains in general are not the effect of volcanoes [...] nor of earthquakes [...]' (p. 120).

But things are not so simple. There are, for example, problems with erratic blocks; there are valleys and lakes[5] between their obvious source rock and their present resting point; and because of topography and the shear size of some of these blocks there is no way that they could have been transported by rivers. However, 'If the Transportation of bowlder-stones [sic] cannot be referred to the agency of Rivers, so neither can the Excavation of Valleys' (p. 138). According to Greenough, there are even more arguments as to why valleys cannot be caused by the present rivers that flow through them: there are, in fact, dry valleys and the river sources often do not sit at the uppermost end of a valley, etc. Consequently, 'It is obvious then, that agents, so circumscribed in their operation as Seas and Rivers, are little calculated either to effect the transportation of the bowlder-stone so often mentioned, or to produce those inequalities of mountain and valley, which the surface of our earth presents' (p. 147). And the 'Plutonists' may say what they want, even if you grant them an eternity of time (Greenough 1819, p. 148), seas and rivers are just too feeble to cause that much erosion. Therefore:

> If Seas and Rivers are, from their feebleness, inadequate to produce the effects which have been produced by the action of water, the only remaining cause, to which these effects can be ascribed, is a Debacle or Deluge.
>
> (Greenough 1819, p. 148)

Having thus established the need for a catastrophic event to produce the ruined state of our present strata, Greenough asks himself whether this Deluge was partial or universal? He comes to the conclusion that 'The universal occurrence of mountains and valleys, and the symmetry which pervades their several branches and inosculations, are further proofs, not only that *a* Deluge has swept over every part of the globe, but probably *the same* Deluge' (p. 155). Satisfied with the argument, he proceeds to the question of timing the event: 'To determine the Age of the world has long been a favourite object with philosophers' (p. 169). Quoting De Luc, Dolomieu and Cuvier, Greenough (p. 170) attempts to give some figures:

> After a patient investigation of the phoenomena [sic] of bays, promontories, deltas, dunes, taluses, seas, lakes, and rivers, they are agreed in thinking that the period of time, which has elapsed since the retreat of the diluvian waters, cannot exceed from five to six thousand years. So much for the positive age of our planet; – let us now consider its age in relation to different events connected with its own history, the history of the solar system, and the history of mankind.

Greenough argues: 'That the order of things, as it existed before the deluge, cannot have differed widely from the present order, will appear from many considerations' (pp. 170–171). As the Earth was already habitable before the Deluge swept the continents with Erratics, we may conclude that '[...] the diluvian catastrophe did not take place till after the establishment of the solar system' (p. 176). In other words, it cannot have been the Deluge of water covering the whole Earth mentioned in Genesis 1, verse 2 (*The Bible: Authorized King James Version*) 'and the spirit of God moved upon the face of the waters', as the rest of the Solar system with sun, moon and the planets (together with the stars) were created only on the fourth day! (Genesis 1, 14–19).

[5] 'The blocks of granite on the Jura attest the non-existence of the Lake of Geneva, at the time of their transportation' (Greenough 1819, p. 177).

As far as I am aware, Greenough (unlike De Luc or Blumenbach) never mentioned his religious convictions in his scientific writings, but here, as we can see, they were obviously at the back of his mind and thus crept into his argument. When he talked about a Geological Deluge, he obviously thought of the second Deluge mentioned in the Bible, i.e. the Noachian Flood (Genesis 7), which at that time would still be within the mainstream of science (cf. Rudwick 2009). On the other hand, Greenough was cautious to equate his Geological Deluge simply with the Biblical Flood, being aware of lacking positive evidence: 'We have no positive evidence to determine whether the deluge took place before or after the Creation of Man: we have only this negative evidence, that neither any part of a human skeleton nor any implements of art have been hitherto discovered, either in regular strata, or in diluvian attritus' (p. 186). Reflecting about the possible causes of the Geological Deluge, Greenough restricts himself completely to natural causes, although he finds the problem to be difficult: 'if we would investigate the means by which this tremendous catastrophe was produced, the mind is easily bewildered in unprofitable conjecture' (p. 189):

> but let a cause be assumed; let us grant that the water was so obtained; how was it afterwards removed? what is become of it now?
>
> Shall we then, fearless of paradox, attribute to the waves constancy, mobility to the land? Shall we say that continents have been submerged, not from the rising of waters, but from their own descent? Extravagant as such a hypothesis may appear, it falls short, very short of that which the Huttonians have long admitted and maintained.[6] [...]
>
> Alas! this expedient, so far from obviating our difficulties, tends only to enhance them. If there were no caverns beneath our continents, how could they sink? [...] The continents having sunk, how have they risen again to their present level? [...]
>
> To the solution of the problem Impetuosity of Motion in the water is indispensable; but an increased Quantity of water is, perhaps, superfluous [...].
>
> (pp. 191–192)

And, indeed, just increasing the amount of water would not be enough for Greenough, as he needed currents violent enough to excavate the valleys and transport the erratic boulders:

> We have seen also, that this order of things, so closely resembling the present order, was suddenly interrupted by a general flood, which swept the quadrupeds from the continents, tore up the solid strata, and reduced the surface to a state of ruin: but this disorder was of short duration; the mutilated earth did not cease to be a planet; animals and plants, similar to those which had perished, once more adorned its surface [...]
>
> (p. 194)

Greenough then asks himself: 'Would a comet fulfil them?' (p. 196). He quickly warms to the idea:

> if we suppose one of suitable dimensions to move in such direction as would allow it only to graze the earth, it is not impossible that the shock of this body [...] might produce the degree and kind of derangement which we are attempting to account for [...].
>
> (p. 197)

> I shall conclude by remarking, that if the hypothesis of a shock derived from the passage either of a Comet or of one of those numerous, important, and long neglected bodies, often of great magnitude and velocity, which occasion meteors, and shower down stones upon the earth, would explain the phenomena of the deluge (a point upon which I forbear to give any opinion,) we need not be deterred from embracing that hypothesis under an apprehension that there is in it anything extravagant or absurd. In the limited period of a few centuries, there is little probability of the interference of two bodies so small in comparison with the immensity of space; but the number of these bodies is extremely great, and it is therefore by no means improbably, says La Place, that such interference should take place in a vast number of years.
>
> (pp. 198–199)

Ironically, this idea, albeit not of a grazing comet but a celestial body actually impacting and causing mega-tsunamis, all sorts of damage not only in the vicinity of the crater and leading to mass extinctions, didn't come into its own before the latter part of the twentieth century (Alvarez *et al.* 1980, Alvarez 1997).

But what about the condition of the Earth prior to the Geological Deluge? 'While its construction was yet going on, is it probable that the surface of our planet exhibited one uninterrupted plane, or that it was diversified then as it is now by protuberances and depressions?' (pp. 200–201). The answer is a definite 'yes' to the latter question, at least for the secondary rocks: 'That the earth was divided into land and water, at a period antecedent to the deluge, is evident, from the remains of land and sea productions so abundantly diffused throughout the secondary rocks; but the situation before this event, appears in many instances to have differed materially from that which has been since assigned to them' (pp. 176–177). 'Unless we suppose that marine moluscae inhabited the land, or timber grew upon the ocean, we must admit that there existed, at the time when these were deposited, land and water; in other words, mountain and valley. Inequalities of surface prevailed then, even

[6] Because the Huttonians, as Greenough always polemicized against them, allowed for any number of cycles of elevation and subsidence, a notion which Greenough utterly despised (Kölbl-Ebert 1999).

during the formation of strata; to what causes are these inequalities to be referred?' (p. 203).

Greenough discussed several possible causes such as crystallization, partial deposition, subsidence, volcanoes and earthquakes; however, volcanoes (and earthquakes associated with them) are but local phenomena: 'It is probable that the effect produced upon the earth by these, was confined in the earliest ages as now, to forming occasional hills by the accumulation of ashes and scoria, or occasional valleys by the falling in of unsupported craters. But the action of running water appears to have been in all ages the principal cause of inequality of surface' (p. 210). And, therefore, Greenough endeavoured to look for evidence of another catastrophe, another Deluge:

> [...] I cannot but consider the almost universal occurrence of conglomerate and greywacke on the confines of what are called primitive rocks, as one of the most important and striking facts yet established in Geology: it seems to prove, that, at the epoch at which these beds were formed, a deluge took place, similar in kind, though perhaps not equal in extent, to that which determined the present outline of the earth. From that period till the work of creation was complete, the mechanical action of running water appears to have been more gentle and partial, but unremitted.
> (pp. 212–213)

To summarize, for Greenough the Earth's history presents itself like this:

- creation of first world: deposition of primitive[7] strata in a submarine environment;
- global deluge, producing surface inequalities: sea and land;
- second world: deposition of secondary and later rocks;
- global deluge: excavation of valleys and deposition of erratics and other 'Diluvium';
- present world, a ruin of the former worlds with only gentle geological action;
- work of creation completed.

But is it completed? It seems that Greenough assumed (in Neptunist fashion) a slow but steady dropping of the sea level from the deposition of the very first rocks, throughout all the geological ages and ongoing at present; and in his notebooks he reflected the possibility of ascertaining this phenomenon:

> Geological Base line.
> is the successive lowering of newer & newer formations asserted by Werner false, or if not to what extent?
> will this enable us to form a baseline.
> Naturalists in many cases can distinguish the depths which different species inhabit. to that extent we can determine what must have been the depth of the water in which they lived when overwhelmed by a new deposit.
> this consideration also might give us a baseline.[8]

Greenough was ready to spot any evidence for this idea in contemporary geological observations. As late as 1834 he deemed it more probable that the sea level of the whole of the Pacific Ocean had dropped as a result of the Valparaiso earthquake in Chile in November 1822, rather than admitting the possibility that the coastal area might have been raised instead.[9]

In his notebook, Greenough writes about the earthquake and the 'cause of the great sea wave', that is, the associated tsunami:

> the sea wave begins with a retirement [...]
> Whenever a communic[ation] is open betw[een] the sea & subterraneous caverns the water will rush into these – hence the backwave which seems to be the first phenom[enon]
> by the heat below, the water so entering, returns in the form of steam & the coast before dry is now immersed some hundred fath[oms] perhaps in water ultimately this water forms new combinations beneath & protanto a permanent diminution of water takes place all over the globe the inrush of the sea traversing caverns produces the rumbling.[10]

Basalt and granite

As a direct consequence of Greenough's bias towards a more 'Neptunistic' Theory of the Earth, he was highly critical concerning all 'Plutonist' interpretations, and thus he could not bring

[7] 'Lehmann appears to have been the first who attempted a chronological arrangement of rocks; he divided them into two great classes – the Primitive and the Secondary.
To these, Werner added a third class, as partaking of the character of both: Was this an improvement? No: unless it would be an improvement to increase the list of primitive colours by the addition of mixed tints, or the list of notes in music by telling in the flats and the sharps. The object of classification is not to enfeeble distinctions, but to strengthen them. Hard lines are indispensable in science, the business of which is not to imitate nature, but make it understood' (Greenough 1819, pp. 232–233). See Guntau 2009 for more on Lehmann's chronology.
[8] UCL: Greenough Papers, 16/4, p. 171 [undated].
[9] GSL: Annual address to the Geological Society of Greenough [21 February 1834], OMI/6. See also [Callcott] (1835); and Kölbl-Ebert (1999, 2002, 2003).
[10] UCL: Greenough Papers, 16/4.

himself to accept any notion that basaltic rocks might be igneous:

> Origin of basalt
> In Geol. Tr. V 3. p. 208 Mr. Wm. Conybeare adduces different reasons to support the igneous origin of basaltic rocks: the reasons are
>
> 1. the identity of chemical composition in basalt & lava.
>
> > to which I answer 1. that both these substances are too heterogeneous & variable to have any fixed chemical composition.
> >
> > 2. that supposing the premises right, they do not justify his conclusion because there is no reason why all water should have been boiled, because that which has been boiled agrees in chemical composition with that which has not; in the same manner trap rocks may assume the same characteristics whether they have been subject to volcanic action or not. Not even of lava is the origin volcanic. Lava is only a modification of rocks previously existing.
>
> 2. the constant occurrence of trap rocks in volcanic districts.
>
> > but granit [sic] is found in these districts almost as frequently as trap.
> > Trap is a most indefinite substance – it means any mixture of those ingredients which are the most common, lime silex aluminium & iron
> > Trap rocks may be the cause of volcanic action as well as the effect of it.
>
> 3. the confession of the Wernerians themselves that the basalt of Auvergne is volcanic.
>
> > Where is this confession recorded? is it the confession of all the Wernerians? if so, is it a confession of what they have witnessed or only of what they have imagined? are there not even Plutonists who doubt that the basalt of Auvergne is volcanic?
>
> 4. the testimony of those best acquainted with districts still exhibiting active volcanoes.
>
> > this is at best very suspicious testimony. but Dolomieu considered basalt to be of doubtful & double origin.[11]

Greenough did not accept a general theory of generating basaltic rocks by means of volcanic action because, in his Neptunistic theory, volcanoes featured only marginally (however, volcanoes were of much concern to the Plutonists). To stress this marginal character of volcanoes and volcanic rocks he very much insisted upon splitting up the nomenclature of rock types (basalt, trap, whinstone and other rock types), which usually were discussed in this context because they resembled true lava; a rock that had been observed to definitely come out of a volcano. Greenough also insisted that basalt, trap, whinstone and others were stratified, and therefore sedimentary, or at least chemical precipitates from water. On the other hand, he tended to lump together various other rocks of similar chemistry, like granites, gneisses and mica schist, stating that: 'In mineralogy it is a convenience to have distinct names for granite & gneiss to mark the difference in their structure [. . .] but the geologist should recollect that the constituents of all these substances are alike and by him therefore they should be considered not in the light of distinct formations but simply of varieties'.[12] This helped him to claim that 'granite' was 'stratified' too, and therefore precipitated from water and not igneous in nature: 'The Huttonians distinguish carefully between stratified rocks and unstratified. The former, according to them, have been deposited by water, and merely hardened by Plutonic heat; whereas the latter have been thrown up in a melted state from beneath, and forcibly injected amid the pre-existing strata. It is a fatal objection to this hypothesis, that consolidated beds often alternate with unconsolidated, stratified with unstratified' (Greenough 1819, p. 266).

In another notebook we find the recipe:

> To form granitic rocks
> A saline solution fully saturated when hot, if cooled rapidly & agitated while cooling will beco[me] confusedly[?] crystalline. In this way nitrate of potass[ium]. is prepared in Sweden: disturbing the solution when crystallizing renders it granular.
> [. . .] – Granic[?granitic] Xls [crystals] may be so developed.
> to form Gneiss. Mica slate
> Gneiss shews diminished agitation – Mica slate still less or more time & the process of Xallizn. interrupted by an imperfect solution – insuffic. heat – or so rapid cooling that the molecules had not time to arrange themselves symmetrically.[13]

The Plutonist concept that granite might rise in a molten state through the Earth was equally quite incomprehensible to Greenough:

> Granite in a state of fusion
> its presumed Ascent
> We live after the Pleistocene period.
> on casting our eyes over a geol. map of Europe we finds extensive tracts of Granic sill open to the day – how happens this?
> what is the date of this Granic – nescio.

[11] GSL: Greenough's notebook, LDGSL 971, p. 138.
[12] UCL: Greenough Papers 16/4, p. 185.
[13] GSL: Greenough's notebook, LDGSL 971, p. 159.

> let us assume an unknown date – still we have no evidence that any granit is of later date than the ~~moun~~ Chalk. see Gryffeths' paper.
> unless we consider Trachit & Granit one & the same
> but be its date what it may – what happened in its early years? was the space it covers protected & fenced in? or what induced not only the sedimentary rocks from forming over it like a counterpane, and the Plutonic rocks from making their way tho' it, as their easiest route?
> or lastly what is become of the [illegible] strata which by the ascent of the Granit were uplifted or thrown right & left to accommodate the newcomer? where is the rock which has been displaced or annihilated?
> if viscous why did the granit not overflow & gutter down into the vallies forming imperfect columns.[14]

This, again, reminds us that Greenough had no confidence in ordinary erosional processes. Granite for him was a rock formed at, or at least very close to, the surface. Imagining fluid granite made no sense because it obviously did not behave like volcanic lava.

Thus, most of the Plutonist world-view was speculation, fruitless theorizing for Greenough, while he was convinced that his own theoretical concept was resting on good, solid empirical evidence; and he endeavoured to collect more evidence to strengthen his own model. Collecting such evidence was all that geology was about and it could be done best in a communal effort, such as the young Geological Society. In order to facilitate Greenough's needs, the geological data to be collected had to be devoid of any theoretical concept but purely descriptive, like circumstantial evidence, which would be collected by impartial investigators in a criminal case and only at the end brought before the judge. He evidently did not ask himself what evidence would falsify his Earth model and then go and look for it, in order to test his Theory of the Earth.

The Geological Society – a school for geology and a research institute

In 1814 Greenough defined geology like this: 'Rational Geology then is the natural History of Mineral beds, of their usual or uncommon characters & appearances, of the substances organic or inorganic which they contain, of their geographical extent, of their continuity redundance, failure, or dislocation & of the order of succession which they severally bear with respect to one another'.[15]

As we can see there is no mention of geological processes and their causes. It is all about pure description. A year later, he emphasized that:

> Geology deserves to be promoted & encouraged either as leading to various improvements in economical industry, or as a curious branch of scientific research. Its general conclusions can only be made by the laborious & expensive[16] comparison of numerous specimens & appearances: for this the artist, the natural historian & the chemist must unite their efforts & these to be constantly supplied must be supported by public establishments Institutions having [...] this object have in some countries been raised in splendour by the powerful protection of the state; but in others the same end may be more slowly attained by the combined & persevering exertions of enlightened & zealous individuals.[17]

And that is obviously, where the newly founded Geological Society came into play. Its purpose was 'of making geologists acquainted with each other, of stimulating their zeal' and of 'inducing [geologists] to adopt one nomenclature, of facilitating the communication of new facts, and of ascertaining what is known in their science and what yet remains to be discovered' (Rudwick 1963, p. 329).

It seems that Greenough at that time believed that geological investigation would reach a final goal, achievable in a foreseeable future, although: 'The early completion of this scheme will depend on the attention with which the members examine the nature & extent of the beds from which they detach their specimens & the relation of those to other neighbouring beds the most distinctly characterized & understood'.[18]

The conviction to work for a goal that might be reached within a limited number of years is quite understandable if we consider that Greenough had his private Theory in mind, which only needed some support from purely descriptive, lithostratigraphic evidence, basically sorting out the laws of deposition in a primordial ocean. One can also imagine that Greenough would have found it interesting from the distribution of erratics or the orientation of valleys to work out the direction of the turbulent currents that swept the continents during the deluge; or that Greenough would have been delighted to get an opportunity to measure sea-level changes and to observe sediment deposition in the seas. But he would have seen no major need to

[14] UCL: Greenough Papers 16/4, p. 183.

[15] UCL: Greenough Papers 5/2, 19; 4 October 1814.

[16] Possibly a spelling error: Greenough certainly meant to write 'extensive'. Perhaps his mind was preoccupied with the financial situation of the Society's collection of specimens (see below).

[17] UCL: Greenough Papers 5/2, 21c; 3 February 1815.

[18] UCL: Greenough Papers 5/2, 20; 4 October 1814.

observe present causes on the continent as he would expect them to be rather different from former times, before 'the work of creation was complete'. His attitude resembled more that of an archaeologist, for whom the experience of the present era is not much help in understanding, for example, extinct illiterate cultures.

Greenough did all in his power to facilitate this endeavour. He and his colleagues quickly succeeded in giving the new Society a national, and later an international, reach – enlisting many so-called Honorary Members out in the countryside to provide the detailed geological information that would be collated in London. In 1808 a brochure entitled *Geological Inquiries* (see Appendix I) was printed with a series of questions relating to the most essential points in geology. It ensured some sort of quality control, but also encouraged the joint effort for a generalized purpose. Planning all of these things and organizing the collection of facts, then later compiling them into the map projects of the Geological Society, must have taken a vast amount of time, therefore much of Greenough's work for the young Society is possibly hidden behind a political and organizational framework.

Some of this effort showed up in the establishment and ordering of what was called the Society's museum.[19] In 1819 Greenough considered[20] 'the Society to be a school for Geology' and the fellows were divided 'into 2 classes, Beginners and Proficients', and therefore the collection was to be also divided into two parts, 'the one elementary tending to the diffusion of Geological Knowledge – the other experimental tending to its advancement'. The elementary part was to be a teaching collection for the self-education of the new disciples, while:

> In the other department of the Museum namely the Experimental, the object should be to accumulate new facts & to test disputed theories When once these facts were generally known & had been recorded the object of such specific collections would have then been attained & therefore it would be unnecessary that they should be permanent. They might then be broken & transferred [...] to other establishments either as presents or in the way of exchange for example.
>
> A collection might be made to ascertain whether the fossils of the older beds belonged to a lower class of animals than those of the newer – Whether in distant districts the same fossils occurred in the same strata &c &c.[21]

The society's museum had temporary exhibitions of specimens, which were later to be given away, partly because of the practical reason that the available space was just not enough to accommodate a large collection. Possibly in 1819, Greenough complained of problems with the foreign collection that was 'packed up for want of cabinets' in about 80 casks in a damp room: 'Each was labelled when first put away but a large proportion of the labels have fallen prey to the rats or to mould'. Specimens containing rock salt or pyrite had been destroyed, and Greenough was of the opinion that this would give a bad impression to the donors.[22] Eventually, the situation became unbearable and the museum had to be enlarged considerably, which proved to be rather expensive, but in 1822 it enabled several thousand specimens and the geographically arranged foreign collection to become accessible '[...] some of them have been presented as illustrative of memoirs published in the Transactions, & others have been collected to elucidate the observations of eminent geologists in different countries'.[23]

Specimens were only interesting in relation to specific questions:

> The name of the substance & its habitat being once recorded there is no great object in preserving the specimen – All such specimens however should be exposed to view for a given time immediately after their presentation. And the attention of the members should be invited to them by conspicuous headings. And the name and habitat of each of these specimens together with the description of it if necessary should be, inserted in the catalogue. After this the specimens might be removed to make way for others. If this plan were adopted members would find new objects of interest as often as they visitted [sic] the museum – The number of cabinets required would be lessened – we should connect ourselves more closely with other similar establishments.
>
> Every contribution presented would be exhibited at the period when it was most likely to excite attention and we should make sensible advances to our main object a knowledge of the geological structure of the globe [...].[24]

Greenough took great care that the arrangement of the specimens was neutral with respect to any preconceived theory: 'Organic remains arranged

[19] See also Herries Davies (2007, p. 31 ff.).

[20] UCL: Greenough Papers 5/2.

[21] UCL: Greenough Papers 5/2, 26 [Dec. 1819].

[22] UCL: Greenough Papers 5/2, 25.

[23] UCL: Greenough Papers 5/2, 33 [1 February1822].

[24] UCL: Greenough Papers 5/2, 29 [1819].

according to their forms without reference to the bed in which they are found ... Collection of specimens illustrating the nature of the rocks & strata arranged according to their ~~ordinary~~ natural position [later addition: '? as far as such positions can be ascertained']'.[25]

This was again the Society's concern in autumn 1814: 'These are the considerations which have guided[?] the council [illegible] arrangement of the Cabinet in a principle independent of any hypothetical view respecting the manner in which mineral beds were formed or their supposed comparative antiquity presenting in combination these facts upon which every Theory must be founded which shall pretend to explain them'.[26] Nevertheless, Greenough did harbour his preference for a Neptunistic world-view and in 1824 the best arrangement for the petrographical collection was still deemed to be that according to the Wernerian system: 'Rocks or aggregates of minerals – This collection consists at present of the rocks of the Wernerian School, arranged & described at Freyberg, & as its importance is felt every opportunity of increasing it will be taken advantage of'.[27]

Fossils and stratigraphy

Stratigraphy also was not safe from Greenough's urge to regulate and teach his fellow geologists, as he felt the necessity to purge it of every 'theoretical' (he always said 'theoretical' but mostly he meant 'Huttonian') notion: 'By the term Formation is meant a series of similar or dissimilar rocks, supposed to have formed in the same manner and at the same period. The idea is therefore purely theoretical' (Greenough 1819, p. 214). On the other hand, it was not easy to reconcile the actual stratigraphic evidence with the prerequisites of a Neptunistic world-view, and Greenough, of course, was aware of this fact: 'Enough had been said to make it evident, that neither any single stratum, nor single rock, nor any imaginable series of rocks can be traced in a continuous line round the globe. Similar strata, similar rocks, similar series of rocks are, however, found in different countries and in different hemispheres' (Greenough 1819, p. 222).

And he very much objected to generalizations made by Werner, which were proclaimed without proper knowledge of the field evidence: 'In the scanty[28] catalogue of rocks with which the Wernerians have furnished us, we find some, as granite, which are common to all climates; some, as primitive gypsum and serpentine, which are confined to a few spots; some, as topaz rock and whitestone, which are peculiar or nearly so, to the neighbourhood of Freyberg. Yet we are told that the primitive, transition, and flötz rocks are almost all Universal Formations' (p. 227).

Greenough freely admitted that things were not simple; nevertheless, he was very concerned with lithostratigraphy and searched for generalizations within the rock record. Again, this preoccupation with lithostratigraphy can only be understood when we bear in mind that he expected all strata to have been precipitated, either chemically or as mechanical sediments, from the primordial ocean. In his 1819 book he therefore invests many pages in discussing whether there are any general laws to be deduced from the evidence: 'Let it be supposed, that certain rocks are known to occur in a certain district; will analogy enable us to predict the order of their occurrence? Do the rocks, of different countries, which resemble each other in external character, resemble each other also in relative position?' (p. 231).

And, indeed, there are certain rules: 'But though every rock alternates with some others it does not alternate with all. Flint alternates with chalk, clay with oolite, red marl with gypsum; but no one, I presume, has seen granite alternating with salt, or serpentine with lias' (p. 232). However, beyond these simple rules, he recognizes that rocks may occur everywhere in geological time and space (p. 259). Nevertheless, there are certain guidelines and trends, albeit not without exceptions: the further up you go, the less the rocks will be crystallized (p. 261), the lower you go, the higher the specific gravity of the rocks will be (p. 264), at the bottom of the stratigraphic column the rocks generally will be more consolidated than above (p. 269).

There is one feature, however, that Greenough would not accept as a stratigraphic criterion: 'Vertical, and highly inclined rocks are reputed older than those which are horizontal' (p. 269). He objects to this notion, because: 'I have shown, in a former essay, that every species of rock assumes, occasionally, every posture; hence, the inclination of rocks forms no evidence as to their antiquity' (p. 270).

Again, from our modern point of view, this is a strange thing to say, especially since Greenough first admitted that the various rocks may occur throughout geological time (Greenough 1819, p. 259), and so we would assume, for example,

[25] UCL: Greenough Papers 5/2, 17c [28 May 1813].

[26] UCL: Greenough Papers 5/2, 19 [4 October 1814].

[27] UCL: Greenough Papers 5/2, 36 [6 February 1824].

[28] Greenough's footnote: 'So scanty that it does not contain even the fresh water beds of Saxony'.

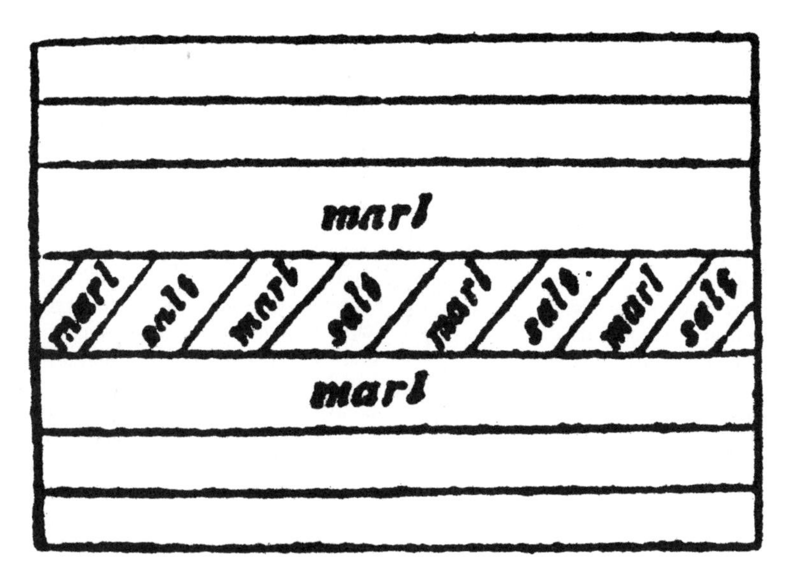

Fig. 2. 'The rock salt of Bocknia is said to lie as represented in the following diagram' (Greenough 1819, p. 273).

that Greenough would accept older, inclined limestone below, and somewhere above this limestone a discordant younger horizontal limestone. However, for Greenough inclined strata and discordances were a core feature of a dynamic Earth in the Huttonian Theory that Greenough so much despised, and therefore he needed to explain them away. He did so by presenting the drawing of an inclined marl–salt sequence, which is supposedly bracketed by horizontal layers (Fig. 2). Greenough also related the story of features that are obviously sills which suddenly leave their concordant, horizontal mode to become dykes. He also observed similar features for quartz veins in 'primitive slate' (p. 273)

And thus he concludes: 'In some situations unconformity of posture, far from proving that the strata, in which it is observed, were formed at different areas, tends rather to prove that these strata were formed simultaneously' (p. 274). So much for the value of lithostratigraphy. But how about fossils and their stratigraphic relevance? 'An opinion has for some time past been entertained in this country, that every rock has its own fossils' (p. 284). But Greenough cannot subscribe to this. For him fossils are of no great stratigraphical value:

> Mr Smith is well known to have embraced this idea at an early period, and, in a table attached to his geological Map of England, has specified a variety of fossils, by which the strata of England may, in his opinion, be identified.
>
> That the fossils contained in secondary strata, are, of all empyrical [sic] and accidental characters, the most useful, in enabling us to follow the direction of these strata, no one can dispute; but their utility has been greatly over-rated. Those who maintain that formations are universal, and produce everywhere the same fossils, must maintain, that these fossils are also universal; in other words, that every part of the world has been peopled, at the same period, by the same animal, which, from the nature of many of those animals, is absurd: for if, in the pre-existing world, as in the present, different animals, inhabited different countries, then are fossils not universal; but each class confined to an area of greater or less extent, and consequently, within and without that area, the fossils of the same formation will not be the same'
>
> (pp. 287–288)

Greenough here is very sceptical of the idea of guide fossils. He grants their usefulness for local geological problems, but as his personal interest lies not in practical questions like, for example, where to find coal, but in the global perspective of a Theory of the Earth, he is convinced that fossils cannot be of much use as they are not available universally. Also, their stratigraphical resolution is not good enough in Greenough's opinion; ammonites, for example, are present throughout the secondary strata.

Much of the well-known conflict between Greenough and William Smith (1769–1839) (Knell 2000; Torrens 2001, 2003; Herries Davies 2007, p. 44 ff.) about their respective geological maps (Smith 1815; Greenough 1820) was probably due to Greenough's preoccupation with a geological Theory for the whole of the globe. Smith's practical concerns as a surveyor went beyond Greenough's horizon of interests. Greenough – at least at that time – neither understood what Smith's map was all about, nor was he interested in it beyond the mere facts that it contained and which could be used in the general scheme of the Geological Society, as Greenough saw it, of producing evidence for a general Theory of the Earth. Greenough simply used Smith's data as he used all the other information provided to him by the Members and Honorary Members of the Society (after all, this was the purpose of the Society). Greenough – with the typical naïveté of the very rich – was probably not even aware of the dire damage he did to Smith's financial situation in producing the Society's map until it was too late to do anything about it.

Greenough's disregard for fossils is obviously not only due to a simple disregard of William Smith's work, as he had already formed this attitude from other considerations regarding the Wernerian Theory:

> It is said in the Wernerian theory, that there is a relation between the nature of fossils, and the age of the rock which contains them [...]
>
> Whether Werner derived this idea from the study of nature, or from the perusal of Telliamed [Maillet 1720], who imagined, that birds and beasts sprung originally from the sea, and that men and women themselves were only an improved breed of fishes, I know not; but there seems as little reason to believe, that a progressive amelioration took place in the nature of the beings called into existence previously to the deluge, as that a corresponding degeneracy can be attributed, with any degree of justice, to those of succeeding ages.
>
> (Greenough 1819, pp. 281–282).

Here Greenough also objects to the notion that species might change or evolve somehow through time.

At least Greenough accepted that fossils are, indeed, remnants of former life forms: 'That fossils are what they appear, and not nature's counterfeits, is a doctrine on which there is no longer any difference of opinion: it is remarkable, however, that amid the vast number of fossils which have been discovered, there is scarcely one exactly similar to any existing plant or animal' (p. 296). And this posed another problem for him. In Bolca, for example, there are fish that at least resemble those we know today, but which do not live in European seas but around several other continents. So what to conclude from this? Did the fish migrate from these different places to meet in Bolca, or did the seas or the continents move? All this appeared unlikely to Greenough and therefore he does not dare to use fossils for geographical interpretations (p. 297 ff.). It is even impossible, Greenough claimed, to say for sure whether a mollusc shell was formed in sea water or freshwater. Elephants today roam the hotter regions of the land, but in Siberia elephants with a warm fur coat (i.e. mammoths) have been found. From all this Greenough concluded that fossils are also of no use for climatic interpretations.

Conclusions

George Bellas Greenough, first President of the Geological Society of London, was an extremely important figure during the formative years of the Society. He himself had a comparatively conservative, but still mainstream, Neptunistic view of the Earth's history, which he hoped to support with abundant empirical evidence. For this purpose he used the facilities of the Geological Society, and the network of Honorary Members who delivered geological data to London – in other words, to Greenough himself. These data he endeavoured to collect and arrange impartially and devoid of theoretical notions. Nevertheless, his strong bias against the 'Plutonists' or 'Huttonians' permeated his writings, notes and speeches to the Society. This bias – strengthened by his exceeding fondness for generalizations, for clear and systematic definitions, and for global views – sometimes became a drag on scientific 'progress' because his preoccupation with his own Theory led him to dismiss important developments, such as Smith's biostratigraphy, when they did not fall within his horizon of interest. Thus, Greenough was slow to recognize their importance and value.

His personal goal of finding evidence to support his own model or Theory was not achieved because, in the course of the ensuing decades after the founding of the Society, those whom Greenough called the Huttonians carried the day with their vision of a dynamic Earth, despite Greenough's best efforts. But it was he who put the young Society on its way and he who gave it the administrative framework in which it could function as a teaching and research facility, making the new science of geology popular. That the Society, at least from the 1830s onwards, rapidly evolved by deviating more and more from Greenough's ideas to follow other interests and views of the Earth only proves the effectiveness of the structures that the efficient politician and organizer Greenough put in place and left to the Society as his legacy.

I wish to thank reviewers Cherry Lewis and Patrick Boylan for their helpful comments and also for correcting my English. Many thanks also to the staff of the archive at University College London, the National Museum of Wales and the Geological Society of London for their friendly and competent help. The research that led to this paper has been financed by the Deutsche Forschungsgemeinschaft (DFG).

Archives

UCL: Greenough Papers at University College London Library.

NMW: De la Beche Papers at The National Museum of Wales.

GSL: Archive of the Geological Society of London.

References

ANON. 1855. G.B. Greenough, F.R.S. [obituary]. *The Literary Gazette, and Journal of Science and Art*, **1995**, April 14, 234–235.

ALVAREZ, L. W., ALVAREZ, W., ASARO, F. & MICHEL, H. V. 1980. Extraterrestrial cause for the Cretaceous-Tertiary extinction. *Science*, **208**, 1095–1108.

ALVAREZ, W. 1997. *T. rex and the Crater of Doom*. Princeton University Press, Princeton, NJ.

BLUMENBACH, J. F. (ed.). 1796. Siebenter und letzter geologischer Brief des Herrn de Luc an Herrn Professor Blumenbach. Aus der französischen Handschrift. Bemerkungen über den Ursprung der organisirten Geschöpfe. *Magazin für das neueste aus der Physik und Naturgeschichte*, **11**(1).

BOULGER, G. S. 1908. Greenough, George Bellas. *In*: STEPHEN, L. & LEE, S. (eds) *Dictionary of National Biography*, Volume VIII. Smith, Elder & Co., London, 524–525.

[CALLCOTT, M.]. 1835. On the Reality of the Rise of the Coast of Chile, in 1822, as stated by Mrs. Graham. *American Journal of Science and Arts*, **28**, 239–247.

EYLES, V. A. & EYLES, J. M. 1955. Two geological centenaries, G. B. Greenough, F. R.S. (1778–1855), and Sir Henry De la Beche, F. R.S. (1796–1855). *Nature*, **175**, 658–660.

GOLDEN, J. 1981. *A List of the Papers and Correspondence of George Bellas Greenough (1778–1855) Held in The Manuscripts Room University College London Library*. University College London.

GREENOUGH, G. B. 1819. *A Critical Examination of the First Principles of Geology in a Series of Essays*. Longman, London.

GREENOUGH, G. B. 1820. *A Geological Map of England and Wales*. Longman, Hurst, Rees, Orme & Brown, London (scale 1 inch: 10 miles).

GUNTAU, M. 2009. The rise of geology as a science in Germany around 1800. *In*: LEWIS, C. L. E. & KNELL, S. J. (eds) *The Making of the Geological Society of London*. Geological Society, London, Special Publications, **317**, 163–177.

HAMILTON, W. J. 1856. [Obituary] George Bellas Greenough. *In*: Anniversary Address of the President. *Quarterly Journal of the Geological Society of London*, **12**, xxvi–xxxiv.

HERRIES DAVIES, G. L. 2007. *Whatever is Under the Earth. The Geological Society of London 1807 to 2007*. Geological Society, London.

KNELL, S. J. 2000. *The Culture of English Geology, 1815–1851: A Science Revealed Through its Collecting*. Ashgate, Aldershot.

KÖLBL-EBERT, M. 1999. Observing orogeny – Maria Graham's account of the earthquake in Chile in 1822. *Episodes*, **22**, 1–5.

KÖLBL-EBERT, M. 2002. Augenzeugin der Gebirgsbildung – Das chilenische Erdbeben von 1822 als wissenschaftshistorischer Kriminalfall. *Geohistorische Blätter*, **5**(1), 73–97.

KÖLBL-EBERT, M. 2003. George Bellas Greenough (1778–1855): A lawyer in geologist's clothes. *Proceedings of the Geologists' Association*, **114**, 247–254.

LEWIS, C. L. E. 2009. Doctoring geology: the medical origins of the Geological Society. *In*: LEWIS, C. L. E. & KNELL, S. J. (eds) *The Making of the Geological Society of London*. Geological Society, London, Special Publications, **317**, 49–92.

MAILLET, B. DE. 1720. *Telliamed ou Entretiens d'un philosophe indien avec un missionnaire français sur la diminution de la mer, la formation de la terre, l'origine de l'homme*. Amsterdam.

MURCHISON, R. I. 1855. Greenough and De la Beche [obituaries]. *The Literary Gazette, and Journal of Science and Art*, **1996**, 251.

ROSE, P. F. 2000. The military service of G. B. Greenough, founder president of the Geological Society. *In*: ROSE, P. F. & NATHANAIL, C. P. (eds) *Geology and Warfare: Examples of the Influence of Terrain and Geologists on Military Operations*. Geological Society, London, 63–83.

ROSE, E. P. F. 2009. Military men: Napoleonic warfare and early members of the Geological Society. *In*: LEWIS, C. L. E. & KNELL, S. J. (eds) *The Making of the Geological Society of London*. Geological Society, London, Special Publications, **317**, 219–241.

RUDWICK, M. J. S. 1962. Hutton and Werner compared: George Greenough's geological tour of Scotland in 1805. *British Journal for the History of Science*, **1**, 117–135.

RUDWICK, M. J. S. 1963. The foundation of the Geological Society of London: Its scheme for co-operative research and its struggle for independence. *British Journal for the History of Science*, **1**, 325–355.

RUDWICK, M. J. S. 1985. *The Great Devonian Controversy. The Shaping of Scientific Knowledge among Gentlemanly Specialists*. University of Chicago Press. Chicago, IL.

RUDWICK, M. J. S. 2002. Jean-André de Luc and nature's chronology. *In*: LEWIS, C. L. E. & KNELL, S. J. (eds) *The Age of the Earth: From 4004 BC to AD 2002*. Geological Society, London, Special Publications, **190**, 51–60.

RUDWICK, M. J. S. 2005. *Bursting the Limits of Time – The Reconstruction of Geohistory in the Age of Revolution*. University of Chicago Press, Chicago, IL.

RUDWICK, M. J. S. 2009. Biblical Flood and geological deluge: The amicable dissociation of geology and Genesis. *In*: KÖLBL-EBERT, M. (ed.) *Geology and Religion: Historical Views of an Intense Relationship Between Harmony and Hostility*. Geological Society, London, Special Publications, **310**, 103–110.

SMITH, W. 1815. *A Memoir to the Map and Delineation of the Strata of England and Wales*. Cary, London.

TORRENS, H. S. 1998. Geology in peace time: An English visit to study German mineralogy and geology (and visit Goethe, Werner and Raumer) in 1816. *In*: FRITSCHER, B. & HENDERSON, F. (eds) *Towards a History of Mineralogy, Petrology, and Geochemistry. Proceedings of the International Symposium on the History of Mineralogy, Petrology, and Geochemistry, Munich, March 8–9, 1996.* Institut für Geschichte der Naturwissenschaften, Munich, Algorismus, **23**, 147–175.

TORRENS, H. S. 2001. Timeless order: William Smith (1769–1839) and the search for raw materials 1800–1820. *In*: LEWIS, C. L. E. & KNELL, S. J. (eds) *The Age of the Earth: From 4004 BC to AD 2002*. Geological Society, London, Special Publications, **190**, 61–83.

TORRENS, H. S. 2003. *Memoirs of William Smith, LL.D., author of the 'Map of the Strata of England and Wales' by his nephew and pupil, John Phillips, F.R.S., F.G.S. first published in 1844 with An Introduction to the Life and Times of William Smith and The William Smith Lecture 2000 by Hugh Torrens*. The Bath Royal Literary and Scientific Institution, Bath.

WYATT, J. F. 1995. George Bellas Greenough: A Romantic geologist. *Archives of Natural History*, **22**, 61–71.

WYATT, J. F. 2004. Greenough, George Bellas (1778–1855). *In*: GOLDMAN, L. (ed.) *Oxford Dictionary of National Biography*. Oxford University Press, Oxford.

Dissenting science: the Quakers among the Founding Fathers

HUGH S. TORRENS

Lower Mill Cottage, Furnace Lane, Madeley, Crewe CW3 9EU, UK (e-mail: gga10@keele.ac.uk)

Abstract: Only three of the 13 founding members of the Geological Society of London were Quakers: William Allen and the brothers Richard and William Phillips. As dissenters, they sought to play a significant part in this new scientific development because they were in many ways excluded from English civil society. Such exclusion had encouraged their entry into trade and commerce, and they saw science as a means of improving the world and their place in it. Their great agitation against slavery, at its height just as the Geological Society of London was founded, significantly enhanced their coherence as a group. One of the first fruits of their interest in science was the Askesian Society, founded in 1796 by Allen and William Phillips, among others. With over half of its membership made up of Quakers, the Askesian was amongst the earlier of the London scientific societies. From its membership, in 1799, grew the British Mineralogical Society, which planned, by survey and analysis, to produce a mineral history of Britain. With Allen, and soon both Phillips brothers, involved in manufacture, analysis and lecturing in the field of chemistry, these interests inevitably led them to want to better understand, and use, mineral resources and to contribute to the founding of the Geological Society.

As a group, the Quakers, or Society of Friends, remain poorly known, and it was estimated, in 1859, that at their peak, in 1680, they seem never to have exceeded a population in Britain of 66 000 (Walvin 1997, p. 92). During the eighteenth century, their number had 'contracted by nearly 50 percent, from a peak of about 35 000 to rather less than 20 000 (Cantor 2005, p. 23) when the Geological Society was founded in 1807. Lacking both creed or paid ministers, with abiding interests in social justice, philanthropy and pacifism, they formed a highly distinctive Christian group, known for a series of remarkable achievements. Some of these had been noted by the Lunar Society's geologist John Whitehurst (1713–1788) in 1781 (Whitehurst 1790), and the French geologist Barthélémy Faujas de Saint Fond (1741–1819) in 1784 (Geikie 1907, vol. 1, pp. 113–120). Both had been equally impressed. The nature and significance of these achievements, briefly described below, provide an important means to understand the Society of Friends and particularly those Quakers who came to contribute to the founding of the Geological Society.

Of these achievements, none is greater than their contribution to the abolition of the trade in slaves in the very year the Society was formed (Anstey 1975). Much of the long Quaker agitation against slavery had been led, from the early 1780s, by a group that included the Cornish-born London printer James Phillips, father of William (1773–1828) and Richard Phillips (1778–1851) (Jennings 1997) who were also Geological Society founders. James had printed their highly influential *Case of the Oppressed Africans* in 1783 ([Dillwyn & Lloyd] 1783) and was a founder member of the Society for the Abolition of the Slave Trade in 1787 and their printer. As G. M. Trevelyan (1923, pp. 51–52) noted, this single movement became the 'model for the conduct of [...] thousands of other movements [...] where every [Englishman] is constantly joining Leagues, Unions or Committees to agitate some question'. Their agitation was recently 'hailed as the greatest achievement of British philanthropy'.[1]

The campaign against slavery reflected a wider concern for the relief of suffering. Here, the London chemist and Geological Society founder William Allen (1770–1843), was highly influential and active. As he wrote in 1812, 'on occasions of public calamity, [the] Friends' post must be the care of the poor and the relief of distress'.[2] The Quakers later work in Ireland after the Great Famine is notable in this regard (Hatton 1993).[3]

[1] *The Times*, 28 December 2006.

[2] *Quaker Faith and Practice* – hereafter *QFP*, 1995, item 24.29.

[3] Quakers remained active in relief work; for example, during, and after, both World War I and World War II (see Davies 1947; Wilson 1952). Of the latter, the French geologist, historian and my friend François Ellenberger (1915–2000), had personal experience after his liberation in April 1945 from OFLAG XVIIA, his Austrian prisoner of war camp. Such work led to the Society of Friends, in America and Britain, being jointly awarded the Nobel Prize for Peace in 1947 (Levinovitz & Ringertz 2001, pp. 175 and 227), after nominations in 1912, 1923, 1924, 1936, 1937 and 1938. The

From: LEWIS, C. L. E. & KNELL, S. J. (eds) *The Making of the Geological Society of London*.
The Geological Society, London, Special Publications, **317**, 129–144.
DOI: 10.1144/SP317.6 0305-8719/09/$15.00

It thus follows that the Quakers were also known for their promotion of peace and their conscientious objection to war. Here, again, James Phillips is an enthusiastic advocate, particularly in 1790 during the French Revolution (Brock 1995, pp. 169 and 179). Similarly, the Quaker-born geologist Robert Bakewell (1767–1843) was active in West Yorkshire, between 1801 and 1808, where he protested against the revolutionary wars with France (Cookson 1982, pp. 201–207).[4] This pursuit of peace is also commemorated in the Society for the Promotion of Permanent and Universal Peace (soon The Peace Society) established in London in June 1816 by a Quaker group led, again, by the extraordinary William Allen (Ceadel 2000, pp. 22–23).[5]

Successes in trade and banking made the Quakers active philanthropists. William Allen and William Phillips were both involved in this respect. William Phillips had acted as tacit agent for his sick father James in distributing thousands of pounds for the Shropshire ironmaster Richard Reynolds (1735–1816) after Reynolds had retired from business in 1789 (Emblen [1939], pp. 33–35; Trinder 2004). Phillips later wrote and published anonymously an important tract on this subject ([Phillips] 1824). William Allen was amongst many Quakers influential in promoting the importance of education and in helping the educationalist Joseph Lancaster (1778–1838) from 1808 (Taylor 1996).[6]

The Quakers also became known for their care for prisoners and for prison reform. This was an initiative led by Elizabeth Fry (1780–1845)[7] after her first visit to Newgate Prison in 1813 (Haan 2004). As historians, we must also appreciate that 'The Society has throughout its history sought to be meticulous in the keeping of records (whatever shortcomings there may have been in practice)' (Milligan & Thomas 1999, Foreword). The oasis that is today's Society of Friends Library in London is sufficient proof of these efforts.

The Quakers have, however, also been seen as possessing foibles. These include their once negative attitudes to music and entertainment;[8] their demand to affirm and not swear oaths; their treatment of men and women as equals; their abhorrence of personal titles; their use of simple gravestones to 'guard against any distinction being made between the rich and the poor' (*QFP* 1994, item 15.19); their need to disown members who had brought Quakerism into disrepute; and their pacifism, which can so easily bring accusations of 'appeasement'.

The Quakers and science in 1807

We have a wonderful view of the situations and attitudes of Quakers in 1807 from the work of the abolitionist pioneer Thomas Clarkson (1760–1846). His work towards the abolition of slavery brought him into close contact with Quakers because of their long and total hostility to slavery, and who then 'differed from their countrymen more than many foreigners in their language, dress and informality' (Wilson 1989, p. 133). So Clarkson decided to undertake their study. This was published in three volumes in 1806, with a second, indexed, edition in 1807 (Clarkson 1807).

> His [*Portraiture of Quakerism*] was the first book to explain the principles and peculiarities of the sect to the world at large. [Clarkson] was admittedly partial to them, as the first friends of abolition, but not blind to their imperfections and he cited criticisms of them and deviations from the purity of their system as he went along.
>
> (Wilson 1989, pp. 103–104)

The Quakers were not the only dissenting group to play a large part of the development of science in Britain in the early nineteenth century. What unified these groups was the extent of their exclusion from English civil and military society. Quakers were

---

work of both Oxfam, founded in 1942, and Amnesty International, founded in 1961, has since been much supported by Quakers.

[4] Bakewell was disowned by the Quakers when he married into a Unitarian family (Torrens 2004*b*), and he was later to befall even greater misfortune when he had to assign all his estates to creditors in July 1810, having been unable to meet their demands.

[5] In 1973 Quakers established Bradford University's unique School of Peace Studies (*QFP*, 1995, items 13.04 and 24.35).

[6] On Quaker schools, see also Stewart (1953). Milligan (2007, vii) has pointed out how the early twentieth century became something of a Quaker watershed. It was a time when many Friends abandoned industry and commerce for education and social work.

[7] Fry's portrait is on the UK's current £5 note. The humorist Gerard Hoffnung (1925–1959) has perhaps become her best known modern disciple (Hoffnung 1988, pp. 156–159).

[8] Their, today inexplicable, attitude toward entertainment and music took 'until 1978 before Ormerod Greenwood could name this attitude an apostasy' (*QFP*, 1995, item 21.31). This change was certainly helped by the most un-Quakerly activities, whether musical or cinematographic from the 1950s, of the trio of Gerard Hoffnung, David Lean (1908–1991) and Donald Swann (1923–1994).

barred by conscience from many professions in the Government, armed services or Anglican clergy, and excluded from English universities. As a result, they became a closely knit group, united by concerns about social inequality, poverty and slavery. As most Quakers had of necessity to be in trade (Milligan 2007), they easily joined forces to help their trade links – as their extraordinary legacy as bankers shows – or their work at the supposed 'birthplace of industry' in Coalbrookdale, Shropshire, demonstrates.[9]

Greenwood called science 'the second "freemasonry" in which Quakers were involved and in which so many of them delighted, and in which they have played a part so much greater than their numbers would suggest' (Greenwood 1977, p. 50). A fine historical introduction to Quakers in the context of the development of science has recently been given by Cantor (2005). Quaker science is also discussed by Raistrick (1950) and Loukes (1960). Clarkson's three volumes dedicate no section to science, but he does report a contemporary belief that Quakers did not excel in learning. However, he saw change afoot:

> I ought, however, to add, that this character is not likely to remain long with the Society; for the young Quakers of the present day seem to me to be sensible of the inferiority of their own education, and to be making an attempt towards the improvement of their minds, by engaging in those which are the most entertaining, instructive, and useful; – I mean philosophical pursuits.
>
> (Clarkson 1807, vol. 3, p. 231)

Loukes suggested that one of the first influences on Quaker attitudes to science had come through medicine, since 'Friends came to look on medicine as the servant of worship'. He recorded how the Quaker John Bellers (1654–1725) (Hitchcock 2004), in 1714, had advocated the need for a national health service in his *Essay towards the Improvement of Physick, by Which the Lives of Many Thousands of the Rich as well as of the Poor may be saved Yearly* (Loukes 1960, pp. 126–128; see also Bolam 1952, pp. 1–7). Loukes continued:

> Friends [...], shared with scientists the belief that a deeper understanding of the world would yield more knowledge of God; but they went beyond the scientists in bringing the sceptical temper into religion. Their rejection of priests and sacraments, theology and 'notions', came from the same sceptical temper, the same determination to understand before believing, that lay behind the scientific movement.
>
> (Loukes 1960, p. 161)

As already noted, three of the Geological Society's 13 founding members were Quakers: William Allen and the brothers William and Richard Phillips. The Society's first historian, H. B. Woodward, wrongly claimed there was another, the chemist William Hasledine Pepys (1775–1856) (Woodward 1907, p. 13; Knight 2009). Cantor noted of Pepys how 'another [Askesian] member, William Haseldine Pepys, [had] adopted Quaker habits, but appears never to have joined the Quakers' (Cantor 2005, p. 138). Kurzer (2003, p. 140) confirmed this:

> Pepys has been repeatedly, but wrongly, described as a Quaker [...], [but] there is documentary evidence that Pepys and his family were in fact members of the Established Church [of England]. The false impression may have arisen from Pepys' association with the Quaker-dominated Askesians, and his early closeness to the prominent 'Friends', William Allen, Richard Phillips and Luke Howard.

Luke Howard was another Quaker who has been claimed as a founding member of the Geological Society (Weindling 1997, p. 264). He was certainly a Quaker in 1807, but he was only a founder member of the Quaker-dominated Askesian Society, not the Geological Society which, perhaps surprisingly, he never joined.

The Founding Quakers

William Allen

William Allen (Fig. 1) was born on 29 August 1770, the eldest son of Job (1734–1800) and Margaret, née Stafford (1748–1830). Job was a silk merchant in Spitalfields, London. According to his obituarist, William Allen's 'inclination for scientific pursuits led him to quit the business [silk weaving] into which his father had introduced him' ([Allen] 1846, vol. 3, p. 460). So 'William entered the pharmaceutical chemistry establishment of fellow Quaker Joseph Gurney Bevan (1753–1814) who offered him a clerkship at his London Plough Court pharmacy and manufacturing chemist's operation [Fig. 2], which he entered in 1792' (Milligan 2007, pp. 8–9).

In 1793 Allen attended chemistry lectures given by Bryan Higgins (1741–1818) ([Allen] 1846, vol. 1, p. 22; Averley 1986, pp. 102–107). The following year Allen was elected a member of the Chemical Society and, in 1795, the Physical Society, both based at Guy's Hospital in London. During this time he had become a physician's pupil at St Thomas Hospital. In 1796 he married for the first time, but his wife died the next year. By 1801 it appears that Allen had become a close friend of one of the future Geological Society's

[9] From 1987, part of the Ironbridge Gorge World Heritage Site.

most significant founders, Dr William Babington (1756–1833) who was at that time an assistant physician at Guy's (Lewis 2009). Babington lectured at Guy's Hospital in London from 1789 on chemistry and mineralogy, as well as on *Materia Medica* (medicine) from 1794 to 1816 (Lawrence 1996, p. 368). Allen's surviving diaries first mention Babington in 1801 ([Allen] 1846, vol. 1, p. 58), whom he joined in giving courses on chemistry the following year (see also Lewis 2009). These lectures were to continue beyond the foundation of the Geological Society: from 1809 to 1816 Babington and Allen were joined by the Geneva-born Alexander Marcet (1770–1822); in 1817 Allen and Marcet alone gave this course; in 1820 Allen was joined by John Bostock (1773–1846) and Arthur Aikin (1773–1854), another founder of the Geological Society (Lawrence 1996, pp. 372–376).

In the year in which Allen joined Babington in lecturing on chemistry, he also began lecturing at the Royal Institution on Road Transport and Agricultural Machinery as Professor of Natural Philosophy (Berman 1978, p. 24). To give such a wide range of lectures it was essential one kept up to date with all scientific and technological advances. It was probably with these aspirations that Allen had helped to form the Askesian Philosophical Society in London in 1796 ([Allen] 1846, vol. 1, p. 26; Inkster 1977; Weindling 1983). However, it is a little ironic, given his status as a founder of both the Askesian and the Geological Societies, that soon after 1808 Allen began to move away from science towards philanthropy. Through his work as a chemical manufacturer he became financially successful (Tweedale 1990) and used his wealth to advance those international causes about which he felt so strongly (Doncaster 1965). His later life was much involved with the continuing international struggle to completely abolish slavery, Clarkson calling him in 1822, 'the greatest man in Europe. He does more good than any man living' (Grahame 1946, p. 51).

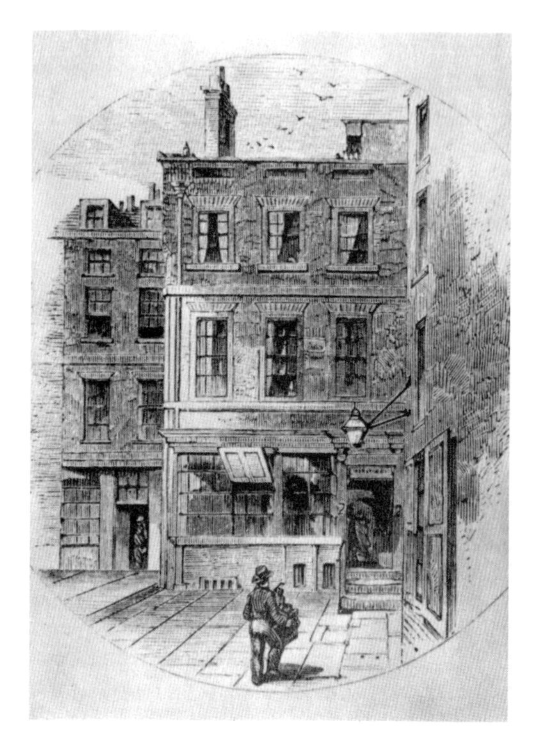

Fig. 2. William Allen's Plough Court Pharmacy (central building) where meetings of the Askesian Society were held (Thornbury [1872] vol. 2, p. 523).

Fig. 1. A likeness of William Allen (1770–1843), in his broad brimmed Quaker hat, taken in Geneva by the Swiss portrait painter Amélie Munier-Romilly (1788–1875), a relative of Sir Samuel Romilly (1757–1818), lawyer and politician, on 11 February 1820, according to Allen's *Diary*. (Original in Friends Library, London).

Allen was equally busy with three new educational initiatives. The first had started in London from 1808, with the Quaker educationalist Joseph Lancaster (1778–1838) (Bartle 2004); the second, from 1813, was at New Lanark, Scotland, with the socialist and philanthropist Robert Owen (1771–1858) (Claeys 2004); and the third was in Sussex from 1824 (Hall 1953; Nicolle 2001). Amid this myriad of concerns, it was notable that in 1820 he still 'seemed literally to have time for everything' (Majolier 1881, p. 41). Allen remained active as a supplier of chemical reagents[10] to members of the Geological Society, but, no doubt overwhelmed with what he now regarded as his more pressing concerns, he resigned from the Geological Society in 1831 (Weindling 1997, p. 264).

The Phillips brothers

The ancestry of the founding Phillips brothers can be traced back to a Quaker couple, Richard Phillips (died 1753) and his wife Esther (1681?–1778) of Swansea. South Wales had long been the leading copper smelting centre in Britain, and since at least 1584 had been using Cornish ores (Grant-Francis 1881, p. 24). Richard's son, William Phillips senior (died 1785), had settled at Trewirgn, near Redruth, in Cornwall before 1743, where he was Cornish agent to an unnamed copper company ([Phillips] 1797, p. 22). He had somewhere married Frances (died 1745) (Milligan 2007, p. 335) and they had two children: Richard, born 18 October 1743, and James, born 24 October 1745;[11] Frances died giving birth to him. William senior remarried Catherine, née Payton (1727–1794), a Quaker minister and writer (Skidmore 2004) in 1772. William senior died on 12 August 1785.[12]

His younger son, James (1745–1799), became a major London printer (Milligan 2007, p. 335).[13] He married Mary Whiting (1747–1808) in 1768 and they had 12 children, two of whom became founders of the Geological Society. James was a vital link in the long chain of Quaker printers based in London from 1680 to 1829 (Muir 1934, pp. 160 ff.). One of the many books he printed was William Pryce's, *Mineralogia Cornubiensis*, in 1778, showing how links continued between the Phillips family in London and Cornish mining. The subscribers to this book included his father William senior and his elder brother Richard, both still 'of Redruth', and himself, now listed as taking 10 copies, of 'George Yard, Lombard Street, London'.

William Phillips junior (Fig. 3), then, was James and Mary's eldest son, and was born on 9 May 1773. He joined his father as a London printer in 1797, when James became ill. One of his first publications, by James Phillips and Son, was [Phillips] (1797) *Memoirs of the Life of Catherine Phillips*. He soon became printer to the Society of Friends and, in 1810, he was chosen to be the Geological Society's printer as well. But he always carefully separated these two roles. His two-page *Catalogue of Books* (Phillips 1820) of his many Quaker publications includes no geological books despite his having dutifully published such works since before the Society was founded.

Fig. 3. William Phillips (1773–1828), late in life, shown holding a large quartz crystal. From a lithograph by M. Ganci, of an original drawing by Bowman, published in 1831 by John and Arthur Arch (Askesian Society member), 61 Cornhill, in 1831. (Woodward 1907 opposite p. 14).

William Phillips proved to be the most significant Quaker member of the Geological Society. In Woodward's (1907, p. 13) opinion 'he was the most distinguished, as a geologist, of the original founders'. William had earlier interests in both minerals and crystallography. He had been active with the British Mineralogical Society from its start, reading it a paper on the use of the divining rod, submitting geological samples and a mineralogical map of, notably in view of his family's connections there, Cornwall. His early interest in crystallography was,

[10] Such as the sample of barium oxide given to Edward Daniel Clarke (1769–1822) of Cambridge (Oldroyd 1972, p. 231) in about 1817 from which Clarke isolated metallic barium with a gas blow-pipe (Otter 1825, **2**, p. 455).

[11] See *Dictionary of Quaker Biography*.

[12] *Bath Chronicle*, 25 August 1785, p. 3. His will (PCC, PROB 11/1135) was proved 19 October 1785.

[13] See *Dictionary of Quaker Biography*.

no doubt, stimulated by Cornish minerals (Smith 1978). His Geological Society obituarist wrote in 1829 how:

> after the invention of Wollaston's reflective Goniometer [William Phillips's] assiduity and success in the use of that [instrument] [...], enabled him to produce his most valuable Crystallographic Memoirs; and the third edition of his elaborate work on mineralogy contains perhaps the most remarkable results ever yet produced in Crystallography, from the application of goniometric measurement, without the aid of mathematics.
>
> (Fitton 1829, p. 113)

William Hallowes Miller (1801–1880), Cambridge mineralogist responsible for the introduction of Miller indices for minerals, helped produce the fifth and last edition of this work in 1852.

William Phillips's greatest scientific influence, unlike that of the other two more chemical Quaker founders, was in stratigraphy. This was a result of his chapter 'The Geology of England and Wales', added to the second edition of his *Outlines of Mineralogy and Geology* (1816). This he had printed separately, for those who already owned the first edition. It took William Smith's new geological map of 1815 as its basis, with a specially permitted reproduction. This chapter Phillips much enlarged in his book *A Selection of Facts to Form an Outline of the Geology of England and Wales* of 1818. This *Selection* was much enlarged through vital, and predominant, contributions from the Rev. William Conybeare (1787–1857) to become their jointly authored classic *Outlines of the Geology of England and Wales, Part 1* (all published).[14]

The President of the Geological Society, William Fitton, noted in 1828 how this book had:

> had an effect, to which nothing since the institution of this [Geological] Society and the diffusion of Geological maps of England can be compared [...] It may be said without fear of contradiction, that no equal portion of the earth's surface has ever been more ably illustrated.
>
> (Fitton 1828, p. 57)

William Phillips' active work as another Quaker philanthropist extended to his undertaking the production of the first five volumes of the Society's *Transactions* (1811–1821), four of this first series (volumes 2–5) at his own risk and at the great cost of 'about £4,500', on which he supported a considerable loss (Woodward 1907, pp. 63–64; Torrens 2004*a*). In 1824 he published anonymously an *Appeal on the Accumulation of Wealth* ([Phillips] 1824[15]). In this he noted, disapprovingly, the recent, post-Napoleonic wars, 'money getting spirit of which the author feared the danger 15 years ago [i.e. from 1809]'. He urged Friends to be more generous with their philanthropy.

William Phillips's own fine collection, 'a rich and valuable cabinet of Minerals, also of a select crystallographical cabinet' was offered for sale in 1829 (Silliman 1829) after his death (Fig. 4). The collection comprised one part, 'made many years ago by a Cornish gentleman', perhaps his Cornish grandfather William senior, and the remaining part which he had collected himself. This was all sold by private contract through the London dealer G. B. Sowerby to Dr John Rutter of Liverpool who bequeathed it to the Medical Institution there. This then passed to the Liverpool Museum in 1887. William Phillips's fine collection of 100 drawers with at least 2500 specimens was destroyed in World War II bombing. His purely geological collection had been sold separately, in 1830 (Cooper 2006, pp. 239 and 244).

William's younger brother Richard (Fig. 5) was born on 21 November 1778. He was also educated as a chemist and druggist under William Allen at Plough Court. He had been elected to the Askesian Society by February 1800 (Weindling 1983, p. 143). Following his later important and largely chemical work, he has been described as the 'first Government chemist in this country' (Russell 2001). Any geological contributions, excluding those on the analyses of minerals, were few, while his chemical contributions, especially those in pharmacy, were enormous. The Royal Society's *Catalogue of Scientific Papers* lists 72 of his papers, dating from 1802 to 1839, as well as others written jointly with his brother William, or Michael Faraday. All are within what we would today class as chemistry, metallurgy or mineralogy. His skill as a chemical analyst enabled him to become the chemist and curator to the Museum of Practical Geology in London in 1839. He was later president of the Chemical Society, between 1849 and 1851.

His later Quaker connections are, however, mysterious. In 1807 he married Anne Rickman, from a Quaker family, but in November 1811 it was recorded that he had been disowned by the Quakers,[16] although Milligan (2007, p. 335) states instead that he had resigned at a date 'not yet traced'. Then the commonest reason for disownment by Quakers was for 'marrying outside the Society'. Brown (1885, p. 175) recorded that between 1699 and 1812 the Evesham Quaker

---

[14] On 30 May 1822 – see the cover of *Philosophical Magazine*, **59**, (289), May 1822.

[15] Available online on Google books.

[16] *Dictionary of Quaker Biography*.

CATALOGUE

OF A

RICH AND VALUABLE CABINET

OF

MINERALS;

AND, ALSO, OF A SELECT

CRYSTALLOGRAPHICAL CABINET.

CONTAINING

A GREAT VARIETY OF CURIOUS CRYSTALS,

TO THE EXTENT OF SOME THOUSAND SPECIMENS, WITH DRAWINGS AND MEASUREMENTS ANNEXED:

THE PROPERTY OF THE

LATE WILLIAM PHILLIPS, F.R.S., F.L.S., F.G.S.,

AUTHOR OF THE "INTRODUCTION TO MINERALOGY;"
AND (JOINTLY WITH THE REV W. D. CONYBEARE) OF THE "GEOLOGY OF ENGLAND AND WALES:"

Now to be disposed of by Private Contract.

Further particulars may be had, by application to G. B. SOWERBY, No. 156, Regent Street, to whom Communications on the subject may be addressed.

Fig. 4. Title page of the 1829 *Sale Catalogue* of the late William Phillips's collections of minerals and crystals. (Library of the Geological Society.)

meeting had disowned 35 people, for the following 'offences'; 'marrying out' (21), immorality and bankruptcy (both four), drunkenness and minor offences (both three). Richard Phillips's disownment for having 'married out' seems impossible in view of the dates, but the exact cause in his case needs further exploration.

The Askesian Society and the British mineralogists

Founded on 23 March 1796, the Askesian was among the earlier general scientific societies in London. Considered by some historians as a precursor to the Geological Society, it is useful to consider

Fig. 5. Photograph of Richard Phillips (1778–1851) late in life. Reproduced courtesy of the Library & Information Centre of the Royal Society of Chemistry (Russell 2001, p. 45).

its social make up and alongside the contribution of the Quakers.[17] The Askesian was founded by seven men who were later joined at various times by at least another 10 others (Table 1). More than half were Quakers at the time of their elections, although at least three were later disowned as Quakers: Samuel Mildred in the late 1790s; Richard Phillips, for unknown reasons in November 1811 (see earlier); and Luke Howard in 1837, on religious grounds. Allen, one of the founding Askesian members, later confirmed how 'the [Askesian] Society, continued for twenty years', that is up to *c.* 1816 ([Allen] 1846, vol. 1, p. 27). However, long before its demise, it produced a mineralogical offshoot:

> In 1799, the British Mineralogical Society was formed from a nucleus of nine members of the Askesian meeting in the Society's rooms at [Allen's] Plough Court, with declared aims 'to assist both the miner and the mine owner in the analysis and reduction of substances'.
>
> (Inkster 1977, p. 18)

This was not the first such society. The Sozietät für die Gesamte Mineralogie had been founded in 1796 in Jena, Germany (Salomon 1990), and it may have been the inspiration for this new British society, as it was for the American one of 1799. By 1801 this German society had 595 members, which rather counters the claim that the Geological Society of London, which the British Mineralogical Society was in turn to inspire, was ever 'possibly the first association ever formed with the idea of studying a distinct scientific discipline' (Whaley 2007).

The British Mineralogical Society planned, by survey and analysis, to produce a mineral history of the nation. Its history, however, is incomplete and has previously only been taken up to 1807. The fullest study, by Weindling (1983), closes its account around 1806. Indeed, all who have discussed the Society have assumed it ceased to exist when its only surviving Minute Book ended in 1805 or 1806 (Watts 1926, pp. 108–109).[18] But the Society must have continued for at least a further eight years, although records are now scarce. In fact, the Askesian and the British Mineralogical Society merged, on 18 December 1806, to become the joint Askesian and Mineralogical Societies ([Allen] 1846, vol. 1, p. 83). The appearance, in April 1986, of a new notebook 'containing the Sessions of the Experimental Committee of the *Askesian Society* from 1801 to 1808' (Smith 1986, item 672)[19] has shed new light on this amalgamated society. This committee, for 1801–1802, is listed in these Minutes as being composed of Pepys, Allen and Samuel Woods (*c.* 1772–1853) and, for 1802–1803, Allen, Richard Phillips and Woods. Most of their experiments concerned galvanism, electricity and heat. On 3 April 1806 the minutes record that 'the Experimental Combustion of the Diamond by means of Galvanism in oxygen gas with a power of four troughs [was] unsuccessful – proposed by Pepys and Allen'. Their session 1806–1807, starting late in 1806, was duly headed 'Askesian and Mineralogical Society, when Mr. Davy's Galvanic Discoveries [were] repeated'.[20]

---

[17] The names listed here follow Weindling (1983, p. 143) but with additions from the Prerogative Court of Canterbury (PCC) and the *Dictionary of Quaker Biography* at the Friends House Library, London.

[18] This minute book is now preserved at the Natural History Museum in London.

[19] Now in Wellcome Library, London, MS 6135, accession 348394, 1990.

[20] Wellcome Library, London, MS 6135, accession 348394, 1990, f. 22.

Further evidence that this society continued after 1806 comes from the work of mineral prospector James Ryan (Torrens 2002). In June 1807 the British Mineralogical Society witnessed trials of Ryan's new rock boring apparatus. Its Secretary, William Allen, noted:

> a New Boring Instrument having been exhibited to the British Mineralogical Society, by Mr. Ryan, and the same being referred to a Committee, the said Committee, after a careful inspection and examination of the Instrument, report – That they are persuaded that it affords results much more satisfactory than any of the methods now in use; since by bringing up the different Strata in solid Cylinders, it supplies the means of a more accurate knowledge of their depth and quality, while it indicates, at the same time, their dip and inclination. Signed by Order of the Society, W. Allen, Secretary, June 1, 1807.

This statement appeared in both Ryan's broadsheets of *c.* 1812 (Ryan *c.* 1812) and *c.* 1817, although it is only dated in the second (Ryan *c.* 1817).

Pepys and Allen's experiments on the diamond, noted above, were successful in 1807. The letter from Davy to Pepys, dated 13 November 1807, that announced the formation of a 'little talking Geological Dinner Club',[21] which became the Geological Society, noted 'if you and Allen had been one person, the Council of the Royal Society would have voted to you the Copleian Medal; but it is an indivisible thing, and cannot be given to two' (Paris 1831, vol. 1, p. 279). The joint paper by Allen and Pepys was read to the Royal Society by Humphrey Davy on 11 and 18 June, and published later in 1807 (Allen & Pepys 1807). The Copley Medal had long been the premier award of the Royal Society, but Allen noted that 'the Royal Society did not know how to manage it with two [authors]. It is however satisfactory to find that they thought it deserved one [medal]' ([Allen] 1846, vol. 1, pp. 87–89).

The last entry in these new Askesian Society minutes is dated January 1808 and perhaps marks the last work of the experimental committee of this joint Society, since the rest of the book is blank. This notes how 'the experiments of Mr. Davy were [then] repeated on a large Galvanic Apparatus of Mr Pepys, consisting of 120 Plates [of] Copper and Zinc, 36 inches square'.[22] Full notices of these same 'experiments on the alkalies before the AMS'[23] were printed in both the *Philosophical Magazine* for January 1808 (Anon. 1808*a*) and the *Monthly Magazine* for May 1808 (Anon. 1808*b*). The alkalies were metals, like potassium and sodium, which Humphry Davy had just discovered in 1807 (Friend 1951, chap. 10).

After this the activities of the new Geological Society seem to have brought an end to many activities of the joint Askesian and Mineralogical Societies, especially after details of the new Society's activities had been issued in March 1808: 'On 13 November last, a society was formed in London under the title of the Geological Society, of which we have avoided giving any notice, till we could announce its objects and constitution' (Anon. 1808*c*). A similar notice appeared in the *Monthly Magazine* for May 1808 (p. 344). However, evidence that joint Askesian and Mineralogical Societies' meetings did continue, in parallel with those of the Geological Society, is provided by the Derbyshire geologist White Watson (1760–1835). He noted in his Commonplace Book that: 'The Mineralogical Society and Askesian Society are now, 1811, combined – forming the Mineralogical Society and meet at Dr Babbington's in Aldermanbury on Thursday once a fortnight in the season (from Watson's Commonplace Book[24]).

So according to Watson it appears that these informal joint meetings at Babington's house – his 'conversazione' as Herries Davies (2007) called them – were still continuing in 1811. At one of these, in 1807, the idea had been put forward that subscribers should be sought to help publish a new treatise on the mineralogy of calcium carbonates, by the French refugee Jacques-Louis, Comte de Bournon (1752–1825) (Bournon 1808*b*; see Herries Davies 2009; Lewis 2009). He had earlier published, by William Phillips, the first part of his major *Traité de Minéralogie*.[25]

The fourteen subscribers to de Bournon's treatise, *Traité complet de la chaux carbonatée et de l'arragonite* (1808*b*), included the Askesians, with all three Quaker founders of the Geological

[21] Davy was taken very seriously ill only 10 days later (*Archives*, 1971, vol. 4, 289 and 3006a–d).

[22] Wellcome Library, London, MS 6135, accession 348394, 1990, f. 32.

[23] AMS, Askesian and Mineralogical Societies.

[24] Now at Alnwick Castle, Northumberland, in the Duke of Northumberland's archives, by kind permission.

[25] The mineral collector Sir Abraham Hume's own copy of this (Bournon 1808*a*) with an additional volume of MSS 'planches de la Chaux Carbonatée', from specimens in Hume's own collection, survives in Cambridge University Library (pressmark 8365.b.15–18).

Table 1. *Members of the Askesian Society**

Name	Askesian Society	Geological Society	Notes
William Allen (1770–1843)	Founder	Founder Subscriber to de Bournon's treatise	**Quaker**. Pharmaceutical druggist and experimental chemist/ geologist and philanthropist (Matthew & Harrison 2004, vol. 1, pp. 833–835; Milligan 2007)
Samuel Woods (*c*. 1772–1853)	Founder First President	Elected 1808. On first Council from 1809 to 1815. Wrote a number of geological papers, one with William Phillips on the geology of Snowdon (1822).	**Quaker**. Eldest son of the woollen draper Joseph Woods (1738–1812), an early and active, abolitionist (Jennings 1997; Milligan 2007, p. 478). Died in Liverpool (*The Times*, 17 March 1853), having relinquished his Quaker membership
Samuel Mildred (1772–?)	Founder		**Quaker**. Druggist and was Allen's business partner from 1795 to 1797 (Milligan 2007, p. 307)
William Phillips (1773–1828)	Founder	Founder Subscriber to de Bournon's treatise	**Quaker**. Printer, geologist and philanthropist (Matthew & Harrison 2004, vol. 44, pp. 158–159; Milligan 2007)
Joseph Fox (*c*. 1758–1832)	Founder		**Quaker**. Cornish-born, then of Finsbury. Physician to the London Hospital (Munk 1878, p. 390; Brett 1979)
Joseph Fox (1775–1816)	Founder		Baptist. Dental surgeon at Guy's Hospital and philanthropist, then of Westminster (Matthew & Harrison 2004, vol. 20, pp. 665–666; PCC will)
Henry Lawson (1774–1855)	Founder		Anglican. Optician of the Spectacle Makers Company and astronomer (Matthew & Harrison 2004, vol. 32, pp. 891–892; PCC will).
Joseph Ball (1759–1831)	Elected 1796		**Quaker**. Iron founder, later of Dulwich (Milligan 2007, p. 30; PCC will).
Luke Howard (1772–1864)	Elected 1796	Subscriber to de Bournon's treatise	**Quaker**. Manufacturing chemist, and Allen's new business partner from 1797 to 1806 (Matthew & Harrison 2004, vol. 28, pp. 400–402; Milligan 2007, p. 247). Now famous as a meteorologist, as a result of the Askesian paper he read in December 1802 (Hamblyn 2002)

William Haseldine Pepys (1775–1856)	Elected 1799	Founder	Anglican. Cutler, scientific instrument maker and chemist (Matthew & Harrison 2004, vol. 43, pp. 652–653; & Knight 2009)
Alexander Tilloch (1759–1825)	Elected 1799	Elected 1809	Independent but influenced by Sandemanian thinking (Cantor 1991, p. 304). Journalist & inventor (Matthew & Harrison 2004, vol. 54, pp. 790–791; PCC will)
Martin Tupper (1780–1844)	Elected 1799		Anglican. Later a student of medicine. Father of the more famous Martin Tupper junior (Hudson 1949)
Richard Phillips (1778–1851)	Had joined by February 1800	Founder Subscriber to de Bournon's treatise	**Quaker**. One of Allen's apprentices and later a famous chemist
Astley Paston Cooper (1768–1841)	Elected February 1800, resigned April 1801		Anglican. Surgeon at Guy's Hospital (Brock 1952; Matthew & Harrison 2004, vol. 13, pp. 227–229; PCC will)
William Babington (1756–1833)	Elected December 1801	Founder Subscriber to de Bournon's treatise	Either a dissenter/Unitarian (Inkster 1977, p. 24) or an Anglican (Weindling 1983, p. 143). Physician to Guy's Hospital and mineralogist (Matthew & Harrison 2004, vol. 3, pp. 89–90, PCC will; Lewis 2009)
Arthur Portsmouth Arch (1768–1839)	Elected December 1801		**Quaker**. London publisher and bookseller. John (1767–1853) and Arthur were 'Quaker brothers at the northeast corner of Bishopsgate Street, where they enjoyed an excellent retail trade' (Rees & Britton 1896, p. 87; Mortimer 1963, pp. 113–114; Milligan 2007, p. 10; PCC wills)
Joseph Woods junior (1776–1864)	Had joined by February 1804	Elected 1810. Member of Committee of Maps and Sections (Minutes of Council 14 June 1810), and published in the Society's *Transactions* in 1824	**Quaker**. Architect, botanist (elected FLS 1807) and geologist (DQB; Matthew & Harrison 2004, vol. 60, p. 216). Younger brother of Samuel above

*The names listed here follow Weindling (1983, p. 143), but with additions from the Prerogative Court of Canterbury (PCC) and the *Dictionary of Quaker Biography* (DQB) at the Friends House Library.

Society, and Howard and Babington.[26] This circle of wealthy and interested subscribers also included a quartet of major mineral collectors. Sir John St Aubyn (1758–1839) made his money from Cornish copper. In 1799 he had purchased the collection of fossils and minerals that had belonged to Richard Greene of Lichfield for £100. The nucleus of his mineral collection, however, was purchased from Babington, who had acquired his material from the famous collection made by John Stuart, the third earl of Bute (1713–1792) (Lewis 2009). The majority of St Aubyn's collection survives at Plymouth City Museum (Bishop 1991). Charles Greville (1749–1809) had served as a Lord of Trade, and was a member of the Board of Trade and Plantations. His crucial role in this quartet, at the Royal Institution and in 'transmitting information and specimens', has been highlighted by Weindling (1997, pp. 251–259). His collection was bought by the British Museum when he died intestate (Smith 1969, p. 242). Sir Abraham Hume's (1749–1838) family fortune had come from the East India Company, which allowed Hume both into politics and to make a major collection of minerals, of which a catalogue of the diamonds was published by de Bournon in 1815, before he returned to France. His collection went to Cambridge University (Cooper 2006). The fourth member of this quartet was Robert Ferguson MP (1767–1840), considered around 1810 to have one of the finest collections in Britain, but which was only rediscovered in 1997 (Lloyd & Lloyd 2000; Lewis 2009).

Late in 1802 the Swedish industrial spy, Eric Thomas Svedenstierna (1765–1825), arrived in London, with a particular interest in improving the technology of high-quality iron manufacture in his native Sweden. He came with letters of introduction and he gives his impressions of this now established mineralogical community. 'One such letter, addressed by a French mine superintendent, was to the royalist French mineralogist and refugee, Count Bournon' who had come to England in 1792. Through Bournon, Svedenstierna 'obtained free entry to several private collections of minerals in London, whose owners not only showed me much kindness, but some of them rendered me specially important services' (Svedenstierna 1973, p. 9). He particularly mentions three of the above quartet – Ferguson was abroad at the time and would soon (on 22 May 1803) become a prisoner of war in France. He remained such in France and Germany until he was paroled home early on 13 October 1804 (ex inf. Brian Lloyd, and see Dawson 1958, pp. 257 and 325). Greville's collection, Svedenstierna thought, was 'perhaps the richest and most complete in Europe', and was under de Bournon's direct supervision. When it was purchased by the British Museum, for £13 727, after Greville's death in 1809, the committee appointed to report on it included Babington, de Bournon, Ferguson and Charles Hatchett (1765–1847), as well as Davy, Richard Chenevix (too often rendered Chevenix – 1774–1830)[27] and William Hyde Wollaston (1766–1828) (Fullmer 1969, p. 59). Mineralogical circles were then tightly knit in London. Svedenstierna continued:

> Besides this collection, Count Bournon has two others under his supervision. One belongs to a Sir John St. Aubin [sic]. It is not so complete as that of Greville, but yet contains a quantity of fine specimens, particularly from Cornwall [...] Sir Abraham Hume owns the other collection and [as] it is situated farther out on his estate some miles from London [Wormleybury, Hertfordshire] I have not seen it.

Svedenstierna also mentions the British Mineralogical Society, to whose meetings he had clearly been introduced, and noted that 'this deserves a distinguished place, although it is less well known. The object of this society extends not only to the scientific characteristics of minerals, but also to the application of the same in agriculture, manufacture, and handicraft' (Svedenstierna 1973, pp. 16–19). He also mentions other mineralogists and dealers in London, and added how:

> among the smaller, but rather good mineral collections was that of Mr. Richard Phillips, which deserves to be seen especially on account of its fine copper specimens, rich tin ores, etc, from Cornwall. Mr Phillips is himself a mineralogist and chemist, and one is received by him with the simple and unassuming courtesy which distinguishes an enlightened Quaker.
>
> (Svedenstierna 1973, pp. 16–19)

Many of these same people had, soon after this in 1804, tried to set up a Society for the Establishment of a National Collection of Mineralogy, with a laboratory, under the auspices of the Royal Institution in London. Charles Hatchett, a rich London coachbuilder and another of Bournon's subscribers, was also involved in this project (Cooper 2006; Raistrick 1967). But financial support for the idea proved insufficient (Berman 1978, p. 89 ff.) and it was abandoned. This failure to form such a national

[26] The others subscribers were: Sir John St Aubyn; Robert Ferguson; Charles Greville; George B. Greenough; Sir Abraham Hume; Charles Hatchett; Richard Knight (1768–1844) (see Hunt & Buchanan 1984); Richard Laird, M.D. (a typographic error for James, see Lewis 2009); and John Williams junior (1777–1849).

[27] Chemist and mineralogist (see Usselman 2004).

collection must have considerably helped precipitate the alternative establishment of the Geological Society in 1807.

Why a Geological Society of London?

The early years of the Geological Society and its origins have been very well covered in the bicentenary history by Gordon Herries Davies (2007). The roles of the mineralogists, the French refugee Comte Louis de Bournon and the Irishman William Babington are there described (also Herries Davies 2009; Lewis 2009). The Society's pre-history had previously been well covered by Weindling (1997), who correctly saw the first President's[28] account as 'too biased'.

A remaining mystery is why a 'geological' society was then founded, amongst a group that had such strong mineralogical interests, and which it demonstrated in the largely mineralogical bias in its own early publications. This must be due to the influence of the two founders who were then most interested in geology, as opposed to mineralogy. These are: the chemist Humphry Davy (see Knight 2009), who made significant early contributions to geology, between 1805 and 1808, at both the Royal Society and Royal Institution (Weindling 1997, pp. 261–263); and the Society's first President, the Unitarian politician, George Bellas Greenough, who also had an early interest in geology (Kölbl-Ebert 2009). James Parkinson, another Society founder had, in 1804, published the first volume of his *Organic Remains of a Former World* (Lewis 2009), and was soon to introduce William Smith's methods in print to the Society in an article very widely reprinted in English, French and German (Parkinson 1811).

Their united geological front was despite the fact that Davy and Greenough argued over the relationship that this new Society should have with the Royal Society, and that each held diametrically opposed views on how such geological science should be applied to mining (which Davy, from Cornwall, supported, but the Londoner, Greenough, opposed) (Weindling 1997, p. 262 ff.). It must have been the remarkably united view that these two shared, of the exciting geological future ahead of the new Society, that encouraged a geological, rather than a mineralogical, society. Michael Cooper has taken the same view in his recent work on British mineral dealers (Cooper 2006, p. 21), but for a rather different view (see Lewis 2009).

The stratigrapher and mineral prospector (but never Society Member), William Smith (1769–1839) remarked, in March 1808, how their 'new Geological Society [had been] purposely established to ascertain the actual stratification of the British Isles'. As I have shown (Torrens 2003, pp. 155–199), there was, initially, clear and considerable opposition within the new Society to the utility and reliability of Smith's stratigraphic results. Their intention was clearly to check Smith's claims and their dependability. This must also explain both why the Society soon undertook its own investigation of these highly geological, rather than mineralogical, problems, and why it later produced a map to rival Smith's (Wigley *et al.* 2007). All this must have influenced the reasoning behind why a geological, as opposed to any continuing mineralogical, society, was founded, just as Smith's results were becoming better known late in 1807.

As is well known, President Greenough long remained convinced that organic remains were not as reliable in identifying strata as Smith claimed (Torrens 2003, pp. 180–185; Knell 2009). Similar, but less well known, is that Davy also shared this view. A well-connected friend of Smith's wrote to him, in October 1809, saying how Davy had:

> said he had never seen your [Smith's] collection [of fossils] but that he had heard much about it. He confesses he is not a believer in a regular succession of strata: composed of the same materials and containing the same fossils through the whole of their course. He says if this discovery is made you have certainly the merit of it [...] I had almost forgotten to add that he said, when you showed your map to the *Board of Agriculture* [in 1806], they thought that your System was not sufficiently proved.
>
> (Torrens 2003, pp. 180–181)

William Phillips, who was not a member of the inner circles of the early Society, wrote on 30 December 1820 to the botanist and geologist Nathaniel Winch (1768–1838) (Boulger 2004):

> Of the concoction of [Greenough's Geological Map 1820] very little is known except by a very few, of which I am not one. In detail it is certainly not quite correct in some places, but on the whole it is a vast improvement on Smith's, to whom nevertheless belongs the great merit of originality - but I fear that his pocket is not so much the heavier for that, as it ought to be.[29]

Phillips, as ever, here saw the philanthropic side of things.

We can see, in conclusion, that the Quaker founders were motivated initially by an interest in supporting such a new science as geology. Allen soon moved on to, in his eyes, more important projects, and all three had to earn their livings; which is

[28] George Bellas Greenough (1778–1855).

[29] Linnean Society archives. Winch MSS, W 104.

why Richard Phillips could become a Government chemist of eminence and William Phillips could publish so significantly within the emerging world of stratigraphy.

I owe many thanks to the late John Fowles (and now Sarah, to whom I owe his copy of Loukes 1960). He, a confirmed atheist (see his *Wormholes*, 1998, p. 346), wrote to me how 'Quakerism was the only religion by which I have ever felt attracted' – *in lit.*, 26 March 1990). I also wish to thank Geoffrey Tweedale, historian of Allen and Hanbury's; Ellen Gibson Wilson for her studies of the Clarkson brothers; and Paul Weindling for his pioneering work on the histories of the early Geological Society and British Mineralogical Society. Three successive librarians at the Society of Friends, Ted Milligan (Reading), Malcolm Thomas (Barry) and Joe Keith (London), have given me all possible assistance. Thanks are equally owed to Michael Bishop (Newport, Isle of Wight), Alan Bowden (Liverpool), Wendy Cawthorne (London), Gina Douglas (London), Margaret Nicolle (Lindfield), Brian Lloyd (Walmer) and Leucha Veneer (Leeds).

References

ALLEN, W. 1846. *Life of William Allen, with Selections from his Correspondence* (3 vols). Gilpin, London.

ALLEN, W. & PEPYS, W. H. 1807. On the quantity of carbon in carbonic acid, and on the nature of the diamond. *Philosophical Transactions of the Royal Society*, **97**, 267–292.

ANON. 1808*a*. Experiments of Mr. Davy. *Philosophical Magazine (Tilloch's)*, **29**, January 1808, 372–373.

ANON. 1808*b*. Experiments of Mr. Davy. *Monthly Magazine*, **25**(1 May), 345.

ANON. 1808*c*. Notice of new Geological Society. *Philosophical Magazine*, **30**, (March), 183–185.

ANSTEY, R. 1975. *The Atlantic Slave Trade and British Abolition 1760–1810*. Macmillan, London.

AVERLEY, G. 1986. The 'Social Chemists': English chemical societies in the eighteenth and early nineteenth century. *Ambix*, **33**, 99–128.

BARTLE, G. F. 2004. Lancaster, Joseph (1778–1838). *Oxford Dictionary of National Biography*. Oxford University, Oxford.

BERMAN, M. 1978. *Social Change and Scientific Organization the Royal Institution, 1799–1844*. Heinemann, London.

BISHOP, M. J. 1991. *The St Aubyn Mineral Collection: 200 Years of Curation*. Thesis for the Diploma of the Museums Association. Plymouth, City Museum.

BOLAM, D. W. 1952. *Unbroken Community: The Story of the Friends School, Saffron Walden 1702–1952*. Friends School, Cambridge.

BOULGER, G. S. 2004. Winch, Nathaniel John (1768–1838). *Oxford Dictionary of National Biography*. Oxford University, Oxford.

BOURNON, J. L. de. 1808*a*. *Traité de Minéralogie, Première Partie* (3 vols). William Phillips, London.

BOURNON, J. L. DE. 1808*b*. *Traité complet de la chaux carbonatée et de l'arragonite* (3 vols). William Phillips, London.

BRETT, R. L. 1979. *Barclay Fox's Journal*. Bell and Hyman, London.

BROCK, P. 1995. Conscientious objection in Revolutionary France. *Journal of the Friends' Historical Society*, **57**, 166–182.

BROCK, R. C. 1952. *The Life and Work of Astley Cooper*. Livingstone, Edinburgh.

BROWN, A. W. 1885. *Evesham Friends in the Olden Times*. West, Newman and Co., London.

CANTOR, G. 1991. *Michael Faraday: Sandemanian and Scientist*. Macmillan, London.

CANTOR, G. 2005. *Quakers, Jews and Science*. Oxford University Press, Oxford.

CEADEL, M. 2000. *Semi-Detached Idealists: The British Peace Movement and International Relations, 1854–1945*. Oxford University Press, Oxford.

CLAEYS, G. 2004. Owen, Robert (1771–1858). *Oxford Dictionary of National Biography*. Oxford University, Oxford.

CLARKSON, T. 1807. *A Portraiture of Quakerism*, 2nd edn (3 vols). Longman, London.

COOKSON, J. E. 1982. *The Friends of Peace: Anti-War Liberalism in England 1793–1815*. Cambridge University Press, Cambridge.

COOPER, M. P. 2006. *Robbing the Sparry Garniture: A 200 Year old History of British Mineral Dealers 1750–1950*. Mineralogical Record, Tucson, AZ.

DAVIES, A. T. 1947. *Friends Ambulance Unit: The Story of the F.A.U. during the Second World War 1939–1946*. Allen and Unwin, London.

DAWSON, W. 1958. *The Banks Letters*. Trustees of the British Museum, London.

DILLWYN, W. & LLOYD, J. 1783. *The Case of our Fellow-Creatures, the Oppressed Africans, Respectfully Recommended to the Serious Consideration of the Legislature of Great Britain, by the People called Quakers*. James Phillips, London (reprinted 2007, Society of Friends, London).

DONCASTER, L. H. 1965. *Friends of Humanity with Special Reference to the Quaker William Allen 1770–1843*. Dr Williams's Trust (Friends Lecture 19), London.

EMBLEN, P. H. [1939]. *Quakers in Commerce*. Sampson Low, London.

FITTON, W. H. 1828. Annual Address. *Proceedings of the Geological Society of London*, **1**, 50–62.

FITTON, W. H. 1829. Annual Address. *Proceedings of the Geological Society of London*, **1**, 112–134.

FRIEND, J. N. 1951. *Man and the Chemical Elements*. Griffin, London.

FULLMER, J. Z. 1969. *Sir Humphry Davy's Published Works*. Harvard University Press, Cambridge, MA.

GEIKIE, A. 1907. *A Journey through England and Scotland to the Hebrides in 1784 by B. Faujas de Saint Fond*, (2 vols). Hopkins, Glasgow.

GRAHAME, J. 1946. James Grahame's diary 1815–1824. *Journal of the Friends Historical Society*, **38**, 51–52.

GRANT-FRANCIS, G. 1881. *The Smelting of Copper in the Swansea District of South Wales from the Time of Elizabeth to the Present Day*. Sotheran, London.

GREENWOOD, J. O. 1977. *Quaker Encounters: Volume 2; Vines on the Mountains*. Sessions, York.

HAAN, F. DE. 2004. Fry (née Gurney), Elizabeth (1790–1845). *Oxford Dictionary of National Biography*. Oxford University, Oxford.

HALL, H. 1953. *William Allen 1770–1843: Member of the Society of Friends*. Charles Clarke, Haywards Heath.
HAMBLYN, R. 2002. *The Invention of Clouds*. Picador, London.
HATTON, H. E. 1993. *The Largest Amount of Good: Quaker Relief in Ireland 1654–1921*. McGill-Queen's University Press, Kingston.
HERRIES DAVIES, G. L. 2007. *Whatever is Under the Earth: The Geological Society of London 1807 to 2007*. Geological Society, London.
HERRIES DAVIES, G. L. 2009. Jacques-Louis, Comte de Bournon. *In*: LEWIS, C. L. E. & KNELL, S. J. (eds) *The Making of the Geological Society of London*. Geological Society, London, Special Publications, **317**, 105–113.
HITCHCOCK, T. 2004. Bellers, John (1654–1725). *Oxford Dictionary of National Biography*. Oxford University, Oxford.
HOFFNUNG, A. 1988. *Gerard Hoffnung: His Biography*. Gordon Fraser, London.
HUDSON, D. 1949. *Martin Tupper [Junior]: His Rise and Fall*. Constable, London.
HUNT, L. B. & BUCHANAN, P. D. 1984. Richard Knight (1768–1844): A forgotten chemist and apparatus designer. *Ambix*, **32**, 57–67.
INKSTER, I. 1977. Science and society in the metropolis: A preliminary examination of the social and institutional context of the Askesian Society of London 1796–1807. *Annals of Science*, **34**, 1–32.
JENNINGS, J. 1997. *The Business of Abolishing the British Slave Trade, 1783–1807*. Cass, London.
KNIGHT, D. 2009. Chemists get down to Earth. *In*: LEWIS, C. L. E. & KNELL, S. J. (eds) *The Making of the Geological Society of London*. Geological Society, London, Special Publications, **317**, 93–103.
KNELL, S. J. 2009. The road to Smith: how the Geological Society came to possess English geology. *In*: LEWIS, C. L. E. & KNELL, S. J. (eds) *The Making of the Geological Society of London*. Geological Society, London, Special Publications, **317**, 1–47.
KÖLBL-EBERT, M. 2009. George Bellas Greenough's 'Theory of the Earth' and its impact on the early Geological Society. *In*: LEWIS, C. L. E. & KNELL, S. J. (eds) *The Making of the Geological Society of London*. Geological Society, London, Special Publications, **317**, 115–128.
KURZER, F. 2003. William Hasledine Pepys FRS: A life in scientific research, learned societies and technical enterprise. *Annals of Science*, **60**, 137–183.
LAWRENCE, S. C. 1996. *Charitable Knowledge: Hospital Pupils and Practitioners in Eighteenth-Century London*. Cambridge University Press, Cambridge.
LEVINOVITZ, A. W. & RINGERTZ, N. 2001. *The Nobel Prize: The First 100 Years*. Imperial College Press, London.
LEWIS, C. L. E. 2009. Our favourite science: Lord Bute and James Parkinson, searching for a Theory of the Earth. *In*: KÖLBL-EBERT, M. (ed.) *Geology and Religion: A History of Harmony and Hostility*. Geological Society, London, Special Publications, **310**, 111–126.
LEWIS, C. L. E. 2009. Doctoring geology: the medical origins of the Geological Society. *In*: LEWIS, C. L. E. & KNELL, S. J. (eds) *The Making of the Geological Society of London*. Geological Society, London, Special Publications, **317**, 49–92.
LLOYD, B. & LLOYD, M. 2000. The journals of Robert Ferguson (1767–1840). *Mineralogical Record*, 31 September, 425–442.
LOUKES, H. 1960. *The Discovery of Quakerism*. George G. Harrap, London.
MAJOLIER, C. R. 1881. *Memorials of Christine Majolier Alsop*. Harris and Co., London.
MATTHEW, H. C. G. & HARRISON, B. 2004. *Oxford Dictionary of National Biography* (60 vols). Oxford University Press, Oxford.
MILLIGAN, E. H. 2007. *Biographical Dictionary of British Quakers in Commerce and Industry 1775–1920*. Sessions, York.
MILLIGAN, E. H. & THOMAS, M. J. 1999. *My Ancestors were Quakers*. Society of Genealogists, London.
MORTIMER, R. S. 1963. Quaker printers, 1750–1850. *Journal of the Friends Historical Society*, **50**, 100–133.
MUIR, P. H. 1934. English imprints after 1640. *The Library*, Series 4, **14**, 157–177.
MUNK, W. 1878. *The Roll of the Royal College of Physicians*, 2nd edn. College of Surgeons, London.
NICOLLE, M. 2001. *William Allen: Quaker Friend of Lindfield, 1770–1843*. Nicolle, Lindfield.
OLDROYD, D. R. 1972. Edward Daniel Clarke, 1769–1822, and his role in the history of the blow-pipe. *Annals of Science*, **29**, 213–235.
OTTER, W. 1825. *The Life and Remains of the Rev. Edward Daniel Clarke*, 2nd edn. Cowie, London.
PARIS, J. A. 1831. *The Life of Sir Humphry Davy* (2 vols). Colburn and Bentley, London.
PARKINSON, J. 1811. Observations on some of the strata in the neighbourhood of London, and on the fossil remains contained in them. *Transactions of the Geological Society of London*, **1**, 324–354.
PHILLIPS, C. 1797. *Memoirs of the Life of Catherine Phillips*. James Phillips and Son [i.e. William], London.
PHILLIPS, W. 1820. *Catalogue of Books Chiefly Written by Members of the Society of Friends and which may be had of [...]* W. Phillips, London (copy in Brotherton Library, Leeds – shelf mark, Birkbeck 1109.323).
PHILLIPS, W. 1824. *An Appeal on the Subject of the Accumulation of Wealth, Addressed to the Society of Friends*. Darton and Harvey, London.
QUAKER FAITH AND PRACTICE. 1995. Yearly Meeting of the Religious Society of Friends, London.
RAISTRICK, A. 1950. *Quakers in Science and Industry*. Bannisdale Press, London.
RAISTRICK, A. (ed.) 1967. *The Hatchett Diary: A Tour through the Counties of England and Scotland in 1796 Visiting their Mines and Manufactures*. Bradford Barton, Truro.
REES, T. & BRITTON, J. 1896. *Reminiscences of Literary London from 1779 to 1853*. Suckling and Galloway, London.
ROYAL INSTITUTION. 1971. *Archives of the Royal Institution of Great Britain. Minutes of Managers' meetings, 1799–1900*. Scholar Press, Menston, Yorkshire.

RUSSELL, C. 2001. In the service of government: Richard Phillips, the first 'government chemist'. *Chemistry in Britain*, **37**(2), 44–46.

RYAN, J. [*c*. 1812]. *Printed broadsheet (3 p.) of 'J. Ryan, Mineralogical Surveyor, Ventilator, and Director of Mines, &c'*. Evans and Ruffy, London (British Geological Survey, Bell Collection, volume 6/162).

RYAN, J. [c. 1817 – *watermark* 1814]. *Printed broadsheet (2 p.) of 'J. Ryan, F.A.S. Mineralogical Surveyor, Director of Mines, &c'*. Smart, Wolverhampton (author's collection).

SALOMON, J. 1990. *Die Sozietät für die Gesamte Mineralogie zu Jena unter Goethe und Johann Georg Lenz.* Böhlau (Mitteldeutsche Forschungen 98), Köln and Wien.

SILLIMAN, B. 1829. Cabinet of the late William Phillips. *American Journal of Science*, **16**, 379–380.

SKIDMORE, G. 2004. Phillips [née Payton], Catherine (1727–1794). *Oxford Dictionary of National Biography*. Oxford University, Oxford.

SMITH, W. C. 1969. A history of the first hundred years of the mineral collection in the British Museum. *Bulletin of the British Museum (Natural History): Historical Series*, **3**(8), 237–259.

SMITH, W. C. 1978. Early mineralogy in Great Britain and Ireland. *Bulletin of the British Museum (Natural History): Historical Series*, **6**(3), 49–74.

SMITH, WILLIAM (BOOKSELLERS) Ltd., READING. 1986. *Catalogue*, **277** (item 672), April 1986.

STEWART, W. A. C. 1953. *Quakers and Education as Seen in their Schools in England.* Epworth Press, London.

SVEDENSTIERNA, E. T. 1973. *Svedenstierna's Tour Great Britain 1802–1803: The Travel Diary of an Industrial Spy*. David and Charles, Newton Abbot.

TAYLOR, J. 1996. *Joseph Lancaster: The Poor Child's Friend.* Campanile Press, West Wickham.

THORNBURY, W. 1872. *Old and New London: A Narrative of its History, its people and its places. Illustrated with Numerous Engravings from the Most Authentic Sources* (2 vols). Peter Cassell and Galpin, London.

TORRENS, H.S. 2002. *The Practice of British Geology, 1750–1850.* Ashgate, Aldershot.

TORRENS, H. S. (ed.) 2003. *Memoirs of William Smith LL.D with Additional Material.* Royal Literary and Scientific Institution, Bath.

TORRENS, H. S. 2004*a*. Phillips, William (1773–1828). *Oxford Dictionary of National Biography*. Oxford University, Oxford.

TORRENS, H. S. 2004*b*. Bakewell, Robert (1767–1843). *Oxford Dictionary of National Biography*. Oxford University, Oxford.

TREVELYAN, G. M. 1923. *British History in the Nineteenth Century (1782–1901)*. Longmans, London.

TRINDER, B. 2004. Reynolds, Richard (1674–1744). *Oxford Dictionary of National Biography*. Oxford University, Oxford.

TWEEDALE, G. 1990. *At the Sign of the Plough: Allen and Hanburys and the British Pharmaceutical Industry 1715–1990.* John Murray, London.

USSELMAN, M. C. 2004. Chenevix, Richard (1774–1830). *Oxford Dictionary of National Biography*. Oxford University, Oxford.

WALVIN, J. 1997. *The Quakers: Money and Morals*. John Murray, London.

WATTS, W. W. 1926. Fifty years work of the Mineralogical Society. *Mineralogical Magazine*, **21**, 108–109.

WEINDLING, P. 1983. The British Mineralogical Society: A case study in science and social improvement. *In*: INKSTER, I. & MORRELL, J. (eds) *Metropolis and Province: Science in British Culture 1780–1850.* Hutchinson, London, 120–150.

WEINDLING, P. 1997. Geological controversy and its historiography: The prehistory of the Geological Society of London. *In*: JORDANOVA, L. J. & PORTER, R. S. (eds) *Images of the Earth*, 2nd edn. British Society for the History of Science Monographs, **1**, 247–268.

WHALEY, J. 2007. Celebrating the world's oldest geological society. *GeoExpro: Geoscience and Technology Explained*, **4**(3), 78–79.

WHITEHURST, J. 1790. Original letter from the late John Whitehurst Esq. *Gentleman's Magazine*, **60**, 1001–1002.

WIGLEY, P. *ET AL*. 2007. *'Strata' Smith: His Two Hundred Year Legacy. Digitally Enhanced Maps and Sections on DVD.* Geological Society, London.

WILSON, E. G. 1989. *Thomas Clarkson: A Biography*. Macmillan, Houndmills.

WILSON, R. C. 1952. *Quaker Relief: An Account of the Relief Work of the Society of Friends 1940–1948.* Allen and Unwin, London.

WOODWARD, H. B. 1907. *The History of the Geological Society of London*. Geological Society, London.

The early Geological Society in its international context

MARTIN J. S. RUDWICK

Department of History and Philosophy of Science, University of Cambridge, Free School Lane, Cambridge CB2 3RH, UK (e-mail: mjsr100@cam.ac.uk)

Abstract: The Geological Society was the world's first formal learned society to be devoted to the earth sciences, but these sciences were already flourishing in other social forms. In Continental Europe, state-supported 'academies of sciences', natural history museums, mining schools and universities all supported many 'savants', who would now be classed as professionals. In Britain and Ireland, in contrast, mineral surveyors and managers of mines worked entirely in the private sector. Throughout Europe, however, all such professionals relied on an infrastructure of 'amateur' observers and collectors (often very far from 'amateurish'), including groups of lower social status such as miners and quarrymen, to provide local information and specimens. The leading figures regarded themselves as belonging, despite the wars, to an informal and cosmopolitan network of savants; they used the international language of French to communicate across national boundaries, and treated Paris as the centre of their intellectual world. The Geological Society modelled itself first on other informal scientific clubs and then on the botanical Linnean Society. It chose to model its periodical on the Royal Society's *Philosophical Transactions* rather than the more utilitarian *Journal des Mines* edited in Paris. It considered adopting the well-established epithet 'mineralogical', but chose instead the rather novel and previously contentious word 'geological', in order to signal its intended focus on careful outdoor fieldwork rather than indoor work with specimens. At the same time it rejected the speculative ambitions of the genre of 'theory of the Earth' in favour of an ostentatious focus on supposedly atheoretical 'facts'.

On 13 November 1807, 11 London gentlemen resolved to turn themselves into a geological *society* (Herries Davies 2007, pp. 1–20).[1] One member of the Geological Society's First Eleven – the oldest and also the most experienced in the sciences of the Earth – was an odd man out. Jacques-Louis, Comte de Bournon (1751–1825), provides a convenient peg on which to hang this brief tour of the scientific world into which the Geological Society was born. The presence of a refugee French count among the English, Scottish and Irish gentlemen was significant on several counts. Bournon was living in London because as a French aristocrat he had had to escape from the turbulence and fanaticism of the Revolution in his native country (Herries Davies 2009). And as a loyal royalist he had remained in exile when the French army officer, Bonaparte, seized power in a *coup d'état* and turned France into a militaristic dictatorship. Under Bonaparte, who in 1805 had crowned himself emperor under his first name Napoléon, France was at war with Great Britain. Shifting alliances with other major powers on both sides had turned the conflict into a pan-European war. And the worldwide commercial and strategic interests of all these nations, together with the involvement of the young United States of America, were making it the first *world* war in all but name.

The long years of war, first Revolutionary and then Napoleonic, certainly constrained those on both sides who were engaged in the natural sciences, but it did not cut them off totally from each other (De Beer 1960). They could not easily visit each other, and the geologists among them could not do fieldwork on enemy territory. But other scientific activities were generally treated by the authorities as irrelevant to the conflict, or as rising above it; and Napoleon's attempted economic blockade could often be bypassed by routing letters, publications and specimens through neutral territory. For example, a few years after helping to found the Geological Society, Bournon received a parcel of specimens from his compatriot Alexandre Brongniart (1770–1847) in Paris; and this prompted Thomas Webster (1772–1844), the Society's first employee, to return to the Isle of Wight and there to discover important new evidence about the history of the Earth (Webster 1814; Englefield 1816, pp. 117–238; Rudwick 2005, pp. 514–521; Heringman 2009).

There was nothing narrowly nationalistic about those who founded the Geological Society. It was not the first 'national' geological society; it was

[1] Two further supporters of the scheme were unable to attend.

From: LEWIS, C. L. E. & KNELL, S. J. (eds) *The Making of the Geological Society of London*.
The Geological Society, London, Special Publications, **317**, 145–153.
DOI: 10.1144/SP317.7 0305-8719/09/$15.00

the first geological body of its kind *anywhere*. But this does not mean it was the first institution to serve the sciences of the Earth; only that it was the first example of a specific kind of institution – the formal learned society – to be devoted to the then new science of geology. This does not diminish its historical importance, but it does pinpoint its significance more precisely.

Academies of the sciences

Contrary to a myth widely believed by modern geologists, the science was not being practised at this time primarily by amateurs. All the sciences – including what came to be called 'geology' – were already deeply professionalized, in the sense that many of those involved were earning their living from such work and spending most of their time on it; Bournon, for example, earned his living in exile by identifying and arranging the mineral specimens of more wealthy collectors. And those who were not professionals were rightly called *amateurs*, because they were *in love with* their science. Many of them practised it with as much professionalism as the professionals, and many were far from 'amateurish' in the pejorative modern sense. If we want to understand the scientific world at the time the Geological Society was founded, we must set aside all our modern prejudices about amateurs and professionals.

However, with respect to paid employment in the sciences, Great Britain was indeed the great exception; it is no coincidence that among the First Eleven only Bournon the foreigner can be counted a professional. In the rest of Europe, all the sciences – including the sciences of the Earth – were relatively well supported by the state. Scientific work was publicly funded for the sake of the cultural prestige that it lent to any nation wanting to be regarded as civilized; and there was also the hope and expectation that it would lead to discoveries of practical economic benefit. In Britain, in contrast, almost all such activities were left to private enterprise.

For example, the venerable Royal Society in London, and its younger sister societies in Dublin and Edinburgh, were purely private bodies that received no state support, although in practice they were often consulted by politicians on scientific and technical matters (Miller 1999). In contrast, every major state in the rest of Europe had its own 'Academy of the Sciences', staffed by a small number of well-qualified and moderately well-paid 'savants' (to call them 'scientists' would be misleading and anachronistic). In effect, these savants were available to give their governments scientific and technical advice when required (Gascoigne 1999). France was acknowledged everywhere as the centre of the scientific world; and the largest and most distinguished academy of the sciences was, unsurprisingly, the one in Paris. The Revolutionary zealots had abolished it as a bastion of elitist privilege, but it had soon been refounded as the 'First Class' of a new and broader Institut de France; its activities had revived; and under Napoleon's regime its members were once again producing some of the finest scientific research in the world (Crosland 1992). For example, it was to the Institut in Paris – just a few months after the Geological Society was founded in London – that Brongniart and his collaborator the zoologist Georges Cuvier (1769–1832) presented their literally epoch-making research on the rocks and fossils of what they called the Paris Basin (Cuvier & Brongniart 1808; Rudwick 1997, pp. 127–156; Rudwick 2005, pp. 471–484); and it was this work that gave Webster, via Bournon, his cue for the Isle of Wight.

Museums, mining schools and universities

Another kind of institution of great importance for the sciences of the Earth was the natural history museum. Cuvier, for example, was one of a dozen well-paid professors at what was then the finest scientific museum in the world, the Muséum d'Histoire Naturelle in Paris (Laissus 1995). Almost every self-respecting European state had its own scientific museum, including many of the small independent states that would later be amalgamated to form modern Germany and modern Italy. Most were on a much smaller scale than the Parisian one, but they did support and often publish the research of their professional curators. In contrast, the British Museum in London, although nominally a public institution, was in practice virtually a private body, and its collections – which at this time included natural history specimens as well as antiquities – were poorly curated. Even two decades after the Geological Society was founded, Charles Lyell (1797–1875) publicly bemoaned the embarrassing contrast between his own country and the rest of Europe, in the quality of its museums and the degree of support that the sciences received from the state ([Lyell] 1826, pp. 154–161).

For the sciences of the Earth, mining schools were, of course, particularly important as breeding grounds of competent experts. Mining schools had been founded in all but one of the major European states, even before the Revolution in France threw the whole of Europe into turmoil. Saxony's school at Freiberg, at the heart of the mining region of the Erzgebirge, counted the great mineralogist Abraham Werner (1750–1817) among its distinguished teachers; and there were other schools, also well staffed and funded, in Prussia, Sweden,

Austria, Spain, Russia and, of course, France. The École des Mines in Paris was the headquarters of the Corps des Mines, which was in effect the world's first state geological survey: it sent trained personnel out into the field to survey the vast French territories for their possible mineral resources. One of these trained surveyors was Brongniart, who later hoped his fieldwork in the Paris Basin and all around Europe would yield new sources of clays and mineral pigments to improve the products of the great porcelain factory at Sèvres, of which he was the director (Préaud 1997). In contrast to all this state-supported activity, there was no mining school in Britain; and the flourishing British mining industry – the world leader in coal production, for example – was entirely in the private sector; although of course there were many highly competent self-employed mineral surveyors, among whom William Smith (1769–1839) is now the best known.

As for universities, most of those thickly dotted around Continental Europe had at least a professor of natural history, who was expected to cover the sciences of the Earth as well as botany and zoology; some had a professor whose chair was defined more specifically as mineralogy; and some professors of medicine did research on fossil bones as well as those of living animals, which also fed into geological debates. For example, the great university at Göttingen, in King George III's Hanoverian realm, was particularly distinguished in all the sciences, both natural and human. Johann Blumenbach (1752–1840), its professor of medicine, was the author of a widely used textbook on all the natural-history sciences, and also contributed importantly to the pan-European debates about the history of the Earth (Blumenbach 1803; Rudwick 2005, pp. 297–300). Even benighted Britain was less of an exception in the academic sphere than in relation to scientific academies, museums and mining schools. In England, John Kidd (1775–1851) and John Hailstone (1759–1847) were the relevant professors at Oxford and Cambridge, respectively; and there were others at the universities in Scotland and Ireland, notably Robert Jameson (1774–1854) at Edinburgh. But their courses were optional and generally formed very small components of the students' education. Often they gave no more than a cultural top-dressing to an education oriented vocationally towards one of the three traditional learned professions of medicine, the law and the church.

Amateurs

All these institutions – academies, museums, mining schools and universities – provided professional employment for many of those who were most active and productive in the sciences of the Earth. But, of course, there were others, not only in Britain but also in the rest of Europe, who were equally active and productive and yet were not dependent on paid employment. These, as already mentioned, were the *amateurs* or lovers of science; many of them made their living in one of the three learned professions, or else in commerce or industry (Knight 2009; Lewis 2009; Torrens 2009). Most of the founders were in this category; for them the sciences were a spare-time interest. Only a few members were also wealthy enough – either by inheritance or by marriage – to have no need to follow any profession or trade; of the Geological Society's founders, only the first President, the young George Greenough (1778–1855), belongs here (Kölbl-Ebert 2009). Some in this category were of great importance because they chose to use their wealth to amass private collections that were as fine as those in many public museums. As mentioned already, many of these amateurs, wealthy or not, worked as competently as any professional: Bournon would have been a case in point, had he not lost his wealth in fleeing from the Revolution in France.

Foreign Members

A much more prominent and productive figure in the same category, Leopold von Buch (1774–1853) of Berlin, used his private wealth to travel more widely around Europe than almost any of his contemporaries, doing fieldwork of outstanding importance for the sciences of the Earth (Buch 1815; Rudwick 2005, pp. 571–585). von Buch was the first prominent foreign geologist to visit Britain after Napoleon's final defeat at Waterloo. On that occasion he became, fittingly, the first of the Society's 'Foreign Members'; it may well have been his visit that prompted the leaders of the Society to create this new category in 1815. It signalled their intention to make the Society fully international in outlook: at the very next meeting seven other distinguished foreigners were added to the list, selected probably on von Buch's advice. Unsurprisingly, they included Brongniart, Cuvier and Werner; all seven were professionals, and five were Parisians. In the next two years 17 further Foreign Members gave the Society a network of prestigious friends spread across Europe from Christiania (now Oslo) to Naples, from Breslau (now Wroclaw) to Lisbon (Woodward 1907, pp. 277–279).

Greenough had already made a trip in the opposite direction, heading of course for Paris, immediately after Napoleon's first defeat and before the

Hundred Days of the dictator's final military comeback. Greenough was surely the most internationally minded of the founders of the Geological Society. He had attended Blumenbach's lectures at Göttingen, and he had travelled very widely, for example as far as Vienna and Sicily, before Napoleon extended his sway and, again, in 1802 during the brief interlude of the Peace of Amiens. As President he would have supported not only the idea of Foreign Members, but also the even earlier decision to imitate the Royal Society by having a Foreign Secretary. This was a position for which Bournon was the obvious first choice, as the only foreigner among the founders; in 1813 he handed it over to the first of a succession of British members, who kept the Society more or less aware of the distinguished work going on beyond the Channel. And the Society's library was soon being enriched with complimentary copies of the publications of foreign savants, including the French, sent by their authors and often reaching London by roundabout routes, until the coming of peace brought a resumption of normal international postal traffic.

Centre and periphery

Leading savants of all nations were, of course, dependent on a varied infrastructure of less prominent people. This was particularly necessary in the sciences of the Earth, because these sciences depended so much (as they still do) on the significance of particular localities and particular specimens. So the geologists based in metropolitan centres such as Paris and London were highly dependent on provincial informants to tell them about relevant local details, and often to guide them in the field to the most fruitful outcrops and the most revealing viewpoints. These local informants ranged in expertise from highly knowledgeable people, such as mineral surveyors and the managers of mines, down to those with merely a sharp eye for something unusual turning up in their locality. And they varied in social position from local nobility and landed gentry, through gentlemanly local doctors, lawyers and clergymen, and ungentlemanly surveyors, down to plebeian quarrymen and the 'fossilists' such as Mary Anning (1799–1847) – unusual only in being a woman – who made a living from finding good specimens to sell to their social superiors (Knell 2000).

Geologists everywhere made good use of this wide range of potentially useful informants. But what distinguished those who founded the Geological Society was their shrewd decision, at the very next meeting after the inaugural dinner party, to recruit many of these provincials as Honorary Members. What was unprecedented was this formal recognition of the potential value of local expertise beyond the metropolitan centre. The convenient ambiguity of the word 'honorary' served to reassure those elected that the honour would not also entail the hefty fees charged to members living in or near London. It was these fees, not social snobbery, that excluded Londoners such as William Smith and John Farey (1766–1826), although they were highly active and accomplished in the sciences of the Earth. In contrast, James Sowerby (1757–1822), although from a similar social stratum, was elected, presumably because he judged that the fees, and the networking that membership made possible, were a good investment for his business as a natural-history artist and dealer.

However, given the middle-class character of the Society's founders, and the social mores of the time, it is not surprising that there was a tacit cut-off in the list of Honorary Members: only informants with the social status of gentleman or above, or at least that of the mineral surveyors, were invited, not lower-class fossilists. But within that limitation the range was very wide indeed, not least geographically (Rudwick 1963, see map on p. 330; Veneer 2009). This innovative move was promptly followed up by the Society's very first publication, the *Geological Inquiries* of 1808. This appealed to the Honorary Members, and others everywhere, to contribute local information of specific kinds. The format of this booklet, and much of its content, were clearly modelled on the famous 'Agenda' (Saussure 1796) with which the great Genevan savant Horace-Bénédict de Saussure (1740–1799) had concluded his four-volume *Voyages dans les Alpes* (Saussure 1779–1796) just over a decade earlier (Rudwick 2005, pp. 342–346).

As an extension from all these provincial sources of information and specimens, leading savants of all nations also benefited from the local knowledge of an even wider network of informants. These were spread right around the globe, along routes determined by the commercial trade, colonial expansion or strategic demands of all the major European powers. The Geological Society recruited a handful of such expatriate informants as Honorary Members – ranging from Nova Scotia in eastern North America to Cape Colony in southern Africa – even before the category of Foreign Member was created; and many others were added later.

Information reaching Europe from the independent United States was in this respect no different in kind from that derived from the French, Dutch and British expatriates who manned trading posts in India and SE Asia, or from the Russian merchants trading across the vast land mass of Siberia. Often what was reported or brought back to Europe in this way was confirmed and improved by state-financed voyages and expeditions, which almost always included scientific as well as commercial and strategic objectives; those that the French and British sent in wartime as far as Australia and the Pacific, and those that were sent out across the Russian empire, are good examples. But in all such cases the scientific information – including much that was relevant to the sciences of the Earth – was treated as factual grist for the theoretical or interpretative mills that were grinding away in Europe. Europe was the centre of the scientific world; the rest of the globe, including the United States, was regarded as peripheral, both literally and metaphorically, except as a valuable source of factual information.

Places of debate

These mills of scientific debate were grinding away in diverse social settings. For example, there were many informal gatherings, ranging from London coffee houses and Parisian cafés to grander occasions such as the *salons* held regularly in the homes of prominent savants such as Cuvier in Paris and Joseph Banks (1743–1820), the President of the Royal Society, in London. On all such occasions new books would be shown around, striking new specimens handled, new maps and instruments displayed, and, above all, new ideas discussed. Then there were small informal clubs and societies where like-minded enthusiasts met regularly to discuss scientific topics of all kinds. The Société Philomathique in Paris, to which many of the younger French savants belonged, was one good example (Mandelbaum 1988); the Askesian Society in London (a partial precursor of the Geological) was another; and the clubbable Dutch had similar groups in almost every city of any size in The Netherlands. This was the model that some at least of the founders had in mind when they made the momentous decision to form a 'geological society'. They wanted to create a quite informal special-interest group, which would be the hub of a wider-ranging circle of informants; together they could advance the sciences of the Earth by pooling their knowledge for the common good.

In contrast, the formal meetings of bodies such as the Institut and the Royal Society were very formal indeed. At the latter, papers were read by one of the secretaries, not by the author even if he was present (the author was, of course, always male). This was intended to prevent any histrionic display by the author from influencing the calm evaluation of what was being reported. For much the same reason, no discussion of a paper was allowed until after the meeting was adjourned and the members moved to another room to talk informally over coffee. A similar formality marked the meetings of the Linnean Society, the one and only specialist scientific society already in existence in England when the Geological Society was founded. The Linnean was not merely tolerated by Banks, but warmly supported, because he regarded its specialized focus on systematic botany as no threat to his hegemonic claim that the Royal Society alone should cover all the major natural sciences in England. So the Linnean Society was the model that the Geological Society adopted, as it transformed itself in its earliest years from an informal club into a formal learned society.

It was not until the 1820s that the Geological Society cautiously brought back a little of its original informality, by allowing discussion of papers within the meeting room and immediately after they had been read. It was only then that it acquired a reputation for lively debates, which made it the envy of other learned societies (Thackray 2003). It was not until 1830 that the Geological Society was taken as a model for a similar body in another country: the Société Géologique de France was founded in Paris just as the brief July Revolution produced a regime, presided over by the 'citizen-king' Louis-Philippe, somewhat similar to Britain's widely envied constitutional monarchy.

Periodicals

Beyond all such gatherings, formal or informal, reports of scientific work reached much wider audiences, of course, if and when they were published. Some of the most impressive publications were produced by the various academies of the sciences spread across Europe, and by other state-supported institutions such as the Muséum d'Histoire Naturelle in Paris. Their nearest equivalent in Britain, the Royal Society's *Philosophical Transactions*, had been the world's very first scientific periodical and remained one of the most prestigious. It provided the model that the Geological Society adopted, as soon as it had weathered its stormy crisis with the Royal Society and established its ambition to be another learned society (Rudwick 1963). The *Geological Transactions*, as they were called informally, mimicked the *Philosophical Transactions* in their lavish format and large plates of copper engravings: impressive but also very expensive. However, in compensation for their high price, all such publications were linked in an efficient international network of exchange, which ensured – even at the height of the wars – that what was published in Paris soon became known in Berlin, Vienna, London and other such centres, and vice versa.

The growing number of privately owned scientific periodicals, which generally offered much quicker publication, were likewise linked in an international network. The monthly *Journal de Physique* (renamed from the pre-Revolutionary *Observations sur la Physique*), edited in Paris, was one of the oldest and the most influential. It covered the whole range of the natural sciences and related technologies: its full title was the *Journal of Physics, Chemistry, Natural History and the Arts* (techniques). A rather pale anglophone imitation, the *Philosophical Magazine* edited in London, covered the same wide range. Such scientific magazines – there were many others in the German- and Italian-speaking countries, but very few beyond Europe – often published summaries or even complete translations of each others' articles and reviews. The *Bibliothèque Britannique* (literally, British Library) edited in Geneva was particularly important because it reported on British scientific and scholarly publications of all kinds for the benefit of readers cut off from Britain by the wars or unable to read English easily. For example, it had earlier carried translations and critiques of the well-known works of both James Hutton (1726–1797) and James Hall (1761–1832); and later, but still in wartime, it reported on what James Parkinson (1755–1824), one of the founders, had told the Geological Society about the rocks and fossils of what he called the 'London Basin' (Parkinson 1811; Rudwick 2005, pp. 512–514).

Only a few periodicals were more specialized in content. For the sciences of the Earth by far the most important was the *Journal des Mines*, which had been founded by the mining corps in Paris at the height of the Revolution. It carried reports and articles ranging from the highly technical to the strictly scientific. For example, it had made Saussure's 'Agenda' (1796) well known, far beyond those who read it in the Genevan's massive volumes on the Alps; and later it published the preliminary version of Cuvier and Brongniart's work on the Paris Basin, only weeks after it had been read at the Institut (Cuvier & Brongniart 1808). And from an early date, even in wartime, it regularly reached the Geological Society, apparently through Bournon. The Society considered the *Journal des Mines* as a possible model when it began to have publishing ambitions of its own, although in the event it adopted instead the more prestigious but less practically oriented model of the *Philosophical Transactions*.

Periodicals, such as the *Journal de Physique*, the *Bibliothèque Britannique* and the *Journal des Mines*, were distributed and read throughout the scientific world, not just in France and other francophone countries. For French was the premier international language of the time, just as much as American English is today, and just as much for the sciences as for politics, diplomacy, literature and the arts. For example, the Russian academy of the sciences in St Petersburg published its original research not in Russian but in French, reserving the much earlier international language of Latin for its more recondite articles such as those on mathematics. The Royal Society in London had published papers in French in the *Philosophical Transactions* if they had been submitted in that form, printing an English translation in an appendix for the benefit of the linguistically challenged. But most readers of the Royal Society's periodical were not so challenged, just as modern scientists in, say, Russia or Japan cannot afford to have any great difficulty in reading research articles in English.

Scientific books

A decade before the Geological Society was founded Hutton had published his *Theory of the Earth* (1795) with page after page of quotations from Saussure's Alpine volumes in their original French (Carozzi 2000): evidently he expected his readers to read them as fluently as his pages in English (good plain English, despite modern myths to the contrary). So when Bournon was publishing his mineralogical monograph in London (Bournon 1808) – the project that led to the founding of the Geological Society – there would have been no question about its being in any language but French: not just because Bournon had written it in his native tongue, but much more because this would ensure that his book could be read throughout the scientific world, not only by the minority of savants who could cope easily with English. As if to make the point, Bournon dedicated his book to the emperor Alexander I in St Petersburg. The typesetters employed by his publisher William Phillips (1775–1828), another of the First Eleven, would have had no difficulty with Bournon's French text: this was nothing out of the ordinary for any British publishing firm that dealt with books of serious non-fiction.

Nor was Bournon's monograph unusual in being financed by the very substantial £50 subscriptions paid by some of those who became the founders of the Geological Society. This procedure was normal throughout the scientific world. Usually an author would publish a leaflet or 'prospectus' inviting subscriptions; and if he received sufficient support the publisher would then go ahead with ordering paper for the text and employing an engraver to make the illustrations (usually the two greatest costs). In due course the subscribers got their copies, and also had the satisfaction of having their names listed in the book, so that all its readers could see that they had been its enlightened patrons. Few substantial scientific books could be published without first getting this guaranteed minimal support; the only exceptions were those published by state-supported institutions such as the Muséum and the Institut in Paris, and those by authors so wealthy that they themselves could bear the financial risk. Only a scientific book written at a fairly popular level could be taken on by a publisher as a purely commercial venture.

Meanings of 'geology'

It has not been possible, within the framework of this paper, to put the *substance* of the science pursued by the Geological Society in its earliest years into its international context; but some comments on the Society's name will serve to point in that direction. By far the biggest surprise at the time was that the founders chose to call their new body 'geological'. 'Geology' was then still a neologism whose meaning was fluid and contentious (which is why this paper has used instead the phrase 'the sciences of the Earth'). Thirty years earlier the Genevan Jean-André de Luc (1727–1817), the tame savant of King George III's German wife Queen Charlotte at Windsor, had publicly hesitated to use the word '*géologie*' because it was almost unknown (Luc 1778, pp. vii–viii). But he had adopted it nonetheless as the terrestrial equivalent of '*cosmologie*'; and he was followed in this usage by Saussure and many others during the rest of the eighteenth century. So the first consistent meaning of 'geology' was to denote a high-level *theory* of the way the Earth worked, and its origin (if it had ever had one). In other words, 'geology' was the name given to the flourishing *genre* of 'theory of the Earth', of which Hutton's work of that title was just one of many examples; a modern neologism, *geotheory*, is useful to denote this kind of all-embracing megatheorizing (Rudwick 2005, pp. 133–139).

By the time the Geological Society was founded, 'geology' in this sense had acquired a bad reputation. It was widely considered far too speculative in its methods and grandiose in its ambitions; too many geotheories had been put forward, more or less inconsistent with one another; too many theories were chasing too few facts. Only a few months before the Geological Society was founded, Cuvier and his colleagues famously used the publication of yet another 'theory of the Earth' as an occasion to criticize the whole genre: as he told the Institut 'it has become almost impossible to pronounce the name of 'geology' without provoking laughter' (Cuvier *et al.* 1807; translated in Rudwick 1997, pp. 98–111). The Geological Society, right from the start, aligned itself de facto with Cuvier's position. Its founders repudiated the kind of bad-tempered wrangling, for example between so-called Neptunists and Plutonists, which in their opinion was disfiguring the savant scene in Edinburgh; in effect they followed Cuvier in calling for a moratorium on grandiose geotheorizing until more of the relevant facts had been established. This underlay the Society's somewhat ostentatious policy that it would not take sides in any such quarrels, and would devote itself instead to the patient collection of reliable observations. And, yet, the founders of the Geological Society chose to call it 'geological'.

The paradox is only apparent, because just at this time the word 'geology' was shifting towards another meaning, which in the long run the Geological Society helped to consolidate. In fact, this new meaning had begun to emerge even before the turn of the century; for example, in the teaching of Déodat de Dolomieu (1750–1801) at the École des Mines in Paris, and in the final work of the great Saussure (Rudwick 2005, pp. 337–348). In the latter's 'Agenda', which (as already mentioned) was the model for the Society's *Geological Inquiries* a decade later, Saussure had begun to use 'geology' to denote his own strongly field-based practice, which aimed at understanding the major physical features of the Earth, such as his beloved Alps. But he had also recognized that the kind of strenuous outdoor fieldwork he had pioneered needed to be complemented by the close indoor study of specimens brought back from the field. So 'geology' was beginning to be used to denote a wide range of studies carried out, above all, in the field but also in the museum and the laboratory. Rather than trying prematurely to construct an all-embracing geotheory, or 'theory of the Earth' in Huttonian style, this new kind of 'geology' would aim at describing and explaining *specific* features of the Earth, such as specific mountains, volcanoes, earthquakes, mineral veins, strata and fossils. This particularistic research programme was what the founders had in mind when they decided to call their new society 'geological'.

The alternative term they considered was 'mineralogical'; there was a clear precedent in the British Mineralogical Society, the informal London group that was their indirect precursor (Weindling 1983). But 'mineralogy', like 'geology', was not what it seems to modern ears; both words illustrate what historians of the sciences have to drum into their students, that familiar words may have carried quite different meanings in the past (the word 'evolution' is a classic example). 'Mineralogy' covered everything in the 'mineral' or non-living part of the natural world: not just minerals in the modern sense, but also rocks and even fossils; all were collected and then stored in the mineralogy cabinets of natural-history museums such as the great Muséum in Paris. But mineralogy, like the botany and zoology studied in the same museums, was primarily an indoor science of specimens. All these three branches of 'natural history' ('Animal, vegetable, or mineral?' in the traditional guessing game) were primarily sciences of description and classification, not sciences of causal explanation. Causes belonged in the quite separate science of 'physics' (sometimes known in Britain as 'natural philosophy'), which in turn tackled the causes of all kinds of natural phenomena, not just the inorganic but also the organic (Rudwick 2005, pp. 48–115).

Thus, when the founders of the Geological Society rejected 'mineralogy' and chose 'geology' instead, they were breaking new ground not just in one respect but in two. They were cutting across the long-standing distinction between 'natural history' and 'physics', between description and causal explanation, and setting out to study the Earth in ways that would integrate the two. They were also aspiring to break out of the indoor setting of mineralogy in order to follow Saussure in 'studying nature on the great scale', as Greenough put it (Rudwick 1963, p. 328). Greenough had acquired a taste for this during his extensive travels in the years immediately before he helped found the Geological Society; and his tour around Scotland in 1805 had also convinced him that the geotheoretical quarrels in Edinburgh were sterile, and that the only way forward was through the more critical collection of factual data, particularly in the field (Rudwick 1962). Greenough's inclination to adopt the new concept of 'geology' would have been reinforced by what Bournon wrote in his mineralogical monograph. For, in effect, Bournon urged his fellow founders to go beyond his own kind of indoor descriptive work on the crystallography of calcite and aragonite, and to tackle the large-scale features that demanded the kind of outdoor fieldwork pioneered by Saussure (Bournon 1808, pp. i–xxxv).

So the new body became the *Geological* Society. It should be clear that its founders intended 'geology' to be equivalent to all that the Society still promotes in the modern world. It was an umbrella term that covered Bournon's minerals, Werner's rock masses, Smith's and Brongniart's strata, Saussure's mountain topography and Webster's fold structures, von Buch's erratic blocks, Parkinson's and Cuvier's fossils, and much else besides. If our present-day colleagues had had a better sense of history, they would not have needed to abandon 'geology' and coin bastard neologisms such as 'geosciences'. 'Geology' is still the best and simplest term for all that our Society exists to promote, now as much as it was two centuries ago.

References

BLUMENBACH, J. F. 1803. *Handbuch der Naturgeschichte*, 7th edn. Heinrich Dieterich, Göttingen. (1st edn published in 1779.)

BLUMENBACH, J. F. 1803. *Specimen Archaeologiae Telluris, Terrarumque in primis Hannoveranarum.* Heinrich Dieterich, Göttingen.

BOURNON, J.-L. 1808. *Traité complet de la Chaux carbonatée et de l'Arragonite, auquel on a joint une Introduction à la Mineralogie en général, une Théorie de la Cristallisation, et de son Application, ainsi que celle du Calcul, à la Détermination des Formes cristallines et de ces deux Substances* (3 vols). William Phillips, London.

BUCH, L. VON. 1815. Ueber die Ursachen der Verbreitung grosser Alpengeschiebe. *Abhandlungen der Königlichen Akademie der Wissenschaften [Berlin], Physikalische Klasse*, **1804-11**, 161–186.

CAROZZI, M. 2000. H.-B. de Saussure: James Hutton's obsession. *Archives des Sciences, Genève*, **53**, 77–158.

CROSLAND, M. P. 1992. *Science Under Control: The French Academy of Sciences 1795–1914.* Cambridge University Press, Cambridge.

CUVIER, G. & BRONGNIART, A. 1808. Essai sur la géographie minéralogique des environs de Paris. *Annales du Muséum d'Histoire Naturelle*, **11**, 293–326 (also *Journal des Mines* **23**, 421–458; translated in RUDWICK 1997, pp. 133–156).

CUVIER, G., HAÜY, R.-J. & LE LIÈVRE, C. H. 1807. Rapport de l'Institut National (Classe des sciences physiques et mathématiques), sur l'ouvrage de M. André, ayant pour titre: Théorie de la surface actuelle de la terre. *Journal des Mines*, **21**, 413–430.

DE BEER, G. 1960. *The Sciences Were Never at War.* Nelson, London.

ENGLEFIELD, H. C. 1816. *A Description of the Principal Picturesque Beauties, Antiquities, and Geological Phaenomena, of the Isle of Wight; with Additional Observations on the Strata of the Island [...] by T. Webster.* Payne & Foss, London.

GASCOIGNE, J. 1999. The Royal Society and the emergence of science as an instrument of state policy. *British Journal for the History of Science*, **32**, 171–184.

HERINGMAN, N. 2009. Picturesque ruin and geological antiquity: Thomas Webster and Sir Henry Englefield on the Isle of Wight. *In*: LEWIS, C. L. E. & KNELL, S. J. (eds) *The Making of the Geological Society of London.* Geological Society, London, Special Publications, **317**, 299–317.

HERRIES DAVIES, G. L. 2007. *Whatever is Under the Earth: The Geological Society of London 1807–2007.* The Geological Society, London.

HERRIES DAVIES, G. L. 2009. Jacques-Louis, Comte de Bournon. *In*: LEWIS, C. L. E. & KNELL, S. J. (eds) *The Making of the Geological Society of London.* Geological Society, London, Special Publications, **317**, 105–113.

HUTTON, J. 1795. *Theory of the Earth, with Proofs and Illustrations* (2 vols). William Creech, Edinburgh.

KNELL, S. J. 2000. *The Culture of English Geology, 1815–1851*. Ashgate, Aldershot.

KNIGHT, D. 2009. Chemists get down to Earth. *In*: LEWIS, C. L. E. & KNELL, S. J. (eds) *The Making of the Geological Society of London*. Geological Society, London, Special Publications, **317**, 93–103.

KÖLBL-EBERT, M. 2009. George Bellas Greenough's 'Theory of the Earth' and its impact on the early Geological Society. *In*: LEWIS, C. L. E. & KNELL, S. J. (eds) *The Making of the Geological Society of London*. Geological Society, London, Special Publications, **317**, 115–128.

LAISSUS, Y. 1995. *Le Muséum d'Histoire Naturelle*. Gallimard, Paris.

LEWIS, C. L. E. 2009. Doctoring geology: The medical origins of the Geological Society. *In*: LEWIS, C. L. E. & KNELL, S. J. (eds) *The Making of the Geological Society of London*. Geological Society, London, Special Publications, **317**, 49–92.

LUC, J.-A. DE. 1778. *Lettres physiques et morales sur les Montagnes et sur l'Histoire de la Terre et de l'Homme, addressées à la Reine de la Grande-Bretagne*. Deture, The Hague.

LYELL, C. 1826. [Review of six provincial English scientific periodicals.] *Quarterly Review*, **34**, 153–179.

MANDELBAUM, J. 1988. Science and friendship: The Société Philomathique de Paris, 1788–1835. *History of Technology*, **5**, 179–192.

MILLER, D. P. 1999. The usefulness of natural philosophy: The Royal Society and the culture of practical utility in the later eighteenth century. *British Journal for the History of Science*, **32**, 185–201.

PARKINSON, J. 1811. Observations on some of the strata in the neighbourhood of London, and on the fossil remains contained in them. *Transactions of the Geological Society*, **1**, 324–354.

PRÉAUD, T. (ed.) 1997. *The Sèvres Porcelain Manufactory: Alexandre Brongniart and the Triumph of Art and Industry, 1800–1847*. Yale University Press, New Haven, CT.

RUDWICK, M. J. S. 1962. Hutton and Werner compared: George Greenough's geological tour of Scotland in 1805. *British Journal for the History of Science*, **1**, 117–135.

RUDWICK, M. J. S. 1963. The foundation of the Geological Society of London: Its scheme for co-operative research and its struggle for independence. *British Journal for the History of Science*, **1**, 325–355.

RUDWICK, M. J. S. 1997. *Georges Cuvier, Fossil Bones and Geological Catastrophes*. University of Chicago Press, Chicago, IL.

RUDWICK, M. J. S. 2005. *Bursting the Limits of Time: The Reconstruction of Geohistory in the Age of Revolution*. University of Chicago Press, Chicago, IL.

SAUSSURE, H.-B. DE. 1779–1796. *Voyages dans les Alpes, précédés d'un Essai sur l'Histoire naturelle des Environs de Genève* (4 vols). Samuel Fauche, Neuchâtel.

SAUSSURE, H.-B. DE. 1796. Agenda, ou tableau général des observations et des recherches dont les résultats doivent servir de base à la théorie de la terre. *Journal des Mines*, **4**, 1–70 (also in *Voyages dans les Alpes*, **4**, 467–529).

THACKRAY, J. C. 2003. *To See the Fellows Fight: Eye Witness Accounts of the Meetings of the Geological Society of London and its Club, 1822–1868*. British Society for the History of Science.

TORRENS, H. S. 2009. Dissenting science: The Quakers among the Founding Fathers. *In*: LEWIS, C. L. E. & KNELL, S. J. (eds) *The Making of the Geological Society of London*. Geological Society, London, Special Publications, **317**, 129–144.

VENEER, L. 2009. Practical geology in the Geological Society in its early years. *In*: LEWIS, C. L. E. & KNELL, S. J. (eds) *The Making of the Geological Society of London*. Geological Society, London, Special Publications, **317**, 243–253.

WEBSTER, T. 1814. On the freshwater formations in the Isle of Wight, with some observations on the strata over the Chalk in the south-east part of England. *Transactions of the Geological Society*, **2**, 161–254, plates 9–11.

WEINDLING, P. 1983. The British Mineralogical Society: A case study in science and social improvement. *In*: INKSTER, I. & MORRELL, J. (eds) *Metropolis and Province: Science in British Culture 1780–1850*. Hutchinson, London, 120–150.

WOODWARD, H. B. 1907. *The History of the Geological Society of London*. Geological Society, London.

Geology beyond the Channel: the beginnings of geohistory in early nineteenth-century France

PHILIPPE TAQUET

Laboratoire de Paléontologie, Département Histoire de la Terre, Muséum National d'Histoire Naturelle, 8 rue Buffon, 75005 Paris, France (e-mail: taquet@mnhn.fr)

Abstract: As Martin Rudwick has emphatically underlined, the beginning of the nineteenth century was marked in France by an intense intellectual awakening that allowed, in the scope of Earth Sciences, new applications of research. Indeed, the joint study of rocks and their associated fossils was made in France in its earliest years by pioneers, afterwards amplified by the endowed work of Georges Cuvier and Alexandre Brongniart on the 'Géographie minéralogique des environs de Paris'. But the integration of the study of fossils into a new geognostic practice was made possible by the combination of a number of favourable circumstances: the presence in France of such new institutions as the Muséum d'Histoire Naturelle and the Ecole des Mines where ambitious and rigorous scientific programmes, backed by a determined political power, were brought together. In these institutions young talented naturalists within premises entirely devoted to research and teaching, coupled with the presence of very diverse collections of natural history, the recruitment of competent staff and significant financial support, led to spectacular results. These studies did, of course, contribute to the rise of geology in France, but they also brought celebrity to their authors, increased the prestige of the institutions and of the authorities in place.

On 1 November 1907 the French geologist, Gustave Dollfus, wrote in a modest monthly natural history journal for amateurs, *La feuille des Jeunes Naturalistes*, a very interesting paper entitled, 'The geology in England one century ago'. He wrote (Dollfus 1907, p. 1):

> The Geological Society of London has just celebrated the centenary of its foundation and has invited for this occasion all its members, its correspondents, delegates from scientific societies, representatives of universities and of geological surveys from the whole world for a friendly assembly which brought together the specialists from all other the countries.
>
> It is indeed a singularly significant anniversary of the creation of such a scientific society at a such period; it was very audacious to put trust in the destiny of a science, which was then just at its beginnings [...], an almost rash confidence in a so uncertain political future; Napoleon I, after the defeat of Prussia, had signed the Treaty of Tilsit with Russia, on 7 July 1807, [... thus making] arrangements for the Continental blockade [of Britain].

On 7 November Russia broke with Britain. However, the British could count on many French friends and among them numerous immigrants who had run away from the chaos of the French Revolution and its guillotine. Amongst them was Jacques-Louis, Comte de Bournon (1751–1816). A disciple of the mineralogist, Romé de l'Isle, and owner of a splendid collection of minerals, he had escaped to England in 1794 in complete ruin. He was obliged to find a job, and mineralogy, which had simply been his hobby, now became his livelihood. In his newly acquired curatorial role, he worked on the arrangement and description of some of the most prestigious English mineralogical collections of the day, including those of Charles Francis Greville (1749–1809) and Sir John St Aubyn (1758–1839), and the diamond collection of Sir Abraham Hume (1749–1838) (Bournon 1815). Bournon's catalogue of Hume's collection, with its angular measurements and its 71 engraved drawings of idealized crystals and actual specimens, made an important contribution to the understanding of diamond morphology. Working with the English chemist Edward Charles Howard (1774–1816) in 1801, Bournon was the first to describe the silicates, sulphides, magnetic metals, strange globules and fine-grained matrices of chondrites. Indeed, he was one of few French scholars convinced of the extraterrestrial origin of meteorites.

In these ways, Bournon made a huge success of his enforced retreat to 'this happy and hospitable land' and, in 1807, he became one of the Founding Fathers of the Geological Society (see Herries Davies 2009; Lewis 2009). As Rudwick (2009) has shown, the nucleus of the Geological Society was made up of mineralogical enthusiasts who had come together to finance the publication of Bournon's *Traité complet de la chaux carbonatée et de l'arragonite* (Bournon 1808) (Rudwick 1963; but see also Knell 2009). The 14 subscribers to this work, each of whom paid £50 towards the

From: Lewis, C. L. E. & Knell, S. J. (eds) *The Making of the Geological Society of London*. The Geological Society, London, Special Publications, **317**, 155–162.
DOI: 10.1144/SP317.8 0305-8719/09/$15.00 © The Geological Society Publishing House 2009.

cost of publication, held a series of meetings to progress the project during 1807. Of that year, Lyell later wrote:

> No measure could be more salutary at such a moment than a suspension of all attempts to form what were termed *theories of the Earth.* A great body of new data were required, and the Geological Society of London, founded in 1807, conduced greatly to the attainment of this desirable end [. . .] Inquiries were at the same time prosecuted with great success by the French naturalists, who devoted their attention especially to the study of organic remains. They shewed that the specific characters of fossil shells and vertebrate animals might be determined with the utmost precision, and by their exertions a degree of accuracy was introduced into this department of science, of which it had never before been deemed susceptible. It was found that, by the careful discrimination of the fossil contents of strata, the contemporary origin of different groups could often be established, even where all identity of mineralogical character was wanting, and where no light could be derived from the order of supposition.
>
> (Lyell 1830, vol. 1, pp. 71–72)

In that turbulent first decade of the nineteenth century, then, an extraordinary success story was taking place in France, as geology replaced its 'geotheories' with 'geohistory', to use Rudwick's (2005) terminology.

An extraordinary decade

In Paris during that first decade, scientific life was intense and, in the emergent science of palaeontology, was dominated by the production of three masterworks: Jean-Baptiste Lamarck's (1744–1829) detailed description of fossil shells from the vicinity of Paris; the long and diligent study of the bones of fossil vertebrates by Georges Cuvier (1769–1832); and Georges Cuvier and Alexandre Brongniart's (1770–1847) geological map of the Paris Basin (Figs 1 and 2).

In 1802 Lamarck published his *Hydrogéologie ou Recherches sur l'influence qu'ont les eaux de surface du globe terrestre* and his *Recherches sur l'organisation des corps vivants*. Between 1802 and 1809 he prepared 39 articles on Tertiary shells, which were published in the *Annales du Muséum* as *Mémoires sur les Fossiles des environs de Paris* (Lamarck 1802–1809). During the preparation of these articles he also published his lectures on invertebrates, *Discours d'ouverture du cours des animaux sans vertebras* (Lamarck 1806), and, in 1809, his famous book *Philosophie Zoologique ou Exposition des considérations relatives à l'Histoire naturelle des animaux* (Lamarck 1809), in which he exposed his ideas on the transformism of living species.

In an 1800 prospectus, Cuvier announced his major scientific project to describe all the bones of terrestrial fossil vertebrates (Cuvier 1800). From 1802 to 1809 he wrote 22 papers for the *Annales du Muséum* (Cuvier 1817). In 1805 five volumes of his *Leçons d'anatomie comparée* (Cuvier 1805) were published and in the following year Cuvier opened his cabinet of comparative anatomy to the public: a thousand skeletons and preparations were exhibited (Taquet 2007). Between 1804 and 1806 he studied and described living and fossil elephants, the fossil bones of mastodons, the *Mosasaurus* skull, the *Megatherium* skeleton, the *Pterodactylus* and many other remarkable vertebrates. These were published in 47 articles in the *Annales du Muséum*, and then under the general title *Recherches sur les Ossemens fossiles de quadrupèdes* (Cuvier 1812).

During this period Cuvier was also working with Brongniart preparing their ground-breaking *Essai sur la géographie minéralogique des environs de Paris* (Cuvier & Brongniart 1808) with its beautiful geological map. At the same time, Brongniart was writing a *Traité élémentaire de Minéralogie avec des applications aux Arts* (Brongniart 1807) for use in the lycées, the new high schools introduced during the regime of Napoleon.

These were not, however, extraordinary or exceptional achievements in French science. At the beginning of the nineteenth century the nation excelled in all areas: in astronomy, with Laplace, Lagrange, Lalande and Delambre; in chemistry, with Lavoisier, Berthollet, Vauquelin and Fourcroy; in mineralogy, with Romé de Lisle and Haüy: in botany, with De Jussieu and De Candolle; and in zoology, with Latreille, Lacepède, Olivier and Savigny. It was at this time that that new system of measurement was introduced that gave us the metre and kilogram, a system that would ultimately take over the world.

Why all this success? And, in particular, why was France so extraordinarily successful in pioneering the new science of geology at the very dawn of the nineteenth century? There were several favourable factors that together help to explain both this success and how insights into the history of the Earth changed. There were six principle reasons: the exceptional geological environment around Paris, rich in quarries and underground galleries, provided considerable supplies of fossils and easy stratigraphy; the Revolution had brought about the reorganization of science, introducing new institutions and societies with interests in the natural sciences; the recruitment of a new generation of talented and open-minded naturalists; the establishment of diverse, focused and long-term research programmes; a rapid and effective programme of publication with international distribution; and the

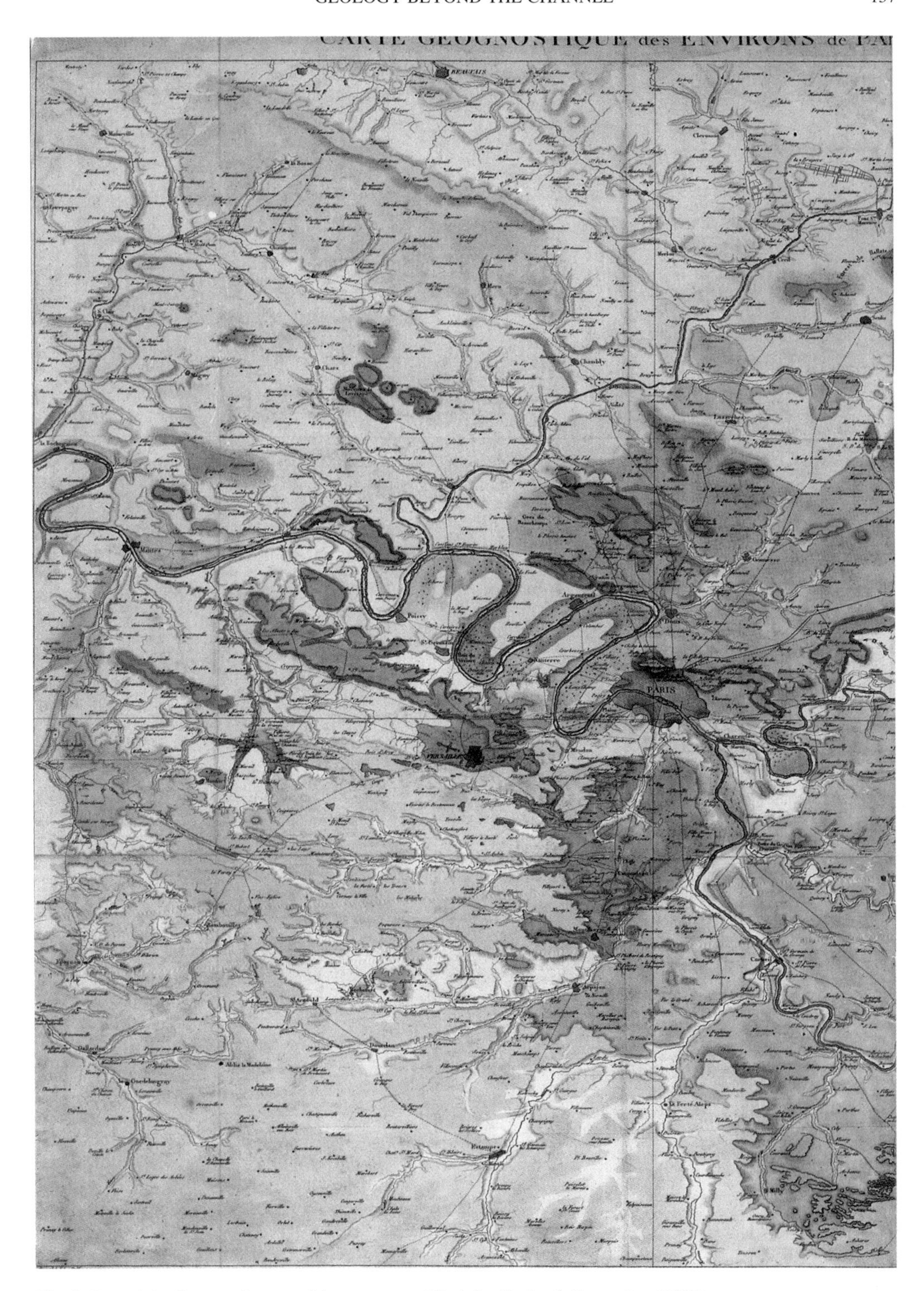

Fig. 1. Part of the Geognostic map of the area around Paris by Cuvier & Brongniart (1808).

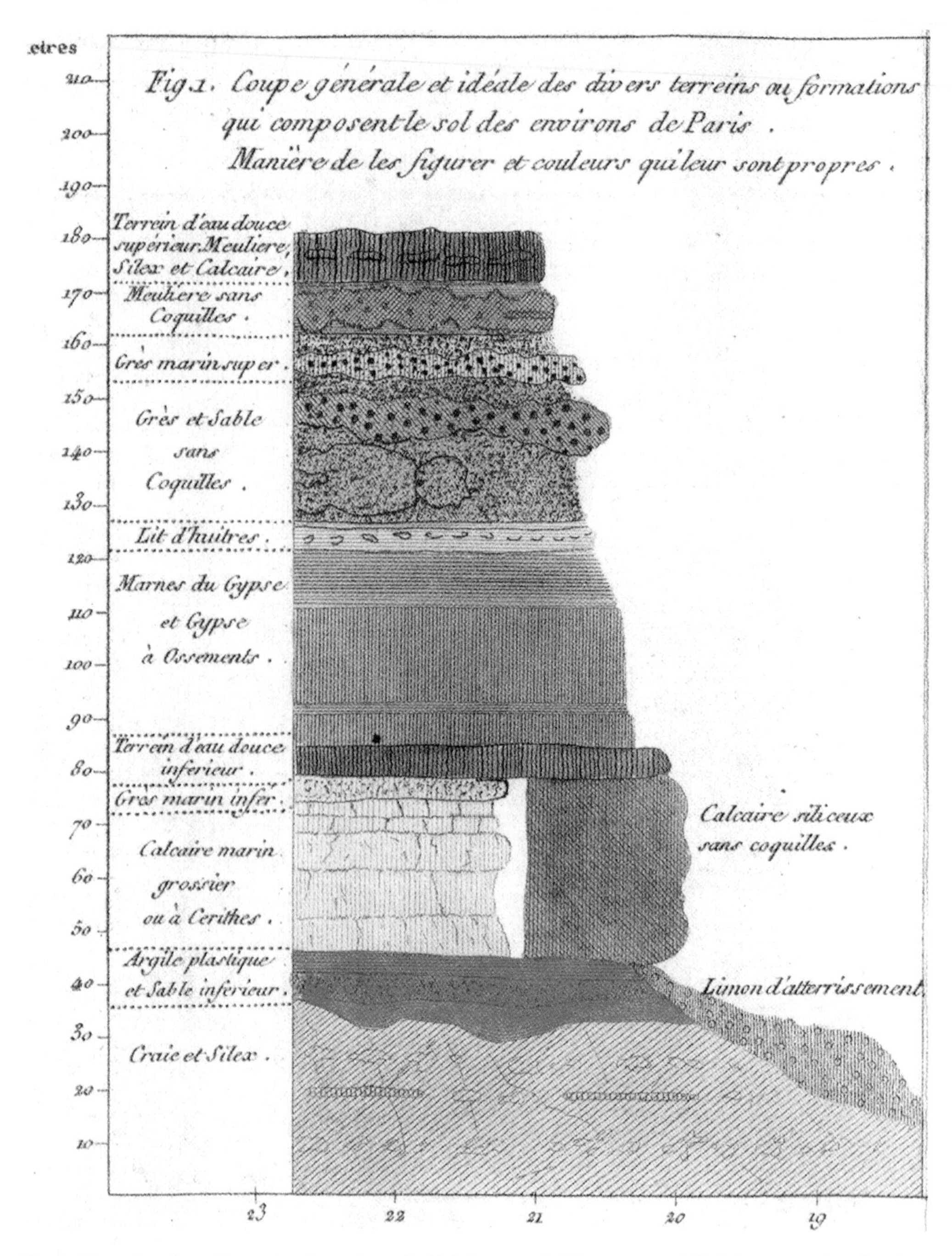

Fig. 2. General section of the various formations of which the ground of the area around Paris is composed as given by Cuvier & Brongniart (1808).

valuing of teaching, which resulted in popular and high-quality lectures in which the speakers clarified and disseminated their ideas. Of course, success in the Earth sciences was not new. Men like Jean-Étienne Guettard (1715–1786) and Antoine Lavoisier (1743–1794), for example, had produced geognostic and mineralogical maps of the Paris Basin, and published on fossil vertebrates of the vicinity of Etampes (south of Paris). These were, however, the products of brilliant individuals who were not solely devoted to natural history.

This was, then, a new era and I shall now discuss in a little more detail the six factors that brought French workers such success in those early years of geology.

An exceptional geological environment

At the beginning of the nineteenth century the rocks beneath Paris and its suburbs were the subject of intensive exploitation. The diversity of its sediments, lying regularly in horizontal layers, allowed for the easy extraction of different materials. At numerous outcrops it became possible to make easy observations of the diverse formations deposited in the Paris Basin.

The building stone for Paris was extracted from the beds of the Lutetian limestone, then better known as the Calcaire grossier, from the Roman times until 1813. Nearly every Parisian house, monument and church was constructed of this rock, which was brought up from subterranean galleries. Today, traces of these workings can still be seen under the Sainte-Geneviève Hill and under the Jardin des Plantes.

The paving stones of the streets of Paris were cut from the Fontainebleau sandstones; the chalk with marl of Meudon was exploited for making the Blanc de Meudon, a powder of chalk used for tinting. Bricks and tiles were made from the plastic clay of Passy or from Vaugirard, and plaster was made from the gypsum of Montmartre. The latter also enclosed the remains of fossil vertebrates, which were put aside by the quarrymen to be sold to the naturalists. Robert de Lamanon had described such bones coming from these quarries in 1782, and had published a map of the Isle de France on which he plotted the ancient evaporate lake. Three years later, the Dutch naturalist Adriaan Gilles Camper (1759–1820), son the famous anatomist Petrus Camper (1722–1789), met Buffon (Bots & Visser 2002) in Paris and examined Monsieur de Joubert's collection of these fossils. He bought some for his father, but owing to the limits of their anatomical knowledge they were unable to interpret them correctly.

Further to the north, siliceous sands were used for making glass, while pyritic lignites, from near Laon and Soissons, were used for the extraction of alum, sulphuric acid or vitriol, and the ashy residues were used to lighten the argillaceous soils of the Aisne department. The constitutional priest Paul-Louis Marie Poiret (1755–1834), an excellent botanist and specialist in the freshwater shells of this region as early as 1800, gave a good interpretation of the nature of the lignites, identifying the freshwater origin of the vegetational remains and correctly interpreting, using his knowledge of conchology, the succession of the deposits in the area between Laon and Soissons. He published in the *Journal de Physique*, from 1800 to 1804, five memoirs on the 'pyritic peat' of the Aisne department and he showed that the sea had invaded this country several times before retiring (Poiret 1800–1803, 1804):

> The different substances found inside the pyritic peat, the elements which constitute it, the successive order of the beds it reveals, the swampy soil on which it lays, the fluviatile shells which are deposited in the lower levels and the marine shells which cover the upper levels, are as many data which can help to discover its formation, the primitive condition of this part of the globe where we find it today, and the different revolutions that took place during the succession of centuries.

Poiret collaborated with Lamarck and he was the principal author of the botanic volumes of the *Encyclopédie Méthodique*. Another constitutional priest, Jacques Michel Coupé (1737–1809), published an embryonic stratigraphy of the Paris Basin in 1805, in the *Journal de Physique* in a memoir entitled *Sur l'étude du sol des environs de Paris* (Coupé 1805).

One final famous outcrop near Paris should be mentioned. This was the sand pit at Grignon, near Versailles. From this quarry thousands of splendid shells from the Tertiary (Lutetian) were collected, and Carl Linné himself named a *Strombus* from here in his *Systema Naturae*.

Science reorganized

The French Revolution of 1789 introduced major changes into the structure and organization of the scientific institutions. The ancient Royal Academy was dissolved on 8 August 1793. The l'Institut National des Sciences et des Arts succeeded it on 25 October 1795, with 144 members, distributed inside three classes: the mathematical and physical sciences; the moral and political sciences; literature and fine arts. The aim of this Institute was to improve science and art by research, and to pursue literary and scientific works that could be useful for all people and for the glory of the Republic. In 1805 the Institute was placed within the buildings of the Collège des Quatre Nations, where it still exists today. In the reign of the First Empire, it played a considerable role in the development of the sciences.

In the same way, the royal garden of medicinal plants was suppressed and succeeded, by a decree of the Convention of 10 June 1793, by a natural history museum with the aim of teaching the subject in all of its scope to the public. This was the Muséum National d'Histoire Naturelle. In 1795 public lectures were delivered by the 13 professors who managed the museum collegially.

In 1790 the Société d'Histoire Naturelle de Paris emerged from the Société Linnéenne de Paris, and numbered among its members all the leading French naturalists and 90 foreign associates. These naturalists worked together every 10 days (decadi) at the Société Philomatique de Paris, which had been founded on 10 December 1788 by six young men who were interested by science. They decided to meet 'for taking as a purpose of emulation the spectacle of the Human spirit'. With the suppression of the Royal Academy of Sciences on 8 August 1793, the Société Philomatique became increasingly important as a venue for the very latest scientific research, its *Bulletin* remaining a reliable, if sometimes irregular, publication even during the Terror. It played a particularly important transitional role between the closure of the old academies and the creation of the Institut National at the end of the 1795. It also organized geological and botanical field excursions.

In July 1794 the Comité de Salut Public laid the foundation of a new École des Mines in the Hôtel de Mouchy. It included a testing laboratory, a mineralogy cabinet and a library.

In a short period of time, then, the scientific scene in Paris was transformed and considerably strengthened. This greatly affected the way that geology would develop in France.

A new generation of naturalists

At the Academy of Sciences at the Institut, the new members were very young. Georges Cuvier, when he was elected in 1795 to the Section of Anatomy and Zoology, was just 26 years old. Napoleon was elected to the Section of Mechanical Arts when only 28 years old. At the Muséum, Etienne Geoffroy Saint-Hilaire (1772–1844) was appointed to the Chair of Zoology when he was only 21. Similarly, Alexandre Brongniart was only 24 years old when he met Cuvier at the museum in Jardin des Plantes in March 1795. These two young naturalists were to become strong collaborators. But even the older generation could be affected by this injection of new blood in Parisian natural history. Lamarck, for example, who was then was 51 years old, abandoned botany to take an interest in invertebrates and, in particular, molluscs.

These naturalists were full-time specialists; they worked, lived and slept in close proximity to the museum. They shared the same patterns of movement, meeting together in the same places: the Société d'Histoire Naturelle, the Société Philomatique and the Institut. They exchanged invitations and established alliances. Alexandre Brongniart married Cecile Coquebert de Monbret, the sister of another naturalist, in February 1800, at the same Lutheran chapel of the Swedish Embassy in Paris where Cuvier would marry four years later.

New directions

The Museum National d'Histoire Naturelle found itself at the hub of a new network transporting natural objects of all kinds into Paris. In June 1795 two elephant skulls arrived at the museum, one originating from Ceylon and the other from the Cape of Good Hope. They came from the Stadhouder collection in The Netherlands, seized by the armies of the young French Republic. In Cuvier's hands these specimens would help him to distinguish the two species of extant elephant: the Asian and the African, and to prove the existence of extinct or lost species, like the mammoth. 'The geologists have been prolific with hypotheses', Cuvier concluded; the facts brought to him in 1795 enabling him 'to realize great progress toward the theory of the Earth' (Cuvier 1796).

From 1800 Cuvier's programme of research into fossil vertebrates benefited greatly from the discoveries of workmen in the gypsum quarries of Montmartre Hill. Feet, skulls and complete skeletons arrived at the museum. Using his extraordinary anatomical knowledge and the laws of comparative anatomy, which permitted him to correlate characters, he was able to reconstruct animals from the past and demonstrate the existence of extinct species; he was able to burst the limits of time in order to tell the history of the Earth and develop the idea of revolutions in the deep past.

At the same time Lamarck was involved with a systematic inventory of the fossil shells of the Paris Basin, particularly those coming from the famous sand pit of Grignon. He demonstrated that the movement of seas was the result of a lengthy, but still active, cause. In 1802 he wrote:

> Comparing the fossils of Courtagnon (near Reims) with those of Grignon, and those collected by Brander in Hampshire, one is pushed to believe that the fossils concerned pertain to the same bed, which allows the association with each geological formation a characteristic fauna; because, apart from a few species, the shells of these three localities are entirely the same.
>
> (*Annales du Muséum* 1802, p. 306)

Cuvier, with his friend Brongniart, began their own cartographic study of the formations of the Paris Basin with the aim of elaborating the succession of faunas through time, each replaced by a revolution of the globe of the kind Cuvier had described. This involved excursions from the centre of the basin to its periphery. All the small communities surrounding Paris were called upon to help. Frédéric Cuvier, Georges's brother, for example, was put in charge of the northern part of the basin. In 1808 the breakthrough came when Georges Cuvier, Alexandre Brongnart and Charles Laurillard travelled to Fontainebleau. As Georges Duvernoy explained later, it was during this excursion that Cuvier demonstrated the existence of alternating marine and freshwater (in fact, a confirmation of the work of Poiret). The result of this study was the publication of a splendid geological map, which – like those British maps that hang in Burlington House (Knell 2009) – form a cornerstone of the development of the geological science. Brongniart in the following years took this exploration of the geology of the Paris Basin still further with a more detailed examination of the succession of invertebrate faunas.

Publication

The three musketeers of geology – Lamarck, Cuvier and Brongniart – took care to publish their work quickly. They were helped by existing journals and in different ways. The *Bulletin de la Société Philomatique* and the *Journal des Mines* permitted rapid publication. In contrast, *Annales du Muséum* offered the detailed presentation of research illustrated with a great number of first-rate plates, made from copper engravings and printed on vellum. Great editors supported all these publications: Levrault, who was the printer of the *Annales*; Déterville, printer of the Memoirs; Baudouin, printer for the Institut.

Teaching

Lamarck, Cuvier and Brongniart were excellent teachers. During these productive years they were active lecturers and attracted audiences of more than 100. Their lessons pushed them to formalize and improve their ideas, version after version before the definitive publication of their discoveries and explanations. Lamarck taught the zoology of invertebrates, and his opening lessons allow us to follow the evolution of his thinking. An audience from across Europe converged on Paris to hear him: the register of students numbered 1200 names, 84 percent were less than 30 years old.[1]

Cuvier gave his lessons on comparative anatomy at the Muséum in front of a large audience. He also gave a course in mineralogical geology at the College de France (two versions of his lectures were written by d'Omalius d'Halloy in 1805 and 1808, and they allow us to follow the evolution of his ideas: Grandchamp 1994). His reports for the Emperor on the progresses of the sciences in 1808 (published in 1810) and the analysis of the works of the class of mathematics and physics published by the Institut once a year give a precise account of the progress of geology and palaeontology.

In 1807 Brongniart published an elementary treatise on mineralogy and gave lectures in geology at the Paris University (Grandchamp 1999).

Conclusion

At the time of the founding of the Geological Society of London – the first in the world and in this regard some 23 years in advance of the French – French naturalists were already actively undertaking research that would leave a significant record in the history of geology. Of particular note are the names of Lamarck, Cuvier and Brongniart. Between 1800 and 1810 this trio of workers put in place the foundations of vertebrate and invertebrate palaeontology and biostratigraphy. They benefited greatly from the physical and intellectual environment of Paris and its surroundings in those years after the Revolution of 1789. Indeed, they also benefited from the Revolution itself.

References

Bots, H. & Visser, R. 2002. *Correspondance, 1785–1787, de Petrus Camper (1722–1789) et son fils Adriaan Camper Gilles Camper (1759–1820).* LIAS Sources and Documents Relating to the Early Modern History of Ideas, **28**. APA–Holland University Press, Amsterdam.

Bournon, J. L. 1808. *Traité complet de la chaux carbonatée et de l'arragonite [sic], auquel on a joint une introduction à la mineralogie en général, ainsi que celle du calcul, à la determination [sic] des formes cristallines de ces deux substances* (3 vols). William Phillips, London.

Bournon, J. de. 1815. *A Descriptive Catalogue of Diamonds in the Cabinet of Sir Abraham Hume*. John Murray, London.

Brongniart, A. 1807. *Traité élémentaire de minéralogie avec des applications aux arts. Ouvrage destiné à l'enseignement dans les lycées* (2 vols). Déterville, Paris.

Coupé, J. M. 1805. Sur l'étude du sol des environs de Paris. *Journal de Physique*, **61**, 363–395.

Cuvier, G. 1796. *Mémoire sur les espèces d'éléphans vivantes et fossiles ... lue à l'Institut national le premier pluviôse an 4 (21 January 1796).* Baudouin, Paris.

[1] P. Corsi (ed.) *Jean-Baptiste Lamark: Works and Heritage* (http://www.lamarck.cnrs.fr/?lang=en).

CUVIER, G. 1800. *Extrait d'un ouvrage sur les espèces de Quadrupèdes dont on a trouvé les ossemens dans l'intérieur de la Terre, adressé aux savants et aux amateurs des Sciences*. Baudouin, Paris.

CUVIER, G. 1805. *Leçons d'anatomie comparée* (5 vols). Baudoin, Paris.

CUVIER, G. 1812. *Recherches sur les ossemens fossiles de quadrupèdes ou l'on rétablit les caractères de plusieurs espèces d'anaimaux que les révolutions du globe paroissent avoir détruites* (4 vols). Déterville, Paris.

CUVIER, G. 1817. *Mémoires pour servir à l'histoire et à l'anatomie des Mollusques. Annales du Muséum. Déterville*, Paris. (Republished in 1969, Culture et civilisation, Bruxelles.)

CUVIER, G. & BRONGNIART, A. 1808. Essai sur la géographie minéralogique des environs de Paris. *Journal des Mines*, **23**, 421–458.

DOLLFUS, G. F. 1907. La géologie il y a cent ans en Angleterre. *La feuille des Jeunes Naturalistes*, IVème série, 38ème année, **445**, 1–6.

GRANDCHAMP, P. 1994. Deux exposés des doctrines de Cuvier antérieurs au 'Discours préliminaire': Les cours de géologie professés au Collège de France en 1805 et 1808. *Travaux du Comité Français d'Histoire de la Géologie*, **3**(8), 13–26.

GRANDCHAMP, P. 1999. Un essai inédit de classification des terrains: Le cours de géognosie professé en 1813 par Alexandre Brongniart à la Faculté des Sciences de Paris. *Travaux du Comité Français d'Histoire de la Géologie*, **3**(13), 99–113.

HERRIES DAVIES, G. L. 2009. Jacques-Louis, Comte de Bournon. *In*: LEWIS, C. L. E. & KNELL, S. J. (eds) *The Making of the Geological Society of London*. Geological Society, London, Special Publications, **317**, 105–113.

KNELL, S. J. 2009. The road to Smith: how the Geological Society came to possess English geology. *In*: LEWIS, C. L. E. & KNELL, S. J. (eds) *The Making of the Geological Society of London*. Geological Society, London, Special Publications, **317**, 1–47.

LAMARCK, J. B. DE. 1802*a*. *Hydrogéologie ou Recherches sur l'influence qu'ont les eaux de surface du globe terrestre*. Paris, chez l'auteur et Agasse, Maillard.

LAMARCK, J. B. 1802*b*. *Recherches sur l'organisation des coups vivants*, Paris, Maillard.

LAMARCK, J. B. 1802–1809. Mémoires sur les fossiles des environs de Paris, comprenant la détermination des espèces qui appartiennent aux animaux marins sans vertébres, et dont la plupart sont figurés dans la collection des vélins du Muséum. *Annales du Muséum national d'Histoire naturelle*, par les professeurs de cet établissement. (Republished in 1978 by the Paleontological Research Institution. Ithaca, NY.)

LAMARCK, J. B. 1806. *Discours d'ouverture du cours des animaux sans vertebras.* (Republished in 1907, in *Bulletin Scientifique de la France et de la Belgique*, **XL**.)

LAMARCK, J. B. 1809. *Philosophie Zoologique. Dentu ou chez l'auteur.* (Republished in 1994, G. F. Flammarion, Paris.)

LEWIS, C. L. E. 2009. Doctoring geology: The medical origins of the Geological Society. *In*: LEWIS, C. L. E. & KNELL, S. J. (eds) *The Making of the Geological Socity of London*. Geological Society, London, Special Publications, **317**, 49–92.

LYELL, C. 1830. *Principles of Geology*, Volume 1. John Murray, London. (Reprinted in 1990, University of Chicago Press, Chicago, IL.)

POIRET, J. L. M. 1800–1803. Mémoire sur la tourbe pyriteuse du département de l'Aisne; sur sa formation; les différentes substances qu'elle contient, et ses rapports avec la théorie de la terre. *Journal de Physique*, **51**, 292–304; **53**, 5–19; **55**, 189–196; **57**, 249–259.

POIRET, J. L. M. 1804. Mémoire sur la formation des tourbes. *Journal de Physique*, **59**, 891–891.

RUDWICK, M. J. S. 1963. The foundation of the Geological Society of London: Its scheme for co-operative research and its struggle for independance. *British Journal for the History of Science*, **1**(4), 325–355.

RUDWICK, M. J. S. 2005. *Bursting the Limits of Time. The Reconstruction of Geohistory in the Age of Revolution*. University of Chicago Press, Chicago, IL.

RUDWICK, M. J. S. 2009. The early Geological Society in its international context. *In*: LEWIS, C. L. E. & KNELL, S. J. (eds) *The Making of the Geological Socity of London*. Geological Society, London, Special Publications, **317**, 145–153.

TAQUET, P. 2007. Establishing the paradigmatic museum: Georges Cuvier's Cabinet d'anatomie comparée in Paris. *In*: KNELL, S. J., MACLEOD, S. & WATSON, S. E. R. (eds) *Museum Revolutions: How Museums Change And Are Changed*. Routledge, London, 3–14.

The rise of geology as a science in Germany around 1800

MARTIN GUNTAU

Zentrum für Logik, Wissenschaftstheorie und Wissenschaftsgeschichte der Universität Rostock, Germany (e-mail: martin.guntau@web.de)

Abstract: Special attention was paid to geology and mineralogy in the German countries around 1800. Following the final decades of the eighteenth century, during which an essential understanding of the natural history of the Earth was gained, geology developed into an independent science. Mining was dependent on geological findings, which in turn promoted geology. This process was driven by lecturers in the mining academies founded at that time, mining civil servants, university professors and also by private scholars. In this process, the Mining Academy of Freiberg, at which German and foreign students took their degrees, was of great importance. Abraham Gottlob Werner worked there as a lecturer who combined geological findings – based on his theory of Neptunism – into one systematic doctrine, imparting his ideas to many students over decades. These students became successful mining and metallurgy officials in the first years of the nineteenth century, and professors of geology and mineralogy at universities in Germany and abroad. During the same period, Leopold von Buch and Alexander von Humboldt contributed to the consolidation of geology as a natural science in Germany. Leopold von Buch had not only recognized the task of developing historical geology; he himself made important contributions to the stratigraphy of the Mesozoic and to palaeontology. The term 'guide fossil' was established by him. His coloured geological map of Western and Central Europe, published in 1826 and with five editions up to 1843, met with great approval.

As in other European countries, the nature of the Earth attracted enormous interest in Germany[1] during the transition from the Enlightenment to Romanticism. The natural scientists of that time were intensively engaged in exploring their natural environment to widen their intellectual horizons and find new utility in Nature for the lives of people. They described the geological conditions and mineralogical deposits in the countries concerned, as well as a variety of ores, fossils and crystals, and many other phenomena. Those interests were also reflected by the goals of contemporary expeditions such as those undertaken by Alexander von Humboldt (1769–1859) or Leopold von Buch (1774–1853). The collecting and systemizing of minerals became a popular and common activity in educated circles. The extensive and beautiful private mineral collections of Adolf Traugott von Gersdorf (1744–1807) in Görlitz, Abraham Gottlob Werner (1749–1817) (Fig. 1) in Freiberg and Johann Wolfgang von Goethe (1749–1832) in Weimar are still in existence. At the end of the eighteenth century, the first geological–mineralogical journals came into existence. As, for example, the *Mineralogische Belustigungen*[2] in Leipzig (Adelung 1768–1771), the *Journal für die Liebhaber des Steinreichs*[3] in Weimar (Schröter 1773–1780) and the *Magazin für die Mineralogie*[4] in Halle (Pfingsten 1789–1790). In addition, numerous articles on mineralogical topics were printed in general periodicals like the *Schriften der Berlinischen Gesellschaft naturforschender Freunde*[5] (1775–1803) and the *Bergmännisches Journal*[6] in Freiberg (1788–1816) (Köhler & Hoffmann 1788–1794).

In 1807 Carl Cäsar von Leonhard (1779–1862) founded the journal *Taschenbuch für die gesammte Mineralogie*[7] in Frankfurt am Main, which published contributions covering the whole spectrum of the natural history of the Earth. This important

[1] During the transition from the eigthteenth to the nineteenth century Germany did not exist as a uniform state as it does at present. Contributions to geological knowledge were mainly developed by countries like Saxony, Thuringia, Bavaria, Prussia and Lower Saxony, which later confederated into the 'German Reich' (1871).

[2] *Mineralogical Amusements.*

[3] *Journal for Lovers of the Realm of Rocks.*

[4] *Magazine of Mineralogy.*

[5] *Writings of the Berlin Society of Naturalist Friends.*

[6] *Mining Journal.*

[7] *Pocket book of Mineralogy.*

From: LEWIS, C. L. E. & KNELL, S. J. (eds) *The Making of the Geological Society of London.*
The Geological Society, London, Special Publications, **317**, 163–177.
DOI: 10.1144/SP317.9 0305-8719/09/$15.00

Fig. 1. Abraham Gottlob Werner (1749–1817).

instrument of the geosciences in Germany has remained in existence (under various titles) right up to the present. Under the name of 'Societät für die ganze Mineralogie zu Jena' a mineralogical society was also founded in Thuringia in 1798 by Johann Georg Lenz (1748–1832), the first President of which was Wolfgang von Goethe. The topicality of founding this society was confirmed by its membership of 1500 men, which included mineralogy and geology scholars, many ministers, professionals working in mining administration, pastors, artists, teachers, and other interested parties such as medical practitioners and apothecaries. A second society of mineralogy was founded in Dresden in 1816. These interests and initiatives are an indication of the interest shown by members of the educated middle and upper classes in studies of the natural Earth, which played a remarkable role in the scientific and cultural life in Germany around 1800.

At that time, mining and metallurgy were of considerable significance for many trades because they constituted a crucial basis for the production of important commodities. The demand for precious metals and more and more raw materials was growing, which was why mining was ascribed a key role in production in many countries. At the end of the eighteenth century initiatives were developed to operate mining more effectively and to explore for more economically viable deposits, especially ores. For this, geological and mineralogical knowledge was necessary: an understanding of minerals, rock and petrifactions, and, above all, of outcrops, structures and deposits, increased in value. In order to impart this knowledge, new mining academies were established alongside the universities and were of great importance, especially in Saxony and Prussia.

The basis of geological knowledge at the turn of the nineteenth century

For practical mining purposes, as well as an element of contemporary education and entertainment among the upper classes, a knowledge of the natural history of the Earth had become both topical and fashionable. To some extent this interest had its roots in the scholarly knowledge of the natural history of the Earth, which had already been acquired and published in the second half of the eighteenth century. Even if at first there were only glimpses of a comprehensive image of the natural history of the Earth, this information contributed significantly to the first interdisciplinary ideas on the structure of the mountains and the history of the Earth. These different kinds of knowledge were summarized under the term 'Mineralogy' in Germany, since recognition of the natural history of the Earth often started with studies of minerals.

The growing body of knowledge about the structure of mountains and the rocks that built them led to ideas about how geological formations had come into being. Such studies of mountains resulted in establishing three or four orders of geological formations, which were described in detail in 1758 by the Italian Giovanni Arduino (1714–1795) (Vaccari 1993; Vai 2009), but they were also being observed by German scholars. Johann Gottlob Lehmann (1719–1767) (Prescher 1969), medical practitioner, chemist and mining official, for example, recognized such 'ordering', particularly in Thuringia, and published his observations in 1756. Still bound by religious beliefs, Lehmann was searching for conformity between the ordering of mountain structures and biblical texts. With the origin of the world the 'primeval mountains' had, said Lehmann, formed from a watery medium during the Creation. A second group of mountains came into being through general changes to the ground caused by the Flood, and he considered the third group, that consisted of horizontal layers of soils and rocks which he called the 'floez formations', to have been generated by the Flood. As an example of the latter, Lehman highlighted a stratigraphic sequence of 'floez rocks' at the southern edge of the Harz Mountains, near Nordhausen, which passed from greywacke, through the New Red and Zechstein formations, into the Bunter Sandstone.

Similarly, the medical practitioner Georg Christian Füchsel (1722–1773) made geological observations in Thuringia, which he published in 1761 (Möller 1963). Typical of the spirit of the Enlightenment, his explanations of Nature were free of biblical texts, which earned him criticism from some of his more orthodox contemporaries. Explicitly, he emphasized that all his ideas were based on personal observations, from which he drew some general conclusions. Füchsel knew 'that the lowermost layers are the first or oldest, the uppermost, however, the last and most recent' ([Füchsel] 1773, p. 23). Assuming the sea to be the main cause of geological events, in 1761 he underlined the historical continuity of natural forces and their effects with the following words, 'The way in which Nature has functioned up to the present time and produced matter is to be considered the norm, we don't know any other' (Füchsel 1761, p. 82). This idea – uniformitarian at its core – enabled Füchsel to develop very realistic concepts regarding sediment formation, fossil genesis and even the geological effects of wind. And for him, all these processes in Nature took place in huge dimensions of time, 'How can you think here of [a] beginning? Nothing less than eternity, and, consequently, an eternal almightiness' ([Füchsel] 1773, p. 140).

Like Lehmann, Füchsel explored the geological rocks in Thuringia and described their characteristics, their deposits, their relations to each other and their historical succession. He was among the first in Germany who also included fossils when characterizing particular layers. By carefully proceeding through the order of the geological layers observed, he managed to bring a historiographical view into his descriptions. For him, the lithological characterizations, together with the remains of organisms, were significant features of particular sediments: on the one hand, they were characterized by the remains of terrestrial fauna or land plants, and, on the other, by the remains of marine organisms. Using these specific features, he was able to differentiate rocks. Füchel's descriptions can be considered the first contribution in Germany to the establishment of stratigraphy, particularly through his development of geological terms. It was in 1761 when he managed to get the first geological map (not in colour) of an area of Thuringia into print (Füchsel 1761). This can be regarded as proclaiming the development of a 'visual language for the geological sciences' in Germany, although it was also developing in other countries.

It was during those years (1750s) that the Prussian 'Berghauptmann' (chief mining inspector), Johann Heinrich Gottlob von Justi (1720–1771) (Koch 1980; Blei 1981, pp. 45, 73 and 331), appeared before the public with interesting ideas about the history of the Earth. Well known as a civil servant and state economist, he also produced publications containing new ideas on mineralogy (1757) and geology (1771). Using concrete arguments, he turned against widespread opinions that the whole surface of the Earth had been covered by a flood (*the* Flood). According to him, it was not acceptable 'that the bible was the absolute judge in the cognition of nature and the sciences' (Justi 1771, p. 277). There had been at least 40 floods in various regions of the Earth which, in the past, had desolated inhabited surfaces of the Earth. These had subsequently become the sea floors and had later been inhabited again. As an explanation for these processes, Justi assumed a drift of the poles, an idea that he propounded. The origin of mountains he mainly ascribed to an elevation of the ground by subterrestrial fire from the centre of the terrestrial body, although he also considered they could have been formed by ocean currents or floods on the Earth's surface (Justi 1771, p. 62 ff.). In addition he thought that all these processes needed long periods of time, and that to estimate their ages 'Millions of years were barely sufficient' (Justi 1771, p. 99). All in all, Justi's ideas were very similar to those of his contemporaries who disassociated themselves from religious interpretations of nature.

Rudolf Erich Raspe (1736–1794), who worked as curator of the so-called 'antiquities cabinet' in Kassel (a coin collection owned by the Landgrave of Hesse), edited Leibniz's philosophical works and became known as the author of the Baron Munchausen's adventurous narratives. He was another engaged in asking geological questions and, as early as 1763, had reported on the mode of origin of the Mediterranean Islands and their fossiliferous layers, with their faults and folds that extend onto the mainland. In his view, these effects were caused by natural events such as earthquakes and elevations of the sea floor, through the force of fire inside the Earth. He also considered this to be the way in which mountains and major coastal areas came into being. In addition, he described the volcanoes in Hesse and explained the basalts as volcanic formations. In his opinion, basalt columns resulted from fast cooling of lava flows in the sea (Raspe 1774).

Backed by geological evidence, support for the petrifaction of once-living beings was winning recognition in the eighteenth century. Johann Ernst Immanuel Walch (1725–1778), Professor at the University of Jena, published a *Naturgeschichte der Versteinerungen*[8] (1755–1773) in nine volumes, continuing the work of Georg Wolfgang Knorr (1705–1761) from Nuremberg. With 275 copper plates in a four-volume folio format, the work provided a survey of all the well-known petrifactions (fossils). In addition, Walch produced comprehensive descriptions and explanations. As early as 1762, he critically disputed widespread opinion that petrifactions were 'sports of Nature or stone' or that they were evidence of organisms that had perished during the Flood. He argued, 'the petrifactions are by no means to be taken as sports of Nature, but they are genuinely petrified bodies of the animal and vegetable kingdoms' (Walch 1762, p. 48). Furthermore, he said, those 'have gone too far that want to make all and everything undeniable evidence of a general Flood' (p. 54). Although the theologian Johann Esaias Silberschlag (1721–1791) in Berlin still defended religious interpretations, including the Flood, in the 1780s Walch's explanations were more and more widely accepted, and became one of the foundations of palaeontology.

In line with this, Johann Friedrich Blumenbach (1752–1840), Professor at the University of Göttingen, a student of Walch's and a teacher

[8] *Natural History of Petrifactions.*

of von Humboldt,[9] referred to the importance of petrifactions for the age determination of rocks in 1790:

> If you view petrifactions from a wide perspective, they are the most infallible documents in the archive of Nature; through them the various revolutions experienced by our planet, its origins and, to a certain extent, its epochs can be revealed, and consequently the respective ages of several of the mountain types can be determined, so it becomes obvious that their history must be considered as one of the most important, most instructive parts of scientific mineralogy.
>
> (Blumenbach 1790, p. 1)

The value of petrifactions was clearly recognized, not only for determining the age of organisms but also for their role in identifying the chronological sequence of strata.

In the last quarter of the eighteenth century, the first reports on the geology of several German countries were published, partially with coloured geological maps. Special attention was paid to the *Mineralogische Geographie der Chursächsischen Lande*[10] (1778) by Johann Friedrich Wilhelm von Charpentier (1738–1805), because of the description of the ore mines, bituminous and brown coal deposits, limestone and granite quarries, and clay pits in Saxony with its traditional and technically well-developed mining facilities. Georg Sigismund Otto Lasius (1752–1833) published his observations on the Harz Mountains (Lasius 1789) and Mathias Bartholomäus von Flurl (1756–1823) did the same with an extensive description of Bavaria and the Upper Palatinate (Flurl 1792), both with coloured lithological maps. Further works, with reports on the geological environment, natural resources, mining, etc., were also published by other scholars in other German countries. At the same time, German scholars went abroad leading expeditions or taking study trips, like Peter Simon Pallas's (1741–1811) long expedition to Russia 1768–1773 (Wendland 1992). Pallas described his geological discoveries and observations in a major three-volume report (1771–1776) and in 1778 he presented his views on the structure of mountains, which were soon accepted by other scholars. According to Pallas, mountains had a structured development: (1) the granite or slate formations that formed the base or core zone; (2) the limestone formations; and (3) the overlying sand and marl of the floez formations. This generalized view of how mountains developed can be considered as a step beyond simple description, and towards a theoretical approach. However, it was soon recognized that mountains have more complicated structures and these ideas were rapidly replaced by more detailed interpretations.

In the last decades of the eighteenth century many geological views evolved that were held by distinguished scholars in various European countries, not least in Germany. They included the following.

- A basic *structure of mountains* had been recognized. This classification was one of the first to be based on numerous observations and descriptions of mountains in Nature. The order of the rocks was subdivided into four epochs, according to their lithological criteria:
  (1) The Primary – this was the basal layer of crystalline rocks, slates, sandstones, etc.; it did not contain petrifactions. Today it would be assigned to the Palaeozoic Era.
  (2) The Secondary consisted of fossiliferous limestone and marbles, which corresponded to the Mesozoic Era.
  (3) The next layers consisted of gravel, clay, sand and also volcanic rocks. They constituted the Tertiary.
  (4) The last epoch largely consisted of alluvial material, which would possibly be assigned to the Quaternary period.
- The *role of fire* – as well as water – was increasingly recognized as one of the causal forces of natural processes in the Earth. The heat from the depth of the terrestrial body was considered the reason for the genesis and activity of volcanoes, the formation of mountains and the genesis of various rocks. The role of fire in the processes mentioned was commented on in Germany by von Justi (1771, p. 62 ff.), Raspe (1774, p. 27), Johann Carl Wilhelm Voigt (1752–1821) (1781, pp. 19–20) and others.
- To an increasing extent, the *history of the Earth* and its chronological past, together with its structure, was given more and more attention by scholars interested in geology. The Earth's biblical age of 6000 years, determined and spread in Europe by religious figures such as James Ussher (1581–1656) and John Lightfoot (1602–1675), was of little consequence in Germany in the second half of the eighteenth century. The philosopher Immanuel Kant (1724–1804) believed (Kant 1755) that 'a number of millions of years [...] passed by before the sphere of shaped nature, in which

[9] Blumenbach also taught the Geological Society's first President, George Bellas Greenough (1778–1855) (Kölbl-Ebert 2009).

[10] *The Mineralogical Geography of Saxony.*

we reside, has prospered to such perfection' (reprinted in Kant 1961, vol. 1. p. 139). Simlarly, von Justi wrote that 'the oldest times of it are at least half a million years away' (Justi 1771, p. 99). Füchsel also spoke of the eternity of the processes in Nature (1773, pp. 139–140).

All these statements appeared before James Hutton (1726–1797) presented his great *Theory of the Earth* (Hutton 1785) to the public with its ideas about geology and the infinity of the existence of the Earth.

In the last decades of the eighteenth century, an understanding the Earth was also enriched by new views in the German countries. Above all, the influence of physicotheology (natural theology) was lessened in geological thinking. Religiously oriented explanations for naturally occurring phenomena clearly lost significance; God as an orientating and enlightening power was less drawn upon, and the explanation of (geological) phenomena by means of observation and logical reasoning had become widely accepted.

The development of geology as a science in Germany

At the end of the eighteenth century, current understanding about the nature of the Earth was taught at universities within the framework of natural history as, for example, in Jena, Göttingen and Leipzig, as well as in Munich and Königsberg. The realm of minerals in natural history became the locale for presentations on the nature of the Earth. In parallel with university education, mining academies were established that mainly imparted technological knowledge about mining and metallurgy, the fundamentals of mathematics and, most importantly, scientific knowledge about the natural history of the Earth. As a prerequisite for the mining and metallurgy industries, the mining academies were founded in the states and countries that had such industries, such as in Freiberg (1765 for Saxony), Schemnitz (1770 for Austria–Hungary), Berlin (1770 for Prussia), Petersburg (1773 for Russia), Almaden (1777 for Spain), Paris (1783 for France), Mexico City (1792 for Spain) and others. In these institutions, scientific knowledge about the natural history of the Earth became disassociated from natural history and was imparted independently due to the practical demands of industry. These changes advanced the growing body of knowledge on the nature of the Earth, and the development of geology and other sciences as autonomous disciplines.

At the turn of the nineteenth century, the Mining Academy in Freiberg received immediate acceptance. The reason for this was the solid education of its students, based on the teaching of excellent scholars such as Christlieb Ehregott Gellert (1713–1795) in chemistry and metallurgy, von Charpentier in mathematics and mechanics, and Werner in mineralogy, geology and mining. Werner (Guntau 1984) was the son of an ironworks inspector in Saxony, who had himself been a student at the Mining Academy from 1769 to 1771; after that, he studied at the University of Leipzig until 1774. In 1775 he was appointed lecturer in mineralogy and geology to the Freiberg Mining Academy, where he stayed until his death in 1817. During those decades, mineralogy and geology in Germany became independent scientific disciplines, in which Werner played an important role.[11]

By 1774 Werner had already attracted attention with his important book *Von den äusserlichen Kennzeichen der Fossilien.*[12] He accurately described the various characteristics of minerals, putting them into a system that developed into a methodology for determining minerals. As collecting minerals was in vogue at that time, Werner quickly became famous. The book also became the basis for his extensive teaching of mineralogy. For many years Werner taught techniques in mining and iron metallurgy. His lectures and seminars on mineralogy (1775–1817), geology (1786–1817), regional geology of Saxony (1781–1782), palaeontology (1799) and the history of mineralogy (1803) were of particular importance, although the subject titles employed here were not used by Werner. In about 1790 he defined the subject of mineralogy as a comprehensive term incorporating all fields of the 'natural body of the Earth'. Accordingly, he subdivided mineralogy into five

[11] Abraham Gottlob Werner's private library and 84 volumes of handwritten papers (hereafter 'Literary remains') are located at the Georgius Agricola Library of the Technische Universität Bergakademie, Freiberg, Germany, where they may be used for research purposes.

[12] *On the External Characters of Fossils [Minerals].* Minerals were called fossils by Werner. Georgius Agricola (1494–1555) introduced the term 'fossil' instead of 'mineral' in his book *De natura fossilium* published in Basel in 1546. The term 'fossil' was widespread in Germany: and, until about 1820, it meant both fossil and mineral. After that only the remains of ancient floral and faunal organisms were called fossils.

disciplines, the terms for which were used in the German literature until about 1820. These were as follows:[13]

(1) *Oryctognosy*:	'Science of the determination of minerals', i.e. the field of present-day mineralogy.
(2) *Mineral Chemistry*:	The chemical composition and analysis of minerals.
(3) *Geognosy*:	The nature of the Earth in its totality, i.e. present-day geology.
(4) *Geographical Mineralogy*:	The regional geology of specific places, particularly with respect to rocks and ore deposits.
(5) *Economic Mineralogy*:	The characteristics of minerals and their processing for the benefit of humanity.

Werner taught geognosy at the Mining Academy of Freiberg for over 40 years, responding to the demands of his students that he should impart practical information for use in the mining industry. However, it was not only industry that he considered in his teaching, but also general geological knowledge. Over the years, he changed and perfected the contents of his lectures, developing a scientific theory about the state of the Earth, as it was in the present and had been in the past. He published several parts of these lectures on geognosy, although not the complete body of work. Nevertheless, a good understanding of Werner's concepts has been handed down to us through the extensive manuscripts of his lectures and the notes taken by his students, as well as through textbooks written by his students in accordance with his doctrines. To illustrate the complexity of his teaching, some key aspects of his lectures and publications on geognosy are presented here (Albrecht & Ladwig 2003).

The basis of Werner's hypothesis was Neptunism, through which he developed a system that guided his observations and interpretations. Although Neptunism was eventually to be replaced as a doctrine, for a time it served as the leading theory. In about 1786 he expressed this belief in Neptunism with statements like the following: 'The solid globe, insofar as we know it, was originally formed entirely from water [...] the work that has been done on it by fire is insignificant in comparison with the whole'.[14] In contrast, the advocates of Volcanism were of the opinion that, above all, heating and melting were the most significant processes that shaped the Earth. The leading proponent of this concept was James Hutton, who wrote 'the power of heat and operation of fusion must have been employed in consolidating strata of loose materials, which had been collected together and amassed at the bottom of the ocean [...] all the solid strata of the globe have been condensed by means of heat and hardened from a state of fusion' (Hutton 1785, pp. 50 and 51 ff.).

Werner mainly propagated his Neptunistic views through his lectures, explaining them to his students for over three decades. The content of his lectures on geognosy (1786–1817) covered a wide range of topics that included the following: the Earth as a planet and the formation of its surface; the structure of mountains; exogenous and endogenous forces that formed and altered the Earth's surface; the various forms of water (floods, rivers, ice, etc.) and their role in the configuration of the Earth's surface; volcanoes and earthquakes; mountains and rock masses in the history of the Earth: primitive, floez, volcanic and alluvial rocks; the characteristics of rocks and their uses; the genesis and tectonics of ore deposits, and deposits of economic minerals. With these lectures Werner imparted a great deal about geology and the nature of the Earth (Schmidt 1999), and the topics covered can be considered the roots of present-day disciplines such as physical and historical geology, petrography and sedimentology.

Oldroyd (1996)[15] points out that Werner published considerably more than is assumed. There are more than 30 scientific publications to be found, a number of which have been translated into various languages (Werner 1774, 1787, 1791, 1818), which explains why they were more widely disseminated.

Through his lectures on petrifactology (in 1799) Werner made an important contribution to the development of palaeontology. In his unpublished work, there is a manuscript of about 50 pages with a cohesive text apparently of his lecture on the petrifaction of animal and vegetation organisms.[16] In his view, the study of petrifactions was aimed at clarifying the history of the living world and, most importantly, at revealing the history of the Earth and its periods (Wagenbreth 1968, pp. 32 ff.).

Based on his Neptunistic hypothesis, Werner developed an elaborate theory about the history of the Earth. In his unpublished work there is a fragment in his handwriting[17] in which he says that the terrestrial body was formed by a precipitation

[13] Literary remains, vol. 10, pp. 68–70.
[14] Literary remains, vol. 2, p. 206.
[15] German edition 1998, p. 149.
[16] Literary remains, vol. 7, pp. 75–126.
[17] Literary remains, vol. 6, pp. 307–312.

of various substances and that this is proven by the content of the different rocks of which the mountains are composed. Water levels dropped gradually. In that early phase there was first chemical precipitation followed by sedimentation. Under these conditions only the first rocks such as gneiss, mica slate or older sandstone could have developed. When areas of the solid surface of the Earth rose above the water, the first organic creatures appeared. Thereafter greasy, combustible and salty substances emerged, followed by the formation of hydrogen, carbon, nitrogen, alkalis, organic acids, etc. The accumulation of combustibles gave rise to the emergence of volcanoes and volcanic mountains. The time Werner allowed for such processes was considerable. He stated, 'If water once totally covered our globe, perhaps 1.000 000 years ago, the present surface of the Earth must be quite different in various ways'.[18] He had an image of immense processes that caused changes to the Earth over long periods of time. Not only did he see an historical succession of incidents, but also an interlocking of events like an evolutionary chain. In all his writings, there are no references to any biblical texts such as the Creation or the Flood. Werner was a deist and consequently separated religion and science from each other, consistent with the Enlightenment of his time.

Werner's theoretical concept was based on findings of an earlier time. The development of knowledge over the years produced new observations and findings, which deprived Werner's Neptunism of its basis. He was not able to recognize this and the Neptunism–Volcanism argument (Wagenbreth 1955, pp. 183–241), which was started by his pupil Johann Carl Wilhelm Voigt (in 1789), merely consolidated his conservative position. Until at least the end of Werner's career, Neptunism served as the 'primary idea' under which his beliefs about geological processes and structures, and the genesis of rocks and deposits, were brought together. Although it was possible for Werner to maintain this position while he was still teaching, quite a few of his pupils – once they left Freiberg – rejected his Neptunistic interpretations and explanations, even though they still appreciated the extensive and effective education they had gained from him. Furthermore, if you wanted to achieve in the field of geology, you attended Werner's lectures.

In addition to his university-style lectures and seminars, Werner also enriched his teaching with excursions into the field or below ground. So under his guidance students were introduced to mapping. By order of the Saxon sovereign, Werner designed and organized a geological mapping programme in Saxony, aimed at providing a basis for searching for economic minerals in the state. Economic deposits, especially coal, were particularly targeted because the state urgently needed combustibles. In 1799 the operation was started as an exercise for the students of the mining academy. The Saxony area to be mapped was subdivided into 75 districts, and each student provided with written and oral instructions for their district. They were then sent into the field to work alone. After about four weeks of reconnaissance on foot they were required to submit the following: a description of the district, a map of their route, a petrography map – along with minerals and rocks collected as evidence – a journal and a statement of their finances. The material handed in was checked and evaluated by Werner, and passed on to the Saxon Board of Mining. For the mapping of the various rocks types, Werner had identified 40 'obligatory' colours, which allowed a uniform 'illumination' (colouring) and labelling among all students. From the results of this collaborative work emerged a picture of the geology of Saxony. The first report on this mapping project was presented in 1811 (Schellhas 1967, pp. 245–278).

Werner's ideas were widespread; there were hardly any publications on mineralogy and geology in which he was not quoted, praised or criticized in one way or another (Ospovat 2003). Famous personalities were students of his as, for example, the natural scientist Alexander von Humboldt, the geologist and palaeontologist Leopold von Buch, the high-ranking mining official Johann Carl Freiesleben (1774–1846), and the statesman and natural scientist Ernst Friedrich Schlotheim (1764–1832). Even foreigners were Werner's pupils, who later filled high government positions or accomplished significant works. The Spaniard Fausto de Elhuyar (1755–1833) was General Director of Mining in Mexico; the Brazilian Manuel Ferreira da Camara (1762–1835) became General Manager of the gold and diamond mines of his country; the Englishman Thomas Weaver (1773–1855) became well known for his geological work; and Fjodor Moiseenko (1754–1781) became – after returning from Freiberg – one of the most significant mineralogists in Russia. In the decades around 1800 Freiberg had about 500 students, most of whom were matriculated; others were registered as private pupils.

Whoever attended Werner's lectures and seminars learned all they needed to know about geology from the classification and description of minerals to the structure of rocks and ores, from the various structures of the Earth's surface to the exogenous

[18] Literary remains, vol. 2, p. 256.

and endogenous formative forces; from the beginning of stratigraphy and palaeontology to his views on the evolution of the history of the Earth. Werner imparted these findings in explanatory texts that were assimilated by his students and enabled them to understand all the new ideas about the nature of the Earth. It was a movement of great significance for the development of geology in the decades around 1800, because through his teachings the genesis of this discipline in the German countries (and beyond) became well known.

The consolidation of geology as a natural science in Germany

As in countries like England and France at the beginning of the nineteenth century, geology in Germany developed as an independent discipline with new questions and answers. A clear structure had been given to the process of observing geology, which had essentially matured. Speculation and theories became unacceptable, facts were wanted. Descriptions of the structure and history of the terrestrial crust were of particular interest. This included questions about the development of mountains, the properties of volcanoes, the relative age of rocks and petrified organisms, the formative forces of the Earth's surface and the geological structure of regions or countries. The Freiberg Professor, Abraham Gottlob Werner, was acknowledged as a successful lecturer. His pupils worked with the knowledge and methods acquired from their master; they had a solid education that also gave them the wherewithal to recognize where his views failed regarding the development of basalts and volcanoes. These scholars climbed the high mountains, travelled across Siberia and South America, and explored other countries to compare their observations with conditions in their homelands.

During his many travels – covering great distances on foot in doing so – through the Alps of Switzerland, the Auvergne in France to the volcanoes in Italy and the Canary Islands, Leopold von Buch (Mathé 1974) was an eager witness to geological conditions. After numerous observations as a result of his fieldwork – most of all in Italy 1805 (Buch 1802–1809), Norway and Lappland in 1806–1808 (Buch 1810), and the Canaries in 1815 (Buch 1825) – Buch rejected Werner's views and arrived at a belief that mountains were elevated above the surface by forces coming from depths within the Earth. From these observations, he developed his theory of the 'elevation crater' to explain the evolution of mountains: ascending magma lifted the strata above into domed structures until the centre collapsed, a vent was formed and the lava broke through. Buch presented this theory at the Academy of Sciences at Berlin in 1818 and published it in 1836 (Buch 1836, pp. 169–190).

Alexander von Humboldt too, had studied the mountains on his long journey through South America (1799–1804) and thought that the huge massifs could only have come into being by active forces of the Earth's interior. He believed he had recognized a relationship between mountain chains and coastlines, and considered that the strike directions of mountain chains bore a relationship to their age. He elaborately studied and presented the properties, development and classification of the various rocks (Humboldt 1823). He also assumed a connection between volcanic activity and earthquakes, comparing the volcanoes in Europe and South America (Humboldt 1825, pp. 137–155). Many of Humboldt's scientific findings, practices and questions were very novel for their time as they belonged to various geosciences that only developed and matured much later. They included disciplines like geomorphology, seismology, volcanology, geothermal activity, hydrology, petroleum geology, environmental geology and speleology.

But it was not only the Earth's current status that constituted geology's subject matter; there was increasing interest amongst geologists in the historical processes that had led to the current state of the Earth's surface. Studies of the deposition of rocks, the structure of mountains and the fossils of organic organisms provided opportunities to answer these questions. In doing so, most geologists in Germany concentrated on their observations and findings without referring to religious ideas or texts in the bible. For example, when in 1806 Leopold von Buch was elected corresponding member of the Academy of Sciences in Berlin, he talked about 'the progression of formations in nature' and drew a picture of the development of the Earth and life on this planet that stretched from the Universe through minerals (inorganic) to flora and fauna (organic) up to the appearance of man (social). For Buch, geology's main objective was to provide a complete history of the Earth: 'If geology has succeeded in leading this great progression of development from a shapeless drop of water into a dominion of man by certain laws, it does not seem unworthy for it to join the prestigious company of the sciences [...]' (Buch 1806, p. 16). Buch belonged to the first of those who clearly emphasized geology as a historically oriented natural science. For geologists, the history of the Earth became an important field of work that initially included lithostratigraphy as a method with which to determine the historical sequence of strata (Oldroyd 1979). This more intensive orientation towards the history of the Earth between

Table 1. *Professors of Mineralogy and Geology, respectively, around 1800*

Names and dates	Attendance at the Mining Academy of Freiberg	Subsequent professional fields and institutions, with dates
Andrada e Silva, José Bonifacio (1763–1838)	1792–1794	Professor of Mineralogy at the University of Coimbra, Portugal, 1801–1807
Breithaupt, Friedrich August (1791–1873)	1811–1812	Professor of Mineralogy at the Mining Academy of Freiberg, Saxony, 1826–1866
Brochant du Villiers, André J.M. (1772–1840)	1795	Lecturer of Mineralogy at the Mining School of Pesey (Savoy), France, from 1802
Engelhardt, Otto Moritz Ludwig von (1779–1824)	From 1805	Professor of Mineralogy at the University of Dorpat, Russia (Estonia), 1820–1842
Esmark, Jens (1762–1839)	From 1792	Reader of Mineralogy at the Mining Seminar of Kongsberg, 1802, and from 1814 Professor of Mineralogy and Geology at the University of Christiania, Norway
Fuchs, Johann Nepomuk von (1774–1856)	1803	Professor of Mineralogy at the University of Landshut, 1807, Custodian 1823 and from 1827 Professor of Mineralogy at the University of Munich, Bavaria
Gerhard, Carl Abraham (1738–1821)	(Not at Freiberg but at the University of Frankfurt/Oder)	Founder of the Mining Academy of Berlin, Prussia, 1770, and Professor of Mineralogy, Geology, Mining and Metallurgy there, 1770–1789
Germar, Ernst Friedrich (1786–1853)	From 1804	Professor of Mineralogy at the University of Halle, Prussia, 1817–1853
Jameson, Robert (1774–1854)	1800–1801	Professor of Geology and Natural History at the University of Edinburgh, Scotland, from 1804
Karsten, Dietrich Ludwig Gustav (1768–1810)	1782–1786	Professor of Mineralogy and Geology at the Mining Academy of Berlin, Prussia, 1789–1810
Leonhard, Karl Caesar von (1779–1862)	(Not at Freiberg but at the Universities of Marburg and Göttingen)	Professor of Mineralogy and Geology at Heidelberg, Baden (Germany), from 1818

Meder, Pjetr Ivanovic (1769–1826)	1793–1794	Professor of Geology and Mineralogy at the Mining Institute of St Petersburg, Russia, 1797–1826
Mohs, Carl Friedrich (1783–1839)	1798–1801	Professor of Mineralogy at Johanneum of Graz, Austria, 1812–1818; at the Mining Academy of Freiberg, Saxony, 1818–1826, and at the University of Vienna, Austria, 1826–1835
Moiseenko, Fedor Petrivic (1754–1781)	1774–1778	Professor of Mineralogy at the Mining Institute of St Petersburg, Russia, 1780
Naumann, Carl Friedrich (1779–1873)	1816–1817	Professor of Geology and Mineralogy at the University of Leipzig, Saxony, 1842–1870
Pusch, Jerzy Bogumil (1791–1873)	1806–1807	Professor of Geology at the Mining Academy of Kielce, Poland, 1816–1826
Rio, Andrés Manuel del (1765–1849)	1791–1792	Professor of Mineralogy at the Mining Academy of Mexico City, Mexico, from 1795
Schreiber, Johann Gottfried (1746–1827)	1771–1773	Director and Mineralogist at the Mining School Pesey/Moutiers (Savoy), France, 1802–1816
Schubert, Gotthilf Heinrich (1780–1860)	1805–1806	Professor of Natural Sciences in Erlangen, Bavaria, 1816, and Lecturer of Mineralogy in Munich, Bavaria, 1827
Steffens, Henrik (1773–1845)	1799–1802	Professor of the University of Halle, Prussia, 1805–1811, and at the University of Breslau, Prussia, 1819–1832
Symonowicz, Roman (1763–1813)	Times unknown	Professor of Mineralogy and Geology at the University of Vilnius, Russia (Lithuania), from 1803
Weiss, Christian Samuel (1780–1856)	1802–1803	Professor of Mineralogy and Geology at the University of Berlin, Prussia, 1810–1856
Werner, Abraham Gottlob (1749–1817)	Freiberg and Leipzig University 1769–1774	Inspector and Lecturer of Mineralogy and Mining at the Mining Academy of Freiberg, 1775–1817
Zimmermann, Johann Christian (1786–1853)	1804–1805	Lecturer of Mineralogy and Geology at the Mining School of Clausthal, Lower Saxony, 1811–1846

1780 and 1815 was called historiography, or, as David Oldroyd has said, 'geology became historicized' (Oldroyd 1996, p. 115).

Leopold von Buch not only recognized the task of developing historical geology, he also made important contributions to the stratigraphy of the Mesozoic and to palaeontology. He originated the term 'guide fossil', when he realized that sedimentary layers of the same type may contain a specific fossil, or combination of fossils. The underlying principle, of course, had been recognized much earlier. In 1796 the English engineer William Smith (1769–1839) observed that the same kinds of bivalves are only found in certain layers, thereby characterizing those strata. The coloured geological map of Western and Central Europe, published by Buch in 1826, was extremely well received, going into five editions until 1843.

In 1811 Georges Cuvier (1769–1832) subdivided the sedimentary sequence of the Paris Basin into 11 sections using lithostratigraphic criteria. He also included the fossils that characterized each horizon, in other words he used biostratigraphic criteria as well (Zittel 1899, p. 149). In 1813 Ernst Friedrich Schlotheim, who had already become known through a publication on palaeobotany (Schlotheim 1804), explained the importance of fossils in stratigraphical research as follows:

> Obviously the occurrence of fossils can furnish us with the most important aid in the closer determination of the relative ages of many kinds of rock, and enables us to distinguish the synchronous origins of these and their subordinate beds. There is an endless variety of Ammonites, Terebratulae, etc. Of these, certain species may occur only in the Transition Series, certain ones in the Alpine limestone, others in the Jura limestone, and still others in our so-called Kupferschiefer, Upper Muschelkalk, and Sandstone formations, etc. Should this supposition be affirmed, we would thus be practically in a position to determine the characteristic fossils for several series of formations and then could reach many unexpected conclusions concerning the rock layers and stratum members belonging to a given formation.
>
> (Schlotheim 1813, pp. 5 and 9)

Schlotheim had been acutely concerned that the biostratigraphic methodology was enforced in Germany in order to determine the relative ages of the sediments. As a result, the significance of palaeontology in the development of historical geology became accepted.

At the beginning of the nineteenth century geological observations, descriptions and knowledge were increasing so considerably that people felt the need to produce overviews and summaries. Along with other authors, Karl Ernst Adolf von Hoff (1771–1837) (Mathé 1985) at Gotha (Thuringia) published several volumes on the alteration of the Earth's surface. Collating information from an enormous quantity of printed sources, he reported changes in the relationship between land and sea, the sinking and uplift of continents, the rising and falling of sea levels, the properties of volcanoes and earthquakes on the Earth's surface, subsidence and elevation of the ground, hot springs, landslides and rock falls, properties of rivers and lakes, ice and glaciers, big pebbles and boulders, sand drift and dunes, and petrified organic creatures. In other words, everything in the history of the Earth that had thus far been recognized. As for the character of natural forces in the history of the Earth, Hoff's ideas were quite similar to those of Charles Lyell (1797–1875). As early as 1822, he wrote: 'that small effects prolonged over huge amounts of time explain a lot in the history of the terrestrial crust' (Hoff 1822–1834, vol. 1, p. 209). Despite the assumption of gentle, non-catastrophic forces in the past and present, he avoided mechanistic uniformity and said that in the history of the Earth 'every now and then, and here and there, obvious effects of natural forces known from our experience could occur faster and more forcefully' (Hoff 1822–1834, vol. 3, p. 237). These ideas corresponded with Lyell's view of uniformity, as well as with Werner's belief that the forces affecting the Earth were the same in its history as in the present: 'As long as the solid terrestrial body is known to us from its surface, it shows itself as a child of its time, as a result of natural effects, as a function of bodies and forces continuing to affect on it'.[19]

By the beginning of the nineteenth century, geology had developed as an independent natural scientific discipline (Guntau 1978, pp. 280–290). A cognitive basis in the form of a large number of findings about the nature of the Earth had accumulated, whose contents – oriented towards central ideas such as Neptunism or Volcanism – were recognized as not belonging to the realm of other (long-established) sciences. But the life of a discipline or science is not only propagated through its cognitive body of knowledge. To survive and have an impact, a discipline needs a social framework such as scientific associations, journals and textbooks, and training at universities or academies. For geology in Germany, the appropriate framework – together with the cognitive processes – came into being in the years around 1800 (Guntau 1985; Prescher 1985). Above all, there were facilities for teaching geology at universities and mining academies. At first, these subjects had been part of courses in other disciplines but now they were independent. Professors of geology who studied in Germany,

[19] Literary remains, vol. 6, p. 128.

and foreign students who became professors in their home countries, then disseminated the geological knowledge and experience that they had acquired there (Table 1).

Conclusion

With the end of natural history as a discipline, geology – along with chemistry, zoology and botany – appeared as a separate subject at German universities and mining academies. Initially Werner's terminology was still used, but once this became outmoded in the 1820s, the term 'geology' was finally accepted in Germany.

In the years shortly before and after 1800 the first manuals and textbooks on geognosy (geology) were edited in Germany, expressing the independence of this science. To some extent, mineralogical and geological works were still to be found formally combined, although separated in terms of contents. Examples of these were the first books on geology, such as *Practische Gebirgskunde*[20] (1792) by Johann Carl Wilhelm Voigt, the two-volume *Lehrbuch der Geognosie*[21] (1805) by Franz Ambros Reuss (1761–1830) and Gotthilf Heinrich Schubert's (1780–1860) *Handbuch der Geognosie*[22] (1813). Still under the influence of Werner's doctrine, publications about geology were also edited abroad, such as Robert Jameson's (1774–1854) *Elements of Geognosy or the Geological System of Werner* (1808) and *Traité de géognosie* (1819) by the Frenchman Jean Francois d'Aubisson de Voisins (1769–1841), which also became available in Germany (in Dresden in 1821) and was met with great approval. Thus geology was introduced to the public as an independent discipline.

Karl von Zittel (1839–1904), in his *Geschichte der Geologie und Paläontologie*[23] (1899) described the period of the development of geological knowledge at the turn of the nineteenth century as the 'heroic age of geology'. Interest in the natural history of the Earth was blossoming in society and mining, which became increasingly dependent on the geological exploration for natural resources. Under these conditions, geological research became organized as a scientific discipline or science, thus the eighteenth century should be regarded as the *prehistory* of geology. In it, fundamental geological discoveries accumulated, and the rate of new information intensified and accelerated. This led to a new quality of understanding and geology *emerging* as an independent discipline around 1800. A significant role in incorporating these observations and findings into the discipline of geology was attributed to the leading ideas such as Neptunism and Volcanism, which were coloured with arguments about the genesis of basalt and the origins of volcanic heat. The first half of the nineteenth century was a time of *consolidation* for geology in Germany, and this was paralleled in several other countries. While establishing geology as a new discipline, it was of the utmost importance to make sure that the previous traditions of scientific geological work were not disrupted. Initially, erroneous and inadequate interpretations of the established body of geological knowledge had to be overcome, and accurate and better explanations found. Then, most importantly, the acquisition of new data had to be secured, new questions followed up and the spread of geological knowledge had to be organized. For this purpose an institutional framework was necessary, such as chairs at universities and opportunities for publication and the scientific exchange of ideas.

In the year the Geological Society of London was established (1807) the founders of geology as a science came from France, England, Scotland, Sweden, Austria-Hungary, Italy, Germany and several other countries, too. It was the work of these many scholars who laid the groundwork for the subject of geology that today we respect as a scientific discipline.

References

ADELUNG, J. CH. (ed.). 1768–1771. *Mineralogische Belustigungen, zum Behuf der Chemy und der Naturgeschichte des Mineralreichs* (6 vols). Heineck & Faber, Leipzig.

ALBRECHT, H. & LADWIG, R. (eds). 2003. *Abraham Gottlob Werner and the Foundation of the Geological Sciences.* Freiberger Forschungshefte D 207. Technische Universität Bergakademie, Freiberg.

AUBISSON DE VOISIUS, J. F. D' 1819. *Traté de Geognosie* (2 vols). Levrault, Strasbourg (1821–1822. *Geognosie*. Arnold, Dresden).

BIELEFELDT, E. 1980. Carl Abraham Gerhard 1738–1821, ein Berliner Geologe der Aufklärung. *Zeitschrift für Geologische Wissenschaften, Akademie, Berlin*, **8**, 207–215.

BLEI, W. 1981. *Erkenntniswege zur Erd- und Lebensgeschichte.* Akademie, Berlin.

BLUMENBACH, J. F. 1790. *Beyträge zur Naturgeschichte der Vorwelt. Magazin für das Neueste aus der Physik und Naturgeschichte*. Ettinger, Gotha.

BUCH, L. VON. 1802–1809. *Geognostische Beobachtungen auf Reisen durch Deutschland und Italien.* Haude & Spener, Berlin.

BUCH, L. VON. 1806. *Über das Fortschreiten der Bildungen in der Natur.* Königliche Akademie, Berlin.

[20] *Practical Orography.*

[21] *Manual of Geognosy.*

[22] *Textbook of Geognosy.*

[23] *History of Geology and Palaeontology.*

BUCH, L. VON. 1810. *Reise durch Norwegen und Lappland* (2 vols). Nauck, Berlin.
BUCH, L. VON. 1818. *Ueber die Zusammensetzung der baltischen Inseln und über Erhebungscratere*. Königliche Akademie, Berlin.
BUCH, L. VON. 1825. *Physicalische Beschreibung der Canarischen Inseln* (3 vols). Königliche Akademie, Berlin.
BUCH, L. VON. 1826. *Geognostische Karte von Deutschland und den umliegenden Staaten in 42 Blättern im Massstab 1:1 000 000*. Schropp, Berlin.
BUCH, L. VON. 1836. Ueber Erhebungskratere und Vulkane. *Poggendorfs Annalen der Physik und Chemie, Leipzig*, **37**, 169–190.
CHARPENTIER, J. F. W. 1778. *Mineralogische Geographie der Chursächsischen Lande*. Crusius, Leipzig.
FLURL, M. B. VON. 1792. *Beschreibung der Gebirge von Bayern und der oberen Pflanz*. Joseph Leutner, München.
FÜCHSEL, G. CH. 1761. Historia terrae et maris ex historia thuringiae, per montium descriptionem. *Actorum Academiae electoralis moguntinae scientiarum utilium quae Erfordiae est, Erfordiae*, **II**, 44–254.
[FÜCHSEL, G. CH.] 1773. *Entwurf zu der ältesten Erd- und Menschengeschichte, nebst einem Versuch, den Ursprung der Sprache zu finden*. Frankfurt (published anonymously).
GUNTAU, M. 1978. The emergence of geology as a scientific discipline. *History of Science*, **XVI**, 280–290.
GUNTAU, M. 1984. *Abraham Gottlob Werner*. Teubner, Leipzig.
GUNTAU, M. 1985. Gedanken zu einhundert Jahren Geologie in Deutschland 1770 bis 1870. *In*: PRESCHER, H. (ed.) *Leben und Wirken Deutscher Geologen im 18. Und 19. Jahrhundert*. Grundstoffindustrie, Leipzig, 9–17.
HOFF, K. E. A. 1822–1834. *Geschichte der durch Überlieferung nachgewiesenen natürlichen Veränderungen der Erdoberfläche* (3 vols). Perthes, Gotha.
HUMBOLDT, A. VON 1823. *A Geognostical Essay on the Superposition of Rocks in Both Hemispheres*. Longmans, London.
HUMBOLDT, A. VON 1825. Über den Bau und die Wirkungsart der Vulcane in verschiedenen Erdstrichen. *Abhandlung der physikalischen Klasse der Koniglichen Akademie der Wissenschaften zu Berlin*. Akademie, Berlin, **4**, 137–155.
HUTTON, J. 1785. *Theory of the Earth, or an Investigation of the Laws Observable in the Composition, Dissolution and Restoration of Land upon the Globe. From Transactions of the Royal Society of Edinburgh*, 1788, pp. 1–96, 4 plates.
JAMESON, R. 1808. Elements of Geognosy. *In*: *III Vol. System of Mineralogy, Comprehending*. Archibald Constable, Edinburgh.
JUSTI, J. H. G. VON. 1757. *Grundriss des gesamten Mineralreichs, worinnen alle Fossilien in einem, ihren wesentlichen Beschaffenheiten gemässen, Zusammenhange vorgestellt und beschrieben werden*. Vandenhoeck, Göttingen.
JUSTI, J. H. G. VON. 1771. *Geschichte des Erd-Cörpers aus seinen äusserlichen und unterirdischen Beschaffenheiten hergeleitet und erwiesen*. Himburg, Berlin.
KANT, I. 1755. *Allgemeine Naturgeschichte und Theorie des Himmels. Petersen, Königsberg & Leipzig. Frühschriften*, Volume 1. KLAUS, G. (ed.), Akademie, Berlin, reprinted in 1961, pp. 35–199.
KOCH, R. A. 1980. Der humanistisch-fortschrittliche Ideengehalt im Lebenswerk des Kameralisten und Berghauptmanns Johann Heinrich Gottlob von Justi. *Zeitschrift für Geologische Wissenschaften, Berlin*, **8**, 181–188.
KÖHLER, A. W. & HOFFMANN, C. A. S. (eds). 1788–1794. *Bergmännisches Journal* (12 vols). Crazische Buchhandlung, Freiberg.
KÖLBL-EBERT, M. 2009. George Bellas Greenough's 'Theory of the Earth' and its impact on the early Geological Society. *In*: LEWIS, C. L. E. & KNELL, S. J. (eds) *The Making of the Geological Society of London*. Geological Society, London, Special Publications, **317**, 115–128.
LASIUS, G. S. O. 1789. *Beobachtungen über die Harzgebirge: nebst einer petrographischen Charte und einem Profilrisse als ein Beytrag zur mineralogischen Naturkunde*. Helwigische Buchhandlung, Hannover.
LEHMANN, J. G. 1756. *Versuch einer Geschichte von Flötz-Gebürgen, betreffend deren Entstehung, lage, darinnen befindliche Metalle, Mineralien und Fossilien*. Klütersche Buchhandlung, Berlin.
LEONHARD, C. C. (ed.). 1807–1829. *Taschenbuch für die gesammte Mineralogie mit Hinsicht auf die neuesten Entdeckungen* (35 vols). Johann Christian Hermann, Frankfurt.
MATHÉ, G. 1974. Leopold von Buch und seine Bedeutung für die Entwicklung der Geologie. *Zeitschrift für Geologische Wissenschaften, Berlin*, **2**, 1395–1404.
MATHÉ, G. 1985. K.E.A. v. Hoff 1771–1837. Verdienste und Grenzen eines Gothanischen Staatsbeamten um die Förderung geologischen Denkens. *In*: PRESCHER, H. (ed.) *Leben und Wirken Deutscher Geologen im 18. Und 19. Jahrhundert*. Grundstoffindustrie, Leipzig, 118–139.
MÖLLER, R. 1963. *Mitteilungen zur Biographie Georg Christian Füchsels*. Freiberger Forschungshefte, **D 43**. Grundstoffindustrie, Leipzig.
OLDROYD, D. R. 1979. *Historism and the Rise of Historical Geology, Part 1. History of Science*, **XVII**, 191–213.
OLDROYD, D. R. 1996. *Thinking about the Earth*. Athlone, London.
OSPOVAT, A. 2003. Why Werner is one of the founders of modern geology. *In*: ALBRECHT, H. & LADWIG, R. (eds) *Abraham Gottlob Werner and the Foundation of the Geological Sciences*, Freiberger Forschungshefte, **D 207**. Technische Universität Bergakademie, *der Koniglichen*, Freiberg, 6–14.
PALLAS, P. S. 1771–1776. *Reise durch verschiedene Provinzen des russischen Reichs in den Jahren 1768–1773*. (3 vols) Kayserliche Akademie, St Petersburg.
PALLAS, P. S. 1778. *Betrachtungen über die Beschaffenheit der Gebürge und Veränderungen der Erdkugel, besonders in Beziehung auf das Russische Reich*. Hartnoch, Frankfurt.
PFINGSTEN, J. H. (ed.). 1789–1790. *Magazin für die Mineralogie und mineralogische Technologie* (2 vols). Gebauer, Halle.

PRESCHER, H. (ed.). 1985. *Leben und Wirken Deutscher Geologen im 18. und 19. Jahrhundert*. Grundstoffindustrie, Leipzig.

PRESCHER, H. 1969. Johann Gottlob Lehmann (1719–1767). Sein Leben und Werk in Dresden, Berlin und Petersburg. *Sächsische Heimatblätter, Dresden*, **15**, 274–277.

RASPE, R. E. 1763. *Specimen historiae naturalis globi terraquei, praecique de novis e mari natis insulis, et Hovkiana telluris hypothesis, de origine montium et corporum petrefactorum*. Schreuder, Amstelodami.

RASPE, R. E. 1774. *Beytrag zur allerältesten und natürlichen Historie von Hessen oder Beschreibung des Habichwaldes und verschiedner andern niederhessischen alten Vulcane in der Nachbarschaft von Cassel*. Cramer, Cassel.

REUSS, F. A. 1805. *Geognosie* (2 vols). Jacobaeer, Leipzig.

SCHELLHAS, W. 1967. *Abraham Gottlob Werner als Inspektor der Bergakademie Freiberg und als Mitglied des Sächsischen Oberbergamtes zu Freiberg*. Freiberger Forschungshefte, **C 223**. 245–278. Grundstoffindustrie, Leipzig.

SCHLOTHEIM, E. F. VON. 1804. *Beschreibung merkwürdiger Kräuter-Abdrücke und Pflanzen-Versteinerungen*. Becker, Gotha.

SCHLOTHEIM, E. F. VON. 1813. Beiträge zur Naturgeschichte der Versteinerungen in geognostischer Hinsicht. *Taschenbuch für die gesammte Mineralogie*, **7**, 2–134.

SCHMIDT, P. 1999. Werners Vorlesung 'Geognosie' nach dessen handschriftlichem Plan von 1816. *Nachrichtenblatt zur Geschichte der Geowissenschaften, Krefeld and Freiberg*, **7/8**, 162–168.

SCHRÖTER, J. S. (ed.). 1773–1780. *Journal für die Liebhaber des Steinreichs und der Conchyliologie* (6 vols). Hofmann, Weimar.

SCHUBERT, G. H. 1813. *Geognosie und Bergbaukunde*. Schrag, Nürnberg.

VACCARI, E. 1993. *Giovanni Arduino*. Olschki, Firenze.

VAI, G. B. 2009. Light and shadow: The status of Italian geology around 1807. *In*: LEWIS, C. L. E. & KNELL, S. J. (eds) *The Making of the Geological Society of London*. Geological Society, London, Special Publications, **317**, 179–202.

VOIGT, J. C. W. 1781. Schreiben an Professor Leske über die Rhönberge, Weimar 3.3.1781. *Leipziger Magazin zur Naturkunde*, **1**, 19–20.

VOIGT, J. C. W. 1792. *Practische Gebirgskunde*. Industrie-Comptoirs, Weimar.

WAGENBRETH, O. 1955. *Abraham Gottlob Werner und der Höhepunkt des Neptunistenstreites um 1790*. Freiberger Forschungshefte, **D 11**, 183–241. Akademie, Berlin.

WAGENBRETH, O. 1968. Die Paläontologie in Abraham Gottlob Werners geologischem System. *Bergakademie. Grundstoffindustrie, Leipzig*, **20**, 32–36.

WALCH, J. E. I. 1755–1773. *Die Naturgeschichte der Versteinerungen der knorrischen Sammlung von Merkwürdigkeiten der Natur* (9 vols). Felsecken, Nürnberg.

WALCH, J. E. I. 1762. *Das Steinreich systematisch entworfen*. Gebauer, Halle.

WENDLAND, F. 1992. *Peter Simon Pallas 1741–1811*. de Gruyter Berlin.

WERNER, A. G. 1774. *Von den äusserlichen Kennzeichen der Fossilien*. Crusius, Leipzig (ed.). 1785 Trattern, Wien; (engl.) 1805. *A Treatise on the External Characters of Fossils* (translated by TH. WEAVER). Mahon, Dublin; 1962. *On the External Characters of Minerals* (translated by A. V. CAROZZI). University of Illinois Press, Urbana, IL.

WERNER, A. G. 1787. *Kurze Klassifikation und Beschreibung der verschiedenen Gebirgsarten*. Walter, Dresden; (engl.) 1971. *Short Classification and Description of the Various Rocks* (translated by A. M. OSPOVAT). Hafner, New York.

WERNER, A. G. 1791. *Neue Theorie von der Entstehung der Gänge, mit Anwendung auf den Bergbau besonders den freibergischen*. Gerlach, Freiberg; (engl.) 1809. *New Theory of the Formation of Veins, with the Application to the Art of Working Mines* (translated by CH. ANDERSON). Encyclopaedia Britannia Press, Edinburgh.

WERNER, A. G. 1818. *Allgemeine Betrachtungen über den festen Erdkörper*. Auswahl aus den Schriften der unter Werners Mitwirkung gestifteten Gesellschaft für Mineralogie zu Dresden. Gleditsch, Leipzig. 1. vol., 39–57; (engl.) 1999. *General Observations About the Earth* (translated by A. M. OSPOVAT). Printed by INHIGEO, University of New South Wales, Sydney (Australia).

ZITTEL, K. VON. 1899. *Geschichte der Geologie und Paläontologie bis Ende des 19. Jahrhunderts*. Oldenbourg, München (engl.). 1901. *History of Geology and Palaeontology to the End of the Nineteenth Century* (translated by M. OGLIBIE-GORDON). Walter Scott, London.

Light and shadow: the status of Italian geology around 1807

GIAN BATTISTA VAI

Dipartimento di Scienze della Terra e Geologiche Ambientali, Università di Bologna – Alma Mater Studiorum, Via Zamboni 67, 40127 Bologna, Italy (e-mail: giambattista.vai@unibo.it)

Abstract: The stratigraphical approach and geological mapping of William Smith in England and Georges Cuvier in France gave birth to modern geology. However, before 1815 neither used the word 'geology', a term first coined by Ulisse Aldrovandi in 1603. At the turn of the nineteenth century most leading geoscientists were based in France and Germany, but those in Britain were poised to take over the lead. After three centuries of dominance in science and geology, was Italian geology in decline? A review of the works of Italian geologists and the role these played in disseminating Italian geological research has been undertaken to examine this question. The French Revolution and the Napoleonic wars shocked the Italian states, disrupted the economic order and discontinued the progress of science. Nevertheless, from 1759 to 1859 over 40 classic papers in geology were published in Italy. Among them, Gian Battista Brocchi's *Conchiologia Fossile* is the most renowned for having inspired Charles Lyell's work. In the middle decades of the nineteenth century Italian geoscientists made up the majority of foreign members of both the French and English geological societies. The Italian Geological Society was not formed until 1881. This was largely due to the earlier political fragmentation of Italy into many small states.

Two different paths can be followed to assess the state of a discipline in a particular country or cultural area[1] at a given time. The first requires analysis of facts, documents and historiographical sources. The second focuses on one or several masterpieces of the discipline, possibly comparing them with equivalent works in other areas or countries. The aim of this paper, therefore, is to explore both paths looking for consistency and reliability of the results obtained. The work of excellence selected to represent the status of geological research in the Italian cultural area at the beginning of the nineteenth century is Gian Battista Brocchi's (1772–1826) *Conchiologia Fossile Subapennina* (1814). This work is not only a masterpiece, but it is also particularly relevant since it contains a famous section on the history of conchology – perhaps the first ever written in such detail – which coincides with the beginnings of modern history in palaeontology and, to a certain extent, geology.

After a detailed chronological review of the literature, Brocchi (1814) was able to make a clear assessment of the status of geology in Italy at the end of the eighteenth century:

> in particular, the Tuscan savants contributed to the advancement of natural history, in particular fossil conchology [palaeontology]: a study which deserves to be called the base of geology.
>
> (Brocchi 1814, p. LXVII)

At the same time, he also identified two peculiarly Italian flaws: self-criticism and xenophilia or an excessive compliance to foreign ideas and fashion (Brocchi 1814, LXVII). Brocchi's general conclusion was:

> My aim was to make a point about the history of Italian science that was not very well known. A wealth of writings came out during three centuries; they did not neglect a single branch of natural science. Since it is not convenient to claim the glory of past merits, I will quote a series of classic authors from the recent past or who are still alive (Olivi, Cavolini, Poli, Renieri, Marsili, Donati, Iano Planco, Breislak, Spallanzani, Targioni, Santi, Savi, Tenore, Viviani, Bertoloni, Brignole). Yet, there are still many gaps.
>
> (Brocchi 1814, pp. LXXIX–LXXX)

This well-documented, proud evaluation that Italian science had held a prominent position is not eulogistic, but shows a realistic picture of Italian geology and an awareness that, however successful they had been in the past, even more could be done.

Charles Lyell (1797–1875) gave full credit to Brocchi's work, directly translating or summarizing most of Brocchi's history of geology in his own history section in the first volume of his *Principles of Geology* (1830–1833). When discussing the status of geology and the role of Italian research in

[1] The term 'cultural area' is used in preference to the country name, since Italy and Germany as we know them today did not exist until the late nineteenth century.

From: Lewis, C. L. E. & Knell, S. J. (eds) *The Making of the Geological Society of London*.
The Geological Society, London, Special Publications, **317**, 179–202.
DOI: 10.1144/SP317.10 0305-8719/09/$15.00

furthering its progress during the previous three centuries, Lyell says:

> It was not till the earlier part of the sixteenth century that geological phenomena began to attract the attention of the Christian nations. At that period a very animated controversy sprung up in Italy, concerning the true nature and origin of marine shells, and other organized fossils, found abundantly in the strata of the peninsula.
>
> (Lyell 1830, p. 23)

This work (Nicholas Steno's *De Solido* (1669)) attests to the priority of the Italian school in geological research (Lyell 1830, pp. 27–28):

> We return with pleasure to the geologists of Italy, who preceded, as we before saw, the naturalists of other countries in their investigations into the ancient history of the earth, and who still maintained a decided pre-eminence [during the eighteenth century].
>
> (Lyell 1830, p. 41)

MacCartney (1976), in a critical analysis of the two histories – Brocchi's and Lyell's – has argued that both have interpreted the sources based on their own geological paradigm. Although commonly true, such an attitude does not alter the validity of Brocchi's dataset, which has never been questioned by any scientist or historian, and certainly not by Lyell who had been able to check Brocchi's data and statements during his repeated visits to Italy.

But only 70 years later things had changed quite drastically, as demonstrated at the centenary celebration of the Geological Society of London in 1907. Sir Archibald Geikie (1835–1924), then President of both the Geological and the Royal Societies, gave a presidential address. He had chosen for his subject 'The State of Geology at the Time of the Foundation of the Geological Society'. In his speech (Geikie 1909), the Italian legacy and its early dominance in geology, so strongly asserted by Lyell, was almost forgotten despite Geikie being a good friend of the Italian geologist Giovanni Capellini (1833–1922) whose handwritten dedication to Capellini can be seen in a copy of Geikie's *Text-book of Geology* (1903)[2] (Fig. 1). The only Italian scientist quoted in Geikie's speech was Scipione Breislak (1748–1826) (Geikie 1909, p. 113, footnote 120). Geikie was an authority on the history of geology and had written a classic, if much-discussed, book on the subject (Geikie 1905) and reviewed others, making his omission even more striking.[3] His speech is notable for its British bias – this 'branch of science is not much older than the Geological Society itself' (Geikie 1909, p. 107) – the recognition of the French and Scottish schools, and the criticism of Werner's geognosy and Neptunism, and yet reference to Italian geology is almost completely absent. How can such a rapid change in the historiography of geology regarding the role of the Italian community be explained?

The obliteration of Italy's leading role in the history of geology by the end of the nineteenth century has both an 'external' and 'internal' rationale. The former is related to the changing approaches to geological sciences in the major European countries, while the latter can be explained by Italian individualism and provincialism, no longer compatible with the changing political and economical conditions in Europe. After three centuries of Italian dominance, the centres of development in the geosciences had moved to Germany and France by the end of the eighteenth century, and to Britain by the beginning of the nineteenth century (Ellenberger 1988, 1994; Oldroyd 1996; Vai & Cavazza 2003; Rudwick 2005; Vai & Caldwell 2006). In addition, the new geology of the nineteenth century, based on palaeontological stratigraphy, was developed mainly in France and Britain. The coincidence of these cultural areas with two long-standing and powerful nations favoured the awareness and popularization of a *national science* that could not be felt in nations such as Italy and Germany, which did not yet exist. At the same time, the international, interdisciplinary and broadly cultural approach to geoscience, culminating in a proliferation of theories of the Earth, came to an end.

After the original and pioneering historicist work done by Brocchi (1814), no new studies on the history of geology in Italy have been undertaken, neither by scientists nor historians, except perhaps for some research on Leonardo da Vinci and a few others.[4] Furthermore, recurrent articles on Italian primacy in geology by leading Italian geologists[5] mostly echoed Brocchi's and Lyell's histories, rather than presenting original historiographical research. Only in the last few decades has there been any attempt to fill this historiographical gap, but these studies mainly cover the golden age from the early sixteenth to late eighteenth centuries (Morello 1979, 1998; Vaccari 1993, 2003*a*; Vaccari & Curi 2003; Vai & Caldwell 2006). A

[2] This is now stored in the historical library of the Capellini Museum in Bologna (A, b, 91).

[3] See, also Vaccari 1993, pp. 5–6 and 16.

[4] See, for example, Baratta (1903), Gortani (1931, 1952, 1963) and Vai (1995*a*).

[5] For example, Bianconi (1862), Capellini (1897), Gemmellaro (1862), Meneghini (1866), Pilla (1840) and Stoppani (1862, 1881).

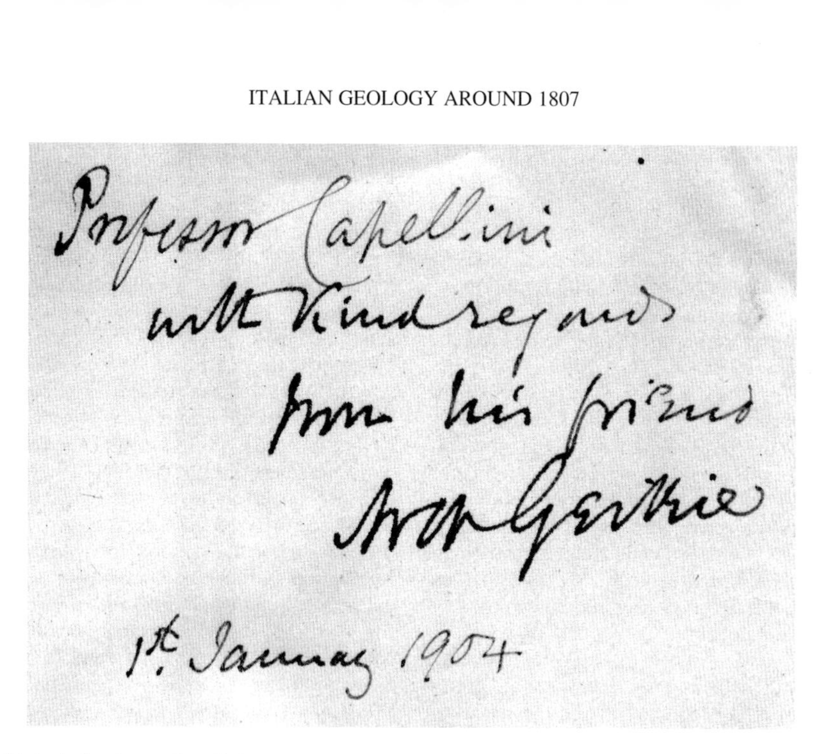

Fig. 1. Geikie's dedication to Capellini.

comprehensive picture of the scientific and cultural status of geology in Italy in the decades around the turn of the eighteenth to the nineteenth century is still not available.

The result of such oblivion for more than a century was silence or, at best, acceptance and propagation of a notion that, after a glorious past, Italian science (including geology) had suffered a decline since the eighteenth century (e.g. Cavazza 1990, 2002). But was this really the case for Italian geology in the crucial decades before and after the founding of the Geological Society in 1807? Factual evidence about the status of geological studies, results and publications in a broad time interval centred on the Geological Society's birth date will be discussed first. The results of this discussion will then be cross-checked, as a case history, by examining how far Lyell's *Principles* (1830–1833) was influenced by Brocchi's *Conchiologia* (1814).

A decline in geology or managerial delay during the Napoleonic turmoil?

The term 'geology' was first used in its modern sense in Bologna in 1603 by Ulisse Aldrovandi (1522–1605) (Vai 2003*a*; Vai & Cavazza 2006). He defined the new discipline as 'la Giologia ovvero de fossilibus' ('Geology, or about what is dug up from the Earth'). In particular, it made reference to his large scientific collection of rocks, fossil organisms and minerals that were to be displayed in a dedicated room bearing this name within his museum. It may be of interest and rather surprising to note here the remarkably similar quotation adopted for the Geological Society's logo; 'quicquid sub Terra est' ('whatever is under the Earth'). Thus, the circular recurrence of both concept and terminology, starting with Aldrovandi in 1603 and ending with the Geological Society in 1807, was closed. During those two centuries, terms such as lithology, petrology, oryctology and geognosy were used more commonly than geology and, when used, geology had several different meanings (Ellenberger 1994, pp. 250–251; Vai 2003*a*, pp. 73–74). In fact, it took almost two centuries of scattered use for the term geology to become accepted after Aldrovandi had introduced it, important examples being Arduino (1759–1760, 1774, p. 232), De Luc (1778), Targioni-Tozzetti (1779), De Saussure (1779–1796), Dolomieu (1794, p. 259) and Hutton (1795).

It is well known that Georges Cuvier's (1769–1832) and William Smith's (1769–1835) pioneering work in stratigraphy and geological mapping around the turn of the nineteenth century gave birth to modern geology. However, before 1815 neither used the term geology. Most of the Founding Fathers of the Geological Society would also not have called themselves 'geologists' before the Society was founded, despite having undertaken many geological activities (Lewis 2009), and its first President, George Bellas Greenough

(1778–1855), was, according to Geikie (1909, p. 112), 'a shrewd pupil of Werner' (but see also Kölbl-Ebert 2009). Geognosy – a knowledge of the Earth – had originated with Abraham Gottlob Werner (1749–1817) at the Freiberg Mining Academy in Germany in the middle of the eighteenth century (Guntau 2009) and its use soon spread all over Europe, including Italy (Morello 1983; Ciancio 1995; Vaccari 2000). Geognosy defined a science based on the recognition of the order, position and relation of the layers forming the Earth. On this basis, geognosists began to draw cross-sections that could represent the inclination of strata (geometry) in deformed areas, but which did not provide a reasonable lateral correlation of the same strata in order to understand their regional setting (structure). Furthermore, for geognosists, fossils were only lithic elements of the rocks and of no great consequence. Geologists, on the other hand, began quite early on to appreciate the time value of fossils, which were recognized as organisms that were independent from the encasing rock. Once this paradigm was established by Smith in 1799 (Smith 1815), Cuvier & Brongniart (1811) and Brocchi (1814), modern geology began its race. It is interesting to note that Cuvier & Brongniart when printing their *geological* map (1811) still used the conventional term *geognostical*, reflecting a still non-critical acceptance of the terminology received during their training. Smith, on the other hand, who had a less formal education, felt free to use the term 'geological' for his map (Smith 1815). Brocchi (1814) was the most critical and rigorous insofar as he desired to 'provide this work with a geognostical map' (Brocchi 1814, p. 56), while establishing his stratigraphical approach 'geologically speaking' (Brocchi 1814, p. 11; Vai 1995*b*, pp. 66–68). (See Table 1.)

But what about the status of Italian geology at this time? If the answer was to be based on the comparative founding dates of national geological societies, the Società Geologica Italiana would not occupy pole position; indeed, it comes behind most of the other European countries (Table 2). Obviously, there is not a direct causal link between the health of a geocommunity in a given country and the founding of its geological society. However, each late arrival in setting up basic geological institutions (such as societies, surveys, museums) has to be carefully assessed as to whether it indicates external constraints or the internal lack of potential. Why, then, did the Italian Geological Society appear 74 years after the founding of the Geological Society in London? Was the alleged Italian scientific and geological decline – suspected since Geikie's time – the reason for the delay? Was there really a true decline?

To answer these questions it is necessary to review the standard and number of naturalists active in 'geology' (geoscientists hereafter) during the eighteenth century, and to have some familiarity with their work – the number of publications they wrote and the questions they raised – as well as having an idea of how many of them took foreign correspondents into the field in Italy, and the ideas they exchanged and exported.

When analysed in this way, a true scientific decline during the period preceding the French Revolution, although commonly accepted and uncritically quoted, is difficult to prove – at least for geology. Works by Marsili, Vallisneri, Moro, Monti, Arduino, Targioni Tozzetti, Donati, Spallanzani and Soldani (Table 1), to name but a few, seem to testify to the contrary, as will be shown. In spite of recent attempts to revive the relevance and works of great Italian scientists,[6] what is still missing is a greater awareness of the wider body of knowledge regarding other Italian geoscientists, their writings and the role they played in the European forum at the time of the 'Italian tours' undertaken by many foreign scientists in the eighteenth century (Vaccari 2007; Wyse Jackson 2007).

A careful survey shows that within an interval of six decades around 1807, from 1777 to 1837, Italian geoscientists actively producing scientific papers exceeded 200 in number (Table 3). Half of these were devoted to the study of volcanoes and/or earthquakes, events of great interest to geoscientists in the second half of the eighteenth century, despite the dominance of Neptunism as a geological model. The works of many of these authors considerably influenced the views and works of the European and North American scientists undertaking the Italian tour. A typical example is the influence Giuseppe Recupero (1720–1778) had on Sir William Hamilton's (1730–1803) views on volcanology (Rudwick 2005).

In the middle of this period of activity came the French Revolution and the Napoleonic wars (1789–1814), which shocked the Italian states and disrupted their social and economic order, severely hampering and even halting, for a time, progress in science (Vai 2003*b*). However, there were local advantages brought by the Napoleonic reforms, such as reproducing French institutions strongly

[6] See, for example, publications on Aldrovandi (Vai & Cavazza 2006), Colonna and Scilla (Morello 2006*a*, *b*), Steno, Kircher and Marsili (Vaccari 2003*a*; Vai 2004*b*, 2006; Yamada 2006), Arduino (Vaccari 2006), Spallanzani (Spallanzani 1994; Vaccari 1998) and Fortis (Ciancio 1995).

Table 1. *A chronograph of leading natural historians from about 1200 to 1800, arranged by cultural and linguistic areas: Italian, German, French and British. Most of the prominent natural historians at the beginning of the nineteenth century were based in France and Germany, but those in Great Britain were poised to take over the lead in a few years time*

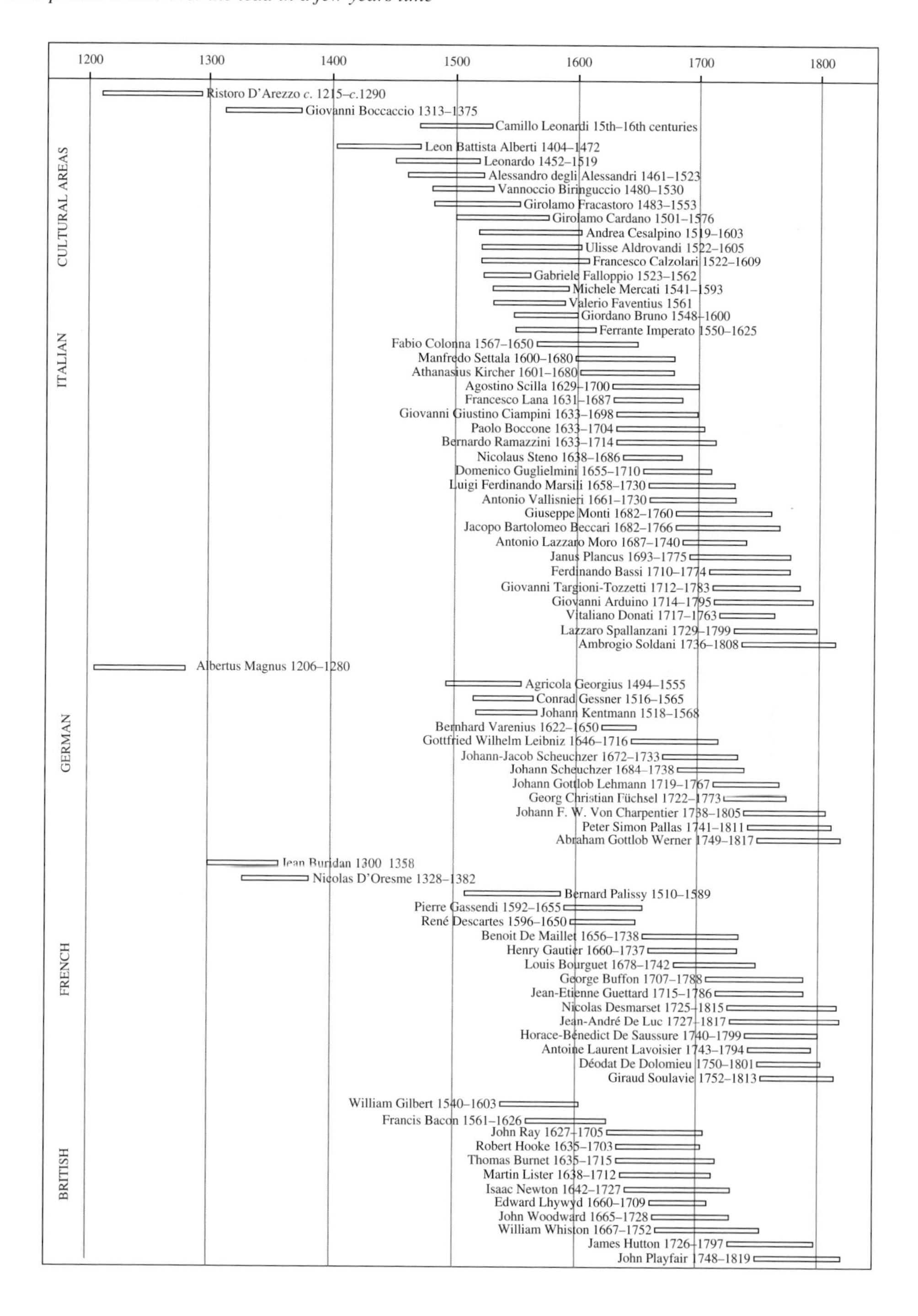

Table 2. *Foundation dates of national geological societies*

England (London)	1807
France	1830
Scotland (Edinburgh)	1834
Germany	1848
Hungary	1848
Sweden	1871
Belgium	1874
Italy	**1881**
Switzerland	1882
America	1888
Denmark	1893
Japan	1893
South Africa	1895
Norway	1905
Austria	1907
The Netherlands	1912
Spain	1985

related to geomineralogical studies in Milan, which became the capital city of the new Napoleonic Repubblica Cisalpina and later the Kingdom of northern Italy. The Istituto Lombardo delle Scienze (1797–1803) and the Consiglio delle Miniere del Regno Italico (1808) were two such institutes (Vaccari 2003*b*). This happened, however, at the expense of the Istituto delle Scienze di Bologna, which had had an excellent reputation for geology since 1711 (Tega 1986). In general, the advantages were too localized and of too short a duration – they did not last beyond the Restoration of 1815 – to outweigh the disadvantages. In some regions, such as the Church States (from Rome to Bologna), Tuscany and southern Italy, the previous scientific and academic structures were damaged or destroyed when, for example, collections were despoiled. These had to be restored, where possible, during a time of contrasts between revolution and restoration, reform and conservation.

A major and long-lasting impact of all this upheaval was on the economy. From having been a rich industrial and trading area, the Italian states were forced to become agricultural satellite areas of France, which gradually induced an increasing state of poverty (Vai 2003*b*, pp. 247–248). The turmoil also weakened the position of research centres such as those in Bologna, Florence, Padua and Siena that had previously been leaders in geology and natural sciences, while the role of Milan, Turin, Genoa, Rome, Naples and especially Pisa, which became the principal geological school of Italy in the third and fourth decades of the nineteenth century, was reinforced. Although there were some benefits, seen in the administrative, technical and political modernization introduced by the Napoleonic reforms, they took decades to become effective. The earlier shift of the scientific lingua franca from Latin to French and later to English had already caused the number of leading Italian scientists with an international visibility to decrease, and this trend was exacerbated by the reforms and the turmoil in the early decades of the nineteenth century. The quite large number of geoscientists of the eighteenth century was facing a gradual contraction.

Alongside the picture of political and economic instability, it may be useful to briefly review the most relevant works of Italian geoscientists printed in the century encompassing the origin of the Geological Society in 1807 and the Napoleonic turmoil in order to have a broader field for unravelling a trend in science development. During the century from 1759 to 1859 more than 40 'classic' papers on geology were published in Italy, with a dominance of authors born in the eighteenth century (Table 4).

The relevance and priority given by Arduino to both chronostratigraphy and lithostratigraphy is well recognized (Vaccari 1993, 2006; Vai 2007*a*). Nevertheless, his maps and geological cross-sections – among the first ever made – as well as his discovery of contact metamorphism in 1782, may come as a surprise to many. As shown in Table 4, there were several other geoscientists active in the Venetian region,[7] reflecting the strong tradition of Padua University and its school of geology. These geologists were well known in Britain and in France, even before Brocchi's works were introduced, and their interests were wide ranging at both the academic and consultant level. Most of them (Arduino, Fortis, Marzari-Pencati, Da Rio) focused on one important area of research, namely the interfingering, and thus interaction, between sedimentary and volcanic strata, and other igneous bodies that are well exposed in the Berici and Euganean hills near Padua. Their recognition of these effects explains why Neptunism and Plutonism are rarely observed in the discussions and works of the Italian geoscientists, and if only these works had been more widely disseminated at the time it might have contributed to solving, or at least mitigating, the largely speculative controversy about the origins of basalt. As it was, this remained a controversial topic for another 100 years. In addition, these geologists were able to provide the first evidence of what we now know to be post-Triassic granitoid intrusion, thereby strengthening the evidence for non-Primitive granites provided by Arduino in 1782. Before this work, granites had

[7] The region a geologist came from is generally indicated by the location given in the titles of their papers.

Table 3. *List of Italian geoscientists active between 1777 and 1837, indicating the main subject or region they worked on*

1	Acerbi	Venetian area	1829*
	Alessi G.	Sicily	1820s*
	Alfano	Naples Kingdom	1823*
	Algarotti M. A.	Venetian	1809–1823*
	Allioni Carlo	Turin	(1728–1804)
	Alvino F.	Vesuvius	1830s*
	Amoretti Carlo	Lombardy	(1741–1816)
	Ancona (d') G.	Calabria	1791*
	Aracri G.	Calabria	1810*
10	Arduino Giovanni	Venetian	(1714–1795)
	Ascoli (d')	Earthquakes	1806*
	Attumonelli M.	Vesuvius	1779*
	Augusti M.	Earthquakes	1789*, 1780*
	Baccio A.	Sicily	1793*
	Balbo P.	Piedmont	1784–1785*
	Baldracco C.	Liguria	1840s*
	Balsamo Crivelli G.	Lombardy & Italy	(1800–1874)
	Barba A.	Vesuvius	1794*
	Barelli V.	Emilia	1827*
20	Bartalini B.	Tuscany	1776, 1781*
	Barzellotti G.	Tuscany	1813*
	Battini D.	Tuscany	1793*
	Beffa C.	Sicily	1828*
	Bellani A.	Vesuvius	1835*
	Bellenghi A.	Sardinia	1832*
	Bernardino F.	Vesuvius	1794*
	Bertoloni Antonio	Bologna	(1775–1869)
	Bonfiglioli Malvezzi A.	Bologna	1777*
	Bonvicino	Turin	1780s–1800s*
30	Borson S.	Turin	1810s–1830s*
	Bottis (de) G.	Naples volcan.	1761*
	Breislak Scipione	Naples Kd	(1748–1826)
	Brocchi Giambattista	Venetian	(1772–1826)
	Bruno G. D.	Turin	1836*
	Calindri G.	Umbria & Marche	1829*
	Camilli S.	Latium	1833*
	Capocci E.	Naples	1835, 1837*
	Cappello A.	Latium	1820s*
	Carena A. P.	Piedmont	1760–1761*
40	Carpi P.	Latium	1829*
	Catullo Tommaso Antonio	Venetian	(1782–1869)
	Cava (la) P.	Vesuvius	1820*
	Ceva Grimaldi G.	Naples K. dom	1821*
	Chiavetta B.	Etna	1809*
	Ciofi A.	Vesuvius	1794*
	Colaci (de) O.	Earthquakes	1783*
	Collegno (di) Giacinto	Turin	(1793–1856)
	Collini C. A.		1776*
	Colosimo V.	Earthquakes	1832*
50	Consigliere (di) C.	Vesuvius	1818*
	Corrao A.	Earthquakes	1783*
	Cortesi G.	Emilia	1809*
	Covelli Nicola	Naples	(1790–1829)
	Cristofori (de) I.	Tuscany	1832*
	Delfico O.	Apennines	1794*
	Donati E.	Vesuvius	1831*
	Donato G.	Earthquakes	1832*
	Dondi Orologio A. Carlo	Venetian	1780s*
	Fantonetti G. D.	Lombardy	1836*

(*Continued*)

Table 3. *Continued*

60	Fasano	Calabria	1788*
	Ferrara M.	Vesuvius	1794*
	Filippi (de) F.	Lombardy	1830s*
	Fortis Alberto	Venetian	(1741–1803)
	Fossombroni V.	Tuscany	1789*
	Galani P.	Earthquakes	1783*
	Galiani F.	Vesuvius	1770s*
	Gallo C. D.	Earthquakes	1783*
	Gallo M.	Vesuvius	1835*
	Galvani C.	Bologna	1780*
70	Gargiolli G.	Tuscany	1837*
	Gemmellaro Carlo	Sicily	(1787–1866)
	Gennaro (di) B. A.	Vesuvius	1779*
	Giannelli B.	Vesuvius	1779*
	Gioeni G.	Vesuvius	1790–1793*
	Gismondi C.	Latium	1817*
	Giuli G.	Tuscany	1835*
	Giusti D. G.	Etna	1819*
	Giustiniani L.	Naples volcanoes	1817*
	Gottardi G.	Emilia	1813*
80	Granata L.	Naples Kingdom	1831*
	Grimaldi F.	Earthquakes	1784*
	Grimaldi L.	Earthquakes	1835*
	Guarini G.	Vesuvius	1830s*
	Guerrazzi	Tuscany	1810s–1830s*
	Guidoni G.	Liguria	1820s–1830s*
	Gusta F.	Earthquakes	1783*
	Inghirami G.	Tuscany	1820s*
	Interlandi	Sicily	1830s*
	Ippolito di Catanzaro	Earthquakes	1783*
90	Jorio (di) A.	Naples	1820*
	Lamarmora A.	Piedmont & Sardinia	1810s–1830s*
	Lancellotti J.	Naples	1812*
	Lanzetta A.	Naples	1796*
	Lavaggiorosso S.	Liguria	1814*
	Lavini G.	Turin	1835*
	Lippi C.	Vesuvius	1816*
	Magalotti L.	Vesuvius	1779*
	Maironi da Ponte Giovanni	Lombardy	(1748–1833)
	Malacarne C. G.	Venetian	1818, 1821*
100	Maraschini P.	Venetian	1810s–1820s*
	Maravigna C.	Etna	1800s–1830s*
	Mascagni P.	Tuscany	1779*
	Mazza L.	Earthquakes	1832*
	Mecatti G. M.	Vesuvius	1777*
	Menabuoni G.	Tuscany	1796*
	Milano M.	Naples Kingdom	1820*
	Minasi G.	Earthquakes	1785*
	Minervino S. C.	Vesuvius	1779, 1794*
	Mirone G.	Earthquakes & Etna	1786, 1787*
110	Monticelli Teodoro	Naples	(1759–1845)
	Mojon G.	Liguria	1806*
	Mongiardini G. A.	Liguria	1809, 1814*
	Morichini Domenico	Latium	(1773–1836)
	Morozzo (de) C. L.	Piedmont	1780s–1790s*
	Napione Galleani G. F.	Piedmont	1780s–1790s*
	Negroni O.	Vesuvius	1779*
	Nesti Filippo	Tuscany	(1780–1847)
	Niccolini A.	Naples	1829*
	Nobili (de) G.	Vesuvius	1822*

(*Continued*)

Table 3. *Continued*

120	Olivi G. B.	Venetian	(1769–1795)
	Onofrio (d') M. A.	Vesuvius	1790s*
	Ortolani G. E.	Sicily	1800s–1810s*
	Pacchi D.	Tuscany	1785*
	Palatino L.	Naples	1826*
	Pareto Lorenzo	Liguria	1830s*
	Pasini L.	Venetian	1820s–1830s*
	Passerini R.	Tuscany	1830*
	Perez Naro P.	Tuscany	1830s*
	Petagna L.	Calabria	1827*
130	Pianciani G. B.	Latium	1817*
	Piccoli Gregorio	Venetian	(1680–1755)
	Pilla Leopoldo	Volcanoes & Calabria	(1805–1848)
	Pini Ermenegildo	Milan	(1739–1825)
	Piombanti C.	Tuscany	1779*
	Picaro A.	Vesuvius	1794*
	Porta L.	Tuscany	1833*
	Procaccini Ricci V.	Volcanoes	1810s–1820s*
	Quattromani G.	Naples Kingdom	1827*
	Ragazzoni B.	Lombardy	1820s–1830s*
140	Ranzani Camillo	Bologna	(1775–1841)
	Re (del) G.	Naples Kingdom	1830*
	Reale S.	Piedmont	1788–1789*
	Recupero Giuseppe	Volcanoes	(1720–1778)
	Repetti E.	Tuscany	1830s*
	Ricca M.	Tuscany	1810*
	Ricci G.	Naples	1821*
	Riccomanni L.	Rome	1782*
	Ridolfi C.	Tuscany	1834*
	Rio (da) Nicolò	Venetian	(1765–1845)
150	Riso (de) B.	Calabria	1810*
	Risso A.	Alps	1810s–1820s*
	Robilant (di) Spirito B.	Turin	(1724–1801)
	Romanelli D.	Naples	1817*
	Sacco G.	Vesuvius	1794*
	Salvadori G. B.	Vesuvius	1823*
	Sanmartino d. Motta	Turin	1784–1785*
	S. Quintino (di) G.	Tuscany	1825*
	Sarconi M.	Earthquakes	1784*
	Sasso A.	Liguria	1827*
160	Savi Paolo	Pisa	1820s–1830s*
	Scacchi Angelo	Naples	1830s*
	Sciciliano A.	Etna	1831*
	Scinà D.	Sicily	1810s–1830s*
	Scortegagna F. O.	Venetian	1820s–1830s*
	Scotti E.	Vesuvius	1794, 1804*
	Serrao E.	Earthquakes	1785*
	Serristori L.	Tuscany	1818*
	Sirugo	Sicily	1835, 1837*
	Sismonda Angelo	Turin	(1807–1878)
170	Sobrero A.	Turin	1835*
	Soldani Ambrogio	Siena	(1736–1808)
	Spadoni P.	Bologna	1793*
	Spallanzani Lazzaro	Pavia	(1729–1799)
	Stoppa G.	Vesuvius	1806*
	Targioni Tozzetti Giovanni	Tuscany	(1712–1783)
	Targioni Tozzetti Ottaviano	Tuscany	1823*
	Tata D.	Vesuvius	1779*
	Tenore M.	Naples	1830s*
	Terzi P. B.	Venetian	1791*

(*Continued*)

Table 3. *Continued*

180	Testa D.	Venetian	1793*
	Tomasi (de) D.	Vesuvius	1794*
	Tondi Matteo	Naples	(1762–1835)
	Torcia M.	Earthquakes	1783*
	Torre (della) P. M.	Vesuvius	1780s–1800s*
	Tramontani L.	Tuscany	1800*
	Turriani	Earthquakes	1784*
	Vagliasindi P.	Etna	1833*
	Vagnone	Piedmont	1816*
	Valenziani M.	Vesuvius	1783*
190	Vasco	Turin	1790–1791*
	Vassalli Eandi	Turin	1805–1808*
	Verri V.	Milan	1830*
	Vetrani A.	Vesuvius	1780*
	Viscardi F.	Vesuvius	1794*
	Vivenzio Giovanni	Earthquakes	(174?–1819)
	Viviani V.	Liguria	1800s, 1814*
	Volta Alessandro	Milan	1784*
	Volta Giovanni S.	Venetian	1780s–1790s*
	Zorda G.	Vesuvius	1806, 1810*
200	Zuccaro A. M.	Earthquakes	1833*
	Zupo N.	Earthquakes	1784*

*Denotes date(s) of publications listed in the *Bibliographie Géologique et Paléontologique d'Italie* (Anon. 1881).

always been considered Primitive in age. An empirical and balanced approach recognizing the action of both fire and water in developing geological processes is also observed in the regional geomorphological and lithostratigraphical monographs of the Tuscan school (e.g. Targioni-Tozzetti, Giorgio Santi). The number of lithostratigraphical names introduced at this time and still used today in the geological literature is remarkably high (Vai 2007*a*).

Many of the works listed in Table 4 are among the first palaeontological and zoological monographs ever written on these subjects in the world. They were the result of the pioneering efforts made by the Bologna school of geology to assemble the first specialized palaeontological and micropalaeontological collections in the Istituto delle Scienze di Bologna in the first half of the eighteenth century (Gortani 1931, 1963; Sarti 1988; Vai & Cavazza 2006, pp. 55–60). These works are also ground-breaking for the early taxonomy and chronostratigraphy they contain (Arduino, Spallanzani, Soldani, Brocchi). As an example, one can quote Volta who, in 1796, was the first to make a complete taxonomic study of the famous Bolca fish fauna of the early Eocene that was to inspire Louis Agassiz's (1807–1873) later fossil fish classification. Innovative research in the field of mineral resources was also carried out by Targioni Tozzetti, Arduino, Di Robilant and Matteo Tondi in the second half of eighteenth century and continued until the beginning of the nineteenth (Vaccari 2003*b*, pp. 265–266). Of major importance were the first surveys of historical volcanic eruptions and earthquakes that were accompanied by geological mapping of these areas (Morello 1998). In addition, the Italian role (Spallanzani, Brocchi, Scarabelli) in developing the prehistorical archaeology and the population of early humans should be noted (Pacciarelli & Vai 1995; Franceschelli & Marabini 2006).

Many of the research highlights of the Italian geoscience community a few decades later, presented at the time of the Congressi degli Scienziati Italiani (1839–1875), have been recorded by Morello (1983). She particularly underlined the importance of the early geological mapping of Italy, which required updated stratigraphic and regional research, accompanied by an agreed set of criteria for the 'semiography' of geological maps and the setting up of a national commission for geological and mineralogical nomenclature by Angelo Sismonda (1807–1878) (Morello 1983, pp. 71–72). The legacy of the Bologna school – original fieldwork and systematic geological mapping of the Apennines – can be seen in the excellent relations that developed between the Museo di Storia Naturale in Bologna and the Muséum in Paris, through the likes of the Reverendo Camillo Ranzani (1775–1841) and Georges Cuvier. Ranzani was Professor of Natural History at Bologna University. Cuvier met him in Bologna in 1810 and was so impressed by the Bologna vertebrate collections that he invited Ranzani to Paris where he stayed for two years, improving comparison

Table 4. *'Classic' Italian papers on geology printed in the century between 1759 and 1859*

Author	Title and date of publication	Highlights
Giovanni Arduino (1714–1795)	1759–1760, *Due lettere del sig. Giovanni Arduino sopra varie sue osservazioni naturali al Cavalier Antonio Vallisnieri* 1769, *Alcune osservazioni Orittologiche fatte nei monti del Vicentino* 1774, *Saggio Fisico-Mineralogico di Lithogonia e Orognosia*	Foundation of stratigraphic chronology (chronostratigraphy and lithostratigraphy); discovery of contact metamorphism (1782); early statement on Actualism; mining maps, proto-geological maps and geological cross-sections among the first ever made; ancient volcanic rock interbedded within marine sedimentary strata
Jacopo Odoardi	1761, *Dei corpi marini che nel Feltrese distretto si trovano* plus French (1762) edn	Major climatic changes and polar shift used to explain marine ingressions; angular unconformity
Giovanni Targioni-Tozzetti (1712–1783)	1768–1779, *Relazione di alcuni viaggi fatti in diverse parti della Toscana*, 2nd edn plus German (1787) and French (1792) reduced edns 1779, *Dei Monti ignivomi della Toscana e del Vesuvio*	Large geomorphological and lithostratigraphical monograph with balanced empirical approach supporting action of both fire and water in geological processes
Giuseppe Recupero (1720–1778)	1755, *Discorso storico sopra l'aque vomitate dal Mongibello e i suoi ultimi fuochi avvenuti nel mese il marzo del corrente anno MDCCLV* 1815, *Storia naturale e generale dell'Etna*	Survey of the 1755 Etna eruption in the Valle del Bove including lahars. He guided several foreign scientists and travellers to visit the Etna volcano
Lazzaro Spallanzani (1729–1799)	1784, *Osservazioni e esperienze da me fatte l'anno 1784 nella laguna di Chioggia e mare vicino* 1792–1797, *Viaggi nelle Due Sicilie e in alcune parti dell'Appennino*, plus French (1795–1797) and English (1798) edns 1797, *Sur les Salses du Modénais*	Foundation of volcanology, water–melt interaction, geochemistry and experimental geology; relation of fossil to living molluscs; physiology of sponges
Spirito Benedetto Nicolis di Robilant (1724–1801)	1786, *Essai Géographique suivi d'une Topographie souterraine, minéralogique*	Inventory of mines and mineral resources
Ambrogio Soldani (1736–1808)	1780, *Saggio orittografico* 1789–1798, *Testaceographia ac Zoophytographia parva et microscopica*	Taxonomy of recent and fossil foraminifera; different species at different depths; pioneering remarks on meteorites
Giovanni Vivenzio 174?–1819	1788, *Istoria de' tremuoti in generale, ed in particolare quelli accaduti nella provincia della Calabria ulteriore e nella città di Messina nell'anno 1783*	Early seismology and seismometers

(*Continued*)

Table 4. *Continued*

Author	Title and date of publication	Highlights
Carlo Ermenegildo Pini (1739–1825)	1790, *Saggio di una nuova teoria della terra* 1790–1792, *Sulle rivoluzioni del globo terrestre provenienti dall'azione dell'acque: memoria geologica* 1810, *Descrizione ed uso di uno stratimetro*	One of the many diluvian Earth's theories. New tools for assessing the amount of mineral resources; scientist-traveller
Giovanni Serafino Volta (2nd Part of 18th century)	1796, *Ittiologia Veronese del Museo Bozziano*	Fossil fishes from Monte Bolca
Giovanni Battista Alberto Fortis (1741–1803)	1774, *Viaggio in Dalmazia* + translations 1778, *Memoria orittografica nella valle di Roncà* 1784, *Viaggio mineralogico nella Calabria e nella Puglia* plus German (1788) edn 1802, *Mémoires pour servir à l'histoire naturelle et principalement à l'oryctographie de l'Italie et des pays adjacens*	First to describe marine shells cemented by a basaltic flow in the Venetian region; scientist-traveller
Carlo Amoretti (1741–1816)	1794, *Viaggio da Milano ai tre laghi, Maggiore, di Lugano e di Como, e dei monti che li circondano* 1811, Della *ricerca del carbon fossile*	Inventory of mines and mineral resources; scientist-traveller
Scipione Breislak (1748–1826)	1801, *Voyages physiques et lithologiques dans la Campanie; suivis d'une mémoire sur la constitution physique de Rome* plus German (1802) edn 1811, *Introduzione alla geologia* plus French (1812) and German (1819) edns and British (1816) review	Role of uplift to explain fossils in the hills; early author of a treatise on geology; magmatic origin of basalts; scientist-traveller
Giovanni Maironi Da Ponte (1748–1833)	1803, *Osservazioni sul Dipartimento del Serio* and *Aggiunta alle* ...	Inventory of mines and mineral resources; origin of concretions
Giuseppe Gautieri (1769–1833)	1804, *Sulla necessità di stabilire una direzione generale per lo scavo delle miniere, e de' fossili e per le manifatture loro relative nella Repubblica Italiana*	Inventory of mines and mineral resources
Giambattista Brocchi (1772–1826)	1807–1808, *Trattato mineralogico e chimico sulle miniere di ferro del Dipartimento del Mella con l'esposizione della costituzione fisica delle montagne metallifere della Val Trompia*	Late Tertiary mollusc taxonomy and stratigraphy; pioneerin species extinction; pioneer in urban geology and urban geological maps and profiles; early attempt at

	1811, *Memoria mineralogica sulla Valle di Fassa in Tirolo* 1814, *Conchiologia fossile subapennina, con osservazioni geologiche sugli Apennini e sul suolo adiacente* 1820, *Dello stato fisico del suolo di Roma*	history of geology; inventory of mines and mineral resources and their origin
Giuseppe Marzari-Pencati (1779–1836)	1822, *Sur les granits dits tertiaires, observes dans le Tyrol* plus Italian (1820) edn	First evidence of post-Triassic granitoid intrusion by contact metamorphism
Teodoro Monticelli (1759–1845 and Nicola Covelli (1790–1829)	1823, *Storia dei fenomeni del Vesuvio avvenuti negli anni 1821, 1822 e parte del 1823* 1825, *Prodromo della mineralogia vesuviana. Orittognosia*	Volcanology and volcanic minerals; covellite was named after Covelli
Nicolò Da Rio (1765–1845)	1833, *Quelques observations sur le gisement des trachytes en general et des Trachytes des Monts Euganéennes en particulier*	Pioneering work on intra-sedimentary laccolithic intrusions
Tommaso Antonio Catullo (1782–1869)	1827, *Saggio di Zoologia Fossile*	Early Wernerian and Brongniartian stratigraphy in the Secondary of the Venetian Succession
Carlo Gemmellaro (1787–1866)	1833, *Sopra la morfologia delle montagne di Sicilia*	Early opponent to the 'crater of elevation' theory
Giovanni Giuseppe Bianconi (1809–1878)	1840, *Storia naturale dei terreni ardenti, dei vulcani fangosi, delle sorgenti infiammabili, dei pozzi idropinici e di altri fenomeni geologici operati dal gas idrogene e dell'origine di esso gas*	Discovered and named the argille scagliose (tectonosomes and olistostromes) in the Apennines; early critic of Darwinian natural selection
Leopoldo Pilla (1805–1848)	1845, *Saggio comparativo dei terreni che compongono il suolo d'Italia* 1847–1851, *Trattato di Geologia*	Contribution to chronostratigraphical classification and correlation of European terrains at the Cretaceous–Tertiary boundary; critical discussion and empirical mitigation of excessive gradualism (also of 'craters of elevation')
Angelo Sismonda (1807–1878)	1845, *Notizie e schiarimenti sulla costituzione delle Alpi Piemontesi*	Jurassic age of the metamorphic rocks in the Western Alps
Alberto Ferrero de Lamarmora (1789–1863)	1858, *Voyage en Sardaigne: troisième partie. Description géologique et paléontologique*	Regional monograph
Abramo Massalongo (1824–1860) and Giuseppe Scarabelli (1820–1905)	1859, *Studii sulla Flora Fossile e Geologia Stratigrafica del Senigalliese*	Foundation of late Tertiary Apennine palaeobotany and stratigraphy

and correlation of the collections. The friendship resulted in an extensive exchange of palaeontological material between the two museums, as indicated by the numerous original samples and casts of French and Italian species studied and donated by Cuvier who left his insignia on them. After having referenced the Italian material in his *Ossemens Fossiles* (Cuvier 1812), the collection is now preserved at the Capellini Museum in Bologna. This tradition of palaeontological studies and geological mapping was continued by the works of Bianconi and Scarabelli (Table 4) who discovered and named the 'argille scagliose' (including what we now call tectonosomes and olistostromes within mélanges), a term still in worldwide use and discussion today (Vai 1995*b*, pp. 91–92 and 98–100; Pacciarelli & Vai 1995, p. 205). A similar geological survey of the Western Alps characterized the works of Sismonda in 1845 (Table 4). It was the first such work to demonstrate the Jurassic age of large masses of metamorphic rocks in the Alps, thus opening the way to the nappes theory. In his 'fixistic' map of the Alps, Sismonda distinguished two units with different lithologies and degrees of metamorphism, using palaeontology; one contained many Jurassic fossils; the other, assumed to be much older or Primitive, finally yielded extremely rare and metamorphosed Jurassic ammonites after tireless fieldwork proving it was, in fact, from Secondary rocks.

Altogether, these works suggest a very active community, still keeping pace with its past, maintaining a critical mass able to produce new ideas and methods, and improving the new science and its applications to the needs of the society before and after 1807, in spite of the political instability. While some of these works were translated into French, German and, sometimes, English, most of these Italian ideas and discoveries were disseminated through personal contacts during guided visits to Italy of distinguished foreign savants. This allowed the foreign visitor an easy, almost effortless, appropriation of Italian geological knowledge. Some of the works listed in Table 4, about 20 percent, are still relevant and quoted today in a scientific context. The importance of the others becomes immediately apparent when they are studied in an historical perspective, which is still to be done for most of them.

A case history: Brocchi's *Conchiologia Fossile* and its influence on Lyell's *Principles of Geology* (1830–1833)

Among the leading scientists of the Italian geocommunity discussed earlier, Giambattista, or Gian Battista, Brocchi (Fig. 2a) (Marini 1987) has been the most continuously quoted, after Arduino. Brocchi's inspirational work, *Conchiologia Fossile* (1814) (Fig. 2b), and its subsequent importance in Lyell's works has already been recognized (MacCartney 1976; Rudwick 1998, 2005, pp. 522 and 583); however, the extent to which this work specifically influenced Lyell's *Principles of Geology* (1830–1833) and, in its turn, *Elements of Geology* (1838) has not previously been qualified or quantified.

Conchiologia is not only a modern palaeontological monograph, but also a treatise on regional geology based on palaeontological stratigraphy, like Cuvier & Brongniart's (1811) and Smith's (1815) memoirs and maps. With these three works, a revolution was achieved. Even the language used by Brocchi is succinct, notwithstanding the length of the work; the style is concise when compared to the convoluted baroque metaphors common in books of the eighteenth century. The full title of the work (Fig. 2b) highlights the contents of the first volume, which is devoted to the stratigraphy and geological interpretation of the entire Apennine mountain chain and adjoining foothills. The second volume is a monograph on the taxonomy, ecology and palaeogeography of late Tertiary molluscs. As with Cuvier and Smith, the aim of the work is not regional but general; it is not limited to description (examples), but looks for concepts and principles derived from new observations. An extract from Manilius' *Astronomica* (first century AD), quoted on page four of *Conchiologia*, indicates Brocchi's farsightedness: 'Emersere fretis montes, orbisque per undas exiliit, vasto clausus tamen undique ponto' (So, by degrees, mountains emerged from the deep, and the round world sprung forth from the waves, but closed in on every side by the vast ocean).

Brocchi's aim was clear and immediate: 'to elucidate the ancient history of the globe' (Brocchi 1814, p. 7), following Arduino's chronological paradigm (four time periods for the history of the Earth), and breaking out of the restrictions caused by 'theories of the Earth'. He strove to unravel the history of the tangible Earth, starting 'from organic remains' and complementing them with 'observations on their setting, the state in which they are formed, the features of the sediments in which they are buried' (Brocchi 1814, p. 7). As equally ambitious as the preceding 'theorists', but taking a different approach, Brocchi expounds the history of the Earth. Using today's terminology, the meaning of which was already clear to him and a few of his contemporaries, he outlines the stratigraphy of the late Tertiary based on the distribution of mollusc (and other) fossils in the Subapennine clay formation, using their palaeoecology and

(a)

(b)

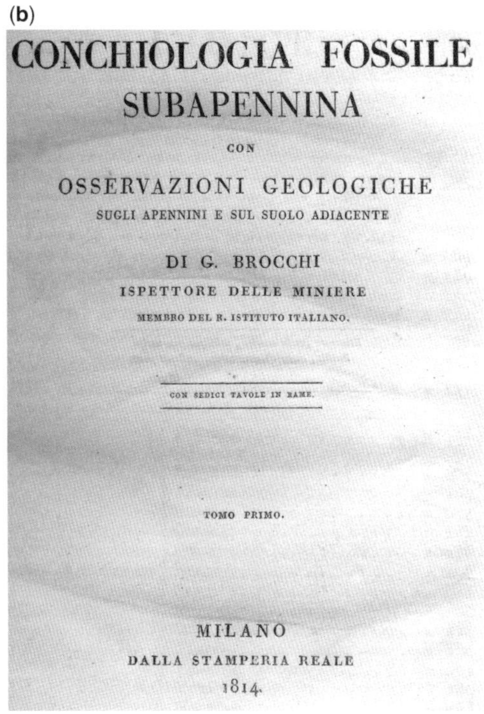

CONCHIOLOGIA FOSSILE
SUBAPENNINA
CON
OSSERVAZIONI GEOLOGICHE
SUGLI APENNINI E SUL SUOLO ADIACENTE
DI G. BROCCHI
ISPETTORE DELLE MINIERE
MEMBRO DEL R. ISTITUTO ITALIANO.
CON SEDICI TAVOLE IN RAME.
TOMO PRIMO.
MILANO
DALLA STAMPERIA REALE
1814.

Fig. 2. (**a**) Gian Battista Brocchi. (**b**) The frontispiece of Brocchi's *Conchiologia Fossile*.

taphonomy, as well as the lithogenesis and diagenesis of the encasing formation (Fig. 3).

Brocchi's basic methodology was to ascertain 'whether we are able to compare with the present what others tell us of the past' (Brocchi 1814, p. 7). It is a much less dogmatic statement than Geikie's equivalent Lyellian aphorism 'the present is the key to the past'. Brocchi was thus calling for a more realistic, not always gradualistic, even partly catastrophic, uniformitarianism, giving no priority either to the present (e.g. Lyell) nor to the past (perhaps Cuvier); he follows a balanced pragmatic rule: 'I have always developed concurrently and relied similarly upon the study of fossil shells and present-day marine molluscs' (Brocchi 1814, p. 8). The difference between the fossil (past) and the living (present) shells is his starting point: 'I did not start this enterprise until I recognized the shells which are dug from underground are totally different from those settling in the present seas' (Brocchi 1814, p. 8).

Brocchi proceeds in a very modern way, providing information on both fossil and living faunal assemblages, and comparing them in both time (biostratigraphy) and space (palaeoecology). He is even aware that palaeontological research can predict results that could be tested by ongoing research (as, for example, trends in global climate change). This is a reverse interpretation of uniformitarianism: that is, the past is a key to the present (Valentine 1973, pp. 14–16) and future. By comparing the very low numbers of living species in Lamark's list of fossils from the Paris Basin to the high species content in the Subapennine fauna, Brocchi establishes the use of the percentage of living forms in a fossil assemblage as a criterion of relative dating of the encasing formations. This approach was later expanded upon in the palaeontological lists and works of Paul Gerhard Deshayes (1797–1875) and adopted in Lyell's chronostratigraphy (Vai 2007*a*). In this way 'fossil conchology becomes a scale for geology' (Brocchi 1814, p. 13), a statement succinctly describing how Werner's geognostic dream died out, being replaced by the stratigraphic paradigm of geology. Unlike Cuvier, who was strongly influenced by his Wernerian education in Germany, Brocchi's scientific consistency and semantic freedom is reflected in the early and well-defined use of the term 'geological' v. 'geognostical' (Brocchi 1814, pp. 11 and 56; see also Vai 1995*b*, p. 68).

Brocchi expressed the concept of a relatively long timescale by comparing the 'much more recent incoherent Tertiary Subapennine formation' with the 'older cemented calcareous Apennine mountains',

Fig. 3. Plates V and XIII from Brocchi's *Conchiologia Fossile*.

and spoke about 'excavated sites of the Lombardy plain hosting bodies buried for many centuries' (Brocchi 1814, p. 25). With his understanding of the distribution of different fossils in formations of different ages, he was confident in formulating a natural law that 'species perish like individuals' (Brocchi 1814, p. 30) (i.e. the 'law of species disappearance' or species senescence), introducing his own view on the evolution of fossil organisms (Corsi 1983; Pancaldi 1983*b*; Berti 1987). Although the concept of species disappearance did not originate with him, Brocchi was the first to elaborate and demonstrate this theory. Brocchi's evolutionary view, although accepting the geological relevance of catastrophes, is based on a general and constant law. However, the rate of disappearance (and, thus, of existence) is different among the various species in a teleological view of 'creation' and 'Nature' (Brocchi 1814, pp. 227–228): 'Species approach their end slowly and gradually as individuals do' (Brocchi 1814, p. 229). Similarly, he was aware that the face of the Earth had undergone changes and catastrophes in very recent times 'compared to the first origin of things'. This applied to the age of emersion of the Subapennine hills (Tertiary), expected to be in the order of thousands of years, and not the thousands of centuries he imagined for the Apennine mountain core (Secondary).

Brocchi's and Lyell's views about many aspects of geology did not coincide. In spite of this, Brocchi's influence on Lyell's work was impressive. Table 5 is a qualitative summary, by chapter, of items in Lyell's *Principles of Geology* (first edition) that have a direct source in Brocchi's *Conchiologia*. Out of a total of 1765 pages in Lyell's *Principles*, 70 are a direct translation or a

Table 5. *Brocchi's (1814)* Conchiologia *as a source for Lyell's (1830–1833)* Principles of Geology, *showing the corresponding pages where the same subject matter or quotation occurred in both**

Lyell 1830–1833 *Principles*	Items considered	Brocchi 1814 *Conchiologia*
Vol. I		**Vol. 1**
Ch. I, p. 1	**Definition of geology**	p. 7; p. I
Ch. I, pp. 1 and 3	**Actualistic approach**	p. 7
Ch. I, p. 4	**Werner**	p. I
Ch. I, p. 4	**Cosmogony**	p. III
Ch. II, p. 16	**Theory of equivocal generation**	p. V
Ch. II, p. 17	**Quotation of Fortis [1802]**	p. LXXII
Ch. II, p. 20	**The ancient authors had 'no purpose of interpreting the monuments left by nature of ancient changes'**	p. I, II
Ch. II, p. 23	**Account of history of geology**. Same title, same approach, same chronologic method, wider scope in Lyell (geology in general), narrower in Brocchi (fossil Subapennine conchology and geology)	p. I, title
Ch. III–V	**History of geology**. According to Lyell, his Ch. III is a shortened version of'Brocchi's Discourse'. More than half the chapter is a literal translation or a summary of Brocchi's text. Additional inserts from Brocchi are in Chs IV and V	part 2
Ch. VI, pp. 92–95	**Organic remains from the Italian strata prove climate formerly hotter**. The point is discussed by Lyell making reference to his visits to Italian collections and outcrops led by Costa (Naples), Guidotti (Parma) and Bonelli (Turin)	p. 14
Ch. VI, p. 100	**Assemblage of fossils in the Secondary rocks are very distinct from those of the Tertiary**	pp. 21–22
Ch. VIII, pp. 135–136	**Rise of the Apennines above the level of the Mediterranean and Adriatic seas**	pp. 52–55 and 78–79
Ch. IX, p. 151	**'State of preservation of fossils in the younger rocks very different than in the older'**	p. 20
Ch. IX, p. 160	**Begins by studying the most modern periods of the Earth's history, attempting afterwards to decipher the monuments of more ancient changes**	p. 22

(*Continued*)

Table 5. *Continued*

Lyell 1830–1833 *Principles*	Items considered	Brocchi 1814 *Conchiologia*
Vol. II		**Vol. 1**
Ch. I, p. 1	**'Duration of each species'**	§ 6, pp. 219–240
Ch. 2, p. 18	**'Constancy of species, Amount of transformation'**	§ 6, pp. 219–240
Ch. 2, p. 20	**'Approach to present system gradual'**	§ 6, pp. 219–240
Ch. 2, p. 23	**'Each species shall endure for a considerable period of time'**	§ 6, pp. 219–240
Ch. VIII, pp. 128–30	**Brocchi on 'the loss of species'**	§ 6, pp. 219–240
Ch. IX, p. 178	**Recent addition of land to both sides of the Italian peninsula**	p. 35
Vol. III		**Vol. 1**
Preface p. viii Ch. II, p. 19	**Proportion of recent species in different Tertiary groups** Brocchi's idea transferred to Lyell through Bonelli, Guidotti, Costa and Deshayes	p. 11
Ch. V, p. 49	**'Necessity of accurately determining species'**	pp. 30–31 ff.
Ch. V, p. 55	**'Recent species found in older tertiary strata are inhabitants of warmer climates'**	p. 14
Ch. V, p. 58	**'Numerical proportion of recent shells in the different tertiary periods'**	p. 11
Ch. V, p. 59	**'Increase of existing species from older to newer formations analogous to fluctuations of individuals in a populations'**	p. 30 ff.
Ch. VI, p. 62	**'Reverse the natural order of historical research' or 'retrospective order of inquiry'**	p. 22
Ch. VIII, pp. 97–101	**'Quantity of past time' much greater than our historical/ human periods**	pp. 22, 32–48, 121, 199–200, 212–214
Ch. XII, pp. 155–157 and 159	**Synthesis of Subapennine formations and Brocchi's opinions**	pp. 82, 143, 147, 148 and 166
Ch. XII, p. 163	**Excellent preservation of shells (pearly lustre, external colour, ligament)**	pp. 26–27

*Ch., Chapter; §, Paragraph.

summary of sections in Brocchi's *Conchiologia* (44, 10 and 16 pages in volumes I, II and III of *Principles*, respectively). More generally, *Principles* deals with Italian geological features and matters compiled by Lyell from both Italian and foreign sources, or derived from his own trips to Italy with local guides, in the following: volume I, 159 pages out of 479; volume II, 20 pages out of 301; volume III, 111 pages out of 385. Although remarkable, this is not surprising. Lyell was fascinated by all aspects of the Italian natural environment and culture. The final decision to write the *Principles* came from his second Italian tour in 1828–1829 (Fig. 4), which overlapped with his honeymoon. Lyell was so excited by his Italian experience that he quoted from Dante's *Divine Comedy* (Lyell 1830, vol. I, p. 63), selecting the verse 'Dinanzi a me non fur cose create se non eterne' ('Before me no things were created unless eternal'). This shows the level of his proficiency in the Italian language and its literature (Rudwick 1998, p. 6), and his intention to benefit from the potential of Italian geological writings of his time. Rudwick (1998, pp. 4–6) in particular stressed the role of Brocchi's works, among others, as 'resources for the *Principles*' and says Lyell was influenced by Brocchi's model of piecemeal faunal change and by his geohistorical thinking (Rudwick 2005, p. 523).

To summarize, Lyell's *Principles* borrowed extensively from Brocchi's *Conchiologia* the ideas of:

- organic remains showing that the climate was formerly hotter;
- state of preservation of fossils in the younger rocks different to that in older rocks;
- assemblage of fossils in the Secondary rocks as distinct from those in the Tertiary;

Fig. 4. A page of Lyell's field notes of northern Italy, 1828. Notebook 14, Kinnordy MSS. Reproduced by courtesy of Lord Lyell of Kinnordy.

- studying the most recent periods of the Earth's history and subsequently attempting to interpret more ancient changes (retrospective order of enquiry);
- proportion of recent species in fossil assemblages;
- increase in percentage of existing species from older to newer formations;
- accurately determining species;
- recent species found in older Tertiary strata are inhabitants of warm climates.

Thus, even the most fundamental tool of Lyell's chronostratigraphic division of the Tertiary into Eocene, Miocene, Pliocene (1833) and Pleistocene (1839) – the percentage of existing or extinct taxa – was borrowed from Brocchi (1814), although Lyell implemented Deshayes' (1830, 1831) faunal lists, updating those originated by Brocchi (1814). Rudwick's (1998, p. 6) statement 'Lyell appropriated a massive and mature body of contemporary geological literature' seems to be especially the case for Brocchi's *Conchiologia*, but there was an additional, perhaps less scientific and more personal, reason for Lyell to have been captivated by it.

Conchiologia reached the British Isles following the Italian tour made by William Buckland (1784–1856) and George Greenough in 1816. Buckland had met Brocchi in Milan where he was given a complimentary copy, which perhaps became the one used by Leonard Horner (1816) for his enthusiastic review and translation of excerpts (Rudwick 2005, p. 604).[8] In fact, the review was a much extended summary and translation of most of the first volume, which is devoted to the geological history of the Apennines – as underlined by the running title of the review: *Geology of the Apennines* (Horner 1816). Most of the translation is excellent and often exhaustive, except for a section (p. 169) dealing with the origin of coarse-grained deposits of the northern Po Plain. Here Brocchi's text is difficult to interpret and the translation fails to understand his idea of the 'catastrophic irruption of the sea'. What Brocchi meant by this is the power of the sea to effect the erosion and re-deposition of large amounts of material to build new sedimentary bodies, but in Horner's summary this concept is lost. Horner's review concluded: 'The description of the shells are illustrated by plates, which we cannot praise too highly; for they are more beautifully executed than any thing of the kind we have ever seen before' (Horner 1816, p. 180) (Fig. 3). However, it is quite surprising that no mention is made by Horner of Brocchi's final chapter on the 'loss of species'. Instead, this concept was popularized by Lyell, crept into Charles Darwin's notebooks and became a constant feature of controversies at the Société Géologique de France, although the name of Brocchi was rarely associated with the idea. Nevertheless, occasional events such as Buckland's and Greenough's Italian tour, and Horner's review and translation, acted as powerful advertisements for Brocchi's work, despite the fact that it was written in Italian. Brocchi was lucky. But he was also aware of having written an unprecedentedly important work. Against friendly advice to publish in French, he preferred to compensate for the limited access to his work, derived from writing in Italian, by having its excellence recognized and extolled by the Italian community. More generally, one has to admit that reception in Europe of a large part of the classic works listed in Table 4 was severely limited due to them being written in Italian. Brocchi's work was also well received in continental Europe. It immediately became a reference in taxonomy for Tertiary and more recent molluscs, and led to the standard recognition of the Subapennine Clay Formation for stratigraphic correlations throughout the Mediterranean countries. In addition, Brocchi's works, although in Italian, were appreciated because his scientific language was consistent with the new European ecumenical language of Cuvier and von Humboldt (Vaccari 1998).

Discussion

At the beginning of the nineteenth century there was still a quite large Italian geoscience community very

[8] Horner (1785–1864) was to become Lyell's father-in-law.

active in the academic theoretical and experimental fieldwork, as well as in the applications of mining and soil science. In general, Italians kept up rather well with the pace of research in Europe. The standard of Italian authors in regional geology, stratigraphy, volcanology and seismology was not less than their French, British and German counterparts, but their visibility was reduced because their works were mostly written in Italian. However, about 10 percent were translated into French and/or German, and, more rarely, into English, signifying their relevance. Only about 15 percent were originally published in French, and less than five percent (mostly taxonomic works) in Latin. All this, plus the successful Congressi degli Scienziati Italiani,[9] would suggest that, at least until the end of the eighteenth century, there was not a decline of interest in geology per se in Italy where the discipline had blossomed in scientific, cultural, institutional and social terms.

Although the potential to establish an Italian geological society was available at about the same time as the Société Géologique de France was set up (in 1830), the political fragmentation of Italy did not facilitate it until the two stages of Italian Unity occurred in 1860 and 1871. Before that, an Italian nation, and even less an Italian state, was only a dream that was fostered by many Italian geologists (Corsi 1995, 1998; D'Argenio 2006). So, the easiest way for the Italian geoscience community to maintain contacts and exchanges with their European counterparts and test their competitive potential was to turn to established disciplinary organizations and to join to the geological societies already active. Paris was near, French the closest language, and France the country of science, culture, progress and vitality, in spite of its ambiguous position towards the challenge of the Italian Unity. In addition, there was a long tradition of exchanges between the Italian academies (e.g. the Istituto delle Scienze di Bologna) and the Académie des Sciences, which was reinforced at the geological and museum level by the action of Cuvier who was one of the supervisors during Napoleon's plundering (Vai 2003*b*). As a result, the best Italian geoscientists transferred their main activities to the geological societies of Paris and London. They used these opportunities not only for scientific but also for political purposes, as shown for example in the cases of Matteo Tondi (1762–1835) – a refugee in Paris during the Revolution and founder of the Royal Mineralogical Museum of Naples in 1801 – and Giuseppe Scarabelli – an active member of the Société since 1846 (Pacciarelli & Vai 1995; D'Argenio 2006). The enforced delay in establishing national scientific institutions that were able to stimulate advancement and competition was an additional problem confronting a fragmented Italy, following the economic crisis caused by the Napoleon turmoil. The total number of scientists decreased as well as the resources available to them for doing research. In addition, the most innovative programmes, such as the Geological Map of Italy that by the early 1860s only was just starting to become a possibility, could not progress (Corsi 2003, 2007). The Italian geological community began to loose confidence and had to wait another two decades, until 1881, before the Società Geologica Italiana was set up. This was due largely to reasons of economic crisis and institutional fragility within the new Italian state. The opportunity was provided by Italy hosting the second International Geological Congress in Bologna. The Congress's Presidents, Giovanni Capellini and Quintino Sella (1827–1884), were also the formal founders of the Società (Vai 2007*b*).

It is worth noting that Italian geoscientists made up the majority of the foreign members in both the Geological Society and the Société Géologique de France in the middle of the nineteenth century, owing to the relatively large size of the Italian geocommunity and, perhaps, a growing interest by European geoscientists wanting to know more about Italian geology. Out of a total membership of 1432 in the Geological Society of London in 1880, there were 40 Foreign Correspondents: eight were from Italy, eight from France, five from Germany, five from the USA, four from Austria, three from Belgium, two from Sweden, two from Switzerland, two from The Netherlands and one from Hungary. Out of a total of 577 members in the Société Géologique in 1883, 99 were Foreign Members: 24 Italians, 18 Belgians, 18 Spaniards, 14 Germans, 13 British and 12 from the USA. The distribution of the Italian members over time in the Société is shown in Table 6. A peak of 10 percent of the total Société Géologique membership was reached by Italians in 1846, with Sismonda and Scarabelli being the leading Italian geologists active in the Alps and the Apennines, respectively.

So it was only when the political conditions of the new Italian state seemed to be favourable, and an opportunity such as the second International Geological Congress Bologna in 1881 provided an ideal launch pad (Vai 2002, 2004*a*), that the Società Geologica Italiana became established; later than in many important countries such as the UK, France, Germany, Hungary, but prior to many

[9] The geology sessions were particularly successful, being joined by renowned foreign geoscientists such as Leopold von Buch, Adolphe Brongniart, Charpentier, Omalius d'Halloy (Morello 1983; Pancaldi 1983*a*).

Table 6. *Number of Italian members at various times within the Société Géologique de France*

Year of election	Italian members	Total members
1838	15	386
1844	25	437
1846	50	500
1859	42	
1868	32	
1880	25	
1883	24	577
1908	12	
1923	9	

others (Table 2). However, it was too late to prevent a major decline in the international relevance of the Italian geological community that had started after the Napoleonic wars and continued for some decades. The decline was felt in the reduction in size of the community, its scientific role and international visibility (except for some prominent individuals), and financial support from government for institutions and major programmes. A good example, derived from ineffective political and institutional support, is the history of the Italian Geological Survey and the Geological Map of Italy well described by Corsi (1995, 2003, 2007) and Brianta & Laureti (2006).

Conclusions and final remarks two centuries later

At the time of the founding of the Geological Society, the Italian geological community was still in good shape, able to supply Lyell with a lot of ideas, and through him many other leading geoscientists such as Charles Darwin. The relatively late founding of the Società Geologica Italiana, seven decades after the Geological Society and half a century after the Société Géologique de France, was not the consequence of a decline in scientific excellence or support from interested participants, but was initiated owing to the turmoil caused by the Napoleonic wars and enforced because an Italian state did not exist prior to 1860. All scientific requirements for establishing an Italian geological society at the time of the foundation of the Société Géologique were in place: there was a reasonably large geological community, trained and equipped with a tradition of discoveries; innovative research blossomed in most aspects of geology; and the leading scientists of this community were active, visible and respected at an international level. This is clearly demonstrated by the fact that the Italian geoscientists were the majority of the foreign members of both the Société Géologique and the Geological Society in the middle decades of the century. The disadvantages were the political fragmentation of Italy into many small states, making it difficult to reach a critical mass for interaction in both field and indoor activity. International visibility was limited by the increasing use of Italian in the publications instead of Latin or French; and the impact of the Napoleonic reforms on the economic and social frame of the country and its scientific structures was only locally compensated for by modernization and the rearrangement of institutions. But even the establishment of the Società Geologica Italiana could not prevent a decline in the international relevance of the Italian geological community that began to be evident in the second half of the nineteenth century and peaked in the first half of the twentieth, in spite of some individuals such as Sismonda, Scarabelli, Capellini, Seguenza, Sacco and Gortani playing an important role at the scientific and organizational level (Vai 2002, 2004*a*, 2007*b*). A recovery began during the second half of the last century.

The fascinating history that started two centuries ago with the founding of the Geological Society is still ongoing. It is partly reflected, and may be represented, in a graph comparing the membership over time of some of the major geological societies (Fig. 5). It stimulates a final remark. In spite of their different ages, the Geological Society, the Société Géologique de France and the Società Geologica Italiana show a relatively similar early development and evolutionary trend through the previous

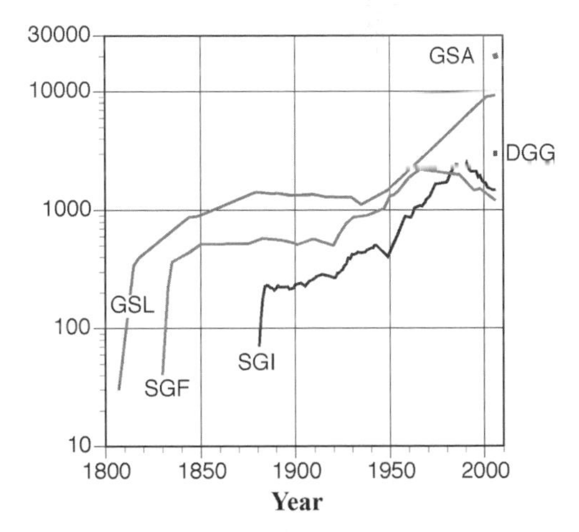

Fig. 5. Changing membership over time for the Geological Society of London (GSL), the Société Géologique de France (SGF) and the Società Geologica Italiana (SGI), compared with total membership of the Deutsche Geologische Gesellschaft (DGG) and the Geological Society of America (GSA) in 2006 (after Vai 2007*b*).

century. However, in the last two or three decades there is a sharp divergence between the still-growing Geological Society of London and the falling Société Géologique France and Società Geologica Italiana. Is this a dangerous side-effect of the 'impact factor', a policy that – as the President of the Société Géologique believes (Brun 2006) – favours papers and publications written in English, or should we be prepared for a major decline in our national geological societies?

I am indebted to the volume editors, to Andrea Candela and to an anonymous reviewer for their useful comments, and to Pietro Corsi, Gilbert Kelling, Giuliano Pancaldi, Hugh Torrens, Ezio Vaccari, the late Steve Walsh and Wendy Cawthorne, the Librarian at the Geological Society of London, for kindly providing information. Figure 4 was passed to me by S. Walsh through L. Wilson. This information is courtesy of Lord Lyell of Kinnordy.

References

ANON. 1881. *Bibliographie Géologique et Paléontologique d'Italie.* Zanichelli, Bologne.

ARDUINO, G. 1759–1760. *Due lettere [...] sopra varie sue osservazioni naturali. Al Chiaris. Sig. Cavalier Antonio Vallisnieri professore di Storia Naturale nell'Università di Padova. Lettera Prima [...] Sopra varie sue Osservazioni Naturali (Vicenza, 30 gennaio 1759). Lettera Seconda [...] Sopra varie sue Osservazioni fatte in diverse parti del Territorio di Vicenza, ed altrove, appartenenti alla Teoria Terrestre, ed alla Mineralogia (Vicenza, 30 marzo 1759).* Letters printed in 1760 in *Nuova Raccolta di Opuscoli Scienti fici e Filologici del padre abate Angiolo Calogierà* (Venezia), **6**, xcix–clxxx.

ARDUINO, G. 1774. Saggio Fisico-Mineralogico di Lythogonia e Orognosia. *Atti dell'Accademia delle Scienze di Siena detta de' Fisiocritici*, **5**, 228–300.

ARDUINO, G. 1782. Lettera Orittologica. *Giornale Letterario*, **3**, 244–248, 253–256, 263–264.

BARATTA, M. 1903. *Leonardo da Vinci ed i Problemi della Terra.* Fratelli Bocca Editori, Torino.

BIANCONI, G. G. 1862. Cenni storici sugli studi paleontologici e geologici in Bologna. *Atti della Società Italiana di Scienze Naturali, Milano*, **4**, 241–267.

BERTI, G. 1987. La formazione di Brocchi. *In*: MARINI, P. (ed.) *L'opera scientifica di Giambattista Brocchi (1772–1826).* Rumor, Vicenza, 13–40.

BRIANTA, D. & LAURETI, L. 2006. *Cartografia, scienza di governo e territorio nell'Italia liberale.* Edizioni Unicopli, Milano.

BROCCHI, G. 1814. *Conchiologia Fossile Subapennina con Osservazioni Geologiche sugli Apennini e sul Suolo Adiacente* (2 vols). dalla Stamperia Reale, Milano.

BRUN, J.-P. 2006. Appel: pourquoi nous avons besoin du Bulletin de la Société Géologique de France. *Géochronique*, **98**, 1–2.

CAPELLINI, G. 1897. Sulla data precisa della scoperta dei minuti Foraminiferi e sulla prima applicazione del microscopio all'analisi meccanica delle rocce, per J. B. Beccari. *Memorie Regia Accademia delle Scienze dell'Istituto di Bologna*, serie 5, **6**, 631–648.

CAVAZZA, M. 1990. *Settecento inquieto. Alle origini dell'Istituto delle Scienze di Bologna.* Il Mulino, Bologna.

CAVAZZA, M. 2002. The Institute of Science of Bologna and the Royal Society in the eighteenth century. *Notes Royal Society, London*, **56**(1), 3–25.

CIANCIO, L. 1995. *Autopsie della Terra: illuminismo e geologia in Alberto Fortis (1741–1803).* Olschki, Firenze.

CORSI, P. 1983. *Oltre il mito: Lamarck e le scienze naturali del suo tempo.* Il Mulino, Bologna.

CORSI, P. 1995. The Pisa School of Geology of the 19th century: An exercise in interpretation. *Palaeontographia Italica*, **82**, iii–vii.

CORSI, P. 1998. Le scienze naturali in Italia prima e dopo l'Unità. *In*: SIMILI, R. (ed.) *Ricerca e istituzioni scientifiche in Italia.* Laterza, Bari, 28–42.

CORSI, P. 2003. The Italian Geological Survey: The early history of a divided community. *In*: VAI, G. B. & CAVAZZA, W. (eds) *Four Centuries of the Word 'Geology': Ulisse Aldrovandi 1603 in Bologna.* Minerva Edizioni, Bologna, 271–299.

CORSI, P. 2007. Much ado about nothing: The Italian Geological Survey, 1861–2006. *Earth Sciences History*, **26**(1), 97–125.

CUVIER, G. 1812. *Recherches sur les ossemens fossiles* (4 vols). D'Ocagne, Paris.

CUVIER, G. & BRONGNIART, A. 1811. *Essai sur la géographie mineralogique des environs de Paris, avec une carte géognostique et des coupes de terrain.* Baudouin, Paris.

D'ARGENIO, B. 2006. Leopoldo Pilla (1805–1848): A young combatant who lived for geology and died for his country. *In*: VAI, G. B. & CALDWELL, W. G. E. (eds) *The Origins of Geology in Italy.* Geological Society of America, Special Paper, **411**, 211–223.

DE LUC, J. A. 1778. *Lettres physiques et morales sur l'histoire naturelle de la terre et de l'homme, adressèes à la Reine de la Grande-Bretagne.* De Tune, La Haye.

DE SAUSSURE, H. B. 1779–1796. *Voyage dans les Alpes, précédés d'un essay sur l'histoire naturelle des environs de Genève* (4 vols). Fauche, Neuchâtel.

DESHAYES, G. P. 1830. Tableau comparative des espèces de coquilles vivantes avec les espèces de coquilles fossiles des terrains tertiaires de l'Europe, et des espèces de fossiles de ces terrains entr'eux. *Bulletin Société Géologique de France*, **1**, 185–187.

DESHAYES, G. P. 1831. [Comment to A. Boué's lecture titled 'Essai pour apprécier les avantages de la paléontologie appliqué à la géognosie et à la géologie']. *Bulletin Société Géologique de France*, **2**, 88–91.

DOLOMIEU, D. 1794. Discours sur l'Étude de la Géologie. *Journal de Physique*, **45**, 256–272.

ELLENBERGER, F. 1988. *Histoire de la géologie.* Lavoisier Tec Doc, Paris.

ELLENBERGER, F. 1994. *Histoire de la géologia*, vol. 2. Lavoisier Tec Doc, Paris.

FRANCESCHELLI, C. & MARABINI, S. 2006. Luigi Ferdinando Marsili (1658–1730): A pioneer in geomorphological and archaeological surveying. *In*: VAI, G. B. & CALDWELL, W. G. E. (eds) *The Origins of Geology in Italy.* Geological Society of America, Special Paper, **411**, 129–139.

GEIKIE, A. 1903. *Text-Book of Geology* (4th edn). Macmillan, London.

GEIKIE, A. 1905. *The Founders of Geology* (2nd edn). Macmillan, London.

GEIKIE, A. 1909. Address of the President. *In*: WATTS, W. W. (ed.) *The Centenary of the Geological Society of London Celebrated September 26th to October 3rd 1907*. Longmans, Green, and Co., London, 107–131.

GEMMELLARO, C. 1862. Sommi capi d'una storia della geologia sino a tutto il secolo XVIII, pe' quali si delegge che le vere basi di questa scienza sono state fondate dagli Italiani. *Atti Accademia Gioenia, Catania*, serie 2, **18**, 5–40.

GORTANI, M. 1931. Bibliografia geologica italiana. *Giornale di Geologia*, **6**, 9–38.

GORTANI, M. 1952. La geologia di Leonardo da Vinci. *Scientia*, **87**, 7–8, 197–208; version 1953 in *Giornale di Geologia*, **23**, 1–18.

GORTANI, M. 1963. Italian pioneers in geology and mineralogy. *Journal of World History*, **7**(2), 503–519.

GUNTAU, M. 2009. The rise of geology as a science in Germany around 1800. *In*: LEWIS, C. L. E. & KNELL, S. J. (eds) *The Making of the Geological Society of London*. Geological Society, London, Special Publications, **317**, 163–177.

HORNER, L. 1816. Art. VII. [Review of] G. B. Brocchi, Conchiologia Fossile Subapennina con Osservazioni Geologiche sugli Apennini e sul Suolo adiacente. Di G. Brocchi, Ispettore delle Miniere, & c. Milano, 1814. 2 vol. 4to. *Edinburgh Review*, **26**, 156–180.

HUTTON, J. 1788. Theory of the Earth, or an investigation of the laws observable in the composition, dissolution and restoration of the land upon the globe. *Transactions of the Royal Society of Edinburgh*, **1**, 209–304.

HUTTON, J. 1795. *Theory of the Earth, with Proofs and Illustrations* (2 vols). Creech, Edinburgh.

KÖLBL-EBERT, M. 2009. George Bellas Greenough's 'Theory of the Earth' and its impact on the early Geological Society. *In*: LEWIS, C. L. E. & KNELL, S. J. (eds) *The Making of the Geological Society of London*. Geological Society, London, Special Publications, **317**, 115–128.

LEWIS, C. L. E. 2009. Doctoring geology: the medical origins of the Geological Society. *In*: LEWIS, C. L. E. & KNELL, S. J. (eds) *The Making of the Geological Society of London*. Geological Society, London, Special Publications, **317**, 49–92.

LYELL, C. 1830–1833. *Principles of Geology* (3 vols). John Murray, London.

LYELL, C. 1838. *Elements of Geology*. John Murray, London.

LYELL, C. 1839. *Eléments de Géologie*. Pitois-Levrault et C., Paris.

MACCARTNEY, P. J. 1976. Charles Lyell and G. B. Brocchi: A study in comparative historiography. *British Journal for the History of Science*, **9**, 177–189.

MARINI, P. (ed.). 1987. *L'opera scientifica di Giambattista Brocchi (1772–1826)*. Rumor, Vicenza, 189.

MENEGHINI, G. 1866. *Del merito dei veneti nella geologia*. Pisa.

MORELLO, N. 1979. *La nascita della paleontologia nel Seicento: Colonna, Stenone, Scilla*. Franco Angeli, Milano.

MORELLO, N. 1983. La geologia nei congressi degli scienziati italiani, 1839–1875. *In*: PANCALDI, G. (ed.) *I congressi degli scienziati italiani nell'età del positivismo*. Editrice Clueb, Bologna, 69–81.

MORELLO, N. (ed.). 1998. *Volcanoes and History*. Glauco Brigati, Genoa.

MORELLO, N. 2006*a*. Agricola and the birth of the mineralogical sciences in Italy in the sixteenth century. *In*: VAI, G. B. & CALDWELL, W. G. E. (eds) *The Origin of Geology in Italy*. Geological Society of America, Special Paper, **411**, 23–30.

MORELLO, N. 2006*b*. Steno, the fossils, the rocks, and the calendar of the Earth. *In*: VAI, G. B. & CALDWELL, W. G. E. (eds) *The Origin of Geology in Italy*. Geological Society of America, Special Paper, **411**, 81–93.

OLDROYD, D. R. 1996. *Thinking about the Earth: A History of Ideas in Geology*. Harvard University Press, London.

PACCIARELLI, M. & VAI, G. B. (eds). 1995. *La collezione Scarabelli. 1. Geologia*. Musei Civici di Imola. Grafis Edizioni, Bologna Casalecchio.

PANCALDI, G. (ed.). 1983*a*. *I congressi degli scienziati italiani nell'età del positivismo*. Editrice Clueb, Bologna.

PANCALDI, G. 1983*b*. *Darwin in Italia: impresa scientifica e frontiere culturali*. Il Mulino, Bologna.

PILLA, L. 1840. *Discorso accademico intorno ai principali progressi della geologia e allo stato presente di questa scienza*. Tipografia Platina, Napoli.

RUDWICK, M. J. S. 1998. Lyell and the *Principles of Geology*. *In*: BLUNDELL, D. J. & SCOTT, A. C. (eds) *Lyell: The Past is the Key to the Present*. Geological Society, London, Special Publications, **143**, 3–15.

RUDWICK, M. J. S. 2005. *Bursting the Limits of Time*. University of Chicago Press, Chicago, IL.

SARTI, C. 1988. *I fossili e il Diluvio Universale*. Pitagora Editrice, Bologna.

SMITH, W. 1815. *A Memoir to the Map and Delineation of the Strata of England and Wales, with a part of Scotland*. John Cary, London.

SPALLANZANI, M. E. 1994. Vom 'Studiolo' zum Laboratorium: Die 'piccola raccolta di naturali produzioni' des Lazzaro Spallanzani (1729–1799). *In*: GROTE, A. (ed.) *Macrocosmos in Microcosmos*. Leske & Budrich, Opladen, 679–694.

STENO, N. 1669. *De solido intra solidum naturaliter contento dissertationis prodromus*. Typografia Stellae, Florentiae.

STOPPANI, A. 1862. *Della priorità e preminenza degli Italiani negli studi geologici*. Bernardoni, Milano.

STOPPANI, A. 1881. Priorità e preminenza degli italiani negli studi geologici. *In*: STOPPANI, A. (ed.) *Trovanti*. Agnelli, Milano, 87–125.

TARGIONI-TOZZETTI, G. 1779. *Relazioni di alcuni viaggi fatti nelle diverse parti della Toscana*, Vol. 12. Cambiagi, Firenze.

TEGA, W. 1986. *Anatomie accademiche. I Commentari dell'Accademia delle Scienze di Bologna*. Il Mulino, Bologna.

VACCARI, E. 1993. Giovanni Arduino (1714–1795): *Il contributo di uno scienziato veneto al dibattito settecentesco sulle Scienze della Terra*. Olschki, Firenze.

VACCARI, E. 1998. Lazzaro Spallanzani and his geological travels to the 'Due Sicilie': The volcanology of the Aeolian Islands. *In*: MORELLO, N. (ed.) *Volcanoes and History. Proceedings of the 20th INHIGEO Symposium*. Brigati, Genoa, 621–652.

VACCARI, E. 2000. Mining and knowledge of the Earth in Eighteenth-century Italy. *Annals of Science*, **57**(2), 163–180.

VACCARI, E. 2003*a*. Luigi Ferdinando Marsili geologist: From the Hungarian mines to the Swiss Alps. *In*: VAI, G. B. & CAVAZZA, W. (eds) *Four Centuries of the Word 'Geology': Ulisse Aldrovandi 1603 in Bologna*. Minerva Edizioni, Bologna, 179–185.

VACCARI, E. 2003*b*. The 'Council of Mines' and geological research in Italy at the start of the 19th century: Research perspectives. *In*: VAI, G. B. & CAVAZZA, W. (eds) *Four Centuries of the Word 'Geology': Ulisse Aldrovandi 1603 in Bologna*. Minerva Edizioni, Bologna, 265–269.

VACCARI, E. 2006. The 'classification' of mountains in eighteenth century Italy and the lithostratigraphic theory of Giovanni Arduino (1714–1795). *In*: VAI, G. B. & CALDWELL, W. G. E. (eds) *The Origins of Geology in Italy*. Geological Society of America, Special Paper, **411**, 157–177.

VACCARI, E. 2007. The organized traveller: Scientific instructions for geological travels in Italy and Europe during the eighteenth and nineteenth centuries. *In*: WYSE JACKSON, P. N. (ed.) *Four Centuries of Geological Travel*. Geological Society, London, Special Publications, **287**, 7–17.

VACCARI, E. & CURI, E. 2003. Quarrying and geology in early 18th century Italy: The lithological column of Gregorio Piccoli (1739). *In*: SERRANO PINTO, M. (ed.) *Geological Resources and History. Proceedings of the 26th INHIGEO Symposium, Universidade de Aveiro, Portugal*, Minerva Central, Aveiro, 417–429.

VAI, G. B. 1995*a*. Geological priorities in Leonardo da Vinci's notebooks and paintings. *In*: GIGLIA, G., MACCAGNI, C. & MORELLO, N. (eds) *Rocks, Fossils and History. Proceedings of the 13th INHIGEO Symposium, Pisa–Padova, 1987*. Festina Lente Edizioni, Firenze, 13–26.

VAI, G. B. 1995*b*. L'opera e le pubblicazioni scientifiche di Scarabelli. *In*: PACCIARELLI, M. & VAI, G. B. (eds) *La collezione Scarabelli. 1 Geologia, Musei Civici di Imola*. Grafis Edizioni, Bologna Casalecchio, 49–104.

VAI, G. B. 2002. Giovanni Capellini and the origin of the International Geological Congress. *Episodes*, **25**(4), 248–254.

VAI, G. B. 2003*a*. Aldrovandi's Will: Introducing the term 'Geology' in 1603. *In*: VAI, G. B. & CAVAZZA, W. (eds) *Four Centuries of the Word 'Geology': Ulisse Aldrovandi 1603 in Bologna*. Minerva Edizioni, Bologna, 64–111.

VAI, G. B. 2003*b*. A liberal Diluvianism. *In*: VAI, G. B. & CAVAZZA, W. (eds) *Four Centuries of the Word 'Geology': Ulisse Aldrovandi 1603 in Bologna*. Minerva Edizioni, Bologna, 220–249.

VAI, G. B. 2004*a*. The Second International Geological Congress, Bologna 1881. *Episodes*, **27**(1), 13–20.

VAI, G. B. (ed.). 2004*b*. *Athanasii Kircheri Mundus Subterraneus in XII Libros digestus*. Editio Tertia, 1678, Arnaldo Forni Editore, Bologna (photostatic edition with foreword by the editor, and presentations by N. Morello and Umberto Eco).

VAI, G. B. 2006. Isostasy in Luigi Ferdinando Marsili manuscripts. *In*: VAI, G. B. & CALDWELL, W. G. E. (eds) *The Origins of Geology in Italy*. Geological Society of America, Special Paper, **411**, 95–127.

VAI, G. B. 2007*a*. A history of chronostratigraphy. *Stratigraphy*, **4**, (2/3), 83–97.

VAI, G. B. 2007*b*. Origine e prospettive della Società Geologica Italiana. *Bollettino Società Geologica Italiana (IJG)*, **126**, 131–157.

VAI, G. B. & CALDWELL, W. G. E. 2006. Preface. *In*: VAI, G. B. & CALDWELL, W. G. E. (eds) *The Origins of Geology in Italy*. Geological Society of America, Special Paper, **411**, vii–xi.

VAI, G. B. & CAVAZZA, W. (eds). 2003. *Four Centuries of the Word 'Geology': Ulisse Aldrovandi 1603 in Bologna*. Minerva Edizioni, Bologna, 327.

VAI, G. B. & CAVAZZA, W. 2006. Ulisse Aldrovandi and the origin of geology and science. *In*: VAI, G. B. & CALDWELL, W. G. E. (eds) *The Origins of Geology in Italy*. Geological Society of America, Special Paper, **411**, 43–63.

VALENTINE, J. W. 1973. *Evolutionary Paleoecology of the Marine Biosphere*. Prentice-Hall, Englewood Cliffs, NJ.

YAMADA, T. 2006. Kircher and Steno on the 'geocosm', with a reassessment of the role of Gassendi's work. *In*: VAI, G. B. & CALDWELL, W. G. E. (eds) *The Origins of Geology in Italy*. Geological Society of America, Special Paper, **411**, 65–80.

WYSE JACKSON, P. N. (ed.). 2007. *Four Centuries of Geological Travel*. Geological Society, London, Special Publications, **287**.

Scientific institutions and the beginnings of geology in Russia

VICTOR E. KHAIN[1] & IRENA G. MALAKHOVA[2]*

[1]*Geological Institute, Russian Academy of Sciences, Pyzhevsky per., 7 Moscow 119017, Russia*

[2]*Department for the History of Geology, Vernadsky State Geological Museum, Russian Academy of Sciences, Mokhovaya st. 11, Building 2, Moscow 125009, Russia*

**Corresponding author (e-mail: malakhova@sgm.ru)*

Abstract: There were three major scientific centres in Russia at the turn of the nineteenth century: the Imperial St Petersburg Academy of Sciences; the Mining Department with its attached Mining School; and the Moscow University. Geology was only then emerging in Russia, where it was mainly focused on mineral prospecting and imported European concepts. However, Alexander I initiated major reforms of science and education, founding new universities in which geological education in Russia found a foothold. The first scientific societies – the Moscow Society of Naturalists and the Mineralogical Society – also played an active role in the consolidation of Russian geology.

Institutions played a fundamental role in the development and consolidation of geology in Russia in the early nineteenth century. These included the Imperial Academy of Sciences at St Petersburg and the Mining Department with its attached Mining School, as well as universities, and particularly the Moscow University. With the accession of Alexander I (1777–1825), in 1801, the country underwent a period of liberal reform of science and education, which had a huge impact on the intellectual development of the nation. It was overseen by a newly instituted Ministry of Education (1802).

This paper maps out an institutional context that has much more in common with scientific developments in Germany and France than in Britain. As will be apparent, in 1807 Russia had no need for a geological society. While Russians engaged in the same geological debates that occupied the founders of the Geological Society of London (Knell 2009; Kölbl-Ebert 2009), these debates found a ready home in institutions already in existence and extended by royal patronage and reform. In this respect, the restructuring of scientific institutions here had much in common with that taking place in Paris in the same period (Taquet 2009). A long-established relationship with Germany also saw considerable German influence on emergent geological thinking in Russia. We should begin, however, with the most significant of these institutions, the St Petersburg Academy of Sciences; an institution respected across Europe.

The internationalism of the St Petersburg Academy of Sciences

At the very beginning of the nineteenth century, geology was present in the affairs of the Academy (established 1724) only in the form of mineralogy (Fig. 1). In this respect, St Petersburg was like many of those other European scientific capitals where a broader discipline of geology was to emerge over the ensuing decades. However, the basis for this emergence did not rely solely on academicians with mineralogical interests. There were others who had been elected for their work in natural history or chemistry who would also contribute to these developments (Table 1).

For three decades – from 1793 to 1826 – Vasily Mikhailovich Severgin (1765–1826) was the only academician elected as a mineralogical specialist (Fig. 2). His influence on the subsequent development of geology in Russia cannot be overestimated. As early as 1789, he advocated the volcanic origin of basalts and became a vocal opponent of Abraham Gottlob Werner (1750–1817). This opposition needs to be understood in the context of a country that felt the constant renewal of interest in Werner owing to its various intellectual exchanges with Germany. Severgin actively pushed the boundaries of Russian mineralogy with his pioneering study of the effects of paragenesis in mineral formation, published in his *First Principles of Mineralogy, or Natural History of Fossils* (1798). Between 1808 and 1809 he published his two-volume

From: LEWIS, C. L. E. & KNELL, S. J. (eds) *The Making of the Geological Society of London*.
The Geological Society, London, Special Publications, **317**, 203–211.
DOI: 10.1144/SP317.11 0305-8719/09/$15.00 © The Geological Society Publishing House 2009.

Fig. 1. The Academy of Sciences in St Petersburg. (Photograph: Russian Academy of Sciences, Department of the History of Geology).

Mineralogy of Russia and, in 1816, his *New System of Minerals Based on Appearance*, in which he introduced a new classification based on crystal structure. He also worked on textbooks for secondary schools and was an advisor for the first Russian circumnavigation of the Earth by Adam (Ivan Fedorovich) Kruzenstern, which took place between 1803 and 1806. Severgin also studied erratic boulders in the Russian NW and proposed their Scandinavian origin. It was this hugely significant body of work by this one man that formed the basis for the expansion of geological study in Russia in the early nineteenth century. For example, Severgin's work on crystallography was followed by Adolf Theodor (Adolf Yakovlevich) Kupffer (1799–1865), who, in 1823, presented his *Preisschrift über genau Messung der Winkel an Krystallen* to the Academy of Sciences in Berlin, where it was published in 1825. That expansion also had much to do with Russia's willingness to engage with the outside world.

Since the time of Peter the Great (1672–1725), Germans had been welcomed to St Petersburg and Moscow. Indeed, in the Baltic provinces of Russia, which lie to the west of St Petersburg, intellectual life was dominated by a wealthy merchant class of German origin. It was natural for young Russian men from these communities to seek their

Table 1. *Full Members of the Academy involved in geological studies in the early nineteenth century (Russian Academy of Sciences 1999)*

Name	Membership	Discipline
Peter Simon Pallas (1741–1811)	1767–1811	Natural History
Ivan Ivanovich Lepekhin (1740–1802)	1771–1802	Natural History
Johann Gottlieb Georgi (729–1802)	1783–1802	Chemistry
Nikolai Yakovlevich Ozeretskovsky (1750–1827)	1782–1827	Natural History
Franz Johann Benedict Herrmann (1755–1815)	1790–1801	**Mineralogy**
Vasily Mikhailovich Severgin (1765–1826)	1793–1826	**Mineralogy**
Christian Heinrich Pander (1794–1865)	1823	Natural history
Adolf Theodor (Adolf Yakovlevich) Kupffer (1799–1865)	1828	**Mineralogy**

Fig. 2. Vasily Mikhailovich Severgin. (Photograph: Russian Academy of Sciences, Department of the History of Geology).

educations in the great German universities, before then gravitating towards the Russian scientific capital. This exchange – so vital to intellectual development in Russia and Germany – is reflected in the number of publications emanating from St Petersburg written in German or in the international scientific language of the period, French.

This unhindered flow of knowledge was further facilitated by the active translation of key foreign works by academicians. Ivan Ivanovich Lepekhin (1740–1802) set the standard for this with his 10-volume translation, with commentary, of Buffon's *Histoire naturelle* (1749). Alexander Sevastyanov (1771–1824) also excelled in this effort. His translation of Linneaus's *Systema naturae* (1735) earned him membership of the Linnean Society of London in 1809. His translation of Werner's lectures, with a supplement by Horace-Bénédict de Saussure (1740–1799), also included his own commentary on the Werner–Hutton dispute. It was effectively the first Russian textbook on geology.

If the Academy was, then, absorbing influences from its European neighbours and, indeed, actively engaging in dialogue with them, it was also expanding its knowledge of its own expansive and expanding territories (Table 2). The eighteenth century had been marked by expeditions into the furthest reaches of Russia and here, too, Russia imported foreign talent. For example, among the last of the official scientific journeys of that century was that by the medically trained German, Peter Simon Pallas (1741–1811), who had acquired the patronage of Catherine the Great (Catherine II) (1729–1796). In his second expedition for her, he travelled south into the Ukraine, Crimea and Caucasus in 1793–1794. Another immigrant was the Swedish cleric Erick Gustav Laxmann (1737–1796), who had spent much of his life in Siberia, where he undertook expeditions for the Academy between 1781 and 1793. Perhaps the most geologically focused of these expeditions, was that led by Austrian-born mining engineer, (Franz Johann) Benedict Hermann, who had become a Corresponding Member of the Academy in 1782, a Full Member in 1790–1801, and Honorary Member 1786–1790, 1801–1815. He was sent to the Urals by Catherine II in 1783, where he founded steel mills. Hermann, like the others mentioned here, became an adopted Russian, living and working in Russia; indeed, residency was required in order to be a Full Member of the Academy.

The Academy became a centre for scientific publishing and in doing so it again demonstrated tremendous internationalism; it embraced the languages of all nations. Its *Nova Acta Academiae Scientarum Imperialis Petropolitanae* became

Table 2. *Results of Academy's expeditions published at the end of the eighteenth century*

Year	Title	Author
1787–1791	Reisen durch Russland und die Caucasischen Gebirge auf Befehl der Russischen Akademie der Wissenschaften Herausgegeben von P. S. Pallas, 2 Th.	J. Güldenstädt
1793	Reise von Kamtschatka nach Amerika mit dem Commandeur-Capitaen Bering (in German)	G. W. Steller
1797–1800	Geographisch-physikalische und naturhistorische Beschreibung des Russisches Reiches, 3 Th., 9 Bd	J. Georgi
1797–1801	Mineralogische Reisen in Sibirien vom Jahr 1783 bis 1796, 3 Th. (in German)	B. Hermann

Table 3. *Foreign geologists who were members of the Academy in St Petersburg at the turn of the nineteenth century (Russian Academy of Sciences 1999)*

Name	Country	Date of election	Membership
Karl von Meidinger (1750–1820)	Austria	1794	Honorary Member
Baron Georges (Jean-Léopold-Nicolas-Frédéric) Cuvier	France	1802	Honorary Member
Karl, or Charles Dietrich Eberhard König (1774–1851)	Great Britain	1805	Corresponding Member
René-Just Haüy (1753–1822)	France	1806	Honorary Member
Carl Cesar Leonhard (1779–1862)	Germany	1811	Corresponding Member
Wilhelm Ludwig von Eschwege (1777–1855)	Portugal	1815	Corresponding Member
Alexander Friedrich Wilhelm Heinrich von Humboldt (1769–1859)	Germany	1818	Honorary Member
Christian Andreas Zipser (1783–1866)	Hungary	1818	Corresponding Member
Jöns Jacob Berzelius (1779–1848)	Sweden	1820	Honorary Member
Sir Humphry Davy (1778–1829)	Great Britain	1826	Honorary Member
Gustav Rose (1795–1864)	Germany	1829	Corresponding Member

Mémoires de l'Académie imp. des Sciences de St.-Pétersburg in 1803 (Ostrovityanov 1964). In 1808 the Academy also began to publish its *Speculative Studies of the Empire at Saint-Petersburg Academy of Sciences*, renamed *Proceedings* in 1821. The *Journal of Technology* was produced between 1804 and 1812. It was in these publications that Severgin published approximately 100 articles, and Hermann reported on his discoveries of gold and coal in the Urals. In 1818 the Academy resolved to establish a special commission, with Severgin as an editor-in-chief, to publish the results of all the expeditions the academy had undertaken in the eighteenth century (Khartanovich 2002, p. 146). The Academy also published Severgin's books including *First Principles* (1798), his *Accounts of Journeys Around the Western Regions of Russia, or Mineralogical, Economic and Other Additions* (1803–1804), his *Dictionary of Mineralogy* (1807) and his *Mineralogy of Russia* (1816) (Ostrovityanov 1964). Hermann's *The Natural History of Copper* (1791) and his three-volume account of mining in Siberia (1797–1801) were also published by the Academy.

At the beginning of the nineteenth century, then, the St Petersburg Academy of Sciences assured itself of an international place for its science by engaging in a liberal exchange of ideas. This internationalism was extended still further by the admission of Honorary or Corresponding Foreign Members (Table 3).

The Mining Department

The Berg Collegium was established in Russia in 1719, under regulations signed by Peter the Great, 'to run a business in ores and minerals' (Loransky 1900, p. 13). Privy Councillor, writer and statesman Andrey Andreevich Nartov (1737–1813), became its head in 1796 until it was reorganized and renamed the Mining Department in 1807. From 1811 it became the Mining and Salt Mines Department.

While the Academy possessed a coterie of talented individuals pushing at the intellectual boundaries of mineralogy and other geological subjects, the Mining Department had more practical objectives and made rapid progress in prospecting for mineral resources. It became a considerable economic force in the country because it was well financed, possessed effective management, and had adopted a military style of education and fieldwork. It led the exploration of new territories, such as the Caucasus which became part of the Russian Empire in 1801. It discovered and exploited the rich goldfields in the Urals in 1814 and produced the first geological map of the Donetsk coal basin in the Ukraine, compiled by Evgraf Petrovich Kovalevsky (1790–1867), in 1828. Indeed, many of these expeditions were mounted to undertake geological mapping, which led to steps, in 1824, to unify these efforts (Soymonov 1829), effectively beginning the geological survey in Russia. With the establishment of its Scientific Committee, in 1825, the role and influence of the Mining Department increased still further, by permitting the involvement of representatives from the Academy and universities in its affairs.

The Department also initiated the publication of the *Mining Journal*, or more precisely, *Mining Journal, or Collection of Data on Mining and Salt-Mines, with Additional New Discoveries in the Applied Sciences*, in 1825. According to

Tikhomirov (1963, p. 377), 'it was the first single-subject professional magazine on the geosciences and mining in Russian and one of the first publications of its kind in the world'. Dmitry Ivanovich Sokolov (1788–1852), a professor of the Military Mining School and University in St Petersburg, was the editor for 15 years. A paper by him entitled 'Progress in geognosy' opened the first issue. Despite its name the journal became an open resource for the discussion of any geological fact, idea or debate. In 1830, for example, it held a lively discussion of the Werner–Hutton debate (Kemerer 1830).

Geological education

From 1773 the Mining School in St Petersburg had trained engineers for the Berg Collegium and it would continue to do so as that organization was reorganized. The School was renamed the Military Mining School in 1804, and was given the rank of an institute in 1806 (Fig. 3). Classes were taught in mineralogy and geognosy, and, from the 1820s, in paleontology, or as it was then known 'petromatognosy'. Both lecturers and students were trained in Germany and came to the School to either teach or to take their studies further. Given this link, it is easy to understand Werner's influence here. The School also boasted a rich museum, founded in 1797, in which there were models of the Earth's interior and a 'mine trainer' (Tikhomirov 1963).

Fig. 3. The Mining Institute, St Petersburg (building constructed in 1806–1811) http://www.spmi.ru/.

Table 4. *The first Russian centres of education*

Name	Foundation date
Moscow University	1755
Mining School in St Petersburg	1773
Dorpat (Tartu) University	1802
Vilna (Vilnius) University	1803
Kazan University	1804
Khar'kov University	1805
Institute of the Corps of Railway Engineers in St Petersburg	1810
St Petersburg University	1819

The Mining School had entered the nineteenth century as the dominant educational body in its field. However, with the educational reforms of the first decade of the century, it soon found itself in the company of a number of new universities, some of which would eventually introduce teaching in geological subjects (Table 4) (Khartanovich 2002, p. 126). These new universities were based on the German model, and Alexander I invited 60 professors from that country to teach Russian students. However, geology was not amongst the subjects at first taught as a specialist discipline, but as Germany was a leader in geological education it was inevitable that it would influence similar developments in Russia.

The Moscow University (Fig. 4) was the only university to predate these reforms and here geology existed only as a small component of natural history. However, the University did possess a few followers of the remarkable polymath Mikhail Vasilievich Lomonosov (1711–1765), who had discussed his ideas on the continuity and variety of geological processes in an era before James Hutton. As Ivan Alexeevich Dvigubsky (1771–1839) noted in his speech to the University in 1806: 'All the processes that we now see also acted many thousand years ago' (Dvigubsky 1806, p. 29). If there were, then, many reasons for the pervasive influence of Werner, there were also local reasons why a counterperspective was popular amongst Russians in early nineteenth century.

Another early geological influence at the University was Johann Gotthelf (Grigory Ivanovich) Fischer von Valdheim (1771–1853). An alumnus of the Freiberg Academy, he was amongst those invited German professors. From 1804, he turned his attentions wholly to Russia where he took up the Chair of Natural History at the Moscow University and headed its Museum of Natural History. In 1811 he published a catalogue of the museum's

Fig. 4. The old building of the Moscow University. (The left wing was the first home of the Moscow Society of Naturalists. It now houses the Society's library and archive). (Photograph: I. Malakhova).

mineral collection based on a chemical classification. He had begun to lecture on orictognosy (mineralogy) in 1805 and published a two-volume monograph, *Orictognosy*, between 1818 and 1820. His scientific talents were, however, broad and he became known as the 'Russian Cuvier' following his palaeontological work in Moscow region. Indeed, it was he who introduced the term 'palaeontology' into the Russian scientific lexicon in 1834 (Milanovsky 2004, p. 33).

Also at the University was Alexander Alexandrovich Iovsky (1796–1854?), who undertook chemical studies of minerals His most notable work, however, was possibly his *Manual for Understanding of the Interior and Origin of the Globe, with Illustrations of its Strata and pre-Deluge Animals* published in 1828.

The new Dorpat (now Tartu) University also became a successful centre for science, drawing in German-speaking Russians from the Baltic provinces. There were a number who soon made a name for themselves in palaeontology: Karl Ernest (Karl Maximovich) von Baer (1792–1876), Christian Heinrich Pander (1794–1865) and Karl Eduard (Eduard Ivanovich) Eichwald (1795–1876). Pander became a pioneer of 'Palaeozoic' palaeontology, known particularly for his work on fossil fishes and celebrated today as the discoverer of the conodont. In 1830 he published a significant book, *Beträge zur Geognosie des Russischen Reiches*, on the stratigraphy and palaeontology of the older rocks around St Petersburg. He had travelled widely before this publication having established himself as an embryologist in Germany under the influence of this friend, Baer. He later supported Roderick Murchison's (1792–1871) attempts to paint Russia in the colours of his Silurian in 1840–1841 (Raikov 1964, p. 64). Both Pander and Baer found education at Dorpat unsatisfactory because of the lack of practical classes, and took the remainder of their studies in Germany. Pander returned to Russia in 1819. Eichwald, a contemporary of Pander, had studied with him at Dorpat. He shared Pander's interests in orictozoology (palaeontology), and came to work on what would become known as Palaeozoic stratigraphy (the Carboniferous System). He would distinguish himself at the universities of Vilna, Kazan and St Petersburg. In

1820, a chair of mineralogy was established and a mineral cabinet founded that would soon develop into a collection of more than 20 000 minerals and fossils. Moritz Fedorovich Engelgard (1779–1842) was the first to hold the chair, and lectured on mineralogy using Severgin's textbooks (Tikhomirov 1963, p. 367).

A mineral cabinet was also created at St Petersburg University just after its foundation in 1819. Sokolov taught mineralogy and geognosy there from 1822, and his colleague, Nickolay Prokofievich Scheglov (1794–1831), published a textbook, *Mineralogy According to the Classification of M. Haüy* (1824), which proved the superiority of a French system of mineralogy over that of the Germans. Scheglov also became known for his work on the organic origin of coal.

A course on mineralogy was introduced at the St Petersburg Institute of the Corps of Transport Engineers in 1816. Lectures here concerned the Earth's interior, the origin of rocks and minerals, raw materials and technology. The building of a new road between Moscow and St Petersburg in 1823 led to the undertaking of special geological studies. In that year the Cabinet of Lithology was founded, and transport engineers compiled the first geological sections for the environs of St Petersburg between 1827 and 1831 (Gumensky *et al.* 1956, p. 5).

Russian scientific societies

If Russia had moved rapidly to put in place a government-sponsored infrastructure for science and done so with unparalleled internationalism, it was also not slow to establish learned societies that would contribute to the new science. The Free Economic Society (1765–1918, 1982–present) was the first public society in Russia. It was founded on 31 October 1765 with the support of Catherine the Great, and aimed to advance agriculture and industry. Nartov became its President between 1797 and 1813. It was during his presidency that the society surveyed the nation by questionnaire in order to obtain regional information. By 1802 a total of 10 regions had reported and their reports included, in addition to information on geographical position, climate and landscape, objects of the 'fossil kingdom', mineral springs, 'stone oils' salt and mineral deposits (Khodnev 1865, p. 53).

Every year the society put forward a list of questions and rewarded those who offered solutions. The society also established an important mineral cabinet, which was begun with a gift from Nartov in 1791. It grew to 4000 specimens of rocks and minerals by 1815, sent in by naturalists from across Russia and Europe (Khodnev 1865, p. 485). The society also issued *Proceedings* and monographs, but of the 56 books it issued before 1830 only a few concerned the geosciences. Among them, however, was the ubiquitous Severgin, who published his *Mineralogical Dictionary in German, Russian and Latin* (1790) and *Manual for the Best Understanding of Literature on Chemistry* (1814) through the society.

Societies also formed in, and affiliated to, the new universities. On 25 July 1807 the Charter of the Moscow Society of Naturalists was signed with the authorities (Fig. 4). The main goal of the new society was: 'To provide information concerning the natural history of the Great Russian Empire' (Varsanofyeva 1955, p. 11). The Society of Naturalists was affiliated with the Moscow University and Valdheim became its first, and life-long, Director. In 1806 the position of Honorary President was introduced, and was occupied by Count Alexey Kirillovich Razumovsky (1748–1822), the Minister of Education, between 1810 and 1816. Through his patronage and politicking, the society was permitted to add the title 'Imperial' to its name in 1807.

From the very beginning this society collected zoological, botanical and mineralogical data from across the Moscow region, enlarging the collections of Moscow University's Museum of Natural History (founded in 1759, and now the Vernadsky State Geological Museum of the Russian Academy of Sciences in Moscow). The society's excursions, public lectures and consultations attracted a large membership. However, both the society and the museum suffered irreparable losses during the Moscow fire of 1812. All the documents and the most historical part of the collections were destroyed. Publications were stopped, which meant that Valdheim's important geological compilation of palaeontological and stratigraphic studies of the Moscow region was not published until 1837.

As was true of scientific endeavour in Russia more generally, it is unsurprising to note that the Moscow Society of Naturalists established a wide international network, which included J. Berzelius (1805), Humboldt (1806), Banks (1806), Haüy (1811), Fox-Strangways (1817), Buckland (1818) and Babington (1818).

The society took the early decision to publish in French. Its *Journal de la Société des Naturalistes de l'Université impériale de Moscou* was published between 1805 and 1812. Collective works and monographs were printed in 1805 as *Mémoires de la Société des Naturalistes de l'Université impériale de Moscou.* The first issue of *Bulletin de la Société des Naturalistes de l'Université impériale de Moscou* appeared in 1829 (Varsanofyeva 1955). These became a vehicle for the work of foreign scientific workers including Haüy, Humboldt, von Buch, Murchison, Verneuil and St Hilaire.

The St Petersburg Mineralogical Society, affiliated to the Military Mining School, was founded 'for science and for the love of the motherland' (Minutes of the Society, cited by Solovyev & Dolivo-Dobrovolsky 1992, p. 15). It came into being when 33 individuals met in the apartments of a Professor Lorenz (Lavrenty Ivanovich) Pansner (1777–1851) on 19 January 1817. A botanist, Christoph Bernard (Boris Ivanovich) von Vietinghof-Scheel (1767–1829) was the first President with Pansner the first Director. Count Alexander Grigoryevich Stroganoff (1795–1891), an alumnus of the Institute of Transport Engineers, was appointed the President of the Society in 1824 and Yakim Grigoryevicn Zembnitzky (1784–1851), a professor of the Military Mining School, acted as the Society's Director between 1827 and 1842. By these means, the Society embraced the diverse mineralogical community in Russia and undoubtedly negotiated the science's institutional politics.

The Mineralogical Society considered the foundation of a museum its main goal, and it soon acquired fossils from the region of St Petersburg, ores and minerals from the Urals, gifts from founders and parcels from enthusiasts. Supporters were found in Germany, Austria and Italy who shared their collections with the Society's museum. Amongst them are fossils from William Buckland in Oxford (Solovyev & Dolivo-Dobrovolsky 1992, p. 24). In 1830 the number of members exceeded 140, and among the first foreign members of the Society were Goethe, Banks and Haüy (Russian Mineralogical Society 2005).

Conclusion

At then end of the eighteenth century Russia was already engaged in data collection, mineral prospecting and the study of rocks and minerals, and it was from this base that geology arose. Palaeontology, stratigraphy and the study of Palaeozoic rocks were early additions in the nineteenth century. Russia rapidly acquired a scientific infrastructure within which the science could develop; the Imperial Academy of Sciences in St Petersburg, the Mining Department with School attached and the Moscow University were particularly important. The educational reforms that led to the founding of new universities extended the possibilities for the new science, and geological education became established. What, however, is particularly distinctive about geological developments in Russia is the nation's internationalism. It welcomed in foreign nationals to lead expeditions and to undertake research, it was open to foreign ideas, and used a range of methods to ensure that there were no linguistic barriers between Russian science and that undertaken elsewhere in Europe. It was, like many of these other nations, rather more infatuated with mineralogy than other aspects of the emergent science, but it also produced a number of world-class palaeontologists. Russia had peculiar advantages at this time. It was a vast and growing country that encouraged exploration. It possessed considerable mineralogical wealth and some fine geology – such as the fairly simple Palaeozoic strata around St Petersburg. It profited from its open-minded internationalism and strong relationships with its scientific neighbours. It also had, as this essay has attempted to demonstrate, a rich culture of old and new scientific institutions. These factors meant that Russia did not need the formation of its own geological society in 1807. Like other parts of Europe, it was essentially capable doing geology without one.

The authors are very grateful for remarks and comments by Dmitry V. Rundqwist and George P. Khomizuri. They are much obliged to Simon Knell for textual improvements.

References

DVIGUBSKY, I. A. 1806. A speech on the present Earth's surface. *Speeches of the Moscow University's Professors Reported in the Public Meeting of the Moscow University on the 30th of June, 1806.* Moscow University, Moscow, 7–29 (in Russian).

FISCHER VON WALDHEIM, G. 1818. *Orictognosy or Short Description of all Fossils, with Dictionary*, Part I. The Printing House of the Medical Surgeon Academy, Moscow (in Russian).

FISCHER VON WALDHEIM, G. 1820. *Orictognosy or Short Description of all Fossils, with Dictionary*, Part II. The Printing House of the Medical Surgeon Academy, Moscow (in Russian).

GUMENSKY, B. M., KOMAROV, N. S. & VORONIN, M. I. 1956. Towards a history of geological studies during road construction in 1817–1870. *Proceedings of the Institute of the History of Sciences and Technology*, **7**, 3–22 (in Russian).

HERMANN, B. F. J. 1791. *The Natural History of Copper or Textbook on Its Study, Processing and Utilization.* The Imperial Academy of Sciences Publishing House, St. Petersburg (in Russian).

HERMANN, B. F. J. 1797–1801. *Essay on Siberian Mines and Works Collected by Academician Ivan Hermann.* Parts I–III. The Imperial Academy of Sciences Publishing House, St. Petersburg (in Russian).

IOVSKY, A. A. 1828. *Manual for Understanding of the Interior and Origin of the Globe, with Illustrations of its Strata and pre-Deluge Animals.* The Reshetnikov Printing House, Moscow (in Russian).

KEMERER, A. B. 1830. A glance at the principle geological theories: Werner's and Hutton's. *Mining Journal*, **10**, 1–18 (in Russian).

KHARTANOVICH, M. F. (ed.). 2002. *Chronicle of the Russian Academy of Sciences. 1803–1860*, Volume II. Spb. Nauka, St Petersburg (in Russian).

KHODNEV, A. I. 1865. *History of the Imperial Free Economic Society from 1765 to 1865*. The Printing House Tovarischestva 'Obschestvennaya pol'za', St Petersburg (in Russian).

KNELL, S. J. 2009. The road to Smith: how the Geological Society came to possess English geology. *In*: LEWIS, C. L. E. & KNELL, S. J. (eds) *The Making of the Geological Society of London*. Geological Society, London, Special Publications, **317**, 1–47.

KÖLBL-EBERT, M. 2009. George Bellas Greenough's 'Theory of the Earth' and its impact on the early Geological Society. *In*: LEWIS, C. L. E. & KNELL, S. J. (eds) *The Making of the Geological Society of London*. Geological Society, London, Special Publications, **317**, 115–128.

KUPFFER, A. 1825. *Preisschrift über genau Messung der Winkel an Krystallen*. The Academy of Sciences, Berlin.

LORANSKY, A. M. 1900. *A Short Historical Account on Institutions of the Mining Department in Russia. 1700–1900*. G. A. Bernstein, St Petersburg (in Russian).

MILANOVSKY, E. E. 2004. *Two Hundred Years of the Moscow University's Geological School in Portraits of its Founders and Prominent People (1804–2004)*. Academical Project, Moscow (in Russian).

OSTROVITYANOV, K. V. (ed.). 1964. *History of the USSR Academy of Sciences (1803–1917)*, Volume II. Nauka, Moscow (in Russian).

PANDER, C. H. 1830. *Beträge zur Geognosie des Russischen Reiches*. Gedruckt bei K. Kray, St. Petersburg.

RAIKOV, B. E. 1964. *Chr. Pander. Der Forscher. Der Mensch*. Nauka, Moscow-Leningrad (in Russian).

RUSSIAN ACADEMY OF SCIENCES. 1999. *List of Members*, Volume I. Nauka, Moscow (in Russian).

RUSSIAN MINERALOGICAL SOCIETY. 2005. *Full Members of the Russian Mineralogical Society*. Russian Mineralogical Society, Petersburg (in Russian).

SCHEGLOV, N. P. 1824. *Mineralogy According to the Classification of M. Haüy*. The Marine Printing House, St. Petersburg (in Russian).

SEVERGIN, V. M. 1790. *Mineralogical Dictionary*. The Imperial Academy of Sciences Publishing House, St. Petersburg (in Russian).

SEVERGIN, V. M. 1803. *Accounts of Journeys Around the West Regions of Russia, or Mineralogical, Economic and Other Additions made in 1802*. The Imperial Academy of Sciences Publishing House, St. Petersburg (in Russian).

SEVERGIN, V. M. 1815. *Manual for the Best Understanding of Literature on Chemistry with Dictionaries: Latin-Russian, French-Russian and German-Russian; Old-fashioned and New lexicons*. The Imperial Academy of Sciences Publishing House, St. Petersburg (in Russian).

SEVERGIN, V. M. 1798. *First Principles of Mineralogy, or Natural History of Fossils*, 2 vols. The Imperial Academy of Sciences Publishing House, St. Petersburg (in Russian).

SEVERGIN, V. M. 1804. *Accounts of Journeys Around the West Regions of Russia, or Mineralogical, Economic and Other Additions made in 1803*. The Imperial Academy of Sciences Publishing House, St. Petersburg (in Russian).

SEVERGIN, V. M. 1807. *Mineralogical Dictionary With all Terms and Recent Discoveries in Mineralogy. Latin, French and German Mineralogical Dictionaries are Attached*, Volumes I–II. The Imperial Academy of Sciences Publishing House, St. Petersburg (in Russian).

SEVERGIN, V. M. 1808–1809. *Experience of Mineralogical Description of the Russian State*, Parts I–II. The Imperial Academy of Sciences Publishing House, St. Petersburg (in Russian).

SEVERGIN, V. M. 1816. *New System of Minerals Based on Appearance*. The Imperial Academy of Sciences Publishing House, St. Petersburg (in Russian).

SOLOVYEV, S. P. & DOLIVO-DOBROVOLSKY, V. V. 1992. *History of the All-Russian Mineralogical Society and its Role in the Geological Sciences*. Nauka, St Petersburg (in Russian).

SOYMONOVM, V. YU. 1829. Directions for mining parties for geognostic description of the Urals chain, ore and gold-fields prospecting. *Mining Journal*, **4**, 1–43 (in Russian).

TAQUET, P. 2009. Geology beyond the Channel: the beginnings of geohistory in early nineteenth-century France. *In*: LEWIS, C. L. E. & KNELL, S. J. (eds) *The Making of the Geological Society of London*. Geological Society, London, Special Publications, **317**, 155–162.

TIKHOMIROV, V. V. 1963. *Geology in Russia During the First Half of the XIX Century*, Volume II. Development of the Main Ideas and Trends in Geology. USSR Academy of Sciences, Moscow (in Russian).

VARSANOFYEVA, V. A. 1955. *The Moscow Society of Naturalists and its Role for the National Science*. Moscow University Publishing House, Moscow (in Russian).

A story of things yet-to-be: the status of geology in the United States in 1807

JULIE R. NEWELL

Social and International Studies Department, Southern Polytechnic State University, 1100 South Marietta Parkway, Marietta, GA 30060, USA (e-mail: jnewell@spsu.edu)

Abstract: In the 1780s observations of geological phenomena by American authors appeared in American publications. Europeans had also begun to explore the American geological landscape, notably the immigrant William Maclure. But an American geological community had not yet formed by 1807. Much of this apparent 'delay' in the development of the geological sciences in the United States resulted from the cultural and political realities of the new nation. In a new and democratic–egalitarian society, it took time to negotiate the nature of the appropriate public support for the practice of science. Individuals with the resources to provide private patronage for scientific undertakings were exceedingly few. The educational institutions that would ultimately be a major factor in the transmission and extension of geological knowledge were only then beginning to multiply and grow. In 1807 Benjamin Silliman completed his first year of science instruction at Yale, but offered only chemistry and mineralogy. Geology would wait several more years. Other institutions and individuals critical to the future of geology in the United States were 'born' in 1807 – including the United States Coast Survey, and Louis Agassiz and David Dale Owen. Roughly another decade would pass before a 'geological community' would emerge in the United States.

As part of the bicentennial celebration of the Geological Society of London in 2007, comparative studies evaluating the state of geology in other countries in the Society's founding year were solicited. Asked to summarize the state of geology in the United States in 1807, an historian's first reaction would be to reply 'not much, not yet'. While accurate, such a reply sheds little light on what was, or was not, happening and why. There was no American centre of science: no London. There were only the very beginnings of geological activity – and that only if the term 'geological' is very broadly defined. And the socio-political context of the early nation is critical to understanding why American geologists, who would achieve so much by mid-century, were barely visible in 1807.

No London

In 1785 observations of geological phenomena by American authors began to appear in the *Memoirs of the American Academy of Arts and Sciences* and the *Transactions of the American Philosophical Society*. The *Medical Repository* included similar articles from its creation in 1797. Each of these publications provided an outlet for a centre of science in the new nation: Boston, Philadelphia and New York City, respectively. But there was no single capital of science, no one centre where the nation's scientific activity was focused. Even once it was created, the nation's *political* capital at Washington, DC, would not function as a major scientific centre; and none of the early centres would predominate over the others. In fact, new centres for geological activity would emerge at New Haven, Connecticut, and in Albany, New York. The American scientific community had no London.

There were, of course, connections of correspondence and acquaintance among the communities at Boston, Philadelphia and New York. But there was nothing even approaching a national community of science in any field, and certainly not in geology.

Geology, very broadly defined

Between 1785 and 1807, 62 articles that might be considered 'geological' appeared in the *Memoirs*, *Transactions* or *Medical Repository*. Three articles dealt with observations of European localities. Five were largely theoretical, including Benjamin Franklin's 1793 'Conjectures concerning the Theory of the Earth'. Three could best be classified as 'miscellaneous'. But 19 provided descriptions of local geological or geographical features, 12 discussed particular fossil finds, seven were clearly mineralogical, another six involved chemical analysis (including agricultural applications of mineral substances, and seven (all in *Medical Repository*) discussed the health effects of various soil and rock types. The 57 articles with named authors were penned by 43 different men.

From: LEWIS, C. L. E & KNELL, S. J. (eds) *The Making of the Geological Society of London*. The Geological Society, London, Special Publications, **317**, 213–217.
DOI: 10.1144/SP317.12 0305-8719/09/$15.00

Among the activities at these various centres, perhaps the American Mineralogical Society deserves special mention. Founded by *Medical Repository* editor Samuel Latham Mitchill in New York in 1799, it represents the earliest known attempt to form any kind of mineralogical or geological society in the United States. It was absolutely characteristic in the utilitarianism of its aims and the rapidity of its dissipation.

Socio-political context

The development and practice of the geological sciences in the United States were directly influenced by the cultural and political realities of the new nation. In an assertively democratic–egalitarian society, it took time to negotiate the nature of such public support for the practice of science as might be deemed at all appropriate.

There was no place in the American ideology of equality and individual effort for American scientists to claim a transcendent right to support for themselves and their science. In addition, individuals with the resources to provide private patronage for scientific undertakings were exceedingly few, and, with rare exception, geologists in America were not men of independent means. Instead, they were required to negotiate support within what historian John Greene has described as 'a context of flamboyant patriotism, political and religious controversy, [and] a practical concern with commerce and industry that alternately inspired and distorted scientific development' (Greene 1984, p. 4). Far too often, at least in the eyes of scientific practitioners, Americans measured all things in terms of utility defined by financial profit. Benjamin Silliman, often called the 'father of American geology', complained in 1805, 'I have been chagrined when conversing with commercial men, to find that friendship–science & all other considerations are immediately brought to the touchstone of loss and gain' (Brown 1989, p. 187).

Even where contributions to the common good might justify government funding for science, constitutional issues virtually precluded direct federal involvement in science before the Civil War (Dupree 1957, pp. 4–5 and 14). Despite the American founding fathers' hopes for generous federal support of learned institutions, it was widely agreed that the Constitution included no provision that specifically empowered the federal government to fund scientific or other learned endeavours directly. Such work could be funded indirectly only when it could be included in projects justified in terms of their contributions to national defence or related to interstate commerce. Such arguments were used to create and protect the United States Coast Survey, the Lewis and Clark expedition, and a number of other military exploring expeditions in the American West.

The Coast Survey, founded in 1807, was justified in terms of its importance to commerce and national defence. In practice, it provided a critical source of research funding and employment for scientists in a number of fields, and was a key institution for the development of many Earth science fields in the United States. But it would take another decade after the 1807 founding for the actual survey work to begin (Slotten 1994).

Financial support for science in the United States, at least until the middle of the nineteenth century, was left primarily to private institutions or individuals, and to state governments. And these entities quite explicitly expected tangible results from whatever they funded. Ultimately, state surveys would be the most important institutions for the support of geology and geologists in the United States in the first half of the nineteenth century. But the earliest state surveys would not be created until the 1820s.

Educational institutions would ultimately be a major factor in the transmission and extension of geological knowledge, but that story, too, was only just beginning in 1807. Samuel Latham Mitchill taught a broad range of geologically related subjects at Columbia College in New York as early as 1792. Parker Cleaveland became professor of mathematics, natural philosophy, chemistry and mineralogy at Bowdoin College in Maine in 1805, but developed most of his knowledge in these fields after his appointment. This was also true for Benjamin Silliman.

In 1798 the President of Yale College, Timothy Dwight, had convinced the college corporation to create a Professorship of Chemistry and Natural History. In 1801, the professorship was still underfunded and unfilled. Dwight forced the issue by appointing recent Yale graduate Benjamin Silliman to the post. Dwight's priorities were clear: Silliman was theologically and politically sound, and he could learn the science. However, Silliman would draw only a tutor's pay while trying to master the knowledge he would need in order to actually begin teaching. He began his studies in Philadelphia, spending two winters there. As early as 1804 he offered mineralogy and geology by including lectures in his chemistry course. In 1805 he travelled to Great Britain to study, and to obtain books and laboratory equipment. Beginning in 1807 Silliman taught a separate course in mineralogy outside the curriculum for interested students, a practice he continued until 1821. But at some time before 1815, he segregated mineralogy and geology from his chemistry lectures. He then taught chemistry first term, mineralogy in the spring and geology in the early summer, ending with the departure of the seniors in July. Students who wanted to work beyond – or entirely outside – the set curriculum often worked as assistants to Silliman. Many of these assistants became important investigators and educators in their own right, including Chester Dewey, Edward Hitchcock, Charles Upham Shepard, James Dwight Dana, Oliver P. Hubbard, and Benjamin Silliman, Jr. But almost all of Silliman's scientific activity at Yale came after 1807.

William Maclure

Benjamin Silliman is often called 'the father of American Geology', largely based on his impact as an educator, and as founder and editor of the *American Journal of Science and the Useful Arts*. Silliman himself assigned that paternal title to William Maclure, largely based on *his* impact as an investigator, author and patron (Silliman 1844, p. 1).

Born in Scotland in 1763, Maclure was a man of business like his father. He travelled widely, including a visit to the United States before he was 20. He moved to the United States in 1796, and retired from business at age 34 to concentrate on his interest in natural history and especially geology. He also became an American citizen. Maclure was an exception to the pattern in early US science: he pursued science as a man of independent means and provided material support – patronage – for the scientific work of others.

Maclure travelled widely in the first decade of the nineteenth century, gathering information and samples and making field observations in both the United States and Europe. In 1807 and early 1808, he visited France and Spain, after which he returned to the US, visiting every state along the Atlantic coast and as many inland areas as he could manage. On 28 January 1809 Maclure read a paper entitled 'Observations on the geology of the United States, explanatory of a geological map' before the American Philosophical Society at Philadelphia. The paper and the map were published in the Society's *Transactions* later that year.

Maclure's map consisted of geological colouring applied to a pre-existing map, labelled using four of Abraham Werner's five categories. The rocks of New England and upstate New York, Maclure indicated, could be assigned to the Primitive, and the Appalachians to the Transition. He assigned the area west of the Appalachians to the Secondary, and traced the Alluvial in a band from Cape Cod down the length of the Atlantic Coast. Despite the contrary opinions of several previous investigators, including four of the 62 papers mentioned above, both Maclure's map and his text were clear about the absence of volcanic rocks. He wrote in the 1809 paper: 'That no volcanic productions have yet been found east of the Mississippi, is not the least of the many prominent features of distinction between the geology of this country and that of Europe, and may perhaps be the reason why the Wernerian system so nearly accords with the general structure and stratification of this continent'.

But Maclure was using Werner's categories as nomenclature for stratigraphic elements with particular lithological, mineralogical and positional characteristics, while explicitly distancing himself from Werner's ideas about the origins of the rocks themselves.

Maclure's 1809 map and paper are characteristic, in three ways, of the work to be done in American geology in the following decades. First, much of the work in at least the next two decades would directly or indirectly refer back to Maclure's map, refining, correcting and extending his characterization of American stratigraphy. Second, American geologists would take a very cautious approach to geological nomenclatures developed in Europe, assessing their utility and appropriateness for application to the stratigraphy of the New World, while explicitly reserving judgement on or openly rejecting any theoretical implications associated with the nomenclature. Third, implicit in the quoted paragraph and explicit in much of the work in American geology that would follow, was a belief that, compared to that of Europe, the American rock record was better – in the sense of being simpler, less corrupted and less likely to mislead.

Together, the faith in the American rocks and the suspicion of the theory associated with various European nomenclatures would lead Americans to propose two kinds of alternatives. Brothers, William and Henry Darwin Rogers, would devise a system using various times of day to reflect the relative order of Palaeozoic strata from earliest to latest. In addition, these and other American geologists would argue that it was absurd to apply the names of *European* 'tribes' and localities to American rocks, and would propose names based on American peoples and places instead.

Maclure influenced the development of American science through his patronage, as well as through his own work. But his interests, and patronage, were not limited to overtly scientific undertakings. In the late 1820s, Maclure joined with Robert Owen to undertake a utopian experiment at New Harmony, Indiana. In 1825 Robert Owen and William Maclure had purchased the town of New Harmony, buildings included, from another utopian group, the Rappites. Maclure was interested in an educational experiment: founding a school based on the Pestalozzian model he had encountered during his travels in Europe. Owen had a long history of interest in education, but his primary objective at New Harmony was the creation of a system of collaborative or cooperative industry. The experiment would ultimately fail, but would have a lasting impact on American geology.

David Dale Owen

Robert Owen's fourth son, David Dale, was born 24 June 1807 at New Lanark, Scotland. Educated at home and then in Switzerland, David Dale developed an interest in mathematics and natural history, especially chemistry. On his return from Switzerland, David Dale spent an additional year studying chemistry in Glasgow.

In 1828 David Dale Owen travelled to New Harmony, which was to be his home for the rest of his life. His father intended for him to take an active role in the cotton spinning industry at New Harmony. David Dale sought to make the case that he would be more useful as a trained chemist. His arguments must have had some effect because in the period from 1831 to 1833 he was in London studying chemistry. Between 1835 and 1837 he studied medicine at Ohio Medical College in Cincinnati, receiving his M.D. in 1837, but he never practiced medicine and never intended to. Medical study was a fairly common path for the study of chemistry and natural history. In Glasgow, in London and in Cincinnati, geology was included, to at least some degree, in David Dale Owen's formal studies. Between academic terms in Cincinnati, he was back in New Harmony with access to Maclure's books and collections, working to identify Maclure's fossil specimens. In the summer of 1836, Owen worked on the Tennessee state survey as an assistant to Gerard Troost. Troost, another scientific immigrant, had also been part of the group of scientific idealists involved in the Maclure–Owen experiment at New Harmony.

By 1833, when David Dale Owen returned to New Harmony from London, the infrastructure of the town was in need of repair and the social experiment and the economy had collapsed entirely. But starting in 1837, New Harmony would emerge as the scientific centre Maclure had hoped for and the centre of industry Robert Owen had desired. The product of this industry, however, was not cotton but geological surveys.

In 1837 David Dale Owen was the right man in the right place when the state of Indiana decided that it, like so many of its sister states during the decade, should have a state geological survey. What Indiana's state legislature, like all state legislatures, really wanted was a quick and accurate accounting of the location, nature and value of its mineral resources. Owen turned out to be phenomenally talented at working quickly and accurately to generate survey reports, with a clear focus on the practical that also reflected current knowledge and developments in the science and had lasting scientific value of their own.

It is impossible to overstate the importance of survey work in conducting and publishing American geology in the period before 1860. And no one headed as many surveys or covered as many states during that period as David Dale Owen. His solo work in Indiana in 1837–1838 led to his appointment to lead the 1839 federal survey of the Dubuque and Mineral Point Districts – rich lead regions in Wisconsin, Iowa and Illinois. Owen oversaw and synthesized the work of 140 individuals surveying 11 000 square miles in two months. The maps were completed in February of 1840 and the report was finished in June (White 1970–1980, p. 258). From 1847 to 1852 Owen conducted another federal survey of mineral lands, this time in the Chippewa Land District of Iowa, Wisconsin and Minnesota. He headed the Kentucky state survey (1854–1860) and also the Arkansas state survey (1857–1860). When the Indiana state survey was reactivated in 1859, Owen was chosen as state geologist to his home state once again.

The arc of David Dale Owen's career coincides with the pattern of government-supported geology and the resulting peaks of discovery, analysis and publication before geological work was suspended for the duration of the American Civil War. There was an incredible burst of activity from the mid-1830s to about 1840, activity increased again starting in the late 1840s and outside forces stopped activity in the very early 1860s. For American geology in general, it was the coming of the war that stopped scientific productivity. For David Dale Owen, it was the culmination of a number of chronic health problems that ended his life three days after he completed the Arkansas report.

Louis Agassiz

Louis Agassiz, like David Dale Owen, was an immigrant. He was born in Switzerland in May 1807, just a month before Owen was born in Scotland. Their shared birth year, although only coincidence, highlights the critical role of immigrants in the American geological community at the very beginning of the nineteenth century. Agassiz's early career also shows just how much the climate for science would change in the United States in the decades after 1807.

Agassiz was well known and well established in European science as he approached his 40th birthday. By the 1840s natural history, including – perhaps even especially – geology, was incredibly popular in the United States. So popular, in fact, that public lectures had been added to the list of ways scientific knowledge could be turned into income in the United States. The wealthy Lowell family had endowed a lecture series in Boston, and, in 1846, Benjamin Silliman helped facilitate an invitation for Louis Agassiz to give a series of lectures at the Lowell Institute.

Silliman, a veteran of public lecture series, including the Lowell Lectures, offered a wealth of advice on how to succeed in this venture and how to be sure one ended up financially ahead. But nothing prepared Agassiz for the crowds he would draw and the adulation he would receive in Boston and well beyond. By the end of 1847 Agassiz had lost his first wife, accepted a new professorship, and begun to build a new and permanent life in the United States. Agassiz's new academic position was in the Lawrence Scientific School at Harvard, the very existence of which provides evidence for the increasingly solid and important place of science in American higher education. Geology benefited significantly from this trend.

One additional institution with which Agassiz would be connected helps to demonstrate just how much had changed by 1847. As the first state geologists struggled in the 1830s to understand how the geological stories they were beginning to unravel connected across state boundaries, they recognized the need for a forum at which they could share, discuss and debate their work. In 1840 they created the Association of American Geologists. By the next year the Association had already begun to expand its membership because not all survey scientists were geologists and not all geologists were survey scientists. The name of the organization was changed to the Association of American Geologists and Naturalists (AAGN), and the AAGN it would remain through the 1847 meeting. But in 1848 the membership would be expanded and the name changed yet again. It now became the American Association for the Advancement of Science.

Conclusion

Only with hindsight can we recognize that the scattering of activity in 1807 really was a promise of things to come. Four decades later, geology would be the most popular and widely supported science in America, where it helped to shape and create institutions that would foster all the scientific disciplines in the decades to come.

References

BROWN, C. 1989. *Benjamin Silliman: A Life in the Young Republic*. Princeton University Press, Princeton, NJ.

DUPREE, A. H. 1957. *Science in the Federal Government: A History of Policies and Activities*. Belknap Press of Harvard University Press, Cambridge, MA (reprinted in 1986).

FRANKLIN, B. 1793. Conjectures concerning the theory of the Earth. *Transactions of the American Philosophical Society*, **3**, 1–5.

GREENE, J. C. 1984. *American Science in the Age of Jefferson*. Iowa State University, Ames, IA.

MACLURE, W. 1809. Observations on the geology of the United States, explanatory of a geological map. *Transactions of the American Philosophical Society*, **6**(2), 411–423.

SILLIMAN, B. 1844. Editor's note. *American Journal of Science and the Arts*, **47**, 1.

SLOTTEN, H. 1994. *Patronage, Practice, and Culture of American Science: Alexander Dallas Bache and the US Coast Survey*. Cambridge University Press, Cambridge.

WHITE, G. H. 1970–1980. David Dale Owen. *In*: GILLISPIE, C. C. (ed.) *Dictionary of Scientific Biography*, Volume 10 (10 vols). Scribner, New York, 257–259.

Military men: Napoleonic warfare and early members of the Geological Society

EDWARD P. F. ROSE

Department of Earth Sciences, Royal Holloway, University of London, Egham, Surrey TW20 0EX, UK (e-mail: ted.rose@virgin.net)

Abstract: At the time the Geological Society was founded in 1807, Europe had entered the latter half of some 23 years of near-continuous warfare, in which the overall scale and intensity were wholly new. Wars from 1792 to 1815 affected the careers of many well-known geologists in France, Germany and the United Kingdom. Influential early members of the Society included a significant number of men with periods of military service or education, or militarily-funded employment: four of its 11 primary founders, Jacques-Louis, Comte de Bournon, James Franck, George Bellas Greenough and Richard Phillips, as well as six of its first 23 Presidents – Greenough, Henry Grey Bennet, John MacCulloch, Roderick Impey Murchison, Henry Thomas De la Beche and Joseph Ellison Portlock. Several councillors, such as Thomas Frederick Colby and John William Pringle, and three of its first five executives – William Lonsdale, David Thomas Ansted and T. Rupert Jones – also had military affiliations. Largely as a consequence of Napoleonic warfare, from 1814 to 1845 national geological mapping in Britain was supported by military funding, and between 1819 and the end of the century geology was a subject taught at various times in all military training establishments within Britain.

When the Geological Society was founded in November 1807, London was the capital of a nation at war with France. It was a long war, arguably a world war, and a war of unprecedented intensity.

War in Europe had begun on 20 April 1792. Fear provoked by the French Revolution of 1789 had stimulated alliances that led to outbreak of the series of intermittent conflicts now known as the French Revolutionary Wars (Blanning 1996). France declared war first on 'the king of Bohemia and Hungary', that is, on Austria, later on Prussia and on the northern Italian Kingdom of Sardinia-Piedmont. On 1 February 1793 France also declared war on Great Britain and the Dutch Republic. On 7 March France declared war on Spain, and that same year the Holy Roman Empire (spread across much of future Germany), and most Italian states, joined the anti-French coalition. A later coalition involved Great Britain, Russia, Turkey, Portugal and the southern Italian Kingdom of Naples. Conflict continued until a peace treaty was signed at Amiens on 27 March 1802. However, in May 1803, after a respite of only 14 months, hostilities against France broke out once more. The Napoleonic Wars began with France as a republic, led by Napoleon as First Consul, but from December 1804 France was transformed into an empire with Napoleon as its Emperor (Esdaile 1995; Gates 1997). By mid 1807, the year of the Society's foundation, only Britain, at war almost continuously since 1793, remained undefeated. After 1807 the wheel of France's fortune gradually turned. Ultimately, Napoleon's army was forced to retreat into France and in April 1814 Napoleon was made to abdicate his throne. Exiled to the Mediterranean island of Elba, he escaped in March 1815 to campaign and briefly reign again, until defeated on 18 June at the Battle of Waterloo. Once more he was forced into abdication and exile. Overall, warfare lasted for some 23 years.

During this period, conflict had blazed at times across almost the whole of Europe, from Portugal in the west across Spain, France and central Europe to Moscow in Russia in the east. It had extended across the Mediterranean, from Italy to Malta, Egypt and the Middle East. It had crossed the Atlantic, to the Caribbean, Canada and the United States. It had reached the Far East, most notably India. French troops had landed in the British Isles, campaigning briefly (and unsuccessfully) both in Wales and Ireland. By the end of hostilities, Britain and its allies had been embroiled in conflicts on land or sea of an almost unprecedented extent.

On 23 August 1793, little more than six months after first declaring war on Britain, the French National Convention proclaimed general mobilization in stark terms:

> Young men will go to battle; married men will forge arms and transport supplies; women will make tents, uniforms and serve in the hospitals; children will pick rags; old men will have themselves carried to public squares, to inspire the courage of the warriors, and to preach the hatred of kings and the unity of the Republic.
>
> (Holmes & Evans 2006, p. 101)

From: LEWIS, C. L. E. & KNELL, S. J. (eds) *The Making of the Geological Society of London*.
The Geological Society, London, Special Publications, **317**, 219–241.
DOI: 10.1144/SP317.13 0305-8719/09/$15.00

An army of over a million men was formed, fired with revolutionary zeal and supported by the nation as a whole. War was to be conducted on a scale and with an intensity that were entirely new.

The dawn of military geology in Europe

There had been an association between military men and mineralogy and mining in parts of Europe from at least the mid-eighteenth century. For example, from 1749 to 1752 Captain Benedetto Spirito Nicolis Di Robilant (1724–1801) and four cadets from the Royal School of Artillery in the Kingdom of Sardinia-Piedmont were sent to Freiberg in Saxony, 'to take courses in metallurgy, mineralogy, chemistry and to study the organization and the productivity of the local mines' (Vaccari 1998, p. 110). However, it was during the French Revolutionary Wars that a general was to take geologists as such on a military operation for the very first time. In July 1798, after victory in late 1797 in command of French forces in northern Italy but lacking sufficient resources to subsequently invade England, Napoleon Bonaparte led an invasion of Egypt. Ostensibly to free its people from their Mameluke overlords, who ruled largely in defiance of Ottoman Turk superiors, this was intended to make France a power in the Levant, to counter British domination in India (Herold 1963).

Napoleon's army was accompanied by a Commission of Sciences and Arts: the 'savants'. Comprising about 150 engineers and other technical experts, their task was to examine almost every aspect of contemporary and ancient Egyptian civilization, and thus provide information about Egyptian society and the country's physical environment and natural resources that would enable the French to govern the region effectively (Gillispie 1989, 1994; Bret 1999). The Commission included four members specifically as '*minéralogistes*' (i.e. geologists) (Bret 1999, p. 61): Déodat de Dolomieu (1750–1801) and three of his former pupils at the redeveloping School of Mines in Paris, Louis Cordier (1777–1861), François-Michel de Rozière (1775–1842) and Victor Dupuy (alias Dupuis: 1777–1861) (Rose 2004*a*, 2005*a*, 2008*a*, *b*).

Dolomieu (after whom the carbonate mineral dolomite is named) and his assistant Cordier (after whom the metamorphic mineral cordierite is named) left Egypt in March 1799, before the French army had created access to the country as a whole. Dolomieu soon came to the end of a long and adventurous life, already well documented (e.g. by Lacroix 1921; Lacroix & Daressy 1922; Gaudant 2005). Cordier, however, developed a distinguished career in the government *Corps de Mines* and as Professor of Geology at the Natural History Museum, Paris, served three times as President of the Geological Society of France and was, ultimately, elevated to the French peerage as Baron Cordier (Jaubert 1862). In 1821 he became a Foreign Member of the Geological Society of London (Woodward 1907), one of the few members resident in France.

However, it was Rozière who was to complete the most significant geological work in Egypt (Drouin 1999; Rose 2008*b*), before the French army was defeated and expelled by a British expeditionary force in the spring and summer of 1801 (Herold 1963; Rose 2005*a*). In addition to four Egyptian articles in the *Journal des Mines* (Rose 2008*b*), Rozière contributed a dozen major items to the *Description de l'Égypte*, the official published record of the 'savants' and a work so massive in its 20-volume first edition (Jomard *et al.* 1809–1828) that a special cabinet could be bought to contain it. Rozière and Dupuy returned to France to lifetime careers in the *Corps des Mines*. Rozière, from 1820 to 1824, was concurrently employed as Professor of Geology and Mineralogy at the Saint-Étienne School of Mines in southern France.

Although the conflicts of 1792–1815 arguably mark the advent of modern warfare, the role of French geologists in Egypt bears closer comparison with that of later colonial explorations in America, Africa and the Far East, than with the work of military geologists in later 'world' wars. There was no need (or means) for a geological appraisal to facilitate a beach landing, as, for example, in the geological preparations for the D-Day landings in Normandy of June 1944 (Rose & Pareyn 1995, 2003; Rose *et al.* 2006) and the German preparations for the invasion of England in 1940 (Rose & Willig 2002, 2004). Although the main responsibility for the young engineers in the expedition was to build or mend fortifications, roads, bridges, canals and public works, the geologists did little to assist them in the ways that military geology was to assist the engineers of both Allied and German forces in the 1914–1918 and 1939–1945 world wars (Rose *et al.* 2000). Moreover, the River Nile and associated waterways plus cisterns and shallow wells provided an adequate water supply, without the need or means to supplement this by deep boreholes sited with geological guidance: a skill developed largely in the twentieth century (Rose 2004*b*; Robins *et al.* 2007). The geologists were deployed to undertake exploration rather than military geology, for there were no geological or, initially, even high-quality topographical maps to guide them. They did record information on raw materials in the expeditionary force area and occasionally notes on water supply: tasks familiar to twentieth century military geologists in both American and European armies. They were not, however, military geologists in the modern sense in that they were not involved in the planning or conduct of military operations, or used to prepare specialist maps (Rose & Clatworthy 2008*a*, *b*).

It is the Prussian general (later Field Marshal) Gebhard Leberecht von Blücher (1742–1819)[1] who has often been credited with first making use of a geologist in military uniform: Professor Karl Georg von Raumer (1783–1865), during the 1813–1814 war of German liberation from French Napoleonic domination. Raumer had been a student of Abraham Gottlob Werner (1749–1817) at the Freiberg Mining Academy in Saxony in 1805–1806 (his fieldwork in 1806 deferred to 1807–1808 to avoid Naploeon's military campaigns in Germany), and became Professor of Mineralogy at the University of Breslau in the then Prussian province of Silesia (now part of Poland) in 1811. Betz (1984), Hatheway (1996), Kiersch (1998) and Kiersch & Underwood (1998) have followed earlier German authors in claiming that Blücher consulted Raumer for information on the terrain of Silesia before triumphing over the French at the second Battle of the Katzbach, on 26 August 1813, and that Raumer was therefore the first geologist to help plan a military operation. However, it is now clear (Linnebach 1912; Rose 2005*c*) that he was used as a normal staff officer and not as a geologist; that he was away from Blücher's army carrying dispatches for some time prior to the Battle of the Katzbach and so was not available for consultation at that critical time; and that he was absent from the entire Battle itself.

Raumer served on the staff of Blücher's Army of Silesia until the allies entered Paris in March 1814, Napoleon abdicated soon afterwards, and the war thus came to an end. In Paris he repeatedly visited the School of Mines, establishing a friendly rapport with an English visitor, the Geological Society's founder President, George Bellas Greenough (1788–1855). Familiar with the geology of the Paris area from pre-war visits, Raumer took Greenough on several local fieldtrips. Raumer was demobilized in early May 1814, awarded the Iron Cross for his military service and returned to a distinguished academic career, initially at Breslau. There in 1816 he was visited by Greenough, accompanied by the Oxford academic William Buckland (1784–1856), the end of the war ending the (arguable) isolation of British from German geologists and the consequent development of separate schools of thought during wartime (Torrens 1998). In 1817 Raumer became one of the earliest Foreign Members of the Geological Society, continuing until he died in 1861 (Woodward 1907).

The first true military geologist was arguably not Raumer, but Johann Samuel Gruner (1766–1824), otherwise known as Johann Samuel von Grouner. A distinguished Swiss mining geologist, Gruner had also studied under Werner at the Freiberg Mining Academy, and later became Director-General of all Swiss mines in the 'Helvetic Republic': briefly a French satellite state (Häusler & Kohler 2003). He participated in military operations during the War of 2nd Coalition, which ended the French Revolutionary Wars, when French troops campaigned in Switzerland against an Austro-Russian army. Later emigrating to Bavaria, when in 1814 Bavaria joined Prussia in the German War of Liberation from Napoleonic rule he commanded a volunteer rifle battalion in the campaign to drive Napoleon's army back to Paris. Drawing on his experience both as a geologist and soldier, he wrote a memorandum in 1820 on the relationship between geology and the science of war. The first such work in its field, this was published posthumously (Grouner 1826), but made no discernable impact either on geology or military training, in Germany or elsewhere. German perceptions of the military applications of geology developed only with the advent of World War I (Rose *et al.* 2000).

In general, geologists were not to be used as such on the field of battle until the twentieth century world wars (Rose & Rosenbaum 1993*a*, *b*, 1998; Rose & Clatworthy 2007*a*, *b*; 2008*a*, *b*). Nevertheless, many influential early members of the Geological Society of London had careers that were affected at some point by the hostilities of 1792–1815. Examples are described here from amongst the founders, early presidents, council members and executives of the Society.

Military affiliations among the Geological Society's founders

Of the 11 Society founders who dined together in London on 13 November 1807 (Herries Davies 2007; Lewis 2009, table 4), in the midst of the Napoleonic Wars, Jacques-Louis, Comte de Bournon (1751–1825), James Franck (1768–1843) and Greenough already had military associations, and Richard Phillips (1778–1851) was soon to acquire them.

Jacques-Louis, Comte de Bournon, from whom the mineral bournonite (a sulphide of copper, lead and antimony) is named, was born into an aristocratic family. His home, the Château de Fabert, lay near the fortress city of Metz in NE France (Herries Davies 2007, 2009), a city in

[1] In 1815 he was joint victor with the Duke of Wellington over Napoleon at the Battle of Waterloo.

which Dolomieu had served as a soldier in his youth (Rose 2008*b*). Although de Bournon had studied mineralogy in Paris under Jean-Baptiste Romé de Lisle, his chosen career initially was that of a professional soldier. He rose to become a captain of artillery, a Lieutenant of the *Maréchaux de France*, and a knight of the Royal and Military Order of St Louis (Bournon 1808, title page). However, as a devoted royalist, he left France when his republican countrymen seized the French royal family in 1791, and fought against the Republic in the opening campaign of the French Revolutionary Wars. By 1793 he was in England: a member of London's émigré community. There his life-long interest in mineralogy helped to provide an income and, as Herries Davies (2007) has described, it was in order to facilitate publication of his mineralogical monograph (Bournon 1808) that had caused most of the group of friends – which founded the Geological Society – to come together initially. De Bournon was thus a professional and veteran soldier whose energies had been diverted from a military career into mineralogy (and so to geology) by the fortunes of war. Between 1810 and 1813 he was to serve on the Council of the Society as its first Foreign Secretary (Woodward 1907).

James Franck was a physician by profession (Lewis 2009); however, he also had a distinguished military career. After graduating from the University of Cambridge as a Bachelor of Medicine in 1792, on 10 June 1794 he was appointed to serve as a 'Physician' with the British army (Hart 1843, p. 418). On 12 November he was assigned for service on the Mediterranean island of Corsica (Peterkin & Johnston 1968, p. 84). At that time, Physicians were commissioned into the Army on the recommendation of the Physician-General, and were always men with higher medical qualifications than the 'Surgeons'. As the élite of the medical profession in the service, they 'were paid at a rate very much higher than that received by the Regimental or Staff Surgeons' (Peterkin & Johnston 1968, p. xxxvii). A Royal Warrant of 12 March 1798 specified that a medical degree of Oxford or Cambridge, or a licence from the College of Physicians in London, were desirable prerequites for appointment, although some latitude was allowed both before and after this date. In wartime, as indeed 1794 then was, a Physician was appointed to the Staff of the Commander-in-Chief of each Army or Expeditionary Force and became his personal medical attendant. Others were appointed to the larger hospitals or garrisons (Peterkin & Johnston 1968, p. xxxviii).

On 4 April 1800 Franck was promoted to the rank of Inspector of Hospitals (Hart 1843; Peterkin & Johnston 1968). Peterkin & Johnston (1968, pp. xxv, xxxi) note that there existed at that time, in addition to the Regimental Medical Officers (surgeons and their 'mates', i.e. warrant officer assistants) who served with regiments of foot guards, cavalry and infantry, a distinct separate medical staff whose duty was concerned with hospitals other than the small regimental hospitals. With the exception of the hospitals in some garrisons both in Britain and abroad, and, later, a few large general hospitals, these hospitals existed only in time of war. Their officers were gathered together on the outbreak of war, as required, typically from the ranks of the general medical profession, or from those who had served on former campaigns and had retired on half-pay.

Franck served as an Inspector of Hospitals during the British campaign of 1801 to expel French forces from Egypt. Napoleon Bonaparte had returned to France in late 1799 to seize power as First Consul, but the survivors of the French army he had led into the country in July 1798 remained as a formidable force: some 25 000 men dispersed at stations throughout Egypt. In March 1801 the British mounted a combined naval and military operation in the eastern Mediterranean to defeat them. The naval force comprised some 180 ships under Admiral Lord Keith; the military comprised about 15 000 men, of 31 regiments or battalions, under Lieutenant-General Sir Ralph Abercromby (Walsh 1803; Daniell 1951; Herold 1963; Ryan 1983). On 8 March Abercromby's troops fought their way ashore at Aboukir Bay, some 20 km from Alexandria. Abercromby was to fight two more battles in the vicinity of Alexandria later in March, being mortally wounded in the second. Nevertheless, French forces in Cairo were compelled to surrender on 27 June, and to evacuate from Egypt between 31 July and 7 August, the 'savants' having been evacuated earlier, in May, as open warfare brought their activities to a close. The remaining French forces surrendered at Alexandria, after siege, on 2 September. Their repatriation began on 14 September, bringing the French invasion of Egypt to an end.

Franck was at first appointed the principal medical officer for the British campaign, but, although regarded as a physician of some distinction, he was 'without war experience and he was superseded by Inspector of Hospitals Thomas Young' (Cantlie 1974, p. 265). Young was an officer who had served Abercromby well on previous occasions. Franck assisted Young in Egypt (Lewis 2009), and seemingly then had no time or inclination to pursue geological interests. Moreover, the British expeditionary force lacked the colonial ambitions of the French, so operated without the support provided by geologists and other 'savants' attached to the French army. There was no British attempt to create an official record comparable with the *Description de l'Égypte* (Jomard *et al.* 1809–1828), or to emulate its extensive description of geology (Rozière 1812*a*) and lavish illustration (Rozière 1812*b*) of Egyptian rocks and fossils.

The campaign over, and war with France ended by the Treaty of Amiens on 27 March 1802, Franck returned to England, where he temporarily retired on half-pay on 25 June 1802. Warfare resumed in Europe in May 1803, but Franck was restored to full pay only on 1 August 1805, and reverted once more to half-pay on 25 December 1806 (Peterkin & Johnston 1968, p. 84). However, on 28 April 1808 Franck was again restored to full pay, prior to service in the Peninsular War which raged across Portugal and Spain (the Iberian Peninsula) between 1808 and 1814. A British expeditionary force landed in Portugal in the summer of 1808, and on 21 August inflicted a sharp defeat on occupying French forces at Vimiera (Holmes & Evans 2006). Of the 30 000 British troops assembled near the start of the campaign, 10 000 had arrived under the command of Sir John Moore after an abortive expedition to Sweden, 'and with him came a substantial number of medical staff officers: Dr. Franck was Inspector of Hospitals, and Dr. William Fergusson Deputy Inspector; there were also two physicians and four surgeons' (Cantlie 1974, p. 301).

The campaign of 1808 soon left the British army in undisputed control of Portugal, with Franck as its principal medical officer (Peterkin & Johnston 1968, p. 84; Cantlie 1974, p. 420). However, in the following winter, a British force that had advanced into Spain under the command of Sir John Moore was compelled to make a poetically celebrated retreat to the coast at Corunna,[2] and there evacuated in January 1809 (Holmes & Evans 2006). Later that year, remaining British forces commanded by Sir Arthur Wellesley (the future Duke of Wellington) campaigned in Portugal and then briefly into Spain. In 1810 the British fought primarily to defend the Portuguese capital, Lisbon, constructing the formidable 'Lines of Torres Vedras', but from March 1811 the character of the war changed, as the French were forced into the first of a series of retreats. However, by 30 October 1811 Franck was so unwell that he was scheduled for repatriation to England (Lewis 2009).

Franck's post in the Peninsula was filled temporarily by Deputy Inspector Bolton, and from 10 January 1812 by Dr (later Sir) James McGrigor. Cantlie (1974, p. 339) was later to comment that 'it is not to decry the ability of Dr. Franck when we say that with McGrigor's appearance on the scene a new era began for the Medical Department in the Peninsula'. Swinson (1972, p. 302) has affirmed that 'It was not until 1812 in the Peninsula that the first organized medical service was instituted by Wellington. On 10 January that year he appointed Dr. James McGrigor as his Principal Medical Officer and consulted him almost daily'. However, it seems that Wellington considered Franck to be a competent Inspector of Hospitals, as, when Franck was invalided to England, Wellington was concerned at his departure, and when later newspaper reports expressed criticism and questioned Franck's ability, Wellington was supportive of Franck rather than his critics (Cantlie 1974, p. 332).

Back in Britain, Franck retired on half-pay with effect from 25 October 1812 (Peterkin & Johnston 1968, p. 84), and was elected to the council of the Geological Society for the two years 1812–1814 (Woodward 1907). The 1812 edition of the *Army List* (War Office 1810–1862)[3] shows that Franck, in his final year on full-pay, served in the Army's Medical Department headed by John Weir as Director-General, with Charles Ker [*sic*] MD and William Franklin MD as Principal Inspectors. Franck himself was listed beneath these as third in seniority of seven Inspectors, above 23 Deputy Inspectors, 24 Physicians, 109 Surgeons, 10 Assistant Surgeons, plus five Purveyors, 25 Deputy Purveyors and 17 Apothecaries.

The 1813 and 1814 editions of the *Army List* record Franck as the first of the few Inspectors 'On the English half-pay', but the 1815 edition lists him as the sole member of the Medical Department categorized amongst 'Retired and Reduced Officers receiving Full Pay', his seniority given as 28 April 1808, the date of his Peninsular War appointment. From 1816 until his death on 27 January 1843 (Peterkin & Johnston 1968, p. 84) annual editions of the *Army List* record him

[2] *The burial of Sir John Moore after Corunna.* Charles Wolfe (1791–1823). This eight-verse poem, which first appeared anonymously in 1817, immediately caught the admiration of the public and has been republished in many subsequent anthologies, correctly attributed to its Irish clergyman author:

> Not a drum was heard, not a funeral note,
> As his cor[p]se to the rampart we hurried;
> Not a soldier discharged his farewell shot
> O'er the grave where our hero we buried.

[3] The *Army List* is a serial publication issued under slightly differing titles, and with frequencies ranging between monthly and annually, from the late eighteenth century to the present day.

as receiving only half-pay (between 1816 and 1818 named incorrectly as James Frank), normally amongst about 20 Inspectors, but from 1825 with seniority credited as from 4 April 1800, his initial date of appointment to the rank. On 29 July 1830 a Royal Warrant changed the rank of Inspector of Hospitals to that of Inspector-General, and accordingly in 'the London Gazette of 11 January 1831 it was notified that the titles of the officers on the half-pay as Inspectors had been altered to that of Inspector-General from 29 July 1830, "but this change is not to be attended with any additional expense to the Public"' (Peterkin & Johnston 1968, p. lxi). Franck is duly shown in the *Army List* as an Inspector-General rather than an Inspector from 1832 to 1843, from 1834 his seniority placing him as first of these.

It is thus clear that the period of extended warfare generated a major career opportunity seized by James Franck, and that Army medical full pay or (more typically) half-pay provided him with an income for nearly 50 years. He was one of the last senior medical officers in the British army before the reforms introduced by his Peninsular War successor James McGrigor pioneered the development of a more modern medical corps.

George Bellas Greenough had a very different background, as described elsewhere in this volume (Kölbl-Ebert 2009). His influence on the Society was considerable and many aspects of his life have already been well documented (e.g. by Wyatt 1995; Kölbl-Ebert 2003; Herries Davies 2007). He was one of the Society's founders, its first treasurer (for one month) and its first President; in 1810 he became one of its first (seven) trustees; he compiled a *Geological Map of England and Wales* (Greenough 1820) largely under the Society's auspices; in 1825 he became one of the first (five) Fellows when the Society gained its Charter (Boylan 2009); and on his death in 1855, he became a significant benefactor to the Society. A lawyer by training but with a lifestyle financed by inherited wealth, he served his country as a Member of Parliament. For 16 years he served also as a soldier in Britain's reserve army (Rose & Rosenbaum 1993*a*; Rose 2000).

On 2 November 1803, barely six months after the onset of the Napoleonic Wars, Greenough had been elected a private soldier in the Light Horse Volunteers of London and Westminster. This regiment formed a corps separate from the regiments of militia, yeomanry and volunteer infantry, which then comprised most of the units in the British reserve army (Rose 2000). Founded in 1779, it was older in foundation, and had diffferent and rather unusual terms of service, as defined in its standing orders by June 1815 (Collyer & Pocock 1843, appendix, pp. iii–xi, and quotes below). The unit was commanded by a colonel rather than a lieutenant-colonel. In peace time its affairs were directed not by its commanding officer alone but by a Committee 'consisting of the Field officers, the Adjutant, Quartermaster, Regimental Sergeant-Major, the Senior Officer present of each troop, and two Privates or Non-Commissioned Officers for each troop, chosen by ballot'. All members had to join as private soldiers and even to enrol as a private: 'Every gentleman must be either proposed or seconded by one of the Committee'. Rather than being paid for their services, members had to pay to join: 'Every Light Horse Volunteer furnishes himself, at his own expense, with uniform and accoutrements and pays a certain sum on his admission, and an annual subscription [and] provides and keeps a horse at his own expense'. Officers were elected and promoted by ballot at a general meeting 'and their names sent up by the Colonel to the Secretary of State for the approbation of His Majesty'. They were thus not subject to the Lord Lieutenant of the county in the manner of other reserve army units. Even private soldiers had the right to be presented to the sovereign when attending court in uniform. Individual commitments were truly voluntary: 'The Light Horse Volunteers ... trust to an *esprit de corps* and point of honour for attendance'. The unit had an obligation to be called out in support of the Civil Power when necessary – but without pay. On 30 June 1808, in the year following his election as a Member of Parliament and also foundation of the Geological Society, Greenough was commissioned as a lieutenant, one of six lieutenants and so subordinate to only nine officers of senior rank in the regiment (Fig. 1). He was to serve as a lieutenant for 11 years.

Greenough resigned his commission in 1819, following an event on 16 August at St Peter's Fields, Manchester, which shocked the country (Rose 2000). Details of the event have proved controversial both in fact and interpretation, then and subsequently. In essence, a political meeting had been called that drew a crowd of 50 000, perhaps 60 000, men, women and children. Fearful of so large a gathering, the local magistrates sent in reserve army cavalry, the Manchester and Salford Yeomanry, to arrest the principal speaker. When these inexperienced horsemen became dispersed and hemmed in, troops of regular army cavalry (the 15th Hussars) were ordered to disperse the crowd by the flat, rather than the edge, of their sabres. The panic as the crowd tried to flee led to 11 deaths and about 400 people being injured. There was an immediate national outcry among both radicals and Whigs against the Tory government, and many members of Yeomanry Corps resigned in protest at a perceived abuse of military power for political ends. Greenough made his

Fig. 1. A squadron officer of the Light Horse Volunteers in review order uniform of the period 1811–1817. Although the illustration is not necessarily of G.B. Greenough himself, it depicts an event in Hyde Park, London, on 20 June 1814, and portrays a type of uniform that he would have worn during his commissioned service in the corps. From a coloured illustration in Collyer & Pocock (1843).

own protest. He resigned from the Light Horse Volunteers and published an identical resignation letter (dated 18 October) plus letter extract (dated 2 November) in national newspapers: the *Morning Chronicle* of Friday 5 November 1819 and *The Times* of Saturday 6 November:

> COPY OF A LETTER FROM G. B. GREENOUGH, Esq. Lieutenant of the Light Horse Volunteers of London and Westminster, to Colonel Bosanquet, the Commander of the Corps.
>
> Dear Sir, – In explanation of the motives which induce me to trouble you with this letter, I beg to refer you to two documents, both of them but too notorious, the one a letter from Lord Sidmouth, his Majesty's Secretary of State for the Home Department, conveying the thanks of the Prince Regent to the Magistrates of Manchester, for their conduct on the 16th of August last; the other a declaration of the Lord Mayor and Aldermen of the City of London, congratulating the lovers of social order on the dignified posture of vigour and resolution which the several authorities throughout the country are assuming, at a period when an unexampled abuse of magisterial authority is from one end of the Kingdom to the other a topic either of regret or alarm.
>
> These two documents have deprived me of that assurance which is obviously indispensable to every honest man who allows his name to continue on the roll of Light Horse Volunteers, an assurance that he will not be called upon as a Member of that Corps to endanger or destroy the lives of the unarmed, the innocent, the unsuspecting, men, women, and children – fellow-subjects, friends and neighbours. In loyalty to the King, in attachment to the constitution, in a desire to counteract, by all legal means, the machinations of those bad men who wilfully impute the distress of the poor to false causes, in order to give effect to their recommendation of desperate remedies, I yield to no man either in the Cabinet or in the Court of Alderman, but I cannot consent to serve any longer in a Corps, in which the only security I have for not being engaged in civil warfare, is the discretion and temper of the Ministers who advised such thanks and the Magistrates who framed such a declaration.
>
> With these sentiments, I beg the favour of you to take whatever steps may be necessary to enable me to resign the Commission which I have the honour to hold in the Regiment under your command. I request also that you will have the goodness to communicate my resignation to the Committee in any way, which, under the circumstances, you may deem expedient.
>
> I have the honour to be, Dear Sir,
> Your obedient faithful Servant, G. B. G.
>
> Parliament St, Oct 18, 1819
>
> Extract of a Letter from Lieutenant Greenough to Colonel Bosanquet, dated London, Nov. 2, 1819.
>
> I beg leave then, with every feeling of respect, both towards you and the corps under your command, to repeat the request contained in my former letter, but I am concerned to say, with one modification. When the letter was written, I desired only that the motives which induced me to resign should be known to you. I am now desirous that they should be made known to the Committee. The dismissal of Lord Fitzwilliam from the Lord Lieutenancy of the West Riding of Yorkshire, for having promoted an inquiry into the unconstitutional proceedings at Manchester, and an addition of 10,000 men to the regular army, for the purpose of intimidating the people, are circumstances which require that the motives of my resignation should be distinctly avowed. The day is come when the friends of a free Government must speak out unless they mean to be silenced for ever. I remain, &c.
>
> G. B. GREENOUGH

These letters provoked both public and private correspondence (Rose 2000), but their acceptance soon ended Greenough's reserve army commitment. It was a commitment additional to geology that had absorbed a minor, but definable, element of his energies over a significant timespan.

Fig. 2. The Royal Military College, Sandhurst, *c*. 1820. From the Sandhurst Collection: published by kind permission of the Curator, Dr Peter Thwaites.

Richard Phillips, the fourth founder with known military associations, was primarily a chemist, and a Quaker until disowned in November 1811 (Torrens 2009). His life has been documented by James (2004). In about 1818 he was appointed Professor of Chemistry at the Royal Military College, Sandhurst (Fig. 2). Thomas Livingstone Mitchell (1792–1855), a Scotsman who was to become famous as a surveyor and explorer in Australia (Branagan 2009), gained a commission as a 2nd lieutenant in the 1st Battalion, 95th Regiment (of Rifles), in 1811, served in the Peninsular War, and was subsequently tasked with making plans of Peninsular sites, based at Sandhurst from 25 September 1819 (Foster 1985). Whilst at Sandhurst, Mitchell attended Phillips' evening lectures from 11 February to 17 March 1820 on chemistry, and shortly thereafter 'the course on Geology' (Foster 1985, p. 84): a course which helped him prepare ultimately for admission as a Fellow of the Geological Society (on 20 April 1827), shortly before his emigration to Australia. It therefore seems that despite his pacifist Quaker origins, by about 1820 Richard Phillips was teaching geology in courses that could be attended by army officers, about three years before such teaching is known to have been instituted at the United States Military Academy, West Point (Rose 2008*a*).

The Royal Military College itself was a product of the 1793–1815 wars (Anon. 1849; Smyth 1961). Its Senior Department was founded at High Wycombe in Buckinghamshire in 1799[4] and transferred to Farnham in Surrey in 1813, moving to Camberley near Sandhurst in 1820–1821, where it transformed into the Staff College in 1858. Its Junior Department[5] was founded at Great Marlow in Buckinghamshire slightly later (1802), moving to Sandhurst in 1812.

The 1793–1815 wars thus created very early opportunities for the teaching of geology in England in a military context (although such teaching at Sandhurst did not become truly significant until 1858, as described below). Introduction of any science teaching was, however, somewhat of a novelty. In 1818 the establishment by subject of professors and masters at the Junior Department comprised a total of 31 teachers in eight subject groupings: fortification (four teachers), French (five teachers), German (two teachers), mathematics (four teachers), arithmetic (three teachers), classics and history (four teachers), landscape drawing (three teachers), military drawing (six teachers); drawing being important since topographical maps sufficiently detailed for military use were still in the process of development (Smyth 1961). The emphasis in the curriculum was not to change greatly for 40 years (Anon. 1849). Early teaching in the Senior Department built largely on the subjects taught in the Junior Department, notably geometry, algebra, fortification and drawing.

Military affiliations among the Society's early Presidents

Amongst the early Presidents of the Society, G. B. Greenough, the Honourable Henry Grey Bennet

[4] In the early years it had about 15 young officers, at least 21 years of age and with at least three years' overseas or four years' home regimental service, who were aspiring to more senior positions of military command.

[5] Initially a school for the sons of military officers, but later for cadets aspiring to commissions in the infantry and cavalry.

(1777–1836), John MacCulloch (1773–1835), Roderick Impey Murchison (1792–1871), Henry Thomas De la Beche (1796–1855) and Joseph Ellison Portlock (1794–1864) had well-known military affiliations, and between them served nine out of the first 26 presidencies, leading the Society for a total of 22 of its first 51 years (Woodward 1907; Herries Davies 2007).

Greenough, whose military activity has been described previously, served as the Society's first President from 1807 to 1813, again in 1818–1820 and yet again from 1833 to 1835, making a unique total of three terms and 10 of the first 28 years.

The second President, Henry Grey Bennet, served from 1813 to 1815. The second son of the 4th Earl of Tankerville, Bennet had been educated at the prestigious Eton School from 1788 to 1792 (Thorne 2004), and then (according to the *Army List* of 1794) joined the army on 16 October 1792. He was commissioned on 25 April 1793 at the age of 16 as an ensign in the 1st Regiment of Foot Guards,[6] barely three months after France had first declared war on Great Britain. He served for five years, being promoted to the list of lieutenants/captains in his regiment on 3 October 1794, before leaving the army in 1798 to study law at Lincoln's Inn, London, and Peterhouse College, Cambridge. Then between 'interludes as assistant to William Drummond, [British] envoy at Naples [from 1801 to 1803], and captain of militia volunteers, he was called to the bar in 1803' (Thorne 2004, p. 105).

Elected Member of Parliament for Shrewsbury in 1806, Bennet shared the political Whig-Radical sympathies of his friend Greenough (Rose 2000). Failing to gain re-election in 1807, he returned to Parliament in 1811 until resignation in 1826 and emigration to Italy. A member of the Geological Society from 1811, he served on its Council from 1812 to 1824, and again from 1825 to 1826, but, apart from this influence in the Society's early affairs, Bennet seemingly made no significant contribution to geology.

The fourth President, John MacCulloch, who served from 1816 to 1818, is described and illustrated elsewhere in this volume (Bowden 2009). A man distinguished as a surgeon and malariologist, as well as a geologist, his life in general has been well documented (Cumming 1980, 1983, 2004; Flinn 1981; Rose 2005*b*; Rose & Renouf 2005). After graduating as a doctor of medicine from the University of Edinburgh in 1793, he remained at Edinburgh and was thus stimulated by geological ideas being widely discussed in academic circles (Porter 1977) until he began a military career as a surgeon's mate in the Royal Artillery on 15 August 1795 (Peterkin & Johnston 1968; Flinn 1981).

At that time the Royal Artillery and the Royal Engineers were commanded by the Master-General of the Ordnance, who held his appointment directly from the Crown and so was not subject to the Commander-in-Chief of the Army (the 'Army' thus comprised principally regiments of guards, cavalry and infantry). The medical departments of the Ordnance and the Army were distinct, until the Board of Ordnance (over which the Master-General presided) was abolished in 1855. Until 1804 'Ordnance' medical officers held only warrants from the Master-General of the Ordnance, not commissions from the Crown, and not until 1814 was a separate heading 'Ordnance Medical Department' inserted in the *Army List*. In the army generally, the 'medical and surgical qualifications of the Mates were not likely to be of a high order. Some were properly qualified men, but the greater number had little other qualification than that they had been apprentices to a surgeon in general practice, and had attended courses in Anatomy, Surgery and Medicine in some University, College or Medical School. At the same time many were highly educated men' (Peterkin & Johnston 1968, p. xxv). In the Royal Artillery, the designation 'Surgeon's Mate' was changed to 'Assistant Surgeon' in May 1797. Peterkin & Johnston (1968, p. 111) cite John MacCulloch (erroneously as James McCulloch) as an Assistant Surgeon of the Ordnance Medical Department from 1 January 1804, but Cumming (1983) states that he was a senior assistant surgeon by 1803.

In addition to performing normal duties as a surgeon, MacCulloch was made assistant to a surgeon, William Cruickshank, who was chemist and assayist to the Board of Ordnance. In 1804 MacCulloch took over Cruickshank's duties. These included lecturing on chemistry at the Royal Military Academy, Woolwich, established since 1741 to educate cadets seeking commissions as officers in the Royal Artillery and Royal Engineers (Smyth 1961). Other duties were primarily responsibility for analysing the purity of sulphur and nitrate shipments imported for gunpowder manufacture, and the efficacy of gunpowder delivered from the mills. His appointment was formalized from 10 October 1806 (Peterkin & Johnston 1968, p. 111), when he retired as a military assistant surgeon and was appointed full-time as the Ordnance chemist.

In the years 1805 and 1807 MacCulloch spent his summer holidays touring parts of the Peak and Lake Districts and the West Country, during which

[6] The regiment was later retitled the Grenadier Regiment of Foot Guards, after defeating the Grenadiers of the French Imperial Guard at the Battle of Waterloo.

he visited mines and made notes about rocks (Cumming 1983). His fieldwork developed into a search for a non-siliceous crystalline limestone suitable for use as millstones for grinding gunpowder and its constituents. The Napoleonic Wars then in progress had prevented importation from Belgium of the 'Namur stone' hitherto used. From 1 September 1809 until the end of the war in 1814 this fieldwork became a task authorized and funded by the Ordnance (Flinn 1981), and MacCulloch thus became the first geologist to be militarily employed in Britain. He was almost pre-empted by William Smith, 'Father of English geology' (Knell 2009), when in 1805 Sir John Sinclair of the Board of Agriculture unsuccessfully proposed that Smith be attached 'as a geologist' to the Ordnance engineers then topographically mapping the country (Torrens 2003, p. xxvii). However, from 1814 to 1821 it was MacCulloch who was appointed by the Board of Ordnance to serve each summer as geologist to the Trigonometrical Survey (Cumming 1983), a project that was ultimately to generate the first geological map of Scotland (MacCulloch 1836). In the winter months he continued to lecture and undertake chemical analyses at Woolwich.

In addition, MacCulloch gained an appointment as a lecturer at the East India Company's military college, at Addiscombe in Surrey (Fig. 3), in chemistry from 1814, and geology also from 1819. Teaching was to cadets aged between 14 and 18, who attended the college for four–six terms, amplifying a curriculum in which prominence was given to mathematics, fortification, Hindustani and military plan drawing, supported by civil drawing, classics and French (Vibart 1894). As such, MacCulloch introduced the teaching of geology to a military curriculum a generation before such teaching at Woolwich and Sandhurst (Rose 1996, 1997*a*), and taught the only course in systematic geology at that time to be taught in England outside the universities of Oxford and Cambridge. His book *A Geological Classification of Rocks* (MacCulloch 1821) deserves credit as the first textbook written specifically to complement an annual British geological lecture course; the first comprehensive, systematic, descriptive, catalogue of rocks to be prepared by a British author (Cumming 1980); and arguably the first British geological textbook to be written primarily for engineers. It was generated (and publication financially supported) as a course book by the East India Company's military college at Addiscombe, following MacCulloch's appointment as part-time lecturer in geology. Its opening dedication is specifically to 'the honourable Court of Directors of the United East India Company', and the directors' names are listed in full. A further potential course book, *A*

Fig. 3. The East India Company's military college (now demolished) at Addiscombe, Surrey. From Vibart (1894).

System of Geology (MacCulloch 1831), drafted in 1821 according to its preface, was less successful. It received more limited financial support from the Court and so lacks a dedication. MacCulloch held his appointment at Addiscombe until his death, from a tragic carriage accident, in 1835.

The 12th President, Roderick Impey Murchison (Fig. 4), served from 1831 to 1832, and then for a second term, from 1841 to 1843. He was the stepson of an army officer, for his mother had married Colonel Robert Macgregor Murray after the death of her first husband (Morton 2004). After education at Durham Grammar School, he gained entry to the Royal Military College, then at Great Marlow, in 1805: at the age of 13 and at the second attempt (being qualified for entry as effectively the 'son' of an army officer).

Before the age of 16, Murchison was commissioned in 1807 as an ensign in the 37th Regiment of Foot but joined his regiment (then based at Cork in Ireland) late in the year, after six months supposedly to complete his education. In 1808 the regiment 'joined a small army of about eight thousand men, later to be reinforced to thirty thousand, under Sir Arthur Wellesley [later to be ennobled as Duke of Wellington], which set sail from Cork to Portugal and the Peninsular War' (Morton 2004, p. 20). The regiment went into action in August, notably at the Battle of Vimiera, where Murchison carried its colours: the flag that formed a rallying point in battle. Promoted to lieutenant, he served in the subsequent campaign into Spain, led by Sir John Moore, and in which Franck served as principal medical officer: a march that began in September but ended under winter conditions of great hardship, a retreat to Corunna where Moore was mortally wounded, and his force being evacuated by sea in January 1809.

Fig. 4. Sir Roderick Impey Murchison, Bt., F. R. S. (1792–1871), veteran infantry officer of the Peninsular War, and second Director-General of the Geological Survey of Great Britain and Ireland. He wears the Military General Service Medal 1793–1814, with clasps awarded for service in the Peninsular War. From Woodward (1907), courtesy of the Geological Society of London.

Back in southern England, Murchison remained with his regiment on home duties for most of 1809, based at Horsham Barracks in Sussex. In the autumn, however, he was appointed '*aide-de-camp*' to his uncle, General Mackenzie of Fairburn, and joined him in Sicily. They returned to England only in 1811, when the General's health deteriorated. The General was soon appointed to a command in the north of Ireland, and after a tedious period of barrack duty with his old regiment in Horsham, Murchison joined him in Armagh. He was promoted captain in January 1812, but saw no further action.

When peace was agreed in 1814, his uncle gave up his staff appointment and Murchison moved to London, temporarily retired on half-pay. When Napoleon escaped from exile in 1815 and initiated the campaign leading to the Battle of Waterloo, Murchison transferred from the infantry into a commission on full pay in the cavalry, the Inniskilling Dragoons, in the hope of action and promotion. But only six troops of his regiment were sent across the Channel, and his was not amongst them. He remained in the depot at Ipswich in Suffolk. When the war finally ended, he married, soon resigned his commission and channelled his considerable energies into other pursuits. In 1824 he became a member of the Geological Society and embarked on a glittering geological career that included his appointment in 1855 as the second Director (in 1866 re-titled Director-General) of the Geological Survey of Great Britain and Ireland (Geikie 1875; Morton 2004; Herries Davies 2007).

The 20th President, Henry Thomas De la Beche, who served from 1847 to 1849, had also attended the Royal Military College at Great Marlow. The only son of Thomas De la Beche (1755–1801), a brevet major (later lieutenant-colonel) in the Norfolk regiment of fencible cavalry (Secord 2004), he was admitted as a cadet in 1810, but seemingly authoritative assertions such as that of Flett (1937, p. 26) that 'He entered the Army after passing through the Military School, but soon retired from the service' and Bailey (1952, p. 21) that 'on the close of the Napoleonic wars he decided to renounce his martial ambition' seem to derive from earlier ambiguous or erroneous biographical statements (Rose 1996). There is no record in the *Army List* that he was ever granted a commission. Rather, McCartney (1977, p. 4) has pointed out that his career at the Military College lasted less than two years and was terminated at a time of unruly behaviour amongst some of the cadets: 'Henry De la Beche, 8 October 1811, removed by order of the Commander-in-Chief'.

De la Beche became a member of the Geological Society in 1817 and later the founding director of what was to become the British Geological Survey. The Survey was founded as the result of a military initiative: 'Early in the spring of last year an application was made by the Master-General and Board of Ordnance to Dr. Buckland and Mr. Sedgwick, as Professors of Geology in the Universities of Oxford and Cambridge, and to myself, as President of this [Geological] Society, to offer our opinion as to the expediency of combining a geological examination of the English counties with the geographical survey now in progress' (Lyell 1836, p. 358). De la Beche had been 'acting under the Direction of the Board of Ordnance' (Greenough 1834, p. 51) from at least 1834, and seemingly from 1832 (Flett 1937, p. 21), and that military direction continued from foundation of the Ordnance Geological Survey until 'It was formally transferred from the control of the Master-General and Board of Ordnance to the First Commissioner of Her Majesty's Woods, Forests, Land Revenues, Works and Buildings' in 1845 (Flett 1937, pp. 45–46). This military financial support is generally credited to the influence of Lieutenant-Colonel T. F. Colby, at that time superintendent of the Ordnance Trigonometrical Survey.

The Society's 26th President, Joseph Ellison Portlock, who served from 1856 to 1858, was a professional soldier throughout his career (Bigent 2004*b*) and the last of the Society's Presidents to have Napoleonic military associations. After passing through the Royal Military Academy at Woolwich, he was commissioned as a second lieutenant in the Royal Engineers in July 1813, and served briefly in England before promotion to lieutenant on 13 December (Rose 1996). In April 1814 he was posted for active service in Canada, serving in August as the junior of two engineer officers in the trenches (Elting 1995) during the siege of Fort Erie, and in September–October helping construct the bridgehead and defensive lines at Chippawa, at which Lieutenant-General Sir Gordon Drummond made a successful stand and 'saved Upper Canada' (Bigent 2004*b*, p. 981) from invasion by forces of the United States that, since 1812, had been allied with Napoleonic France. Following the close of hostilities he took part in numerous expeditions of exploration within Canada before returning to England in October 1822.

In 1824 Portlock was selected for employment with the Ordnance Survey as its work extended into Ireland: first at the planning headquarters in London, and from 1825 with T. F. Colby to initiate fieldwork in Ireland itself. There he remained attached to the trigonometrical branch of the work, of which he soon became the senior and ultimately the sole officer. An account of the origin and early progress of the geological department of the Ordnance Survey of Ireland is given in the preface to its 'first' (and only) report, by Portlock (1843, pp. iii–xi, quoted below). He acknowledges the 'most promising beginning' to geological study made by J. W. Pringle, on Colby's initiative, but its discontinuance due to pressure of basic topographic work. Subsequently geology was 'permitted, but not commanded' under Portlock's nominal charge, until 1832, when he 'commenced the formation of a geological department'. In 1834 he began work on a Memoir of Londonderry intended to describe not only the geology of the region, but also its natural history and 'Productive Economy'. In 1837 he directed his attention more exclusively to the geological department. He formed at Belfast not only a geological and statistical office, but also a museum for geological and zoological specimens, and a laboratory for the examination of soils. 'It is only from this time, then, that the geological branch of the Survey [...] can be considered as an organised work'. However, in 1840 plans to continue the Londonderry Memoir were abandoned, the office, museum and laboratory at Belfast 'broken up, and everything connected with the department removed to Dublin'. There Portlock was directed 'to prepare for publication all the geological data' collected for Londonderry and adjacent areas. The work was completed and the Preface signed in January 1843, and it was published later that year as a substantial, extensively illustrated report 'under the authority of the Master General and Board of Ordnance' (Portlock 1843, title page).

Promoted to second captain on 22 June 1830, Portlock was later promoted to first captain in September 1839, and in 1843 he returned to the ordinary duties of a Royal Engineer officer (Rose 1996). Further promotions followed: to brevet major 9 November 1846, lieutenant-colonel 13 December 1847, colonel 28 November 1854 and major-general (on retirement) 25 November 1857. He served variously in Corfu, England and again, briefly, in Ireland, maintaining an active geological influence. He wrote an elementary textbook on geology (Portlock 1849) that was reprinted annually in 10 editions, and contributed substantial articles on geology (Portlock 1850) and palaeontogy (Portlock 1852) to the massive three-volume *Aide-Mémoire* (Lewis *et al.* 1846–1852) published to equip engineer and artillery officers of the British and East India Company's armies with technical information. His book was noteworthy for making the point that 'the Soldier [...] may find in Geology a valuable guide in tracing his lines both of attack and defence' (Portlock 1849, p. 14).

Military careers of some early councillors

Thomas Frederick Colby (1784–1852), who had sent Portlock to Ireland and used his influence to provide financial support for De la Beche's geological mapping in England, was another career soldier (Portlock 1869; Bigent 2004*a*). Commissioned as a second lieutenant in the Royal Engineers in December 1801 at the age of 17, and soon assigned to the Ordnance Survey (Rose 1996), he was promoted first lieutenant on 6 August 1802, second captain 1 July 1807, captain 5 March 1812, brevet major 19 July 1821, lieutenant-colonel 29 July 1825, colonel 10 July 1837 and, finally, major-general 9 November 1846 on retirement.

In 1814, whilst a captain, Colby was elected a member of the Geological Society and was influential in arranging the appointment of MacCulloch to assist the Ordnance by geological work in Scotland. He served on the Society's Council 1815–1818, 1819–1820 and 1822–1825 until his departure from England to direct the Trigonometrical Survey in Ireland.

The Trigonometrical Survey of Great Britain was extended into Ireland from 1824. As plans were made in London, Colby lived in an atmosphere of science. He belonged to nearly every scientific institution in the metropolis, attended their public meetings and dined three or four days a week at their clubs. He was appointed superintendent of the whole of the Ordnance Survey in 1826. One of his first actions was to establish a geological branch in Ireland, and entrust this to a fellow Royal Engineer officer, Captain J. W. Pringle (although the first memoir including geology was to be published under his own name (Colby 1837) after reactivation of the branch by Portlock).

John William Pringle (*c.* 1791–1861) (known as John Watson Pringle from 1836, *vide* Rose 1997*b*), a colonel's son, was also a professional soldier: a cadet at the Royal Military Academy Woolwich from 24 March 1807, commissioned into the Royal Engineers as a second lieutenant on 23 August 1809, promoted to first lieutenant 1 May 1811 and second captain on 21 July 1815 (Rose 1999).

Pringle served at Chatham from September 1809 until January 1811, before posting to serve in the Peninsular War from January 1811 to July 1814. There he served initially in Lisbon's defensive 'Lines of Torres Vedras'. However, in 1813 and 1814 he served with Wellington's army as it advanced across Spain and into France, being wounded in battle at the River Nive but fit again to serve at Orthes, where a decisive victory over the French was won on 27 February 1814. In April Napoleon abdicated and the war was over. The war ended, Pringle was posted in March 1815 to 'The Netherlands' (at that time comprising both Holland and Belgium), and consequently was one of 11 Royal Engineer officers to participate in the Battle of Waterloo: the only one to be injured.

After the war, as manpower in the British army was reduced to a peace time establishment, Pringle transferred to half-pay from 1 May 1817. He travelled to France and later Freiberg in Saxony, where he is reported to have received training in geology under the distinguished mineralogist Friedrich Mohs (famous for proposing in 1822 the 10-point scale of mineral hardness subsequently widely adopted). After further travels, he was appointed to the Trigonometrical Survey in June 1826 and assigned to the post of 'Superintendent of the Geological Survey of Ireland' from 14 November 1826: a role well documented by Herries Davies (1974, 1983, 1995) and Wyse Jackson (1997). By 1828 Pringle had only one subordinate fully trained in geological survey (Lieutenant J. E. Portlock) and four partly trained (Lieutenants G. F. Bordes, R. S. Fenwick, W. Lancey and A. W. Robe), and the Geological Survey was terminated abruptly on 1 September 1828 to return the officers to their priority task: topographic survey. Little of its work survives: notably a draft map prepared in 1828 for the parish of Aghanloo, County Londonderry (Fig. 5), and two geological crosssections, also from County Londonderry – the first by Lieutenants Fenwick and Lancey RE dating from 28 November 1827; the second by Lancey alone dating from 6 January 1828 (Rose 2008*c*, fig. 1).

Pringle returned to England in March 1829, and served at Woolwich until August 1831, from 1830 as Public Examiner for Commissions at the Royal Military Academy and with the regimental rank of captain from 16 March 1830. He remained on the active list until 1832, but then retired: on full pay because of wartime wounds. Pringle had been elected to Fellowship of the Geological Society on 2 March 1827 on return to England from Ireland, and served on its Council from 1831 to 1832, and again from 1833 to 1834. He resigned on marriage and took up permanent residence in Bath in 1847: coincidentally the year in which the Military General Service Medal 1793–1814 was instituted (33 years after its last qualifying action). Pringle was awarded the medal with three clasps, for

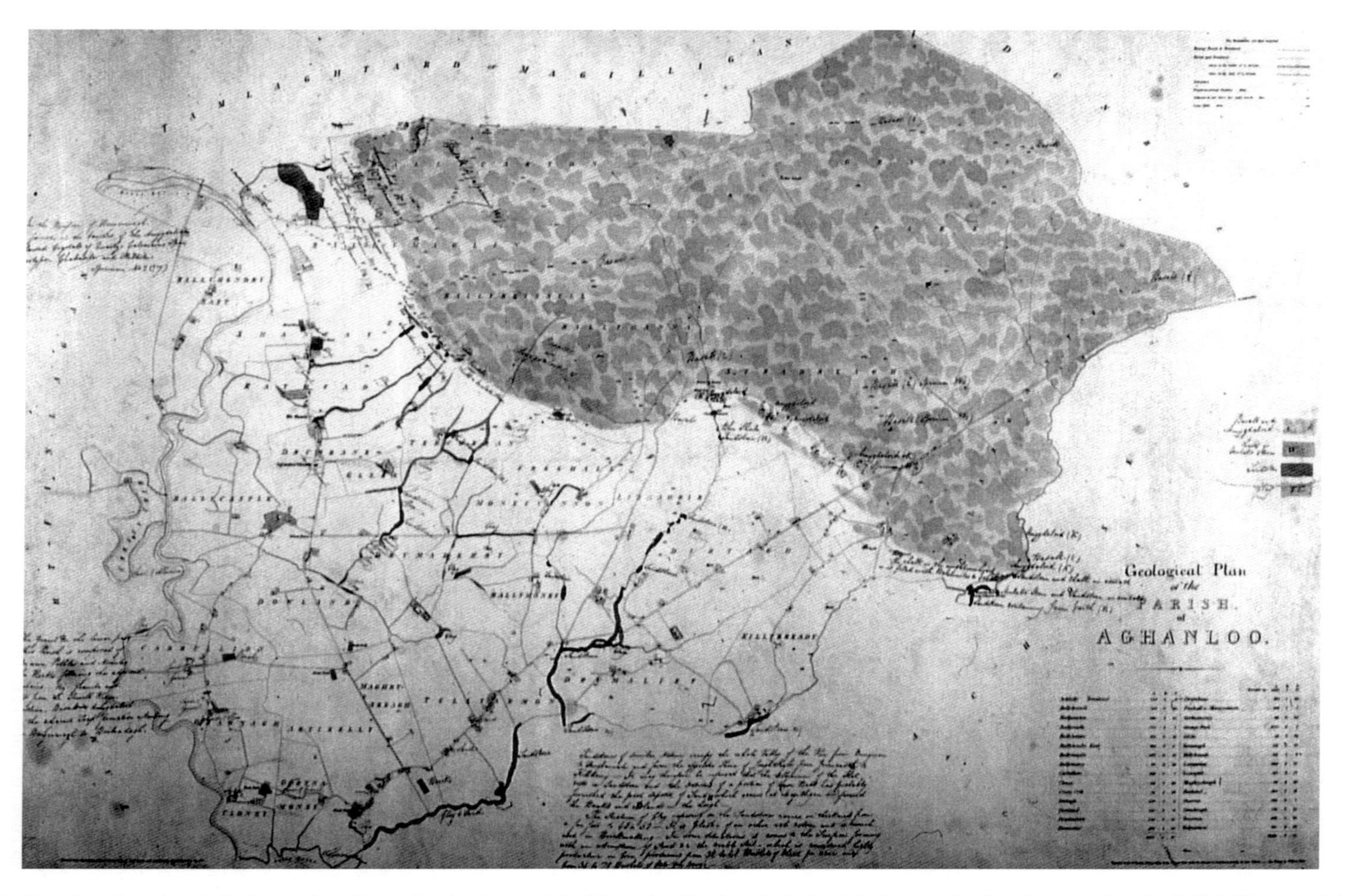

Fig. 5. Part of a coloured geological map of the County Londonderry parish of Aghanloo, Northern Ireland, original at a scale of six inches to the mile (1:10 360) – one of the first government-sponsored geological maps of any part of the United Kingdom, made by Royal Engineer officers in 1828. Geological annotations (black on the figure) have been added in red ink to a printed base map, and a blotched carmine wash has been added to indicate the extent of plateau basalts in the NE. From Rose (1999), after [Herries] Davies (1974); reproduced by permission of the National Archives of Ireland.

service in the Peninsular War and actions at Nivelle, Nive and Orthes, adding this to his Waterloo medal awarded earlier (the first British medal to be awarded to all ranks serving in a military campaign). He was promoted to the rank of major with effect from 28 November 1854, and his death certificate records that he died as a retired major, on 12 October 1861.

Military affiliations among the Society's first executives

Of the first five executives of the Society who served in its early years, from 1829 to 1862 (as listed by Herries Davies 2007, p. 319), three with military connections contributed 28 out of those 33 years of service: W. Lonsdale, D.T. Ansted and T. Rupert Jones.

William Lonsdale (1794–1871) was the Society's curator and librarian from 1829 to 1838 (Pierce 2004; Woodward 1907, p. 308; *non* Herries Davies 2007, p. 319), and as his health failed, assistant secretary and librarian from 1838 to 1842. Initially, he too was destined for the life of a professional soldier: he was commissioned as an ensign in the 4th (or Royal) Regiment of Foot on 1 February 1810, joining two brothers serving in the same regiment (Rose 1996). Promoted lieutenant on 15 May 1812, he served in the Peninsular War and was present at the battle of Salamanca, for which he much later received the Military General Service Medal with one clasp. He also served in the Waterloo campaign and so earned the Waterloo medal. However, with peace-time reduction of the army following the end of the war in 1815, he retired on half-pay on 25 March 1817 and remained in that status until his death. Co-founder with Adam Sedgwick (1785–1873) and Murchison of the Devonian System, his contributions to geology were recognized by the Society by the award of its Wollaston Fund in 1833, 1844, 1846 and 1849, and in 1846 the Wollaston Medal. Yet, a relative commented (Rose 1996, pp. 135–136):

> Though he did not apply to be again [militarily] employed, he always appeared to have liked the life he had in the army, and had always a strong interest in military matters, and especially in his old regiment. He worked hard for Geology, but with us he was the old soldier to the last.
>
> Whether it was an innate love of method, or whether it was the result of military discipline [....] no one can fail to be struck with the regularity and order impressed on everything to which Lonsdale put his hand.

David Thomas Ansted (1814–1880), a Fellow of the Society since 1838 and late in life perceived by some of his contemporaries as a geologist of eminence (Reeve 1864; Thackray 2004), was employed as its vice-secretary from 1844 to 1847. This newly-defined post made him head of the Society's permanent staff, with responsibilities (further defined in 1845) to edit the publications of the Society, have custody of all manuscript papers and, in effect, to be present at the Society's apartments for three hours per day on three days per week (Rose & Renouf 2005). He was responsible for the library catalogue and the meetings programme, and for preparing the first issues of the *Quarterly Journal of the Geological Society* as this replaced the *Transactions* as the Society's premier publication. However, the Society's Council was never satisfied with his performance, and when in January 1847 a Council resolution criticized him for a 'lack of zeal and diligence' he speedily resigned his appointment.

A graduate of the University of Cambridge, Ansted's interest in geology had been kindled there by attending lectures given by Professor Adam Sedgwick. He received his MA in 1840, and was elected a Fellow of Jesus College. Also, in 1840, he was appointed Professor of Geology at King's College London: a post he held until 1853. In 1844 he was elected a Fellow of the Royal Society of London and published his first book, a textbook on geology (Ansted 1844). The next year he published a short, simplified summary of it (Ansted 1845). These became standard reference books in Britain. Duties at the Geological Society in London caused Ansted to give up residence in Cambridge early in 1845, and later that year he was additionally appointed lecturer in geology at the East India Company's military college at Addiscombe: thus, directly succeeding John MacCulloch, after the lectureship had effectively lain vacant for 10 years. He joined an institution founded initially to help meet the high demand for trained military officers during the 1793–1815 wars, in a post initiated by MacCulloch because he was able to convince its Directors from wartime experience of the potential value of a knowledge of natural resources to a military profession. One of his books was awarded as a class prize at Addiscombe, and presumably formed the basis of at least his early lectures at the college. Ansted taught geology at Addiscombe until 1861, when as a result of the war following the 'Indian Mutiny' of 1857, the East India Company lost its sovereign powers, its army was united in direct Crown allegiance with the British army, and the college at Addiscombe closed as cadet training became focused at Woolwich and Sandhurst (Rose 1996, 1997*a*). An Addiscombe graduate published a noteworthy article on the military applications of geology (Smith 1849) during Ansted's tenure of the lectureship, but seemingly without any influence from the Addiscombe lectures (Rose 1997*a*).

Fig. 6. Staff College, Camberley: the Professors in 1874. The professor of geology, T. Rupert Jones, is standing bareheaded at the left rear. Others are, from the left: (rear row) Lieutenant-Colonels Parsons, Farrell, Schaw and Barker; (front row) Professors Dowson and Charante, Colonel Hamley, Dr Atkinson, the Revd J. F. Twisden and Dr Overbeck. From Rose (1997*a*), courtesy of the Joint Services Command and Staff College, and the Institution of Royal Engineers.

Thomas Rupert Jones (1819–1911) was the Society's assistant-secretary, curator and librarian from 1850 to 1862, and served on its Council for the four periods 1865–1870, 1876–1877, 1878–1881 and 1883–1889. Rupert Jones had a civilian medical background: he was apprenticed as a surgeon from 1835 to 1842, and worked as a medical assistant from 1842 to 1850 (Rose 1996; Woodward & (revised by) Cleevely 2004). However, he was an enthusiastic amateur geologist and palaeontologist, hence his appointment to the Society's staff. The post was not well paid, and in 1858 he achieved an additional appointment: lecturer in geology at the Royal Military College, Sandhurst, a post for which he must surely have been recommended by the Society's then President, Major-General Portlock. Like Ansted, he thus joined an institution founded as a consequence of the 1793–1815 wars, but in an appointment whose creation probably owed much to the timely advice of war veteran Portlock, an Inspector of Studies at the Royal Military College, and from May 1857 until 1862 a member of the newly formed Council of Military Education (Rose 1997*a*). Rupert Jones taught at Sandhurst from appointment as Lecturer on Geology on 26 March 1858, and Resident Lecturer from May 1862 (when he resigned his Society posts) until he retired on reduction of the Royal Military College on 31 December 1870.[7]

Rupert Jones also taught at the adjacent Staff College (Fig. 6), re-established in 1858 and with new buildings from 1862, and he continued to do so until retirement in 1882. He supposedly taught the relations of geology to topography, to questions of sanitation, and to water supply; in other words, the practical applications of the science (Woodward 1907). Certainly he seems to have inspired an early paper by a Staff College graduate (Hutton 1862) that

[7] In 1871 the purchase system for commissions in the cavalry and infantry was abolished, with consequent change in the policy of optional attendance at the College for gentleman cadets.

was eminently practical in its focus. But the geology examination paper amongst those printed for the Royal Military College in 1859 (Anon. 1859) provided 20 questions that tested only basic knowledge of mineralogy, petrology, palaeontology and stratigraphy, e.g. 'How do you distinguish the ammonite, goniatite, ceratite and nautilus?' and 'Name the specimens of minerals, rocks and fossils marked 1 to 12'. The mineralogy and geology paper printed in the set for the Staff College had a very similar non-applied focus, its 16 questions including 'Mention some fossils by which you would be able to distinguish a silurian from the carboniferous limestone'. Such questions were seemingly a fair reflection of the teaching provided. Printed notes for the geology course at the Royal Military College in 1866 (Anon. 1866) list 14 lectures given weekly between 2 August and 16 November (each lecture given on two successive days) and justify the teaching as the final part of an introductory lecture in terms of its use in searches for water and coal, and choice of routes for roads. However, the following lectures provided an introduction to sedimentary, igneous and metamorphic rocks, to stratigraphy and structure, and to the geology of England and Wales, Scotland, Ireland and areas of British military interest overseas, such as Gibraltar and Malta. Most are based on *Page's Introductory Textbook* (Page 1854). The syllabus was similar in succeeding years and at the Staff College at that time (Morris & Jones 1870). Few army officers perceived a direct military relevance in such teaching, and when Rupert Jones retired he was not replaced.

Conclusion

It is thus clear that the Geological Society of London contained amongst its early influential members a significant number of veterans of military operational service: de Bournon had served against revolutionary France in Europe in 1792; Cordier for Napoleon in Egypt in 1798; Franck against the French in Egypt in 1801; Franck, Murchison and Pringle against the French in Portugal and Spain at times between 1808 and 1814; von Raumer in the 1813–1814 war of German liberation from French Napoleonic domination; Portlock against invading Americans in Canada in 1814; and Pringle and Lonsdale in 1815 against Napoleon at Waterloo.

Several members were trained for a military career. That of De la Beche never progressed beyond education at a military college. De Bournon's career was cut short by the fortunes of war; Franck retired to half-pay thorough illness; Pringle retired ultimately on account of earlier war wounds; Lonsdale and Murchison, as well as Franck and Pringle, gave up military careers when the wars ended and British armed forces were consequently reduced to a peace-time establishment. The aristocrat Henry Grey Bennet abandoned a nascent military career in the Grenadier Guards for one in the law and politics; whereas Colby and Portlock completed their careers as distinguished senior officers of the Royal Engineers. Those like De la Beche and Murchison who had been trained in landscape drawing as cadets, or who like Colby and Portlock were highly experienced in military topographic survey, no doubt found these aspects of their military background a helpful basis for developing geological map work.

The Society's founder President, Greenough, found time to serve as an officer in the British reserve army whilst guiding the Society in its early years and preparing his *Geological map of England and Wales* (Greenough 1820).

The wars consequent on the French Revolution of 1789 and the threat of invasion of Britain stimulated detailed topographic mapping of the nation as a whole, for military purposes, funded and directed by the Board of Ordnance. The Ordnance Survey was founded (as the Trigonometrical Survey of the Board of Ordnance) in 1791 under the direction of Major Edward Williams and Lieutenant William Mudge[8] of the Royal Artillery (Owen & Pilbeam 1992). The first topographical map published by the new Survey (in 1801) was a one-inch-to-one-mile map of Kent, the county most likely to face invasion from France. The Board of Ordnance, advised by Colby, complemented topographic mapping by also financing pioneering national geological mapping projects in the UK: MacCulloch (from 1814 to 1821) in Scotland, Pringle (from 1826 to 1828) and Portlock (from 1830 to 1843) in Ireland, and De la Beche (from about 1832 to 1845) in England and Wales – leading to establishment of the Geological Survey of Great Britain and Ireland in 1835, a Survey funded under civilian auspices since 1845.

The war also impacted on military training establishments in Britain (Fig. 7), for all except the Royal Military Academy at Woolwich were founded in an attempt to professionalize the British army in the face of Napoleonic threat, and the Academy itself was reorganized as a response to that threat. The Senior Department of the Royal Military College (later transformed into the Staff College) trained young officers aspiring to positions of higher command; the Junior Department of the Royal Military College (later amalgamated

[8] Mudge was elected a member of the Geological Society, in 1810, when a lieutenant-colonel.

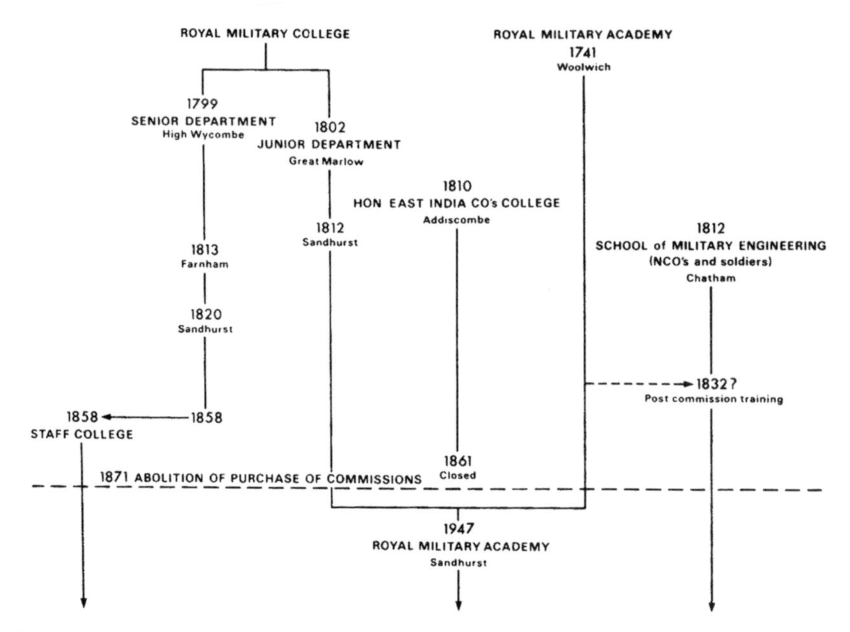

Fig. 7. Military training establishments in nineteenth-century Britain. After Jones (1974), courtesy of the author and the British Cartographic Society.

to form the Royal Military Academy Sandhurst), founded as a school for the sons of army officers, such as Murchison and De la Beche, soon came to train cadets seeking commissions as officers in the infantry and cavalry. The Honourable East India Company's College at Addiscombe trained cadets seeking commissions in the Company's armed forces, while the Royal Military Academy provided the more technical training required by cadets seeking commissions as officers in the 'Ordnance' corps, the Royal Artillery and the Royal Engineers. The School of Military Engineering (Ward 1909) provided training, in the early years for Non-Commissioned Officers (NCOs) and soldiers but soon also for engineer officers on graduation from the Royal Military Academy, in engineering techniques (e.g. skills in topographical cartography, taught to NCOs and soldiers from an earlier date than commissioned officers).

Geology was among the subjects taught at various times during the nineteenth century in each of these institutions, and military funding thus created employment opportunities for several early members. Geology was taught at the Royal Military College in about 1820 by Richard Phillips; at the Staff College, Camberley, from 1858 to 1882 and at the Royal Military College, Sandhurst, from 1858 to 1870, by Rupert Jones, and later at the Staff College by a Major Mitchell and then Lieutenant-Colonel Charles Cooper-King. It was also taught at the Honourable East India Company's College, Addiscombe, by John MacCulloch from 1819 to 1835, and by David Ansted from 1845 to 1861; at the Royal Military Academy Woolwich by James Tennant from 1848 to 1868; and at the School of Military Engineering Chatham from about 1888 to 1896 by A.H. Green (Rose 1997*a*). Of these, only Rupert Jones held a full time appointment to teach geology at a military institution; others were visiting lecturers or taught geology only as a subsidiary subject.

Warfare delayed geological fieldwork (e.g. that of Karl von Raumer in Germany in 1806) and somewhat isolated continental geologists from those of England, arguably leading to separate schools of thought (Torrens 1998). However:

> The British Government became increasingly aware during the Napoleonic Wars of the contrast between the organization of science teaching on the Continent under state patronage and the paucity of official interest and encouragement in this country [Britain]. When it was seen how quickly Napoleon was able to mobilize the French scientific and technological teaching for the war effort, successive British administrations during the first decades of the nineteenth century came to realise that some attempt must be made through the use of central funds to encourage both teaching and research in these neglected fields.
>
> (Edmonds 1979, p. 33).

As a consequence of this wartime awareness, William Buckland, Reader in Mineralogy at the

University of Oxford since 1813 (and later a President of the Geological Society), was appointed also to a new Readership in Geology in 1818, coincidentally as Adam Sedgwick (another future President) was appointed to the Woodwardian Professorship of Geology at the University of Cambridge. As Herries Davies (2007) has eloquently described, in the peace that followed the long war, geology and the Geological Society thrived.

I am grateful to the owners of copyright for permission to use illustrations, as indicated in figure captions, and to Peter Doyle, Mike Taylor and Cherry Lewis for their comments on the article as first submitted.

References

ANON. 1849. *Complete guide to the Junior and Senior Departments of the Royal Military College, Sandhurst.* C. H. Law, London.

ANON. 1859. *Examination Papers for Admission to the Staff College, Royal Military College, and for Direct Commissions.* Harrison, London. [British Library shelfmark 837.e.16.].

ANON. 1866. *Notes of Lectures on Geology [at the] Royal Military College [Sandhurst].* London [British Library shelfmark Ac 4356/2].

ANSTED, D. T. 1844. *The Geology, Introductory, Descriptive & Practical* (2 vols). Van Voorst, London.

ANSTED, D. T. 1845. *A Geologist's Text Book, Chiefly Intended as a Book of Reference for the Geological Student.* Van Voorst, London.

BIGENT, E. 2004*a*. Colby, Thomas Frederick (1784–1852). *In*: MATTHEW, H. G. C. & HARRISON, B. (eds) *Oxford Dictionary of National Biography. Vol. 12. Clegg–Const.* Oxford University Press, Oxford, 491–493.

BIGENT, E. 2004*b*. Portlock, Joseph Ellison (1794–1864). *In*: MATTHEW, H. G. C. & HARRISON, B. (eds) *Oxford Dictionary of National Biography. Vol. 44. Phelps-Poston.* Oxford University Press, Oxford, 981–982.

BAILEY, E. B. 1952. *Geological Survey of Great Britain.* Murby, London.

BETZ, F., JR. 1984. Military geoscience. *In*: FINKL, C. W. (ed.) *The Encyclopedia of Applied Geology.* Van Nostrand Reinhold, New York, 355–358.

BLANNING, T. C. W. 1996. *The French Revolutionary Wars 1787–1802.* Arnold, London.

BOURNON, [J.-L.] COMTE DE. 1808. *Traité de minéralogie. Première partie. Renfermant l'introduction à la minéralogie en général, la théorie de la cristallisation, l'étude de la chaux carbonatée proprement dite, et de l'arragonite, avec application du calcul cristallographique à la détermination des formes cristallines de ceux substances.* William Phillips, London. [English translation of the 'Discours préliminaire'. *In*: LEWIS, C. L. E. & KNELL, S. J. (eds) *The Making of the Geological Society of London.* Geological Society, London, Special Publications, **317**, Appendix II, 457–464.

BOWDEN, A. J. 2009. Geology at the crossroads: aspects of the geological career of Dr John MacCulloch. *In*: LEWIS, C. L. E. & KNELL, S. J. (eds) *The Making of the Geological Society of London.* Geological Society, London, Special Publications, **317**, 255–278.

BOYLAN, P. J. 2009. The Geological Society and its official recognition, 1824–1828. *In*: LEWIS, C. L. E. & KNELL, S. J. (eds) *The Making of the Geological Society of London.* Geological Society, London, Special Publications, **317**, 319–330.

BRANAGAN, D. F. 2009. The Geological Society on the other side of the world. *In*: LEWIS, C. L. E. & KNELL, S. J. (eds) *The Making of the Geological Society of London.* Geological Society, London, Special Publications, **317**, 341–371.

BRET, P. (ed.). 1999. *L'expédition d'Égypte, une entreprise des Lumières, 1798–1801. Actes du colloque. Paris 8–10 Juin 1998.* Académie des Sciences – Technique et Documentation, Paris.

CANTLIE, N. 1974. *A History of the Army Medical Department. Volume 1.* Churchill Livingstone, Edinburgh.

COLBY, T. F. 1837. *Memoir of the City and North Western Liberties of the County of Londonderry: Parish of Templemore. Ordnance Survey of Londonderry. Volume 1.* Hodges & Smith, Dublin.

COLLYER, J. N. & POCOCK, J. I. 1843. *An Historical Record of the Light Horse Volunteers of London and Westminster; With the Muster Rolls From the First Formation of the Regiment, MDCCLXXIX, to the Relodgement of the Standards in the Tower, MDCCCXXIX.* Wright, London.

CUMMING, D. A. 1980. John MacCulloch, F. R. S., at Addiscombe: The lectureships in chemistry and geology. *Notes and Records of the Royal Society of London*, **34**, 155–183.

CUMMING, D. A. 1983. *John MacCulloch, Pioneer of 'Precambrian' Geology* (2 vols.). PhD thesis, University of Glasgow.

CUMMING, D. A. 2004. MacCulloch, John (1773–1835). *In*: MATTHEW, H. G. C. & HARRISON, B. (eds) *Oxford Dictionary of National Biography. Vol. 35. Macan–Macpherson.* Oxford University Press, Oxford, 182–184.

DANIELL, D. S. 1951. *Cap of Honour. The Story of the Gloucestershire Regiment, the 28th/61st Foot, 1694–1950.* George Harrap, London.

DROUIN, J.-M. 1999. Récolter, décrire et raconter: Delile et Rozière. *In*: BRET, P. (ed.) *L'expédition d'Égypte, une entreprise des Lumières, 1798–1801.* Académie des Sciences – Technique et Documentation, Paris, 261–277.

ELTING, J. R. 1995. *Amateurs to Arms!: A Military History of the War of 1812.* Da Capo Press, New York.

EDMONDS, J. M. 1979. The founding of the Oxford Readership in Geology. *Notes and Records of the Royal Society*, **34**, 33–51.

ESDAILE, C. 1995. *The Wars of Napoleon.* Longman, London.

FLETT, J. S. 1937. *The First Hundred Years of the Geological Survey of Great Britain.* HMSO, London.

FLINN, D. 1981. John MacCulloch, M. D., F. R. S. and his geological map of Scotland: His years in the Ordnance. *Notes and Records of the Royal Society of London*, **36**, 83–101.

FOSTER, W. C. 1985. *Sir Thomas Livingstone Mitchell and his World 1792–1855; Surveyor General of New South Wales 1828–1855*. The Institution of Surveyors, Australia, Sydney.

GATES, D. 1997. *The Napoleonic Wars 1803–1815*. Arnold, London.

GAUDANT, J. (ed.). 2005. *Dolomieu et la géologie de son temps*. École des Mines, Paris.

GEIKIE, A. 1875. *Life of Sir Roderick I. Murchison* (2 vols). Murray, London.

GILLISPIE, C. C. 1989. Scientific aspects of the French Egyptian expedition: 1798–1801. *Proceedings of the American Philosophical Society*, **133**, 447–474.

GILLISPIE, C. C. 1994. The scientific importance of Napoleon's Egyptian campaign. *Scientific American*, **271**(3), 64–71.

GREENOUGH, G. B. 1820. *A Geological Map of England and Wales*. Scale 1 inch: 10 miles, 4 sheets (2nd edn, 1839, 6 sheets). Longman, Hurst, Rees, Orme and Brown, London.

GREENOUGH, G. B. 1834. Address delivered at the Anniversary meeting. *Proceedings of the Geological Society of London*, **2**, 42–70.

GROUNER, J. S. VON. 1826. Verhältnis der Geognosie zur Kriegs-Wissenschaft. *Moll's neue Jahrbücher der Berg- und Hüttenkunde, Nürnberg*, **6**(2), 187–233.

HART, H. G. 1843. *The New Annual Army List for 1843, Containing the Dates of Commissions, Together with a Statement of the War Services and Wounds of Nearly Every Officer in the Army, Ordnance and Marines*. John Murray, London.

HATHEWAY, A. W. 1996. Conference report: 1994 Seattle Symposium reveals complexity and value of military geology. *Engineering Geology*, **44**, 245–253.

HÄUSLER, H. & KOHLER, E. 2003. Der Schweitzer Geologe, Oberberghauptmann und Major Johann Samuel Gruner (1766–1824): Begründer der Militärgeologie. *Minaria Helvetica*, **23a**, 47–102.

HEROLD, J. C. 1963. *Bonaparte in Egypt*. Hamish Hamilton, London.

[HERRIES] DAVIES, G. L. 1974. First official geological survey in the British isles. *Nature*, **249**, 407.

HERRIES DAVIES, G. L. 1983. *Sheets of Many Colours; The Mapping of Ireland's Rocks 1750–1890*. Royal Dublin Society, Dublin.

HERRIES DAVIES, G. L. 1995. *North from the Hook: 150 Years of the Geological Survey of Ireland*. Geological Survey of Ireland, Dublin.

HERRIES DAVIES, G. L. 2007. *Whatever is Under the Earth: The Geological Society of London 1807 to 2007*. Geological Society, London.

HERRIES DAVIES, G. L. 2009. Jacques-Louis, Comte de Bournon. *In*: LEWIS, C. L. E. & KNELL, S. J. (eds) *The Making of the Geological Society of London*. Geological Society, London, Special Publications, **317**, 105–113.

HOLMES, R. & EVANS, M. M. 2006. *Battlefield: Decisive Conflicts in History*. Oxford University Press, Oxford.

HUTTON, F. W. 1862. Importance of a knowledge of geology to military men. *Journal of the Royal United Service Institution*, **6**, 342–360.

JAMES, F. A. J. L. 2004. Phillips, Richard (1778–1851). *In*: MATTHEW, H. G. C. & HARRISON, B. (eds) *Oxford Dictionary of National Biography. Vol. 44. Phelps-Poston*. Oxford University Press, Oxford, 136–137.

JAUBERT, H. F. 1862. *Notice sur la vie et les travaux de M. Cordier*. L. Martinet, Paris.

JOMARD, E.-F., LELORGNE DE SAVIGNY, CÉSAR, M. J. & FOURIER, J. B. J. (eds). 1809–1828. *Description de l'Égypte ou Recueil des observations et des recherches qui ont été faites en Égypte pendant l'expédition de l'armée française* (20 vols). L'Imprimerie Impériale, Paris.

JONES, Y. 1974. Aspects of relief portrayal on 19th century British military maps. *The Cartographic Journal*, **11**, 19–33.

KIERSCH, G. A. 1998. Engineering geosciences and military operations. *Engineering Geology*, **49**, 123–176.

KIERSCH, G. A. & UNDERWOOD, J. R. 1998. Geology and military operations, 1800–1960: An overview. *In*: UNDERWOOD, J. R., JR. & GUTH, P. L. (eds) *Military Geology in War and Peace*. Geological Society of America, Reviews in Engineering Geology, **13**, 5–27.

KNELL, S. J. 2009. The road to Smith: How the Geological Society came to possess English geology. *In*: LEWIS, C. L. E. & KNELL, S. J. (eds) *The Making of the Geological Society of London*. Geological Society, London, Special Publications, **317**, 1–47.

KÖLBL-EBERT, M. 2003. George Bellas Greenough (1778–1855): A lawyer in geologist's clothes. *Proceedings of the Geologists' Association*, **114**, 247–254.

KÖLBL-EBERT, M. 2009. George Bellas Greenough's 'Theory of the Earth' and its impact on the early Geological Society. *In*: LEWIS, C. L. E. & KNELL, S. J. (eds) *Geological Inquiries. The Making of the Geological Society of London*. Geological Society, London, Special Publications, **317**, 115–128.

LACROIX, A. 1921. *Déodat de Dolomieu, membre de l'Institut National (1750–1801). Sa correspondance – sa vie aventureuse – sa captivité – ses oeuvres* (2 vols). Libraire Académique Perrin et Cie, Paris.

LACROIX, A. & DARESSY, G. 1922. *Dolomieu en Égypte (30 Juin 1798–10 Mars 1799)*. L'Institut français d'Archéologie orientale, Cairo, Mémoires présentés à l'Institut d'Égypte **3**.

LEWIS, C. L. E. 2009. Doctoring geology: The medical origins of the Geological Society. *In*: LEWIS, C. L. E. & KNELL, S. J. (eds) *The Making of the Geological Society of London*. Geological Society, London, Special Publications, **317**, 49–92.

LEWIS, G. G., JONES, H. D., NELSON, R. J., LARCOM, T. A., DE MOLEYNS, E. C. & WILLIAMS, J. (eds). 1846–1852. *Aide-mémoire to the Military Sciences* (3 vols). Weale, London.

LINNEBACH, K. (ed.). 1912. *Erinnerungen aus den Jahren 1813 und 1814, von Karl von Raumer, herausgegeben und eingeleitet von Karl Linnebach*. Voigtländers Verlag, Leipzig.

LYELL, C. 1836. Address to the Geological Society. *Proceedings of the Geological Society of London*, **2**, 357–390.

MACCULLOCH, J. 1821. *A Geological Classification of Rocks, With Descriptive Synopses of the Species and Varieties, Comprising the Elements of Practical Geology*. Longman, London.

MACCULLOCH, J. 1831. *A System of Geology, With a Theory of the Earth, and an Explanation of its Connection, With the Sacred Records* (2 vols). Longman, London.

MACCULLOCH, J. 1836. *Geological Map of Scotland*. Arrowsmith, by order of the Lords of the Treasury, London, scale 4 inches: 1 mile, 4 sheets, hand coloured.

MCCARTNEY, P. J. 1977. *Henry De la Beche: Observations on an Observer*. Friends of the National Museum of Wales, Cardiff.

MORRIS, J. & JONES, T. R. 1870. *Geology: Heads of Lectures on Geology and Mineralogy in Several Courses, from 1866 to 1870, at the Cadet College, Royal Military College, Sandhurst, Together with Synopses of Lectures used at the Staff and Cadet Colleges, Sandhurst*. Van Voorst, London.

MORTON, J. L. 2004. *King of Siluria: How Roderick Murchison Changed the Face of Geology*. Brocken Spectre Publishing, Horsham.

OWEN, T. & PILBEAM, E. 1992. *Ordnance Survey: Map Makers to Britain Since 1791*. Ordnance Survey, Southampton, and HMSO, London.

PAGE, D. 1854. *Introductory Text-book of Geology*. Blackwood & Sons, Edinburgh [12th edn, 1888].

PETERKIN, A. & JOHNSTON, W. 1968. *Commissioned Officers in the Medical Services of the British Army 1660–1960. Volume 1*. Wellcome Historical Medical Library, London.

PIERCE, S. W. 2004. Lonsdale, William (1794–1871). *In*: MATTHEW, H. G. C. & HARRISON, B. (eds) *Oxford Dictionary of National Biography. Volume 34. Liston–McAlpine*. Oxford University Press, Oxford, 422–423.

PORTER, R. 1977. *The Making of Geology: Earth Science in Britain, 1600–1815*. Cambridge University Press, Cambridge.

PORTLOCK, J. E. 1843. *Report on the Geology of the County of Londonderry, and of Parts of Tyrone and Fermanagh*. Milliken, Hodges and Smith, Dublin.

PORTLOCK, J. E. 1849. *A Rudimentary Treatise on Geology: For the Use of Beginners*. Weale, London.

PORTLOCK, J. E. 1850. Geognosy and geology. *In*: LEWIS, G. G., JONES, H. D., NELSON, R. J., LARCOM, T. A., DE MOLEYNS, E. C. & WILLIAMS, J. (eds) *Aide-mémoire to the Military Sciences*, Volume 2. Weale, London, 77–182, pls 1–14.

PORTLOCK, J. E. 1852. Palaeontology. *In*: LEWIS, G. G., JONES, H. D., NELSON, R. J., LARCOM, T. A., DE MOLEYNS, E. C. & WILLIAMS, J. (eds) *Aide-mémoire to the Military Sciences*, Volume 3. Weale, London, 1–39, pls 1–2.

PORTLOCK, J. E. 1869. *Memoir of the Life of Major-General Colby*. Seeley, Jackson & Halliday, London.

REEVE, L. A. 1864. David Thomas Ansted. *In*: REEVE, L. A. (ed.) *Portraits of Men of Eminence in Literature, Science and Art, With Biographical Memoirs*, Volume 3. Reeve, London, 13–16.

ROBINS, N. S., ROSE, E. P. F. & CLATWORTHY, J. C. 2007. Water supply maps for northern France created by British military geologists during World War II: Precursors of modern groundwater development potential maps. *Quarterly Journal of Engineering Geology and Hydrogeology*, **40**, 47–65.

ROSE, E. P. F. 1996. Geology and the army in nineteenth century Britain: A scientific and educational symbiosis? *Proceedings of the Geologists' Association*, **107**, 129–141.

ROSE, E. P. F. 1997*a*. Geological training for British army officers: A long-lost cause? *Royal Engineers Journal*, **111**, 23–29.

ROSE, E. P. F. 1997*b*. 'John W. Pringle (*c*. 1793–1861) and Ordnance Survey geological mapping in Ireland' by Wyse Jackson (1997): Biographical comment. *Proceedings of the Geologists' Association*, **108**, 157.

ROSE, E. P. F. 1999. The military background of John W. Pringle, in 1826 founding superintendent of the Geological Survey of Ireland. *Irish Journal of Earth Sciences*, **17**, 61–70.

ROSE, E. P. F. 2000. The military service of G. B. Greenough, founder president of the Geological Society. *In*: ROSE, E. P. F. & NATHANAIL, C. P. (eds) *Geology and Warfare: Examples of the Influence of Terrain and Geologists on Military Operations*. Geological Society, London, 63–83.

ROSE, E. P. F. 2004*a*. Napoleon Bonaparte's Egyptian campaign of 1798: The first military operation to be assisted by geologists? *Geology Today*, **20**, 24–29.

ROSE, E. P. F. 2004*b*. The contribution of geologists to the development of emergency groundwater supplies by the British army. *In*: MATHER, J. D. (ed.) *200 Years of British Hydrogeology*. Geological Society, London, Special Publications, **225**, 159–182.

ROSE, E. P. F. 2005*a*. Napoleon Bonaparte's invasion of Egypt 1798–1801 – the first military operation assisted by both geographers and geologists – and its defeat by the British. *Royal Engineers Journal*, **119**, 109–116.

ROSE, E. P. F. 2005*b*. British military geology in India: Its beginning and ending. *Royal Engineers Journal*, **119**, 46–53.

ROSE, E. P. F. 2005*c*. Karl von Raumer: A pioneer of German military geology. *Geology Today*, **21**, 182–186.

ROSE, E. P. F. 2008*a*. Military applications of geology: A historical perspective. *In*: NATHANAIL, C. P., ABRAHART, R. J. & BRADSHAW, R. P. (eds) *Military Geography and Geology: History and Technology*. Land Quality Press, Nottingham, 17–34.

ROSE, E. P. F. 2008*b*. The French invasion of Egypt led by Napoleon Bonaparte in 1798: Its pioneers of military and Egyptian geology. *In*: NATHANAIL, C. P., ABRAHART, R. J. & BRADSHAW, R. P. (eds) *Military Geography and Geology: History and Technology*. Land Quality Press, Nottingham, 45–60.

ROSE, E. P. F. 2008*c*. World wars: a catalyst for British geological innovation. *Open University Geological Society Journal*, **29**(2), Symposium Edition, 10–17.

ROSE, E. P. F. & CLATWORTHY, J. C. 2007*a*. Specialist maps of the Geological Section, Inter-Service Topographical Department: Aids to British military planning during World War II. *The Cartographic Journal*, **44**, 13–43.

ROSE, E. P. F. & CLATWORTHY, J. C. 2007*b*. The Sicilian and Italian campaigns of World War II: Roles of British military geologists in Allied engineer 'Intelligence' and 'Works'. *Royal Engineers Journal*, **121**, 94–103.

ROSE, E. P. F. & CLATWORTHY, J. C. 2008*a*. Fred Shotton: A 'hero' of military applications of geology during World War II. *Quarterly Journal of Engineering Geology and Hydrogeology*, **41**, 171–188.

ROSE, E. P. F. & CLATWORTHY, J. C. 2008*b*. Terrain evaluation for Allied military operations in Europe and the Far East during World War II: 'Secret' British reports and specialist maps generated by the Geological Section, Inter-Service Topographical Department. *Quarterly Journal of Engineering Geology and Hydrogeology*, **41**, 237–256.

ROSE, E. P. F. & PAREYN, C. 1995. Geology and the liberation of Normandy, France, 1944. *Geology Today*, **11**, 58–63.

ROSE, E. P. F. & PAREYN, C. 2003. *Geology of the D-Day landings in Normandy, 1944*. Geologists' Association Guide, **64**.

ROSE, E. P. F. & RENOUF, J. T. 2005. John MacCulloch (1773–1835), Richard Nelson (1803–1877) and David Ansted (1814–1880): Pioneers of geological studies on Jersey and military geology. *Annual Bulletin of the Société Jersiaise*, **29**, 71–98.

ROSE, E. P. F. & ROSENBAUM, M. S. 1993*a*. British military geologists: The formative years to the end of the First World War. *Proceedings of the Geologists' Association*, **104**, 41–49.

ROSE, E. P. F. & ROSENBAUM, M. S. 1993*b*. British military geologists: Through the Second World War to the end of the Cold War. *Proceedings of the Geologists' Association*, **104**, 95–108.

ROSE, E. P. F. & ROSENBAUM, M. S. 1998. British military geologists through war and peace in the 19th and 20th centuries. *In*: UNDERWOOD, J. R., JR. & GUTH, P. L. (eds) *Military Geology in War and Peace*. Geological Society of America, Reviews in Engineering Geology, **13**, 29–39.

ROSE, E. P. F. & WILLIG, D. 2002. German military geologists and terrain analysis for Operation Sealion: The invasion of England scheduled for September 1940. *Royal Engineers Journal*, **116**, 265–273.

ROSE, E. P. F. & WILLIG, D. 2004. Specialist maps prepared by German military geologists for Operation Sealion: The invasion of England scheduled for September 1940. *The Cartographic Journal*, **41**, 13–35.

ROSE, E. P. F., HÄUSLER, H. & WILLIG, D. 2000. Comparison of British and German military applications of geology in world war. *In*: ROSE, E. P. F. & NATHANAIL, C. P. (eds) *Geology and Warfare: Examples of the Influence of Terrain and Geologists on Military Operations*. Geological Society, London, 107–140.

ROSE, E. P. F., CLATWORTHY, J. C. & NATHANAIL, C. P. 2006. Specialist maps prepared by British military geologists for the D-Day landings and operations in Normandy, 1944. *The Cartographic Journal*, **43**, 117–143.

ROZIÈRE, [F.-M.] DE. 1812*a*. De la constitution physique de l'Égypte, et de ses rapports avec les anciennes institutions de cette contrée. *In*: JOMARD, E.-F., LELORGNE DE SAVIGNY, M. J. C. & FOURIER, J. B. J. (eds) *Description de l'Égypte. Histoire naturelle. Tome second.* L'Imprimerie Impériale, Paris, 407–682.

ROZIÈRE, [F.-M.] DE. 1812*b*. Minéralogie, pls. 1–15. *In*: JOMARD, E.-F., LELORGNE DE SAVIGNY, M. J. C. & FOURIER, J. B. J. (eds) *Description de l'Égypte. Histoire naturelle, planches. Tome III.* L'Imprimerie Impériale, Paris.

RYAN, B. P. 1983. Aboukir Bay, 1801. *In*: BARTLETT, M. L. (ed.) *Assault From the Sea: Essays on the History of Amphibious Warfare*. Naval Institute Press, Annapolis, 69–73.

SECORD, J. A. 2004. Beche, Sir Henry Thomas De la (1796–1855). *In*: MATTHEW, H. G. C. & HARRISON, B. (eds) *Oxford Dictionary of National Biography. Vol. 4. Barney–Bellasis.* Oxford University Press, Oxford, 686–688.

SMITH, R. B. 1849. Essay on geology, as a branch of study especially meriting the attention of the Corps of Engineers. *Corps Papers, and Memoirs on Military Subjects; of the Royal Engineers and the East India Company's Engineers*, **1**, 27–34.

SMYTH, J. 1961. *Sandhurst: The History of the Royal Military Academy, Woolwich, The Royal Military College, Sandhurst, and the Royal Military Academy Sandhurst 1741–1961.* Wiedenfeld & Nicholson, London.

SWINSON, A. 1972. *A Register of the Regiments and Corps of the British Army*. Archive Press, London.

THACKRAY, J. C. 2004. Ansted, David Thomas (1814–1880). *In*: MATTHEW, H. G. C. & HARRISON, B. (eds) *Oxford Dictionary of National Biography. Vol. 2. Amos–Avory.* Oxford University Press, Oxford, 268–269.

THORNE, R. 2004. Bennet, Henry Grey (1777–1836). *In*: MATTHEW, H. G. C. & HARRISON, B. (eds) *Oxford Dictionary of National Biography. Vol. 5. Belle-Blackman.* Oxford University Press, Oxford, 105–106.

TORRENS, H. S. 1998. Geology in peace time: An English visit to study German mineralogy and geology (and visit Goethe, Werner and Raumer) in 1816. *In*: FRITSCHER, B. & HENDERSON, F. (eds) *Towards a History of Mineralogy, Petrology and Geochemistry. Proceedings of the International Symposium on the History of Mineralogy, Petrology and Geochemistry, Munich, March 8–9, 1996, Algorismus 23.* Institut für Geschichte der Naturwissenschaften, Munich, 147–175.

TORRENS, H. S. 2003. *Memoirs of William Smith, by his Nephew and Pupil John Phillips, With Additional Material by Hugh Torrens.* Bath Royal Literary and Scientific Institution, Bath.

TORRENS, H. S. 2009. Dissenting science: The Quakers among the Founding Fathers. *In*: LEWIS, C. L. E. & KNELL, S. J. (eds) *The Making of the Geological Society of London.* Geological Society, London, Special Publications, **317**, 129–144.

VACCARI, E. 1998. Mineralogy and mining in Italy between the eighteenth and nineteenth centuries: The extent of Wernerian influences from Turin to Naples. *In*: FRITSCHER, B. & HENDERSON, F. (eds) *Towards a History of Mineralogy, Petrology and Geochemistry. Proceedings of the International Symposium on the History of Mineralogy, Petrology and Geochemistry, Munich, March 8–9, 1996, Algorismus 23.* Institut für Geschichte der Naturwissenschaften, Munich, 107–130.

VIBART, H. M. 1894. *Addiscombe: Its Heroes and Men of Note*. Constable, Westminster.

WARD, B. R. 1909. *The School of Military Engineering, 1812–1909*. Royal Engineers Institute, Chatham.

WAR OFFICE. 1810–1862. *A List of all the Officers of the Army and Royal Marines on Full and Half-pay: With an Index*. War Office, London.

WALSH, T. 1803. *Journal of the Late Campaign in Egypt. Including Descriptions of that Country and of Gibraltar, Minorca, Malta and Macri*. Cadell & Davies, London.

WOODWARD, H. B. 1907. *The History of the Geological Society of London*. Geological Society, London.

WOODWARD, H. B. & (REVISED BY) CLEEVELY, R. J. 2004. Jones, Thomas Rupert (1819–1911). *In*: MATTHEW, H. G. C. & HARRISON, B. (eds) *Oxford Dictionary of National Biography. Vol. 30. Jenner–Keayne*. Oxford University Press, Oxford, 655–656.

WYATT, J. F. 1995. George Bellas Greenough: A Romantic geologist. *Archives of Natural History*, **22**, 61–71.

WYSE JACKSON, P. N. 1997. John W. Pringle (*c*. 1793–1861) and Ordnance Survey geological mapping in Ireland. *Proceedings of the Geologists' Association*, **108**, 153–156.

Practical geology and the early Geological Society

LEUCHA VENEER

Division of History and Philosophy of Science, Department of Philosophy, University of Leeds, Leeds LS2 9JT, UK (e-mail: phllv@leeds.ac.uk)

Abstract: Historical accounts of the interests and activities of the Geological Society of London have often portrayed the Society as indifferent to practical geology and as being focused instead on fieldwork and gentlemanly debate. However, the Society's founders displayed a wide range of interests in the applications of geology and mineralogy. The British Mineralogical Society had been established in 1799 to elucidate the mineral history of Britain and to amass practical information for the benefit of British mining, and the members of this society later brought with them into the Geological Society their practical and mineralogical interests. Close study of the early projects of the Geological Society shows that the membership at this time had a great interest in the practical applications of geology, and especially mining. However, it is clear that there was a significant change in the Society's programmatic activities in the late 1810s and 1820s, when the Society moved away from practical geology. These changes reflected a changing membership.

Historical accounts of the interests and activities of the Geological Society of London have often portrayed the early Society as focused on fieldwork and thriving on gentlemanly debate, with little concern for practical geology or the application of mineralogical knowledge to mining. Roy Porter, in particular, helped to establish this as the standard interpretation in his now classic study (Porter 1973, 1977). While this conception of a geology driven mainly by theoretical considerations has been helpful in assessing aspects of the work of individual geologists, many of them members of the Geological Society (Rudwick 1985; Secord 1986; Thackray 2003), it fails to account for the wider interest in the practical applications of geology, particularly amongst many of the Geological Society's founding and early members.

Porter argued that geology throughout the first half of the nineteenth century was unaffected by practical considerations, and that geological endeavour in Britain centred on London and the gentlemanly practitioners of the Geological Society (Porter 1973). His analysis was one of a number of groundbreaking studies of the mid-1970s in which historians of science sought to replace crude economic accounts of the development of science with more sophisticated social and cultural arguments (Shapin 1972; Thackray 1974). This trend has been continued by Simon Knell (2000) in his cultural history of early nineteenth-century geology. This acknowledged other kinds of geology, including, as Jack Morrell (1983, 2005) and Hugh Torrens (1998, 2001, 2002) have elaborated, the economic and practical motivations of various practitioners of nineteenth-century geology.

The key questions to which Porter devoted himself were, first, can mining in the early nineteenth century be said to have stimulated geology, and, second, did geology aid mining at all? His claim that geologists were not interested in the mining industry, except as a potential source of valuable geological data, focused on a few of the leading geologists of the period 1780–1830, and half of those mentioned were leading members of the Geological Society in the 1820s (Porter 1973). It is certainly the case that many leading geologists of the 1820s were not particularly interested in the mining industry, and especially not when considered as individual geologists with fieldwork to do and papers to publish. However, there were some in the Society with economic interests, even in this period; albeit that by this time they were in something of a minority. In the Society's earliest years, however, I believe there was a major interest in the application of mineralogy and geology to mining that has largely been overlooked.

Of the 13 founder members of the Geological Society, seven had been core members of an earlier scientific society, the British Mineralogical Society. Founded in 1799, it had been established in an attempt to elucidate the mineral history of Britain in order to amass practical information for the benefit of British mining. The members of the Mineralogical Society brought with them into the Geological Society their practical and mineralogical interests, and these were reflected in some of the Society's initial programmatic aims. However, the Geological Society grew very quickly in the 1810s and 1820s, resulting in rapid change to its aims and methods. Examining some of these transformations and uncertainties in the Society's early work reveals how the gentlemanly orthodoxy emerged from more diverse beginnings. In what follows, I examine the aims and membership of

From: LEWIS, C. L. E. & KNELL, S. J. (eds) *The Making of the Geological Society of London*. The Geological Society, London, Special Publications, **317**, 243–253.
DOI: 10.1144/SP317.14 0305-8719/09/$15.00

the British Mineralogical Society, and provide an account of the early interests of the Geological Society and its members, demonstrating that earlier practical interests were carried over into the Society's early programme. I also argue that the Society's interests in the late 1810s and early 1820s represented the collective interests of a group of individuals, suggesting that this altered as the membership of the Society itself changed.

The British Mineralogical Society

The British Mineralogical Society was founded in 1799 with the aim of gathering mineralogical data in order to attain a better practical knowledge of the mineral history of Britain. The society's members aimed to achieve this first by increasing their own knowledge of mineralogy, and especially of new or little-known mineral substances, and then by disseminating that knowledge. Through the chemical analysis of mineral specimens sent to the society from all over the country, and particularly from British mining districts, they hoped to gain a more thorough knowledge of British minerals. They then intended, through publication of the results and through networks of correspondence, to make this knowledge more widely available to those for whom it would be practically useful, possibly eventually with the publication of a practical work on mineralogy.

The importance of these aims to the small society is clear in the stipulation that members were to be highly competent in mineralogy, as well as in the chemical analysis of mineral specimens, and it is also particularly noteworthy that this group chose to call themselves the *British* Mineralogical Society. It was an attempt to form a national body in Britain to advance the study of the Earth for overwhelmingly practical ends. However, in the event, the society's programme of mineral analysis proved very difficult to promulgate and maintain, and the society declined fairly rapidly, merging with the Askesian Society, another scientific society based in London, in 1806 (Torrens 2009). However, the activities of its members did not immediately cease, but became instead a part of the work of the Askesians (Torrens 2009). The society had originally emerged from the Askesian Society to focus particularly on mineralogy and in spite of its apparent failure, its activities and intentions merit further attention, as do the reasons behind the failure.

The founding meeting of the British Mineralogical Society was held in April 1799 in the room used by the Askesian Society in Plough Court.[1] The Askesian Society was a small philosophical and learned society centred on the Dissenting community, which lived and worked in the Lombard Street area of the City of London (Inkster 1977). The name was derived from the Greek ἀσκησις, originally meaning physical exercise or training, although for this particular group it had the connotation of philosophical self-discipline and mental training. Although the British Mineralogical Society was formed by members of the pre-existing Askesian, it was not a breakaway group that felt the Askesian Society as a whole no longer fulfilled their philosophical aims. Rather it was a group of common-minded individuals who had a deep interest in, and were highly competent in, chemistry and mineralogy, and who had particular mineralogical aims that they felt would be better served by a small, more specialized, society.

The founding proposals for the British Mineralogical Society were made by William Haseldine Pepys (1775–1856), a scientific instrument maker who had extensive chemical and mineralogical interests. Other chemists and mineralogists present at the first meeting included William Allen (1770–1843), a chemical manufacturer, Alexander Tilloch (1759–1825), a commercial scientific publisher, and Richard Knight, also a scientific instrument maker. More mineralogists quickly joined, beginning with Arthur Aikin (1773–1854), who made a living from his scientific writings, and his brother Charles (1775–1847), a surgeon, later in the same year. They were followed by the physician, William Babington (1756–1833), the Irish mineralogist, Richard Kirwan (1733–1812), and the chemist, Richard Phillips (1778–1851), in 1800. Phillips's brother William (1773–1828) also attended the society's meetings as a visitor on several occasions before himself joining. The society also had contacts amongst mining engineers, although it was never able to recruit the surveyor, William Smith (1769–1839), who at this time was already engaged in making mineralogical and geological maps (Weindling 1983).

Others residing in provincial mining districts and who had deep interests in the practical applications of mineralogy and geology also joined, including David Mushet (1772–1847) and the Rev. William Turner (1761–1859) in 1800, and John Taylor (1779–1863) in 1801. These became Corresponding Members, having interests in local practical mineralogy and possessing financial interests in mining or related industries. Mushet was manager of the Clyde Ironworks, and later moved to the Calder Ironworks near Airdrie. Turner was a Unitarian minister and a lecturer on natural philosophy in

[1] NHM: Minute Book of the British Mineralogical Society.

Newcastle, and a leading member of the Newcastle Literary and Philosophical Society. He had a keen interest in mapping the local coal-field and improving local mining generally. Taylor, also a Unitarian, was a mining entrepreneur and engineer living near Tavistock in Devon, who also had mining interests in Cornwall. These provincial figures were to form an important part of the network of correspondence and specimen exchange that the society intended to establish.

There was widespread public interest in the application of scientific principles to industry at this time, and leading members of the British Mineralogical Society used public forums such as the commercial scientific press to publicize their activities. The *Philosophical Magazine*, under the editorship of Tilloch, had already emphasized the need for a greater understanding of mineralogical subjects in the mining districts. In its first volume, in June 1798, a contributor described the discovery of a highly valuable vein of cobalt in a mine in Cornwall that had been dismissed as useless until the intervention of a knowledgeable mineralogist. The writer emphasized the importance of applying mineralogical and geological knowledge to mining:

> It is to be lamented that there are few or no skilful mineralogists in Cornwall; and that we have no good *practical* work in the English language to enable them to apply to further use, what little knowledge they have acquired from working the rich tin and copper mines in that county [...] For want of proper books of this kind in the English language, there is reason to believe that many valuable mineral products are every day lost in Cornwall; for every substance that appears not to possess the characteristics of the tin or copper, of which they are in search, is thrown away among the rubbish.
>
> (Anon. 1798)

The waste of highly valuable mineral substances through the ignorance not only of the working miners but of their captains and managers was paraded before the journal's readers as an indictment of the failure of mining proprietors to apply scientific principles to the art of mining. This same point, more generally applied, was soon reiterated by David Mushet in the first of many contributions that he made to the *Philosophical Magazine*. Here, he widened the issue to include other branches of British industry, emphasizing that, in order to apply scientific principles to industry in general, dialogue was required between the practical man and the man of science. Only with co-operation and mutual respect would anything useful be achieved:

> inaccurate and fallacious principles are often brought forward by men of science, even the best intentioned, from a want of that practical knowledge, which can only be acquired by a long and personal acquaintance with the processes carried on in the large way of manufacture. The mischiefs hence occasioned are incredible: it tends to separate the man of science and the manufacturer; it shackles the latter with increasing prejudice; makes him view the former with a suspicious eye; is the principle reason why science has been so long excluded from our manufactories; and why the accurate results of the laboratory have so long been despised by the practical artist, and been deemed undeserving of experiment on an extended scale. The artist and the man of science should mutually inform each other: principles will then, and not till then, acquire consistence and correctness, and their values will be established on the surest foundation.
>
> (Mushet 1798, pp. 9–10)

The need to found British industry on sure scientific and practical principles was to become a familiar theme in the writings of Mushet in particular, as well as in the editorials of Tilloch's journal, and in the work of other writers with similar interests. Such writings also appeared in the rival to Tilloch's *Philosophical Magazine*, the *Journal of Natural Philosophy, Chemistry, and the Arts*, which was published by William Nicholson (and consequently was often simply known as *Nicholson's Journal*). It was in this context, and amongst men with an interest in mineralogy and a belief in the importance of useful practical scientific knowledge, that the British Mineralogical Society was formed in 1799 to collect mineralogical data and improve knowledge of the mineral history of Britain. The society quickly published its own statements, in which concern regarding the efficiency of the mining industry is clearly to be seen:

> To assist in preventing a waste of the bounties afforded by Providence to this country, THE BRITISH MINERALOGICAL SOCIETY has been instituted; but the landholders and proprietors of mines can alone overcome the obstacles which arise from the prejudices of the miners, by causing them to produce samples of the various strata they pass through, and by giving them premiums for every new substance of which they may be able to produce specimens. – The country will be benefited exactly in the proportion that such a practice shall become general.
>
> The British Mineralogical Society will analyse, *free of expense*, package and carriage excepted, for the proprietors of mines or landed estates, whatever substance they may meet with [...] In return, the Society expects that singular or curious specimens [...] will be sent for the improvement of the Society's Cabinet, which will thus in time become a national ornament.
>
> (Pepys 1799, pp. 111–112)

Here again we can see the concern regarding the prejudice of miners against scientific expertise and the certainty that a national body could have a beneficial national effect, assuming that professionals could be encouraged to contribute. This announcement of the foundation of the society was written by Pepys and appeared in *Nicholson's Journal* in

1799, after Nicholson requested further information from the British Mineralogical Society about its activities.[2]

The background against which the core members of the British Mineralogical Society appear is highly significant, both in understanding the society's aims and in accounting for its lack of success in achieving them. The society was centred on the same Dissenting community as the Askesian Society. There were a number of Dissenting groups in Britain at this time, including Quakers (the Society of Friends), Unitarians and others, all of whom dissented from at least one of the articles of faith required by the established Church of England. All these groups had in common a certain isolation from society as a whole, since they were barred from the military and from public office, and they were also usually effectively prevented from receiving a university education in England, since only Anglicans were admitted to Oxford and Cambridge.

Located in the eastern part of the City of London was a business and manufacturing district with a large Dissenting community. There were Quakers with family businesses including printing (such as the Phillips family) and chemical and pharmaceutical manufacture (such as the Allens). Many Quakers were in business (often banking or manufacturing), and as a group Quakers were particularly renowned for their philanthropy and their association in this period with causes such as the abolition of the slave trade and conscientious objection to war in Europe (Torrens 2009). Most of the members of the Askesian Society and the British Mineralogical Society were Dissenters, for example, Allen and the Phillips brothers were Quakers, and Richard Knight and the provincial members, William Turner and John Taylor, were Unitarians, although Pepys was an Anglican. Although Weindling suggests that there were no particular unifying religious or political characteristic in the nucleus of British Mineralogical Society membership (Weindling 1983), nevertheless in the wider context of the general practical and philanthropic interests of such a Dissenting community, the aims of the society towards the practical improvement of mineralogical knowledge can be readily understood. The problems faced by the society, however, also stemmed partly from its somewhat limited circle: they had difficulties in communicating through wider scientific and industrial networks precisely because of this. Overwhelmingly Dissenting and London-based, they had little opportunity to forge links with other groups such as men of science in the universities, although through members such as the printer William Phillips and Alexander Tilloch, editor of the *Philosophical Magazine*, they did have a close association with the commercial scientific press. They also began with few connections with practical men in mining districts, although they did endeavour to establish such links through their system of corresponding membership.

Although the difficulties the British Mineralogical Society faced in achieving its aims did, in part, arise from its somewhat limited circle, it also stemmed from the methodological nature of the society's programme, which was both more difficult and less fruitful than had been expected. The minute book shows that in its seven-year existence as an independent society it received about 25 specimens for analysis, although it also received many donations of specimens for its cabinet.[3] It seems probable that few landowners and mine owners were actually willing to send specimens, for a variety of reasons, including the industry's natural, and economically motivated, suspiciousness and secrecy. There was also a prejudice against practical scientific assistance, or rather a belief that it could result in no particular benefit – a difficulty that both the society as a body and its members as individuals discussed in print. However, the British Mineralogical Society itself was also at fault, for it seems not to have publicized its activities very widely. And, finally, there was the problem of mineral analysis itself, as it was generally time-consuming and did not always produce satisfactory results.

The society's other planned projects never advanced beyond the proposal stage. These included gathering mining knowledge and terminology from across the country so that eventually a dictionary of British mining terms could be produced, thus achieving better understanding and communication within the industry at a national level. Nevertheless, although the society was unable to wholly achieve its aims or complete its projects, later reflections of the intentions and aims of the members can be seen in the early work of the Geological Society.

The Geological Society

Seven of the 13 founding members of the Geological Society had been part of the core membership of the British Mineralogical Society. These were the Phillips brothers, Babington, Aikin, Pepys, Allen and Knight. One more at least, Humphry Davy, had had close connections with the British

[2] NHM: Minute Book of the British Mineralogical Society.

[3] NHM: Minute Book of the British Mineralogical Society.

Mineralogical Society, having attended meetings and contributed to discussions. Many British Mineralogical Society members from provincial districts, such as Mushet, Turner and Taylor, also joined the Geological Society very early on, as Honorary Members; some of the 40 such members proposed at the Society's second meeting. John Taylor served as Treasurer for over 20 years, and many former British Mineralogical Society members were prominent on the Society's early committees and Council. Five of the eight members on the committee that considered which papers should be read at meetings, for example, had belonged to the earlier society.

The papers communicated to the Geological Society in its first five years show that there was a great deal of interest in mining and mineralogy. The minute book covering this period reveals that, on average, about half of the communications concerned mining or some other practical aspect of geology. Specimens and sections from mines were also exhibited. These early meetings contained two or three communications, although sometimes longer papers took up an entire evening or were serialized.[4] An early practical paper by Taylor concerned improvements to the blasting process in mines, read in February 1808. Just a few months later, in April 1808, the mining engineer James Ryan exhibited a core of solid rock drilled out using his patent boring machine (a machine he had already exhibited to a meeting of the British Mineralogical Society in 1806). Maps intended to be of practical use were also highly interesting to the Society and, in December 1808, Society member and President of the Royal Society, Sir Joseph Banks, presented a mineralogical map of Derbyshire by the surveyor John Farey. A year later, in November 1809, William Rashleigh presented a paper on the stream-works of Pentewan in Cornwall.

Throughout this early period communications on these practical aspects of geology were received by and laid before the Society on a regular basis. They came from Ordinary Members of the Society in London, from Honorary Members who resided in mining districts and from practical men, including mineral surveyors and engineers such as Farey, Ryan and Robert Bald, some of whom were members of the Society. Those who were not members communicated with the Society either through the Honorary Members in their local area or through the patronage of wealthy landed gentlemen with scientific interests, such as Sir Joseph Banks, or directly through the Secretaries and President. It was networks such as these, functioning on a variety of levels, that the British Mineralogical Society had not been able to wholly establish, and which, by contrast, gave the Geological Society such strength and depth from its earliest years. The early membership comprised a wide professional and social range: Anglicans and Dissenters; university men and medical practitioners; metropolitan and provincial leisured gentlemen; noble patrons; and, of course, those with interests in geology and its application to mining.

This influx of papers, maps, sections and specimens was greatly encouraged by the Geological Society's programmatic endeavours. The Society at this time was actively engaged in gathering geological data from across the country, and beyond. In 1808 it produced *Geological Inquiries* – its first publication, intended initially for its Honorary Members. This pamphlet, of about 20 pages in length, begins with an introduction to the methods and purposes of geology: although a 'sublime and difficult science [...] it is susceptible of division into many different departments, several of which are capable of being extended by mere observation'. The science of geology itself 'can be performed only by Philosophers,' but the necessary data can be collected by anyone: 'the Miner, the Quarrier, the Surveyor, the Engineer, the Collier, the Iron Master, and even the Traveller' (Geological Society 1808*a*, p.2). The nature, high status and importance of geology, and therefore of the Society itself, are clear to see here, and the pamphlet reflects a concern with status and position that had not really been relevant or important to the British Mineralogical Society in presenting a public image. However, the need to involve as many people as possible on the national level is also clear, and the Geological Society was offering itself as a repository for the collection of geological information so that, amongst other things, 'Mineralogical maps of districts, which are now so much wanting, may be supplied [...] and, above all, a fund of practical information obtained applicable to purposes of public improvement and utility' (Geological Society 1808*a*, p. 3). This is, of course, a general statement of interest in practical geology, but the mention of mineralogical maps, which are 'so much wanting' is worthy of note.

Throughout 1808 the Society continued with its active programme of gathering geological and mineralogical data and at its May meeting it again seemed to echo an ambition of the British Mineralogical Society:

> the present state of the Nomenclature of Geology and Mineralogy renders it highly necessary that this Society should adopt some measures which shall tend to remove the confusion that now prevails, in order

[4] GSL: First minute book of the Geological Society.

> that its own members may have one uniform language for the relation of Geological and Mineralogical facts.[5]

A Committee of Nomenclature was established, and was to lay before the Society for approval a printed list of terminology. The Society also intended to form a collection of provincial mining, geological and mineralogical terminology, and to print it for the use of its members. Although it appears that nothing was ever actually printed, there are two handwritten notebooks in the archives of the Society, which show that some progress in this particular project was made. The first is a ledger entitled 'Names of Mining Instruments and Apparatus and the Provincial Names of Mineral Substances', and is a dictionary listing mining terminology from across the country. Each entry has a definition and an indication of the district in which it was used, and also refers to an authority, from which the definition had been taken. These authorities were usually, but not exclusively, members of the Society, and included George Bellas Greenough, Aikin, Turner, Farey and Banks. However, publications, such as William Hooson's *The Miners Dictionary* (1747) and Richard Kirwan's *Elements of Mineralogy* (1784) were occasionally also referenced. The second notebook is a quarto volume entitled 'Mineralogical Nomenclature: Reports of the Committee, Appendix', and appears to be entirely in Greenough's handwriting. It lists mineralogical terms, mostly the names of minerals, with a description of the substance and a derivation of the name. These volumes apparently never appeared in any published form, and this seems to be due to the fact that they were never completed. This is probably accounted for by the changing focus of the Geological Society's activities in the early 1810s. However, the similarity of the contents of this ledger to the earlier dictionary project intended but never undertaken by the British Mineralogical Society is striking.

In 1810 the Society began to discuss the possibility of commencing publication of a series of transactions, and the first volume appeared in 1811. The Society could have produced a practically oriented periodical – something similar to the French *Journal des Mines*, for instance – and this was indeed suggested by some members. Instead, however, the Society finally decided to produce a more sumptuous and impressive set of *Transactions*, printed on the best-quality paper and expensively bound. My analysis of the contents of the first series of *Transactions*, which ran from 1811 to 1821, suggests that over this period interest within the Society both in mineralogy and in practical geology and mining decreased.

The first volume shows that the Society in 1811 was continuing to pursue the same general approach as it had taken from the beginning, emphasizing its interests in mineralogy and the practical applications of geology, combined with the need to continue to gather geological data and to avoid speculation and controversy. Thus, the preface to the first volume was intended to suggest very clearly to the readership that the science of geology was still not very far advanced, therefore requiring controlled co-operation, such as had already been instigated by the Geological Society:

> It may indeed be asserted that there is no object of research in which [...] co-operation is more necessary than in Geology. In this science [...] more that is important remains to be ascertained by future inquirers, than in any other branch of natural knowledge; while the variety of attainments, and the degree of leisure requisite for the prosecution of it, can seldom fall to the share of one individual.
>
> (Geological Society 1811, p. vi)

The preface also emphasized that geology, being somewhat labour-intensive, required a great number of observations to be collected, confirming that the Society was acting as a repository for specimens, observations, maps, sections and books. It further mentioned that the Society had had a recent, rapid increase in membership, and that this would therefore allow it to fulfil this function perfectly. The editors also made sure to remind the reader that geology was to be considered both respectable and useful:

> [...] it is sufficient to observe, that [geology is] rich in the beautiful and sublime productions of nature; and that, practically considered, its results admit of direct application to purposes of the highest utility.
>
> (Geological Society 1811, p. viii)

To turn to the articles that appeared in the first series, it seems that many, although not all, papers read at meetings were later published in the *Transactions*, although only a selection of the very earliest papers ever appeared in print. An overview of the first series of *Transactions* shows that over the 10-year period 1811–1821 approximately five percent of the papers directly engaged subjects related to mining, with another five percent including some discussion of the practical benefits of their subject matter, such as in Aikin's discussion of the Shropshire coal-field (Aikin 1811), Taylor's comments on metal mining in Devon and Cornwall (Taylor 1814) and Nathaniel Winch's observations on coal mining in his geology of the Durham area (Winch 1817).

In terms of the changing content of the *Transactions*, there are two unmistakable trends over the

[5] GSL: First minute book of the Geological Society.

period. The gradual increase in papers relating to stratigraphy and palaeontology was coincidental with a decrease in papers concerning mineralogy. In the first volume, almost half the papers were strictly mineralogical in their focus, but by 1821 the proportion of strictly mineralogical articles had fallen to approximately one-tenth of the whole. The number of papers with content concerning the practical applications of geology was more variable throughout the series, but it approximately halved over the same period. The increase in stratigraphical investigations and the growing interest in organic remains are even clearer: papers devoted entirely to stratigraphy increased from less than one-twentieth to more than one-tenth, and, although there was no interest in organic remains in the first volumes, they accounted for one-fifth of the papers in the fifth volume in 1821 (this transformation is discussed by Knell 2009). In the early years of the Society, there seems to have been little interest generally in fossils and organic remains, apart from the work of James Parkinson, who had already written on the subject and who contributed to the *Transactions* on the subject again in 1819 (Parkinson 1819). By 1821, however, such papers were a regular feature in the *Transactions*.

The decrease in the proportions of practical geological and mineralogical publications does not appear to be due to any alteration in the aims and interest of professional men such as Farey and practical mineralogists such as Aikin. Their work on such subjects continued to appear in the *Philosophical Magazine* and other commercial scientific publications, as they had long done. However, the trends here described continued to develop in the second series of *Transactions*, with the further development of the Society's focus on wider geological considerations, stratigraphy and palaeontology. The reasons for beginning a new series in 1824 had more to do with the publishing history of the journal than any sudden change in content. The first series had been published by William Phillips, and the final two volumes had actually been produced at a loss (which Phillips himself, rather than the Society, had covered), so it was decided in 1824 to begin anew with a slightly less sumptuously produced series.

The range of papers delivered to the Society, both verbally and in publication, reveal that there was a strong practical element in its early concerns. These interests broadened, however, rather rapidly. It is, however, noteworthy that the Society, as a body, seems never to have been disposed to give direct assistance to anyone seeking practical advice on geological matters, as an incident in 1812 shows. The Secretary had received a letter from a correspondent in Dorset, requesting the opinion of the Society on the likelihood of finding coal in the neighbourhood, based on the geological and other indications that he described in the letter. The Council resolved that the Society did not as a body give such opinions, and the Secretary was therefore instructed to respond to this effect.[6] By contrast, this was exactly the kind of matter the British Mineralogical Society had been trying to tackle. The changing interests of the Geological Society, illustrated in particular by the changing contents of the *Transactions*, are due mainly to the changing pattern its membership. This conclusion is supported by a prosopographical analysis of the membership from its foundation until 1840, compiled using data from the Society's membership lists taking the years 1808, 1810, 1819, 1829 and 1839 as samples.

The Society's initial membership, a group comprising the founders and those from around the country who were admitted to the Society during its second meeting in December 1807, is interesting and diverse. Of the 40 Honorary Members admitted at this second meeting, eight were from Cornwall or had interests in Cornish mines, and there were also members from other mining districts or with practical interests as well, such as Joseph Townsend, a friend of, and advocate for, William Smith. There were also mineralogists or geologists from several universities, including John Hailstone (Woodwardian Professor at Cambridge), John Kidd (Professor of Chemistry at Oxford) and Robert Jameson (Professor of Natural History at Edinburgh). In this group there were fairly equal numbers of Anglicans and Dissenters, and about half the members had had some university education. In addition to university men, there were clergy and medical men. The largest group, however, making up nearly one-fifth, comprised businessmen, and particularly manufacturers and mine managers. Indeed, many of the provincial figures had practical interests.

The membership of the Society grew rapidly, from the 13 founders in November 1807 to 112 members by the end of its first session in 1808. By 1829 the Society boasted over 500 members, from Londoners to correspondents in the farthest-flung reaches of the British Empire (see Branagan 2009). In particular, there was a great influx of metropolitan gentlemen and university men after the first wave of provincial honorary members. The rapid incursion of university men with geological interests, particularly John Kidd, William Buckland and William and John Conybeare at Oxford, had an important effect in the 1810s

[6] GSL: First minute book of the Geological Society.

Table 1. *Members residing in different areas of Britain, 1808–1839**

	1808	1810	1819	1829	1839
London	13	66	167	179	294
East	2	3	12	33	48
West	5	6	21	22	24
SW	13	22	31	31	47
Rest of England	7	34	73	124	208
Rest of GB	14	40	59	63	83

*These data were compiled from the Geological Society membership lists for 1808, 1810, 1819, 1829 and 1839.

Table 2. *Percentage of membership residing in different areas of Britain, 1808–1839*

	1808	1810	1819	1829	1839
London (%)	24	39	46	40	42
East (%)	4	2	3	7	7
West (%)	9	4	6	5	3
SW (%)	24	13	9	7	7
Rest of England (%)	13	20	20	27	30
Rest of GB (%)	26	23	16	14	12

(Rupke 1983; Knell 2009). Into the 1820s Cambridge men became more numerous, although only Adam Sedgwick took on a particularly dominant role on the Society's controlling committees and Council. The high proportion of businessmen and manufacturers in the very early days was fairly quickly overwhelmed by landowners and gentlemen of the clergy, the law and the military. The early balance of Anglicans and Dissenters also tipped in favour of the established Church.

Table 1 shows the number of Geological Society members resident in different parts of Britain for each sample year. It is important to note that the categories 'West' and 'East' are dominated by men at the universities of Oxford and Cambridge, respectively, and it is to illustrate this point that these two categories are listed separately. 'SW' is also included separately, rather than being incorporated into the 'Rest of England', because of its immense statistical significance in the first few years – nearly one quarter of the membership resided there in 1808. These regional categories were defined using counties, chiefly in order to discover and highlight concentrations of Geological Society members, such as that of the university men, and are constituted thus: 'West' includes Oxfordshire, Worcestershire and Gloucestershire; 'East' includes Essex, Cambridgeshire, Suffolk and Norfolk; and 'SW' includes Cornwall, Devonshire, Somersetshire and Dorsetshire (as they were usually referred to in the early nineteenth century). The high proportion of members from the SW in the earliest days is a particular conundrum, possibly attributable to pre-existing networks of individuals with mineralogical interests in the region. This may not be so surprising considering that Cornwall in this period boasted several well-known mineralogists and naturalists, and, in the area around Bath, the work of William Smith had also attracted attention. The growing dominance of the metropolis and the university towns over the period is apparent from Table 2, where the same data are represented proportionally, and these data are also graphically represented in Figure 1. By 1819 half of the membership resided in London or in the university towns of Oxford and Cambridge, and the concerns of the practically minded mineralogists were of far less importance in the Society's plans.

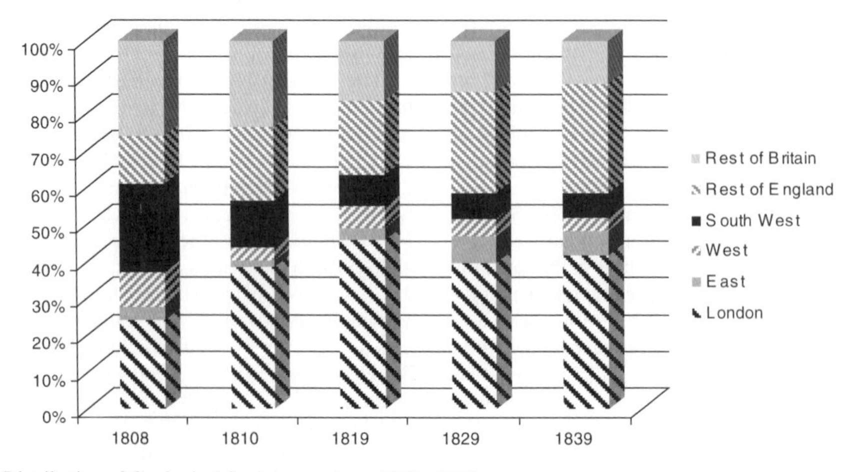

Fig. 1. Distribution of Geological Society members, 1808–1839.

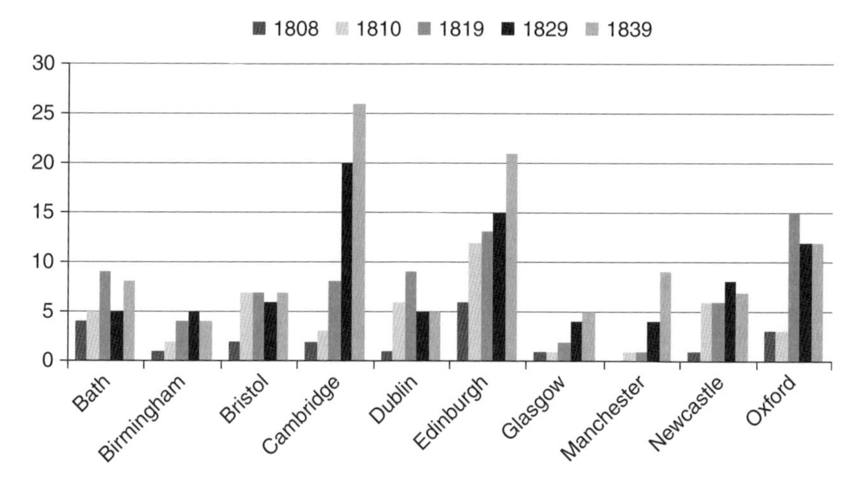

Fig. 2. Geological Society members residing in major British towns.

The final part of the analysis, the results of which are displayed in the graph in Figure 2, reinforces these findings. The number of members residing in the university towns compared to the number of members in the best represented of Britain's other major towns again shows the dominance of the university men by the 1820s.

This changing pattern of membership was at the heart of the Society's change of direction, and also, of course, had wider implications for the discipline of geology as a whole, and perhaps most particularly for the highly significant uptake of palaeontological stratigraphy in the Society in 1820s (Knell 2009). However, although the Society's stance altered somewhat during this period, the loss of interest, as a body, in economic and practical geology was an effect of dilution rather than complete displacement, and those geologists and mineralogists with practical and economic interests were able to continue their own work within the Society. It was simply that as the Society grew – so considerably and so rapidly – there were suddenly proportionally far fewer members with overwhelmingly practical interests.

Conclusion

During its first few years, the Geological Society had a deep interest in economic mineralogy and geology, and particularly in mining. The members who promoted these aspects of geology within the Society were mainly practically minded mineralogists who had been associated with the British Mineralogical Society. Although the programmatic interests of the Society diverged from these early practical aims quite rapidly as the Society grew and its membership profile changed, this did not have a great detrimental effect on the efforts of individuals to continue their work. Members with practical interests mostly retained their membership and were, in general, able to pursue their interests, both within the Society and, later, in the provincial scientific and geological societies that began to form across Britain in the 1820s and 1830s. Moreover, mineral surveyors such as Farey and Smith, and later others such as Thomas Sopwith and William Henwood, continued to work and to publish on geology, on both a national and local level, producing papers, sketches, plans, sections and maps. Mining engineers and colliery viewers such as John Buddle and mine managers such as John Taylor also took a very active interest in geology and published papers, both in provincial journals and national periodicals. Sopwith, Henwood and Buddle were all members of the Geological Society in the 1830s and 1840s. Although economic geology did not maintain its early important position within the Geological Society, mining did stimulate geology in Britain in the nineteenth century, and many practitioners were also able to apply their geological knowledge to benefit the mining industry.

The Geological Society's involvement in practical and applied geology waned as the membership profile altered in the 1810s, diluted by a growing interest in stratigraphy. The changing nature of the disciplinary project of geology was thus, in part, produced by the changes in the community of practitioners forming one of its leading institutions. While the standard view of the Geological Society as indifferent to economic and practical geology may be broadly accurate for the mid-nineteenth-century period, it is not so for the Society's foundational years. Furthermore, practical geology maintained a place in the Society in the work of individual practitioners, and, perhaps even more notably when considering the growth of British geology as a whole, was renewed in the provinces with the rise of provincial geological societies.

I owe thanks to Cherry Lewis for her kind invitation to speak at the 'Talk with the Founding Fathers' meeting at which this paper was presented, and I am also extremely grateful to John Christie, Melanie Keene, Jack Morrell, Martin Rudwick, Jonathan Topham, Hugh Torrens and the referees, Simon Knell and Cherry Lewis, for comments on various versions of this paper. Finally, without access to the Geological Society's library and archives, and without the kind assistance of Wendy Cawthorne, this paper would not have been possible.

Archives

GSL: Archives of the Geological Society of London.

NHM: Earth Sciences Library, Natural History Museum, London.

References

AIKIN, A. 1811. Observations on the Wrekin, and on the great coal-field of Shropshire. *Transactions of the Geological Society*, **1** (first series), 191–212.

ANON. 1798. Scientific Intelligence. *Philosophical Magazine*, **1**, 111–112.

BRANAGAN, D. F. 2009. The Geological Society on the other side of the world. *In*: LEWIS, C. L. E. & KNELL, S. J. (eds) *The Making of the Geological Society of London*. Geological Society, London, Special Publications, **317**, 341–371.

GEOLOGICAL SOCIETY. 1808*a*. *Geological Inquiries*. The Geological Society, London.

GEOLOGICAL SOCIETY. 1808*b*. *List of the Geological Society*. The Geological Society, London.

GEOLOGICAL SOCIETY. 1810. *List of the Geological Society*. The Geological Society, London.

GEOLOGICAL SOCIETY. 1811. Preface. *Transactions of the Geological Society*, **1** (series 1), v–ix.

GEOLOGICAL SOCIETY. 1819. *List of the Geological Society*. Geological Society, London.

GEOLOGICAL SOCIETY. 1829. *List of the Geological Society of London*. Geological Society, London.

GEOLOGICAL SOCIETY. 1839. *List of the Geological Society of London*. Geological Society, London.

HOOSON, W. 1747. The Miners' Dictionary. Explaining not only the terms used by miners, but also containing the theory and practice of that most useful art of mineing, more especially lead-mines. Wrexham.

INKSTER, I. 1977. Science and society in the Metropolis: A preliminary examination of the social and industrial context of the Askesian Society of London, 1796–1807. *Annals of Science*, **34**, 1–32.

KIRWAN, R. 1784. Elements of Mineralogy. London.

KNELL, S. J. 2000. *The Culture of English Geology, 1815–1851: A Science Revealed through its Collecting*. Ashgate, Aldershot.

KNELL, S. J. 2009. The road to Smith: how the Geological Society came to possess English geology. *In*: LEWIS, C. L. E. & KNELL, S. J. (eds) *The Making of the Geological Society of London*. Geological Society, London, Special Publications, **317**, 1–47.

MORRELL, J. 1983. Economic and ornamental geology: The Geological and Polytechnic Society of the West Riding of Yorkshire, 1837–1853. *In*: INKSTER, I. & MORRELL, J. (eds) *Metropolis and Province: Science in British Culture, 1780–1850*. Hutchinson, London, 231–256.

MORRELL, J. 2005. *John Phillips and the Business of Victorian Science*. Ashgate, Aldershot.

MUSHET, D. 1798. Strictures on Mr. Joseph Collier's 'Observations on Iron and Steel'. *Philosophical Magazine*, **2**, 9–10

PARKINSON, J. 1819. Remarks on the fossils collected by Mr. William Phillips near Dover and Folkstone. *Transactions of the Geological Society*, **5i** (series 1), 52–60.

PEPYS, W. H. 1799. British Mineralogical Society. *Journal of Natural Philosophy*, June 1799, 138–140.

PORTER, R. 1973. The Industrial Revolution and the rise of the science of geology. *In*: TEICH, M. & YOUNG, R. (eds) *Changing Perspectives in the History of Science: Essays in Honour of Joseph Needham*. Heinemann Educational, London,

PORTER, R. 1977. *The Making of Geology: Earth Science in Britain, 1660–1815*. Cambridge University Press, Cambridge.

RUDWICK, M. J. S. 1985. *The Great Devonian Controversy: The Shaping of Scientific Knowledge among Gentlemanly Specialists*. University of Chicago, Chicago, IL.

RUPKE, N. A. 1983. *The Great Chain History: William Buckland and the English School of Geology (1814–1849)*. Clarendon, Oxford.

SECORD, J. A. 1986. *Controversy in Victorian Geology: The Cambrian–Silurian Dispute*. Princeton University, Princeton, NJ.

SHAPIN, S. 1972. The Pottery Philosophical Society, 1819–1835: An examination of the cultural uses of provincial science. *Science Studies*, **2**, 311–336.

TAYLOR, J. 1814. On the economy of the mines of Cornwall and Devon. *Transactions of the Geological Society*, **2** (series 1), 309–327.

THACKRAY, A. 1974. Natural knowledge in cultural context: the Manchester model. *American Historical Review*, **79**, 672–709.

THACKRAY, J. C. (ed.). 2003. *To See the Fellows Fight*. British Society for the History of Science, Monograph, **12**.

TORRENS, H. S. 1998. Coal hunting at Bexhill 1805–1811: How the new science of stratigraphy was ignored. *Sussex Archaeological Collections*, **136**, 177–193.

TORRENS, H. S. 2001. Timeless order: William Smith (1769–1839) and the search for raw materials 1800–1820. *In*: LEWIS, C. L. E. & KNELL, S. J. (eds) *The Age of the Earth: From 4004 BC to AD 2002*.

Geological Society, London, Special Publications, **190**, 61–83.

TORRENS, H. S. 2002. *The Practice of British Geology, 1750–1850*. Ashgate, Aldershot.

TORRENS, H. S. 2009. Dissenting science: the Quakers among the Founding Fathers. *In*: LEWIS, C. L. E. & KNELL, S. J. (eds) *The Making of the Geological Society of London*. Geological Society, London, Special Publications, **317**, 129–144.

WEINDLING, P. 1983. The British Mineralogical Society: A case study in science and social improvement. *In*: INKSTER, I. & MORRELL, J. (eds) *Metropolis and Province: Science in British Culture, 1780–1850*. Hutchinson, London, 120–150.

WINCH, N. J. 1817. Observations on the Geology of Northumberland and Durham. *Transactions of the Geological Society*, **4** (series 1), 1–101.

Geology at the crossroads: aspects of the geological career of Dr John MacCulloch

ALAN J. BOWDEN

Earth Sciences, National Museums Liverpool, William Brown Street, Liverpool L3 8EN, UK (e-mail: alan.bowden@liverpoolmuseums.org.uk)

Abstract: Dr John MacCulloch MD was a pioneer of geological cartography. Prior to his surveys there had been few attempts to map and survey Scotland. Of these few, only the student efforts of Louis-Albert Necker de Saussure and the published map of Aimé Boué have attempted to show the whole country. MacCulloch's geological map of Scotland, published posthumously in 1836, remains one of the great cartographic milestones in the history of geology. Earlier, MacCulloch was the first government-appointed geologist through his work on the Board of Ordnance whilst carrying out the Millstone, Meridian and Mountain surveys. MacCulloch's work often generated controversy after 1820 when others began to take an interest in Scottish geology. MacCulloch was an active member of the Geological Society and was quickly recognized as a geologist of rare ability and influence, and was appointed to several committees. He was made Vice-President in 1815 and became its fourth President in 1816. From 1820 his influence and activity in the Society began to wane as the demands of his Survey work and ill health began to take their toll. He finally resigned from the Society in 1832.

In the frontispiece to the 1836 memoir of MacCulloch's Geological Survey of Scotland that accompanied the map, Samuel Arrowsmith wrote:

> The following 'Memoirs' have been printed *verbatim* from the manuscripts of the late Mr. MacCulloch, without note or comment. His sudden and lamented death has prevented my remarking on certain of his opinions and statements; but these, it is hoped, will be received as he meant them to be, by such as are competent judges of the real errors and deficiencies in the unfortunate Map, that became so fertile a source of his indignation.

These rather negative lines hide a 20-year story of one man's involvement in a line of arduous investigation that culminated in one of the greatest achievements in geological cartography during the early to mid-nineteenth century. Variously referred to as MacCulloch or M'Culloch, his work has been discussed by Judd (1898), Eyles (1937), Boud (1974), Cumming (1977, 1983, 1985) and Flinn (1981). Of these, the unpublished PhD thesis of David A. Cumming (1983) remains the single most important and detailed account of MacCulloch's life, work and influence.

Guernsey-born, John MacCulloch M.D., F.R.S. (1773–1835) (Fig. 1) was a geologist of exceptional ability, whose observational skill and dedication paved the way for the establishment of the first professional geological survey of Scotland. By 1795 he had completed five years of study at Edinburgh, having graduated with an MD in 1793. Then just 20 years old, he found his first appointment as a surgeon's mate with the Royal Artillery Medical Department, possibly stationed in a unit near Edinburgh. In 1803 he was promoted to the post of chemist and assayist, and took on lecturing duties at the Royal Military Academy at Woolwich. He continued in this post until 1806 when he accepted a civilian appointment as chemist at the Board of Ordnance. By 1814 he was lecturing in chemistry and in 1819 lecturing in geology at an East India Company seminary at Addiscombe, near Croydon (Addiscombe was a country house and estate, turned into a Military Seminary in 1809). This places MacCulloch amongst a small number of paid public communicators of geology in the early nineteenth century. MacCulloch's medical training also gave him the opportunity to set up a private practice at Blackheath and he became a Licentiate of the College of Physicians of London in 1808. His time spent practising medicine became severely reduced once he began his geological survey work for the Board of Ordnance from 1809 onwards, finally terminating around 1820.

MacCulloch stood at a crossroads in the development of the geological sciences and it is here that the tide of history has, perhaps, been unkind to him. A Wernerian by initial training and a Huttonian by principle, he eschewed theories for hard observational facts. In an age where it was unfashionable not to take sides (Neptunist or Plutonist), MacCulloch pursued a geology that relied upon acute field observational skills. His ability as an artist, chemist and mineralogist enabled him to see field relationships with a greater precision than many of his critics. For example, MacCulloch observed that

From: LEWIS, C. L. E. & KNELL, S. J. (eds) *The Making of the Geological Society of London*. The Geological Society, London, Special Publications, **317**, 255–278.
DOI: 10.1144/SP317.15 0305-8719/09/$15.00 © The Geological Society Publishing House 2009.

Fig. 1. MacCulloch in his prime, painted by Benjamin R. Faulkner, Ordnance Geologist, lecturer on geology at the East India's Academy at Addiscombe. Portrait given to the Royal Society by Louisa Margaretta White. Reproduced by permission of the President and Council of the Royal Society. © Royal Society.

the Red Sandstone (Torridonian) in the NW Highlands rested unconformably upon gneiss and was subsequently overlain by schists. He recognized that this violated one of the Wernarian doctrines that a mechanical sediment could not occur between 'Primitive' chemical precipitates such as gneiss and schist. For a brief discussion on how such observations affected the opinions of geologists mapping the NW Highlands of Scotland during the early to mid-nineteenth century see Oldroyd (1990).

MacCulloch's field observations culminated in his highly acclaimed work *A Description of the Western Islands of Scotland* (1819). The maps in the Atlas of this three-volume work formed the basis, with later revisions, for the western seaboard of his 1836 Geological Map of Scotland and subsequent issues (see Cumming 1977 for a detailed analysis of this work).

In an assessment of his scientific work up to the publication of *A Description of the Western Islands of Scotland* William Henry Fitton (1780–1861) remarked:

> The geological publications of Dr. MacCulloch have now become so voluminous, as to justify criticism upon their manner as well as their substance. The author is too fond, we think, of animadverting upon the clumsiness and insufficiency of systematic geology; and he alludes occasionally to the doctrines of Werner, in a tone too much like that of Sarcasm [...] A second fault of this accomplished geologist is the extreme diffuseness of his papers, with a certain want of unity, in part, arising from the inclusion of too many collateral statements. We expect, and have received so much from Dr. MacCulloch, combining as he does, the qualifications of a chemist, a draughtsman, and a geological observer of singular enterprise and activity, that he will forgive our pressing upon his attention a defect, which a little additional labour will very easily remove.
>
> (Fitton 1818, p. 87)

From 1811 until 1818 MacCulloch published 17 papers in the *Transactions of the Geological Society of London*. His writing style is occasionally verbose and rather discursive, but not without a sense of humour, unlike the clinical and lucid prose of Lyell. MacCulloch was an individual who had a passion for learning in numerous scientific and artistic disciplines. He was especially fond of the classics and literature, fostered by his schooling at Lostwithiel and later during his studies in Edinburgh. MacCulloch's love of learning is shown in his publications, which reflect his zest for knowledge and polymathic interests. However, the downside to this polymathic approach to publication was his tendency to become discursive and later to produce rather acerbic rhetoric in some of his reviews (Cumming pers. comm.).

MacCulloch's geological map of Scotland

MacCulloch's Geological Map of Scotland (Fig. 2) was the result of many seasons spent in the field whilst assigned to the Trigonometrical Survey. The Survey was a department of the Board of Ordnance commanded by Colonel William Mudge (1762–1821) of the Royal Artillery and run by Captain Thomas Colby (1784–1852) of the Royal Engineers. Colby took a keen interest in the emerging science of geology and warmly supported the inclusion of a geologist on survey teams. This led Colby in later years to be instrumental in establishing geological surveys throughout the British Isles, including Ireland (Fig. 3). Cumming (1983) states that the impetus for employing a geologist on the survey probably came from the Office of the Master General of the Ordnance, Lord Mulgrave (1755–1831) with Mudge recommending MacCulloch for the appointment, possibly on the strength of his work on the Millstone Survey conducted from 1809 to 1815. This had involved MacCulloch in a survey of limestones suitable for making the millstones used in gunpowder manufacture. It is possible that the map's origins lie in 1811 when he was engaged in this work, for he later recalled:

> The first journeys which I undertook in Scotland were for a partial purpose required by the Ordnance, of a very limited practical nature. But as it thus became necessary to traverse many tracts of country though somewhat widely dispersed, I felt that I might, without any diversion of time, or any additional expense to the Government, note the other geological facts which fell my way, and thus produce some sort of partial surveys of the country in general.
>
> (MacCulloch 1836, pp. 18–19)

The idea of surveying the region as a whole gradually came to mind after being appointed in 1814 to begin two investigations. One was a geological investigation eight miles either side of the zenith sector stations lying along the meridian from Balta in Shetland to the English Border. Sir Charles Close in his history of the early years of the Ordnance Survey states:

> The idea was, that Dr. MacCulloch (as his name is usually spelt), should be able to point out where and what abnormal deflections of the plumbline were likely to occur; with a view to avoiding stations for the zenith sector where large deflections might be expected. The zenith sector observations being required for the determination of the meridian curvatures, which were themselves required in the construction of the one-inch map.
>
> (Close 1926, p. 61)

The second survey (the Mountain Survey) was to find a mountain more suitable than Schiehallion to calculate the Earth's mean density – a repeat of Maskelyne's experiment (see Smallwood 2009). This arose from a question posed by the mathematician Charles Hutton (1737–1823) who stated:

> What is the density of the matter in the hill? Is its mean density equal to that of water, of sand, of clay, of chalk, of stone or of some of the metals? [...] A considerable degree of inaccuracy in this point could only be

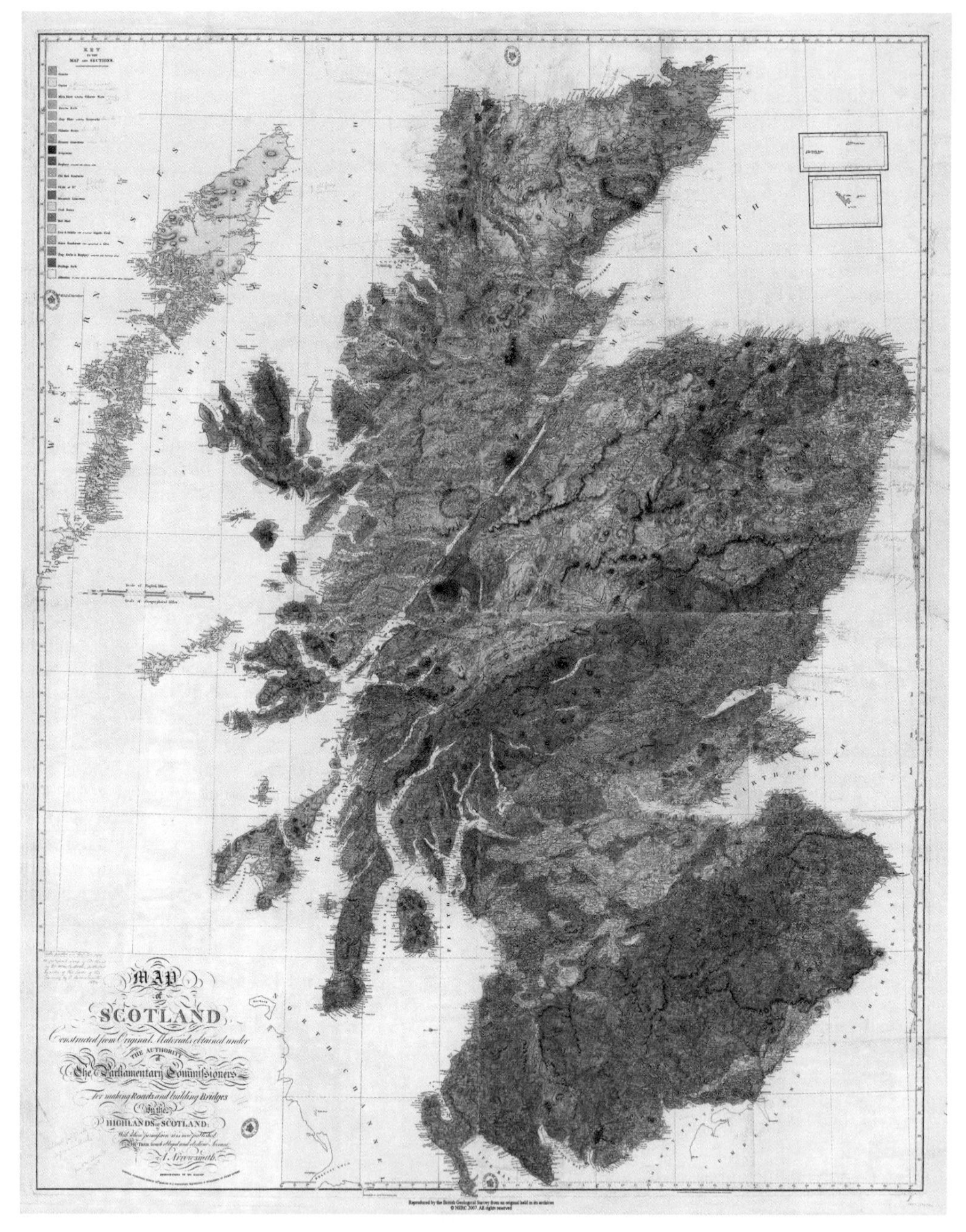

Fig. 2. MacCulloch's Geological Map of Scotland. Reproduced by permission of the British Geological Survey.

> determined by a close examination of the internal structure of the hill.
>
> (Hutton 1778, p. 781)

In an attempt to answer this question the Edinburgh Professor of Mathematics, John Playfair (1748–1819), subsequently undertook a lithological examination of Schiehallion in 1801 in order to gain a more precise knowledge about the specific gravity of the varied lithologies, which he published in 1811. MacCulloch was critical of Playfair's

Fig. 3. Major-General Thomas Colby (1784–1852) of the Royal Engineers. Longest serving Director of the Ordnance Survey 1820–1846. Commemorated in St James' Cemetary, Liverpool. © Crown copyright Ordnance Survey.

results, pointing out that Playfair's work was flawed by faults in computational technique arising from an uninformed knowledge of the local geology, in particular the interpretation of the position and extent of rocks of dissimilar specific gravity. He wrote:

> If the form and situation of this mountain were such as to offer great conveniences to the mathematicians engaged in this problem, it is, unfortunately, deficient in uniformity and simplicity of structure. Not only is it a matter of extreme difficulty to ascertain the true distribution of the strata throughout the whole mass, but the specific gravities of the different materials vary in such a manner as to vacillate between 2.4 and 3, or even more. All the strata are, at the same time, elevated at high, but unequal angles; while it is also difficult to discover their relative proportions on the surface, and, still more, the positions which the rocks of different specific gravities assume in the interior of the mountain; a circumstance of considerable importance, on account of the angular differences of the action of the several columns on the plummet. The different strata, it must be remarked, consist of quartz rock, micaceous schist, hornblende schist and limestone. Under these uncertainties, a true mean specific gravity could neither be determined nor applied; that element could only have been perfect, in this case, when the actual specific gravity of each attracting part of the hill, and quantity of each, were ascertained.
>
> (MacCulloch 1831, vol. 1, p. 22)

Perhaps we see here an early indication of the arguments between the geological field observer and the geophysical modeller.

Each of these surveys, coupled with only verbal guidance from Colonel Mudge, allowed MacCulloch the freedom to indulge in more general mapping and surveying activities. This gave him the opportunity, in his own time, to collect field data with the intention of eventually publishing a geological map of Scotland. To this end, in 1821, he suggested that a mineralogical survey of Scotland could be conducted without extra expense as a by-product of other duties. MacCulloch had been collecting data on the geology of Scotland since 1811 and these data could readily be incorporated in the proposed map.

He approached Colby, now a Major steeped in London scientific society and successor to Major-General William Mudge's superintendence of the Trigonometrical Survey.[1] In 1818 Colby had written to Mudge about the difficulties of surveying in Scotland:

> The country is so extremely wild and destitute of accommodations of every kind and the mountains are so high and difficult of access [...] In the western part of Scotland, from the want of roads and carts, and the extreme height of the mountains, no station can be visited without very considerable expense.
>
> (in Close 1926)

MacCulloch asked for a meeting, which took place on 3 April 1821. It was a heated exchange in which MacCulloch, possibly suffering from illness that heightened his tetchy nature, was apparently not very coherent and threatened to hang or shoot himself (Fig. 4a). He was evidently hoping for official orders for his geological map but Colby stated that it was of no concern of his as there were no formal written instructions concerning it. Perhaps Colby was reflecting his own misgivings about the expense of such a work. However, on 27 July 1821, MacCulloch was given permission to continue with his map at his own expense. In July 1826 MacCulloch eventually received a Treasury endorsement to continue with the map after he

[1] Mudge was promoted shortly before his death in 1821 and Colby took over full responsibility for the Trigonometrical Survey in 1826 (Rose 1996). Colby was Director of the Ordnance Survey from 1820 to 1846 and later promoted to Major-General. A memorial was erected to him in St James' Cemetery, Liverpool.

Fig. 4. MacCulloch's sketches during illness. (**a**) is a self portrait and (**b**) appears to represent a hallucinatory demon, elements of which are to be found in other sketches from late 1821–1825. From a sketchbook presented to the Geological Society in 2002 by Dr Henry Oakley. Reproduced courtesy of the Geological Society of London.

was pensioned off (at £248 per annum) from his post as Chemist with the Board of Ordnance on 13 January 1826 (Flinn, 1981). The instruction being that he could 'transfer his undivided attention and services to the geological survey under the Treasury'. Here the Treasury appeared to have seen the utilitarian need for such a survey as they declared on 4 July 1826 that:

> A complete mineralogical map of Scotland must prove, not only highly interesting in a scientific point of view, but of considerable practical utility to many branches of industry connected with the mineral kingdom.
> (cited in Eyles 1937, p. 120)

Payments for MacCulloch's two surveys with the Board of Ordnance were a vexatious issue, as were the payments made by the Treasury. Flinn (1981) and Eyles (1937) analysed the payments and raised some uncertainty about the mileages claimed by him. However, MacCulloch was a workaholic using every available daylight hour in gainful activity whilst on survey work. In response to a Treasury query regarding his submitted 1828 fieldwork expenses, MacCulloch replied in a Parliamentary paper for 1831 concerning the *Mineralogical Survey of Scotland*:

> This kind of survey, though 'scientific', as you properly remark, does not require anything demanding rest and delay; on the contrary, the more ground that can be traversed, the faster it is done. It consists in the sight, in contact, or near at hand, of so much ground; geological knowledge and experience learning to determine the tracts of rocks by single masses, analysis and inference, and the physiognomy of ground.[2]

Altogether, MacCulloch spent a total of 15 seasons in the field between 1814 and 1830. Six of these were paid by the Board of Ordnance and five were fully paid by the Treasury. Four further field seasons were paid by the Ordnance, but MacCulloch had to fund all of his expenses from his own resources. On average, MacCulloch was in the field for 154 days a year covering around 2530 miles in each season on foot or horseback. For each of the paid Ordnance surveys he received £1 pay, £1 subsistence per day and two shillings per mile travelling allowance. During the Treasury surveys he was away for 184 days for each field season and travelled an average of 8000 miles each year, during which he received £2 pay, £1 subsistence and two shillings per mile travelling expenses each day. This equates to travelling 45 miles each day, which is not an insurmountable figure bearing in mind that Cumming (1983) estimated that between the years 1814 and 1820 MacCulloch had travelled about 18 000 miles overland on foot or horseback. What is not recorded are the thousands of miles he sailed, surveying from small open boats, often taking the helm himself in some of the most treacherous waters, weather conditions and in areas that were heretofore unsurveyed with any pretence to accuracy.

MacCulloch completed his map and memoir in 1832, not, however, without controversy. The fact that he had been paid to complete his map enraged his former friend Robert Jameson (1774–1854) who heard about the payments in 1830. In an unsigned statement from the *Edinburgh*

[2] House of Commons Command Paper No. 140. Mineralogical Survey of Scotland, MacCulloch to J. Stewart on 12 July 1829, pp. 23–24.

Philosophical Journal there is a clue as to where this enmity may have arisen:

> Professor Jameson [...] has now [...] collected so extensive a series of facts and observations, that he will soon be able to present to the public a Map of the Mineralogy of Scotland. Dr MacCulloch, who has had the good fortune to be employed in mineral researches in Scotland, at the expense of Government, has it also in agitation to publish a Map illustrative of the geology of this country.
>
> (1819, p. 418, cited in Eyles 1937)

Jameson never presented a mineralogical map of Scotland to the public and his earlier work on Hebridean mineralogy, published in 1800, which he accused MacCulloch of plagiarizing, was very sketchy in comparison with MacCulloch's detailed observations, as published in his *A Description of the Western Islands of Scotland* (1819). MacCulloch had referred to Jameson's work but only in disparaging terms as mineralogically exact but of little use due to its limited coverage and lack of structural analysis (Cumming 1977).

MacCulloch and Jameson had been very friendly during their student days when attending the inspirational lectures of Rev. John Walker (1731–1803), Professor of Natural History in Edinburgh. During this period the study of natural history, and the Earth sciences in particular, became very popular in Edinburgh – perhaps as a result of the growing debates there concerning the theories of Hutton and Werner. The friendship became progressively more strained from 1809 as MacCulloch's work led him to adopt a more Huttonian stance. They finally broke contact in 1824 when Jameson took over as the editor of the *New Edinburgh Philosophical Journal* to which MacCulloch had previously been contributing (Cumming 1985). Jameson felt that he and the Wernerian Natural History Society had sole responsibility for the co-ordination of geological activities in Scotland. Therefore he attempted, without success, to get the payments to MacCulloch stopped. As a result MacCulloch justifiably complained of a vendetta and conspiracy against him and the publication of the map. Further evidence of difficulties during the Treasury Survey is mentioned by MacCulloch who resorted to subterfuge by travelling under an assumed identity in the Highland region. Was this a consequence of his book on the Highlands and Western Islands, so criticized by Browne in 1825, or was it a result of Jameson turning influential Scottish opinion against him (Cumming 1983)?

A memorial presented by the Highland and Agricultural Society of Scotland, of which Jameson was a committee member, tried to get the Treasury to gift all of MacCulloch's notes, drafts and materials for their own use in promoting a mineralogical map of Scotland. This audacious attempt at the 'theft' of a lifetime's work and an individual's intellectual property occurred as the result of a misunderstanding about the level of detail of MacCulloch's surveys. The Highland and Agricultural Society were under the misapprehension that MacCulloch's materials were of a more sketchy and unfinished nature. Perhaps Murchison was, in part, responsible for the memorialists' approach to the Treasury regarding MacCulloch's work. At the time Murchison was the President of the Geological Society of London (1831–1833) and, not having been a party to MacCulloch's materials, stated that 'he was not [...] by any means sanguine of the value of Dr. MacCullochs notes, but a sketch of a map might be useful' (Eyles 1937, p. 126).

Murchison's misleading statement probably led the members of the Highland and Agricultural Society to view MacCulloch's efforts as being of less value than they actually were. This misinterpretation could also have been compounded by an earlier comment from Colby, which indicated that MacCulloch's materials were unfit for any purpose (Eyles 1937). Unfortunately, this appeared to have been taken out of context and Colby was probably referring to the accuracy of Arrowsmith's topographic base map rather than the quality of MacCulloch's geological survey.

Despite a clash of personalities, Colby supported MacCulloch's work at a time when attempts were being made by the Highland and Agricultural Society to take it from Treasury control. He advocated the publication of the map and accompanying memoirs as well as indicating that MacCulloch was the person best suited to complete the map as he knew more about Scottish geology than any of his contemporaries. Furthermore, he inferred that MacCulloch was very unlikely to produce anything that was not of value as he had devoted almost his entire professional life to the survey. In this, Colby was showing his appreciation of the difficulties of mapping and surveying in Scotland borne out from his own extensive experience on the Board of Ordnance triangulation surveys conducted from 1815 to 1819.

Treasury delays and MacCulloch's persistent ill health meant that the map's publication was delayed until 1836 when it was privately printed by Arrowsmith after MacCulloch's death. This had the unfortunate consequence of credit being given to the Highland and Agricultural Society and the Treasury. MacCulloch's role was so subsumed that his name did not appear on the first printed edition of the map. This was subsequently rectified in the later issues of the map in 1840 and 1843.

MacCulloch's great undertaking represents one of the supreme feats of geological surveying conducted in some of the most difficult terrains in the British Isles. MacCulloch utilized 18 colours in the key to delineate lithology, although only 17 actually appear on the base map. It is noticed that the key colours appear brighter than those used on the actual map. The blackness of the original engraving of Aaron Arrowsmith's base map resulted in the colours appearing slightly muddy in areas of dense hachuring – a criticism that MacCulloch was keen to point out. An accomplished and talented artist, MacCulloch was well aware of the effect of hachuring in attempting to delineate topographic features and he employed his skills to great effect in preparing the geological map plates in his description of the *Western Islands* (1819) much to the vexation of his publisher (Cumming 1977). His artistic bent is summed up in the following passage from the Edinburgh Review:

> If geology is not so purely concerned with the descriptions of visible objects as some other branches of natural history, it is still very conversant in these. There are innumerable cases in which no powers or minuteness of description can convey to the reader clear ideas of the subject under review; when three or four strokes of a pencil are more descriptive than many pages of letter-press. We have no hesitation in saying, that it is impossible to describe a large proportion of geological facts without drawing.
>
> (MacCulloch 1823, p. 418)

In the memoir, MacCulloch gave full instructions to the colour copyists of the map as to the right pigment mixes and tints to achieve the effect he desired. As an example the schists and gneisses (Moine and Lewisian) are coloured yellow using gamboges. The Torridonian (Old Red Sandstone as mentioned above) is in pink using pure lake in a diluted form, whilst the green representing quartzites (MacCulloch's quartz rock) is made from Antwerp blue and gamboges. Prussian blue is used for the trap rocks and the colourists are instructed to use this as a weak tint over large areas as the poor quality of the engraving would render it too muddy in appearance if applied in too pure a form.

MacCulloch was highly aware of the problem of using colourists for providing adequate copies of the map as he noted:

> I trust that, in case of publication, the colourist copiers, thus forewarned, will attend to this, as to all else which I have here noted for their instruction. But I cannot hold myself responsible for the accuracy of their copies: experience has long shown me how very uncertain those are, even to gross mistakes or utter obscurity. The original must remain the point of appeal in cases of doubt or criticism.
>
> (MacCulloch 1836, p. 106)

MacCulloch's perfectionism resulted from his earlier experiences of using colourists during the production of the Atlas volume in his *Western Isles*, described by Cumming (1977). He continues:

> It were better that this Map were not published at all than published at a price insufficient to command the very best colouring that can be procured in that market place of labour.
>
> (MacCulloch 1836, p. 106)

MacCulloch expounds at length his criticisms of the inaccuracies of the base map in terms of its physical geography/topography as well as political geography (roads, villages, etc.): although he was careful not to be critical of Arrowsmith's reputation as a geographer and compiler of distinction. Much of this is due to the problem of using old military field survey sheets compiled by General William Roy (1725–1790) between 1747 and 1755. These were subsequently utilized by the geographer Aaron Arrowsmith (1750–1823) in his four-sheet quarter inch to the mile base map, later used by MacCulloch for plotting the distribution of rock types and boundaries. This map used the shading devised by the map maker and landscape artist Paul Sandby (1730–1809) to distinguish hill and mountain features on the original military survey sheets (Boud 1974). MacCulloch expressed indignation that the increase of access in terms of new roads and expanded settlements had rendered the accurate charting of geological data more difficult since the positions of the new roads could not be plotted with any level of accuracy on the old base map. MacCulloch mapped at a time when there was no standard reference system, and he was reliant on the accurate mapping of both topographical and settlement features in order to delineate his geological boundaries.[3] He also criticized the representation of topographic features such as hills and mountains on the map as bearing, in many cases, little relation to reality. Certainly, the printing of such features made marking geological data difficult. MacCulloch had hoped for topographic maps to be made available by a survey team at the Ordnance. Unfortunately, the first 6″ sheets were not made available until 1846, long after his death. MacCulloch must have looked with envy upon the Ordnance Survey maps that were being made available to his contemporaries in England. One wonders what the result would have been if MacCulloch had had the cartographic advantages of William Smith using John Carey's excellent specially prepared base map. A particular irony is

[3] A shortened version of his criticisms appears in the appendix to his *System of Geology* along with advice for the student field surveyor. Here he implies that the military surveyor often has recourse to more accurate information than that of the civilian worker.

that Aaron Arrowsmith's early employment was with John Carey prior to his founding his family firm of cartographers in 1786 (Boud 1974). Indeed, MacCulloch rather waspishly states:

> It would require but a small geologist indeed to lay down the rocks of any part of England on the Ordnance maps; as he is to be envied on whom such a duty may hereafter fall: while the reputation which the public will probably assign to him, ought in justice to be transferred to the geographical surveyors of the splendid work.
>
> (MacCulloch 1836, p. 14)

MacCulloch's attitude to geological survey mapping is perhaps best summed up in an article penned for the *Edinburgh Review*:

> Let the geologist who means to instruct, who is desirous of real reputation, treat his survey as if he were a geographer also; let him ascertain the exact geographic boundary of every rock on the surface, and determine the true nature and relations of that, be it simple or complicated. Let him to that map add sections – real, if they can be obtained, and deduced from fair comparison of positions, if they cannot be actually examined. Let these sections be as numerous as the subject demands; and, if he has done his duty, his reader will be able, not only to find everything to which he may be directed, and find it right, but will be enabled to construct a model of the country as deep as a geological eye can pierce. Till that plan be adopted, we, at least, shall be equally well pleased to see no more geological maps.
>
> (MacCulloch 1823, cited in Cumming 1977, p. 273)

Another point that MacCulloch had issue with is one familiar to all field geologists, namely the marking of small-scale features such as thin beds of limestone, serpentine or small-scale intrusive features ('trap veins'). He apologizes that for the sake of visibility in colour, these have been made larger than they actually appeared in the field thus unavoidably introducing another source of error. However, despite his criticisms of Arrowsmith's base map MacCulloch remarked:

> The general distributions and places of rocks which form Scotland will now be known; whereas, before this, with the exception of my own work on the Western Islands, and Dr. Hibbert's Map of Shetland (a survey of great merit), not a mile of land in all Scotland had been surveyed and recorded. The whole country, indeed, islands and all, were so absolutely unknown when I commenced this work that I was unable to borrow the description of a single mile in aid of it; as, throughout its whole progress, I have not derived even a hint, far less a fact, or an acre from any other hand.
>
> (MacCulloch 1836, p. 16)

In this he is, perhaps, unjust as he fails to acknowledge the work of either Jameson on the Scottish Isles (1798, 1800) or the agriculturalist and mineralogist Rev. James Headrick (1758–1841) in the Isle of Arran (1807). MacCulloch also neglects to mention the work of William Smith (1769–1839) whose mapping extended into the lowlands of Scotland. The cursory mapping of the Genevese Louis-Albert Necker de Saussure (1786–1861) whilst he was a student in Edinburgh from 1806 to 1808 is similarly unacknowledged. Necker's fieldwork was based on Thomas Kitchen's base map of 1778, and a copy of the map was presented to the Geological Society on 4 November 1808 and would have probably been known to MacCulloch as he was on the Committee for Maps from 1809. This was the earliest representation of geological cartography covering the whole of Scotland at a scale of $12\frac{1}{2}$ miles to the inch. According to Herries Davies (2007) Necker's map displays a youthful approach to geological cartography and also betrays a lack of knowledge of Scottish geology based on limited field experience. This lack of geological experience might have been recognized by MacCulloch who could have dismissed the map as it would have failed to match up to his own critical standards.

The Austrian geologist Aimé Boué (1794–1881), one of the founders of the Société Géologique de France in 1830, used and acknowledged MacCulloch's work in his *Éssai géologique sur l'Écosse* (Boué 1820). His geological map at a scale of 40 miles to the inch was, in part, generalized from MacCulloch's published data up to 1819.

Despite these criticisms MacCulloch would be correct to state that he had surveyed and depicted almost the whole of the country in as great a detail as it is possible for one person to achieve within the constraints allowed. Perhaps MacCulloch should have the last word on his endeavours:

> Let it hereafter prove what it may, it is the first and only work of such a nature and extent that has been executed by a single hand, and with even an attempt after accuracy.
>
> (MacCulloch 1836, p. 56)

Reaction to the map

MacCulloch's geological map stands alongside William Smith's 1815 map of England and Wales, and Richard Griffith's 1838 map of Ireland, as one of the foundation stones upon which all later work has been built. In 1833 at the start of his final presidential term (1833–1835) George Bellas Greenough (1778–1855) remarked that 'whatever the intrinsic excellence' of MacCulloch's geological map of Scotland it would be 'immediately useful'.[4] James Nicol (1810–1879) in drawing up

[4] GSL: Ordinary Minute Book 6, 21 February 1833, p. 205.

his 1844 geological map of Scotland utilized MacCulloch's map as a base for his own investigations, stating his indebtedness to MacCulloch and others who pioneered such work. MacCulloch's map was not generally superseded until 1860 when the Geological Survey began to issue maps at one inch to the mile based on newly available Ordnance Survey six inch to the mile topographic sheets. Thereafter, MacCulloch's map only remained useful for those more remote areas of Scotland, which were not surveyed until the 1870s, and later, in 1880, during the surveys of Cromarty and Caithness.

Others, however, were less than charitable. Sir Roderick Impey Murchison (1792–1871), for example, stated that the map 'was so replete with errata that it would be a waste of time to attempt to enumerate them' (Murchison 1851, p. 139). However, Murchison's opinion should be treated with caution as he was often highly critical of MacCulloch. John Wesley Judd (1840–1916), in 1898, attempted to portray MacCulloch's map in a more benign light and particularly praised the accuracy aspect of which Murchison had been critical. In particular, Judd reasserted MacCulloch's original observations about the true field relations and age of the Torridonian Sandstone and its relation to the Durness Limestone. Murchison and Adam Sedgwick (1785–1873) had opposed MacCulloch's interpretation, and in deference to them MacCulloch, in 1828, appeared to allow his views to be replaced by Murchison's insistence that the Torridonian strata were nothing more than down-faulted Old Red Sandstone. Hence, they appear as a uniform pink colour on the map and are marked up as Old Red Sandstone in the key. This reversal of MacCulloch's original ideas possibly occurred as a result of Sedgwick & Murchison's paper of 1828 arising out of their seaborne superficial examination of both the western and eastern sandstones during 1827. Here Sedgwick, who was inclined to use lithological criteria more than Murchison, conflated both of the sandstones as Old Red Sandstone. The resulting paper appeared at a time when MacCulloch's perceived influence on Scottish geology was in decline amongst his Geological Society peers. Sedgwick told Murchison that 'he fairly threw down the gauntlet to old Mac' (Clark & Hughes 1890, p. 322). Despite this unwarranted criticism, MacCulloch, in his *System of Geology* (1831), restated his views that the red-brown sandstones in the west, first observed by the Welsh naturalist and topographer Thomas Pennant (1726–1798) in 1774 (Pennant 1774), were distinct from the Old Red Sandstone in the east. Unfortunately, this reassertion of his observations did not appear in the final posthumous first edition of the map in 1836.

MacCulloch and malaria

From 1821 MacCulloch suffered from indifferent health, and this no doubt affected his perceptions and professional work. Many seasons spent in arduous fieldwork, often in inclement conditions and over inhospitable ground, led to a progressive weakening of a less than robust constitution. Between 1821 and 1822 MacCulloch suffered from a severe illness, which might have been a protracted bout of the recurrent malaria that had occurred during earlier seasons of fieldwork and may have existed in a passive state since his childhood in France. The next three years were spent in convalescence during which he indulged in extensive writing projects such as completing his text book on the *Geological Classification of Rocks* (1821), the four-volume *Highland and Western Islands of Scotland* (1824) and his *System of Geology*, largely written during 1821 but published in 1831. He also wrote 33 papers on a range of subjects and produced a guide book, *A Description of the Scenery of Dunkeld and of Blair in Atholl*, during this period.

The extent of MacCulloch's ill health is revealed in a letter to the Duke of Wellington, January 1826:

> At present, I have entirely lost my health, &, as far as I really believe, forever; from a disorder produced by the hardships of that unlucky survey; scarcely having been able to leave my room these last few months, and never having had a week's real health these five years, since my first attack in Scotland.[5]

MacCulloch was introspective and revealed little of his inner self to friends and colleagues. This may have contributed to a chronic state of depression when suffering the effects of malaria. Evidence for this is shown in a series of sketches made from 1821 to 1825. Figure 4a is a self-portrait made during a particularly depressed episode, which may have been related to the difficulties MacCulloch had with Colby regarding his survey. Colby had a reputation for being slightly less than tactful, although generally supportive of his officers (Close 1926); a personality trait that would have clashed with MacCulloch's mercurial temperament. The iconography of this image may indicate a hallucinatory state with thoughts of suicide by either hanging or shooting. MacCulloch's state of mind as possibly represented by these drawings suggest both fury and despair concerning his health and

[5] PRO: MacCulloch, J. 1826. Memorial to his Grace the Duke of Wellington, Master General of the Ordnance, w.a. 44/642:257–264. Also cited in Cumming (1983, p. 216).

work commitments. Other sketches from this period also have hanging as a common theme, as well as hallucinatory demons (as in Fig. 4b). During this period many of his landscape and other sketches were executed in a looser style.

MacCulloch's depressed state of mind probably contributed to his gradual withdrawal from the company of other geologists as he concentrated on the completion of his Scottish geological surveys (Cumming 1983). Against a background of increasing criticism, MacCulloch found support in Lyell who assisted him in mapping Forfarshire during 1822–1824 when MacCulloch was laid up with one of his frequent bouts of 'ague'.

MacCulloch's frequent illness from 1821 onwards no doubt coloured his views and perceptions whilst he was laid low. Swings of mood and sensitivity to criticism began to cause problems in both personal and professional relationships. Lyell commented in his obituary address:

> his spirits were much depressed by bodily sufferings – his imagination was then haunted with the idea his services in the cause of geology were undervalued and it was in vain to combat this erroneous impression.
>
> (Lyell 1836, p. 359)

This perception combined with a perfectionist attitude and a quest for cartographic precision contributed to many difficulties with those who dealt with him, particularly publishers and his geological peer group.

During the winter of 1830–1831, MacCulloch suffered a near-fatal illness having already been laid low by malarial attacks in 1825–1826 and again in 1828–1829. This particular illness led him to temporarily lose the use of his right arm. The privations as a result of his illness caused delays in the processing of his field observations and maps for the completion of his survey work as well as unavoidable delays in producing reports for the government.

However, MacCulloch's initial training as a doctor and personal experience of malaria encouraged him to write the first treatise on the subject in Britain, which appeared in 1827. He introduced the term malaria as a cause rather than the disease itself (Cumming 1983). A supplement to the work was produced in 1828 and, taken together, these books provide the most extensive contemporary review of the geographical occurrence, symptoms and treatment of intermittent fevers. The works were a milestone in their time that did much to encourage the development of the discipline of malariology and the study of tropical medicine in Britain.

Although it is difficult to fully assess the impact MacCulloch's indifferent health had upon his interpersonal relationships and work commitments, malaria alone may not have been the cause of many of the problems experienced after 1821. Speculative indications of a possible underlying physiological problem can be gleaned from his self-portrait (Fig. 4a). The blown pupil and later references to temporary loss of use in his arm may indicate that he had suffered a series of micro-strokes, not in themselves sufficient to cause actual physical disability but may be of sufficient magnitude to distort personality and hence his dealings with friends and colleagues.

MacCulloch and the Geological Society

John MacCulloch became a member of the newly fledged Geological Society of London in 1808, just a few months after it was inaugurated. His request to be admitted as the 67th member of the Geological Society was considered and recorded in the minutes of the Ordinary meeting held on 1 January 1808, with his election as a Member occurring at the following meeting. MacCulloch was active from the start and read his first paper over the course of two meetings on 1 and 6 April 1808. This paper, 'An Account of Guernsey and the other Channel Islands', was the first to be published by the Geological Society and appeared in the first volume of the *Transactions* published in 1811. In it he already displayed considerable ability as a mineralogist and as an observer of field relationships in a complex crystalline rock topography. He provided rudimentary mineralogical spot sketch maps demonstrating the utilitarian usefulness of geological mapping, and his artistic abilities were put to good use in providing plates that illustrated geological and topographical features (a feature that later figured significantly in his work on the Western Isles) (Fig. 5).

From 1808 to 1821, when illness curtailed many of his activities, MacCulloch was a leading light and one of the driving forces in the Society. He was increasingly looked upon as the expert in Scottish geology, and the minute books reveal the extent of MacCulloch's involvement and later disillusionment with the Society after his illness of 1821. He was elected to the Committee on Nomenclature on 6 May 1808, made a permanent member of the Committee for Maps on 1 April 1809, appointed to the Committees for Chemical Analysis and the Investigation of Extraneous Fossils on 14 June 1810, and finally elected to the Committee on Papers on 15 February 1811. He was also instrumental in moving forward a recommendation that led to the appointment of Thomas Webster (*c.* 1772–1844) on 24 June 1812 as Keeper of the Museum and Draughtsman to the Society (Herries Davies

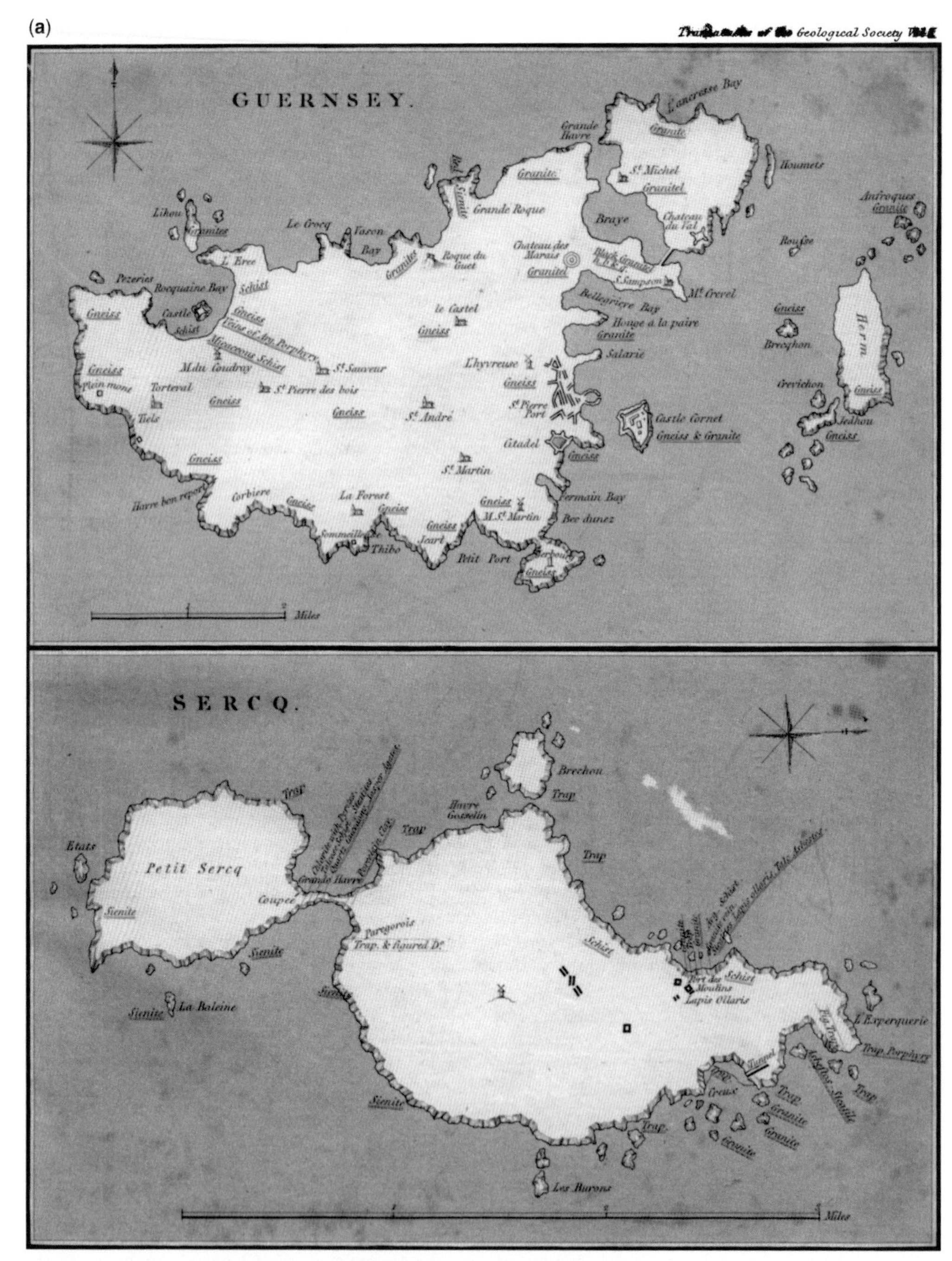

Fig. 5. MacCulloch's first paper in the *Transactions* of 1811. (**a**) and (**b**) are his generalized mineralogical spot sketch maps of the Channel Islands.

(b)

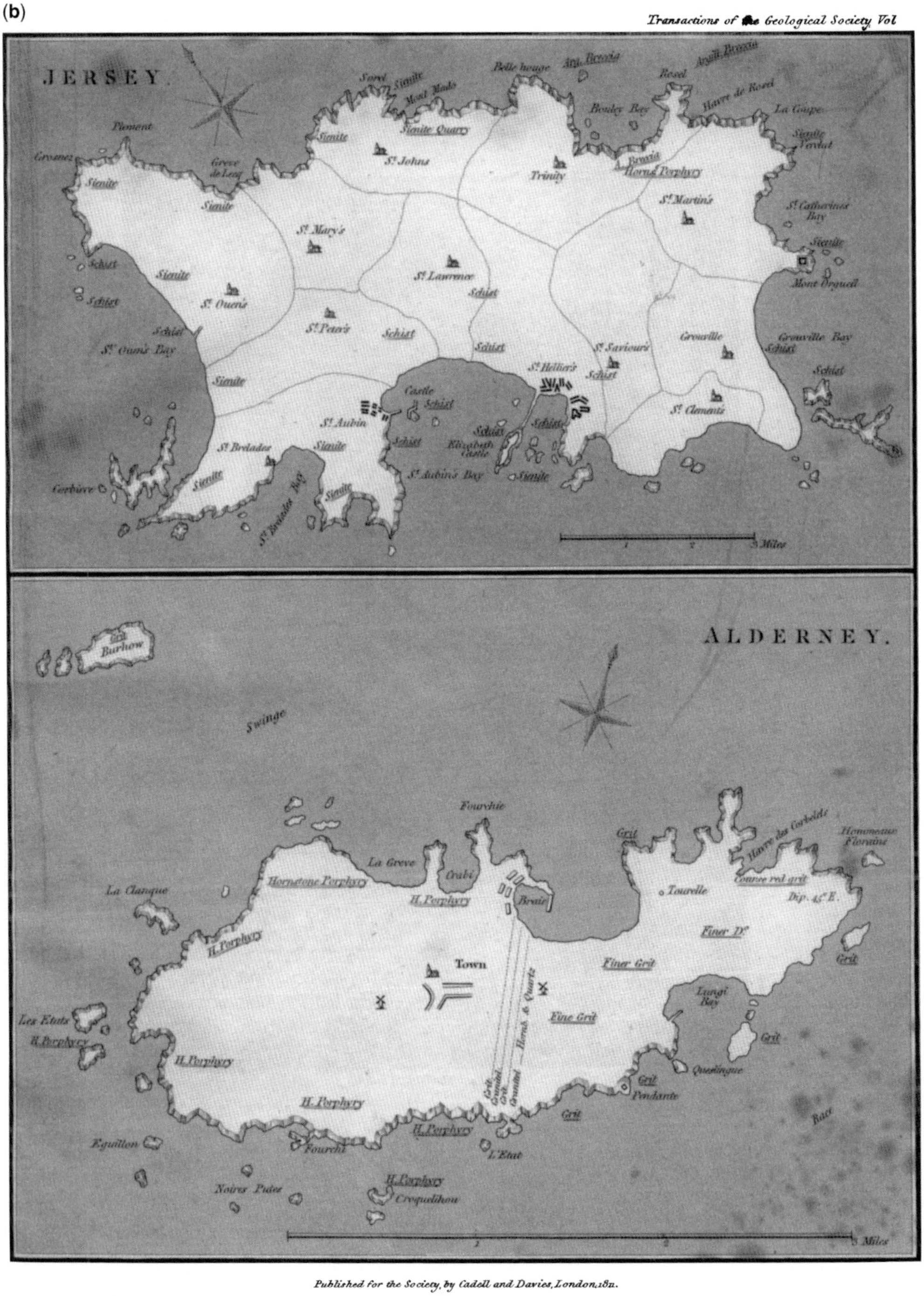

Fig. 5. MacCulloch's first paper in the *Transactions* of 1811. (**a**) and (**b**) are his generalized mineralogical spot sketch maps of the Channel Islands.

(c)

Transactions of the Geological Society Vol.I.
J. Mac Culloch del
No III. A GRANITE VEIN in PORT DES MOULINS in the ISLE of SERCQ.
Published for the Society, by Cadell and Davies, London, 1811.

Fig. 5. (**c**) This sketch displays MacCulloch's awareness of geology and topography whilst examining marine erosion upon crystalline rocks. His observational skills would be further developed in succeeding published papers, culminating in the Atlas volume of the Western Isles, see Figure 7.

2007). Cumming (1983) provides a more detailed analysis of MacCulloch's Society affairs.

He was elected Vice-President in 1815 and, as his influence grew as a result of his work on the Board of Ordnance Mountain and Meridian surveys, he became more involved in Society affairs. Upon completion of his vice-presidential term MacCulloch was immediately elected a member of Council. The quality of his papers read to the Society, as well as his personal enthusiasm, created a most favourable impression amongst his peers and he was subsequently elected as the fourth President at the Annual General Meeting held on 2 February 1816. During this term of office, MacCulloch strongly supported the move to purchase premises at 20 Bedford Street where the Society was to remain for the next 12 years (Herries Davies 2007). Despite being involved in administrative affairs, teaching and fieldwork, MacCulloch continued to read lengthy papers to the Society, updating the results of his Scottish survey work. As a result, he read to the Society his seminal paper on the Parallel Roads of Glen Roy early in 1817 (Fig. 6) in which he postulated a lacustrine origin for the feature. This firmly cemented his reputation as a first-class field geologist and may have helped him to secure his re-election as President in February 1817.

However, during this second presidency his attendance at Council meetings had declined, no doubt forced by the demands of the survey work and other duties. However, upon completion of his second term of office MacCulloch was again selected as a Council member and re-elected as Vice-President in 1820–1821. This probably reflected the high esteem felt for his *A Description of the Western Islands of Scotland* (Fig. 7). This book also led to his election as an FRS, proposed by Society member, mineralogist, physicist and chemist, William Hyde Wollaston (1766–1828), and supported by, amongst others, Colby and Sir William Congreve (1772–1828), comptroller of

(d)

No. V. View of FOURCHI POINT in the ISLE of ALDERNEY.

Published for the Society, by Cadell and Davies, London, 1811.

Fig. 5. (**d**) MacCulloch's sketch illustrating the effect of marine erosion upon 'hornstone prophyry'. He states 'it is the broad and perpendicular fracture of the rock, which causes the picturesque appearance at the western extremity of the island' (MacCulloch 1811, p. 6).

the Royal Laboratories at Woolwich. All this approbation led MacCulloch to become active in Society affairs again, reading two papers and attending Council meetings. These papers were the last he presented before finding other outlets for his work. However, his attendance soon fell off again and from 1821, affected by personal difficulties, he became increasingly estranged from the Society. He did become a Fellow of the newly Chartered Society in 1825, but eventually resigned in 1832 after Murchison's harsh criticisms of his *System of Geology*.

During his later years MacCulloch became critical of those 'Society Poseurs' who were moved to become Fellows because they saw that an FGS could be worn as the badge of a gentleman savant. He protested that such poseurs were travelling the land with geological hammers on ostentatious display (Herries Davies 2007). Paradoxically, such a criticism was specifically levelled at MacCulloch by Dr John Browne, who, in a vituperative book written as a critical response to MacCulloch's *Highlands and Islands of Scotland*, published in 1824, retorted: 'plodding antiquaries, crazy sentimentalists, silly view-hunters, cockney literati and, worst of all impudent 'stone doctors' armed with their hammers' (Browne 1825).

Yet, MacCulloch also had his supporters. He was well educated and descended from landed gentry. Despite this, he had to earn his own living but could clearly afford the Society's high fees. He enjoyed considerable patronage from influential sections of the Scottish aristocracy, in particular Thomas Douglas, fifth Earl of Selkirk (1771–1820), from his Edinburgh student days, and Lord John Murray FRS, the fourth Duke of Atholl (1755–1830) who was also a member of the Society until he resigned in 1827. MacCulloch was also friendly with Sir Walter Scot and dedicated his 1824 four-volume work on the *Highland and*

Fig. 6. The parallel roads of Glen Roy (from MacCulloch 1817, Plate 16).

(a)

J. Mac Culloch del.
Chas. Heath sculp.
ENTRANCE OF FINGAL'S CAVE, STAFFA.
Published June 1, 1819, by Constable & Co. Edinburgh.

Fig. 7. (**a**) and (**b**) Topographic plates of Staffa from the Atlas volume of MacCulloch's description of the western islands, 1819. This is of interest as MacCulloch provides a justification for the portrayal of topographic features in his works. In his *Highland and Islands* he writes:

> The difficulty of drawing these columns is such that no mere artist, be his general practice what it may, is capable of justly representing any point upon this island. It is absolutely necessary that he should have an intimate mineralogical acquaintance, not only with rocks in general, but with all the details and forms of basaltic columns; since no hand is able to

(b)

Mc Culloch del
Byrne Sculpt
VIEW OF STAFFA FROM THE SOUTH WEST.
Published June 1.1819. by Constable & Co Edinburgh

Fig. 7. (*Continued*)
copy them by mere inspection; so dazzling and difficult to develop, are all those parts in which the general, as well as the particular character consists. This is especially the case in attempting to draw the curved and implicated columns, and those which form the causeway; where a mere artist loses sight of the essential part of the character, and falls into a sort of mechanical or architectural regularity. That fault pervades every representation of Staffa, except one, yet published; nor are there any of them, which might not have been produced in the artist's workshop at home.

(MacCulloch, 1824, vol. 4, p. 387)

Islands as a series of essay letters to him. MacCulloch's entry into the world of Scottish aristocratic society led to the criticism from his paymasters that he was guilty of spending too much time enjoying the hospitality of great country houses, although some of this time was spent as recuperation from illness. Also his connections led to a suspicion amongst the west coast Scots that he sided with the lairds who were pressing for land reforms.

He was increasingly remote from the Society at that time when the full force of palaeontological stratigraphy swept through the geological brotherhood (Knell 2009). MacCulloch was now deemed a conservative. In his isolation he began to write anonymous and rather scathing reviews of geology and medicine works. In a single page he dismissed George Poulett Scrope's (1787–1876) work on volcanoes; Scrope was then Secretary to the Geological Society. MacCulloch then followed this dismissal with a long tract expounding his own opinions.

However, according to Cumming (1983), MacCulloch inadvertently set the mould for what was to follow with regard to professional survey work. The professionalism of geology was slow to take off in Britain, but MacCulloch set an early benchmark for conducting government surveys. At a time when there was little guidance or, indeed, expectation or understanding of the role of the geologist, MacCulloch was instrumental in helping to lay the foundations for the current geological establishment in Britain. The use of carefully planned survey work with meticulous attention to detail, the sharing of factual information gathered via surveys, reports, maps and sections, and the dissemination of such knowledge in publication helped to professionalize the science and the role of the Geological Society within it.

His Geological Society peers who were so critical of his conservative approach towards understanding the role of fossils in stratigraphy were also ultimately helped by the path he trod. One major problem arising out of MacCulloch's survey work was a lack of perception of the role of the geologist by the government bureaucracy of the day. During his Board of Ordnance Surveys and the last 'Treasury Survey' MacCullcoch was paid by the mile, which led to a dispute over expenses. Official memories of this led to difficulties for Henry De la Beche (1796–1855) as the founder of the Geological Survey of Britain (Flinn 1981). When the time came to initiate the official Government Geological Surveys of England it was decided to commission a number of small areas with budgets tied to the production of maps and reports before another area was undertaken. This ensured a measure of control over the Survey geologists who became charged with targets, budgets and timescales.

MacCulloch and palaeontology

MacCulloch worked primarily in hard crystalline metamorphic topography and was critical of some of the new thinking concerning the use of fossils in stratigraphy as a cure for all geological problems in correlation. Part of the problem lay in a perceived dichotomy between the proponents of stratigraphical palaeontology and those of mineral geology who were linked by association to the old Wernerian doctrines. Perhaps due to his blunt and occasionally waspish writing style, an example of which is found in a letter to Leonard Horner (1785–1864) sent on 2 June 1820 where he described those who were applying the new ideas as 'namby pamby cockleologists and formation men' (cited in Herries Davies 2007, p. 71), his remarks on palaeontology were often taken out of context. This led to his becoming increasingly embittered with the geological establishment of the time. Indeed, he asked whether the proponents and practitioners of 'fossil conchology' brought up on the soft, southern, relatively unaltered, highly fossiliferous young sediments would be able to unravel the complexities of the regions in which he worked. Paradoxically, it was the Smithian fossil-based stratigraphy that formed the 'cutting edge' of the science at that time and therefore MacCulloch's apparent isolation in the north gave him a short-sighted perspective of how this development would ultimately affect the work he undertook.

MacCulloch reinforced his view on the usefulness of fossils, with increasing charges of conservatism being made against him, by writing an article in Brewster's *Edinburgh Encyclopaedia*. Here he states that:

> In geological reasonings, the nature of fossils affords evidence with respect to the identifications of certain strata, whether these are distant, or whether near to each other but interrupted and disturbed. If it does not in these cases do all that has been fondly expected of it, it is still an assistant and collateral evidence.
>
> (MacCulloch 1830, p. 685)

In this comment MacCulloch was alluding to his belief that fossils were too vulnerable to destruction by geological change to serve as the sole means of correlating rocks over large distances. One of MacCulloch's critics also expressed a similar view when he wrote:

> Good physical groups are the foundations of all geology; and are out of all comparison the most remarkable monuments of the past history of our globe, so far as it is made out in any separate region. Organic remains are in the first instance, but accessories to the information conveyed by good sections.
>
> (Sedgwick 1848, p. 218)

Sedgwick wrote this, upholding the validity of a lithological and structural approach to mapping, whilst working on the Skiddaw Slates where fossils were scarce and distorted, just as MacCulloch encountered in the Highland region.

It is often forgotten that MacCulloch is credited with the first published faunas in several areas of Scotland, notably in the 'Primary' Rocks of Sutherland (MacCulloch 1819). Most notable is the observation that MacCulloch made of strange cylindrical structures in the quartz rock, described as 'imbedded cylindrical bodies' (MacCulloch 1814, p. 461). He attributed these pipe-like structures to the remains of infilled worm burrows indicating former animal life in this rock series: an interpretation that still holds today. He can also be credited for recognizing the Brora coalfield as being of Liassic or Oolitic age based upon palaeontological criteria, unlike John Farey (1766–1826) who deemed the deposit to be of Carboniferous age in 1812.

MacCulloch's tree

MacCulloch had thought sufficiently enough about palaeontology to be able to recognize and differentiate strata using palaeontological criteria. In particular he recorded instances of fossil material when they were found in the most unlikely of geological locations, although he characteristically kept to observational details and did not allow himself to get involved in speculation as to their origin. An illustration of MacCulloch's approach is the observation he made of the trunk of a fossil conifer, now assigned to the genus *Cupressinoxylon*, discovered at Rubha na h-Uamha, Bearraich whilst working on the SW part of the Ardmeanach Peninsula on Mull (Fig. 8)

> This substance is contained in a perpendicular vein, about fifty feet in height, and five feet in breadth, the lower end which reaches to the shore; the upper termination abruptly in a mixture of solid and columnar basalt by which it is everywhere surrounded [...] Towards the bottom it varies and becomes mixed with a black substance, which on examination is found to consist of minute fragments and a fine powder of carbonised wood, the vegetable organization still being visible in the former. Amongst this black matter a portion of the trunk of a tree appears, in a direction parallel to the side of the vein, and therefore erect, being entire and unbroken for the space of at least six feet [...] On careful examination of this wood it appears to be fir, which at least it perfectly resembles in its anatomical structure [...] The phenomena of wood in basalt has been often quoted as an argument against the igneous origin of that substance. Whatever conclusions are to be drawn from this fact, it is at least necessary to be accurate in stating it; and I believe that in all the instances hitherto described, the wood has, as in the present case, been found in a conglomerate or in some other rock, either lying under, or entangled in the basalt, and not in the basalt itself. In none of these cases does it appear to prove anything either for or against that theory.
>
> (MacCulloch 1819, pp. 568–569)

He reiterated his observation in 1830 by stating that the 'trap conglomerate' does occasionally contain carbonized fragments of wood. However, he also notes that, in his experience, there are no instances of animal remains or 'sea shells' to be found in such rocks:

> It is in these indurated shales [covered by masses of trap or penetrated by veins, as on Sky], or siliceous schists, that the shells are found, and not in the superincumbent basalt; a circumstance perfectly intelligible, because the common soft shales with which these are connected, and of which they are modifications, are among the ordinary repositories of these shells.
>
> (MacCulloch 1830, p. 685)

The final straw: reactions to MacCulloch's *System of Geology*

Murchison's stinging attack on MacCulloch when Charles Lyell (1797–1875) published his *Principles of Geology* might have been the final cause of MacCulloch estrangement from the Geological Society. This, unfortunately, coincided with the eventual publication of MacCulloch's two-volume student text, *System of Geology* (1831), most of which had been written a decade earlier and which he had attempted to get published in 1824. Murchison wrote:

> If you wish to study geological science as it is, in the writings of your own countrymen, you will naturally consult the works of Lyell and De La Beche. But for a knowledge of what it was, I may request you to peruse these volumes of Dr MacCulloch.
>
> (Murchison 1833, p. 376)

Sir Archibald Geikie (1835–1924) further compounded the criticism in historical terms by stating:

> MacCulloch's system made its appearance like the sullen protest of this last high-priest of a supplanted religion [...] in 1831, he threw this system at the heads of his rivals, and in the face of the geological world. The book may be looked upon as almost the last expiring effort of the old mineralogical school of geology in Britain. In perusing it, the reader might suppose himself to be in the midst of the literature of the end of the previous century. Fossil remains are ignored, together with all the new lines of inquiry which they had opened, and the rocks are described according to their mineral characters precisely as if William Smith had never lived.
>
> (Geikie 1875, vol. 1, pp. 202–203).

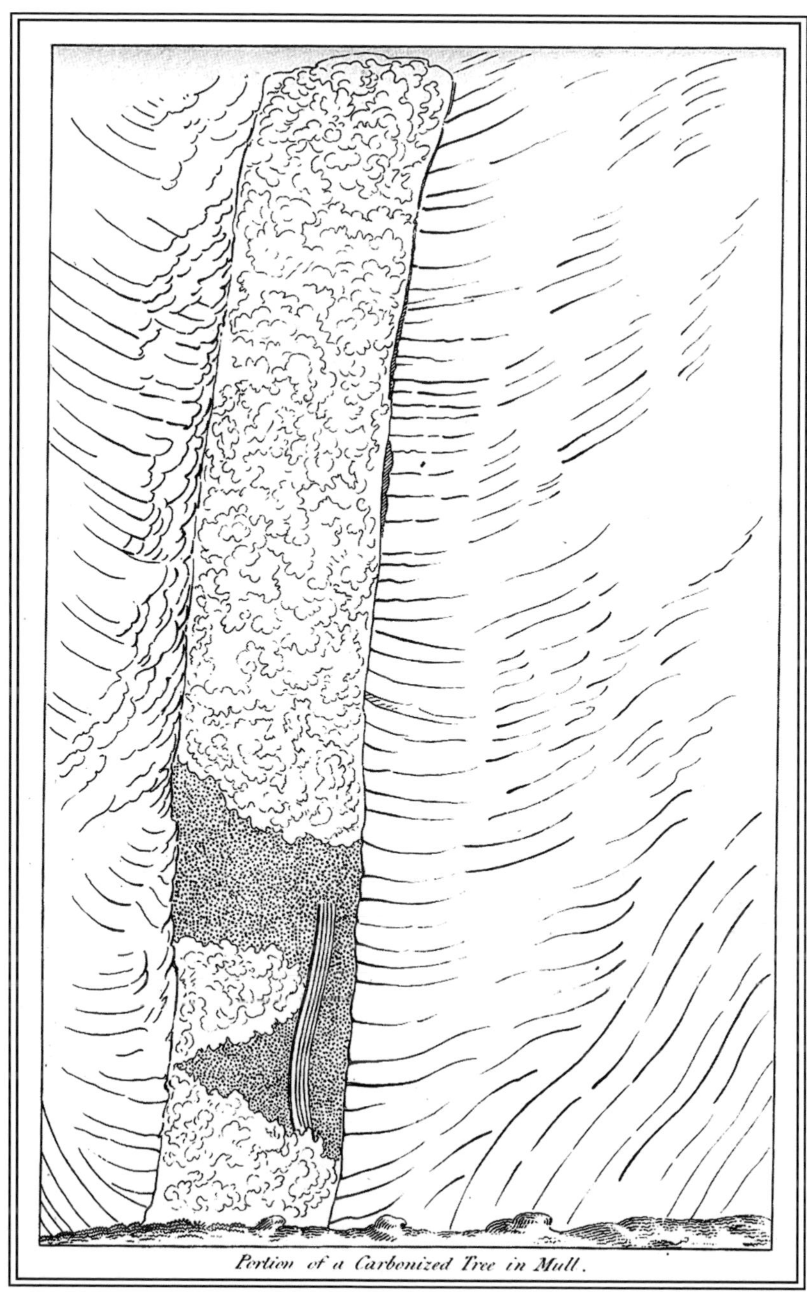

Portion of a Carbonized Tree in Mull.
I. Mac Culloch del.

Fig. 8. MacCulloch's depiction of the tree described in his *Description of the Western Islands of Scotland.* Atlas volume, plate XXI . He states that 'the proportional altitude has been materially reduced, for the purpose in showing the manner in which this vein terminates above. The surrounding rock is an irregular congeries of imperfect columns' (MacCulloch 1819, vol. III. p. 29).

Murchison's and Geikie's ill-considered statements arose from a misunderstanding of the purpose of the book. MacCulloch hadn't set out to make an attack on the proposals outlined in the first volume of Lyell's *Principles*, published 1830. He merely presented his ideas and observations derived from a lifetime of studying crystalline rocks. In effect, MacCulloch wrote one of the first geological textbooks for engineers, specifically designed for training potential serving officers in the East India Company. His first textbook was the critically acclaimed, *A Geological Classification of Rocks*, published in 1821; *System* was basically a follow-up volume containing a synthesis of information that would be of value to serving engineering officers in India. The appendix on mapping and field observation is still of value today. MacCulloch was at heart a salaried military geologist (see Rose 1996, 2009). In this respect, his intellectual agenda in writing textbooks differed from that of his peers in the Society. However, Murchison's criticisms effectively ruined the remaining year of an already outspoken and unpopular geologist. It is partly due to Murchison's withering criticisms of the *System of Geology*, given at an Annual General Meeting of the Geological Society, that MacCulloch resigned from the Society in 1832. It is noteworthy, however, that Lyell expressed his debt to MacCulloch and his *System* in subsequent volumes of *Principles* (1832–1833). Indeed, MacCulloch was Lyell's main authority in his discussion of 'Primary Rocks'.

Postscript: marriage, death and legacy

MacCulloch appears to have lived alone for most of his life. However, in 1835 he married Louisa Margaretta White (1811–1858) who came from East Bergholt, Suffolk. The ceremony was registered at St Martin-in-the-Fields Church, London, on 6 July. MacCulloch probably knew Margaretta for some years prior to the marriage, as her family resided near Addiscombe where he had been a lecturer at the East India Company Military Academy from 1814 to 1835. During the honeymoon in Cornwall, MacCulloch was driving a pony phaeton and, whilst turning in a narrow lane, locked the wheels and was thrown from the carriage. As a result he suffered a compound fracture of the right leg which necessitated amputation. MacCulloch was carried to the house of his friend Capt. Giddy RN of Poltair, near Penzance, where he prepared himself for the operation. MacCulloch's health was delicate. He delayed the operation for several hours preparing himself mentally for the ordeal that he was sure he wouldn't survive. He died on 21 August 1835 and was buried in Gulval Churchyard.

A memorial brass to MacCulloch was recorded in Lake's Parochial history of the church archives to 1868. The inscription read:

> Spe Beatæ Resurrectionis Johannes Macculloch, M.D., R.G.L.S.S. Obiit. 21. die mens: Aug. 1835. Ætat. 62. Dei Optimi Maximi Creatoris Parentis Gubernatoris Protectoris Omnipotentiam Benignitatem Iustitiam & Vigilantiam Ingenio Scientiâ & Literis probavit & illustravit.

In translation it reads:

> In the hope of a blessed resurrection John MacCulloch MD RGLSS [of the Royal, Geological and Linnean Societies] died on the 21st day of the month of August in 1835, age 62. By his talent for knowledge/science and literature he tested/proved and made famous/clear the omnipotence, kindness/bounty, justice and watchfulness of God the best and greatest, the creator and originator, director and protector.
>
> (Taylor pers. comm.).

MacCulloch's grave is still visible in Gulval churchyard with nothing to indicate that here lies one of the great originators of field geology and geological cartography in Britain and the fourth President of the Geological Society. MacCulloch left an important legacy as the first government-supported surveyor in Britain in terms of developing survey work and the gradual professionalism of geology. His unique blend of multidisciplinary skills opened up the geology of a largely unexplored tract of country for others to follow and interpret. As a result of this he produced one of the classic geological texts of the early nineteenth century, *A Description of the Western Islands of Scotland including the Isle of Man*. MacCulloch recognized and described several new rock types: chert, quartz rock (renamed quartzite), hypersthene rock and augite rock (two pyroxenites). He discovered and described the mineral chlorophaeite, and stratigraphically differentiated the three major sandstone groups, now known as the Torridonian, Old Red Sandstone and New Red Sandstone. He also correctly inferred the true nature of the Tertiary volcanic centres and examined their relationship with the older sedimentary sequences. Here he correctly deduced that the Cuillin granites and gabbros of Skye, along with the associated felsite and basalts, were younger than the Jurassic sedimentary succession. MacCulloch's observational skills noted that, although seemingly interbedded, the alteration of sediment at the contacts demonstrated that the igneous rocks were intrusive and therefore younger. Although this may seem obvious to modern eyes, it is perhaps sobering to note that Geikie examined the same succession more than 40 years later and came to the erroneous conclusions that the gabbros were pre-Cambrian, the bulk of the igneous rocks were Jurassic (Bathonian) and that some of the basalts were Tertiary in age. MacCulloch may also be credited with the first description of igneous layering in an intrusive body on the isle of Rùm.

MacCulloch's contributions to the understanding of Scottish geology were fundamental, and he deserves a better place in history than has been ascribed to him by Murchison and Geikie. Lyell gave him a more fitting tribute during part of his obituary address:

> Having expressed myself unreservedly on some of the peculiarities and defects of his style, I may affirm that as an original observer, Dr. MacCulloch yields to no other geologist of our times, and he is perhaps unrivalled in the wide range of subjects on which he displayed great talent and profound knowledge. For myself, I may acknowledge with gratitude that I have received more instruction from his labours in geology than those of any living writer".
>
> (Lyell 1836, p. 359)

A less welcome legacy is that he is probably the only geologist whose image has appeared on the base of a set of chamber pots. These were commissioned by MacKinnon of Coirechatachan after MacCulloch's disparaging remarks concerning the habits of some of the west coast Scots (Cooper 1973).

With the passing of MacCulloch, a controversial and complex pioneering spirit on hard rock petrology and its structural complexities was lost. Although field mapping techniques and surveys developed apace, the advancement of petrology was held back for several decades while British geologists became enamoured with palaeontology and stratigraphy. Workers such as Buckland, Mantell, Owen and Conybeare brought to the public consciousness the discovery, reconstruction and interpretation of the great reptiles of the Mesozoic, a legacy that exists today with museum collections. Why bother with intractable hard rock regions and their attendant structural complexities that were largely inaccessible in the 1830s–1840s when the great fossil discoveries were opening up a window on the past in the public's imagination. With a couple of notable exceptions such as James Nicol (1810–1879) in Aberdeen and Henry Clifton Sorby (1826–1908) in Sheffield, a British hard rock petrographic tradition did not fully resurrect itself until after the pioneering thin section work carried out in Germany during the 1860s–1870s; a tradition that survived until another paradigm shift occurred with the discovery of North Sea Oil.

When the occupations of the Founding Fathers of the Geological Society are examined it is noticed that there is a distinct group of medical men, chemists and others of independent means as well as members who possessed a military background (see Lewis 2009; Rose 2009). John MacCulloch MD was both a salaried military geologist, via his lectureship at Addiscombe, and a physician, as well as the first paid professional government sponsored survey officer in Britain. Today, a bust of MacCulloch and a copy of his *Geological Map of Scotland* are located on the stairway up to the Library and Council room at the Geological Society Apartments in Burlington House.

The author is indebted to the valuable assistance offered by Wendy Cawthorne at the Geological Society and her cousin J. Stephen Gould who kindly located MacCulloch's grave. Professor Richard J. Howarth is thanked for his reading of an initial draft of the manuscript that helped to correct many errors of style. Any remaining errors are my own. I thank the referees Leucha Veneer and Dr Mike Taylor for their helpful comments without which this contribution would have been the poorer. Finally, I wish to acknowledge a special debt to Dr David A. Cumming whose work on MacCulloch still remains the seminal source of reference and who freely granted permission to refer to his unpublished PhD thesis. In particular, I am grateful for the kind comments he made on a revised draft of this manuscript.

Archives

GSL: Geological Society of London archive.
PRO: Public Record Office.

References

BOUD, R. C. 1974. Aaron Arrowsmith's topographical map of Scotland and John MacCulloch's Geological Survey. *The Canadian Cartographer*, **11**(1), 24–34.

BOUÉ, A. 1820. Essai Geologique sur L'Ecosse. Paris, Mme Ve Courcier, Librarie pour les sciences. 519pp, 7pl, 2 folding maps.

BROWNE, J. 1825. *A Critical Examination of Dr. MacCuilloch's work on the Highlands and Western Islands of Scotland.* D. Lizars, Edinburgh.

CLARKE, J. W. & HUGHES, T. M. 1890. *The Life and Letters of the Reverend Adam Sedgwick* (2 vols). Cambridge University Press, Cambridge.

CLOSE, C. 1926. *The Early Years of the Ordnance Survey.* Institution of the Royal Engineers, London; reprinted 1969, HARLEY, J. B. (ed.), David & Charles, Newton Abbott and Augustus M. Kelley, New York.

COOPER, D. 1973. *Road to the Isles: Travellers in the Hebrides 1770–1914.* Routledge & Keegan Paul, London.

CUMMING, D. A. 1977. A description of the western Islands of Scotland: John MacCulloch's successful failure. *Journal of the Society for the Bibliography of Natural History*, **8**(3), 270–285.

CUMMING, D. A. 1983. *John MacCulloch, Pioneer of 'Precambrian' Geology.* Unpublished PhD thesis, University of Glasgow.

CUMMING, D. A. 1985. John MacCulloch, blackguard, thief and high priest, reassessed. *In*: WHEELER, A. & PRICE, J. H. (eds) *From Linnaeus to Darwin: Commentaries on the History of Biology and Geology.* Society for the History of Natural History, London, Special Series, 77–88.

EYLES, V. A. 1937. John MacCulloch, F. R. S., and his geological map: An account of the first geological survey of Scotland. *Annals of Science*, **2**(1), 114–129.

FITTON, W. H. 1818. Reviews of the *Transactions of the Geological Society*, volume III. *Edinburgh Review*, **29**, 87–88.

FLINN, D. 1981. John MacCulloch M. D., F. R. S., and his geological map of Scotland: His years in the Ordnance 1795–1826. *Notes and Records of the Royal Society*, **36**(1), 83–101.

GEIKIE, A. 1875. *Life of Sir Roderick Impey Murchison* (2 vols). John Murray, London.

HEADRICK, J. 1807. *View of the Mineralogy, Agriculture, Manufactures and Fisheries of the Island of Arran*. A. Constable, Edinburgh, 396pp.

HERRIES DAVIES, G. L. 2007. *Whatever is Under the Earth: The Geological Society of London 1807–2007*. Geological Society, London.

HUTTON, C. 1778. An account of the calculations made from the survey and measures taken at Schehallien, in order to ascertain the mean density of the Earth. *Philosophical Transactions of the Royal Society*, **68**, 781.

JAMESON, R. 1798. *An Outline of the Mineralogy of the Shetland Islands, and of the Island of Arran*. Wm. Creech, Edinburgh.

JAMESON, R. 1800. *Mineralogy of the Scottish isles: With Mineralogical Observations made in a tour Through Different Parts of the Mainland of Scotland, and Dissertations upon the Peat and Kelp* (2 vols). B. White & Son and W. Creech, Edinburgh.

JUDD, J. W. 1898. The earliest geological maps of Scotland and Ireland. *Geological Magazine, Decade IV*, V, 145–149.

KNELL, S. J. 2009. The road to Smith: how the Geological Society came to possess English geology. *In*: LEWIS, C. L. E. & KNELL, S. J. (eds) *The Making of the Geological Society of London*. Geological Society, London, Special Publications, **317**, 1–47.

LEWIS, C. L. E. 2009. Doctoring geology: the medical origins of the Geological Society. *In*: LEWIS, C. L. E. & KNELL, S. J. (eds) *The Making of the Geological Society of London*. Geological Society, London, Special Publications, **317**, 49–92.

LYELL, C. 1830–1833. *Principles of Geology* (3 vols). John Murray, London.

LYELL, C. 1836. Address to the Geological Society, delivered at the Anniversary on the 19th February. *Proceedings Geological Society of London*, **2**(44), 358–359.

MACCULLOCH, J. 1811. Account of Guernsey, and the other Channel Islands. *Transaction of the Geological Society of London*, **1** (series 1) 1–22.

MACCULLOCH, J. 1814. Remarks on several parts of Scotland which exhibit quartz rock, and on the nature and connexions of this rock in general. *Transactions of the Geological Society of London*, **2** (series 1) 450–487.

MACCULLOCH, J. 1817. On the Parallel roads of Glen Roy. *Transactions of the Geological Society, London*, **4**, 314–392.

MACCULLOCH, J. 1819. *A Description of the Western Islands of Scotland including the Isle of Man: Comprising an Account of their Geological Structure; with Remarks on their Agriculture, Scenery and Antiquities*. A. Constable & Co., London.

MACCULLOCH, J. 1821. *A Geological Classification of Rocks with Descriptive Synopses of the Species and Varieties, Comprising the Elements of Practical Geology*. Longman, London.

MACCULLOCH, J. 1823. French Geology of Scotland: Review of Boué, A. Essai Géologique Sur l'Ecosse. *Edinburgh Review*, **38**, 418.

MACCULLOCH, J. 1824. *The Highlands and Western Isles of Scotland, Containing Descriptions of their Scenery and Antiquities*. (4 vols). Longman, London.

MACCULLOCH, J. 1830. Organic remains. *In*: BREWSTER, D. (ed.) *Edinburgh Encyclopaedia*, Vol. 15, 681–756. William Blackwood, Edinburgh.

MACCULLOCH, J. 1831. *A System of Geology, with a Theory of the Earth and an Explanation of its Connections with the Sacred Records* (2 vols). Longman, London.

MACCULLOCH, J. 1836. *Memoirs to His Majesty's Treasury respecting a Geological Survey of Scotland*. Samuel Arrowsmith, London, **ii**, 142.

MURCHISON, R. I. 1833. Address to the Geological Society, delivered on the evening of 17th February 1832. *Proceedings of the Geological Society of London*, **1**, 362–386.

MURCHISON, R. I. 1851. On the Silurian rocks of the South of Scotland. *Quarterly Journal of the Geological Society*, **VII**, 137–178.

OLDROYD, D. R. 1990. *The Highlands Controversy: Constructing Geological Knowledge through Fieldwork in Nineteenth century Britain*. University of Chicago, Chicago, IL.

PENNANT, T. 1774. *A Tour in Scotland and Voyage to the Outer Hebrides*. John Monk, Chester.

PLAYFAIR, J. 1811. Lithological survey of Schichallien. *Philosophical Transactions of the Royal Society*, **101**, 347.

ROSE, E. P. F. 1996. Geologists and the army in nineteenth century Britain: A scientific and educational symbiosis? *Proceedings of the Geologists' Association*, **107**, 129–141.

ROSE, E. P. F. 2009. Military men: Napoleonic warfare and early members of the Geological Society. *In*: LEWIS, C. L. E. & KNELL, S. J. (eds) *The Making of the Geological Society of London*. Geological Society, London, Special Publications, **317**, 219–241.

SEDGWICK, A. 1848. On the organic remains found in the Skiddaw slate, with some remarks on the classification of the older rocks of Cumberland and Westmorland. *Quarterly Journal of the Geological Society, London*, **4**, 216–225.

SEDGWICK, A. & MURCHISON, R. I. 1828. On the old conglomerates, and other secondary depositis on the north coasts of Scotland. *Proceedings Geological Society, London*, **1**, 77–80.

SMALLWOOD, J. R. 2009. John Playfair on Schiehallion, 1801–1811. *In*: LEWIS, C. L. E. & KNELL, S. J. (eds) *The Making of the Geological Society of London*. Geological Society, London, Special Publications, **317**, 279–297.

John Playfair on Schiehallion, 1801–1811

JOHN R. SMALLWOOD

Hess Ltd, Level 9, Adelphi Building, 1–11 John Adam Street, London WC2N 6AG, UK
(e-mail: john.smallwood@hess.com)

Abstract: John Playfair first visited the Scottish mountain, Schiehallion, during Nevil Maskelyne's 1774 plumbline deflection experiment, which was conducted to measure the density of the Earth. The mathematician Charles Hutton analysed the survey data from the experiment, reporting the mean Earth specific gravity as 4.5 in 1778. Playfair undertook a lithological mapping exercise in 1801, to improve the accuracy of Hutton's estimate, and reported a range of 4.56–4.87 in 1811. The computation of the gravitational effect of topography with variable subsurface density effectively made him the creator of the first geophysical model. As such, not only was Playfair's Schiehallion contribution pioneering in itself, but it was representative of his more significant works in both mathematics and geology, in that he built on existing benchmark work with novel and valuable additions of his own. Although Playfair's map of the extent of the Schiehallion quartzite was quite accurate, the Society's fourth President, John MacCulloch, having visited Schiehallion, was dismissive of Playfair's representation of the subsurface density variation. MacCulloch spent several years searching Scotland for a more favourable site for a plumbline experiment, travels that allowed him to compile the data for his 1836 geological map of Scotland.

The Scottish mathematician and natural philosopher, John Playfair (1748–1819) was born at Benvie, Angus, and went up to university at St Andrew's to train for the church at the age of 14. His mathematical ability became apparent and he delivered lectures on natural philosophy prior to graduating in 1765. Following completion of his theological studies in St Andrews, in 1769, Playfair spent much time in Edinburgh in the company of intellectuals: James Hutton (1726–1797), Dugald Stewart (1753–1828), Adam Smith (1723–1790), Joseph Black (1728–1799) and Robert Adam (1728–1792). He finally obtained an academic post with his appointment to the position of Joint Professor of Mathematics at Edinburgh in 1785 (Fig. 1). Playfair published an edition of the six books of Euclid's *Elements of Geometry*, intended for his students. To these six books he added volumes on the circle, planes, solids, trigonometry and the arithmetic of sines: 'The chief peculiarity of it is, the introduction of algebraic signs in the fifth book, in order to render the proportions more compact, and consequently more easily followed by the eye' (Playfair 1822, p. xviii). Playfair's print run extended to five editions of 1000 copies each, four prior to his teaching from it at Edinburgh (Playfair 1822). His legacy to the mathematical world also includes an eponymous axiom motivated by seeking a solution to Euclid's contentious 'parallel' postulate (Wilson 1935).

In the spring of 1797, with the death of his friend the geologist, naturalist, chemist and agriculturalist, James Hutton, Playfair's thoughts turned to writing a biographical memoir. This evolved to become his *Illustrations of the Huttonian Theory of the Earth* (1802). Playfair argued his qualification to 'explain [...] Dr. Hutton's Theory of the Earth in a manner more popular and perspicuous than is done in his own writings' as 'having been instructed by Dr. Hutton himself in his theory of the earth; having lived in intimate friendship with that excellent man for several years, and almost in the daily habit of discussing the questions [...] treated of' (Playfair 1802, p. iii). He had accompanied Hutton, for example, with James Hall (1761–1832), in observing the famous unconformity between the Old Red Sandstone and Silurian slates at Siccar Point in 1788.

Playfair's accessible writing style ensured that he met his aim in correcting 'the obscurity of [Hutton's writing which has] often been complained of [...] so little attention has been paid to the ingenious and original speculations which they contain' (Playfair 1802, p. 3). He was voted an Honorary Member of the Geological Society of London at its first meeting in December 1807.

In the latter part of his life he was largely occupied as the first President of the Astronomical Institution of Edinburgh (1811) and with writing his textbooks on natural philosophy. However, he wished to produce a second edition of his *Illustrations*, enhanced with more field observations from the Continent, and the cessation of hostilities permitted this in 1815. On route to what turned out to be his last field trip, Playfair met Alexander von Humboldt (1769–1859), Georges Cuvier

From: LEWIS, C. L. E. & KNELL, S. J. (eds) *The Making of the Geological Society of London*.
The Geological Society, London, Special Publications, **317**, 279–297.
DOI: 10.1144/SP317.16 0305-8719/09/$15.00

Fig. 1. John Playfair, from an engraving of James Thomson, published 1 September 1819 by Henry Colburn, Conduit Street. Reproduced with permission. © The Royal Society.

(1769–1832) and Alexandre Brongniart (1770–1847)[1] in Paris (Playfair 1822). Humboldt later wrote that he sought to emulate Playfair in his own writing: 'Among those who were the first, by an exciting appeal to the imaginative faculties, powerfully to animate the sentiment of enjoyment from communion with nature, and consequently, also, to give impetus to its inseparable accompaniment, the love of distant travels, we may mention [...] in Great Britain, the intellectual Playfair' (Humboldt 1850, pp. 74–75).

This paper discusses the contribution of Playfair to Nevil Maskelyne's 1774 Schiehallion experiment. In some ways this contribution has similarities to his major mathematical and geological works in that in each he built on an existing landmark contribution with novel and valuable additions of his own. These works then became stepping stones to further major landmarks.

The Schiehallion experiment

In 1772 Nevil Maskelyne (1732–1811), the Astronomer Royal, proposed to measure the deflection of the vertical near some suitable mountain, an experiment suggested a century earlier by Sir Isaac Newton (1642–1727). Maskelyne wished to 'greatly illustrate the general theory of gravity' and to 'make the universal gravitation of matter palpable', and specifically to 'serve to give a better idea of the total mass of the Earth and the proportional density of the matter near the surface compared with the mean density of the whole Earth' (Maskelyne 1775*a*, p. 496). The principle of such an experiment is that a hanging plumbline is deflected towards a nearby mountain by gravitational attraction, as the mountain's mass attracts the plumb-bob or plummet. The estimation of the Earth's mean density from the deflection of the plumblines depends on the relative gravitational attractions of the mountain (horizontally) and the Earth (vertically, towards the Earth's centre). Such deflections are typically very small, as in spite of its relative proximity, the mass of a mountain is very small relative to the mass of the Earth.

Maskelyne's experiment was to be the second such attempt, the French astronomer and mathematician Pierre Bouguer (1698–1758) having been arguably unsuccessful on Mount Chimborazo in the Andes due to difficult conditions and instrumentation 'too small and imperfect for the purpose' (Bouguer 1749; Maskelyne 1775*a*, p. 497). The astronomer and mathematician Charles Mason (1728–1826) conducted a reconnaissance trip in 1773 to find a suitable site in England or Scotland. He identified Schiehallion, in Perthshire, a large, steep and isolated mountain elongated in the east–west direction as the preferred experimental venue (Fig. 2).

Astronomical and topographic measurements

In the summer of 1774 Maskelyne had built and occupied two observatories on the flanks of Schiehallion (Fig. 2), taking astronomical measurements referenced to a vertical defined by a free-hanging plumbline at each location (Maskelyne 1775*b*). A total of 337 observations were taken of the apparent zenith distance at the meridian of 43 stars. Leadstone (1974) reviewed the instrumentation and methods used by Maskelyne, and described the meticulous care, rigour and ingenuity applied during the taking of the astronomical observations.

To calculate the attraction of the mountain, and measure the latitudinal separation of the two observatories, a detailed topographic survey was also

[1] Following his 1812 review of their paper, Playfair took the opportunity to visit the Paris Basin with Cuvier & Brongniart in person.

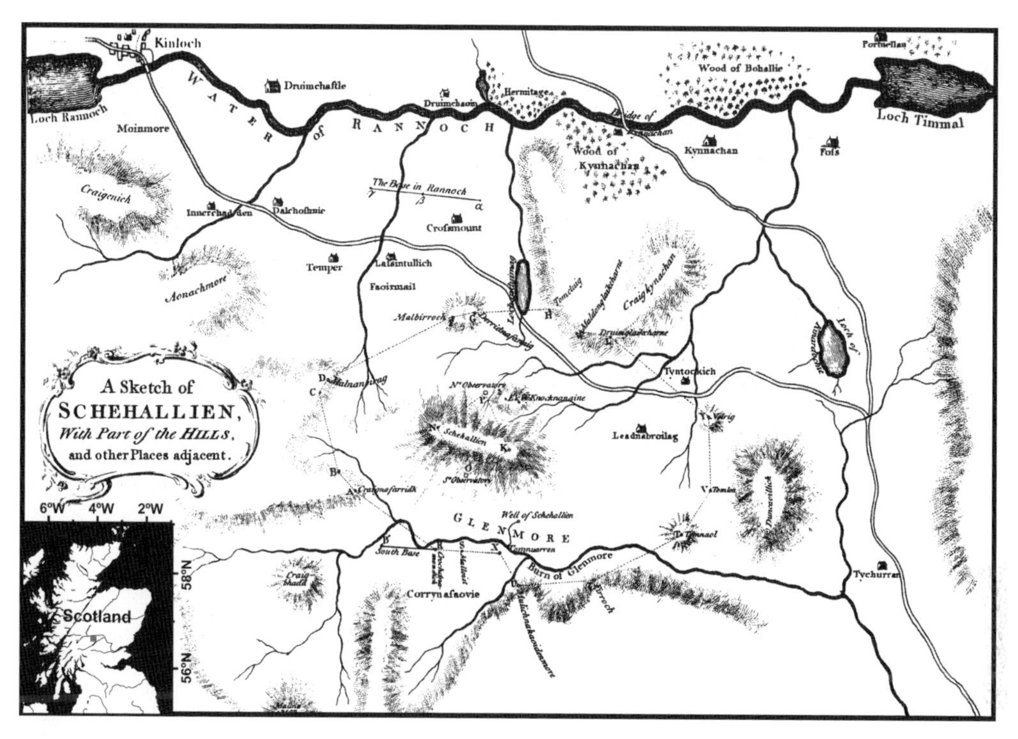

Fig. 2. Sketch map of the Schiehallion area, from Hutton (1778). Reproduced with permission. © The Royal Society. The survey of the mountain used triangulation and reference to two measured baselines, in Rannoch and Glenmore. The positions of north (P) and south (O) observatories on the flanks of Schiehallion are marked. Cairn N marked the summit and Cairn K was built on the summit ridge. Inset map shows the location of Schiehallion (square) in central Scotland.

undertaken, largely by the astronomer and mathematician, Reuben Barrow (1747–1792), and local surveyor, William Menzies. This consisted of triangulation between cairns on and around the mountain (Fig. 2). Theodolitic measurements were made along transects between the base station cairns with assistants holding vertical poles at series of points along each transect. The continuation of the topographic survey over three field seasons in 1774, 1775 and 1776 is reflected in Maskelyne's comment that the name 'Schehallien [...] signifies in the Erse language, Constant Storm' (Maskelyne 1775*b*, p. 503).

Maskelyne's experiment had created 'so great a noise' that 'many friends did it meet with [were] interested in the success of it', and he received many visitors while he was on the mountain (Maskelyne 1775*b*, p. 525). Among these were 'Mr. [William] Ramsay [1806–1865], professor of natural history at Edinburgh' and 'Mr. [Patrick] Copland [1748–1822] and Mr. Playfore of the university of Aberdeen' (Maskelyne 1775*b*, p. 525). This appears to be an erroneous attribution of affiliation from Maskelyne, as although Playfair had trialled for the position of the Chair of Mathematics at Marischal College, Aberdeen in 1766, he was beaten in an 11-day contest by the mathematician, William Trail (1746–1831).[2] In fact, when meeting Maskelyne in 1774, Playfair was the Minister of Liff and Benvie, a post in which he had succeeded his father. His visit to Maskelyne on Schiehallion 'partaking of his hardships and labours [...] for a few days' initiated 'a friendship [...] which terminated only with the life of the Astonomer Royal', as 'an acquaintance contracted among wilds and mountains is much more likely to be durable than one made up in the bustle of a great city' (Playfair 1822, p. lxxviii).

[2] The competition does not appear to have concluded with any bad feeling, as Trail and Playfair were two of the founder members of the Royal Society of Edinburgh in 1783 along with Copland, Benjamin Franklin (1706–1790), James Hutton and others.

The precision of Maskelyne's instruments was sufficient to yield a measurement of the mean sum of the two plumbline deflections towards Schiehallion of 11.6 arc-seconds (0.00322°). The determination of a tangible total deflection allowed Maskelyne to report success in meeting one of the aims of the experiment, which was to demonstrate the operation of gravitational attraction between objects on the Earth's surface. He also reported as one of his preliminary conclusions following the experiment that the Earth's mean density was about double that of the mountain, 'totally contrary to the hypothesis of some [...] who suppose the Earth to be only a great hollow shell' (Maskelyne 1775*b*, p. 533). In this conclusion he reiterated what Bouguer had stated nearly 30 years earlier (Bouguer 1749).

Hutton's analysis of results

The mathematician, Charles Hutton (1737–1823), computed that Maskelyne's angular deflection of 11.6″, 'after allowing for the centrifugal force arising from the spin of the Earth', suggested a ratio of 17804 : 1 for the 'attraction of the earth [...] to the sum of attractions of the hill' (Hutton 1778, p. 780). It only remained to compute the expected gravitational effect of the mountain, so that the gravitational effect of the Earth could be estimated.

Hutton devised a means of numerical integration to calculate the mountain's attraction, with 'some hints' from the chemist Henry Cavendish (1731–1810) (Hutton 1778, p. 750). He constructed 20 concentric rings in stepwise radial increments of 666.67′ around each observatory, and divided each of the annular rings into 24 sectors, increasing in angular width with regular steps of the sine of the angle from the meridian (Fig. 3). Ranalli (1984) gave more detail on the advantages of the radial grid geometry devised and the approximations Hutton made in order to make the laborious computations manageable: 'All the matter in [... each] pillar may be supposed to be collected into its axis [...] the length of which axis [... is] the mean altitude of the pillar' (Hutton 1778, p. 748). While the height of each pillar was determined from the survey observations, many of the sectors had no direct observation points within their perimeter. This problem was solved by 'connecting together by a faint line all the points of the same relative altitude' on a large-scale plan (Hutton 1778, pp. 756–757). This technique led to Hutton being credited variously (for example, Leadstone 1974) with the invention of the contour line. Although, in fact, this was not the first appearance of contour lines in the literature, he may have been the first to display topographic data in this form. Unfortunately, he did not publish a copy of the topographic contour map he constructed.

Hutton initially selected a constant subsurface specific gravity of 2.5 (density of 2500 kg m^{-3}) and suggested a mean Earth specific gravity of 4.5 (4500 kg m^{-3}), a value much higher than that of surface rocks, and therefore requiring a still higher density at depth within the Earth, like Maskelyne's initial estimate. The estimate of a specific gravity of 4.5 fell somewhat below both Newton's earlier guess of between 5 and 6, and the currently accepted mean density for the Earth of 5515 kg m^{-3} (Lowrie 1997).

Improvement to the result

In 1782 Maskelyne hosted Playfair's first visit to London. During this visit Maskelyne introduced Playfair to many of the scientific elite.[3] There may well have been discussion of the Schiehallion experiment and the density of the Earth, as in addition to Maskelyne and Playfair's shared personal experiences on the mountain, Henry Cavendish was among Playfair's new acquaintances. Cavendish was involved in assisting Charles Hutton to calculate the mountain's attraction. Playfair noted afterwards that 'most of the members of the Royal Society seem to look up to him as to one possessed of talents confessedly superior; and; indeed they have reason to do so, for Mr. C. so far as I could see, is the only one among them who joins together the knowledge of mathematics, chemistry and natural philosophy [...] [his] gleam of genius breaks often through [... his] unpromising exterior' (Playfair 1782, p. lxxxv).

Cavendish subsequently went on to make his own measurement of the mean density of the Earth using a torsion balance (Cavendish 1798). He expressed his surprise that the difference between his and Hutton's result was outside his own estimated error 'it seems very unlikely that the density of the earth should differ from 5,48 [5480 kg m^{-3}] by so much as 1/14 of the whole' (Cavendish 1798, p. 522). Playfair had many conversations, some no doubt spurred by Cavendish's result, with both Maskelyne and Hutton over the possible improvement in the Schiehallion experiment.

[3] Playfair initially found the Royal Society Club rather inhospitable. John Smeaton (1724–1792) and Alexander Aubert (1730–1805) appear to have been a 'great consolation to a stranger, amid the inattention of the English philosophers' (Playfair 1782).

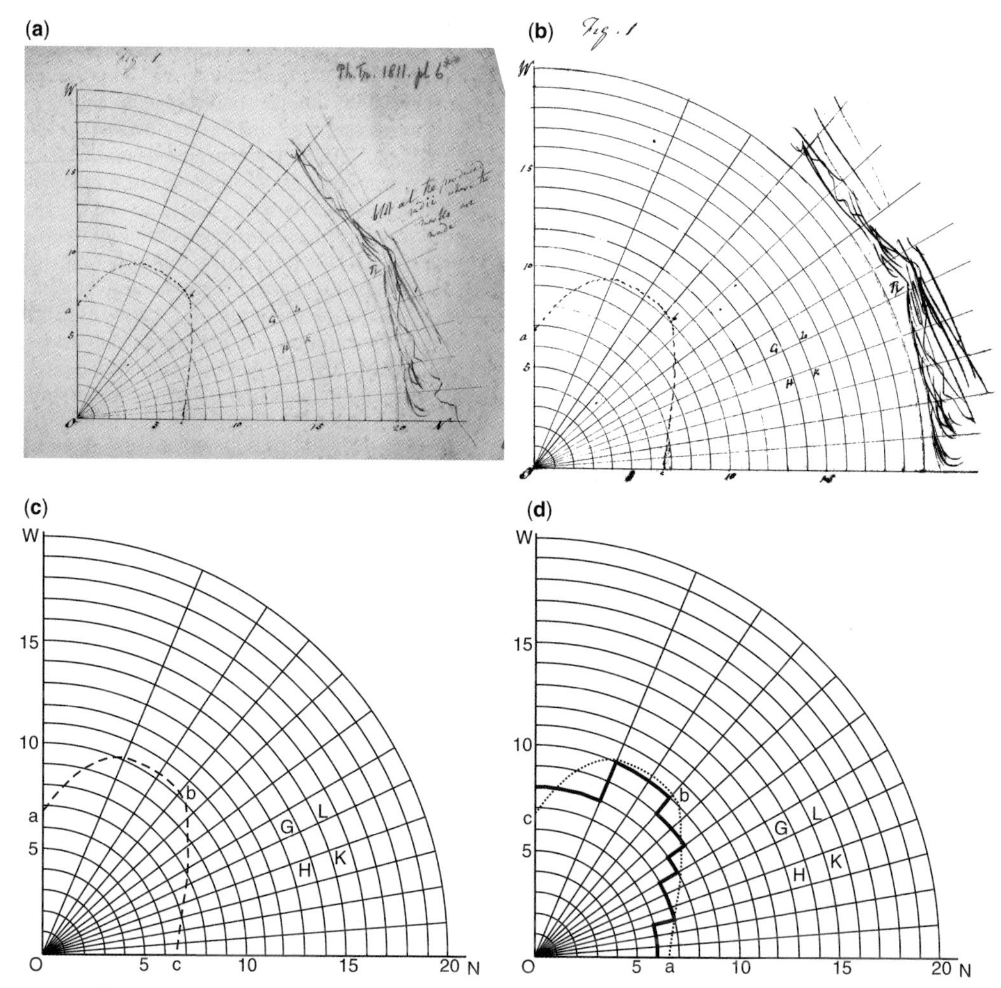

Fig. 3. Northwest quadrant of Playfair's (1811) map of Schiehallion, omitted from contemporary publication. Reproduced with permission. © The Royal Society. (**a**) Original figure as submitted to *Philosophical Transactions.* The text instruction to the engraver reads 'Blot out the produced radii where the marks are made'. (**b**) Drafted version of (a). It appears that Playfair started to sketch a panoramic view from the observatory around the edge of the figure, but decided not to include it in the publication and scored it out. (**c**) Drafted version of (a) as should have been published as figure 1 of Playfair (1811). (**d**) Version of the figure as it appeared in Playfair (1822). The bold solid line appears to be an attempt to follow segment boundaries of the grid to separate quartzite from schistus, but is not explained in the caption and appears to be an irrelevant editorial addition, as Playfair (1811) used a fraction of quartzite to schistus along the contact to calculate a proportional density, not whole cells of one or the other.

They recognized that the Schiehallion density estimate for the Earth of 4500 kg m^{-3} was likely to be an underestimate, as the density of the rocks composing the mountain was 'considerably above the mean' [of 2500 kg m^{-3}] used by Hutton (Playfair 1811, p. 348). Playfair's field experience and attendance during the original experiment made him the obvious candidate to undertake the necessary lithological and density survey of the mountain.

Playfair on Schiehallion

When reading his paper, 'Account of a lithological survey of Schehallien', in 1811, Playfair said that he had 'long wished to attempt such a survey of the mountain [...] and having mentioned the circumstances to the Right Hon. Lord Webb Seymour, he entered readily into a scheme' (Playfair 1811, p. 348). Playfair had been instructing his friend, Lord John Webb Seymour (1777–1819), for a number of years in geology and mathematics. Webb Seymour displayed a notable dedication to whatever the subject of his current interest[4] and the Schiehallion excursion served a useful purpose in diverting Webb Seymour from a period when devotion to his mathematical studies was in danger of proving injurious to his health (Hallam 1843). He travelled with Playfair on several occasions over Scotland and sometimes in England, assisting with his geological observations.[5]

So, in June 1801, Playfair together with his protégé Webb Seymour travelled from Edinburgh to 'a small village as near as [... possible] to the bottom of the mountain' from where they undertook their excursions over Schiehallion (Playfair 1811, p. 348). They mapped a boundary between the 'granular quartz[ite ...] hard, compact and homogeneous' that composed 'a large proportion of the mountain [...] from about the level of the two observatories up to the summit', and the 'schistus containing much mica and hornblende' on the lower slopes (Figs 3 and 4) (Playfair 1811, p. 353). Thirdly, they found 'in several places toward the base of the mountain [...] a granular and micaceous limestone highly cristallized'. Playfair noted that 'in general the strata [... were] nearly vertical' (p. 353), and so inferred that 'the rock which breaks out any where at the surface, continues the same through the interior of the mountain, in the direction of a perpendicular plane, down to its base, or perhaps to an indefinite depth' (p. 355), particularly for the schist (Fig. 5a). Playfair and Webb collected a selection of hand specimens (pieces between 1000 and 4000 grains (0.065–0.26 kg), mostly 2000–3000) for which specific gravities were obtained by Dr Kennedy: 2.64 for the quartzite (the standard deviation of Kennedy's measurements is calculated to be 0.01, 13 samples), 2.83 for the mica–hornblende slate (standard deviation 0.13, 10 samples) and 2.77 for limestone (standard deviation 0.07, five samples).

To revise Hutton's calculation of the mountain's attraction, Playfair used the topography tabulated by Hutton and the same grid layout, 20 rings and 48 azimuthal segments, over a circular area extending about four kilometres from each observatory (Fig 4). His computations differed from Hutton's in that he could not simply sum the contributions from topography above and lack of it below the observatories, as he wished to incorporate the effects of the different subsurface densities of the rocks he had mapped. He first had to employ his skill as a mathematician to calculate the gravitational attraction of various geometric solid bodies (Playfair 1809; Howarth 2007). The new Schiehallion model utilized his half-cylinder result, with a datum 1440′ below the north observatory (level R, Fig. 5), and summed the effects of columns of rock of variable density above this datum. The similarity in density between the slate and the limestone meant that these were treated together, and a proportional density was assigned to cells traversed by the mapped quartzite–schist boundary.

Although initially of the view that the quartzite cropping out above the observatories extended to depth, Playfair 'subsequently had opportunity to examine other of the Grampians where granular quartz is found at the summit [and found it] certain that the same rock does not go down into the interior' (Playfair 1811, p. 356). He therefore carried out two calculations, the first assuming a vertical boundary between quartzite and schist (Fig. 5a), and the second assuming that the rock beneath the level of the observatories was schist rather than quartzite. These calculations resulted in a revision of the estimate of Earth specific gravity to lie between 4.5588 (vertical boundary case) and 4.867 (horizontal boundary case). Recognizing the remaining discrepancy with Cavendish's result, Playfair concluded his 1811 paper by urging the repetition of experiments on the same lines so that the mean density of the Earth could be known more accurately.

An update on Playfair's calculations

Hutton and Playfair's published data allow their calculations to be repeated and investigated for sensitivity to density and datum changes (Table 1).

[4] Henry Hallam (1777–1859) commented 'Lord Webb Seymour soon adopted a plan, which even the reading men at Oxford seldom thought necessary to pursue. He resolutely declined all invitations, and during the whole remainder of his stay at Christ Church was never seen at a wine party' (in Hallam 1843, pp. 473–474).

[5] William Conybeare (1787–1857) and William Phillips (1775–1828) reported some of the observations made by Playfair and Webb Seymour in Yorkshire in their *Outline of the Geology of England and Wales* (Conybeare & Phillips 1822), and Webb Seymour scripted the publication of his and Playfair's observations in Glen Tilt (Webb Seymour & Playfair 1815). Lord Webb Seymour was later to be invited to become a visiting member of the Edinburgh 'Friday Club' at which Playfair, Walter Scott (1771–1832), Francis Horner (1778–1817), James Hall (1761–1832), Duguld Stewart, Francis Jeffrey (1773–1850) and other 'senior literati' would dine. Jeffrey transferred his affection to Leonard Horner (1785–1864), President of the Geological Society of London, who also joined the club following his brother Francis' death (Cockburn 1852).

(a)

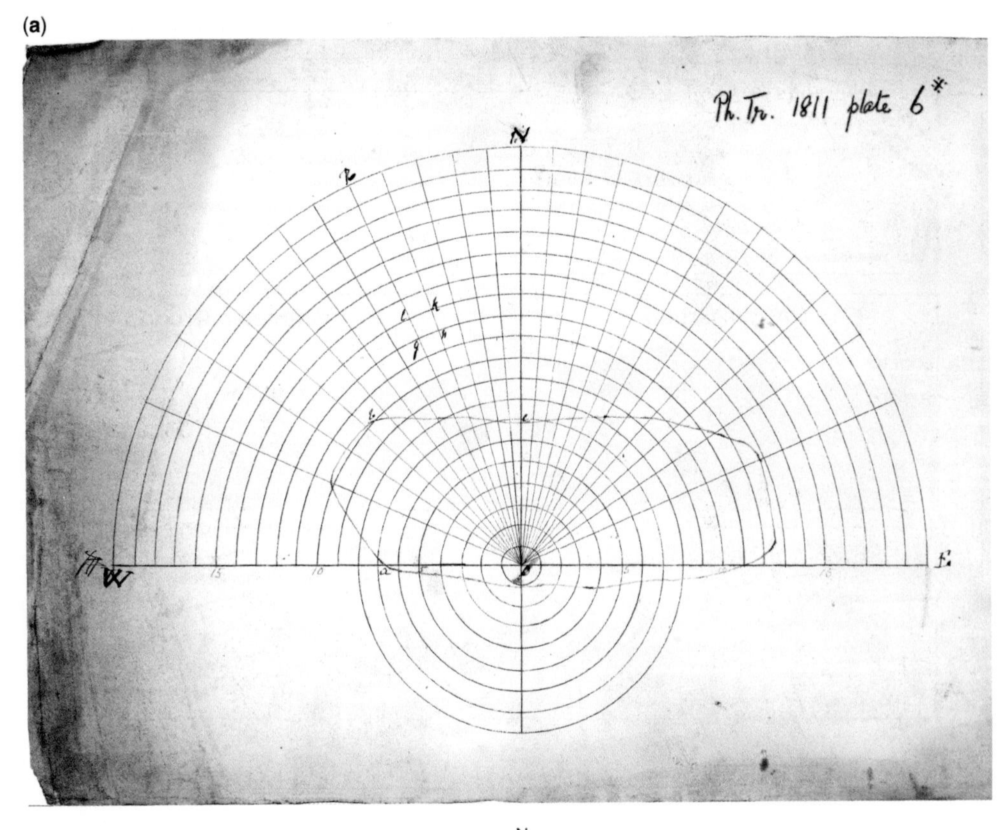

(b)

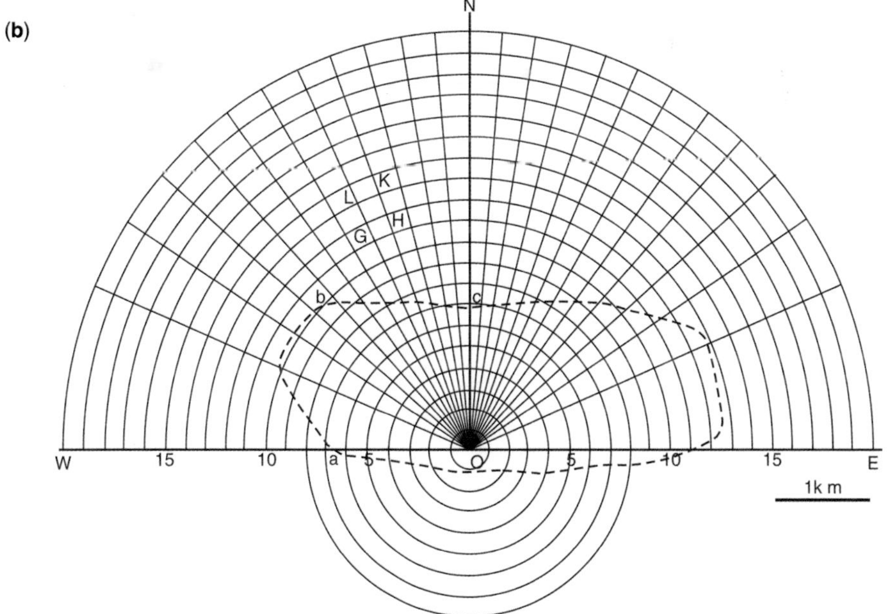

Fig. 4. Extent of Schiehallion quartzite as mapped by Playfair (1811), with part of Hutton's topographic calculation grid. Reproduced with permission. © The Royal Society. (**a**) Original figure submitted to Philosophical Transactions. (**b**) Drafted version of (a), as it should have appeared as figure 4 of Playfair (1811).

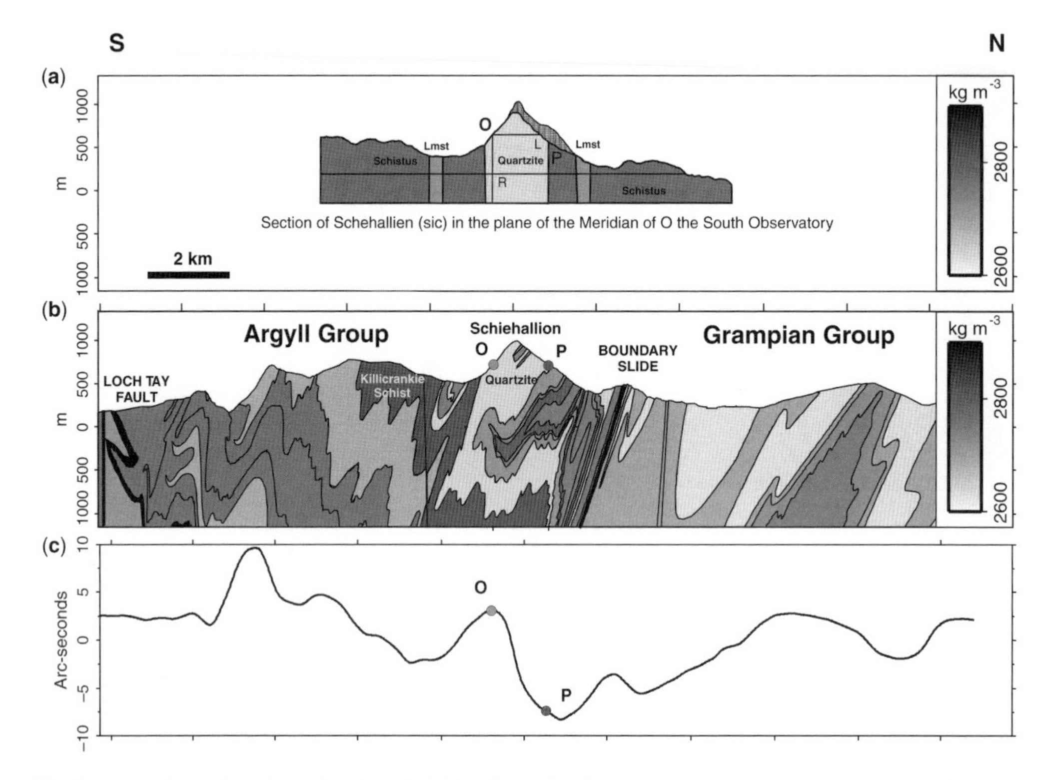

Fig. 5. (**a**) North–south cross-section through Schiehallion, after figure 3 from Playfair (1811) (published as fig. 2). Shaded by density; Playfair allocated a specific gravity of 2.64 to the quartzite and 2.81 to the surrounding schistus. (**b**) Geological cross-section after British Geological Survey (2000), converted to a density model using densities assigned from specimen measurements (table 2 of Smallwood 2007). (**c**) Northward deflection of the vertical (black) extracted along line of section from constant (2640 kg m^{-3}) density topography model, 100 m grid of topography, 10 km radius from each field point (Fig. 7a). The stations O and P were positioned almost optimally to capture the maximum inward deflection due to topography.

While Maskelyne quoted the cumulative inward plumbline deflection as 11.6″, recomputation of the observatory positions suggests that this would be better reported as 11.3″ (Smallwood 2007). This deflection correction changes the ratio of the gravitational attraction of the Earth to that of the mountain, including the correction for centrifugal effect, from 17 804, reported by Hutton (1778) to 18 256. The effect on Earth density is to add approximately 100 kg m^{-3} (Table 1) over Hutton's original value[6] of 4480 kg m^{-3}. Furthermore, as inferred Earth density scales linearly with mountain density in a constant subsurface density case, had Hutton used a more realistic 2640 kg m^{-3} he would have initially suggested 4840 kg m^{-3} (Table 1).

Compared with this figure, the inclusion of a lower density between the observatories than outside them (Playfair 1811) has the effect of reducing the modelled inward plumbline deflection (Table 1). So although Playfair's mapping of quartzite and surrounding denser 'schistus' was seeking to narrow the gap between Hutton's (1778) figure and that of Cavendish (5480 kg m^{-3}), the vertical quartz–schist boundary case, favoured initially by Playfair, did not really accomplish this. Only by decreasing the amount of quartzite between the observatories, by surmising the horizontal boundary, was an improvement seen. Similarly, deepening the datum used has an effect opposite to that desired (Table 1). What would have helped Playfair's result would have been to have the benefit of the recomputation of the observatory positions (leading to an Earth density of 4600 kg m^{-3}), or even including a higher density between the observatories reflecting the mix of lithologies (a 60 kg m^{-3} increase leads to an increase in inferred Earth density of 200 kg m^{-3}, Table 1).

[6] Hutton (1778) reported an Earth specific gravity of $4\frac{1}{2}$. Resumming his sines yields 4480 kg m^{-3}.

Table 1. *Earth Density determinations from Schiehallion Experiment with Hutton (1778) topographic grid*

	Hutton 1778 Original	Hutton 1778 Variant	Hutton 1778 Variant	Playfair 1811 Original	Playfair 1811 Variant	Playfair 1811 Variant	Playfair 1811 Variant	Hutton 1821 Original
Measured ratio Earth : mountain attraction[1]	17 804	18 256	18 256	17 804	18 256	18 256	18 256	18 256
Datum (ft, relative to observatories)[2]	0	0	0	−1440	−1440	−2440	−1440	0
Mountain density (kg m^{-3})	2500	2500	2640	Quartz 2640 Schist 2810	Quartz 2640 Schist 2810	Quartz 2640 Schist 2810	Quartz 2700 Schist 2810	2750
Model ratio Earth : mountain attraction	9933[3]	9939[3,4]	9933[3]	3905/3965[4]	3967	4074	3807	9960[3,4]
Earth density (kg m^{-3})	4480[4,5]	4590	4840	4560/4490[4]	4600	4480	4800	5040

[1]Original measured ratio of Earth : mountain attraction of 17 804 based on a cumulative inward plumbline deflection of 11.6″. Revised relative observatory positioning gives a ratio of Earth : mountain attraction of 18 256 based on a deflection of 11.3″ (Smallwood 2007).
[2]Although not explicit, several lines of evidence support Playfair's use of a datum 1440′ below the south observatory as well as the north.
[3]Assuming the Earth and the mountain have the same density.
[4] Recalculated, difference due to an increased computing accuracy since the original.
[5]Quoted by Hutton (1778) as $4\frac{1}{2}$.

Charles Hutton later recomputed the mountain's attraction at the age of 84. With no new useful field data he spurned Playfair's mapping and reverted to a constant density subsurface approach for a final pass at the analysis. This single subsurface mountain density, the mean density from Kennedy's quartzite and schist sample measurements (2750 kg m^{-3}) led Hutton to an Earth density estimate of 4950 kg m^{-3} (Hutton 1821), recalculated here as 5040 kg m^{-3} (Table 1), still some way short of the currently accepted value.

MacCulloch's geological map

Playfair had no useful extant geological maps of Schiehallion to refer to in 1801 during his fieldwork on Schiehallion. Although preceding publication of Playfair's results, Louis Albert Necker's (1786–1861)'s 1808 map of Scotland,[7] coloured according to the rock formations, grouped the whole area between Loch Rannoch and Loch Tay as 'Primitive rocks stratified as Gneiss, Mica Slate, Clay Slate' (Fig. 6).

The Schiehallion experiment turned out, however, to play a crucial role in the production of MacCulloch's landmark geological map of Scotland (Bowden 2009). The opportunity for Dr John MacCulloch (1773–1835) to create his famous 1836 map arose as he gathered data during fieldwork initially, at least, partially aimed at locating a more favourable site to repeat the Schiehallion experiment (MacCulloch 1836; Cumming 1983; Bowden 2007). MacCulloch had travelled widely from 1811 to 1814 seeking a suitable domestic source of millstones for the Ordnance powdermills (Flinn 1981; Rose & Rosenbaum 1988), and subsequently he was tasked with seeking alternative sites at which the Schiehallion experiment could be repeated, in line with Playfair's suggestion. An appropriate site had to display suitable form (a conical or ridged mountain) with suitable isolation and simple subsurface density distribution. Although he sought to promulgate his appreciation of the 'majestic [...] beautiful simple and conical form of Schehallien' with self-publication of a field guide, written whilst convalescing from a bout of malaria (MacCulloch 1823, pp. 261 and 268), he regarded it as quite unsuitable for the plumbline deflection experiment.

MacCulloch found the mountain 'easy to climb from Kinloch', but, 'in vain [... he] sought for the remains of Dr. Maskelyne's observatories; for time seems to have performed its appointed duty towards them' (MacCulloch 1824, vol. 1, pp. 434–435). His investigations of Schiehallion rapidly led him to the conclusion that the subsurface was too complicated to render detailed remapping worthwhile, at least for the purposes of revisiting the Maskelyne experiment. He was disturbed to find the structure of the mountain complex: 'I discovered what I had long before suspected: the error of this celebrated experiment, and the consequent wreck of its conclusions' and was also quite dismissive of Playfair's mapping: 'Nor did our late friend Playfair succeed in effectually correcting them by his geological investigation; since that itself was insufficiently conducted; having proceeded on an incorrect and superficial view of the structure of the mountain' (MacCulloch 1824, vol. 1, p. 435).

In his 1829 report to the Ordnance on suitable mountains for plumbline deflection experiments, MacCulloch therefore wrote off Schiehallion simply for 'Geology' (Cumming 1983). He elaborated in his *System of Geology* (1831, vol. 1, p. 22):[8]

> If the form and situation of the mountain [Schiehallion] were to offer great conveniences to the mathematicians engaged in this problem, it is, unfortunately, deficient in uniformity and simplicity of structure. Not only is it a matter of extreme difficulty to ascertain the true distribution of the strata throughout the whole mass, but the specific gravities of the different materials vary in such a matter as to vacillate between 2.4 and 3, or even more. All the strata are, at the same time, elevated to high but unequal angles; while it is also difficult to discover their relative proportions at the surface, and, still more, the positions which the rocks of different specific gravities assume in the interior of the mountain; a circumstance of considerable importance, on account of the angular difference of the action of several columns on the plummet [...] Under these uncertainties, a true mean specific gravity could neither be determined nor applied.[9]

Although Playfair's recognition of the quartzite on Schiehallion may have stimulated him to include that division in his mapping (Fig. 6), MacCulloch

---

[7] This map was presented to the infant Geological Society of London on 4 November 1808 (Woodward 1907; Eyles 1948).

[8] This topic occupied MacCulloch considerably. He later again summarized 'I pointed out numerous and serious errors, arising from a false view of the extent and positions of strata of highly different specific gravities' (MacCulloch 1836, p. 19).

[9] MacCulloch's comment is borne out by the result of more recent mapping (Figs 5–7), justifying his accolade from Conybeare & Phillips (1822, p. xviii) as being among 'the best geological observers [along with Von Buch and Necker]'.

(**a**) Necker 1808 map (1939)

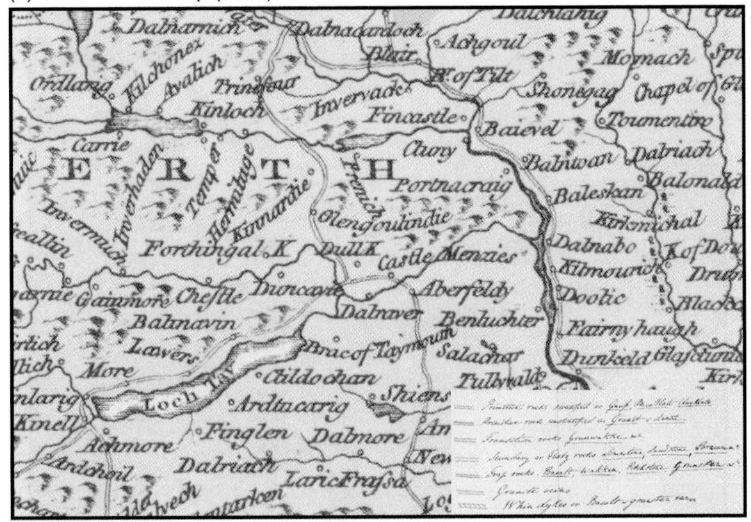

(**b**) MacCulloch 1840

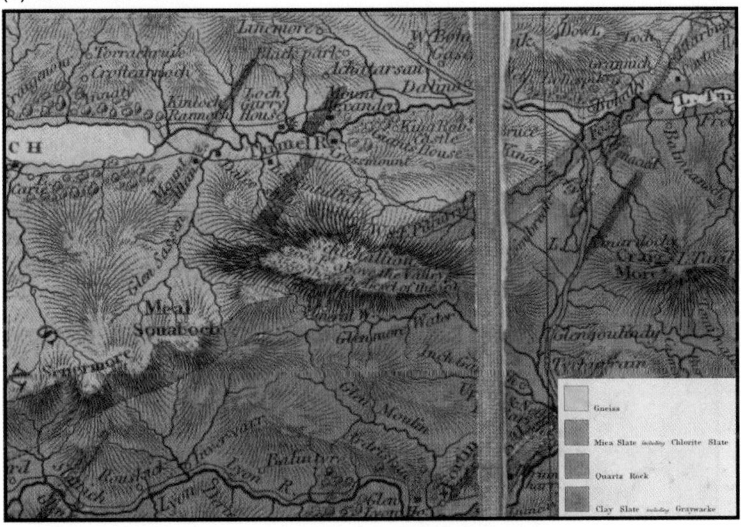

Fig. 6. Historical succession of geological maps across the Schiehallion area. The east end of Loch Rannoch and the west end of Loch Tummel are shown in all maps, separation distance, currently 11 km, was 14 km prior to 1986. (**a**) Necker 1808 map (1939) – the whole area termed 'Primitive rocks stratified as Gneiss, Mica Slate, Clay Slate' (**b**) MacCulloch 1840 – a band of 'Quartz rock' recognized between 'Gneiss' to the NW and 'Mica slate' to the SE. (a)–(d) Reproduced by permission of the Trustees of the National Library of Scotland.

(1836, p. 16) made the statement accompanying his geological map of Scotland that:

> before this, with the exception of my own work on the Western Islands [Macculloch 1819], and Dr. [Samuel] Hibbert's [1819] Map of Shetland [...] not a mile of land in all Scotland had been surveyed and recorded [...] I was unable to borrow the description of a single mile in aid of it. I have not derived even a hint, far less a fact, or an acre, from any other hand.

This statement may be one of those which map-maker, Aaron Arrowsmith (1750–1823), had in mind when penning the preface to MacCulloch's (1836) posthumous Memoirs, 'The following [...] have been printed verbatim [...] these, it is hoped,

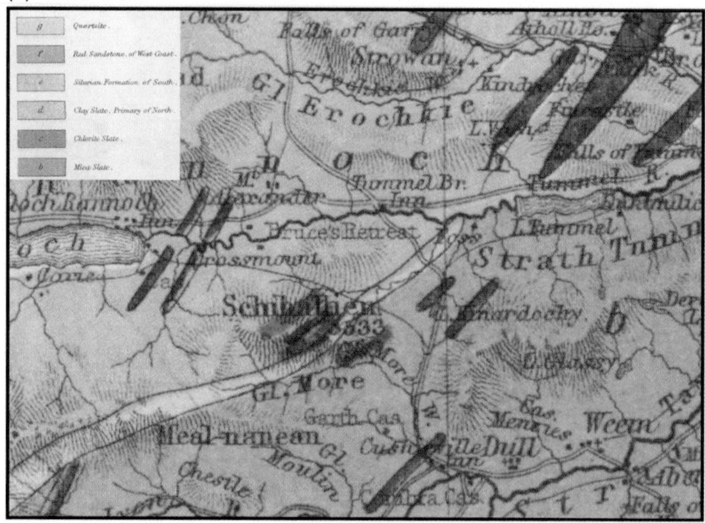

(d) Geikie 1876

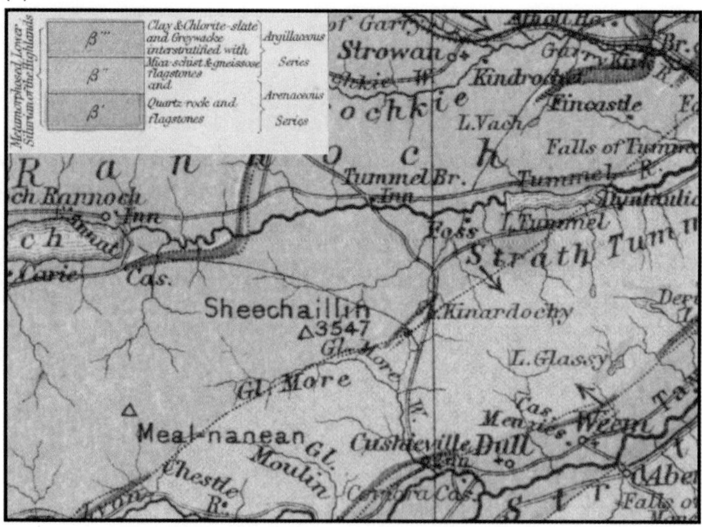

Fig. 6. Nicol (1846) derived partly from MacCulloch, but with an improved basemap, and more felsic and basic intrusives indicated on Schihallien (sic). (**d**) Geikie (1876), the first map shown here to attempt a stratigraphic approach, Scheehaillin (sic) composed of 'Metamorphosed Lower Silurian of the Highlands' (later Dalradian: Oldroyd 1990) with 'Mica-schist & gneissose flagstones' (later Argyll Group) to the SE and 'Quartz-rock and flagstones' (later Grampian Group) to the NW. (a)–(d) Reproduced by permission of the Trustees of the National Library of Scotland.

will be received [...] by such as are competent judges of the real errors and deficiencies'. Bowden (2007, 2009) argues that this sentiment is rather harsh,[10] but one such judge was George Bellas Greenough (1778–1855), the first President of the Geological Society of London, who noted that

[10] Harsh, but perhaps not surprising, since, in the Memoir, MacCulloch says he was 'condemned' to use Arrowsmith's basemap.

(e) BGS 1:25 0 000 1986; 1987

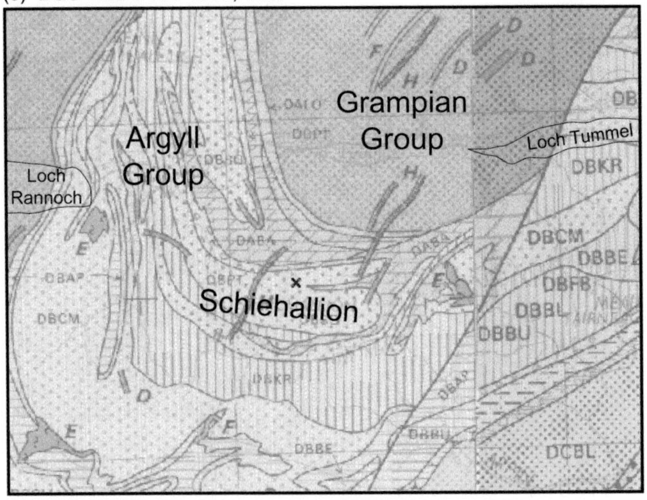

(f) BGS 1:50 000 2000

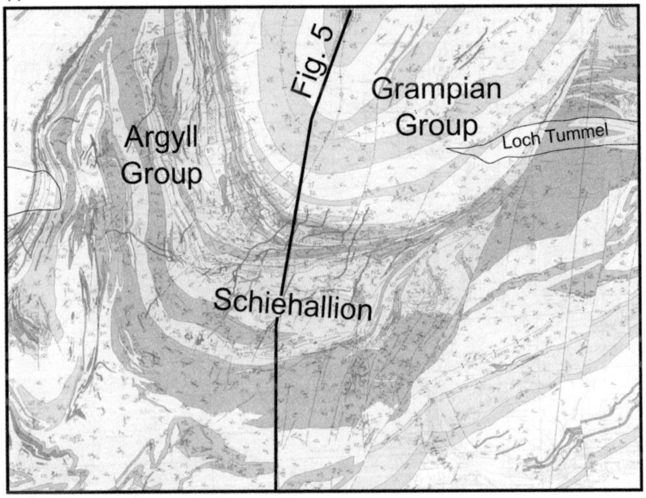

Fig. 6. (**e**) British Geological Survey 1:250 000 series (1986, 1987), the large-scale deflection of the Grampian–Argyll Group boundary recognized (see Stephenson & Gould 1995; Treagus 2000 for discussion), the outcrop of the Schiehallion Quartzite (green with pink dots) mapped to strike NW–SE. (**f**) British Geological Survey 1:50 000 (2000), the detail including subdivision of the Grampian group. This was converted to the density model in Figures 5 and 7 (Smallwood 2007). (e) and (f) Reproduced by permission IPR/99-54CA, British Geological Survey. © NERC. All rights reserved.

MacCulloch could have said the 'same of [Louis Albert] Necker, [Leopold Von] Buch [1774–1853] and Lord Webb [Seymour]'.[11]

Playfair observed that lithology appeared to exert control on the form of Schiehallion (Playfair 1811), the hard, homogeneous quartzite defining

[11] In a marginal note in Greenough's copy of MacCulloch's (1836) Memoir (p. 20, Geological Society of London Library). Lord Webb is presumably being credited for his Memoir on James Hutton's classic Glen Tilt locality (Webb Seymour & Playfair 1815). It is likely that MacCulloch was familiar with Necker's map as he was elected to the Geological Society of London in 1808.

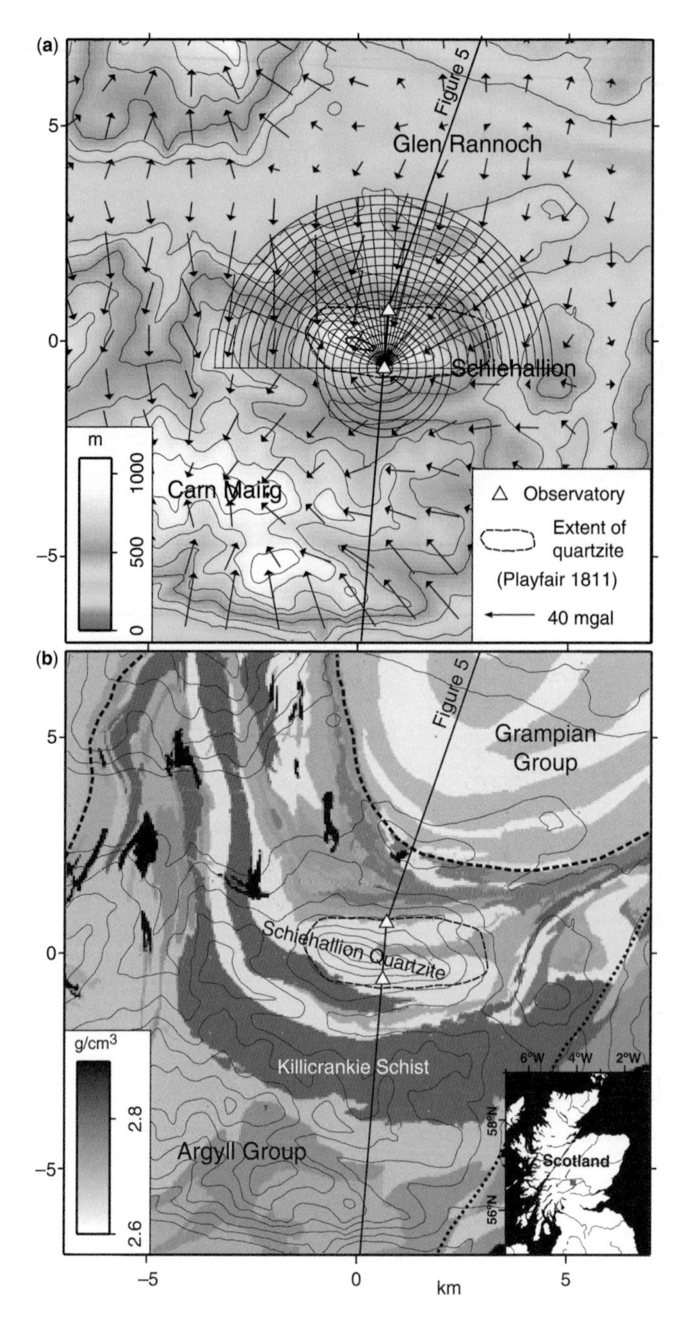

Fig. 7. Maps centred on Schiehallion's summit. (**a**) Topography. Maskelyne's observatories (triangles) were located on the flanks of the mountain. Figure 4 from Playfair (1811) superimposed to indicate extent of Barrow's survey data and Hutton's 1778 modelling (Fig. 4). Vectors show the approximate direction and magnitude of horizontal component of gravity due to topography within a 10 km radius of each field point (constant 2640 kg m^{-3} subsurface density assumed). There is a strong gravitational attraction towards Schiehallion, particularly on the north flank (Fig. 5c). (**b**) Surface density from digitized geological formations (British Geological Survey 2000) each allocated a density according to the mean of density measurements from that formation (table 2 of Smallwood 2007). Thick dotted line indicates the Loch Tay Fault; thick dashed line indicates the boundary between the Argyll and lower density Grampian groups of the Dalradian Supergroup. The thin dashed line shows Playfair's (1811) boundary between quartzite and schistus, which shows broad agreement with contacts from more detailed recent mapping.

the east–west orientation of the mountain. In spite of MacCulloch's attention to Schiehallion and his correct recognition of structural complexity, Playfair's quartzite boundary concurs more closely with modern mapping than that of MacCulloch (Figs 6 & 7). MacCulloch's difficulty may have arisen owing to the regional nature of his mapping and his trouble reconciling his fieldwork onto Arrowsmith's base map (MacCulloch 1836). These difficulties in recording observations on an imperfect base map (MacCulloch 1836; Boud 1974) were not unique. James Nicol's (1810–1879) 1846 (partly derivative) map misplaces the quartzite even further away from Schiehallion, and Geikie (1876) chose to group it as part of the arenaceous Silurian metamorphic succession that he later termed Dalradian (Oldroyd 1990) (Fig. 6).

MacCulloch was somewhat overcome with a melancholy spirit while considering the challenge of representing the subsurface density structure of Schiehallion:

> While the very words are falling from my pen, Dr. [Charles] Hutton is gone where, we trust, all the labyrinths of the universe will be revealed to him; leaving, to mathematicians, a name seldom equalled for science, for utility, never; and, to his friends, the memory of a character adding to that science an unwearied fund of knowledge and conversation, a cheerful and kind disposition, and the simplicity of a child. Smeaton, Maskelyne, Burrowes, Playfair, all are gone. My turn is next. While I write, my pen threatens to stop for ever. It will remain for another to determine the attractions of Schihallien. He too must follow: but the mountain will remain; a monument to its mathematicians, to terminate only with the great globe itself.
>
> (MacCulloch 1824, p. 435)

In this MacCulloch echoed Hutton's final published words, a challenge to successors: 'let any person [...] look over and repeat the calculations [...] and try if he can find any inaccuracy in them' (Hutton 1821, p. 283).

A modern view on the Schiehallion experiment

Two centuries on and it has been possible to overcome MacCulloch's paralysing resignation over geological complexity and to take up Hutton's challenge on accuracy, with the benefit of more detailed and extensive topographic measurements, surface gravity anomaly measurements, geological mapping and a comprehensive set of sample density data around Schiehallion. A fine-scale computer model with Maskelyne's astronomical measurements yields a mean Earth density of $5480 \pm 250\ \mathrm{kg\ m^{-3}}$ (Smallwood 2007; a range largely reflecting standard error on Maskelyne's astronomical observations) matching the currently accepted value ($5515\ \mathrm{kg\ m^{-3}}$). The reasons for Hutton and Playfair's estimates falling low were mainly in the inaccuracies and coarse representation of the topography, and the limited range that was surveyed. The subsurface density variations are very much a secondary factor (table 3 in Smallwood 2007).

In fact, if Playfair's quartzite mean density ($2640\ \mathrm{kg\ m^{-3}}$) is used as a single subsurface density, with properly sampled modern topography data, Maskelyne's observations suggest a mean Earth density of $5620\ \mathrm{kg\ m^{-3}}$. Only surveying resource and calculation time, not technology or theory, were lacking, to prevent this result appearing in 1778. Had such a result been published, it would not have been compelling to search for a Schiehallion experiment follow-on site, with possible implications for the appearance of MacCulloch's map. However, MacCulloch had been seeking a suitable geological project since 1811 when he left Wales, and appeared to be determined to map Scotland (Bowden 2007).

Relations between Playfair, Hutton and MacCulloch

The Schiehallion experiment was a significant part of the scientific careers of Playfair, MacCulloch and Charles Hutton, and it played a role in shaping their relationships.

Hutton's view of Playfair appears to have been strained. His last publication (1821, p. 276) refers to the 'Schehallien experiment [... of] Dr. Maskelyne and myself' and complains that 'it was customary among certain persons to withhold the mention of my name, with regard to the great share that I had in the experiment' (p. 277). While referring to Playfair as an 'excellent philosopher and geologist' (p. 281), that should be taken in the context of Hutton's questionable defence of the Schiehallion experiment over Cavendish's method. Hutton, possibly piqued by Playfair's improvement to his result, and with a somewhat chequered Royal Society history, did not join Maskelyne in supporting Playfair's election to the Royal Society, which was granted in 1807.[12] It appears that Hutton's perception of being frozen out from receiving credit for the Schiehallion experiment may have soured his relationship with Maskelyne in particular, after

[12] Recognizing Playfair's 'eminent knowledge in Mathematics and Natural Philosophy'.

earlier cordial relationships.[13] Perhaps Playfair, a close friend of Maskelyne (Playfair 1822), was caught on Maskelyne's side of the rift; hence Hutton's decision to drop Playfair's techniques and results for his final publication on the issue in 1821. Playfair wrote defending Maskelyne 'he has been accused of sometimes detracting from the discoveries of others when they interfered with his own; I must say, however, that I could never observe any thing of this kind, though I saw him placed in one of those critical situations where envy and jealousy, had they lurked anywhere within him, could scarcely have failed to make their appearance' (Playfair 1782, p. lxxviii).

Conversely, MacCulloch's posthumous eulogy to Charles Hutton, quoted earlier, bears the hallmarks of a deep admiration. This was a tribute indeed, as in his Memoir (1836, p. 19) he uses the term 'mathematician' with implied mild scorn when referring to those on the Board of Ordnance. MacCulloch's first lectures, as chemist at the Ordnance at the Royal Military Academy in Woolwich, in 1804, made him a colleague of Hutton's, who was nearing the end of his tenure as Professor of Mathematics at the time. Notably, Playfair, also recently deceased at the time of MacCulloch's writing (1824), only merited mention by name in the same passage.

The state of the relationship between Playfair and MacCulloch is less clear. MacCulloch studied medicine at Edinburgh from 1790 to 1793 and, although he studied his chemistry under Playfair's friend Joseph Black (Cumming 1983), as a medical student he may not have interacted with Playfair, then the Joint Professor of Mathematics. MacCulloch's criticism of Playfair's attempts to improve the Schiehallion experiment results are typically robust, but fair, and are made in the light of detailed first-hand observations. Similarly, his decision to ignore Playfair's fieldwork in his 1836 Memoir extended to others, and Necker could have felt more justifiably affronted. This should not therefore be considered personal. If nothing else, MacCulloch cannot have been ungrateful for the opportunity afforded to him by Playfair's argument for a repeat mountain experiment to pursue his passion in the creation of a geological map of Scotland.

Playfair's legacy

Although he is remembered primarily as a geologist (for example, Guicciardini 1989), and his *Illustrations* are now considered one of the most conspicuous landmarks in the progress of British geology[14] (for example Challinor 1954), after Playfair's death Francis Jeffrey (1773–1850)[15] commented, 'it is perhaps to be regretted that so much of his time, and so large a proportion of his publications should have been devoted to the subjects of Indian astronomy and the Huttonian theory of the Earth' (Jeffrey 1819, in Playfair 1822, p. lxiii). Playfair maintained a stream of publications throughout his life and left his mark on all the fields in which he had worked: for example, Playfair's axiom, in mathematics, Playfair's law of accordant fluvial junctions in geomorphology (for example, Kennedy 1984), glaciology (Seylaz 1962) and physics,[16] as well as in geology, his style 'a model of purity of diction, simplicity of style, and clearness of explanation' (Playfair 1822, p. xx). Humboldt's acclamation of Playfair's inspirational qualities and writing reassures that these expressions were not merely nepotistic hyperbolae. Humboldt also quoted Playfair's work on isothermal lines as of the greatest value (Playfair 1822). In this distinguished list, what part has the Schiehallion contribution to play in his legacy? I suggest three areas in which this contribution was significant: as the first geophysical model; as a picture of Playfair's research *modus operandi*, reflecting his character; and as one of the springboards for MacCulloch's ground-breaking geological map of Scotland.

The first geophysical model

The face value of Playfair's work on Schiehallion was in improving Hutton's earlier Earth density estimate by characterizing the mountain more accurately, and this it achieved, although the improvement was rather small. In addition to the quartzite, schistus and limestone, Playfair recognized the presence of other lithologies on Schiehallion, for example 'veins and dykes of porphyry and greenstein' (Playfair 1811, p. 353), but was unable

[13] Maskelyne had been on the appointment board for Hutton's professorship in 1773 and the two had obviously worked together well during the initial analysis of the Schiehallion experiment in 1775–1778.

[14] His second edition of the *Illustrations* was not completed at the time of his death. This promised to be a definitive modern-style textbook of observations, including those from his continental trip, deductions and synthesis of geology.

[15] John Playfair's nephew James included this post-obituary appreciation in his 1822 *Works of John Playfair.* The article appeared in *The Times*, and is attributed there and by James Playfair to Francis Jeffrey, long-serving editor of the *Edinburgh Review* and Playfair's friend from the Edinburgh Friday Club.

[16] He left the third volume of his 1814 *Outlines of Natural Philosophy* incomplete. The first covered dymanics, mechanics, hydrostatics, hydraulics, aerostatics and pneumatics, and the second astronomy.

to include them in his mapping and calculations. Limited by manual mathematical calculations and the level of detail of the 1774–1776 topographic survey, his analysis was correspondingly simple, by today's standards, and this hindered the value of the quantitative results (Table 1). However, the real significance of Playfair's work was the represention of distinct areas of lithology linked to a physical property, density, and his calculation of the resulting gravitation response. As such, this renders his work in effect the first gravity survey with a corresponding geophysical model.

A model of Playfair's modus operandi

Playfair's two most substantial contributions to the literature were his edition and extension of Euclid's *Geometry* and his *Illustrations* of Hutton. In both cases, Playfair built on a substantive existing work and added to it through his own insights and style. In that sense, the Schiehallion experiment contribution followed a similar pattern. Maskelyne won the Royal Society's Copley medal for his ingenuity and diligence, and the success of the Schiehallion experiment. Playfair's addition to it certainly progressed the understanding from the experiment, and required innovation and skill to do so, but fell short of equalling the significance of the original results. This may be understood to be in keeping with Playfair's character. He had 'no pleasure in controversial writing' (Playfair 1802, p. 342), which paradigm-shifting work often requires. His nephew wrote of his character that he possessed the 'highest powers [...] kindness of heart, and [...] admiration of genius' (Playfair 1822, p. xxi). His desire to avoid confrontation is exemplified by his treatment of the civil engineer and physicist, John Smeaton (1724–1792), during his 1782 London visit. Although sharing an interest in geology[17] and in the siting of the Schiehallion experiment (Maskelyne 1775*a*), Playfair was obliged to extend the view to Smeaton that his work on loss of 'mechanical power' during collision of bodies was 'by no means new'. Playfair records his attempts at sensitivity in breaking this news: 'This I told him with all the softness I could, imagining that he might be hurt by it, and knowing well, that to tell an author his discoveries are not new is the next thing to telling him they are not true' (Playfair 1782, in Playfair 1822, p. lxxxiv).

A springboard for MacCulloch

Without Playfair's improvement to Hutton's analysis, the inclusion of subsurface density variation, there would have not have followed the step of looking for a more suitable site for the repeat of the plumbline experiment in Scotland. Mason had already examined, and ruled out, several sites in England (Maskelyne 1775*b*). It seems Bouguer's (1749) suggestion of carrying out a plumbline experiment in a valley was not seriously entertained[18] and the search was undertaken for an isolated Scottish mountain with simple density structure. Consequently, MacCulloch was afforded the opportunity to travel the length and breadth of Scotland. This was the perfect activity for him to indulge in his passion for creating a geological map of Scotland (Cumming 1985; Bowden 2009).

MacCulloch was a very early member of the Geological Society of London and following election in 1808 served on the Nomenclature Committee (Woodward 1907) and later as the Society's fourth President (1816–1817). He would have been intimately familiar with the mapping of England and Wales by William Smith (1769–1839) and George Greenough, and was inspired to undertake the mapping of Scotland. It seems unlikely that he was ignorant of Necker's map,[19] presented to the Geological Society in November 1808.

Although the millstone survey was discontinued by order from the Treasury to the Ordnance after the defeat of Napoleon in 1815, Flinn (1981) and Cumming (1984) suggested that MacCulloch's preoccupation with creating the Scotland map appeared to have eclipsed the application of urgency to his official duties in seeking limestone for millstones. Focus on the map also drove him to extend the geological surveys around the meridian line measurement sites beyond the suggested range (Cumming 1983) and to undertake fieldwork at his own expense after 1821 (Cumming 1980). But it was the search for the perfect plumbline experiment peak that afforded him the widest opportunity to compile observations.

Conclusions

Playfair's qualifications to contribute to the Schiehallion experiment were that he had participated in Maskelyne's observations in 1774 and possessed

[17] Smeaton, best known for building the Eddystone Lighthouse, had traced the continuity of the Lias through England and Wales as part of his interest in locating lime that possessed the property of forming a good cement for works exposed to the sea (Smeaton 1791).

[18] Hutton's (1821) suggestion of repeating the experiment at the great pyramid in Egypt was also never pursued.

[19] Greenough annotated MacCulloch's claim of originality for his map with several exclamation marks in his copy of MacCulloch (1836).

the skill both in geological fieldwork and mathematics to carry out both the lithological survey and the complex calculations required to predict the gravitational effect of topography of varying density. The resulting improvement to Hutton's (1778) calculation of Earth density was necessary and tangible, but unfortunately insufficient. Adjusting the Hutton model could only go so far towards understanding why the Maskelyne experiment on Earth density fell short of Cavendish's accurate determination. This was partly because the geological structure of Schiehallion is more complex than Playfair was able to represent in his model, as recognized by MacCulloch, but mainly because the topographic survey was of limited area and insufficiently sampled in the model (Smallwood 2007). Playfair's computations were far from futile, though, as his output was the production of the first geophysical model, in that a physical property, density, was assigned to mapped lithological units and their geophysical, gravitational, response computed for the first time.

I argue here that the Schiehallion experiment contribution was characteristic of much of Playfair's work in that it reflected a desire to build on an existing landmark work with new practical techniques, ideas and conclusions of his own. As his *Illustrations* of Hutton produced a popularization of those views, and led to Playfair's law of branching fluvial valleys, which is still referred to, the spin-off from his Schiehallion work was a significant paper on the gravitational attraction of solids (in 1809). Playfair's approach illustrated that progress in science does not necessarily need confrontation.

Finally, Playfair's earnest request that the plumbline experiment be repeated was taken up by the Board of Ordnance and, in consequence, MacCulloch's geological map of Scotland, the first government-commissioned national geological map, was produced in 1836. Playfair's visit to Schiehallion had significant consequences.

Figures 1, 2 and parts of Figures 3 and 4 are reproduced by permission of the Royal Society. Parts (a)–(d) of Figure 6 are reproduced by permission of the Trustees of the National Library of Scotland. Parts of Figures 5, 6 and 7 are reproduced by permission of the British Geological Survey.
Thanks are due to Alan Bowden and Richard Howarth for their constructive reviews, and to Hess for permission to publish, although the opinions expressed herein do not necessarily represent theirs.

References

BOUD, R. C. 1974. Aaron Arrowsmith's topographical map of Scotland and John MacCulloch's geological survey. *Canadian Cartographer*, **11**, 24–34.

BOUGUER, P. 1749. *La Figure de la Terre, déterminée par les observations de MM. Bouguer et La Condamine*, in 4°. Joubert, Paris.

BOWDEN, A. J. 2007. Book review of MacCulloch's 1840 Geological Map of Scotland, facsimile edition, BGS Edinburgh. *Scottish Journal of Geology*, **43**, 181–184.

BOWDEN, A. J. 2009. Geology at the crossroads: Aspects of the geological career of Dr John MacCulloch. *In*: LEWIS, C. L. E. & KNELL, S. J. (eds) *The Making of the Geological Society of London*. Geological Society, London, Special Publications, **317**, 1–18.

BRITISH GEOLOGICAL SURVEY. 1986. *Tay-Forth Sheet 56°N–04°W. 1:250 000 Series, Solid Geology.* British Geological Survey, Keyworth, Nottingham.

BRITISH GEOLOGICAL SURVEY. 1987. *Argyll Sheet 56°N–60°W. 1:250 000 Series, Solid Geology.* British Geological Survey, Keyworth, Nottingham.

BRITISH GEOLOGICAL SURVEY. 2000. *Schiehallion, Scotland Sheet 55W, Solid Geology, 1: 50 000.* British Geological Survey, Keyworth, Nottingham.

CAVENDISH, H. 1798. Experiments to determine the density of the Earth. *Philosophical Transactions of the Royal Society of London*, **88**, 469–526.

CHALLINOR, J. 1954. The early progress of British geology; III, From Hutton to Playfair, 1788–1802. *Annals of Science*, **10**, 101–148.

COCKBURN, H. T. 1852. *Life of Lord Jeffrey, with a Selection of his Correspondence*, Volume 1. Lexden, London.

CONYBEARE, W. & PHILLIPS, W. 1822. *Outlines of the Geology of England and Wales.* Phillips, London.

CUMMING, D. A. 1980. John MacCulloch, F.R.S., at Addiscombe: The lectureships in chemistry and geology. *Notes and Records of the Royal Society of London*, **34**, 155–183.

CUMMING, D. A. 1983. *John MacCulloch, Pioneer of 'Precambrian' Geology.* PhD dissertation, University of Glasgow.

CUMMING, D. A. 1984. John MacCulloch's 'Millstone Survey' and its consequences. *Annals of Science*, **41**, 567–591.

CUMMING, D. A. 1985. John MacCulloch: Precursor of the British 'Geological Survey'. *Geology Today*, **1**, 124–125.

EYLES, V. A. 1948. Louis Albert Necker of Geneva, and his geological map of Scotland. *Transactions of the Edinburgh Geological Society*, **14**, 93.

FLINN, D. 1981. John MacCulloch M.D. F.R.S. and his geological map of Scotland: His years in the Ordnance, 1795–1826. *Royal Society of London Notes and Records*, **36**, 83–101.

GEIKIE, A. 1876. *Geological Map of Scotland, the Topography by T. B. Johnston.* W. & A. K. Johnston, Edinburgh.

GUICCIARDINI, N. 1989. *The Development of Newtonian Calculus in Britain 1700–1810.* Cambridge University, Cambridge.

HALLAM, H. 1843. Biographical notice of Lord Webb Seymour. *In*: HORNER, L. (ed.) *Memoirs and Correspondence of Francis Horner MP*, Volume I. John Murray, London.

HOWARTH, R. J. 2007. Gravity surveying in early geophysics II: From mountains to salt domes. *Earth Sciences History*, **26**, 229–261.

HUMBOLDT, A. VON. 1850. *Cosmos, A Sketch of a Physical Description of the Universe*, Volume 2. Translated by E. C. Otte. Harper & Collins, New York.

HUTTON, C. 1778. An account of the calculations made from the survey and measures taken at Schehallien, in order to ascertain the mean density of the Earth. *Philosophical Transactions of the Royal Society of London*, **68**, 689–788.

HUTTON, C. 1821. On the mean density of the Earth. *Philosophical Transactions of the Royal Society of London*, **111**, 276–292.

JEFFREY, F. 1819. Some account of the character and merits of the late Professor Playfair (from *The Times*). *In*: PLAYFAIR, J. G. (ed.) *The Works of John Playfair, Esq.*, 1822, Volume 1. Archibald Constable, Edinburgh, lx–lxxvi.

KENNEDY, B. A. 1984. On Playfair's law of accordant junctions. *Earth Surface Processes and Landforms*, **9**, 153–173.

LEADSTONE, G. S. 1974. Maskelyne's Schiehallion experiment of 1774. *Physics Education*, **9**, 452–458.

LOWRIE, W. 1997. *Fundamentals of Geophysics*. Cambridge University Press, Cambridge.

MACCULLOCH, J. D. 1819. *A Description of the Western Islands of Scotland*. A. Constable & Co. London.

MACCULLOCH, J. D. 1823. *A Description of the Scenery of Dunkell and of Blair of Atholl*. J. Mallet, London.

MACCULLOCH, J. D. 1824. *The Highlands and Western Isles of Scotland [...] Contained in Six Letters to Sir Walter Scott* (4 vols). Longman, London.

MACCULLOCH, J. D. 1831. *A System of Geology* (2 vols). Longman, London.

MACCULLOCH, J. D. 1836. *Memoirs to His Majesty's Treasury respecting a Geological Survey in Scotland*. Samuel Arrowsmith, London.

MACCULLOCH, J. D. 1840. *A Geological Map of Scotland by Dr. MacCulloch*. G. F. Cruchley, London.

MASKELYNE, N. 1775*a*. A proposal for measuring the attraction of some hill in this kingdom by astronomical observations. *Philosophical Transactions of the Royal Society of London*, **65**, 495–499.

MASKELYNE, N. 1775*b*. An account of observations made on the mountain Schehallien (sic) for finding its attraction. *Philosophical Transactions of the Royal Society of London*, **65**, 500–542.

NECKER, L. A. 1939. *Scotland, Coloured According to the Rock Formations, Presented to the Geological Society 1808*. Edinburgh Geological Society, Edinburgh.

NICOL, J. 1846. *Geological Map of Scotland*. W. Blackwood & Sons, Edinburgh.

OLDROYD, D. R. 1990. *The Highlands Controversy: Constructing Geological Knowledge Through Fieldwork in Nineteenth-Century Britain*. University of Chicago, Chicago, IL.

PLAYFAIR, J. 1782. Journal. In: PLAYFAIR, J. G. (ed.) *The Works of John Playfair, Esq.*, Volume 1. Archibald Constable, Edinburgh, Appendix No. 1. lxxvii–lxxxviii.

PLAYFAIR, J. 1802. *Illustrations of the Huttonian Theory of the Earth*. Cadell & Davies, London and William Croack, Edinburgh.

PLAYFAIR, J. 1809. Of the solids of greatest attraction, or those which, among all the solids that have certain properties, attract with the greatest force in a given direction. *Transactions of the Royal Society of Edinburgh*, **6**, 187–243.

PLAYFAIR, J. 1811. An account of the lithological survey of Schehallion. *Philosophical Transactions of the Royal Society of London*, **101**, 347–378.

PLAYFAIR, J. 1822. Biographical account of the late Professor Playfair. *In*: PLAYFAIR, J. G. (ed.) *The Works of John Playfair, Esq.*, Volume 1. Archibald Constable, Edinburgh, xi–lix.

PLAYFAIR, J. G. 1812. Review of Essai sur la géographie mineralogique des environs de Paris by G. Cuvier and Alex. Broigniart, Paris 1811. *Edinburgh Review*, **20**, 369–386.

RANALLI, G. 1984. An early geophysical estimate of the mean density of the Earth: Schehallion, 1774. *Earth Sciences History*, **3**, 149–152.

ROSE, E. P. F. & ROSENBAUM, M. S. 1998. British military geologists through war and peace in the 19th and 20th centuries. *In*: UNDERWOOD, J. R., JR & GUTH, P. L. (eds) *Military Geology in War and Peace*. Geological Society of America, Reviews in Engineering Geology, **13**, 29–39.

SEYLAZ, L. 1962. A forgotten pioneer of the glacial theory: John Playfair (1748–1819). *Journal of Glaciology*, **4**, 124–126.

SMALLWOOD, J. R. 2007. Maskelyne's 1774 Schehallion experiment revisited. *Scottish Journal of Geology*, **43**, 15–31.

SMEATON, J. 1791. *A Narrative of the Building and a Description of the Construction of the Eddystone Lighthouse*. H. Hughs, London.

STEPHENSON, D. & GOULD, D. 1995. *British Regional Geology: The Grampian Highlands* (4th edn). HMSO for the British Geological Survey, London.

TREAGUS, J. E. 2000. *Solid Geology of the Schiehallion District*. Memoir of the British Geological Survey, Sheet 55W (Scotland).

WEBB SEYMOUR, J. & PLAYFAIR, J. 1815. Account of Observations [...] upon some geological appearances in Glen Tilt. *Transactions of the Royal Society of Edinburgh*, **7**, 302.

WILSON, D. K. 1935. *The History of Mathematical Teaching in Scotland to the End of the Eighteenth Century*. University of London, London.

WOODWARD, H. B. 1907. *The History of the Geological Society of London*. Geological Society, London.

Picturesque ruin and geological antiquity: Thomas Webster and Sir Henry Englefield on the Isle of Wight

NOAH HERINGMAN

Department of English, 107 Tate Hall, University of Missouri, Columbia, MO 65211, USA (e-mail: HeringmanN@missouri.edu)

Abstract: Thomas Webster published his earliest geological observations by commission in Sir Henry Englefield's *Description of the Isle of Wight* (1816), a work concerned as much with historic architecture and picturesque landscape as with geology. This paper shows how Englefield's broad three-part agenda fostered the development of Webster's specifically geological competence and sensibility. As a professional draftsman and architect, Webster was especially well equipped to translate Englefield's architectural and picturesque idiom into a more geological register. Their collaboration also illustrates how well the style and content of local history – a traditional literary and learned genre – could be applied to geology. For Webster in particular, the image of ruins was essential for representing the historicity of geological phenomena. This paper adduces close readings of numerous passages from Englefield and Webster's work to show how strongly the traditional language and research questions of antiquarianism continued to shape geology even as it became a professional specialization.

In May 1811 a budding field geologist wrote to his employer from the Isle of Wight: 'the whole of the country between [the Undercliff] and the sea, had, apparently, at some remote period, been one immense mass of ruin, though now covered with woods, cornfields, and villas. It was impossible to view this scene, and to contrast its present with its former state, without feelings of interest' (Englefield & Webster 1816, p. 129).[1] This aesthetic discourse on a landscape, eloquent as it is, would be unremarkable for its time were it not that the author, Thomas Webster (1772–1844), was one of the first to refer to himself as a 'professional geologist'.[2] Webster's major geological publications rely upon the picturesque and the sublime, as well as an antiquarian discourse of ruin, to establish analogies for geological processes and geological time. As Ralph O'Connor (2007) has recently argued, these and other literary devices served geology well at least up to the 1850s. Webster's research, in context, illustrates very clearly the continued importance of aesthetic categories even as geology parted ways with Romanticism and conjectural history. His geology profits as well from its close affiliation with archaeology.

Thomas Webster was a member, a salaried employee and eventually a Fellow of the Geological Society of London from 1809 to 1826. His 'On the freshwater formations in the Isle of Wight' (1814*a*) stands out among early contributions to the *Transactions of the Geological Society* for several reasons. One is the high degree of scientific rigour that makes Webster's analysis still compelling today.[3] Another is his delicate handling of the war-born tensions accompanying any geological comparison across the English Channel during the Napoleonic Wars, which were still at their height when he began his research in 1811 (Heringman 2009). This paper examines another striking feature of Webster's writing: its strong affiliation with antiquarianism and local history.

His influential research on the Isle of Wight was commissioned by the author of an antiquarian and topographical work on that locale by Sir Henry Englefield (1752–1822). Webster's contribution to the book, actually completed in 1813 and before

[1] This work will be cited by page number only for the remainder of the essay.

[2] Webster identified himself as 'Thomas Webster, F. G. S. &c. Professional Geologist, and Lecturer on Geology' in his *Proposal for Publishing (by Subscription) A Description of the Isle of Wight*, a unique copy of which was recently discovered by Hugh Torrens in a collection made by Dawson Turner (Prospectus, p. 1). Webster's prospectus is bound in Vol. 4 of this collection, which carries the manuscript title 'Prospectuses of Books, Engravings, Lithographs, &c. collected, with the view, not only of showing the state of the literature, arts, & sciences of the passing day, but as too often illustrative of the vanity of human wishes, etc'.

[3] A field trip organized to celebrate the Society's bicentenary in 2007 visited the Isle of Wight and, in particular, the key sites of Webster's fieldwork.

From: LEWIS, C. L. E. & KNELL, S. J. (eds) *The Making of the Geological Society of London*. The Geological Society, London, Special Publications, **317**, 299–318.
DOI: 10.1144/SP317.17 0305-8719/09/$15.00

his essay in the *Transactions*, shows how deeply his geology was influenced by the broader concerns of Englefield's *Description of the Principal Picturesque Beauties, Antiquities, and Geological Phaenomena, of the Isle of Wight* (1816). As a professional draftsman hired by Englefield, the Geological Society and other patrons, Webster brought to his geological work a different intellectual background and different objectives from those of the paying members of the Geological and other learned societies. Webster's unique abilities and his socioeconomic position visibly shape both the 1814 essay and his extensive contributions to Englefield's book.

As a 'knowledge worker' whose various fields of expertise – architecture, agriculture, drawing – were called for on different occasions by different employers, Webster was apt to draw on other literary forms of local history for his geological essays, just as he took notice of geology and antiquities simultaneously when working as a draftsman on the Isle of Wight.[4] Sir Henry Englefield was a long-time Vice-President of the Society of Antiquaries as well as a Fellow of the Royal Society and had correspondingly wide-ranging interests, although he published only two short geological essays.[5] When Englefield decided to turn his notes and sketches from his earlier journeys to the Isle of Wight (1799–1801) into a handsome illustrated folio he called on the talents of Webster – an established artist who had been developing a particular interest in geology via commissions from Humphry Davy at the Royal Institution (Siegfried & Dott 1980, pp. xxxii–xxxiii and p. 156 n.13). As an experienced architect and topographical draftsman, Webster had a knowledge of architectural antiquities that would have appealed to Englefield, who supervised the Society of Antiquaries' programme of publishing engravings of the English cathedrals (Sweet 2004, pp. 302–303). At the same time Webster's growing geological expertise made him an ideal contributor who could confirm and develop the connections between human and geological antiquity implied by the title of Englefield's book. Martin Rudwick has recently argued that the emergence of primary-source-oriented antiquarianism or 'erudite history' strongly influenced the evidentiary turn in geohistorical science (Rudwick 2005, p. 182). The appearance of path-breaking geological research in a fashionable antiquarian work of local history shows how this symbiosis continued to operate in the nineteenth century.

By comparing Webster's writing with Englefield's in the context of local history, I hope to show how the relationship between geology and antiquity persisted in the era of advancing professionalization. At the same time, I will examine Webster's own professionalism: by way of his influential research on the Isle of Wight, Webster capitalized on his employment as Keeper of the Museum and Draftsman to the Geological Society to reinvent himself as a professional geologist.

Local history, fieldwork and collaboration

The pre-historian Stuart Piggott has argued that a new empiricism infused the English county histories of the period 1790–1820. Displacing the 'irrational beliefs of Gothick romance' that had, in his view, corrupted antiquarian research in the mid-eighteenth century (Piggott 1976, p. 120), it renewed the rigour of early Baconian investigators such as Robert Plot and Edward Llwyd. Piggott particularly praises Richard Colt Hoare's *Ancient Wiltshire*, published from 1806 to 1821 under the epigraph 'we speak from facts, not theory.' Piggott's archaeological bias leads him to distinguish

[4] The biographical literature on Webster is limited, although there is no shortage of primary materials, including numerous manuscript letters, some still unpublished, and a manuscript autobiography housed at the Royal Institution, utilized so far only by Ironmonger (1958). Other sources on Webster's life and career include the articles by Challinor (1961, 1963–1964), Edwards (1971, 1972, 2004) and Kirk (1996), most of which consist largely of the texts of Webster's letters. Some 46 of these have now been published, together with 124 letters to him. Discussions of his geological work include Rudwick (2005, pp. 512–521) and Torrens (1983, pp. 138–139). Webster continues to be of interest to geologists because, as Kirk (1996) has put it, 'he made major contributions both to the understanding of Jurassic and Cretaceous stratigraphy, and to the interpretation of the structure of the area [SE], which was well ahead of its time' (pp. 311–312).

[5] The scholarship on Englefield is even more limited. Apart from the *ODNB* entry by Bernard Nurse, there are only passing references in Sweet (2004) and other histories of antiquarianism. Englefield's essay on flints embedded in chalk and a brief continuation appeared in the *Transactions of the Linnaean Society* (1800, 1801) and arose from his observations on the Isle of Wight. Englefield published occasionally in the *Philosophical Transactions*, more regularly in the *Archaeologia*, and wrote the commentary for six numbers of the Society of Antiquaries' cathedral series. The work with which this essay is mainly concerned, the *Description*, was reviewed by John Playfair in the *Edinburgh Review*, and anonymously in the *Augustan Review*, which places great (pejorative) emphasis on Englefield's Catholicism. There is a brief but useful discussion of the *Description* in Freeman (2004, pp. 14–17).

excavation and material culture categorically from manuscripts and archival sources of evidence, and he even suggests that Earth science went into decline between Hooke and Whitehurst because of the way in which texts came to dominate historical study (Piggott 1976, p. 117). Piggott seems somewhat too ready to dismiss textual and archival research, and Rudwick, by contrast, has emphasized its importance for geology. In fact, both archaeological and philological research began to display new levels of empirical rigour in the later eighteenth century, as richly illustrated by the journal *Archaeologia*, founded in 1771 by the Society of Antiquaries. The founding of the journal 64 years after the Society's revival itself suggests an increasing scientific orientation. A variety of early contributions to the journal register these changing standards of evidence and authentication. In his career with the Antiquaries, Englefield was closely associated with Richard Gough, the society's director and the main force behind the journal. They were allies in the embattled cause of preserving the original architectural features of England's Gothic cathedrals. The Society's research on cathedrals – much of it performed by hired researchers similar to Webster – promoted excavation and structural investigation. Piggott's emphasis on the new county histories helps to explain the connection between this kind of archaeological work and the new geology. Geological fieldwork *is* local history, and it makes sense that Englefield's concern with local archaeological evidence in particular would have led him to stray into the precincts of 'that part of natural science lately called Geology', as he terms it in his preface (p. i). The terminology is new, but the *Description* also inherits the synthesis of natural history and antiquarianism that characterized the literary genre of local history from at least the 1670s (Piggott 1976, chap. 6).

Englefield's unexpected detour into geology was – or might have been – the making of Webster's career. Webster's Isle of Wight research made him a recognized authority and led to further publication, but not to the financial success that he must reasonably have expected (Kirk 1996, pp. 313 and 315). Englefield (Fig. 1) was a connoisseur who published extensively on Gothic architecture, as well as on astronomy and other subjects; he

Fig. 1. *Vases from the Collection of Sir Henry Englefield* (1820), part of frontispiece. Portrait of Englefield engraved by Henry Moses from his own drawing. Courtesy of Ellis Library, University of Missouri-Columbia.

was also Secretary of the Society of Dilettanti and had his vase collection illustrated by another draftsman, Henry Moses, in 1820 (Fig. 2). His recognition, in 1811, that 'that part of natural science' might nonetheless be beyond his purview, was by no means inevitable and neither was his decision to employ Webster to complete the geological part of the book. Englefield explicitly credits Webster as co-author on the title page and cites Webster as an authority repeatedly in his own part of the book.[6] By contrast, John Carter and Jacob Schnebbelie, who did much of the fieldwork and visual documentation for the Society of Antiquaries' series on cathedrals, received credit for their illustrations but only rarely for their research and never as authors (although both eventually published their own work). Webster, like Carter, was trained as an architect and architectural draftsman, but his professional engagements had led him to natural science rather than antiquities: Webster designed the lecture hall of the Royal Institution in

[6] Here is the full text of the title page: 'A Description of the Principal Picturesque Beauties, Antiquities, and Geological Phoenomena of the Isle of Wight. || By Sir Henry Englefield, Bart. || With Additional Observations on the Strata of the Island, and Their Continuation in the Adjacent Parts of Dorsetshire. || By Thomas Webster, Esq. || Illustrated by Maps and Numerous Engravings by W. and G. Cooke, from Original Drawings by Sir H. Englefield and T. Webster. || London: || Printed by William Bulmer and Co. Cleveland-Row, St. James's, for Payne and Foss, 88 Pall-Mall. 1816.' No information is available concerning Webster's financial arrangement with Englefield. Hugh Torrens has suggested (pers. comm.) that newly discovered Webster letters at Southampton University may shed some light on this question.

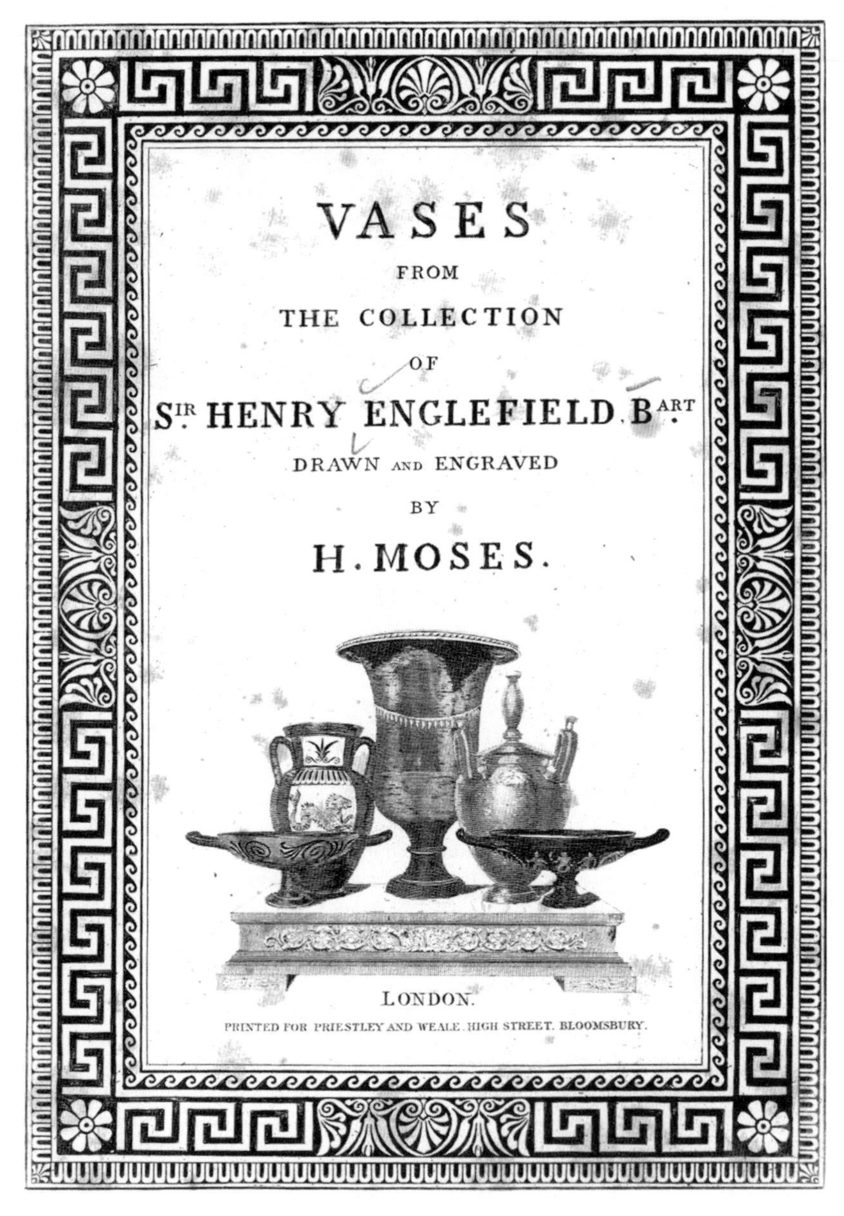

VASES
FROM
THE COLLECTION
OF
Sr. HENRY ENGLEFIELD Bart.
DRAWN AND ENGRAVED
BY
H. MOSES.

LONDON.
PRINTED FOR PRIESTLEY AND WEALE. HIGH STREET. BLOOMSBURY.

Fig. 2. *Vases from the Collection of Sir Henry Englefield* (1820), title page. Engraved and drawn by Henry Moses. Courtesy of Ellis Library, University of Missouri-Columbia.

Albemarle Street and he supplied panoramic paintings for Humphry Davy's 1805 lectures in that theatre, the first English course of public lectures in geology (Siegfried & Dott 1980, p. 156 n. 13; reproduced in O'Connor 2009). While at the Royal Institution, he came into contact with Englefield and with Arthur Aikin, who, along with Davy, was one of the founders of the Geological Society. Hugh Torrens has suggested that Aikin probably gave Webster his first exposure to geological fieldwork (Torrens 1983, pp. 138–139). In 1812 Webster became Draftsman to the Society and Keeper of the Museum. As the society's first salaried officer, Webster embraced geology both as a vocation and as a plausible professional path. In 1826 he elected to pursue that path outside the

Geological Society when they refused him a raise and promotion after 14 months of negotiations (Edwards 1971, p. 472 n. 58–59).

Webster's research on the Isle of Wight, directed and subsidized by Englefield, led to two publications in his own name in the second volume of the Society's *Transactions*: the very substantial essay on the Freshwater formation and a shorter piece on a new species of fossil Alcyonium, published with Englefield's consent. At the same time, Webster's geological expertise helped to justify the publication of Englefield's book, making it more competitive in a rather saturated corner of the literary marketplace. Englefield acknowledges as much in his preface, citing Webster's expertise, along with his own originality, to justify the publication in 1816 of observations made 15 years before. Englefield warms gradually to Webster's praise, noting at first only that his own manuscript work from the turn of the century, especially the geological observations, still has not been 'superseded by the labours of later travellers', prompting him to extend and publish the research:

> As I felt my own strength and activity unequal to the exertions required, I requested Mr. Webster, who was not unacquainted with natural history, and who, to the talents of an expert draftsman, added the more valuable qualities of most scrupulous accuracy and patient investigation; to revisit those parts of the Island as I had examined imperfectly [...] and to extend his researches into that part of Dorsetshire where it was evident that the great chalk range of the Isle of Wight reappeared, after being cut off by the sea for near 12 miles.
>
> (p. ii)

Having begun in this somewhat patronizing tone, Englefield goes on to credit Webster with 'unparalleled' discoveries and declares that Webster's observations 'constitute the most valuable part of the book', fully deserving of the explicit authorial credit reflected in the title page and organization of the book (Webster wrote just over half of it). Finally, Englefield notes that he has given 'the whole of [Webster's] researches in his own words' and recommends Webster's 'masterly descriptions' to his readers (p. iii), having prepared them carefully to accept the authority of a mere draftsman.[7]

Although Webster's geological essay has justifiably become the standard for geological readers, it is important to read this paper side by side with Englefield's book, both to recover the antiquarian spirit that motivated Webster's research and to analyse the changes that Webster made to adapt his work to a geological audience. Both works display a variety of literary and visual conventions. Englefield's book, as denoted by its price of 10 guineas, occupied the niche of the antiquarian plate book. In this work he does not aspire to the archaeological rigour extolled by Piggott in the case of Hoare's *Ancient Wiltshire*, but does evince a well-schooled scientific curiosity in both archaeological and geological matters, and the overall difference in style and content between Englefield's and Webster's parts of the book is less than one might imagine. The most important difference is Webster's ability to enumerate and correlate the strata (and observe their dip), which he does in verbal as well as tabular and cartographical form. Webster also made extensive observations by boat, which, as he notes in his first letter, was essential for observing 'the true dip of the strata' (p. 120). These features, and his concluding table, 'Order of the Upper Strata', in particular, anticipate the larger scale of Webster's conclusions in their final form in the *Transactions* (Webster 1814*a*). Here, he applies his observations on the local history of the Isle of Wight and the corresponding parts of Dorset to the Chalk formation as a whole. These habits of correlation and comparison, and, following Englefield, simultaneous attention to geology, antiquities and the picturesque, encourage the conceptual movement from the local observations to regional structures. Both antiquarianism and picturesque theory promote attention to recurring visual patterns, relative dating and geological situation.

Webster's ruined arches: geological ruin in its architectural and social contexts

Webster himself makes one particularly explicit analogy between archaeological and geological forms:

> In connecting together, and as it were restoring (to use the language of the antiquary) this series of strata, which, though now in ruins, was probably much more entire in some former condition of the earth, a singular conclusion presents itself to the imagination [...] with nearly the same certainty which we feel in putting together the fragments of an ancient temple. The conclusion to which I allude is, that the chalk of the middle range, together with the clay of Alum Bay

[7] It is a little surprising that Englefield does not mention Webster's credentials as a geological draftsman. Webster had done significant illustration work for Humphry Davy's geology lectures and contributed illustrations to John Pinkerton's *Mineralogy: A Treatise of Rocks* (1811). Pinkerton was, like Englefield, primarily an antiquary. Englefield might not have known of Webster's work for Aikin and could not have known that many of these illustrations would eventually be published in Roderick Murchison's *The Silurian System* (Torrens 1983, pp. 138–139). Webster also 'draft[ed] the topographical base' of G. B. Greenough's 1819 *Geological Map of England and Wales* (Kirk 1996, p. 312).

> and the inferior strata, once formed a sort of arch or vault, by which it joined to the horizontal strata of the hills of St. Catherine and Shanklin.
>
> (p. 201)

This passage clearly illustrates the difference between the older antiquarian premise of continuity between natural and human history, still defended by John Whitehurst in 1786, and the self-conscious use of historical references as not continuous with but analogous to pre-human events whose antiquity is of a higher order of magnitude. Rudwick (2005, p. 517) has noted that this passage and the accompanying section focus attention on geohistorical sequence in place of conjectural causes, and especially on the astonishingly recent date of this 'arch' or fold structure. The antiquarian context of Webster's analogy thus merits careful attention. The 'denudating causes' responsible for removing the top of Webster's imagined arch (Fig. 3) are directly analogous to the weathering that has removed and obscured the vestiges of beacons, for example, formerly set atop early medieval structures along the coast, and are linked in a more general way to the poor state of preservation of many of the region's architectural antiquities. Englefield encouraged Webster to return to the Isle of Wight in the summer of 1812, primarily to extend his correlation of strata between the Isle of Wight and the Dorset coast west of Purbeck. Twenty of the *Description*'s 50 plates are devoted to the Dorset coast, and four of them depict an antiquarian site that Englefield gave Webster particular instructions to examine in 1812, because it had not been illustrated before (p. 191).

Webster's writing on this site, St Adhelm's (now Aldhelm's) Chapel, beautifully exemplifies his stylistic amalgamation of antiquarian, geological and picturesque elements. The chapel resides amidst the 'grand' and 'immense' ruins of the two lowest strata in Webster's column, not exposed at all on the Isle of Wight:

> St. Adhelm's Head, [. . .] whose dangerous rocks have been so often fatal to mariners, is a vertical section of the oolite, and of the argillaceous strata below it. [. . .] deep fissures in the upper part of the cliff predict the impending ruin of many places, and the agitation of the sea for above a mile from the shore, shows the nature of its rocky bottom, the remains, no doubt, of land which has gradually been covered by the ocean.
>
> (pp. 188–189)

The chapel itself (Fig. 4), displaying formal properties of 'the simplest kind of Saxon architecture' (p. 189), bears marks of such 'high antiquity' that the unusual forms of the arches supporting the structure initially appeared to Webster to be a result of deformation over time ('high antiquity' is another phrase that recurs in Webster's geological descriptions). On closer inspection he realized that these arches were not ruined, but were originally 'composed of portions of circles having their centres a little above the springing of the arch, which gives it something of the Moorish character, a circumstance rare, though not unique, in Saxon architecture' (p. xxiv; cf. p. 190). The cross vaults supported by these arches also made an impression on Webster with their 'extremely elegant' simplicity, and in his conclusion he relied on the term 'vault' (p. 205) to underscore the three-dimensional

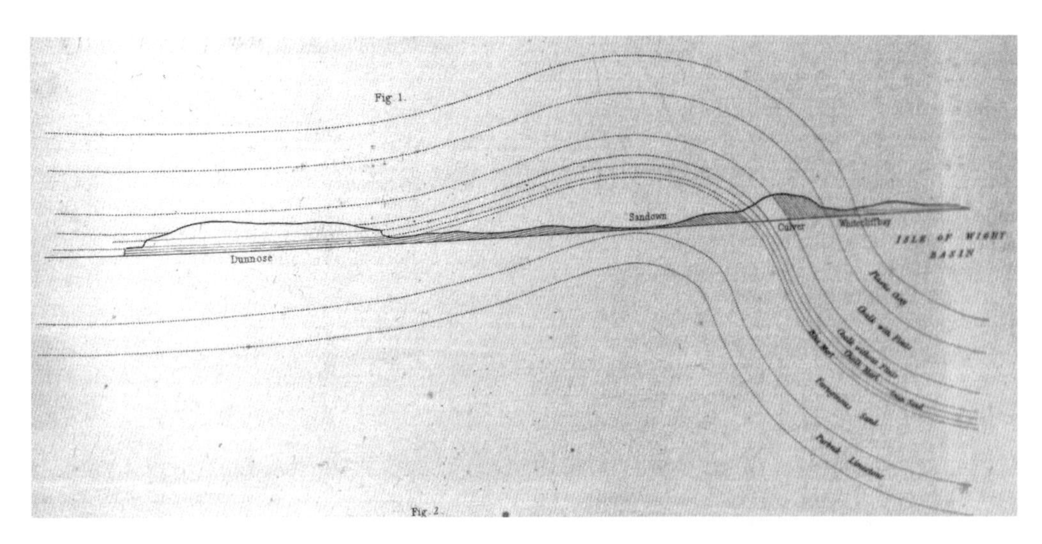

Fig. 3. *Description of the Isle of Wight* (1816), plate XLVII, Fig. 1: theoretical section or 'ruined arch'. Engraving by W. and G. Cooke from a drawing by T. Webster. Courtesy of Howard-Tilton Memorial Library, Tulane University.

Fig. 4. *Description of the Isle of Wight* (1816), plate XLIII. Engraving by W. and G. Cooke from a drawing by T. Webster. Courtesy of Howard-Tilton Memorial Library, Tulane University.

character of his hypothetical ruined arch or fold structure.

The recent discovery by workmen of a cylindrical base – presumably used to hold a warning beacon for vessels – on the roof of the chapel especially fired Webster's imagination. Echoing Englefield's description of a similar structure on the Isle of Wight, Webster imagined the 'religious inhabitant' of this 'chauntry', as he called it, praying for the safety of distressed mariners as he shuddered at the 'dismal scene' visible through the structure's one window (p. 190). Webster's initial description of the cliff, then, evokes the perceptions of this early medieval anchorite, while the oddly melancholy contemplation of the lone fisherman lounging in the converted chapel – now a storehouse – preserves the historical mood and at the same time provides a visual echo of the labouring figures featuring so prominently in many of Englefield's drawings (engraved as plates I–XII) and in other illustrations of Webster's as well.

These figures must be understood as both picturesque and antiquarian elements. In some cases, where the figures indicate the scale of the cliffs, they are geological elements as well. In fact, the picturesque convention, deriving primarily from the landscapes of Salvator Rosa, of rural labourers or banditti dwarfed by craggy landscapes, is likewise an important art historical legacy for geological representation (Gilpin 1792, pp. 20–21; Bermingham 1986, pp. 81–83). The antiquarian aspect of these figures derives less from their scale than from their costume, and their illustration of customs and manners. While the mood of the lounging figure may evoke the melancholy life of Webster's imagined anchorite, the baskets and other fishing gear speak clearly of this occupant's worldly vocation. Many of the illustrations, by both Englefield (Fig. 5) and Webster (Fig. 6), feature fishermen, their boats and their equipment, although they are rarely mentioned in the text. Webster mentions them twice as sources of data (pp. 147 and 156), and he also includes quarrymen and their technology in two plates (for example, Fig. 7), citing their methods more than once for the light they shed on the stratigraphy or lithology of a particular locale.

Richard Gough's preface to his monumental *Sepulchral Monuments of Great Britain*

Fig. 5. *Description of the Isle of Wight* (1816), plate VI. Engraving by W. and G. Cooke from a drawing by H. Englefield. Courtesy of Howard-Tilton Memorial Library, Tulane University.

(1786–1799) offered one vivid defence of the value of dress, and material culture more broadly, for the study of the past. Gough valued the effigies in Gothic cathedrals particularly, as almost the only sources from which 'we can deduce such parts of the history of past times' as fashion, customs and manners (Gough 1799, vol. I, p. 1). Englefield brought a similar emphasis to the Society of Antiquaries' series on the cathedrals, and the letters of Gough's draftsman, Jacob Schnebbelie, document his attention to sartorial detail in drawing these effigies.[8] The closely parallel emphasis on native costume in the ethnographic drawings of Sydney Parkinson on the *Endeavour* (1768–1771), in the very different arena of the South Pacific, points to the still underrecognized importance of antiquarianism as a framework for early ethnography, which in its broadest sense also includes the labouring figures in Englefield's *Description.*

Like Parkinson in the South Pacific, Webster had to resist picturesque convention to portray his object of study more accurately, and his figures become more than merely conventional in this light. At Culver Cliff (Fig. 8) he reported that his point of view 'was chosen rather with a view to shew the real dip of the strata than as the most picturesque' (p. xi) and the same logic is implicit in many of the views, although at times the two objectives coincide. Unlike Parkinson and Schnebbelie, Webster did not die young of the hazards attending his profession, although he might easily enough have been swept off the stump of rock on which he sat for two hours, becoming soaked to the skin, while taking both his spectacular view of the Needles and the view of Alum Bay that was included on the volume's only colour plate (Fig. 9). The Needles image in particular is picturesque as well as accurate, and taken from the only station available. In defending Webster's scientific merits, Englefield rightly insists that 'in no instance has accuracy been sacrificed to the effect of the engraving' (p. iii), but effect was nonetheless crucial to Webster's art. He made use of picturesque conventions quite freely both in his prose and in his

[8] SAL: Antiquarian Society MS. 267, fol. 6: Letters from and to Mr Jacob Schnebbelie, Draughtsman to the Society of Antiquaries.

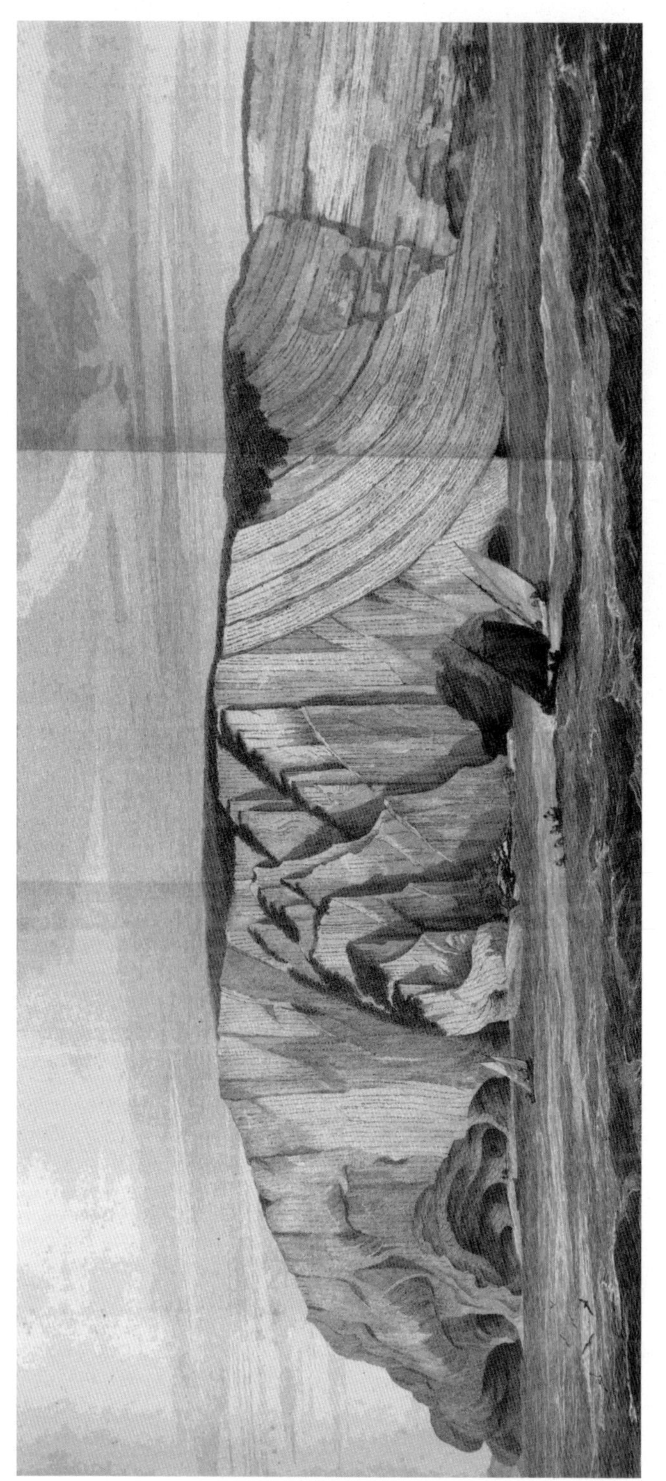

Fig. 6. *Description of the Isle of Wight* (1816), plate XXVII. Engraving by W. and G. Cooke from a drawing by T. Webster. Courtesy of Howard-Tilton Memorial Library, Tulane University.

Fig. 7. *Description of the Isle of Wight* (1816), plate XXXIII. Engraving by W. and G. Cooke from a drawing by T. Webster. Courtesy of Howard-Tilton Memorial Library, Tulane University.

drawings, as we have seen in the case of the lounging fisherman, when they did not interfere with a stated scientific objective. Lulworth Cove in Dorset, for example, 'unit[es] great picturesque beauty with singularly instructive geological phenomena' (p. xxiii), while the columns of erosion-resistant chalk off Handfast Point 'render it somewhat dangerous to pass, but add to the picturesque effect of the place' (p. xviii). Englefield, however, was more apt to note technical 'defects' in the island's picturesque beauty, which result in part from a deficiency in landed property, but he too waxes eloquent on its sublime scenes such as Alum Bay: 'the Needle rocks rising out of the blue waters, continue the cliff in idea, beyond its present boundary, and give an awful impression of the stormy ages which have gradually devoured its enormous mass [...] all is wild ruin' (p. 81).

Knowledge work among the ruins

Englefield's portion of the book is in conventional chapters, unlike Webster's letters, which appear more or less in their original form, in part surely to reinforce the authenticity of the fieldwork reported in them. He devotes his longest chapter to picturesque beauties, but in doing so he creates a fairly consistent line of demarcation between the aesthetics of landscape and the science of antiquities, which (like geology and other more objective domains) is reserved for a separate chapter. His chapter on the picturesque is full of allusions to and borrowings from the classic authors of picturesque theory and travel: William Gilpin, Richard Payne Knight, Sir Uvedale Price and others. Although he names none of these precursors explicitly, he shares the technical vocabulary of 'vistas', 'perspectives' and 'terminations' (p. 56) with Gilpin and his disdain for the landscape designs of Capability Brown (pp. 87–88) with Price and Knight. Webster, by contrast, uses the terms of the picturesque in more colloquial and empirical ways. One interesting and probably unintended effect of the epistolary form is Webster's synthesis of the three elements at least partially segregated by Englefield, a synthesis that shows how often geology and the picturesque, and/or antiquarianism, reinforce, rather than conflict, with each other.

Englefield displays his substantial expertise in architectural history in the chapter on antiquities, but only when he feels it is called for. There is a

Fig. 8. *Description of the Isle of Wight* (1816), plate XVIII. Engraving by W. and G. Cooke from a drawing by T. Webster. Courtesy of Howard-Tilton Memorial Library, Tulane University.

curious emphasis on the island's deficiencies that unites all the chapters and helps to explain, in my view, his delay in publishing the book as well as the need for Webster's contribution to justify its publication. Englefield continually reminds his readers that one of his objectives is to demystify the Isle of Wight, to correct the enthusiasm of previous accounts, and to bestow praise on the landscape and antiquities only where it is due – in some cases on sites that previous authors have not noticed: 'So much has been within these few years written on the beauties of the Isle of Wight, that it may appear presumptuous to rank it lower in a picturesque view than these authors have uniformly placed it' (p. 55). The interior of the island, Englefield observes, is 'as destitute of beauty, as any tract of the same extent in England'; even when its historic buildings are taken into consideration, 'scarcely a subject for the pencil can be found' (p. 58).

The chapter on antiquities thus begins in very unpromising fashion: 'The remains of antiquity in the Isle of Wight are very few in number, and inconsiderable in point of size and beauty' (p. 89). Englefield finds only about a dozen structures deserving of extended analysis, primarily churches, and five buildings are engraved from his drawings. One of these is Yaverland Church (Fig. 10), which provides the occasion for some of Englefield's most detailed architectural history and also anticipates Webster's careful attention to the arches of St Adhelm's Chapel. The arch here displays the 'very rude Saxon zig-zag ornaments' that repeatedly draw Englefield's attention (p. 101), and he notes that this arch 'is the only entrance into the Chancel, from the Nave of the Church. So solid a wall of separation, and so small an opening, are not common, and mark an high antiquity' (p. viii). The broken staircase to the left of the arch 'appears coeval with the wall, and it is a very curious and perhaps singular remnant of antiquity', recalling the reading desks attached to the choirs 'of the earliest Christian churches' (p. 102). This analysis displays Englefield's knowledge of ecclesiastical history, but the reasoning is also closely analogous to the reconstruction of stratigraphic sequence and deformation adumbrated in Englefield's geological discussion and elaborated fully in Webster's letters. At the same time, Webster's geological observations confirm Englefield's feeling that the island – although not insignificant in a picturesque or an antiquarian view – is most curious for its geology, which ultimately justifies the publication of his work.

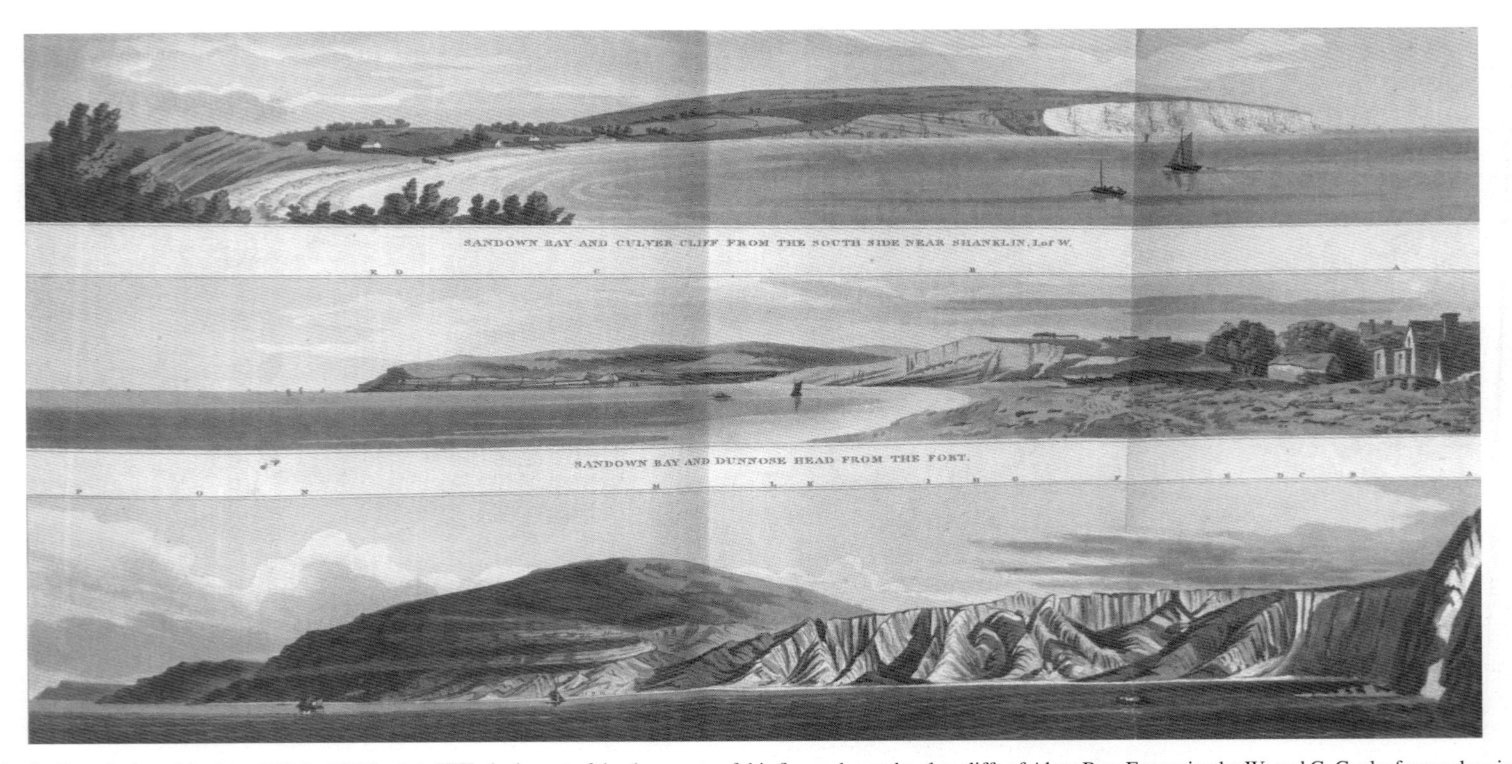

SANDOWN BAY AND CULVER CLIFF FROM THE SOUTH SIDE NEAR SHANKLIN, I. of W.
SANDOWN BAY AND DUNNOSE HEAD FROM THE FORT.

Fig. 9. *Description of the Isle of Wight* (1816), plate XIX: the lowest of the three parts of this figure shows the clay cliffs of Alum Bay. Engraving by W. and G. Cooke from a drawing by T. Webster. Courtesy of Howard-Tilton Memorial Library, Tulane University.

Fig. 10. *Description of the Isle of Wight* (1816), plate IX. Engraving by W. and G. Cooke from a drawing by H. Englefield. Courtesy of Howard-Tilton Memorial Library, Tulane University.

Another structure noticed by Englefield resembles St Adhelm's Chapel in function, if not structure: it is a small freestanding turret on St Catherine's Hill, with an octagonal upper story containing windows suitable for the projection of a light. Archaeological evidence suggests that this lighthouse was manned by a hermit who lived in a cell close by, and for Englefield, 'standing on this airy summit, it is impossible not to picture to the mind the venerable inhabitant of this cell lifted almost out of the habitable world, and only recalled to it by the friendly care of his charitable lamp' (p. 94). The 'religious inhabitant' of St Adhelm's imagined by Webster (p. 190) arises from the same intimate relationship between coastal geology and navigation. These figures clearly mark the affinity between antiquarian local history and the new 'geo-historical science' (to borrow Rudwick's term).

This motif of marine piety is an essential feature of the local history of almost any rocky coastal landscape. Charlotte Smith's learned local poem *Beachy Head* (1807) provides a good analogue to both Englefield and Webster in its rendition of the legendary Parson Darby as a hermit residing in a cave under Beachy Head, piously pursuing his vocation of saving drowning mariners until he and a portion of the cliff are consumed by the raging sea. Webster's essay 'On the freshwater formations', which extends his observations eastward along the coast, tells of the narrow escape of Parson Darby's successor, the then current vicar of East Dean, who leaped to safety across a widening fissure just before a 300-foot × 80-foot mass of the chalk plunged into the Channel (Webster 1814a, pp. 191–192 n.). Smith's interest in the natural history of Beachy Head certainly extends to the qualities of the chalk, although, she does not develop a structural understanding of the formation as Webster does.[9] Webster, even in this more formal paper, is far from indifferent to the sublime, whether in its coastal manifestations or in the form of 'mysteries on which all the speculations of

[9] Smith's description of the Isle of Wight in her second novel, *Emmeline* (1788), resembles Englefield more than Webster in its literary polish (Smith 1788, 302–303). For a sustained comparison of Smith and Webster see Heringman (2009).

geologists have not thrown any certain light' (Webster 1814*a*, p. 245).

The picturesque, the sublime and the elegiac theme of ruin all converge in the tales of shipwreck that belong to the description of many sites of geological interest. Webster happened to make his hard-won sketch of the Needles the day after the wreck of the *Pomone*, a 50-gun frigate coming from Persia (p. 161). Webster mentions several other shipwrecks (cf. pp. xvii, xix–xx and 191) and Englefield is moved by his discovery of an impromptu burial ground, stemming from a 1782 shipwreck, to include a quite powerful original verse elegy (p. 61), the only piece of his *Description of the Isle of Wight* ever to be reprinted.[10] The *Pomone* provides an especially compelling case because Webster actually witnessed the aftermath of the wreck, which sets the tone for his description of the Needles and underscores the difficulties of his own practice as a draftsman and geologist. The letter in question (VI) begins with a tone of heightened interest and wonder, conditioned by the astonishment that Englefield and his predecessors reserve for Alum Bay: on the day described in this letter, Webster took his views both of the colourful clay cliffs of Alum Bay (Fig. 9) and of the Needles (plate XXV) from the same rock off the western tip of the island (pp. xv and p. 163). He points out that the shipwreck inevitably reinforced the natural wonder of this landscape:

> Viewing this scene as merely picturesque, and independently of the feelings of regret naturally excited by the loss of so valuable a ship, it was one of the grandest I have ever witnessed. [...] such a wreck, on a spot so extraordinary, formed a combination, which, though not strictly accordant with the rigid laws of picturesque composition, yet was in nature highly sublime.
>
> (p. 161)

Webster notes that the wreck – together with the spectacle of feverish salvage work – has attracted many other 'spectators, in pleasure yachts and other boats' (p. 162); ironically, the same swell that precipitated the wreck now 'made it almost impossible to approach the *Pomone* without danger, but increased very much the picturesque effect of the groupes'.

An unusually large portion of this letter is devoted to aesthetic response – the clay cliffs, too, turn out to be more sublime than picturesque (p. 158) – and it also records Webster's important conclusion that the same strata not only run all the way across the island but also reappear (albeit horizontally) at Hordwell Cliff in Hampshire (pp. 159–160). Webster's willingness to view the coast from the water – which Englefield apparently had not done – was a prerequisite for the accuracy and comprehensiveness of his observations, as he indicated already in his first letter to Englefield (p. 120). Webster also added a lengthy commentary on the aesthetic appeal of the coast as viewed from the Needles in the 'Explanation of the Plates' included in the front matter to the volume, a rhapsody on its 'romantic' character that merits attention in its own right:[11]

> The wonderfully coloured cliffs of Alum Bay, the lofty and towering chalk precipices of Scratchell's Bay, of the most dazzling whiteness and the most elegant forms, the magnitude and singularity of the spiry insulated masses, which seem at every instant to be shifting their situations, and give a mazy perplexity to the place, the screaming noise of the aquatic birds, the agitation of the sea, and the rapidity of the tide, occasioning not unfrequently a slight degree of danger, all these circumstances combine to raise in the mind unusual emotions, and to give to the scene a character highly singular and even romantic.
>
> (p. xvi)

Webster's literary skill in managing the sublime may well owe something to his time with Humphry Davy at the Royal Institution, since Davy's geology lectures frequently invoke the sublime as well (e.g. Siegfried & Dott 1980, pp. 12–13 and 77–78).

Webster's own adventures by boat thus had a different purpose from those of the spectators, although often enough they afforded aesthetic pleasure and occasionally antiquarian data as well. The wrecked vessel, for example, 'afforded [him] a scale' for measuring the height of the Needles, which appear deceptively small from a distance (p. 161). When the tide was out he ventured closer, not to view the wreck, but simply to find a station for his drawing, which he finally did, appropriately, on a ruin, 'the stump of the lofty spire of chalk which chiefly gave the name [Needles] to these rocks, and which being worn through by the waves, fell down in 1764' (p. 163). Webster reports that he had about two hours on this six-foot wide rock to make his sketches, 'though completely wetted through by the spray', before the tide began to

[10] The poem, with a paragraph of Englefield's prose, was included in an account of a new technology (the 'Cliff Waggon') for rescuing 'shipwrecked persons' in the evangelical *Saturday Magazine* (Parker 1834, p. 93). The article – including Englefield's explanatory paragraph but not the poem – was then reprinted the next year in New York (Bacheler 1835, p. 51).

[11] The spectacular plate described here (plate XXV) is reproduced in Freeman (2004, p. 14), Kirk (1996, p. 311) and Rudwick (2005, p. 515).

come in again. In this case Webster draws on an antiquarian work, Sir Richard Worsley's *History of the Isle of Wight* (1781), to historicize the geological monument. Worsley's book, referred to elsewhere by both Webster and Englefield, includes an engraving of the rock when it was still standing.[12]

The antiquarian framework

Many other episodes in Webster's letters document the intersection of human and geological antiquity, and specifically the utility of antiquarianism as a disciplinary framework for geological observation and induction. He recognized this kinship most explicitly in the passage cited above comparing the Isle of Wight [now Hampshire] Basin to an 'arch or vault,' in which he self-consciously decides 'to use the language of the antiquary' (p. 201). His observations concerning building stone sometimes proved crucial for determining the order of the strata, and his longest antiquarian digression concerning St Adhelm's Chapel is followed by an account of a more strictly geological antiquity, the 'coal money' found at Kimmeridge, just west of St Adhelm's Head. Having already described the 'Kimmeridge coal' (oil shale) as an effective but foul-smelling local fuel, Webster then touches on specimens that bear obvious signs of human workmanship and appear to be 'some ancient sort of token money' (p. 191).[13] In exploring the coast between Luccombe and Bonchurch on the east side of the island, where Englefield had been bewildered by a massive landslip, Webster uses the language of the antiquary more routinely – the word 'ruin' appears 10 times in Letter III. Although the specimens in this case are natural, the practice of the antiquary echoes in Webster's assurance that he 'climbed over the ruins' and 'searched, among the fallen masses, for some specimens of what you had seen' (p. 133). Here is a practical basis for the convergence, long noticed by intellectual historians, between Gothic ruins and geology as newly fashionable themes beginning in the eighteenth century (e.g. Aubin 1934).

Webster's account amplifies the idea of sublime confusion from Englefield's description of this site (cf. p. 38), and yet his recognition that 'the whole of this coast is composed of ruins' (p. 138) facilitates some effective geological reasoning: based on his observation of the dip in the remaining horizontal mass, he identifies the massive fault running through the chalk on this part of the coast (p. 137). Along the same stretch of coast, while noting the location of a sandstone bed that was evidently used for 'carved work' in the island's Gothic churches (p. 141), Webster observes 'singular forms in relief' that turn out to be natural (pp. 141–143): a new species of fossil Alcyonium (Webster 1814*b*). In successive letters the antiquarian framework continues to facilitate the inductive process while also leaving room occasionally for bolder conjectures (e.g. p. 150) that come to fruition in the final letters reconstituting the chalk formation across southeastern England as a whole. These two letters (XI–XII), although maintaining a strict insistence on 'deductions from the facts themselves' (p. 198), also defend the antiquarian mode of reasoning by analogy. Webster acknowledges here that 'the fabric, whose ruins we have been contemplating, is of a nature too vast to be completely conceived as a whole', but he maintains that nonetheless 'the mode at least' of 'great changes' may be 'inferred' from the geological fragments 'by a reasoning not unphilosophical' (pp. 215–216).

Renewed comparison suggests that the elegiac tone is much more pervasive in Englefield, perhaps to the extent that the antiquary predominates over the geologist in him. At the same time, Englefield is deeply invested in an ideology of the picturesque derived from the privilege attached to landed property, an ideology that might be presumed to have a weaker hold on Webster.[14] As a Catholic and an ardent medievalist, Englefield found cause to lament the decline of architectural and moral standards in both his picturesque and antiquarian chapters. For Webster, 'ruin' as a geological metaphor – apart from the theme of shipwreck – does not carry these connotations. Like other writers of picturesque tours and prospect poems throughout the eighteenth century, Englefield looked first to gentlemen's seats for instances of the picturesque (cf. Barrell 1972, chap. 1). He was disappointed both in the number and the quality of great houses with large landscaped grounds. The few seventeenth-century manor houses did not have significant grounds, and 'those of modern erection on the beautiful parts of the coast, scarcely rise above the character of villas; none being seated in anything like a park. Indeed in the whole island', Englefield continues, 'there is but one seat which can rank with what is generally called an improved place' (p. 87). This is the seat of Sir Richard Worsley, the author of the history cited later by Webster, but is also the very place that was

[12] In addition to Worsley's history, other histories of the Isle of Wight referred to in the *Description* include books by Henry Windham (1793), the Rev. Richard Warner (1795) and, especially, Charles Tomkins (1796).

[13] Later found to be the central disks from Romano-British armlets (Calkin 1955).

[14] For Webster's controversial project of educating 'mechanics' at the Royal Institution, see Ironmonger (1958).

spoiled (in Englefield's view) by the celebrity designer Capability Brown (p. 88, cf. p. 104). To make matters worse, at least two of the villas are in a Gothic revival style, one of them designed by James Wyatt (p. 57), the architect who became the arch-enemy of Englefield, Gough and other medievalists by carrying out radical renovations in Salisbury (1787–1793) and other cathedrals (Evans 1956, pp. 207–214). Englefield's contempt for this shallow Gothicism is readily apparent as he laments 'the infelicity of attempts to imitate the style of our despised ancestors, whose works, while modern science strives in vain to rival, modern insolence brands with the appellations of dark and barbarous' (p. 58).

Although lacking in picturesque credentials, Worsley's property, Appuldurcombe, got high marks from Englefield for its fine collection of painting and classical sculpture. More importantly, Worsley was responsible for the restoration of the hermitage and lighthouse on St Catherine's Hill that prompted one of Englefield's most enthusiastic flights in his chapter on antiquities (pp. 94–95). Worsley's antiquarian labours thus benefitted both Englefield's and Webster's portions of the book, if in rather different ways. Englefield identified with him as a gentleman landowner and classically inspired antiquary, but dismissed the decadent modern architecture associated with his class in favour of Saxon ecclesiastical buildings (such as the lighthouse, now known as St Catherine's Oratory).[15] These structures display a 'rude' vigour (p. 101) and suggest a noble morality more in keeping with the 'rude and majestic' coastal scenery (p. 76). Englefield's antiquarianism is a moral discourse to the extent that his encounter with ruins – both geological and architectural – led him to lament a perceived moral and social decline, as when he elaborated a contrast between the 'venerable inhabitant' of this early Christian lighthouse and modern lighthouse keepers motivated by 'love of idleness' in place of 'sublime benevolence' (p. 95). Webster, however, was less apt to associate ruin with a narrative of decline because it had a primarily geological significance for him. His correspondingly greater emphasis on natural scenery – and thus in a loose sense on public land – also suggests a weakening of the association between the picturesque and landed property. Both men, however, stand outside the patriotic system of values commonly at work in neoclassical discourse about ruins, according to most scholarship on the subject (Goldstein 1977; Janowitz 1990).

Englefield does have his geological moments, and it would be a mistake to overdraw the distinction between his brand of antiquarianism and Webster's. One historian has pointed out that 'antiquarianism [...] provide[d] a language within which people from very different backgrounds could communicate and exchange information' (Sweet 2004, p. 60). The epistolary form used by Webster, a basic form for scientific communication since the beginning of the *Philosophical Transactions*, also tended to promote the common language and values shared by the correspondents. When Englefield admires the 'naked and stupendous ruin' presented by the south front of St Catherine's Hill (p. 11) he is merely setting a scene. But when he notes that Hurst Castle (opposite the NW shoulder of the island) is set 'on the brow of a submarine cliff' (p. 14) he is also making a significant observation about the contours of the Channel. In at least one case his attention to scenery and topography leads to an original geological observation and a credible hypothesis. Reflecting on 'the shattered state of the flints in the chalk' (p. 30), Englefield presents evidence that they could not have been deposited alternately with the chalk, and then concludes that they must have been altered by the same event that 'subverted the strata of chalk' as a whole. This problem is the subject of his two brief geological articles (Englefield 1800, 1801) published separately after his initial observations on the island were completed. In the *Description* he illustrates this hypothesis in the best antiquarian fashion by comparing this hypothetical 'immense concussion' to the impact on a wooden pier of a 'vast stone' landed at St Petersburg to form the base for a 'colossal statue' of Peter the Great (p. 33). Both Englefield and Webster, in their different ways, perceived that geological monuments too – however slowly – were being gradually 'effaced from the earth,' like the shallow graves of the 1782 shipwreck victims (p. 60).

Conclusion

While the elegiac mode is latent rather than manifest throughout this book, the motif of ruin must be

[15] In his description of the eastern outcrop of the vertical clay strata above the chalk at Whitecliff Bay, Englefield makes a connection that expresses his affiliation with elite neoclassical antiquarianism much more than any true geological resemblance: 'Some of the views in Sir William Hamilton's Campi Phlegrei, have a considerable resemblance to these very extraordinary cliffs' (p. 79). Hamilton was a fellow member of the Society of Dilettanti. Webster does, at times, share Englefield's aesthetic prejudices, as when he decries the want of taste in new construction at Shanklin at the beginning of Letter VIII.

central to any understanding of how it became the forum for Webster's path-breaking geological investigations. Many of Webster's descriptions are initially ambivalent, and could apply equally to geological or antiquarian research: 'I climbed, therefore, over the ruins' – to quote his narrative of the Undercliff again – 'and searched, among the fallen masses, for some specimens of what you had seen' (p. 133). Englefield's antiquarian vocabulary provides the medium required by Webster to express the historicity of these geological ruins. It facilitates the reconstruction of stratigraphic sequence for which these letters, sections and his subsequent essay on the chalk are so justly celebrated. Webster, however, with his strong background in applied science, concluded his portion of the book by distinguishing geology (as practical) from antiquarian and aesthetic reverie. His penultimate letter differs in tone and content from the others: it is long and boldly speculative, but then seeks to justify its theoretical reach – and, perhaps, also to question the values of the leisured elite that dominate Englefield's chapters and colour many of his own observations – by underscoring the utility of his model of an 'imaginary vault' or arch, the Isle of Wight Basin, produced by 'wonderful revolutions' (pp. 205 and 215):

> Now that they are known to be the remnants of a great whole, the original of which can be traced only by its ruins, the subject appears in a more interesting point of view: every detached fact, though by itself of little value, becomes important, as forming a link in the great chain: and Geology, instead of being a collection of the dreams of philosophers, affords among its other important purposes, a basis upon which may be founded the improvement of Agriculture, and hence becomes a science of the highest practical utility.
> (p. 225)

Although agricultural improvement was certainly important to gentleman landowners, Englefield follows picturesque convention in objecting on aesthetic grounds to agricultural land use (e.g. p. 58), and his part of the book omits this connection between geology and agriculture.[16]

Webster is optimistic in this letter for good reason. He was summing up two very substantial summers of fieldwork on the island (1811) and the opposite coast of Dorset (1812), and writing from London, where he had had significant time to elaborate the hypothesis of a regional fold structure (the 'arch or vault' depicted in Fig. 3) and test its merit by reconstructing the sequence of strata across a much larger region. He wrote one final, unexpected letter from the island the next winter (February 1813) explaining to Englefield that he had returned once more – evidently at his own cost – to investigate more concretely the relationships between the Isle of Wight Basin and the Paris Basin described by Cuvier and Brongniart (p. 226).[17] Webster had evidently read Cuvier and Brongniart's memoir before he began his work for Englefield, but according to this letter, he had now seen for the first time (winter 1812–1813) the specimens that Brongniart sent to London to illustrate the memoir. He was evidently embarking on the more extended fieldwork along the coast that would lead to his 1814 essay 'On the freshwater formation'. This last letter contains the kernel of that paper's argument and also fleshes out into a narrative the hypothetical abstract of deposition and 'denudation' included in the previous letter (pp. 223–224); and it includes a detailed table showing 'The Order of the Upper Strata' (p. 235–237). Webster's research was beginning to bear scientific and professional fruit, instilling a confidence also apparent in the 'Explanation of the Plates', composed last, as when he argues for the originality and importance of his observations at Handfast Point in Dorset (Fig. 6): 'no one had yet noticed strata of chalk *quite* vertical', he writes, and this unparalleled phenomenon seems to demand an entirely new order of explanation (p. xvii).[18]

Although the letters collected by historians, especially Challinor, abundantly demonstrate the respect that Webster accrued from geologists such as William Henry Fitton and William Conybeare, they also tell a more melancholy story. As Keeper of the museum, librarian, house secretary and de facto editor of the Geological Society's *Transactions*, Webster had little opportunity to pursue additional fieldwork or publication (Challinor 1961*a*, p. 175). Many of his letters between 1812 and 1826 (when he resigned) express his frustration with these time constraints, his inadequate salary

[16] Here Englefield is probably following William Gilpin, the father of picturesque theory, who dismissed marks of 'cultivation' in the landscape as 'disgusting' (Gilpin 1808, vol. I, pp. 6–7).

[17] In March 1822 Webster wrote to T. R. Underwood concerning the cost of his return trips to the island, beginning in 1813 and continuing periodically at least to the early 1820s when his views were under attack by George Sowerby and others: 'I have expended no less than 200 pounds out of my own pocket' in 'the Isle of Wight affair'. A letter to Brongniart, dated November of the same year, affirms that Brongniart and Cuvier's memoir on the Paris Basin originally inspired this work. Both letters are printed in Challinor (1961*b*, pp. 154 and 168).

[18] A reference to the 'upper freshwater formation' (p. xv) and a correlation with the Paris Basin (p. xi) in the 'Explanation of the Plates' prove that it was written after the 1813 letter.

and what he perceived to be his subordinate position in the Geological Society. As Wendy Kirk has put it, 'his time at the Geological Society was not a particularly happy one' (Kirk 1996, p. 311). In Webster's own words from an 1822 letter to T. R. Underwood, 'I do not think meanly of Englishmen generally, but I have got among a bad lot' (Challinor 1961*b*, p. 155). New evidence indicates that he prepared and attempted to publish his own book on the geology of the Isle of Wight at this stage. An uncatalogued publisher's prospectus recently discovered by Hugh Torrens in the British Library, dated 1827, attempts to promote the publication of this book, but there is no further record of its existence. In this prospectus Webster describes himself as a 'professional geologist', as noted earlier, and emphasizes his priority as discoverer of the island's tertiary formations equally with his qualifications as an educator: the volume, he says, would be aimed primarily at the 'young geologist' and priced at no more than ten shillings.[19] Webster lectured on geology until his death in 1844, but evidently continued to struggle financially even after he was named Professor of Geology at the University of London (now UCL) in 1841 (Edwards 1971, p. 469).

In 1822 he wrote to Adam Sedgwick gracefully but somewhat bitterly pleading for the respect that he felt eluded him as 'a professional man dependent on his own exertions for the means of existence' (Challinor 1961*b*, p. 148). After the heady days of Webster's Isle of Wight research, being a professional geologist turned out to mean administrative work, editing and lecturing for a return of not much more than £100 a year. This was Webster's salary at the Geological Society, and, although his initial contract only called for a three-day work week, all sources indicate that his manifold duties there increasingly took up all his time (Edwards 1971, p. 472 n.51). According to one recent authority, the minimum weekly income of a gentleman in the early nineteenth century was £5, and the 'large constituency of potential readers' in the next-lowest bracket of 50–100s. (£125–250 per annum) could afford to buy books only occasionally (St Clair 2004, pp. 193–195). More research is needed on Webster's finances, especially during 1811–1812, but it is clear that he would have entered the ranks of the book-buying public only in a very good year, based on what we know about his income. In 1841, when Fitton testified that Webster's works '[were] referred to by all the geologists in Europe', Webster applied for and received a government pension of £50 per annum (Kirk 1996, p. 313). Only during his most successful year at the University of London, based on figures provided by Kirk (1996, p. 322), would he have equalled his income during his Geological Society days. Yet, Webster was chosen over other candidates by the University because of his professional experience and his bent for applied science: he was clearly the best man to teach geology to future engineers.

In light of these professional ambitions, it is thoroughly ironic that a stint of old-fashioned aristocratic patronage in 1811–1812 allowed Webster to do his only substantial research. As a largely self-taught geologist with a professional background in art and architecture, he was able in this context to adapt the languages of antiquarianism and the picturesque to scientific ends and set new standards for an emerging geology. This accomplishment would have been impossible at a later stage when geologists' curricula became more standardized and their social status more secure. For the same reasons, however, Webster's career in geology never fulfilled the promise of his path-breaking research on the Isle of Wight and the Chalk formation.

I am especially indebted to Martin Rudwick and Hugh Torrens for their superb leadership on the November 2007 field trip to the Isle of Wight that constituted part of the History of Geology Group's celebration of the bicentenary of their parent society. I learned more about the present topic in two days on the island than I had in three months of reading Webster and Englefield in Missouri. I would also like to thank the other participants in the memorable November meeting at Burlington House, especially Cherry Lewis for inviting me to speak and Simon Knell for his care as an editor. Thanks are also due to the University of Missouri-Columbia Research Council for travel funding.

Archives

BL: British Library, Dawson Turner prospectus collection, pressmark 1879.b.1.

SAL: Society of Antiquaries of London.

References

AUBIN, R. A. 1934. Grottoes, geology, and the Gothic revival. *Studies in Philology*, **31**, 408–416.

BACHELER, O. (ed.). 1835. The cliff wagon. *The Family Magazine*, **2**, 51–52.

[19] BL: Webster's prospectus is bound in Vol. 4 of the Dawson Turner prospectus collection, pressmark 1879.b.1. p. 2. Webster's letters make it clear that he was occupied for some years with defending his views of the Isle of Wight and collecting more evidence (see footnote 17 earlier); presumably, he intended to include this material in the book, which must be more or less identical with the work in progress alluded to in his letters.

BARRELL, J. 1972. *The Idea of Landscape and the Sense of Place, 1730–1840: An Approach to the Poetry of John Clare*. Cambridge University Press, Cambridge.

BERMINGHAM, A. 1986. *Landscape and Ideology: The English Rustic Tradition 1740–1860*. University of California, Berkeley, CA.

CALKIN, J. B. 1955. 'Kimmeridge Coal-Money': The Romano-British shale armlet industry. *Proceedings of the Dorset Natural History and Archaeological Society*, **75**, 45–71.

CHALLINOR, J. 1961*a*. Some correspondence of Thomas Webster, geologist (1773–1844) – I. *Annals of Science*, **17**, 175–195.

CHALLINOR, J. 1961*b*. Some correspondence of Thomas Webster, geologist (1773–1844) – II. *Annals of Science*, **18**, 147–175.

CHALLINOR, J. 1963*a*. Some correspondence of Thomas Webster, geologist (1773–1844) - III. *Annals of Science*, **19**, 49–79.

CHALLINOR, J. 1963*b*. Some correspondence of Thomas Webster, geologist (1773–1844) - IV. *Annals of Science*, **19**, 285–297.

CHALLINOR, J. 1964*a*. Some correspondence of Thomas Webster, geologist (1773–1844) - V. *Annals of Science*, **20**, 59–80.

CHALLINOR, J. 1964*b*. Some correspondence of Thomas Webster, geologist (1773–1844) - VI. *Annals of Science*, **20**, 143–164.

EDWARDS, N. 1971. Thomas Webster (Circa 1772–1844). *Journal of the Society for the Bibliography of Natural History*, **5**, 468–473.

EDWARDS, N. 1972. Some correspondence of Thomas Webster (circa 1772–1844), concerning the Royal Institution. *Annals of Science*, **28**, 43–60.

EDWARDS, N. 2004. Webster, Thomas (1772–1844). *Oxford Dictionary of National Biography*. Oxford University, Oxford.

ENGLEFIELD, H. 1800. Observations on some remarkable strata of flint in a chalk-pit in the Isle of Wight. *Transactions of the Linnaean Society*, **10**, 103–109.

ENGLEFIELD, H. 1801. Additional observations on some remarkable strata of flint in the Isle of Wight. *Transactions of the Linnaean Society*, **11**, 303–308.

ENGLEFIELD, H. & WEBSTER, T. 1816. *Description of the Principal Picturesque Beauties, Antiquities, and Geological Phaenomena, of the Isle of Wight*. Printed by William Bulmer for Payne and Foss, London.

EVANS, J. 1956. *A History of the Society of Antiquaries*. Printed for the Society of Antiquaries, Oxford.

FREEMAN, M. 2004. *Victorians and the Prehistoric: Tracks to a Lost World*. Yale University Press, New Haven, CT.

GILPIN, W. 1792. A poem on landscape painting. *In*: *Three Essays*. Printed for R. Blamire, London, 2–44 (appendix).

GILPIN, W. 1808. *Observations on Several Parts of England, Particularly the Mountains and Lakes of Cumberland and Westmoreland, Relative Chiefly to Picturesque Beauty*, 3rd edn (2 vols). Cadell and Davies, London.

GOLDSTEIN, L. 1977. *Ruins and Empire: The Evolution of a Theme in Augustan and Romantic Literature*. University of Pittsburgh Press, Pittsburgh.

GOUGH, R. 1799 [1786–1796]. *Sepulchral Monuments in Great Britain. Applied to Illustrate the history of Families, Manners, Habits, and Arts, and the Different Periods from the Norman Conquest to the Seventeenth Century* (2 vols in 5). Printed by John Nichols, London.

HERINGMAN, N. 2009. 'Very vain is Science' proudest boast': The resistance to geological theory in early nineteenth-century England. *In*: ROSENBERG, G. D. (ed.) *The Revolution in Geology from the Renaissance to the Enlightenment*. Geological Society of America Memoir, **203**, 247–257.

IRONMONGER, E. 1958. The Royal Institution and the teaching of science in the 19th century. *Proceedings of the Royal Institution*, **37**, 139–158.

JANOWITZ, A. 1990. *England's Ruins: Poetic Purpose and the National Landscape*. Blackwell, Oxford.

KIRK, W. 1996. Thomas Webster (1772–1844): First professor of geology at University College London. *Archives of Natural History*, **23**, 309–326.

NURSE, B. 2004. Englefield, Sir Henry (1752–1822). *Oxford Dictionary of National Biography*. Oxford University, Oxford.

O'CONNOR, R. J. 2007. *The Earth on Show: Fossils and the Poetics of Popular Science, 1802–1856*. University of Chicago, Chicago, IL.

O'CONNOR, R. 2009. Facts and fancies: the Geological Society of London and the wider public, 1807–1837. *In*: LEWIS, C. L. E. & KNELL, S. J. (eds) *The Making of the Geological Society of London*. Geological Society, London, Special Publications, **317**, 331–340.

PARKER, J. W. (ed.). 1834. Preservation of life from shipwreck. *The Saturday Magazine*, **4**, 92–93.

PIGGOTT, S. 1976. *Ruins in a Landscape: Essays in Antiquarianism*. Edinburgh University Press, Edinburgh.

RUDWICK, M. J. S. 2005. *Bursting the Limits of Time: The Reconstruction of Geohistory in the Age of Revolution*. University of Chicago, Chicago, IL.

ST CLAIR, W. 2004. *The Reading Nation in the Romantic Period*. Cambridge University, Cambridge.

SIEGFRIED, R. & DOTT, R. H. (eds). 1980. *Humphry Davy on Geology: The 1805 Lectures for the General Audience*. University of Wisconsin, Madison, WI.

SMITH, C. 1788. *Emmeline: The Orphan of the Castle*. Broadview, Peterborough. (Republished in 2003.)

SWEET, R. 2004. *Antiquaries: The Discovery of the Past in Eighteenth-Century Britain*. Hambledon & London, London.

TOMKINS, C. 1796. *A Tour to the Isle of Wight, Illustrated with Eighty Views, Drawn and Engraved in Aquatinta*. 2 vols. Printed for G. Kearsley, London.

TORRENS, H. S. 1983. Arthur Aikin's mineralogical survey of Shropshire 1796–1816 and the contemporary audience for geological publications. *British Journal for the History of Science*, **16**, 111–153.

WARNER, R. 1795. *The History of the Isle of Wight; Military, Ecclesiastical, Civil, & Natural: To Which is Added a View of its Agriculture*. Printed for T. Baker, Southampton.

WEBSTER, T. 1814*a*. On the freshwater formations in the Isle of Wight, with some observations, on the strata over the Chalk in the south-east part of England. *Transactions of the Geological Society of London*, **2**, 161–254.

WEBSTER, T. 1814*b*. On some new varieties of fossil Alcyonia. *Transactions of the Geological Society of London*, **2**, 377–387.

WORSLEY, R. 1781. *The History of the Isle of Wight*. Printed for A. Hamilton, London.

WYNDHAM, H. 1794. *A Picture of the Isle of Wight: Delineated upon the Spot*. Printed by C. Roworth for J. Egerton, London.

The Geological Society and its official recognition, 1824–1828

PATRICK J. BOYLAN

School of Arts, City University London, Northampton Square, London EC1V 0HB, UK (e-mail: P.Boylan@city.ac.uk)

Abstract: Under the 1824–1826 presidency of William Buckland, the still young Geological Society negotiated with the Government a very important advance in terms of the official recognition of the Society and the emerging science of geology that the Society represented, by obtaining a prestigious new legal status in the form of a Royal Charter of Incorporation. Then, under William Fitton's presidency, in 1828 the government granted the Society rent-free accommodation for both its meetings and its rapidly growing library and museum in the government offices in Somerset House, London. The objectives behind these two related moves are considered, although it is unfortunate that little of the detailed background documentation to these developments seem to have been preserved within either Society or government records. A brief account of what might have been – the possibility of seeking a Coat of Arms for the newly Chartered Society – concludes the story.

In 1824 only five national learned societies had the official recognition and benefits of a Royal Charter. Three of these had been created by Royal Charter – the Royal Society in 1660, the Royal Society of Edinburgh in 1783 and the Royal Irish Academy in 1785. The remaining two had been founded as unincorporated private members' societies (like the Geological Society), but had subsequently been granted a Royal Charter – The Society of Antiquaries, founded in 1707 and chartered in 1751, and the Linnean Society, founded in 1778 and similarly chartered in 1802.

While it remained unincorporated, the Geological Society had no legal identity or status and so could not enter into any contract such as the lease on its rented apartments or have a bank account in its own right. It could not even be recognized as the legal owner of the Society's rapidly growing collections. Instead, individual members had to act as trustees on behalf of the other members of the Society. A Royal Charter would thus give the Geological Society a legal personality in its own right, allowing it to own or lease property, to have bank accounts, to enter into contracts and to employ staff, all in its own name.

The direct and indirect advantages to a society that flow from chartered status in terms of both administrative convenience and public recognition were summarized in an authoritative mid-nineteenth century study of learned societies, quoted by Woodward in his centenary history of the Geological Society:

> a society that is "incorporated by Royal Charter" is an official body publicly and legally recognised; it has perpetual succession and a common seal; and the statutes or bye-laws, which are framed for the ordinary guidance of the members, must be in perfect accordance with the stipulations or principles of the Charter. Societies of this kind naturally take precedence of all others, and where several are in other respects (or are assumed to be) of equal importance, priority of incorporation is a reasonable ground of distinction.
>
> (Woodward 1907, pp. 70–71)

For many such as William Buckland (1784–1856) (Fig. 1), a strongly monarchist Tory and Anglican clergyman (Gordon 1894), a Royal Charter would be seen to carry with it very considerable status within the prevailing social system of the day, which was still very much based on the Court. It could perhaps be compared to the granting of a peerage to an individual in terms of advancement within 'society'. Being only the sixth learned society to achieve such distinction, incorporation would guarantee the Society, and, by implication, the science of geology it represented, a very high status in terms of the recognized order of national and social precedence. More radical members, who perhaps cared little for such considerations in relation to a Court that was widely held in low esteem because of the excesses of the Prince Regent (by then George IV), would nevertheless at least recognize the practical advantages of gaining Chartered status.

Seeking and obtaining a Royal Charter of Incorporation

In 1824 Buckland was elected to the first of his two terms as President of the Geological Society (1824–1826 and 1839–1841). Immediately afterwards, the Society set out to advance its status and, through this, official recognition of geology as a science.

From: LEWIS, C. L. E. & KNELL, S. J. (eds) *The Making of the Geological Society of London*. The Geological Society, London, Special Publications, **317**, 319–330.
DOI: 10.1144/SP317.18 0305-8719/09/$15.00

Fig. 1. William Buckland lecturing. Lithograph by George Rowe, 1823 (from Boylan 1970).

From the very limited documentation that survives, it is not clear exactly when, or by whom, this process was initiated. However, in his Anniversary Address to the Society at the end of the first year of his second presidency in 1840, Buckland seemed to claim much of the credit for this, saying: 'fifteen years have passed since I was placed, by your kindness, in the honourable position of filling this chair, at that important period in our history when we received the national recognition of a Royal Charter. I shall never cease to consider it one of the brightest rewards of my labours in geology [...]' (Buckland 1840, p. 211). One of the senior Council members involved, or at least consulted, was Henry Warburton (1784–1858). One of the earliest members of the Society, joining in 1808, and an early Secretary (1814–1816), Warburton was a wealthy and very influential amateur, who was already active in radical politics and had a powerful network of metropolitan connections. He was to serve as an MP between 1826 and 1847. In politics at least he was well known as a 'fixer' (Matthews 2004). The last part of a long letter of 12 March 1824 from Warburton to Buckland, which was mainly about urgent matters relating to illustrations for papers in the forthcoming volume of the *Transactions*, concludes with urgent advice to the Oxford-based Buckland on tactics and timing in relation to soundings about a Charter application, particularly with regard to what he terms the '*Dei majores*' (literally, the Great Gods) of the Society:

> I am desired to remind you of the subject of the charter. That from the present time to the middle of May is the season when persons of consequence are to be in town, and are to be consulted; And that if any thing at all is to be attempted, you had better take the opportunity of your visit to town, next Friday fortnight, to talk over the matter with the grand men of your acquaintance. Could you be in town a few days before Friday, you would be able to give the Council some information at its meeting on that day. – Or if you could stay in town some days after Friday, a special Council meeting might be summoned to receive your report. The best way will be to request the favour of an interview with the Dei majores expressly for the purpose of opening to them the subject.[1]

The move to seek a Charter was potentially both controversial and risky. Warburton's letter implies that there could be opposition within the Society, at least among the most senior members, hence the suggestion that Buckland should consider calling on at least some of these personally in advance of raising the matter more formally with the whole Council. Perhaps more significantly, in order to obtain a Charter it was first necessary to prepare and submit a formal Petition to the Privy Council, seeking the grant of a Charter. This would require the expenditure of an estimated £300[2] (nearly half the Society's liquid assets) on legal and ther expenses at a time when the Society was struggling under the financial burden of a heavy publications programme and other commitments. Since one of the Privy Council's own tests in relation to the granting or refusing of a Charter was to ensure that the organization was financially secure, this aspect of the Geological Society would have to be taken into account. Also, the Petition would be a public document available for public consultation and any interested organization could object to the granting of a Charter. In this respect there was particular concern about the attitude of the Royal Society.

When the Geological Society had been formed in 1807, a few founders such as Humphry Davy (1778–1829) believed that they were forming nothing more than a small geological dining club and several of the most senior members of the Royal Society had been happy to join on that basis (Knight 2009). However, a year later the Royal Society became alarmed to find the Geological Society rapidly becoming a respected learned society, developing library and museum collections, and leasing its own accommodation. It was argued within the Royal Society that as such, the Geological Society could become a potential competitor; the Royal therefore confronted the geologists. In the event, Sir Joseph Banks (1743–1820), President of the Royal Society, failed to get the Geological Society to agree to restrict its activities and become an 'Assistant Association', wholly subordinate to the Royal Society (Lewis 2009), so urged Royal Society Fellows to resign from the Geological Society. Accordingly, Sir Joseph Banks (1743–1820), Humphry Davy and Sir Charles Greville (1749–1809), respectively, President, Secretary and Vice-President of the Royal Society, immediately resigned from the Geological Society, along with Charles Hatchett, a wealthy mineralogist.

Thus in 1824 it was clearly felt there was a serious risk of further confrontation with the Royal Society over the ambition to obtain a Royal Charter. This time a fight-back could have been much more serious than a call for its Fellows to resign from the Geological Society: the Royal Society had the right to petition the Privy Council, either objecting to the granting of any Charter to the geologists, or at least demanding major restrictions on the powers requested by the Geological Society, if a Charter was granted. The legal costs of responding to either kind of counter-petition from the Royal Society would have added greatly to the Geological Society's expenses. Furthermore, there was a risk that they would still come away empty-handed in the end, with the reputation of the Society seriously damaged. Alternatively, they might be granted a Charter, but one incorporating conditions that would have the effect of restricting rather than enhancing the Society's activities. In either case, the Society could end up much worse off than before. As the new President, Buckland was embarking on a high-risk strategy.

[1] Buckland (Oke/Gordon) Archive formerly deposited in the Devon Record Office, catalogue ref. DRO 138M/F71. Henry Warburton to William Buckland 12 March 1824.

[2] Professor Arthur Lucas has pointed out to me that at least some of the geologists probably knew that more than 20 years earlier the Linnean Society's Royal Charter had cost it much more than this (£450 5s 6d), even though two lawyer members had given their services free of charge.

In addition to the published accounts of the Geological Society's negotiations and lobbying for the grant of a Royal Charter reported by Woodward (1907, pp. 68–73), the Society's archives include a series of both official files and more ephemeral items relating to the Society's Charter and Byelaws.[3] This includes a series of formal notices of Council meetings of 1824–1825 addressed to William Clift (1771–1849), anatomist, palaeontologist and illustrator, and Curator of the Royal College of Surgeons. At that time, Clift was an active member of the Society's Council and a Fellow of the Royal Society. These notices are in the form of pre-printed pro formas in which the details of the meeting had been added by hand by one of the Society's Secretaries.[4] Beyond these and the formal Minutes of meetings, almost nothing of the detailed documentation and working papers that might be expected survive in the Society's own records, and no other records generated and held by the Society's lawyers have been traced, despite extensive searching.[5] The indications are that much of the consultation and discussion that must have been necessary took place in informal meetings.

The first of these printed notices relating to the Royal Charter proposal, addressed to 'William Clift Esq. College of Surgeons Lincoln's Inn Fields' and with postal frankings for 20 April 1824, was for the first meeting of the new Council, that had been elected at the annual meeting at which Buckland had been elected President. The young lawyer and recent Oxford student of Buckland's, Charles Lyell (1797–1875) had become one of the two Honorary Secretaries in 1823 with the long-serving Thomas Webster (1773–1844) continuing as the permanent Secretary, Librarian and Curator. This notice suggested that Buckland had followed the strategy recommended by Warburton, and that it was the President who was initiating consideration of seeking a Royal Charter:

> A meeting of the Council of the Geological Society will be held on Friday the 23rd instant at Three o'Clock, at which the favour of your company is requested.
>
> The President will submit to the Council the propriety of applying to the Government for a Charter for the Geological Society.
>
> House of the Geological Society
> 20, Bedford Street, Covent Garden
> the 20th day of April 1824.[6]

At that meeting, on 23 April, the Council accepted Buckland's proposal and appointed nine of its members as a Special Committee to prepare the draft of a Charter for submission to the Crown. The members were: Buckland as President; the joint Secretaries Thomas Webster and Charles Lyell; the Treasurer John Taylor (1779–1863), a civil and mining engineer and entrepreneur; together with Henry Thomas Colebrooke (1765–1837); Henry Warburton; William Henry Fitton (1780–1861); Daniel Moore (*c.* 1760–1828), a well-known Lincoln's Inn solicitor and very active in most London scientific societies; and William Babington (1756–1833), a founder member of the Society (Lewis 2009), and as such no doubt one of the *Dei majores* that Warburton was referring to in his 13 March letter to Buckland.

Soon afterwards, Joseph Fitzwilliam Vandercom (*c.* 1754–1850), a member of the Society who was a leading London solicitor, experienced in Chancery and Privy Council work, was added to the Special Committee. He was to play a leading role in both the drafting of the Charter and the necessary consultations and negotiations with Government and the Privy Council, without any payment for his own professional services. However, the Council was advised that even with this very generous offer from Vandercom in relation to his own work and advice, the cost of drafting and petitioning the Crown for a Royal Charter would still be an estimated £300 for legal charges and administrative and printing expenses. The stamp duty payable on the Charter Deed alone would be £50. With only £643 2s 9d available in cash at the bank, plus £378 10s in investments, the Society had little by way of working capital in relation to its current commitments, let alone significant reserve funds. In the circumstances, applying for a Charter (with no certainty of success) would represent a very substantial financial risk and burden to the Society (Woodward 1907, p. 68).

The following week the Council deputed Henry Warburton, as a member of the Councils of both the Royal Society and the Geological Society, to inform the Council of the Royal Society what was being proposed. Behind the scenes, Buckland

[3] GSL: Business Papers. GS 1–22: Charter and Byelaws 1810–1993.

[4] GSL: GS 4: Bound volume of documents covering Regulations, printed reports of Council, Charter, and notices of Council and Special General Meetings, 1810–1836.

[5] The Society's solicitors for the Charter application, Vandercom & Co., remained independent until 1978 when they merged with Fladgate & Co. However, enquiries of the Archivist of the current firm, Fladgate LLP of London, have failed to produce any relevant records.

[6] GSL: GS 4/6.

was actively seeking support in both scientific and political circles, particularly within Lord Liverpool's (1770–1828) Government where Robert Peel (1788–1850), a personal friend of Buckland's, had recently been appointed Home Secretary.[7] A General Meeting of the Geological Society, held on 21 May 1824 and attended by 39 members, supported the proposal and empowered the Council 'to take such measures as shall appear to them to be the most efficient for obtaining a Charter of Incorporation for the Society'. A first draft of a Charter was quickly prepared and the responses from the informal consultations with the Council of the Royal Society were positive, so a special meeting of the whole membership of the Society was called for 2 July 1824. William Clift's copy of the notice of the meeting reads:

> A Special General Meeting of the Geological Society will be held at the House of the Society, on Friday, the 2nd July, at Eight o'Clock in the Evening, for the purpose of considering the Draft of a Charter of Incorporation for the Geological Society.
>
> By Order of the Council
> T. Webster, Sec.
>
> House of the Geological Society
> 20, Bedford Street, Covent Garden
> the 25th Day of June, 1824.[8]

This meeting endorsed the proposed text, although the series of drafts and notes of amendments[9] shows that consultations and negotiations with the Privy Council had been protracted and very detailed, much of the detailed work apparently being carried out by a Charles Batley of Lincoln's Inn.[10] Changes and additions proposed in the course of the negotiations were then marked in red ink by Batley on a draft dated 10 June 1824. For example, the original draft gave the Objects of the Society as forming 'a Society for investigating the Mineral Structure of the Earth'. At some (apparently unrecorded) point in the negotiations, the Society had decided that it wanted to broaden this by the addition of the words 'and for promoting such investigation',[11] and this was inserted in the draft in red ink, but this small and apparently innocuous addition was not included in the final text as approved by the Privy Council. There does not seem to be any evidence as to when or why this proposed addition was rejected. It seems most likely that during the face to face negotiations that Vandercom and/or Batley had with representatives of the Government, some concern was expressed by these individuals who did not want the Geological Society to have an explicit power in its Royal Charter that would permit it to campaign and lobby Government in support of its science. While it was acceptable for the Society and its members to carry out such investigations themselves, they did not wish the already high-profile young Geological Society to be given free rein to campaign actively in support of geological research.

The final version of the Society's formal Petition seeking a Charter and the draft text for the requested Charter[12] were printed by the Society's printer (and one of its founder members), William Phillips (1773–1828) (Torrens 2009). These were then lodged with the Home Office on 14 January 1825 and are preserved in the National Archives at Kew.[13] It is known that Vandercom and Warburton then had a meeting with the Attorney General in person, as they reported on this at the 18 March 1825 Council meeting. The only recorded reservation was a technical point about the permitted limit on the power of the proposed Chartered Society to possess real estate. As the Attorney General was reported to have stated that he had no objection to any of the other clauses of the submitted draft (Woodward 1907, p. 69), the request to have the power also to 'promote' geological investigation had presumably been dropped at some date between June 1824 and March 1825.

---

[7] Peel quickly became Buckland's most important patron, first supporting Buckland's nomination as a Canon of Christ Church in 1826, and then, as Prime Minister, Peel finally persuaded a very reluctant Queen Victoria to appoint the notoriously eccentric Buckland to the Westminster Deanery in 1845 when Peel nominated him to the office. The Peel Papers in the British Library, Add. MS. 40355–40555, include over 150 communications from Buckland, many of then soliciting Peel's support for various scientific causes. For correspondence, regarding the proposed Charter see BL Add. MS. 40363 f. 259 (March 1824).

[8] GSL: GS 4/7.

[9] GSL: GS 3, 4, 5 and 6.

[10] It has not been possible to find any biographical information on Batley.

[11] GSL: GS 18. This manuscript version is marked 'Finally settled in Committee 5 June. I have perused and settled and approved this draft [signed] Chas. Batley, Lin. Inn 10 June 1824'. However, at least two versions of this draft were later printed by William Phillips.

[12] GSL: GS 6/8. Headed 'Petition for Charter by the Geological Society' and dated 1824.

[13] National Archives, PRO: HO 72/1/24 (Home Office Records: Charters and Related Papers, Geological Society of London).

George the Fourth by the Grace of

Fig. 2. The original sealed copy of the Royal Charter, dated 23 April 1825 (GSL: Business Records GS/1).

The Special General Meeting of the full Society on 21 May 1824 that first considered the Royal Charter proposal of the Council had learned that at the Royal Society's Council meeting its President, Sir Humphry Davy (himself a member of the Geological Society), had expressed his opinion in favour of 'the propriety of such an application'.[14] However, Lyell reported a potential setback to Council in November, in that Davy had tendered his resignation from the Geological Society. No explanation was given and there must have been concern that this was a sign that the Royal Society had reversed its position and might in due course petition against the granting of a Royal Charter.

The Society was under considerable financial pressure. It had recently taken over the financing of the *Transactions* – which were running several years in arrears – from the printer William Phillips, and the cost of renting its Covent Garden premises and maintaining its growing Library and Museum was stretching its resources. At worst, it could easily waste half or even two-thirds of its limited funds for nothing. In the face of such a gamble it was probably a good thing that some of the key players, including Buckland and Warburton, were not against the odd bet, according to the records of the Geological Society Club (Woodward, 1907, pp. 66–67). But whatever may have been said behind the scenes, the Geological Society was in no mood for debate or compromise and the Council accepted the resignation of the President of the Royal Society without comment. Fortunately, although Davy's resignation may have been an indication of some hostility within the Royal Society to the granting of a Royal Charter to the Geological Society, if this was the case its reservations were not carried forward as a formal objection to the granting of a Charter. Accordingly, on 23 April 1825, the King in Privy Council approved and sealed with the Great Seal a Royal Charter creating the Geological Society of London as a Body Corporate. It now had its own seal and, among other things, the right to own property and to grant Fellowships (Fig. 2). On 6 May 1825 Vandercom wrote officially to the Council saying that the Charter was in his possession, and detailing the legal and other costs incurred in obtaining it.[15] A payment of £385 14s 6d was made to Vandercom, not as a fee but in repayment of his disbursements of expenses in connection with the Charter application.

[14] GSL: Council Meeting Minutes 21 May 1824, Min. 9, CM 1/1.

[15] The original sealed copy of the Royal Charter, dated 23 April 1825, is on display in the Society's Apartments in Burlington House, but is registered as the Geological Society Archives Business Records GS/1. There are further printed copies in the Charter and Byelaws series in the Archives, the first being an edition printed by William Phillips in 1827 (GSL: GS 6/9). The full text was published by Woodward (1907, pp. 263–267).

Under the terms of the Charter, William Buckland and the four Vice-Presidents on the existing Council–Arthur Aiken (1773–1854), John Bostock (1773–1846), George Bellas Greenough (1778–1855) and Henry Warburton – were appointed the first five Fellows of the Society. It was, no doubt, significant that they included the founder President, Greenough, and another founder member, Aiken, confirming the active support of what Warburton had termed the Society's *Dei majores*. Buckland was appointed President to serve until not later than February 1826, and any three of the five designated founder members were authorized to serve as a quorum for the purposes of electing members of the unincorporated predecessor body (the existing Society members) and others worthy of membership, as either Fellows or Foreign Members.

Buckland, Warburton and Greenough held the first formal meeting of the interim 'Council of Five' under the Charter, and appointed as Fellows the remaining 14 members of the existing Council and the five Trustees of the Society who had not been named in the Charter. Of the original founders, only William Hasledine Peyps (1775–1856) and William Babington were among this group, although all the founders were still alive (Lewis 2009, table 4), except James Parkinson (1755–1824) who had died the previous year. Further appointments of members under the Charter quickly followed and by the end of a meeting of the new Council, held on 20 May 1825, 338 Fellows and 48 Foreign Members had been formally elected.

From its early days, the Geological Society had appointed a wide range of Honorary Members who were exempt from paying membership dues (Woodward 1907, pp. 268–291; Herries Davies 2007, pp. 16–17). At the end of the first year, 1807, the Society, then less than two months old, still had just the 13 founder members, but 42 Honorary Members had been appointed at the first formal meeting on 4 December. By the end of 1808 the Ordinary Members totalled 48, compared with 87 Honorary Members. Some of these were distinguished scientists or academics, others men of influence in politics and society, but the list seems to be predominantly drawn from the growing number of non-member provincial correspondents providing British and foreign 'geological intelligence' to the Society, particularly in relation to the geological map project being led by Greenough (Knell 2009; Kölbl-Ebert 2009). The last Honorary Member appointments were made in 1810, by which time 120 had been elected. By May 1825 the Society still had 48 of these original Honorary Members on the books. Each was given the option of becoming Fellows without ballot on application and payment of the standard admission fee, but, in the event, only six of these applied to become subscription-paying Fellows. These included the engineer James Watt (1736–1819) and two of those elected as local correspondents almost two decades earlier: Edward Mammat of Leicestershire and Thomas Meade of Somerset (Woodward 1907, pp. 70 and 285).[16]

Finally, on 3 June 1825, the new Royal Charter was formally presented to Buckland as President at a special meeting of the Society in the 20 Bedford Street apartments, after which Buckland hosted a symbolic and celebratory dinner at the nearby Freemasons' Tavern in Covent Garden, where the inaugural meeting of the Society had been held on 13 November 1807.

On 17 February 1826 Buckland, on behalf of the 'Committee of Five' appointed in the Charter, reported formally to the first annual general meeting on the measures taken to implement the transitional provisions of the Charter, including their appointment of Fellows and Foreign Members, and the accounts of the expenses incurred. The annual meeting passed two resolutions of thanks, one to Vandercom for his 'liberal conduct in gratuitously bestowing his attention to [...] the progress of the Charter', and the second to the Committee of Five:

> for the zeal and judgement with which they have discharged the important duties entrusted to them for their prudence in adhering, as far as possible, to the system according to which the proceedings of the Geological Society have hitherto been conducted.[17]

At the end of his two years in office, Buckland handed over the presidency to the first elected President and Council of the new Geological Society of London Incorporated by Royal Charter, John Bostock FRS. He did so in the knowledge that the very considerable expense, efforts and risk involved in both the formal submissions and behind-the-scenes lobbying (in which Buckland himself had taken the leading role and excelled) had paid off, and that the public standing and influence of the Society, and through it English geology, had never been higher.

[16] The others were Lieutenant General Sir James Affleck, Sir Richard John Griffith and Professor Thomas Thomson, MD, FRS.

[17] GSL: GS 6.

The Somerset House Apartments, 1828

In 1874 the Society moved into its present apartments in Burlington House. In greatly welcoming this move, the then President, John Evans (1823–1908), summarized the history of the Society's accommodation in his 1875 Anniversary Address, the first to be given in the new Burlington House Meeting Room. Early in 1809 the Society had rented accommodation at No. 4 Garden Court, Temple, moving the next year to 3 Holborn Row, Lincoln's Inn Fields, which they shared for five years with the Medical and Chirurgical Society (Lewis 2009). From 1816 the Society rented a fairly substantial building in Covent Garden at 20 Bedford Street (Evans 1875, pp. lvii–lx).[18] However, apart from the cost, the Bedford Street Apartments were less than ideal in terms of the accommodation offered. For example, when William Conybeare (1787–1857) demonstrated and reported on the almost complete Plesiosaur found by Mary Anning (1799–1847) at Lyme Regis (Conybeare 1824; North 1956), the specimen proved too large to get much beyond the front door.

As Buckland had recognized, obtaining a Royal Charter was very important in terms of public recognition and the growing status of the Society and its science, but it did not pay the bills. Even with an increasing membership, the Society had an increasing gap between its income and its expenditure, and the substantial costs of obtaining the Charter had depleted the already limited reserves. The accounts submitted to the 20 May 1825 annual meeting showed that the Society's invested capital was down to just £188 5s 7d. The Society would have been very well aware that three of the older chartered bodies – the Royal Society, the Society of Antiquaries and the Royal Academy – had been provided by the Government with free accommodation in Somerset House when it was built in 1780. The original Somerset House, an old Tudor palace, had been demolished in 1775 so the new building could house many of the Government's offices and the principal learned societies under one roof in order to promote greater efficiency among the Government bureaucracies – large parts were occupied by the Inland Revenue, for example. Having gained the Charter, could the Geological Society at the same time both ease its financial situation and enhance its status even further, by obtaining Government accommodation as well?

Buckland's successor as President, the elderly retired physiologist John Bostock, held office for only one year rather than the usual two years, and was followed in February 1827 by William H. Fitton, also a medical man until he married a wealthy heiress and was able to devote all his time to his scientific interests, particularly geology. Like Buckland, Fitton was a progressive and reforming President, introducing, for example, the publication of the Society's *Proceedings* and the tradition of the President's annual Anniversary Addresses, which quickly became a highlight of the English scientific calendar.

One immediate issue of concern for Fitton was the adequacy and cost of the Bedford Street accommodation, thus the newly chartered Society began to look at the possibility of obtaining free accommodation in Somerset House. Although the building had been fully occupied, things changed in October 1826, the date laid down by Parliament for the final draw of the scandal-ridden National Lottery, under the Lottery Abolition Act of Parliament of 1823.[19] The Lottery Office had occupied part of the eastern wing of Somerset House, but these rooms were now vacant.[20]

The Society began enquiring, or probably more accurately lobbying, about the possibility of the Geological Society moving into this accommodation rent free on the same basis as the Royal Society.[21] It seems clear from the, again, very limited surviving documentation, that the Royal Society and Robert Peel, then the Home Secretary, were mobilized in support. The Lords

[18] Although 3 Holborn Row, Lincoln's Inn Fields, survives (in the same block on the north side of Lincoln's Inn Fields as the Sir John Soane Museum), the other two do not. Garden Court, Temple, was rebuilt in 1884–1885 (Bradley & Pevsner 1997, p. 347), and 20 Bedford Street was rebuilt *c.* 1879: the rebuilding plans and specification are in the London Metropolitan Archives, Duke of Bedford Papers, catalogue ref. E/BER/CG/05/05/004.

[19] Parliamentary Acts, 1 and 2 Geo 4 *c.*120.

[20] In their research for the Society's 2007 bicentenary celebrations, Professor Gordon Herries Davies and Wendy Cawthorne, the Society's Assistant Librarian, identified the location of the former Lottery Office and later the Society's Somerset House Apartments on the east side of the quadrangle (Herries Davies 2007, pp. 58–61).

[21] The 'Solicitors' Bundle' of files of documents prepared by the Society for the recent High Court action against the Deputy Prime Minister over the terms of the occupation of the Burlington House Apartments brings together copies of potentially relevant documents going back to this original allocation of free accommodation in Somerset House, including Peel's March 1828 letter. However, these files have not yet been catalogued and added to the Society's Archives. See also Woodward (1907, pp. 73–75) and Herries Davies (2007, pp. 58–61) for more details of the negotiations relating to the move to Somerset House.

Fig. 3. The Geological Society in session in the first Somerset House meeting room with its 'parliamentary' layout, *c.* 1830. (GSL: GSL/312).

Commissioners of the Treasury appear to have been unwilling to grant a lease or occupation licence to the Geological Society as such (perhaps as they were afraid of establishing a precedent), but an ingenious compromise was reached. The former Lottery Office at Somerset House was allocated by Treasury Minute to the Royal Society as an addition to its own accommodation and occupation agreement, but on the understanding that this would then be made available to the Geological Society for its use. This decision was communicated to both the Royal Society and the Geological Society by Robert Peel in a letter dated 19 March 1828.[22]

The distinguished neo-classical architect Decimus Burton (1800–1881), a Fellow of the Geological Society, was engaged to design and supervise the adaptations of the premises (and later of additional rooms allocated in 1834), and a subscription was opened to raise the £1500 cost of the works. The Bedford Street lease was surrendered from 12 August 1828 and the Society met officially for the first time in Somerset House the same day. However, it was some months before the works were fully completed and the Society could start using the new Meeting Room. The first meeting of the 1828–1829 winter session, on 7 November 1828, had to be held at the Harley Street house of the President, William Fitton.

But it had all been worth the effort. The Geological Society was one of the few scientific organizations singled out for praise by Charles Babbage (a frequent guest at the Society's meetings) in his notorious 1830 *Reflections on the Decline of Science in England and on some of its Causes.* Babbage particularly praised the open debates on the papers presented in the Society's potentially confrontational 'parliamentary' room layout (Fig. 3), one of Buckland's innovations during his presidency (discussed by, among others, Thackray 2003, pp. v–ix). Babbage commented:

> [The Geological Society] possesses all the freshness, the vigour, and the ardour of youth in the pursuit of a youthful science, and has succeeded in a most difficult experiment, that of having an oral discussion on the subject of each paper read at its meetings. To say of these discussions that they are very entertaining is the least part of the praise which is due to them.
>
> (Babbage 1830, p. 45).

Conclusion

By the end of 1828 the Geological Society of London, still only 21 years old, had in less than

[22] A copy of Peel's letter is now in the 'Solicitor's Bundle' referred to previously.

Fig. 4. Henry De La Beche's proposal for a coat of arms for the Geological Society, 13 June 1825. (Buckland (Oke/Gordon) Archive, collection of Roderick Gordon and Diana Harman, formerly Devon Record Office DRO 138M/F73.)

five years achieved two major advances in official recognition and support for itself and for the science of geology more widely, despite the traditional laissez-faire hostility to any form of public support for science (or, indeed, the arts). The Society had obtained the Royal Charter of Incorporation and all that this signified in terms of official and social status and recognition in the prevailing social and political systems and values of the early nineteenth century. It had also been granted highly prestigious and rent-free accommodation in Somerset House, albeit indirectly through the device of allocating this through the Royal Society. The Geological Society was henceforth to be found alongside the country's leading cultural and scientific bodies: the Royal Society (founded in 1660), the Society of Antiquaries (founded exactly 100 years before the Geological Society in 1707) and the Royal Academy of Arts (founded in 1768).

However, one of the other ambitions Buckland had for the Society during his 1824–1826 Presidency was not fulfilled. As noted earlier, the Royal Charter gave the Society a legal identity or personality of its own and one of the perhaps lesser-known privileges that came with this was the right to a Coat of Arms. It is clear that Buckland carried out, privately at least, soundings about applying to the College of Heralds for a Grant of Arms to the Society in 1825. A letter to Henry De La Beche (1796–1855) dated 7 July 1825 shows that Buckland had been consulting about this possibility, no doubt among others, with three of the leading Royal Academicians of the day: Sir Francis Chantrey (1781–1841), the most fashionable and successful British sculptor of the first half of the nineteenth century; Sir David Wilkie (1785–1841), genre painter; and Sir Robert Ker Porter (1777–1842), who was Painter-in-Ordinary to the King and then the Queen from 1830 to his death (McCartney 1977, pp. 57–59 and 70). This letter to De La Beche was apparently in response to one in which De La Beche, a noted geological illustrator and caricaturist, had offered sketch designs for a Coat of Arms for the Society, as well as two different designs (plus one variant) for a Seal for the Society.[23] The Coat of Arms proposal had been 'admired exceedingly by Chantrey, Wilkie and Sir R. K. Porter', according to Buckland (McCartney 1977, p. 58). De La Beche proposed that the coat of arms (Fig. 4) should include in the main shield a simplified section through a bone cave, apparently based on one of the plates in Buckland's *Reliquiae Diluvianae* (1823), in the upper left-hand quarter.[24] The other upper quarter would have included three ammonites, beneath which would be a full-width idealized geological section of the northern flanks of the Alps. Rampant skeletons of an ichthyosaur (on the left) and a plesiosaur (on the right) would serve as heraldic supporters of the shield. De La Beche proposed that the crest above the shield should be the commonly found heraldic device of the *bras armé* – an arm wielding a

[23] Buckland (Oke/Gordon) Archive formerly deposited in the Devon Record Office, catalogue ref. DRO 138M/F73. Henry De La Beche to William Buckland 13 June 1825.

[24] Contrary to the interpretation of McCartney (1977), the cave section does not seem to be of Paviland Cave (Goat Hole): it seems to match more closely *Reliquiae Diluvianae* plate 14, which shows Scharzfeld, one of the German caves studied by Buckland.

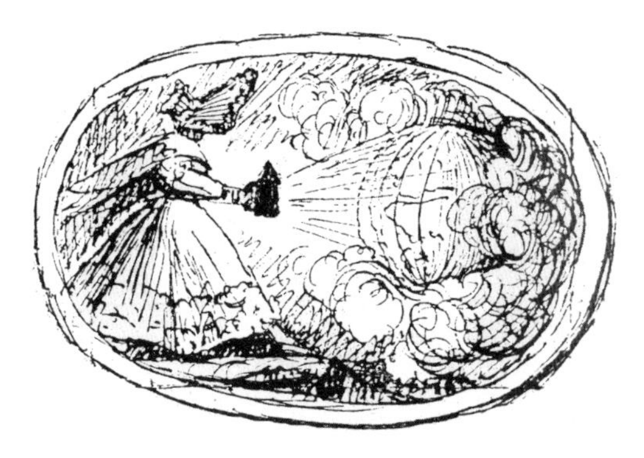

Fig. 5. Henry De La Beche's proposal for a seal for the Geological Society, 13 June 1825: 'Science dispelling the darkness that covered the Earth'. (Buckland (Oke/Gordon) Archive, collection of Roderick Gordon and Diana Harman, formerly Devon Record Office DRO 138M/F73.)

weapon, although in this case the 'weapon' would be a geological hammer rather than the usual dagger or arrow. In the same letter De La Beche had also proposed a sketch of a female figure shining a light on all the Earth as a design for an official seal for the Society (Fig. 5). In his own words, this aimed to show 'Science dispelling the darkness that covered the Earth!!'

It seems that Buckland was himself exploring the possibilities for a coat of arms, as there is a small undated sheet of paper in his distinctive hand filed with the 15 June 1825 letter from De La Beche with another possible design (Fig. 6). In this, the upper half of the shield has sketches of six fossils including a large ammonite. A key to the lower half of the shield indicates that the left-hand side would include four images representing geological surveying – '1 Quadrant', '2 bottle of acid', '3 Barometer', '4 Specific gravity balance', while the fifth image opposite these, '5 crystals', would represent mineralogy. The supporters of the shield would be two ichthyosaurs, but represented here by life reconstructions rather than De La Beche's ichthyosaur and plesiosaur skeletons, while a quill pen crossed with a miner's or geologist's hammer would be displayed above the shield.

In the event, no further progress was made with what would have been the ultimate public expression of the Society's newly attained status as a Chartered Society. However, in 1849, at a time when there seems to have been some discussion about seeking a 'Royal' title for the Society,

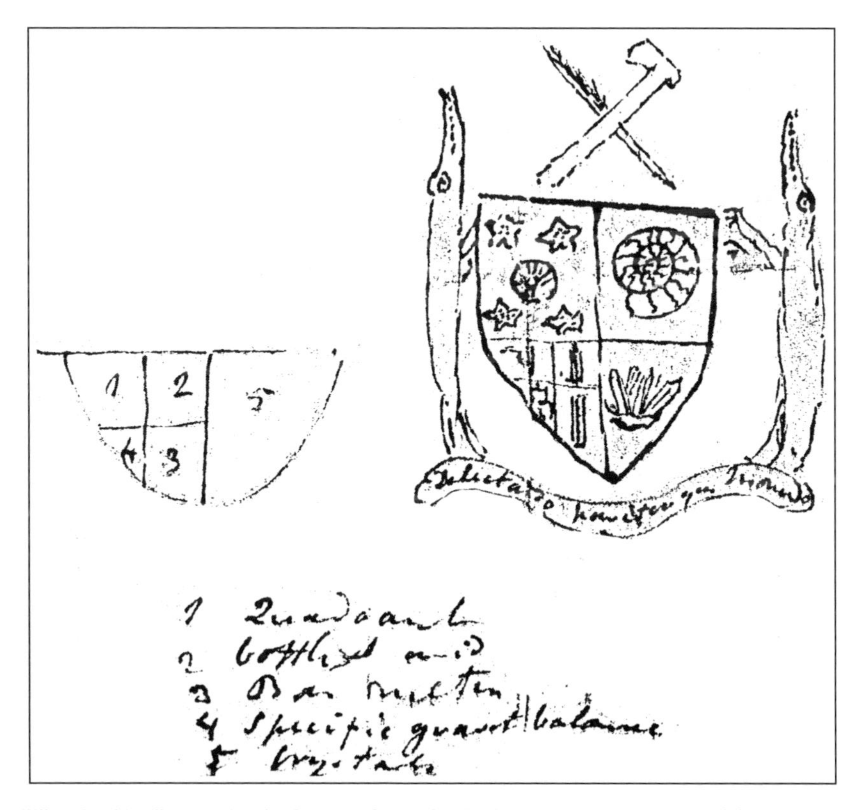

Fig. 6. William Buckland's own sketch of a coat of arms for the Geological Society (undated, but presumably June 1825, since it was filed with the De La Beche letter of 13 June 1825). (Buckland (Oke/Gordon) Archive, collection of Roderick Gordon and Diana Harman, formerly Devon Record Office DRO 138M/F73.)

Fig. 7. The Geological Society's historic logo, based on William Phillips' typography for the Society's 1811 *Transactions*, still in use as its corporate identity today.

William Hellier Baily (1819–1888), a draughtsman and later a geologist with the Geological Survey, returned to the subject with a proposed coat of arms for the 'Royal Hammerers', which adapted some elements of De La Beche's 1825 design, but again no progress was made in seeking a Grant of Arms (McCartney 1977, p. 59). Consequently, in its bicentenary year the Geological Society of London continues to rely on the plain 'GS' black lettering monograph (Fig. 7) used by William Phillips in 1811 on the original copy of the Society's *Transactions*, for its logo and visual identity.

I was particularly indebted to Arthur Lucas for his very detailed editing advice and recommendations. Many thanks are also due Alan Bowden and Cherry Lewis, editor of this paper. Any remaining errors or weaknesses are, of course, my own. I am also very grateful to Roderick Gordon for permission to reproduce the three sketch designs for a Society coat of arms from the Buckland and De La Beche correspondence.

Archives

GSL: Archives of the Geological Society of London.

References

BABBAGE, C. 1830. *Reflections on the Decline of Science in England and on some of its Causes*. Fellowes, London.

BOYLAN, P. J. 1970. An unpublished portrait of Dean William Buckland, 1784–1856. *Journal of Society for the Bibliography of Natural History*, **5**, 350–354.

BRADLEY, S. & PEVSNER, N. 1997. *The Buildings of England. London I. The City of London*. Penguin Books, Harmondsworth.

BUCKLAND, W. 1823. *Reliquiae Diluvianae: Observation on the Organic Remains Contained in Caves Fissures and Diluvial Gravel, and on Other Geological Phenomena, Attesting the Action of an Universal Deluge*. John Murray, London.

BUCKLAND, W. 1840. Anniversary Address of the President. *Quarterly Journal of the Geological Society*, **3**(68), 210–267.

CONYBEARE, W. 1824. On the discovery of an almost perfect skeleton of the Plesiosaurus. *Transactions of the Geological Society*, (Series 2), **1**(2), 381–389.

EVANS, J. 1875. Anniversary Address of the President *Quarterly Journal of the Geological Society*, **31**, xxxvii–lxxvi.

GORDON, MRS. [E. O.] 1894. *The Life and Correspondence of William Buckland, DD, FRS*. John Murray, London.

HERRIES DAVIES, G. L. 2007. *Whatever is Under the Earth. The Geological Society of London 1807–2007*. Geological Society, London.

KNELL, S. J. 2009. The road to Smith: how the Geological Society came to possess English geology. *In*: LEWIS, C. L. E. & KNELL, S. J. (eds) *The Making of the Geological Society of London*. Geological Society, London, Special Publications, **317**, 1–47.

KNIGHT, D. 2009. Chemists get down to Earth. *In*: LEWIS, C. L. E. & KNELL, S. J. (eds) *The Making of the Geological Society of London*. Geological Society, London, Special Publications, **317**, 93–103.

KÖLBL-EBERT, M. 2009. George Bellas Greenough's 'Theory of the Earth' and its impact on the early Geological Society. *In*: LEWIS, C. L. E. & KNELL, S. J. (eds) *The Making of the Geological Society of London*. Geological Society, London, Special Publications, **317**, 115–128.

LEWIS, C. L. E. 2009. Doctoring geology: the medical origins of the Geological Society. *In*: LEWIS, C. L. E. & KNELL, S. J. (eds) *The Making of the Geological Society of London*. Geological Society, London, Special Publications, **317**, 49–92.

MATTHEWS, H. C. G. 2004. Henry Warburton, (1784–1858). *Oxford Dictionary of National Biography online*. Oxford University Press.

MCCARTNEY, P. J. 1977. *Henry De La Beche: Observations on an Observer*. Friends of the National Museum of Wales, Cardiff.

NORTH, F. J. 1956. W. D. Conybeare. His geological contemporaries and Bristol associations. *Proceedings of Bristol Natural History Society*, **29**(2), 133–146.

THACKRAY, J. C. 2003. *To See the Fellows Fight: Eye Witness Accounts of Meetings of the Geological Society of London and its Club, 1822–1868*. British Society for the History of Science, Faringdon, Oxfordshire.

TORRENS, H. S. 2009. Dissenting science: The Quakers among the Founding Fathers. *In*: LEWIS, C. L. E. & KNELL, S. J. (eds) *The Making of the Geological Society of London*. Geological Society, London, Special Publications, **317**, 129–144.

WOODWARD, H. B. 1907. *The History of the Geological Society*. Geological Society, London.

Facts and fancies: the Geological Society of London and the wider public, 1807–1837

RALPH O'CONNOR

Department of History, University of Aberdeen, Crombie Annexe, Meston Walk, Aberdeen AB24 3FX, UK (e-mail: ralph.j.oconnor@gmail.com)

Abstract: The leading lights of the Geological Society announced the birth of a newly scientific form of Earth science by claiming to dissociate geology from the grand theories, theological controversialism and flights of fancy that they felt had dominated eighteenth-century practice. For these gentlemen, geology was to comprise strict empirical induction. They cultivated a historical myth according to which their predecessors had been hopelessly romantic theory-mongers with overactive imaginations, while they themselves were sensible, sober men of science.

But if this was so, how did geology succeed in winning such an enormous middle- and upper-class public by the late 1830s? Public support required public interest, and public interest in this period was most easily stirred by the romantic, the speculative and the poetical. Older theories of the Earth remained popular for this very reason, as did biblically literalist reconstructions of Earth history. The challenge for the new school of geology was to dissociate Earth science from the content and methodology of such theories while retaining their accompanying sense of excitement, wonder and pleasure. This paper explores how members and allies of the Geological Society negotiated (or ignored) their own suspicions about the deceptive power of the 'imagination' when promoting geology as a science worth the public's attention.

One of the chief means by which the Geological Society's founders presented themselves as leading figures was to redefine 'geology' as a strictly empirical science, devoted to fact-gathering rather than speculation. This was not merely a bid to avoid the undignified cosmological polemics that they felt marred geological debate in Edinburgh. Speculation was a dangerous business in early-nineteenth-century London. The ruling classes were still anxious about the possibility of revolution breaking out in Britain; many of them blamed the recent French Revolution on French Enlightenment intellectuals, whose subversively secular speculations had supposedly encouraged the masses to rise up against Church and State. Geology was one of the new sciences that ended up being tarred with this brush (Porter 1978, pp. 435–436) because its best-known literary medium – the 'theory of the Earth' – was an extremely speculative genre, often dealing in grand narratives of Earth history and often ignoring (or being very creative with) the biblical text of Genesis 1. Geology, according to conservative commentators throughout the early nineteenth century, was a symptom of 'the *malaria* of French philosophy' (Brown 1838, p. 38).[1]

How then to cure geology of French malaria? The Geological Society's official solution was to purge their science of any vestige of grand theory, theological controversy and, above all, an excess of imagination (Laudan 1977; Porter 1977, pp. 204–208; compare Daston 2001). Writing theories of the Earth was indefinitely postponed; for now, geologists were to gather and arrange facts in line with the empirical philosophy advanced by Francis Bacon two centuries earlier.[2] The first volumes of the Society's *Transactions* bore a Latin epigraph taken from Bacon's *Novum Organum* including the injunction 'non belle et probabiliter opinari, sed certo et ostensive scire' ('not to produce attractive and plausible conjectures, but certain and demonstrable knowledge').[3] The Society's spokesmen presented this Baconianism as fully in line with developments in other sciences, all of which were beginning 'to assume a character of strict experiment or observation, at the expense of all hypothesis, and even of moderate theoretical speculation' ([Fitton] 1817*a*, p. 175). If 'systems'

[1] On the effects of the conservative backlash in the years following the French Revolution, see Morrell (1971) and Brooke & Cantor (1998, pp. 195–200).

[2] A similar redefinition of geology was being undertaken by savants in Napoleonic France, less narrowly empiricist than the English equivalent but employing precisely the same strategy of disparaging more speculative approaches as romantic nonsense (Rudwick 2005, pp. 456–463).

[3] Title page. *Transactions of the Geological Society*, **2**, i. (1814).

From: Lewis, C. L. E. & Knell, S. J. (eds) *The Making of the Geological Society of London*. The Geological Society, London, Special Publications, **317**, 331–340.
DOI: 10.1144/SP317.19 0305-8719/09/$15.00

of any kind were to be avoided, philosophical ignorance could even be trumpeted as a gentlemanly virtue. William Fitton noted without any irony that:

> the members of the Geological Society have derived great benefit from their want of systematical instruction [. . . .] They were neither Vulcanists nor Neptunists, nor Wernerians nor Huttonians, but plain men, who felt the importance of a subject about which they knew very little in detail.
>
> ([Fitton] 1817*b*, p. 70)

By remaining ostentatiously unbiased by any theoretical allegiance, this new breed of philosophers could reinforce their status as disinterested and 'sober-minded' gentlemen in the conservative political climate of the early nineteenth century ([Fitton] 1817*a*, p. 177).

If the geologists saw themselves as 'plain men', it did not necessarily follow that all plain men could hope to become geologists. In their pamphlet, *Geological Inquiries*, first issued in 1808 and reprinted in various forms well into the 1810s, George Greenough and his colleagues urged the public to take their place in the hierarchy of knowledge production by sending specimens and observations to the Geological Society, envisaged as a 'repository for any facts that may be communicated to them'. Only the gentlemanly philosopher, with an independent income to guarantee his 'total devotion of time' to his science, could hope some day to 'reduce Geology to a system', but such a system was perceived as (at best) a distant future promise rather than something attainable or even desirable at the present time (Anon. 1817, pp. 421–422). Fitton announced to readers of the *Edinburgh Review* in 1817 that:

> matter-of-fact methods have lately been gaining ground in Geology [. . .] hypotheses are now scarcely listened to; and even the well-organized theories which, a short time since, created so much controversy, receive in this day little attention or comment.
>
> ([Fitton] 1817*a*, p. 177)

Fitton was certainly overstating the case: theories of the Earth continued to be read and enjoyed by the intelligentsia. But his statement reveals the kind of philosophical practice that the Geological Society wished to promote among the wider public in the 1800s and 1810s as part of its programme to redefine the scope and nature of geology.

As with many disciplines emerging in the early nineteenth century, geology's self-styled elite forged a historical myth in which they represented enlightenment liberated from a benighted era of fable and superstition, or as a science that had outgrown its 'infancy of speculation and conjecture' ([Fitton] 1817*a*, p. 176; see Porter 1977, p. 148). For the Geological Society, rebranding 'geology' usually meant being rude about their predecessors on whose shoulders they were standing. Men like Thomas Burnet, the Comte de Buffon and even Jean André Deluc were publicly castigated for having let their imaginations run away with them. Members of the Society used widely circulating periodicals and public lectures to ensure that the British intelligentsia got the message. In his 1805 geology lectures at the Royal Institution, for instance, Humphry Davy called Burnet's work 'highly interesting and entertaining' as a 'poetical romance', but from a 'philosophical' viewpoint 'wholly unworthy of attention' (Siegfried & Dott 1980, p. 43).

Other members and associates of the Geological Society wrote admiring reviews of early issues of the Society's *Transactions* in leading literary periodicals such as the *Edinburgh Review*. Such reviews typically began by alluding to the dark ages of unbridled imagination against which the new Baconians stood. John Playfair, who since the death of James Hutton had reinvented himself as a thorough empiricist (Porter 1977, p. 207), opened his review of the first volume in 1811 by looking back to:

> the time when the vague and cursory information that every man might glean from the objects that were perpetually before him, when combined and magnified by a powerful imagination, was sufficient for all the purposes of geological speculation. According to this view of the matter, a man [. . .] stood no more need of the assistance of others, than if he had been at work in the regions of Poetry or Romance.
>
> ([Playfair] 1811–1812, p. 207)

The result had been a rash of 'visionary and fantastic theories', or what Fitton (when introducing his review of the third volume in 1817) called 'easy and tempting speculations which, till of late, it has been the fashion to dignify with the name of Geology' ([Playfair] 1811–1812, p. 207; [Fitton] 1817*b*, p. 74; compare [Fitton] 1823–1824, p. 196). The contrast between these sober Baconians and their romantic predecessors was sometimes stretched to the point of caricature, as when Playfair referred to early geology as 'a species of mental derangement, in which the patient raved continually of comets, deluges, volcanos and earthquakes' ([Playfair] 1811–1812, p. 207). All this was set in sharp relief by a sensible chap like Greenough, whom one could rely on to gather facts or take one out for expensive dinners. As John Ayrton Paris put it, 'The fabulous and romantic age of geology may be said to have passed away' (Paris 1818, p. 168).

In the 1800s and 1810s, then, various members of the Geological Society of London were engaged in disseminating a flattering image of themselves as scientific reformers and architects of a collaborative research programme, set against a backdrop of bad practice depicted in terms of the romantic and imaginative excesses of solitary would-be geniuses. This picture has since become embedded in popular accounts of the 'heroic age' of geology. It needs to be taken with more than a pinch of salt, on several levels. First, it completely misrepresents the aims, achievements and practices of previous investigators into the structure and history of the Earth; this point is too obvious to need defending here (see Porter 1977; Rudwick 2005).

More importantly, its characteristic trope of setting 'truth' against 'imagination' is a blunt polemical instrument and one of the oldest tricks in the rhetorical armoury of Western literature for discrediting an opponent (see Heringman 2004, pp. 269–271). It should not be taken as evidence that science and the imagination, or science and speculation, were necessarily seen as mutually exclusive in all circumstances by the person wielding this rhetorical instrument. For instance, in a book published in 1831 but drafted long before, John MacCulloch (1773–1835) cast aspersions on the diluvial theory by alluding to 'that love of the marvellous which so often loses sight of science, and of truth also'; but he made this remark in a book that defended the practice of speculation and was itself subtitled *A Theory of the Earth* (MacCulloch 1831, vol. I, p. 448). He was also renowned for the poetic quality of his geological writings on the Hebrides, which were often excerpted by later popularizers wishing to exploit their readers' love of the marvellous (e.g. Mantell 1838, p. 644). For MacCulloch, the romantic imagination had its place in scientific writing as long as it was not felt to intrude on the sacred ground of scientific reasoning. For the more severely empirical Greenough, romantic effusions about nature were kept at bay in his published writings but indulged to the full in his private diaries (Wyatt 1995, pp. 172–174).

Furthermore, although empiricism continued to play a defining role in its public image, in practice the Geological Society did not long hold to the extreme empiricism favoured by Greenough. In the 1810s the Society's more recent recruits, such as Thomas Webster, William Buckland and the Conybeare brothers, began serious attempts to correlate strata across large tracts of Britain and (later) Europe (Rupke 1983, pp. 118–123). Theory and generalization were central to this enterprise: hard-line empiricists such as Greenough and John Kidd rejected the idea of a standard stratigraphy for this very reason (see, for example, Greenough 1819), but by the late 1810s research of this kind was taking up a sizeable proportion of the Society's *Transactions*, along with William Smith's method of using characteristic fossils to identify strata. For geologists like Buckland (as for his Continental colleagues) this opened the door once again to the possibility of reconstructing a history of the Earth and its inhabitants, a still more speculative and imaginative pursuit.

In various ways, then, the Geological Society's embargo on speculation and imagination operated in much more limited ways than might be thought from the prominence of this rhetoric in the Society's public statements. Even here the Society's position was not unambiguous. For every assertion of sober empiricism in review articles, popular treatises or public lectures, one may find at least one counterbalancing assertion of geology's significance in terms of its appeal to the speculative imagination, and such assertions often used explicitly 'visionary' or romantic language. The reason for this was simple. It was all very well for Greenough and his colleagues to advise their projected regional network of investigators to act as docile fact-gatherers, but in order to maintain this network (and support from the wider public) it was necessary from the very beginning to ensure that geology retained its traditional appeal as a romantic, sublime, imaginatively inspiring endeavour. Even *Geological Inquiries* conceded this point, referring to geology as 'a sublime and difficult science', a juxtaposition that nevertheless implies that only those with leisure to comprehend the many difficulties of the science would be able to attain its 'sublime' heights (Anon. 1817, p. 421).

The old theories of the Earth had won much of their readership and cultural prestige not because of their economic utility, but because their romantic visions of former worlds and apocalyptic cataclysms had caught their readers' imaginations, somewhat along the lines of the huge historical landscapes being painted in the late eighteenth and early nineteenth centuries by such artists as Philip James de Loutherbourg, J. M. W. Turner and John Martin (Paley 1986; Boime 1990, 2004). Burnet's Restoration-era *Sacred Theory of the Earth* was still admired in the Regency period as a great work of literature (Jackson 1985, p. 318), and deservedly so. If the kind of geology being promoted by the Geological Society was to gain the cultural prestige it needed, it had to provide this kind of interest for a wider audience. For those who engaged in popularization (by which word I mean simply 'making something more widely known'), this would mean cultivating imaginative excess and encouraging speculation among their public, in the teeth of the Society's official party line. In what remains of this paper I shall outline some of the ways in which this was done in the Society's first three decades.[4]

[4] For a detailed account of how geology was popularized in early nineteenth-century Britain, see O'Connor (2007*b*).

Popularization

Today we often use the word 'popularization' to refer to information handed out relatively cheaply to ordinary people, but very few readers of geology books or audiences at geology lectures were ordinary people: they were the wealthiest and most prestigious sector of society. The disparity in incomes across the social spectrum was far greater in the early nineteenth century than it is today. The top slice of church livings earned their incumbents over £1000 per annum.[5] Many curates earned as little as £50 per annum; clerks, tradesmen and servants would be lucky to earn half that. To make this point more concretely, Table 1 gives some examples of the real-life cost of a few geological essentials for a typical lower-middle-class earner, a lawyer's clerk in London in the 1830s.

It comes as a shock to realize that the classic fossil treatise of the Napoleonic years, James Parkinson's *Organic Remains of a Former World* (1804–1811), would have cost our lawyer's clerk almost a third of his annual earnings. Books before the 1840s were prohibitively expensive, cheap periodicals were in their infancy and, in any case, most of the British (or at least English) population were illiterate or only semi-literate.[6] As for museums, it may be true that before the 1830s many cabinet collections charged no entry fee, such as Gideon Mantell's museum in Lewes. But most museums were private-membership institutions: you had to know the right people. You could collect fossils for free (depending on the locality); but keeping up to date with the latest developments in geology was a different proposition. Geology in early nineteenth-century Britain was an exclusive science, unless you happened to be practising it for a living like William Smith.

Let us now examine how geologists and their allies sought to awaken these audiences' excitement, wonder and pleasure at the new science. We glance first at two of the Geological Society's founders, Humphry Davy and James Parkinson. Both men brought poetry back into the science by promoting it well outside a Geological Society context. By 1807 Davy was one of the most celebrated and flamboyant lecturers in London (Golinski 1992; Knight 1992), famous (if contemporary caricatures are to be taken literally) for inducing fits of chortling and farting with his newly discovered gas, nitrous oxide. Contemporary journalists found this hilarious. Davy was the star performer at the prestigious Royal Institution, where his research was sponsored by aristocrats who, besides agricultural utility, wanted some fun for their money. Between 1805 and 1811 Davy also lectured on geology, promoting it as an extension of the Grand Tour:

> The imagery of a mountain country, which is the very theatre of the science, is in almost all cases highly impressive and delightful, but a new and a higher species of enjoyment arises in the mind when the arrangements in it [...] are considered.
>
> (Siegfried & Dott 1980, p. 13)

Davy's procedure is typical of scientific popularizers in his day, as in our own: he latches onto something his audience already takes pleasure in (going on expensive holidays) and channels this pleasure in a new direction. His *Elements of Agricultural Chemistry* (1813) contains a striking visual parallel to this process, incorporating a picturesque engraving of mountains by Thomas Webster, designed to give both pleasure and mineralogical information (Fig. 1). As he put it in his lectures:

> Every mountain chain offers striking monuments of the great alterations that the globe has undergone. The most sublime speculations are awakened, the present is disregarded, past ages crowd upon the fancy, and the mind is lost in admiration of the designs of that great power who has established order in which at first view appears as confusion.
>
> (Siegfried & Dott 1980, p. 13)

This does not sound much like the hard-nosed empirical programme sketched out in *Geological Inquiries*, but Davy knew that to win over his audience he had to capture their imaginations.

---

[5] To gain a crude sense of what this figure might signify in present-day terms, one might compare today's median average British full-time annual salary (*c*. £21 840 in 2006) with a typical 'comfortably middle-class' British income in the 1830s, 48 shillings per week (see Topham 1992, p. 400 n. 19). This would deliver a ratio of 1:175 for pounds sterling, making the church living cited worth £175 000 today. However, this ratio would represent only a 'comfortably middle-class' perspective and cannot be applied across the board. Rudwick (1985, p. 461) has suggested 1:40 from a more genteel perspective, while Taylor & Torrens (1986, pp. 145–146) have suggested 1:200 from a farm labourer's perspective. Even in this limited domain, such ratios are useful only as *very* general pointers, because of serious discrepancies across time in the cost of living, taxation policy and other important variables (on the costs of living in this period see Burnett 1969).

[6] T. W. Heyck, drawing on a range of statistical accounts relating to England and Wales, suggests 'that in the early Victorian years more than a third of the population was illiterate, and that perhaps another third to a half was only semi-literate' (i.e. unable to read connected prose) (Heyck 1982, p. 26). This would leave between one-sixth and one-third of the population fully literate. On working-class literacy see Vincent (1989).

Table 1. *The cost of geology. Costs of some items for a London clerk earning 10s a week*

Item	Cost	Cost in real terms
First year's membership of the Geological Society of London in the 1830s (admission fee plus annual subscription)	9 guineas (£9 9*s*)	$4\frac{1}{2}$ months' wages
Parkinson's *Organic Remains* (1804–1811)	8 guineas	16 weeks' wages
Buckland's *Geology and Mineralogy* (1836)	£1 15*s*	3 weeks' 2 days' wages
'Cheap' edition of Lyell's *Principles of Geology* (1834)	24*s*	2 weeks' 2 days' wages
Annual subscription to Mudie's circulating (i.e. lending) library	1 guinea	2 weeks' wages
Entry fee to many commercial museums and exhibitions	1*s*	0.6 day's wages
Samuel Clark's educational book *The Little Geologist* (*c.* 1840)	1*s*	0.6 day's wages
Quartern loaf of bread (*c.* 2 kg, i.e. four times as much bread as today's supermarket loaf)	10*d*	$\frac{1}{2}$ a day's wages

A similar strategy was used by Parkinson, whose three-volume magnum opus *Organic Remains* was written in epistolary form for an audience of genteel collectors (Thackray 1976). Collectors at this time often treated fossils as curious and attractive objects, and the fictional protagonist of Parkinson's book has just realized how little he knows about such objects:

> How mortifying will it be to have objects presented daily to my view, whose form alone renders them highly interesting; and whose history is probably fraught with entertainment; and to find myself totally ignorant of their origin.
>
> (Parkinson 1804–1811, vol. I, p. 5)

The protagonist writes in this vein to a learned friend, who promises to send him 'a regular, and systematic history' of organic remains. The rest of the trilogy is taken up with the friend's letters, lavishly illustrated with coloured engravings. Entertainment is duly promised in spectacular descriptions of monsters:

> you will behold the bones of an animal, of which the magnitude is so great; as to warrant the conviction, that the bulk of this dreadful, unknown animal, exceeded three times that of the lion; and to authorise the belief, that animals have existed, which have possessed, with all the dreadful propensities of that animal, its power of destroying, in a three-fold degree. In a word, you will be repeatedly astonished.
>
> (Parkinson 1804–1811, vol. I, p. 11)

Organic Remains, published between 1804 and 1811, had only limited relevance to the mainly mineralogical focus of the Geological Society around this time, but it was important in tapping into a genteel interest in fossils as aesthetically pleasing curiosities, and moulding that interest into something more philosophically focused. Like Davy, then, Parkinson presented geology as something that could satisfy, and indeed outdo, an existing enthusiasm among the upper classes.

The importance of the romantic, speculative dimension to the popularization of geology becomes still clearer when we look at what happened when popularizers ignored it in favour of a rigid empiricism. In MacCulloch's *Geological Classification of Rocks*, intended to impart 'the Elements of Practical Geology' to his public (MacCulloch 1821, title page), the author confined himself to describing and classifying rock types without drawing any causal inferences from them, let alone speculating on Earth history. The *Eclectic Review* was not impressed. Geology, its reviewer stated, was by its very nature 'romantic' and speculative, and MacCulloch's allusion to geology in his title seemed to promise more than mere mineralogy:

> The mere study of rocks, as it is here brought before us, is dry and revolting, apart from the inferences and discussions which belong to Geology. [...] there are few persons, we imagine, who would sit down with pleasure to such a catalogue as is here given of rocks, when it is made the ultimatum. [...] What are we to do with the elementary knowledge derived from Dr. Macculloch? Does he intend, that we should commit to memory his descriptions, and his divisions, and subdivisions, and then rest in quiet suspense [...]? This is certainly too much for the patience of ordinary inquirers.
>
> (Anon. 1821, p. 432)

It is worth pointing out that this reviewer had no problem with the Geological Society's redefinition of geology against their overly imaginative

Fig. 1. Keyed diagram of rock types, painted by Thomas Webster and engraved by F. C. Bruce for Davy's (1813) *Elements of Agricultural Chemistry*. This diagram represents a striking blend of picturesque and scientific visual conventions (Davy 1813, fig. 16). Reproduced by kind permission of the Syndics of Cambridge University Library.

predecessors, as long as imagination were not allowed to gain the upper hand entirely. Geology was still, the reviewer insisted, a thoroughly romantic science:

> In calling Geology *romantic*, we refer not to the tissues of wild conjecture and descriptions of unreal events, which Burnet and Buffon dignified with the name of "Theories of the Earth." Such were not science, but an exercise of fancy, an embodying of waking dreams into a beautiful, perhaps, but an improbable system of world-making. We say that Geology is a romantic science independently of its fabulous historians.
>
> (Anon. 1821, p. 431)

Geology was romantic because it was 'a new and interesting branch of Antiquities', leading to speculations about such topics as 'the original chaos' and 'the effects of the Deluge'. In the mind of this and many other readers, popularizers of geology (at least from the late 1810s and later) who omitted such grand historical speculations altogether from their work were not doing their job properly. MacCulloch's later defences of geological theorizing suggest that he had taken the point: he introduced his own provisional 'Sketch towards a Theory of the Earth' by observing that 'To omit such a view' in a work intended for 'the mere student' would be 'a dereliction of duty, as it would be to disappoint reasonable expectation' (MacCulloch 1831, vol. II, p. 410).

This was certainly clear enough to Buckland, who by the early 1820s was expounding the grand historical aspects of geology at Oxford University, the prime training ground for England's elite clergymen and an important base for generating wider support for the science (Rupke 1983). Mixing Genesis with geology also went against the Geological Society's party line, but Britain's future intelligentsia needed to be taught how and why geology represented a valuable support for Scripture, rather than a threat. More than this, Buckland brought the science to life for his students. With the help of visual aids, private jokes, outdoor field trips and his own impersonations of extinct animals, he helped them to *imagine* the former worlds he was talking about, speculating in the most spectacular way imaginable and making full use of the comic potential of fossil hyena droppings along the way (O'Connor 2007*b*, pp. 71–116). It was around Buckland that the practice of composing occasional verses on geology really took on a life of its own, and these verses – circulated privately among fellow dons and geologists – served as a useful testing ground for imaginative techniques that would soon be launched before a wider public, with their tongue-in-cheek tales and songs about heroic geologists and the wonders of the distant past they explored.

By the mid-1820s mammoths and hyenas were old news, and a new range of fossil monsters lurched into the limelight: the saurians.[7] These gave rise to some of the most entertaining papers among the normally rather dry *Transactions of the Geological Society*: Buckland's 1829 paper on pterodactyles even contained a lurid quotation from *Paradise Lost* (Buckland 1835, p. 219). But the fossil saurians also offered a golden opportunity to take up where Parkinson had left off. This opportunity was seized by the Sussex surgeon Gideon Mantell, a key figure in the identification of the great lizards later to be termed 'dinosaurs'. Like Parkinson, Mantell targeted his first two books (1822, 1827) at wealthy landowners with an interest in local history – and the local history of Sussex now stretched back into the pre-human period. Also like Parkinson, Mantell stressed the enormous size of these extinct creatures:

> If we attempt to pourtray the animals of this ancient country, our description will possess more of the character of a romance, than of a legitimate deduction from established facts. [...] The gigantic *Megalosaurus*, and yet more gigantic *Iguanodon*, to whom the groves of palms and arborescent ferns would be mere beds of reeds, must have been of such a prodigious magnitude, that the existing animal creation presents us with no fit objects of comparison.
>
> (Mantell 1827, p. 83)

Here the official rhetoric of the Geological Society, disparaging the exercise of fancy and imagination as unhealthy 'romance', is apparently turned on its head. Mantell not only admits that an accurate geological description is bound to sound like a romance, but he uses this as a positive selling-point for the science. His stance is not far from that of a circus ringmaster, a strategy that enabled ambitious geologists like Mantell to enhance their own prestige and authority in front of an awestruck audience.

That same year (1826), the young Charles Lyell performed a similar feat for the readers of the *Quarterly Review*. The *Quarterly* was the most prestigious of the more conservative review periodicals: this was where the Tory intelligentsia came for current opinion on key cultural issues. Like Playfair and Fitton in the Whiggish *Edinburgh Review*, Lyell used the *Quarterly* as a platform to canvass support for the Geological Society by trumpeting its recent

[7] Ichthyosaurs and pterodactyles had been known to fossilists and geologists long before Buckland's hyena den research, but they were not widely publicized in Britain until the early 1820s in the wake of discoveries of the iguanodon, megalosaurus and plesiosaurs.

discoveries. But whereas Playfair and Fitton had used the concepts of 'romance' and 'poetry' as sticks with which to beat eighteenth-century theorists, contrasting them with the sober and narrower practices of modern geologists, Lyell used these same concepts to play up the spectacular appeal and visionary possibilities of present-day geology. In a long review of the Society's 1824 *Transactions*, Lyell presented palaeontology as a realization of the fictitious monsters of mediaeval romance:

> We cannot wonder that geologists have been suspected of a love for the marvellous. [...] The Pterodactyls [...] or flying lizards, described by Cuvier, recal still more forcibly to our recollection the winged dragons of fabulous legends. [...] the size of some of them, their long jaws armed with sharp teeth, and the hooked nails of their claws, would render them truly terrific were they to revisit Christendom, now no longer under the shield of the Seven Champions.
> ([Lyell] 1826, pp. 523–524).

The 'love of the marvellous' is here invited back in to the discourse of geology as a selling point for outsiders, albeit at the level of metaphor and simile rather than causal explanation. Lyell even uses the word 'terrific', a staple adjective of contemporary Gothic novels. His readers liked reading romances, so he advertised geology to them by presenting it as a romance. This analogy became a mainstay of polite geological popularization in the Victorian period, even when print became cheap enough for the working classes. Romance was a capacious literary form, embracing everything between Edward Bulwer-Lytton's silver-fork novels and the penny dreadfuls, and appealing right across the class spectrum; accordingly, romance was an ideal touchstone for popularizers.

The idea of geology as a romance was also a mainstay of biblically literalist geological popularization, which enjoyed a considerable vogue in the 1820s – much to the annoyance of the Geological Society, whose members felt that this threatened 'to take us back again to the darker ages of geology' ([Fitton] 1823–1824, p. 206). Part of the problem was that the literalists often pre-empted the self-styled geological elite in bringing the new science before a wider public. One popularizer of literalist geology, James Rennie, wrote an attractive and successful series of introductory *Conversations on Geology* between a mother and her two children, in which the mother states:

> I [...] call Geology romantic, because it not only leads us to travel among the wildest scenery of nature, but carries the imagination back to the birth and infancy of our little planet, and follows the history of deluges and hurricanes and earthquakes.
> ([Rennie] 1828, p. 8)

From the 1820s onwards, old-Earth and young-Earth geologies were promoted in competition with each other (O'Connor 2007*a*). Both sides often had harsh things to say about their competitors, but they were also quick to latch onto and cannibalize each other's innovations in spectacular rhetoric, attractive analogies, pictorial techniques, publication formats and display media. Geology rose to celebrity in the 1830s on the back of a rhetorical and iconographic 'arms race' in which new audiences were targeted and new genres colonized. Buckland, for instance, recycled his Oxford lecturing techniques before the mixed audiences of the British Association for the Advancement of Science, leading field trips on horseback and delighting onlookers with his saurian impersonations (Morrell & Thackray 1981, pp. 158–159). Mantell, meanwhile, commissioned George Scharf to paint a huge pictorial representation of his iguanodon description to hang in his museum in Brighton, bringing the infant genre of 'scenes from deep time' (Rudwick 1992) out from the realm of privately circulated caricature into the public spotlight.

Lyell's own public-relations efforts in the 1830s were so successful that they won him a lasting place in the geological pantheon. The historical importance of his *Principles of Geology* (1830–1833) lies less in any scientific breakthrough he achieved than in the masterly way in which he synthesized and dramatized existing knowledge before a wider public, first among the upper classes, then (with the cheaper edition of 1834) among the wealthier members of the middle classes. Laced with attractive poetry quotations and framed as a reflective philosophical treatise rather than a simple textbook, the *Principles* also helped to accelerate the ongoing shift in Geological Society practice away from hard-line empiricism towards a climate in which theoretical speculation once again had a central role to play. Not everyone approved of Lyell's imperialistic approach to other people's work, but his choice of genre enabled him to go public on many issues which the Geological Society had largely kept quiet about before, such as the immensity of past time. By bringing the depth of geological time fully before his readers' imaginations, Lyell's literary gifts played a major role in cementing the public status of geology as a discipline in its own right (Secord 1997, p. xxiii).

But the most distinctive feature of Victorian geological popularization was something Lyell strenuously denied, the evidence for progression in Earth history. This evidence enabled other geological writers to engage the literalists head-on, recasting and amplifying Genesis 1 as a stately pageant from primitive plants to humans, a cosmic drama spanning millions of years to match the vast tracts of space revealed by astronomy. Buckland's 1836 Bridgewater Treatise, *Geology and Mineralogy*, was one of the earliest and best-known embodiments of this new creation-myth, which spawned countless imitations, successors and competitors. These grand narratives, infused with a sense of divine providence and the inevitable progress of civilization, inhabited numerous different genres and drew on a bewildering range of literary and mythological reference-points in order to harness the dramatic appeal of the old 'theories of the Earth' for the benefit of modern science. In the hands of Mantell and Hugh Miller, this kind of storytelling entered the realms of high art (O'Connor 2007*b*, pp. 357–432). It proved extremely durable, forming the bedrock of the 'evolutionary epic' as expounded by later popularizers such as Robert Chambers, Thomas Huxley and Richard Dawkins (Secord 2000, pp. 77–110; Lightman 2007, pp. 219–294; Dawkins 2004).

The return to speculation

Thirty years after its foundation, then, the Geological Society had come full circle. Theoretical speculation was back at the heart of its practice; the Society's apparent equation of poetry with falsehood had been flaunted by countless quotations from poems both old and new, not only in popular books but even in its *Transactions*; 'love of the marvellous' was no longer a byword for delusion; and the providential momentum of Protestant biblical scholarship was put to new use in telling the story of life before man. Nor was the imagination invited back in at a merely decorative level. Geology's propensity for grandiose cosmic speculation and the wild imaginings of romance, while simultaneously giving these visions the dignity of 'fact', pushed it to the top of the hierarchy of the sciences in Victorian Britain. Its capacity for reconstructing former worlds in all their glory, surely one of the most speculative endeavours in any branch of physical science, made it an important conceptual reference point by which other disciplines claimed 'scientific' status (such as anthropology, psychology and history). Credit for its prestige deserves to go to those who reconnected it with its roots. Despite John Ayrton Paris's confident claim (Paris 1818, p. 168), the 'romantic age of geology' was not dead and buried in 1818. It had only just begun.

I am grateful to Simon Knell and Noah Heringman for their detailed and helpful comments on this paper. Any remaining errors and overstatements for dramatic effect are mine alone.

References

ANON. 1817. Geological Inquiries, proposed by the Geological Society. *Philosophical Magazine*, **49**, 421–429.

ANON. 1821. Macculloch on Rocks. *Eclectic Review*, 2nd series, **15**, 430–441.

BOIME, A. 1990. *Art in an Age of Bonapartism 1800–1815*. University of Chicago, Chicago, IL.

BOIME, A. 2004. *Art in an Age of Counterrevolution 1815–1848*. University of Chicago, Chicago, IL.

BROOKE, J. & CANTOR, G. 1998. *Reconstructing Nature: The Engagement of Science and Religion*. Clark, Edinburgh.

BROWN, J. M. 1838. *Reflections on Geology*. James Nisbet, London.

BUCKLAND, W. 1835. On the discovery of a new species of Pterodactyle in the Lias at Lyme Regis [read 1829]. *Transactions of the Geological Society of London*, 2nd series, **3**(1), 217–222.

BUCKLAND, W. 1836. *Geology and Mineralogy Considered with Reference to Natural Theology* (2 vols). William Pickering, London.

BURNETT, J. 1969. *A History of the Cost of Living*. Penguin, London.

DASTON, L. 2001. Fear and loathing of the imagination in science. *In*: GRAUBARD, S. R., MENDELSOHN, E., WILSON, E. O. & GALISON, P. (eds) *Science in Culture*. Transaction, New Brunswick, NJ, 73–95.

DAVY, H. 1813. *Elements of Agricultural Chemistry*. Longman, London.

DAWKINS, R. 2004. *The Ancestor's Tale: A Pilgrimage to the Dawn of Life*. Weidenfield & Nicolson, London.

[FITTON, W. H.] 1817*a*. Transactions of the Geological Society. Vol II. *Edinburgh Review*, **28**, 174–192.

[FITTON, W. H.] 1817*b*. Transactions of the Geological Society, Vol. III. *Edinburgh Review*, **29**, 70–94.

[FITTON, W. H.] 1823–1824. Geology of the Deluge. *Edinburgh Review*, **39**, 196–234.

GOLINSKI, J. 1992. *Science as Public Culture: Chemistry and Enlightenment in Britain, 1760–1820*. Cambridge University Press, Cambridge.

GREENOUGH, G. B. 1819. *A Critical Examination of the First Principles of Geology; in a Series of Essays*. Longman, London.

HERINGMAN, N. 2004. *Romantic Rocks, Aesthetic Geology*. Cornell University, Ithaca, NY.

HEYCK, T. W. 1982. *The Transformation of Intellectual Life in Victorian England*. Croom Helm, London.

JACKSON, H. J. (ed.). 1985. *The Oxford Authors: Samuel Taylor Coleridge*. Oxford University Press, Oxford.

KNIGHT, D. M. 1992. *Humphry Davy: Science and Power*. Blackwell, Oxford.

LAUDAN, R. 1977. Ideas and organizations in British Geology: A case study in institutional history. *Isis*, **68**, 527–538.

LIGHTMAN, B. 2007. *Victorian Popularizers of Science: Designing Nature for New Audiences*. University of Chicago, Chicago, IL.

[LYELL, C.] 1826. Transactions of the Geological Society of London. *Quarterly Review*, **34**, 507–540.

LYELL, C. 1830–1833. *Principles of Geology, Being an Attempt to Explain the Former Changes of the*

Earth's Surface, by Reference to Causes Now in Operation (3 vols). John Murray, London.

MacCulloch, J. 1821. *A Geological Classification of Rocks, with Descriptive Synopses of the Species and Varieties, Comprising the Elements of Practical Geology*. Longman, London.

MacCulloch, J. 1831. *A System of Geology, with a Theory of the Earth, and an Explanation of Its Connexion with the Sacred Records* (2 vols). Longman, London.

Mantell, G. A. 1822. *The Fossils of the South Downs; or Illustrations of the Geology of Sussex*. Lupton Relfe, London.

Mantell, G. A. 1827 [published 1826]. *Illustrations of the Geology of Sussex: Containing a General View of the Geological Relations of the South-Eastern Part of England*. Lupton Relfe, London.

Mantell, G. A. 1838. *The Wonders of Geology*. Relfe and Fletcher, London.

Morrell, J. 1971. Professors Robison and Playfair, and the *Theophobia Gallica*: Natural philosophy, religion and politics in Edinburgh, 1789–1815. *Notes and Records of the Royal Society of London*, **26**, 43–63.

Morrell, J. & Thackray, A. 1981. *Gentlemen of Science: Early Years of the British Association for the Advancement of Science*. Clarendon, Oxford.

O'Connor, R. J. 2007*a*. Young-Earth creationists in nineteenth-century Britain? Towards a reassessment of 'scriptural geology'. *History of Science*, **45**, 357–403.

O'Connor, R. J. 2007*b*. *The Earth on Show: Fossils and the Poetics of Popular Science, 1802–1856*. University of Chicago, Chicago, IL.

Paley, M. D. 1986. *The Apocalyptic Sublime*. Yale University, New Haven, CT.

Paris, J. A. 1818. Observations on the geological structure of Cornwall. *Transactions of the Royal Geological Society of Cornwall*, **1**, 168–200.

Parkinson, J. 1804–1811. *Organic Remains of a Former World* (3 vols). John Murray, London.

[Playfair, J.] 1811–1812. Transactions of the Geological Society. *Edinburgh Review*, **19**, 207–229.

Porter, R. 1977. *The Making of Geology: Earth Science in Britain 1660–1815*. Cambridge University Press, Cambridge.

Porter, R. 1978. Philosophy and politics of a Geologist: G. H. Toulmin (1754–1817). *Journal of the History of Ideas*, **39**, 435–450.

[Rennie, J.] 1828. *Conversations on Geology*. Samuel Maunder, London.

Rudwick, M. J. S. 1985. *The Great Devonian Controversy: The Shaping of Scientific Knowledge among Gentlemanly Specialists*. University of Chicago, Chicago, IL.

Rudwick, M. J. S. 1992. *Scenes from Deep Time: Early Pictorial Representations of the Prehistoric World*. University of Chicago, Chicago, IL.

Rudwick, M. J. S. 2005. *Bursting the Limits of Time: The Reconstruction of Geohistory in the Age of Revolution*. University of Chicago, Chicago, IL.

Rupke, N. A. 1983. *The Great Chain of History: William Buckland and the English School of Geology (1814–1849)*. Clarendon, Oxford.

Secord, J. A. 1997. 'Introduction'. *In*: Lyell, C. *Principles of Geology*. (J. A. Secord (ed.)). Penguin, London, ix–xliii. (Elibron Classics facsimile reprint of the 1837 edition by John Murray, London.)

Secord, J. A. 2000. *Victorian Sensation: The Extraordinary Publication, Reception, and Secret Authorship of Vestiges of the Natural History of Creation*. University of Chicago, Chicago, IL.

Siegfried, R. & Dott, R. H. Jr. (eds). 1980. *Humphry Davy on Geology: The 1805 Lectures for the General Audience*. University of Wisconsin, Madison, WI.

Taylor, M. A. & Torrens, H. S. 1986. Saleswoman to a new science: Mary Anning and the fossil fish *Squaloraja* from the Lias of Lyme Regis. *Proceedings of the Dorset Natural History and Archaeological Society*, **108**, 135–148.

Thackray, J. C. 1976. James Parkinson's *Organic Remains of a Former World* (1804–1811). *Journal of the Society for the Bibliography of Natural History*, **7**, 451–466.

Topham, J. R. 1992. Science and popular education in the 1830s: The role of the *Bridgewater Treatises*. *British Journal for the History of Science*, **25**, 397–430.

Vincent, D. 1989. *Literacy and Popular Culture: England 1750–1914*. University of Cambridge, Cambridge.

Wyatt, J. 1995. *Wordsworth and the Geologists*. Cambridge University, Cambridge.

The Geological Society on the other side of the world

DAVID F. BRANAGAN

School of Geosciences, The University of Sydney, Sydney, NSW 2006, Australia
(e-mail: dbranaga@mail.usyd.edu.au)

Abstract: From its earliest years the Geological Society of London attracted the attention of scientifically- and technically-minded men in Australia and New Zealand. Members 'at home' in Britain were also eager for geological information about the antipodes. The publications of the Society acted as a major source of information about the geology of these southern lands, from vertebrate palaeontology and modern glaciation at sea level to ancient glaciations and mineralization (particularly of gold). At least 360 members were active in Australasia in the nineteenth century. Strong antipodean Society membership continued through the twentieth century. What is noteworthy is the number of mining figures, of varying scientific competence, who boasted of their membership. There were significant contributions to the Society's journals on Australasian geology from the 1820s to the early 1900s. Many topographic features on the maps of both Australia and New Zealand are named for Members and Fellows of the Geological Society. The lists of elected and 'would-be' antipodean members include a few enigmatic examples of chicanery, fraud and disappearance.

From the earliest European exploration of the Antipodes, there was interest in the natural character of these lands. First observations of zoology and botany indicated considerable differences from the 'known world', and it seemed likely that the geology might also be different, or at least 'out of step' with accepted knowledge. There were discoveries of enigmatic fossil vertebrates and, later, evidence of active glaciation at sea level, followed by the discovery and confirmation of ancient glaciations. For the more practically- and financially-minded there were the discoveries of coal and gold. Discovery of the latter caused a social revolution in Australia. It also generated global study and discussion of the origin of ore deposits and particularly gold. These matters have been discussed by various authors (Vallance 1975; Branagan 1994, pp. 118–119; Moyal 1993), and they set the scene for this paper.

Because of its relatively unknown geology, the continent of Australia (and nearby islands) was of great interest to members of the Geological Society of London from its inception in 1807. Some of its earliest members visited Australia, while others studied rocks collected there or wrote about the geological data, which were accumulating. The Society's embryonic museum received early donations of these minerals, rocks and fossils (Moore *et al.* 1991). And while little about Australasia was presented at the earliest meetings of the Society, or published in the pages of its sporadically appearing *Transactions*, the publications of the Society were to be a major source of information about the geology of these southern lands.

The years to 1840

Australasia might be considered to have entered the mindset of the Society with the admission Sir Joseph Banks (1743–1820), the first 'patron' of Australian science, as a Ordinary Member, in 1808, only a few months after its founding. Banks had gained some familiarity with New Zealand and even noted aspects of the geology of Cape Turnagain (on the far northeastern coast of the North Island) in his diary: 'before dinner we were abreast of another cape which made in a bluff rock the upper part of a reddish couloured stone or clay the lower white beyond this the Country appeared pleasant with little smooth hills like downs…'.[1] While Banks did not publish these remarks, the same location was noted in the first publication to be rushed out when Cook's expedition, of which he was a part, returned to England. This was published prior to Hawkesworth's 'official' work:

> Cape Turnagain is remarkable for a stratum of a bright brown colour; its prominence gradually diminishes towards the north-side, but to the southward its descent is more sudden. The soundings opposite to it, at the distance of a mile and a half are about thirty-two fathom, with coarse yellow gravel at the bottom.
>
> ([Magra] 1771–1772)

A little later it was the eastern coast of the Australian continent that drew Banks's attention and where he encountered aspects of the geology in the landfalls at Botany Bay and Cooktown. He also captured views of The Brothers, Mt Warning and the

[1] Banks, 16 October 1769 (Beaglehole 1962).

From: LEWIS, C. L. E. & KNELL, S. J. (eds) *The Making of the Geological Society of London.*
The Geological Society, London, Special Publications, **317**, 341–371.
DOI: 10.1144/SP317.20 0305-8719/09/$15.00

Glasshouse Mountains (all of volcanic or shallow intrusive origin and named by Cook) as they sailed along the coast and in the hazardous passage through what came to be called the Great Barrier Reef. It is a region that has linked past and present through intensive studies, which continue to the present day.

Banks' direct connection with the Society lasted less than a year. As President of the Royal Society he felt he could not support what had so rapidly become a rival, and he resigned on 3 March 1809. Banks wrote 'these new-fangled Associations will dismantle the Royal Society and not leave the old lady a rag to cover her' (Cameron 1952, p. 177; Herries Davies 2007, p. 24). However, three weeks later, perhaps to clear the air and as 'a good friend', Banks sent a valuable collection of minerals (Cameron 1952, p. 178).

The first direct mention of Australasian materials in the Society's possession is the record of the donation by Dr John MacCulloch (1773–1835), on 3 April 1812, of a sample of iron ore from Botany Bay. The term 'Botany Bay' encompassed, at that time, the whole of a region then known as 'European Australia'. While the location named for the specimen was not particularly specific, iron-rich laterite does occur close to what is today called Botany Bay, and might have been the source of the donated specimen. How MacCulloch obtained the specimen is not known.

This was followed in 1824–1825, by plant fossils, ironstone and limestone, maps and a 'drawing of Cape Pillar', Tasmania.[2] The largest donation, on 3 November 1825, came from Australian-born Captain Phillip Parker King (1791–1856). These were 'Specimens from New Holland and New South Wales', almost certainly the collection made during his Australian circumferential voyages of 1818–1822, when assisted by Allan Cunningham (1791–1839) (King 1826; Branagan 1985; Branagan & Moore 2008). A little earlier, on 4 May 1825, Dr William Henry Fitton (1780–1861), a joint Secretary of the Society, presented the Library with his 'An account of some geological specimens . . .', which reported on King's specimens and was published by the Society (Fitton 1825, 1826). He also presented 100 copies of his printed 'Instructions for collecting . . .' (from Fitton 1825), which were to be given to members. This booklet was on board HMS *Beagle* and apparently also influenced Charles Darwin (Herbert 2005, pp. 102–103).

Rather more tenuous is a link to Society founder member, Humphry Davy, who is commemorated in Mount Davy near Greymouth on the west side of the South Island, New Zealand. Among the rock collections transferred from the Geological Society to the British Museum (Natural History) in 1911 was a suite of 36 specimens from the Firth of Thames on New Zealand's North Island, which was listed as 'collected by Sir Humphry Davy and presented by him to the Geological Society in 1822'. The presentation certainly occurred, but, of course, the specimens were not collected by Davy, who was never in New Zealand. Tee (1998, pp. 93–102) attributes the collection to Captain James Love of HMSS *Coromandel*, in 1820. Love and Davy, both from Penzance, Cornwall, were almost certainly friends.

The 'honour' of the first Society paper on Australian matters goes to the Reverend William Buckland (1784–1856), who, on 5 May 1820, presented the results of his examination of two collections of rocks, one from Madagascar and one from New South Wales, the latter sent to him by the surveyor/explorer John Oxley (*c.* 1785–1828) (Buckland 1821). Although Oxley's locations were not specified, only being inland 'on the west side of the Blue Mountains', Buckland thought they showed 'not only an identity in the older formations of rocks that constitute the earth's surface, but also a strong resemblance in the leading features of many of the secondary strata that follow and repose upon them'. Buckland noted 'a strong analogy between the coal formation of the Hunter's River and River Hawkesbury in New South Wales and that of England, which well deserves to be accurately investigated'. That investigation, in fact, proved to be a long drawn out affair (Vallance 1981).

The earliest antipodean members

Amongst the early members of the Society were four 'Australians': James Bicheno (1785–1851), who joined in 1823; Edward Abbott (dates unknown), who joined in 1825 (see Appendix 1); the Venerable Archdeacon Thomas Hobbes Scott M.A. (1783–1860), who was elected on 4 June 1824; and the Reverend William Branwhite Clarke (1798–1878) (Fig. 1), who was elected on 19 May 1826.

Bicheno and Abbott made no particular contribution to the Society. Bicheno, who went to Van Diemen's Land (Tasmania) in 1842, did, however, publish on botany in the Linnean Society, and is remembered in the name of a small town on the east coast. He is one of the earliest 'Australian' members to be put on the map of Australia.

Scott was in New South Wales for two years (1819–1821), as a layman clerk accompanying his brother-in-law, John Thomas Bigge (1780–1843), who had been sent out by the British Government

[2] *Transactions of the Geological Society, London*, second series, **2**, 1829.

Fig. 1. Rev W. B. Clarke, pioneer Australian geologist and enthusiastic Fellow for more than 50 years.

to undertake a 'Royal' Commission into the affairs of the Colony (Border 1967; Ritchie 1971). Scott had matriculated at the age of 30, gaining a B.A. degree at Oxford in 1817 and a M.A. the following year. Oxford was then a linchpin in the Geological Society's development of geology and it is very likely that Scott was exposed to it there (Knell 2009). His geological interests may have developed still further when the Bigge Enquiry came to investigate the coal mining being undertaken some 150 km north of Sydney, at what later became known as Newcastle, and which utilized convict labour. Whatever the influence, Scott began to note the geology of the areas he visited and collect specimens, both through personal effort and by exchange. On the basis of his colonial experience he published a useful, but in parts misleading, short paper on the geology of New South Wales and Van Diemen's Land (Tasmania), which gained wide circulation (Scott 1824).

On his return to England in 1821 Scott took Holy Orders. Then, in October 1824, he was appointed Archdeacon of New South Wales (at that time within the Diocese of Calcutta) with wide responsibilities and powers. When Scott returned to New South Wales his interest in geology continued. In 1829 he was 'stranded' for some time in the newly established colony of Western Australia (Scott 1831) and, again, he collected geological specimens. He amalgamated his collections and, in 1831, donated them to the Society. Scott's geological collection became part of the basis for his early interpretation of Australian geology, but some of his misleading remarks were unfortunately picked up by other writers and spread more widely (Vallance 1975; Branagan & Moore 2008). No pictorial presentation of Scott has been located.

Clarke became a deacon at Norwich Cathedral in 1821, the same year he graduated from Cambridge with a Bachelor of Arts. He was already attracted to geology, having attended lectures by Adam Sedgwick. Sir Charles Bunbury (1809–1886) regarded Clarke's paper on Suffolk geology, presented to the Society on 8 March 1837 (Clarke 1837*a*), as 'very dull'; the meeting only becoming 'some fun' later when an animated debate began between Sedgwick and Greenough (Woodward 1907, p. 130). There were three other offerings from Clarke to the Society between 1837 and 1839 (Clarke 1837*b*; Organ 1994, pp. 95–97). When Clark departed for Australia, early in 1839, he was already in rather poor health and planned only to spend a 'season' or two in the antipodes. He told the Society's then Secretary, Robert Hutton (1784–1870), that he proposed to promote the Society in the south. This he certainly did, but he was destined never to see England again.

Although Clarke does not, perhaps, deserve the commonly applied accolade of 'Father of Australian Geology' (Woodward 1907, p. 189; Jervis 1944; Grainger 1982), from his arrival in the colony in 1839, until his death in 1878, he made major contributions to Australian and, indeed, international geology (Organ 1997; Moyal 2003). Clarke contributed some 13 further papers to the Society: five between 1843 and 1848; one in 1852; three in 1855; two in 1861–1862; and two in 1866–1867 (Etheridge & Jack 1881, pp. 24–34; Organ 1994, pp. 90–134). He maintained contact with both Sedgwick and Murchison during their heydays in the Society but, in the early 1850s, gained the ire of Murchison over claims concerning his part in the discovery of gold in Australia, which were contra to Murchison's highly publicized claims (Moyal 2003, vol. 1, pp. 23–25).

Other members from this period include Sir John Franklin (1786–1847) and (Sir) Thomas Mitchell (1792–1855) (Fig. 2), both of whom became members in 1829. It was in Australian waters (1801–1804) that Franklin was a midshipman with Matthew Flinders (1774–1814), who stirred his interest in geography and surveying, and, no doubt, also in geology – for which Flinders showed some curiosity. Franklin went to Tasmania as Governor in 1837 with his second wife (Lady Jane (1791–1875)), and the two were active in setting up what later became the Royal Society of Tasmania, as well as a natural history museum.

Fig. 2. (Sir) T. L Mitchell, who published important early papers and first placed the names of Society members on the map of Australia.

However, Sir John apparently did not contribute to the Geological Society's publications or its collections.

Mitchell, a surveyor–explorer, on the other hand, presented a significant paper to the Society in April 1831, describing the Wellington Caves region of New South Wales; the site of important discoveries of fossil vertebrates (Mitchell 1831). This was the first paper from a 'resident' Australian. Mitchell's manuscript geological map of the Nineteen Counties (New South Wales) – coloured on a base dated London 1834 (re-engraved), with manuscript annotations: 4, Shale; 3 & 2, Sydney Sandstone and sandstone and shales; 1, coal; Trap – identifying the variations in rock types, is in the Society's Archives.[3] It was presented to the Society by Mitchell in 1838 (it is reproduced in Branagan (1990*a*) and, in colour, in Oldroyd (2007)).

From 1840 to 1874

From 1840 membership of the Society became of importance to a variety of people who had interests in Australia and New Zealand. The identified antipodean memberships from the earliest days to the early 1900s are shown in Appendices 1 (Australian) and 2 (New Zealand). Known details of these members and other significant non-member geologists who worked in the antipodes can be found in Branagan & Branagan (1997 – based on a card index by T. G. Vallance, 1928–1993 see the Archives later in this paper). Many have entries in the *Australian Dictionary of Biography*, the *Dictionary of New Zealand Biography*, McCarthy (1991) and the *Oxford Dictionary of National Biography*. Only a few members can be discussed in this paper, so a rather eclectic selection has been chosen from the deservedly famous to the relatively unknown. Those from the 1840s include the Reverend Richard Taylor (1805–1873) elected on 11 March 1840; and three who became members on the same day, 1 December 1841: Samuel Stutchbury (1798–1859), (Sir) Charles Nicholson (1808–1903) and Owen Stanley (decease noted Committee Meeting, 30 April 1851) (West 1967, pp. 470–471).

Taylor's geological work, carried out in both New South Wales and New Zealand in conjunction with his missionary work, has only recently been assessed (Mason 2007, 2008). Owen (1990, pp. 437–438) gives details of his life but little on his geology, naming him as a Fellow of the 'Royal Geological Society'. Taylor made many good observations and astute interpretations, particularly on the Sydney Basin structures, the coalfields of New South Wales and the Wellington Caves in the 1830s[4] (Osborne 1991, pp. 16–17), and on aspects of volcanic geology in New Zealand. His work, including many fine sketches, is in several unpublished journals in the Alexander Turnbull Library in Wellington.

Sixteen years prior to his membership, Stutchbury had visited New South Wales and New Zealand, and made important observations in 1825–1826 en route to the Pacific Islands with a pearl fisheries expedition. He left behind his position as assistant to William Clift at the Royal College of Surgeons, which was filled by Richard Owen (1804–1892) (Branagan 1993, 2000). Stutchbury's description of geological features in the Bay of Islands, on the North Island of New Zealand, was preceded only by Banks and the Forsters. At the time of his joining the Society, Stutchbury was at the Bristol Institution. He returned to eastern Australia in 1850 to leave his mark on Australian geology as Mineral Surveyor for the Colonial Government of New South Wales, mapping an area of about 80 000 square miles (approximately 208 000 km^2) in the ensuing five years (Branagan 2005*a*).

[3] GSL: Item 994/4.

[4] ATL: Richard Taylor, Unpublished Journal of Life in New South Wales, 1833–1873, pp. 54–59.

Ten years younger than Stutchbury, Nicholson made his name as a 'statesman, landowner, businessman, connoisseur, scholar and physician'. A graduate in medicine from Edinburgh University, he was awarded a doctorate in 1833, on the basis of a thesis in Latin on the causes and treatment on asphyxiation. Nicholson helped to found the University of Sydney and was Chancellor from 1854 to 1862, when he returned to England to live. However, he maintained his links with the University, advising on professorial appointments and contributing to the University's library and museum. His main scientific/cultural interests were in archaeology, gaining fame as an antiquarian travelling to, amongst other places, eastern Russia and Egypt, where he was involved in archaeological excavations (MacMillan 1967, pp. 283–285). He published nothing on geology, as far as is known, although Ludwig Leichhardt (1813–?1848) refers to a geological map Nicholson had prepared for him on part of the Hunter Valley region north of Sydney. Leichhardt thought it incorrect (Aurousseau 1968, vol. 2, pp. 621 and 808; Branagan 1994). Aurousseau, a geologist, surmised that Nicholson possibly used Mitchell's Survey Map of the Colony of New South Wales (1834) and put some geology on it, although he might indeed have had a copy of Mitchell' own manuscript geological map mentioned earlier (Branagan 1990*a*; Oldroyd 2007).

Owen Stanley (1811–1850) was a naval officer, who came under the influence of Phillip Parker King, and was involved in the 1830s in Arctic exploration, carrying out magnetic and astronomical observations. His work in command of HMS *Rattlesnake* (1846–1850), surveying parts of north Australia and New Guinea, made his name, although his two scientific officers, T. H. Huxley (1825–1895) (elected in 1856) and J. MacGillivray (1821–1867), did not think too highly of him. He is commemorated on the world map by the Owen Stanley Range in New Guinea, which he observed in 1849, perhaps the largest monument to a member of the Society, being several hundreds of kilometres long, with elevations of over 3000 m.

Two who did not join

While one might have expected both Phillip Parker King and the explorer Paul Edmund Strzelecki (1797–1873) to have joined the Society this did not happen. Instead, while certainly interested in geology, they seem to have preferred the Royal Geographical Society, by whom Strzelecki was awarded its prestigious Gold Medal in 1846, not long after his book on Australia had been published (Strzelecki 1845). Strzelecki's exploratory work had earlier been referred to by Greenough, in his capacity as President of the Royal Geographical Society (Greenough 1841, pp. lxii–lxiii). Although Strzelecki never joined he was certainly helped with his book by the Society's then Assistant Secretary and Librarian, William Lonsdale (1794–1871), and by Professor John Morris (1810–1886), who brought Strzelecki's manuscript more 'up to date' geologically, by examining fossils and placing them stratigraphically ((Strzelecki 1845; Branagan 1995). Morris was later the recipient of the first Lyell Medal (in 1876) (Herries Davies 2007, p. 125).

Governor Franklin and Lady Jane became friendly with Strzelecki and encouraged his geological work (Strzelecki 1845). During the preparation of his very large geological map of SE Australia he received both moral and practical assistance from Franklin. Writing to Roderick Murchison, on 19 November 1842, and informally introducing Strzelecki, Franklin wrote of Strzelecki's 'very interesting Geognostical Map which he has Constructed to elucidate the course of the Mountain Ranges – in NSW & V[an] D[iemen's] Land which he has examined. I have some hope that he may present that Map to the Geological Society – but am not certain ... '(Paszkowski 1997, p. 205). Franklin later gave Strzelecki a formal introduction to Murchison, who waxed lyrical about the map in his annual address to the Royal Geographical Society, on 27 May 1844, saying that 'M de Strzelecki has prepared a most valuable and colossal map ... which he cannot publish at his own expense; it is well worthy of the patronage of the British government' (Paszkowski 1997, pp. 156 and 177). The Government did not take up the suggestion, however, which was unfortunate for Strzelecki, as, at the time, he was not well-off and had hoped to gain financially from the sale of the map and the rock collection he had also made. In the event neither was sold (apart from a few specimens bought by John Morris).

However, Strzelecki recovered financially and following his work in Ireland during the potato famine of 1846–1847, he was awarded a knighthood. His Australian geological material was then donated to the British Museum and the Geological Survey of Great Britain. While little remains of the rock collection, the map, which lay apparently unexamined for almost a hundred years, is now in the Library of the British Geological Survey (Branagan 1974, 1986). Strzelecki's book received a number of reviews, perhaps the most notable of which was one, some 33 pages long in the *London Quarterly Review* (June–September 1845), which Havard (1940, p. 84) attributed to Buckland. Around this time Fitton was one of the three supporters for Strzelecki's application for British citizenship which was achieved in November 1845 (Paszkowski 1997, p. 217). Strzelecki had been introduced to Fitton through Governor Franklin (Paszkowski 1997, p. 174), and also possibly through Phillip Parker King.

Although Joseph Beete Jukes (1811–1869) joined the Society in 1836, his work on the Australian coast, and particularly the Great Barrier Reef, belong to the 1840s. His perceptive observations and classification remain, in many senses, the best ever made (Jukes 1847, 1850; Branagan 2008*b*). His geological map of Australia had been preceded by the Frenchman Boué (Branagan 2008*b*) and by Strzelecki, the latter having based his work to a large extent on personal field observations on SE Australia (Strzelecki 1845). Jukes, however, included material on other parts of the Australian coast and nearby regions. Jukes and Strzelecki met in 1842 in Sydney and then at the home of Phillip Parker King in Port Stephens, where they discussed geology and particularly the palaeontology described in Murchison's *Silurian System* – Jukes had a copy and tried to identify Strzelecki's fossils.

Economic interests

Robert Marsh Westmacott (*c.* 1801–1870), elected in 1852, was an army officer, but best known for his sketching, particularly (from a geological point of view) of landscapes of the Illawarra region south of Sydney. He also had an interest in coal mines in that region and made a failed attempt to open a mine there in 1840 as the Australian Agricultural Company (AA Co.), operating north of Sydney at Newcastle, had been assigned sole mining rights by the Government. However, after a visit to England, he returned to take up the position, being vacated by King, of superintendent to the AA Company. He was friendly with both the Reverend Richard Taylor and the Franklins (Proudfoot 1992, pp. 850–851).

Edward Hammond Hargraves (1816–1891), elected on 16 May 1855, is noted in his application for Fellowship of the Society as 'the First Practical explorer of the Australian Gold'. The controversial story of Hargraves' alleged responsibility for the discovery of Australian gold, including clashes with Clarke, has been explored in great depth (for example, Blainey 1969; Silver 1986) so will not be discussed here. However, Hargraves' relatively brief exploration for gold in Western Australian in 1862 is often forgotten. The collection of rocks he made at this time were donated to the Society in 1863, together with letters and a map, and are now among the collections held in the Natural History Museum (BM 1911 1560). The Society seems to have had trouble keeping in touch with him towards the end of his life.

Fig. 3. A. R. L. Selwyn. Murchison Medallist, 1876, and successively Director of the Geological Surveys of Victoria and Canada.

Interest in Australian gold and mineralization in general was maintained throughout the 1850s–1860s. Phillip Parker King ensured that the Society was kept informed about the geological surveys carried out by Stutchbury and Clarke, sending the official New South Wales Government published reports, although Stutchbury complained that he did not have a chance to check the proofs for accuracy. A measure of the interest is shown by papers in the *Quarterly Journal* by A. R. C. Selwyn (Fig. 3) (1854, 1858, 1860), H. Rosales (1855), J. Phillips (not a member) (1858) and G. M. Stephen (1812–1894) (1854) (an administrator), which helped to keep up the interest.

Putting geologists on the map

Vallance (1975, p. 18) commented on the surprising number of names of European geologists on the map of Australia. Here I will look solely at Society members who have such memorials. In fact, the vast expanse of Australia, or at least the maps depicting such a space, are indeed liberally sprinkled with the names of Geological Society members, some famous, many rather less so. There is a sole name, Scott Strait, Western Australia, for Archdeacon Scott, given, as early as 8 September

1820, by Phillip Parker King. Then all was quiet for 15 years. The explorer and surveyor, (Sir) Thomas Mitchell really set the ball rolling in 1835 (Mitchell 1838) on his expedition into northwestern New South Wales. It was perhaps appropriate that the first such name he bestowed was that of the energetic Dr John MacCulloch (1773–1835) (on 24 June 1835), President (1816–1818) and producer of a fine map of Scotland, who joined the Society in 1808 (Bowden 2009). Mitchell proceeded to spread the names, Mounts Murchisson [sic], Daubeny (1818), Lyell (1819) and Scrope's (1824) Range (now sadly reduced to Scope's), all on 26 June, 1835. A month later Mitchell also put George Bellas Greenough's name on the map as Greenough's Group (26 July 1835) having earlier (21 June 1835) observed 'two summits of a distant range' for 'a gentleman who has done so much to uniting geology with geography'.

(Sir) George Grey (1812–1898) was not far behind Mitchell in this naming game. Exploring the west coast of the continent he was liberal with his associates. He started with Lyell (12 March 1838), putting his name to 'the most remarkable hill in this part of the country'. But that wasn't enough, for he had to add Lyell Range, a 'remarkable range of dunes ... in compliment to the distinguished geologist of that name'. Murchison again received due notice, by a river (2 April 1839), with a somewhat lesser stream named for Greenough (8 April 1839). But Grey had earlier (24 March 1839) named Greenough Point 'after George Bellas Greenough, President of the Royal Geographic Society'. Mt Horner, south of Geraldton (Western Australia), 'after my friend Leonard Horner' (1785–1864) was of the same period (11 April 1839) (Grey 1841). Some time later (8 October 1854), the surveyor, Robert Austin (1825–1905), named a Mt Murchison, presumably after Sir Roderick, at the head of the eponymous river (Feeken *et al.* 1970).

In his third expedition in 1836, Mitchell ventured south from his earlier work in the Darling River region of New South Wales. Only two namings on this expedition are notable. They are Mt Darwin (10 October 1836), given a month before Charles Darwin joined the Society, and Mt Nicholson (19 September 1836). The latter was acknowledging his Sydney friend, Dr (later Sir) Charles Nicholson, who had also to join the Society. Nicholson later received another topographic name, after he joined, this time in Queensland: Leichhardt wrote, on 27 November 1844, 'to [a] bell-shaped mountain bearing N.68°W., I gave the name Mount Nicholson in honour of Dr. Charles Nicholson, who first introduced into the Legislative Council of New South Wales, the subject of an overland expedition to Port Essington' (Roderick 1988, p. 255).

Charles Darwin, by then a member, became immortalized in 1839 when J. C. Wickham (on 9 September 1839) named Port Darwin (Northern Territory) after him. This is according to Feeken *et al.* (1970, p. 124), but Hordern (1989, pp. 168–169) attributes its naming 'after Charles Darwin, naturalist [. . .] just for friendship's sake' to J. Lort Stokes and C. C. Forsyth. Wickham (or Stokes) added that the locality consisted of 'fine-grained sandstone: a new feature in the geology of this part of the continent, which afforded us an appropriate opportunity of convincing an old shipmate and friend, that he still lived in our memory'. Mitchell (1848) returned to the geologists again in his fourth expedition of 1846, but only three were Society members, among names for 'such individuals of our own race as had been most distinguished or zealous in the advancement of science, and the pursuit of human knowledge' (quoted in Foster 1985, p. 392). He wrote this on 18 June 1846 when atop of his newly-named Mt Owen, after Richard Owen who joined in 1837. The next member noted was Buckland with a Tableland. It is somewhat surprising that Mitchell had not named something earlier for Buckland, as he had followed up Buckland's cave exploits in Britain by looking for bones in caves at Bungonia, south of Sydney, as early as 1829 (Foster 1985, p. 135). Mt Faraday (1824) was also 'put down' (18 June 1846). Three months later, on 10 September 1846, Mitchell named Mts Hutton and Playfair for the two great Scottish pioneers, Playfair being one of the 42 Honorary Members elected at the first meeting of the Society (Herries Davies 2007, p. 17). By this time the first, almost full-time, resident geologist, W. B. Clarke, had been honoured by Leichhardt by the naming of Clarke River in Queensland 'in compliment to the Rev. W. B. Clarke of Parramatta' (22 April 1844).

Another spate of names followed in Western Tasmania in February 1862 given by the then Government Geologist of Tasmania, Charles Gould (1834–1893), son of the zoologist John Gould (1804–1881). They consist of some of the old favourites: Mounts Jukes, Darwin, Lyell (site of a famous mine), Owen and Sedgwick (his first and only time). A Mt Murchison was named in the same region a little later.

Two servants of the Society

Two nineteenth century employees of the Society – Assistants in the Library and Museum – later served Australian geology. They were Ralph Tate (1840–1901) (Fig. 4) and Sydney B. J. Skertchly (1850–1926). Tate was employed between 1864 and 1867, resigning to 'join an exploring expedition in Nicaragua' according to Woodward (1907).

Fig. 4. Professor Ralph Tate, a Society employee who made major contributions to Australian geology.

Alderman (1976), however, has him in central America and Venezuela working on mining prospects for two years, before returning to England to teach at various mining schools and, then, 'undertaking a major investigation of the Yorkshire Lias'. It was this last that saw him awarded a moiety of the Murchison Fund of the Society in 1874. In that year he was appointed Professor of Natural Sciences at the newly established University of Adelaide, beginning his work there the following year, and becoming a nationally recognized geological expert in a number of fields, but particularly in invertebrate palaeontology, as well as contributing considerably to zoology (Alderman 1976). In all, Tate published 10 papers in the Society's journals, a number exceeded only by Clarke. However, like many antipodean geologists, he supported the societies in the region, whose publications were more readily accessible in Australasia. Tate makes several appearances on the map of Australia.

Skertchly seems to have been appointed following Tate's resignation, but he resigned the next year. Skertchly was apparently encouraged by Alfred Tylor (1823–1884), the geologist and anthropologist, and had been a Royal School of Mines student and a pupil of Tate, which probably explains the succession at the Society. Skertchly had a wide range of scientific contacts, including Charles Lyell, Charles Darwin, A. C. Ramsay and T. H. Huxley. He later had a fascinating professional life in Egypt, the British Geological Survey (studying Post-Tertiary deposits and the gun-flint industry, *inter alia*), and publishing his controversial ideas about human antiquity in *Nature*. He and his wife travelled through Europe before going to California, Borneo and then China, where he was appointed Professor of Botany at the College of Medicine for Chinese in Hong Kong, with Sun Yat-Sen as a student. The Sino-Japanese war saw his final move to Queensland where he later joined the Geological Survey and carried out a wide range of fieldwork particularly related to mining. A colleague called him 'a most remarkable and many-sided man with a distinct touch of genius ... quick-brained, silver-tongued, with a retentive memory, he had a marvellous range of knowledge and a distinct literary style' (A. H. Longman, quoted by Marks 1988, p. 622). However, he didn't get on so well with his boss, Robert Logan Jack, who felt Skertchly lacked the energy required for the work in an enervating climate. Elected in 1871, Skertchly was 'removed' after a membership of 22 years, but, perhaps because of his former Society work, is probably one of few 'removed' members, who gained a Society obituary.[5] However, the obituary kindly says 'although long resigned his variegated career demands notice' (Bather 1927, pp. lx–lxi). Although missing from the map of Australia, there are two Mounts Skertchly, one in Borneo and one in Canada.

Mid-nineteenth century achievement and recognition

Walter Mantell (1820–1895), the son of Gideon Mantell (1790–1852) (member in 1818), who joined the Society in 1858, was an important geologist in his own right. During (Sir) James Hector's (see later) absences, Mantell was Acting Director of the Geological Survey of New Zealand. Having left home at the age of 20, Walter became reconciled with his father in later life. Walter corresponded with Lyell concerning earthquakes, with Darwin about glaciers, and with Owen on vertebrate palaeontology, having published a paper in the Society's journal in 1852 on the moa. Indeed, the fossil remains of birds collected by Walter Mantell were discussed in a paper given by his father on 2 February 1848. Gideon Mantell's diary noted

[5] See later for a discussion of the process of 'removal'.

'The Dean of Westminster [Buckland] unfortunately indulged more than usual in buffoonery, and completely marred the discussion. Which consequently was utterly unworthy the subject' (Herries Davies 2007, p. 102). Walter was not particularly complimentary to Owen, who treated his father very shabbily (Dean 1999): 'he [Owen] has made considerable blunderings and flounderings in his search after renown rather than truth' (Sorrenson 1990, p. 268). Vallance (1984) and Tee (1998) point to many important documents of Gideon Mantell kept by Walter and still held in New Zealand, including letters from Michael Faraday (1791–1867) (who amongst so many other things was a Society Council Member 1828–1830), indicating Faraday's abiding interest in geology.

The Catholic priest, Fr Julian Edmund Tenison Woods (1832–1889) elected on 30 November 1859, although English-born, was influenced by his time studying in France, particularly concerning geology. Woods had an extraordinary career as missionary, educator and founder of religious Orders of nuns, while still pursuing geological pursuits and aspects of biological science in his spare time (Branagan 2009).

His report on metamorphism in the Mount Lofty Ranges (Woods 1858) is one of the earliest on the subject in Australia (Vallance 1975, p. 35). Woods benefited greatly from advice he received from Charles Lyell, particularly in his publications on the subdivision of the Tertiary strata of South Australia and Victoria (Woods 1860, 1862, 1865), which used Lyell's percentage ideas. He made an important study of the volcanic region around Mt Gambier (South Australia), and, of more significance, carried out pioneering work on the origin of the karst systems of the Mt Gambier–Naracoorte region.

By 1862 Woods had put together a volume that set out his conclusions on many aspects of the region, and which appeared in London the same year. It is still a valuable resource (Woods 1862). In fact, Woods' ideas on the formation of caves were far in advance of those held in Europe and much closer to present understanding (Hamilton-Smith 1996). Woods moved to New South Wales in 1871. Numbered among his students at St Charles Seminary Bathurst, was John Milne Curran (1859–1928), who joined the Society in 1884. In 1879–1880 Woods was elected President of the Linnean Society of New South Wales, and the Royal Society of that same colony awarded him its prestigious Clarke Medal in 1882. In his last years he carried out important work in Malaysia, Japan and other parts of Asia (Branagan 1996).

Captain Frederick Wollaston Hutton (1836–1905) was elected in 1860, while still in England. In 1862 he published *The Use of Geology to Military Officers*, the beginning of a long history of geological publications. Moving to New Zealand in 1866 he was closely associated with Professor George Ulrich (1830–1900), (who had moved from Victoria) at Otago, where Hutton was Provincial Geologist before moving on to a professorship at Christchurch in 1880. Hutton died returning to New Zealand from leave in Europe, and was buried at sea in the South Atlantic, near Cape Town. He was, perhaps, the most original geological thinker in New Zealand in the nineteenth century.

Sir Julius von Haast (1822–1887), elected on 18 February 1863, was an influential figure in New Zealand geology, based in Christchurch, where he raised the profile of the provincial Canterbury Museum to considerable heights. He published papers on his moa discoveries and on regional geology in the Society's journals. Like Mitchell before him in Australia, Haast was responsible for the placing of numerous member's names on New Zealand topographic features. His son recalled that his father named more than 100 topographic features, particularly in the glacial region of the South Island (Haast 1948; Burrows 2005). They include the Clarke, Darwin, Lyell, Murchison, Mueller and Ramsay glaciers. There are eight or so features with Haast's name. The most notable is the 3138 m peak, named by Robert von Lendenfeldt (1858–1913), near New Zealand's highest mountain, Mt Cook, named by Lort Stokes in 1851. Tee (2007) suggests that the South Island of New Zealand contains 'more places named after scientists than any place on earth', and a considerable proportion of the names consist of members of the Geological Society of London. Haast's marriage to Mary Dobson linked several members of the Society: Haast, the engineer Edward Dobson, Mary's father and his son Arthur (1841–1941).

Alfred William Howitt (1830–1908) was from a Quaker family, which spent some time in Germany. In Australia he became a regional administrator (police magistrate and warden of goldfields), but devoted his spare time to geology. An explorer of note, pursuing metamorphism in considerable detail, he joined the Society in 1874. Vallance (1986, p. 46) suggested that in many ways he anticipated the better-known work of George Barrow (1853–1932). He studied the metamorphic progression in the mountainous region of eastern Victoria, a region ranging from folded early Palaeozoic strata to crystalline schists and gneiss, intruded by granitic bodies. He was almost certainly the first in Australia to use thin sections to discover the secrets of metamorphism. Howitt's paper (1879) in the Society's journal is a forerunner for Howitt's work, published in Victoria over the next 13 years. Vallance (1986) placed his achievements in this field of research very highly.

Sir James Hector (1834–1907), elected on 10 April 1861, owed his appointment in 1864 as first Director of the Geological Survey of New Zealand to the urgings of several Society members, including Walter Mantell and James Coutts Crawford (elected in 1865), a geologist then working from Wellington, New Zealand. Hector was already in New Zealand as Provincial Geologist for Otago in the South Island. He went on to create not only a national geological survey but also the Dominion Museum (Fleming 1986, pp. 5–8, 1987; Mason 1990, p. 7). Hector had some geological disagreements with Sir Julius Haast (see earlier) in 1872 and Captain Hutton in the 1880s. Hector's name appears on New Zealand maps in about equal quantity with Haast (McKenzie 1987). In hindsight, perhaps his most important memorial is the Royal Society of New Zealand, which he managed to form, and which remains the umbrella society of New Zealand science, giving it a continuing link with politicians and the national purse strings (Burnett 1936; Fleming 1987; Dell 1990).

Hector was one of a number of antipodean geologists to be officially recognized by the Society in the 1870s, being awarded the Lyell Medal in 1876. However, by a small margin, he was not the first Australasian to receive such recognition. That honour goes to Alfred Selwyn, who was awarded the Murchison Medal in that same year; although by this time he had moved to Canada, after 18 years in Victoria (1851–1868) as Director of the Geological Survey (Branagan 1990*b*). As noted above, Selwyn published several important papers in the Society's journal prior to joining. The award, for Selwyn's contributions to 'Silurian Geology', was accepted on his behalf by his mentor and former field companion in Wales, Sir Andrew Ramsay (1814–1891). In the following year, 1877, the by then 'venerable' Clarke became the second Australian to awarded the Murchison Medal, just a year before his death. Two years later the palaeontologist Frederick M'Coy (1823–1899), elected in 1852, became the recipient of the same award. The highly prestigious Murchison Medal, close to the top of recognition by the Society, had first been awarded in 1873 and it is remarkable that so may contributors to Australian geology were recognized early on. Together with Tate's more minor award, this was, indeed, a vintage decade for antipodean members.

The late 1800s

In 1880, north Queensland was the site for the appearance of Scottish geologists' names on the map of Australia. This was the work of Robert Logan Jack (1845–1921), which commemorated his former Scottish geological colleagues – Peach River, Geikie, Horne and Croll creeks, and, later, Mt Croll. Possibly realizing his error in giving Geikie a minor stream he named Geikie Range (Jack 1921, vol. 2, 502–507), although Jack might, indeed, have intended to honour both James and his more famous brother, Archibald. While Jack himself has apparently missed out on having a topographic name, his predecessor in Queensland, Richard Daintree (Bolton 1965), is now well remembered by Daintree National Park and Daintree Rain Forest, covering a large area of coastal north Queensland.

There had been naming through the 1860s and 1870s, particularly by the explorers/geographers J. McDouall Stuart (Stuart 1864), Ernest Giles, W. C. Gosse, Peter Egerton Warburton and others. Naming continued into the 1900s, associates W. G. Woolnough and T. Griffith Taylor both being named in New South Wales, Taylor again in the Australian Capital Territory and Woolnough for a seamount off the New South Wales coast. See also Foster (1985, pp. 501–504) for numerous memorials to Sir Thomas Mitchell.

Alexander Mackay (1841–1917), a Scot (elected in 1888), reached New Zealand in 1863 and was a gold-seeker for five or six years, including time in New South Wales and Queensland. Through contact with von Haast he began to undertake geological mapping and fossil collecting, and was then appointed by Hector in 1872 to the Geological Survey of New Zealand. Mackay was one of the pioneers. Essentially self-taught, he made many important surveys, but illness marred his final years (Cooper 1993, pp. 290–291; [Mason] 1995, pp. 5–13; Bishop 2008). Mackay opposed Haast in believing (correctly) that the 'moa-hunters' were the Maoris and not a vanished pre-Maori entity.

Cooper (1993, p. 291) suggests that Mackay's greatest achievement was 'freeing New Zealand Sciences from the strictures of a European-based "received wisdom" enabling them to see, interpret and report the uniqueness of New Zealand Geology', opening the way 'for the development of new theory with worldwide importance, brilliantly exploited by later workers'. McKay's significant work as 'an original thinker' on block-faulting in the evolution of the mountain-system of New Zealand was particularly noted by Alfred Harker (1918), although Mackay might have been aware of some North American work in this field. Another member, at a distance, was involved in McKay's geology. This was W. J. Sollas (1849–1936) who was highly paid to carry out a petrological examination of rocks from the Cape Colville Peninsula, the results of which were published in an expensive two-volume work, but which, despite good reviews by various international colleagues, was poorly received in New Zealand (Sollas 1905; Mason & Watters 1996, pp. 19–24).

Patrick Marshall (1869–1950), elected in 1894, was English-born. Watters (1996, 1997) gives an expansive statement of Marshall's life and achievements. In his student days, he was influenced particularly by F. W. Hutton and G. H. Ulrich, gaining a DSc in 1900 for his thesis on the Auckland volcanic field, carried out while he was a teacher at Auckland Grammar School. Interestingly, his first research, in the 1890s, was in the field of entomology, an indication of the influence of Hutton, who was prolific in this field as well as in geology and other fields of natural history. In 1906 Marshall published a long paper in the Society's *Quarterly Journal* on the geology of Dunedin, a region that was studied in even more detail by his successor, the Australian-educated W. N. Benson (1885–1957), the adopted Kiwi.

Not unlike arguments in Australia about the age of Late Palaeozoic rocks, but, in fact, somewhat more complex, Marshall was involved in a long controversy about the age and classification of much of the older sedimentary rocks of New Zealand, differing in particular from Alexander Mackay's ideas, which were later shown to be closer to the mark (Watters 1997, p. 7). Marshall also clashed with Professor James Park (1857–1946) of Otago University (1901–1931) on various matters, but particularly on the extent of Pleistocene glaciation in New Zealand in which Marshall proved correct. Despite his achievements, Marshall was badly treated at the University of Otago, and for a time went back to secondary school teaching, while still carrying out much important research on the Pacific region. Marshall is remembered for his recognition of the fundamental 'andesite line' of the western Pacific, essentially separating continental and oceanic crustal regions. He also recognized the specific origin and coined the appropriate name for the volcanic rock type 'ignimbrite'. However, there were many other contributions, such as his studies of the geology of many Pacific islands and some important palaeontological work (Benson 1950; Hocken 2002, 2003).

Robert Arthur Buddicom, elected on 22 June 1899 ('Gentleman', B.A. Oxford), became Robert Bedford (1874–1951). Buddicom graduated with a B.A. (Oxford) in 1897 in chemistry and biology, worked at the marine biological station in Naples (1897–1898), and was later (1906–1914) a demonstrator and lecturer at London Medical College. After a court case, in which he was found guilty of misrepresentation in a prospectus, he moved to Australia under the name Robert Bedford, with his second wife and two children. Bedford settled at Kyancutta, an isolated locality in South Australia, growing wheat. There he undertook a wide variety of public services, including acting as the local doctor despite lacking an official medical degree; Douglas Mawson being one who voted against him receiving the equivalent of an 'ad eundum' degree from the University of Adelaide. Bedford had clashed with Mawson over Bedford's visits to the Henbury meteorite craters in Central Australia (obtaining evidence that the impact was a relatively recent one) and his subsequent sale of specimens to the British Museum. Bedford also carried out extensive research on *Archaeocyatha*, identifying 32 new species and eight new genera, publishing a series of monographs between 1934 and 1939 under the imprint of the Kyancutta Museum, which he founded in 1929 (Brett-Crowther 1979; Laube 1990).

What is noteworthy in the latter half of the nineteenth century is the number of mining figures, of a variety of competence, in both Australia and New Zealand who 'boasted' of their membership. In fact, it was a matter of some prestige in the mining industry and among local geologists to be able to append the letters F.G.S. to their names. In these years, there was a constant movement of members between the Australian colonies and New Zealand, which continued into the twentieth century.

James Malcolm Maclaren (1873–1935), elected in 1901, was born on the Thames mining field on the North Island of New Zealand, being the first native-born geologist. A student, first at the Thames School of Mines, Maclaren suffered by the loss of his exam papers that were sent to England for marking, a process which was sensibly abandoned shortly after! Nevertheless, he had a brilliant academic career, gaining a DSc at Auckland. He worked first for the Geological Survey of Queensland and then with the Geological Survey of India. He was offered but declined the post of Director of the New Zealand Geological Survey in 1904, the position being taken up by the Canadian, James Mackintosh Bell (1877–1934). In 1906 Maclaren embarked on an outstanding international career as a consultant mining geologist. His consulting career was largely associated with the Goldfields Group, founded by Cecil Rhodes. Maclaren published little, but he was frequently in the mining news. He was 'a man whose work must be judged not by what he published but by the influence he exercised in the counsels and on the policy of the leaders of the industry whom he advised' (Anon. 1935, p. 195).

Maclaren's extraordinary life has hardly been written about, but he travelled over much of the world, as recorded by his wife, who often went with him and documented how he travelled by various means, including 780 miles by mule in Peru and Ecuador, 510 miles by dog sledge in Canada and Alaska, 210 miles by elephant in Burma, 610 miles by camel in Persia, and 190 miles in the comparative comfort of a houseboat in Korea! (Mason 1994, pp. 28–35, 2003, pp. 3–11).

Celebrating the Society's centenary?

The centenary of the Society was very much a 'Top of the Town' affair, as told by Watts (1909). There were dignitaries (many, but not all, geological) from most European countries. There were long speeches galore, in a variety of languages, numerous functions and hosts of honorary degrees awarded by the universities. There were also field excursions. The celebrations extended from 26 September to 3 October 1907. Watts (1909) records the speeches, and the celebratory letters and telegrams, *in extenso*. However, the congratulations from the antipodes were, perhaps, relatively muted. The most fulsome came from McIntosh Bell, on behalf of the officers of the Geological Survey of New Zealand, who 'desire to extend their most hearty congratulations ... and to express their great indebtedness for the inspiration they have received from the Society's widespread endeavour in the science of geology' (Watts 1909, p. 75).

From Australia only the Royal Society of Tasmania wrote, claiming its position as the longest established scientific society of the Australian Commonwealth, noting that the 'important services rendered by representatives of the [Geological] society to all parts of the Commonwealth ... in promoting geological research, and aiding in the development of their material resources' were duly appreciated. It was hoped that the Society's 'career in future may be as prosperous as it has been in the past' (Watts 1909, pp. 74–75). What is more, the Tasmanian body was represented at the celebrations by R. M. Johnston (1843–1918), Registrar-General, and in his own right a distinguished geologist, but sadly not a member. However, Johnston apparently claimed also to represent the Geological Society of Tasmania, a body whose existence is open to considerable doubt!

Edgeworth David, a dedicated member (see later) of Sydney University, was in the throes of going south to Antarctica with Shackleton, and had been in England the previous year en route to the International Geological Congress in Mexico, so was unable to attend. However, no doubt it was his doing that saw the University of Sydney represented by his friend the embryologist James P. Hill (1873–1854), who had only just left Sydney to take up the Jodrell Chair of Zoology and Anatomy at University College, London. The University of Melbourne was represented by J. W. Gregory (1864–1932), a member who had moved from Melbourne to Glasgow University several years previously. Arthur Smith Woodward (1864–1944), another member, from the British Museum (Natural History), represented the National Museum of Victoria (the great monument to Frederick McCoy). Making up the numbers was the Honourable C. H. Rason (1858–1927), Agent-General for Western Australia, a former brief and 'lightweight' State Premier, and noted for his comment 'I'm not a model of virtue and fine feeling, old man' (Bolton 1988, p. 334), who attended on behalf of the Geological Survey of that state. New Zealand was represented by the chemist H. G. Denham (1880–1943), then studying at Liverpool University, on behalf of the Canterbury College, Christchurch and the New Zealand Geological Survey.

Maclaren was in London at the time of the centenary celebrations and was, perhaps, only one of three antipodean members present at the Society's celebratory dinner held at the Hotel Metropole (he had seat 258 of the '278 ladies and gentlemen' present). Others were Rudolf de Salis and S. H. Cox (*c.* 1852–1920) (Watts 1909, p. 136). But Maclaren was not just there to celebrate the Society. It is not quite clear from the wording, but Maclaren seems to have been a supporter of womens' liberation. A special General Meeting of the Society was held on 17 June 1908 to consider the resolution he proposed: 'that Fellows non-resident in the United Kingdom be invited to express an opinion concerning the admission of women to Fellowship or Associateship of the Geological Society of London' (Mason 2003, p. 9). This seems to have been a move to encourage those 'Home' members who had moved over the past several years to allow women to become Society members (Herries Davies 2007, pp. 158–163; Burek 2009).

Antipodean intrigue

The lists of antipodean members (see Appendices 1 and 2) contain a number of applicants who were refused membership, or whose elections were later declared void. It has not been possible to follow up these matters, but they are clearly items for further research. In some cases the refusals seem contrary to the evidence presented, while other members accepted, about the same time, seem to have had an 'easy ride' to election, with little evidence of expertise.

About 1877 the Society changed its application form and began to require, perhaps, somewhat more of a geological background than had been the case previously. However, both before and after this date there seem to have been some problem applications.

The earliest of such 'problems' was the case of John (Frederick) Calvert (1825–1897), who applied for membership in 1853. Calvert has always been somewhat of a mystery man, but he gained a reputation as a 'Mining Scoundrel' (Vallance Data Base, 1997, No. 536 in Branagan & Branagan 1997) and he is one of the few 'Australian' applicants of which there appears to have been a ballot for his membership (in fact, there might have been two ballots). However, the numbers went against him, 26 to 6. Calvert had just published his extraordinary book on gold (Calvert 1853), including his alleged travels, which appear to be rather more fiction than fact, and was also, about this time, involved in brawling with Murchison about gold, so the circumstances were very probably against him. Calvert's extraordinary claims, documented by Coninsby (1895), Wallace (1890) and Calvert himself (1853), make him an interesting 'character', but unravelling truth from fiction is extraordinarily difficult. He seems to have given himself an extra 10–14 years of life, variously claiming a birth date of 1811, 1813 or 1815, but Vallance (in Branagan & Branagan 1997), who 'chased' Calvert for some years, gives 9 March 1825. Calvert almost certainly 'appropriated' Stutchbury's mineral collection textually, if not actually, and claimed to be the author of some maps, which he certainly was not (Branagan 1992, pp. 107–108).

Although the treatment of Calvert is not surprising, the reasons for the rejections of Alfred James Hodgkinson-Carrington (1897) and Henry John Wolveston Brennand (1904–1907) are not at all clear. A Victorian government employee, and a member of several mining institutions, Hodgkinson-Carrington's election of 10 March 1897 was declared void, without explanation, almost exactly a year later (9 March 1898). Brennand, a physician and surgeon of Sydney, and a member of the Chemical Society of London and Royal Society of NSW was nominated by respected members Lawrence Birks, T. W. Edgeworth David and Herbert Jevons (1875–1955), and was apparently elected on 3 February 1904. However, for some unknown reason his election was declared void at a Committee Meeting just three years later (13 March 1907). Whether Birks' removal from the Society's membership list in 1905 had any significance is unknown. Brennand seems to have had rather better qualifications (excepting a less famous surname) than Charles James Buckland, a bank manager, elected in 1887.

Albert Spencer Ellam (November 1896) and Benjamin Pherson Ekberg (February 1897) did not get as far as a brief membership, both being balloted out. Ellam, a 'geologist and metallurgist', was then working in New Zealand, but claimed extensive experience in four continents, and was a member of several mining associations, with respectable nominators. Ekberg, a civil engineer, was then working in Ballarat, Victoria, had studied with von Haast, and was nominated by F. M. Krause, A. W. Howitt and R. A. F. Murray, all very respectable members. The Rev. Granville Ramage, a Melbourne graduate and a member of the Geological Society of Australasia, was elected in December 1900, but his membership was merely cancelled the following December. On the other hand, there is Charles A. V. Butler, who seems to have been elected twice (1889 and 1895)!

While from the late 1870s there was a period of 'joining' from the Antipodes, it also proved a period of 'removal', often after just a few years of membership (see Appendices 1 and 2). Removal meant, of course, that the members drifted away, and did not pay their fees. There were, naturally, varied reasons for this, changes of jobs and omitting to notify change of address, inability to afford the fees as economic circumstances changed, retirement or a feeling that the fees did not offer value for money. Such would be likely for those well away from London and unable to attend meetings, and not particularly interested in the papers that were being published. As there were many involved in mining, the formation of other societies, both local and international, no doubt also played a part. In this period were founded, for instance, the Society of Economic Geologists, petroleum geology societies, the Australasian Institute of Mining Engineers, the Institution of Engineers Australia, local 'Royal societies' or the equivalents, and even the rapidly growing Geological Society of America. However, Elliott Moore Cairnes is perhaps the only member to have been removed twice (1893, 1905).

The removal of Sydney Skertchly, former servant of the Society, after a membership of 22 years has been noted earlier. Another removed member to have the 'rare privilege' of an obituary was Andrew Gibb Maitland (1864–1951), removed in 1922.[6]

Only one example of apparent fraudulent use of the letters F.G.S. is known to me. These letters are appended to the name Captain E. de Hautpick, a Russian-born 'oil expert' who visited Australia twice during the 1920s (Branagan 2008*a*). On at least one of several pamphlets he published in Adelaide he adds F. G. S. (de Hautpick 1928). There is no evidence that de Hautpick had been a member, or had applied and been refused (Wendy Cawthorne, pers. comm. 2006). However, de Hautpick might just conceivably have used his then membership of the Geological Society of France to award himself an F.G.S.!

[6] See his obituary *Quarterly Journal of the Geological Society, London*, **79**, p. xxv (1953).

The decease of three members deserves noting. They are Harold Rutledge (1920–1954), killed in an air crash at Singapore on 13 March 1954, aged 34. I myself heard the news on an old portable radio while working on a drill site in NW Queensland. The others were much earlier. On 9 February 1871 Edward Heydelbach Davis (1845–1871) was drowned on the west coast of the South Island of New Zealand, when his horse failed to negotiate a swollen creek (Johnston 2007). The disappearance on 12 October 1880 of Lamont Young (1851–?1880) (elected 1875) remains the great unsolved mystery of Australian geology. Did he drown, was he murdered or did he just escape for his own reasons? The last seems very unlikely. There is no space to follow this intriguing story, which has been told in various ways (not always quite accurately) (see Pearl 1978; Branagan 2005*b*, pp. 25–26; Branagan & Packham 2000, p. 319). Whatever the details, if it had not been for Young's disappearance it is unlikely that T. W. Edgeworth David would have gone to Australia and the history of geological endeavour in Australia would have been very different.

Fig. 5. (Sir) T. W. Edgeworth David, Australia's most famous geologist. A coral reef driller, Antarctic explorer, World War I hero and eminent Fellow of the Society.

Edgeworth David: our man in Australia

While there were numerous later Australasian members who added lustre to the Society, the activities of T. W. Edgeworth David (1858–1934) (Fig. 5) probably marked the peak of antipodean involvement in the Society. David became a member in January 1883, having made his application just prior to his departure from Wales for New South Wales to take up his appointment on the Geological Survey of New South Wales, essentially replacing Lamont Young. At that very moment, David had his first paper on glaciation in South Wales for the *Quarterly Journal* in the pipeline (David 1883), having been encouraged by his mentor, the former President, Sir Joseph Prestwich (1812–1896).

Through a meeting in the Hunter Valley of New South Wales in August 1885 with Richard Oldham (1858–1936), who was visiting from India, David became convinced that there was an older (Late Palaeozoic) glaciation in Australia. He worked quickly on this problem and, in 1887, sent a paper to the Society proclaiming the evidence for a (probable widespread) event. It was presented to a sceptical Society audience, who felt the evidence was unconvincing. Nevertheless, the paper was published by the Society (David 1887). It was nearly 10 years before David attacked the Society sceptics. David had been hard at work, *inter alia*, on glacial evidence. He put this together for part of his Presidential address to the Australasian Association for the Advancement of Science (David 1895), before paring it down to present it personally to the Society on his first visit 'home' in 16 years. This time around he took care to load his luggage up with numerous striated boulders (some of considerable size and weight, but it was pre-air travel!). It proved a major triumph. David, with his personality and oratorical skills, accompanied by the specimens, swayed the audience (David 1896). It was an 'admirable paper' commented W. T. Blanford (Herries Davies 2007, p. 149). J. A. Watt (*c*. 1868–1958), David's student and 1851 Exhibiter, was also in the audience, and told Professor Baldwin Spencer (1860 – 1929) that the paper was 'exceedingly well received'.[7] This effort, in addition to news of David's Funafuti 'success', the relatively deep drilling to 'prove' Darwin's theory of coral atoll formation (Branagan 2005*b*, pp. 85–105), undoubtedly led to David being awarded the Bigsby Award in 1899.

Although David was to publish only two more papers in the Society's journal (David 1926; David & Woolnough 1926) he was, nevertheless,

[7] ML: Letter from Watt to Baldwin Spencer, 12 February 1896, Spencer papers, Ms 29, vol. 7.

constantly involved in Society activities, in nominating various colleagues or students for membership, speaking, off the cuff, when attending meetings in London, and using the Society's rooms virtually as his office when in London. There were visits: in 1900, when wrapping up his Funafuti report of the Royal Society, which gained him a Fellowship in that august body; and in 1906 en route to Mexico for the International Geological Congress, loaded with even more large specimens of glacial boulders, this time mainly evidence for a Late Precambrian event, based on the work of his friend Walter Howchin (1845–1937) in South Australia (David 1906). David was briefly in London early in 1911, discussing his Antarctic work, particularly the extraordinary journey with Mawson and Dr A. Mackay to the vicinity of the South Magnetic Pole. He was in England again early in 1914 finalizing arrangements for the British Association meeting in Australia scheduled for the following August. David's work for the BAAS meeting in 1914 (David 1914), including his discussion on the Talgai Skull at that meeting (David & Wilson 1914), following his writing up of aspects of the geology of the Shackleton Antarctic Expedition (Shackleton 1909; Priestly [sic] & David 1912) were the probable reasons for the award of the Society's most prestigious Wollaston Medal to David in 1915, the first Australian to receive it. It was accepted, on David's behalf, by the Agent General Sir George Reid. However, Herries Davies (2007, p. 227) suggests that it might have been awarded during the war 'as a modest addition to the cement of the Allied alliance'. Just the year before, Walter Howchin (1845–1937) gained a moiety of the Lyell fund to help his fieldwork.

World War I interrupted the long-awaited antipodean British Association for the Advancement of Science meeting, and it was not long before David was involved in that earth-shattering event. He arrived in France in May 1916 with the Australian Tunnellers, having been largely responsible for their foundation. Shortly after he met the only geologist with the British army, Captain W. B. R. King (1889–1963), Society President many years later (1953–1955), and they did considerable work together in the years to 1919 (Branagan 2005*b*; Herries Davies 2007, pp. 223–227). David's survival from a severe accident caused by hurtling to the bottom of a dry well when the winding mechanism failed in September 1916 took him to London for convalescence, where he insisted on being put in a hospital in Park Lane, not too far from the Society's rooms (and the Geological Survey) so that he could catch up on necessary maps and references as he convalesced (Branagan 2005*b*, p. 292). He was back to the front in November, again working at a frantic pace. In the next several years he had brief visits to London and apparently always went to Burlington House. Certainly, he spent most of a three-week period of leave in March 1917 in the Society's library. A particularly poignant moment for David was his re-meeting with the distinguished Foreign Member Professor Charles Barrois (1851–1939) in Lille, a few weeks before the war ended (see Herries Davies 2007, p. 225 for a photograph of a field trip November 1918).

During the war years few papers were presented for reading and publication, and the Society was 'pushed' to find satisfactory material for meetings. One antipodean caught for the purpose was Sir Douglas Mawson, who was called upon to give a two-lecture presentation on the Antarctic ice-sheet. Mawson took part, through his mentor, Edgeworth David, in Shackleton's 1907–1909 Antarctic Expedition, and attended a Society dinner and meeting in 1910 (Ayres 1999, p. 32), but it was not until 1918 that he joined the Society, probably after the lectures he had given. In general, Mawson was never one for scientific societies and conferences. However, he was awarded the Society's Bigsby Medal in 1919, David being in attendance at the ceremony (Branagan 2005*b*, p. 321).

Herries Davies (2007, p. 230) notes that David spoke about the use of geology on the Western Front at a Society meeting on 26 February 1919, but the War Office thought what he said was highly confidential and the Society 'was refused permission to publish even a summary of his lecture'.

Nevertheless, during 1919 David laboured hard with King to write a book on military geology, based largely on their experiences in the previous few years. None of the publishers seemed to have been interested (possibly warned off by the War Office). This work, entitled *The Work of the Royal Engineers in the European War. 1914–1919: Geological Work on the Western Front* was eventually published, in 1922, by the Institute of Royal Engineers, at Chatham (David & King 1922). However, only a careful reading of the introduction gives an indication of who were the authors.

In 1926, after retiring from his Chair of Geology at the University of Sydney, David returned a final time to the comfort of the Geological Society's rooms, which he used as his London address, hoping to complete his long-planned *Geology of the Commonwealth of Australia*, a project that continued to grow to massive proportions and eventually overwhelmed the struggling author, hit by age and increasing ill health. He spent nearly two years in London, but couldn't say 'no' to invitations to lecture at University College London, joining his former student Frank Debenham (1883–1965) at the opening of the Scott Polar Research Institute in Cambridge. He also accepted various honorary degrees around the British Isles, and took a field trip with Sir E. B. Bailey (1881–1965) and former student C. E. Tilley (1894–1973) to Schiehallion, and attended the funeral of his former Sydney friend Archibald Liversidge (1846–1927). Consequently, the 'book' made little progress. However, his large map of the *Geology of the Commonwealth of Australia*, first presented in 1931, earned great respect. David was to make a final effort concerning Society recognition of Australasian work. Through his friendship with Thomas Holland (1868–1947), then President, David reminded the Society of Walter Howchin's outstanding achievements and he was awarded the Lyell Medal in 1934, just at the time of David's death.

Conclusion

This paper has attempted to show the place of the London Geological Society at the very extreme of its scientific network. Clearly, the Society has played an important role in the culture of Australasian science. Indeed, there are many other Fellows who had a huge impact on geology in Australia, including (Sir) Frederick McCoy, Baron F. von Mueller, Henry Yorke Lyell Brown, Sir John Forrest, Richard Daintree, Archibald Liversidge and Robert Logan Jack. Most of these have received particular study (see, for example, Bolton 1965; Zeller & Branagan 1994; Grey *et al.*, 2001–2002; Jack 2008; Macleod 2009). Australian achievements in vertebrate palaeontology, ancient glaciations and the formation of mineral deposits were particular areas of important research by Australian Fellows.

Similarly, the New Zealand Society story as told above has been equally selective. The 50 or so New Zealand members identified (Appendix 2) contain a roll-call probably more prestigious than the Australian list. In addition to those briefly discussed above, J. A. Thomson, H. T. Ferrar, Robert Speight and W. N. Benson should be noted. Of the numerous members in New Zealand, Haast and Hector made significant studies of Pleistocene glaciation, still active close to sealevel, while, later, Marshall recognized the aforementioned 'Andesite Line' in the Pacific region, and the nature of ignimbrites. For these Fellows and others, the F.G.S. became a 'badge of honour', and particularly so amongst those who had few other qualifications. The Society's publications offered an important two-way conduit for the exchange of information and ideas, the Society's journals finding a ready home in libraries, companies and homes across the antipodes.

In all, at least 360 members were active in Australasia, many joining in the early years. The strength of this membership continued into the twentieth century, but has not been developed in this paper. The Geological Society of London had a real presence in the Antipodes. The sense of this is embodied in the message sent by J. Mackintosh Bell to the 1907 Centenary Celebrations which could be repeated today expressing 'hearty congratulations ... and ... great indebtedness'.

Many people contributed to the presentation of this paper. The Geological Society of Australia (Federal Executive), the New South Wales Division of that Society and HOGG provided generous financial support, for which I am most grateful. The notebooks of T. G. Vallance (1928–1993), passed to me by Mrs Hilary Vallance, were the source of much of the information on membership. Dr Sue Turner, Queensland Museum, provided information on Sydney Skertchly, while Dr Barry J. Cooper, South Australian Department of Mineral Resources provided information on Tenison Woods. Wendy Cawthorne, Society Librarian, answered queries on membership and archives. The Earth Sciences Department, Natural History Museum, London kindly gave me access to the collections passed from the Geological Society to the Museum in 1911. I was greatly assisted by the New South Wales State Library (including the Mitchell Library) and the Fisher Library, University of Sydney. I am grateful for the enthusiastic work on the history of New Zealand Geology fostered by the founder of the New Zealand Geological Society Historical Studies Group, Alan Mason. The interest in the project by the Earth Sciences History Group, Geological Society of Australia, is also acknowledged. Positive comments and advice by referees D. Oldroyd and T. Darragh and the editors are gratefully acknowledged.

Appendix 1. *Australian members of the Geological Society of London up to the early twentieth century*

Surname	First names	Application no.	Election date	Additional information
Abbott	Edward	632	21 Jan 1825	Of Van Diemen's Land
Aldridge	Edward George	3496	1886	d. 28 Dec 1917
Alford	Charles John	3481	16 Dec 1885	Sydney 1896–1898
Allhusen	Ernest Lionel	4120	1 Dec 1897	Western Aust. Geol. Survey
Anderson	William	4210	12 Apr 1899	India, then in Natal, prev. Sydney Univ.
Ayers	(Sir) Henry	2569	7 Dec 1879	S. Australia
Ball	Lionel C.	4481	7 Dec 1904	Queensland Dept Mines
Banks	(Sir) Joseph		1808	Res. April 1809
Barnard	W. H.	2890	1 Dec 1875	Victoria 1875–1901
Baron	Rev. Richard	3663	17 Apr 1889	London Missionary Soc. Madagascar & ?Aust
Bates	Thomas	3393	17 Apr 1889	
Bather	Francis Arthur	3512	1 Dec 1886	BM (Nat. Hist.), visited Australia
Bell	Stanislaus Napier	4572	11 Nov 1906	Derby, Tasmania
Belt	Thomas	2324	7 Feb 1866	Only UK addresses?
Benton	W. E.	3223	1 Dec 1880	Res. 1891; Sydney 1889
Bicheno	James, F. L. S.	582	7 Mar 1823	Tasmania from 1842
Biden	William Downing	1753	19 Nov 1856	Aust. 1864–1875
Birks	Lawrence	4060	2 Dec 1896	Adelaide,1900–1901, Sydney 1902–1904
Blatchford	Torrington	4173	7 Dec 1898	W. Aust. Geol. Survey; Rem. CM4 2 Jul 1902
Bleasdale	Rev John, D. D.	2405 [?2408]	6 Nov 1867	Melbourne to 1877 only
Bolton	Arthur Jodrell	4311	6 Feb 1901	Victoria Assayer/ metallurgist, Rem. 1904
Booker	Frederick W	7595	21 Oct 1953	NSW, Res. CM, 22 Feb 1967
Boyle	Arthur Robert	3146	3 Dec 1879	Queensland 1884–1889, NSW, Victoria to 1900
Brain	Thomas Henry	1482	11 Mar 1846	Principle Sydney Coll., declared void ?Jan 1854
Branagan	David Francis	10964	7 Jun 1972	Mostly NSW
Brennand	Henry John Wolveston	4457	3 Feb 1904	Physician & surgeon, Sydney, member Chem. Soc. Lond., Roy. Soc. NSW, declared void CM 13 Mar 1907
Breton	Lieut William Henry, R. N.	1026	7 May 1834	d. *QJGS* 1888
Brown	Richard		1864	Poss. related H. Y. L. Brown
Brown	Henry Yorke Lyell	3357	20 Jun 1883	Australia, Canada, Aust., member to 1912
Brown	Robert	?	?	Obit., *QJGS* 1859
Browne	Arthur Richard	3670	22 May 1889	Australia visiting mines, 1889
Buckland	Charles James	3554	22 Jun 1887	Bank Manager Sydney, *c.* 1862–1896
Buddicom (Bedford)	Robert Arthur (1874–1951)	4227	22 Jun 1899	Gentleman, B.A. Oxford; palaeontologist

(*Continued*)

Appendix 1. *Continued*

Surname	First names	Application no.	Election date	Additional information
Bufton	Rev. John (PhD)	4284	5 Dec 1900	Duanalley Tasmania; Bunbury, W. Aust. [more botany], d. *c.* 1912
Burgon	John Alfred	1105	13 Apr 1836	Aust. 1858–1864; d. 8 Nov 1876
Burton	Samuel James	4067	2 Dec 1896	Author of *The Western Australian Goldfields*
Busby	Alexander	1353	23 Feb 1842	Cassilis, NSW
Butler	Charles A. V.	3665	17 Apr 1889	
Butler	Charles A. V.	3986	1895	UK; Same?
Cairnes	Elliott Moore	3775	17 Dec 1890	Rem. 30 Jun 1893
		4147	2 Feb 1898	Rem. 28 Jun 1905 Melbourne; Geol. Soc. Australasia
Calvert	John		balloted 22 Mar 1853, not elected	Ballot 5 Apr 1854 (26 neg., 6 affirmations)
Cameron	John Macdonald	4363	24 Jun 1885	Sydney 1893–1898
Cameron	William Dugald	2693	12 Mar 1873	Civil Engineer
Cameron	W. M.	2661	4 Dec 1872	Rem. 1878 (*QJGS*, **35**, 1879, *PGS* p. 18)
Campbell	Rev. Joseph	3512	1 Dec 1886	Elect. in UK, Aust. to 1900; Rem. 27 Jun 1900
Card	George William	3887	11 Jan 1893	Geol. Survey, NSW
Carne	Joseph Edmund	3674	19 Jun 1889	Geol. Survey NSW to 1920
Chapman	Frederick (1864–1943)	5703	15 Jun 1927	d. 19 Dec 1943, *QJGS*, **100**, p. lxvi–viii
Chewings	Charles	3826	9 Dec 1891	Heidelberg 1892–1893; Ph.D. 1894, S. Aust., some W. Aust. 1894–1932
Churchward	William Gould	4158	6 Apr 1898	W. Aust. from 1895
Clarke	Arthur Walter	3684	4 Dec 1889	Charters Towers, Queensland
Clarke	James	1847	1858	Roy. Soc. Vict., 1864–1907 continues to 1918
Clarke	William Branwhite M. A.	669	19 May 1826	NSW from 1839
Cobbe	Hervie Nugent Grahame	4285	5 Dec 1900	Near Coolgardie, W. Aust. to 1908, then UK
Collins	Arthur Launcelot	3875	7 Dec 1892	Afghanistan, Australia & elsewhere
Cottle	Victor MacDonnell		Feb 1946	Mostly Tasmania
Cotton	Leo Arthur	5430	22 Mar 1922	Univ. Sydney
Couchman	Lieut-Col. T.	3680	20 Nov 1889	Melb. 1890–1892
Cruttwell	Alfred Cecil	2674	18 Dec 1872	Res. 1890, *QJGS*, **47**, 1891, *PGS*, p. 22
Curran	Rev. John Michael Milne (1859–1928)	3388	9 Jan 1884	NSW (incl. Dubbo) to 1890, rem. 24 Jun 1891
Curtis	Alfred Harper	3823	25 Nov 1891	Mining Engineer
Daintree	Richard	2614	6 Dec 1871	d. CM 15, 10 Jul 1878
Dana	James Dwight	1592	8 Jan 1851	? for foreign membership
Davey	Thomas G.	3896	22 Mar 1893	Harrietville, Vict. to 1899, then UK, Rem. 3 Jul 1912
Davey	Thomas Henry	4573	7 Nov 1906	Ballarat, Bulong, W.A.

(*Continued*)

Appendix 1. *Continued*

Surname	First names	Application no.	Election date	Additional information
David	T. W. Edgeworth	3325	10 Jan 1883	Aust. from 1882
Davidson	Allan Arthur	4624	17 Apr 1907	UK, but explored Cent. Aust.
Davies	Joseph	4563	13 Jun 1906	Bulong, W. Aust.? Rem. 1917
Davies	William ?Atwood Tregaskis	4561	8 May 1906	Tasmania; Bulong [West. Aust.]
Darwin	Charles	1127	30 Nov 1836	Aust. 1830s
De Latour	Edward Arthur	4567	13 Jun 1906	Tasmania
De Müller	W. J. E.	3427	3 Dec 1884	Lond.; Port Darwin Gen. Mangr. NT Goldfields of Aust.1897–1898; to 1916
Dennant	John	3498	7 Apr 1886	Victoria to 1906
De Salis	Rudolf	3367	5 Dec 1883	Civil Engineer
De Salis	William Fane, M. A.	1256	22 May 1839	Aust 1843–1845 or later? to 1860
Dixon	John	3876	7 Dec 1892	Newcastle, NSW Collieries Inspector
Dunn	Edward John	3392	6 Feb 1884	Cape Town; Victoria 1886–1936
Dunstan	Benjamin	3811 re-elected 5704	8 Apr 1891	Rem. 1915, 15 Jan 1927; NSW to 1896; Queensland 1897–1914
Dutton	?	4390	1902	Rem. 1925, Cardiff, Wales
East	Joseph James	3777	17 Dec 1890	Adelaide to 1900
Eastaugh	Frederick Aldis	4601	9 Jan 1907	Metallurgist & Assayer, Sydney Univ.
Edmonston	James	1083	16 Dec 1835	d. 19 Mar 1852
Edwards	George Maitland	4383	18 Jun 1902	Rem. 1915; in Queensland
Ekberg	Benjamin Pherson	4116B	balloted 24 Feb 1897, not elected	Ballarat & NZ
Ernst	Fritz Joseph	4471	8 Jun 1904	Gentleman, Hobart, Tas.
Etheridge	Robert Jnr	2358	5 Dec 1866,	Vict. & NSW Res. 1882
Evans	John	3850	6 Jan 1892	Bulli; Rem. 1 Jul 1898; West. Aust. 1901
Fawns	Sydney	4033 4484	11 Mar 1896 7 Dec 1904	Rem. 1902 Coolgardie to 1900
Fischer	Carle, M.D, M.R.C.S. L.R.C.P., F.L.S.	3115	8 Jan 1879	Sydney 1880–1893
Fitton	Edward	2037	7 May 1862	Res. 21 Dec 1864
Ford	Henry William	3830	9 Dec 1891	Northcote, Vict. to 1901 Res. 18 Jan 1905
Forrest	John (Sir)	3369	5 Dec 1883	W. Aust. Surveyor/Explorer
Fowler	Thomas Walker	3930	10 Jan 1894	Melbourne initially; Res. 6 Nov 1912
Franklin	Captain John (Sir)	769	3 Apr 1829	Hobart, Tas., 1840s
Fryer	William	1903	15 Jun 1859	Possibly not the Queenslander, Fryar
Frecheville	William	2888	17 Nov 1875	d. 30 Jul 1940
Gibson	James	4402	3 Dec 1902	Ballarat, Vict; Transvaal
Glauert	Ludwig	4251	7 Feb 1900	Obit. *QJGS PGS*, 1611, 26 Nov 1963

(*Continued*)

Appendix 1. *Continued*

Surname	First names	Application no.	Election date	Additional information
Glenny	Henry J. P.	3588	28 Mar 1888	Victoria to 1891
Gould	Charles.	1891	1859	Tas to 1876, Hong Kong 1886–1895
Gregory	John Walter	3548	11 May 1887	Aust 1900–1904. d. 1931
Griffiths	George Samuel	3517	1 Dec 1886	Victoria to 1892
Guppy	Henry Brougham	3434	14 Jan 1885	Rem. 8 18 Jun 1890
Halligan	Gerald Harnett (1856–1942)	4258	7 Mar 1900	Rem. 4 Dec 1929, back 1934 (*QJGS*, **90**); obit. *QJGS*, **99**, 1943 p. lxxxiv
Hargraves	Edward Hammond	1715	16 May 1855	'First Practical explorer of the Australian Gold'
Harlin	Thomas	1908	16 Nov 1859	1832–1913; Rem. 1871
Harman	Frederick Edwin	3504	21 Apr 1886	Argentina; Aust.
Harper	Leslie Frank	4403	17 Dec 1903	NSW Geol Survey. Res. 20 May 1926
Harrop	William Nobbs	4431	18 Nov 1903	Kanowna, W. Aust. Rem. 1907
Hart	Godfrey Stephen	4940	18 Dec 1912	Mt Morgan, Queensland to 1917; Portland, NSW 1918–1947
Hart	Thomas Stephen	4343	4 Dec 1901	Melb. & Ballarat to 1914
Hawker	Edward	2758	11 Mar 1874	Adelaide, 1888–1904
Hay	James Douglas	4038	15 Apr 1896	10 years Aust. pre-1896; Res. CM 23, 8 Nov 1916
Heathcote	Charles Francis	3637	19 Dec 1888	Tasmania to 1917; S. India; Vict. 1934
Herman	Hyman	4214	26 Apr 1899	Vict. & Tas. (1904-?), Vict. 1908 to '28
Heussler	Christian Adolph	4108	28 Apr 1897	Kalgoorlie; Rem. 27 Jun 1900
Higgs	?Samuel, Jnr	2051	3 Dec 1862	Mostly Cornwall
Hodgkinson-Carrington	Alfred James	4100	10 Mar 1897	Melbourne, Victorian Govt; election declared void 9 Mar 1898
Holland	Thomas H. (Sir)	3789	4 Feb 1891	Mostly India, UK
Hooker	Dr Joseph	1488	6 May 1846	Geol. Survey GB
Hopkins	Evan	1386	1 Feb 1843	To 1866, but only UK addresses
Howchin	Walter, Rev.	3094	6 Nov 1878	S. Aust.
Howitt	Alfred William	2760	11 Mar 1874	Public Servant, Victoria; explorer
Humble	William	3918	6 Dec 1893	Newcastle to 1908. Res. 13 Apr 1910 (still alive 1922)
Hunt	Robert	1673	4 Jan 1854	Australian Mint
Hurst	Henry Ernest	4043	29 Apr 1896	Kalgoorlie, W. Aust.
Huxley	Thomas Henry	1748	9 Apr 1856	Sch. Mines, Lond.; Aust. 1830s
Jack	Robert Lockhart	4237	6 Dec 1899	Obit. *PGS*, 1628, 31 Dec 1965, p. 203
Jack	Robert Logan	2550	27 Apr 1870	Scotland, Queensland, etc.

(*Continued*)

Appendix 1. *Continued*

Surname	First names	Application no.	Election date	Additional information
Jackson	Clements Frederick Vivian	4414	25 Feb 1903	Rem. CM 5, 3 Jul 1912. Retired 1939 State Min. Engr., Queensland
James	John William	2948	10 May 1876	S. Africa to 1878, NSW 1879 to 1904
Jaquet	John Blockley	3794	25 Feb 1891	NSW to 1908
Jenkins	Henry Michael	2058	7 Jan 1863	d. 12 Jan 1887
Jennings	William E, B.A.	3048	1877	Res. 1882; Sydney to 1891
Jevons	Herbert Stanley	4164	20 Apr 1898	d. *PGS*, 1534, 30 Mar 1956
Johnson	Joseph Colin Francis	4029	26 Feb 1896	d. noted CM 21 Jun 1905; S. Aust.
Josephson	Joseph [?Joshua)] Trey [?Frey]	1861	5 Jan 1859	d. noted CM 17, 24 Mar 1897; Sydney to 1896
Jukes	Joseph Beete	1090	20 Jan 1836	d. noted 10 Nov 1869
Junner	Norman Ross	5972	5 Dec 1934	d. noted 10 Oct 1973; Lyell Medal 1944
Keene	William	2285	21 Jun 1865	Obit. *QJGS*, **29**, 1873, pp. xli–ii
Kekovich	George Ormond	3462	27 May 1885	UK; Ceylon 1887–1888; Queensland 1889–1892
Kitson	Albert Ernest	4127	1 Dec 1897	d. 8 Mar 1937, Vict. Dept Mines
Krausé	Ferdinand Moritz	3335	7 Feb 1883	Vict. to 1899, W. Aust. 1900, Transvaal to 1911
Lakeland	William John	4590	5 Dec 1906	Burma; Ballarat Grad
Langtree	C. W.	3626	5 Dec 1888	Victoria
Larcombe	Charles Oswald George	4662	4 Dec 1907	Rem. 1919; NSW Geol Surv to 1908, Sch. Mines, Kalg., W. Aust. to 1918
Lautour	Edward Arthur		1906	Magnet, Tas. to 1918
Lawder	Arthur William	2646	1872	Mostly India
Lawrence	Henry Lakin	3791	4 Feb 1891	Queensland & NZ
Leake	George	604	1824	Swan Rr [Perth], W. Aust. to 1845
Lindon	Edward Byron	3739	30 Apr 1890	d. CM, 27 Jan 1892; Queensland to 1891
Litton	Robert Tuthill	3519	1 Dec 1886	d. noted 4 Nov 1896; founded Geol. Soc. Australasia
Liversidge	Archibald (1847–1927), F.R.S.	2702	9 Apr 1873	d. 26 Sep 1927
Lock	Charles George Warnford	4265	4 Apr 1900	d. CM 14, 3 Nov 1909
Lowles	John Isaac	4114	9 Jun 1897	Rem. 1904, Coolgardie
Lucas	Arthur Henry Shakespeare	3254	27 Apr 1881	Res. CM 11 Jan 1888
MacAndrew	Harold	3565	7 Dec 1887	Aust. 1889–1894
MacArthur	Leslie William Alexander			Sir A. Geikie intimated he be proposed for honorary fellowship, 1892
McConnel(l)	David Cannon	1832	26 May 1858	d. CM 16, 6 Nov 1889

(*Continued*)

Appendix 1. *Continued*

Surname	First names	Application no.	Election date	Additional information
McCoy	Frederick, (Sir)	1635	1 Dec 1852	UK; Melb; d. CM 16, 24 May 1899
MacDonald	Simon		1954	Obit. in Yearbook 1971
Mackenzie	John	2684	5 Feb 1873	Newcastle, Sydney to 1903
McKnight	Frederick	3851	6 Jan 1892	Melb. (amateur geol.)
McMath	John	5962	21 Nov 1934	Obit. *PGS*, 1529, 20 Sep 1955, p. 145
Mactear	James Andrew	3968	5 Dec 1894	W. Aust. to 1899, d. 3 Jun 1903 London
Madigan	Cecil Thomas	5763	5 Dec 1928	Obit. *QJGS*, **103**, 1947, p. lv
Maitland	Andrew Gibb	3645	23 Jan 1889	Rem. CM 15 Nov 1922; Obit. *QJGS*, **79**, 1923, p. xxv
Markes	James Francis	3934	24 Jan 1894	Vict. & W. Aust.
Marks	Edward Seaborn	4069	2 Dec 1896	d. noted 21 Jan 1951
Marshall	Charles Edward	5939	6 Dec 1933	Res. CM, 10 Dec 1981
Marshall	William	3864	23 Mar 1892	Sydney, Geol. Soc. Australasia
Masey	Thomas Adair	2542; 3576	23 Feb 1870 25 Jan 1888	Rem. CM 24 May 1876; S. Aust.; d. *PGS*, 50, 1894
Maxwell	Walter	4191	21 Dec 1890	Hawaii
Mellors	Paul	4274	20 Jun 1900	2 years near Coolgardie, W. Aust.
Mennell	[?Philip Dearman]	4347	4 Dec 1901	Rem. 30 Nov 1927; Res. 25 Jan 1928; d. 1 Jun 1966
Miles	Keith Rodney	6264	14 Apr 1943	Obit. Ann. Rep. Geol. Soc., 1976, 31–32
Millen	John Dunlop	4510	22 Feb 1905	Res. CM 11, 20 Mar 1918
Milles	Robert Sydney	3375	5 Dec 1883	Tas. to 1894
Milligan	Joseph M.D., F.L.S.	1973	20 Feb 1861	d. CM 10, 28 May 1884, Tas.
Mitchell	Thomas Livingstone (Sir)	694	20 Apr 1827	d. CM 7, noted 5 Nov 1856
Molesworth	Francis Hylton [Hilton]	4364	5 Feb 1902	Rem. 1926; Sydney Tech College, d. 1934
Montgomery	Alexander	4296	5 Dec 1900	Obit. *QJGS*, **90**, 1934, p. lxiv; Aust. & NZ
Moore	Charles	1672	4 Jan 1854	d. *QJGS*, **38**, 1882, *PGS*, p. 51–52
Moreing	Charles Algernon	3587	14 Mar 1888	Res. CM 14, 5 Nov 1913
Morris	Alfred	3306	7 Jun 1882	Obit. *QJGS*, **44**,1888, *PGS*, p. 47
Morris	John Fossbrook	4381	11 Jun 1902	Korea, d. CM 15, 17 Nov 06
Moulden	John Collett	4021	8 Jan 1896	South Australia
Mundy	[? Godfrey] Charles	828	16 Apr 1830	?Artist/author
Murchison	John Henry	1701	1 Nov 1854	d. 1891
Murray	Reginald Augustus Frederick	3628	5 Dec 1888	Rem. 1904
Nicholas	William	2789	2 Dec 1874	Vict. to 94, W.A. to 1920
Nicholson	Charles, M.D. (Sir)	1342	17 Nov 1841	Sydney to 1858, then UK
Niven	George	3487	13 Jan 1886	Res. CM 1, 20 Dec 1899
Noetling	Fritz (Freidrich Wilhelm)	3903	10 May 1893	Rem. CM 6, 28 Jun 1905, xxii; Calcutta

(*Continued*)

Appendix 1. *Continued*

Surname	First names	Application no.	Election date	Additional information
Oddie	James	3602	9 May 1888	Ballarat, Vict. to 1913
Officer	Charles Myles (Major)	3979	6 Feb 1895	Melb., Geol. Soc. Australasia
Ogilvie	Alexander Grant	4202	22 Feb 1899	Tas. to 1902; UK & Chile to 1907
Olden	Charles	4098	24 Feb 1897	Rem. CM 3 Dec 1930; Perth W. Aust. 1889–1977
Oldham	Robert Dixon	3521	14 Dec 1886	Obit. *QJGS*, **93**, 1937, p. ciii
Osborne	George Davenport	7192	14 Mar 1951	Obit. *PGS*, 1541, 26 Sep 1956, pp. 140–141
Owen	Frank	3770	10 Dec 1890	W. Aust. only 1895
Owen	Richard, F.R.S.	1169	31 May 1837	Vert. Palaeontologist
Parker	Thomas	4072	2 Dec 1896	Queensland to 1918
Parkes	James Villiers	3771	10 Dec 1890	d. noted CM 5 Jul 1927; S. Aust. & W. Aust.
Parkinson	James	3648 [Vallance says also 2536]	23 Jan 1889	Mostly British Columbia
Parsons	Cyril Edward	3989	24 Apr 1895	RSM; Newfoundland; ?Aust.
Paul	Frederick Parnell	4542	6 Dec 1905	d. CM 15, 9 Nov 1917
Payne	Edward	4451	16 Dec 1903	Rem. 1911; Aust & S. Africa
Phillips	David Watkins	6353	21 Jun 1944	Obit. *PGS*, 1611, 28 Nov 1963, pp. 154–155
Plews	Henry T.	1853	1858	Queensland to 1873, UK to 1885, d. 1885
Poole	William	4299	5 Dec 1900	Civil Engineer
Potter	Edward	4276	20 Jun 1900	Res. CM 16, 2 Jan 1909 [Gawler Sch of Mines]
Power	Frederick Danvers	3575	11 Jan 1888	Obit, *PGS*, 1541, 26 Sep 1956, p. 141
Priestley	Raymond Edward	4813	25 May 1910	Res. CM 11, 9 Apr 1919
Pringle	Henry Arthur	4083	6 Jan 1897	Res. CM 14, 20 Feb 1929; W. Aust.
Pritchard	Edward Cook	3273	7 Dec ?1881	Rem. CM 20 Jun 1891
Pritchard	George Baxter	4512	22 Mar 1905	Rem. CM 21 Jun 1934
Ramage,	Granville H.	4300	5 Dec 1900	Cancelled CM 4 Dec 1901
Ramsay	Edward Pierson	3377	5 Dec 1883	Rem. *QJGS*, **52**, 1896, *PGS*, p. cxxvii.
Rands	William Henry	3378	5 Dec 1883	d. CM, 20 Jan 1915
Rickard	Thomas Arthur	3853	6 Jan 1892	France; Aust.
Ridge	Harry Mackenzie	5372 5993	20 Apr 1921, re-elected 20 Nov 1935	Res. 29 May 1929 but FGS at Broken Hill 1899–1905
Robe	Captain Alexander	913	28 Mar 1832	Royal Eng.; ?brother of F. H. Robe.
Robertson	J. R. M., M.D.	3290	8 Feb 1882	Scotland to 1888; NSW to 1934
Rosales	Henry	3021	9 May 1877	d. CM 9, 10 Jan 1917, Vict.
Rosewarne	David Davey	3734	26 Mar 1890	S. Aust. (& Darwin) to 1900; NZ 1901–1902; London 1903–1912; NSW 1913–1914
Ross	William J. Clunies	3282	11 Jan 1882	B.Sc London; NSW

(*Continued*)

Appendix 1. *Continued*

Surname	First names	Application no.	Election date	Additional information
Rutledge	Harold	6798	15 Dec 1948	Obit. *PGS*, 1515, 18 Sep 1954
Rylands	Thomas Glazebrook	2115	20 May 1863	Also F.L.S.
Salter	William	2586	11 Jan 1871	Rem. CM 9 Jul 1879; Victoria
Sawkins	James	1728	21 Nov 1855	World, include. Australia
Scott	Joseph	3921	6 Dec 1893	Rem. CM 28 Jun 1905; NSW
Scott	Thomas Hobbes	616	4 Jun 1824	d. noted CM 6 Mar 1867
Seal	Albert E.	4320A	Not elected 20 Feb 1901	10 years mining Aust.
Seaver	Jonathan Charles Billing Pockerage	3495	10 Mar 1886	Rem. 29 Jun 1892; Gulgong NSW
Selwyn	Alfred Richard Cecil	2596	22 Mar 1871	d. CM 15 3 Nov 1902
Shaw	John	1430	26 Jun 1844	d. CM 13, 29 Feb 1888
Shaw	J. S.	2736	3 Dec 1873	Rem. 9 Jul 1884
Shipman	James	3458	29 Apr 1885	d. CM 17, 4 Dec 1901
Skeats	Ernest Willmington	4186	7 Dec 1898	d. CM 18 Mar 1953
Skertchley	Sydney Barber Joseph	2623	6 Dec 1871	Rem. CM 6, 30 Jun 1893; obit. *QJGS*, **83**, 1927, p. lx
Slee	W. H. J.	3630	5 Dec 1888	Res. CM 7, 7 Feb 1906
Smeeth	William Frederick	3886	21 Dec 1892	Obit. *PGS*, 1502, 14 Sep 1953
Smith	Henry	1578	13 Mar 1850	Gov. Geol. Surv., NSW
Smith	R. Neil, M.A.	3987	20 Mar 1895	NSW; W. Aust.; Tas.; Victoria
Smyth	Robert Brough	1752	5 Nov 1856	Victoria to 1880
Sollas	William Johnson	2875	9 Jun 1875	d. 20 Oct 1936
Spalding	Henry Arthur	3781	17 Dec 1890	Res. CM 19, 25 Jan 1893
Stanley	Evan Richard	5045	4 Nov 1914	d. CM 11, 4 Feb 1925
Stanley	Owen	1346	1 Dec 1841	d. CM 12, 30 Apr 1851
Stephen	George Milner	1682	1 Feb 1854	Melb. to 1893
Stephens	Thomas	2715	25 Jun 1873	Obit. *QJGS*, **71**, 1915, p. lxi
Stephens	William John	3363	7 Nov 1883	Sydney Uni. 1884–1890
Stirling	James	3428	3 Dec 1884	Rem. CM 3, 29 Jun 1892
Stokes	Henry Gilbert	3846	23 Dec 1891	Queensland to 1905; Adel. to 1920
Stonier	George Alfred	3808	8 Apr 1891	Sydney 1898; India 1903; London 1904–1943
Storey	[Charles] Blades Coverdale	4579	7 Nov 1906	d. CM 25 ?Jan 1953
Streich	Victor	3923	6 Dec 1893	S. Aust.–W. Aust., Elder Expedition
Stutchbury	Samuel	1343	1 Dec 1841	Obit. *QJGS*, **16**, 1860, p. xxix
Süssmilch	Carl Adolph	4505	1 Feb 1905	Obit. *QJGS*, **103**, 1947, p. lxi
Sutherland	Henry Hubert	3774	10 Dec 1890	d. noted CM, 6 Nov. 1918
Sweet	George	3734	12 Mar 1890	Melbourne; Funafuti
Tate	Ralph	1994	20 Nov 1861	d. CM 16, 6 Nov 1901

(*Continued*)

Appendix 1. *Continued*

Surname	First names	Application no.	Election date	Additional information
Tattam	Charles Maurice	5766	5 Dec 1928	Melb. Univ.
Taylor	Richard Cowling	680	2 Mar 1827	d. CM 15, 15 Dec 1858
Taylor	Thomas Griffith	4732	10 Feb 1909	Res. CM 19, 4 Dec1921
Thomas	David James	7649	25 Nov 1953	Vict. Geol. Survey
Thomas	David John	4971	23 Apr 1913	Obit. *QJGS*, **91**, 1935, pp. xcix–c
Thomas	William Roberts	3783	17 Dec 1890	d. CM 17 Dec 1952
Thompson	William	3612	20 Jun 1888	Queensland to 1893; London to 1905
Thureau	G. A. H.	2885	3 Nov 1875	Victoria; Tas.
Tilley	Cecil Edgar	5429	8 Mar 1922	d. CM 31 Jan 1973
Tom	Isidore	4506	1 Feb 1905	Res. 21 Jan 1920
Townley	Kenneth Allison	6129	16 Nov 1938	Rem. 16 Jan 957
Tremenheere	Hugh Seymour	2436	5 Feb 1868	d. CM 20, 8 Nov 1898
Tremenheere	Seymour	1414	31 Jan 1844	Declared void CM 18 Dec 1844
Turton	Albert Henry	4143	5 Jan 1898	Rem. CM 2 Jul 1902; Tas.
Twelvetrees	William ?Harper	3135	9 Apr 1879	Res. *QJGS*, **72**, 1916, p. xix
Vallance	Thomas George	7363	30 Jan 1952	d. 1993
Van Waterschoot van der Gracht	Willem	4176	7 Dec 1898	E. Indies; Aust.
Von Mueller	(Baron) Ferdinand	3292	22 Feb 1882	Victoria
Wade	Arthur	4559	21 Mar 1906	Aust. 1920s
Walcott	Richard Henry	3883	7 Dec 1892	NZ; Melbourne
Washington-Gray		5621	2 Dec 1925	d. CM 9 Jan 1963
Wathen	G. Henry	1715	4 Apr 1855	*Geology of the Gold Provinces of Australia*
Watkins	Artur Octavius	3712	8 Jan 1890	Res. CM 18, 13 Jan 1932
Westmacott	Captain Robert Marsh	1625	5 May 1852	?Res. *c.* 1863
White	William	3973	5 Dec 1894	d. CM 18 Nov 1911; Queensland
Wilkinson	Charles Smith	2950	10 May 1876	Obit. *QJGS*, **48**, 1892, *PGS*, p. 54ff
Williams	Thomas David	3751	18 Jun 1890	Mostly S. Africa
Williams	Luke	4350	4 Dec 1901	Res. CM 13, 4 Mar 1931; Tas.
Wilson	Charles Herbert	4446	2 Dec 1903	Port Darwin
Winch	Nathaniel			Obit. *PGS* 3, 67.
Witherden	Offen Charles	4447	2 Dec 1903	d.17 Sep 1924; N. Terr.
Woodhouse	Alfred		1887	W. Aust. 1895–1897
Woods	Julian Edmund Tenison	1914	30 Nov 1859	d. CM 15, 7 Nov 1889 Aust; Asia
Woodward	Bernard Henry	3659	6 Mar 1889	W. Aust. Analyst; Mus. curator
Woodward	Harry Page	3381	5 Dec 1883	d. CM 15, 18 Apr 1917
Woolnough	Walter George	4261	7 Mar 1900	d. CM 24 Feb 1960
Young	Henry Graeme Lamont	2909	15 Dec 1875	Disappeared October 1880, NSW

Abbreviations: d., deceased; Rem., removed; Reinst., reinstated; CM, GSL Council Minute; *PGS*, *Proceedings of the Geological Society, London*; *QJGS*, *Quarterly Journal of the Geological Society, London*.

Appendix 2. *New Zealand members of the Geological Society of London up to the early twentieth century*

Surname	First names	Application no.	Election date	Additional Information
Adams	Charles Edward	4387	3 Dec 1902	Victoria College, NZ
Andrew	Arthur R.	4571	7 Nov 1906	Univ. NZ
Binns	George Jonathan	3364	5 Dec 1883	Inspector of Mines, Dunedin
Bonwick	Edwin Walter	4366	26 Feb 1901	Rem. *PGS*, 1908
Buller	Walter, F.R.S.	2515	8 Dec 1869	Wanganui
Campbell	Rev Joseph	3512	1 Dec 1886	Rem. 27 Jun 1900
Campbell	William Dugald	2693	12 Mar 1873	Res. 16 Dec 1914
Clarke	William Wild	3594	11 Apr 1888	Civil/ Mining Eng.
Cochrane	Neil Macdonald	4306	9 Jan 1901	Grey Rr & other coalfields, NZ
Cox	Samuel Herbert	2798	2 Dec 1874	NZ, later NSW
Crawford	James Coutts	2280	24 May 1865	Wellington
Daniel	Peter Francis	4062	2 Dec 1896	Greymouth
Davies	John	2615	6 Dec 1871	Eng. Canterbury, NZ
Davis	Edward Heydelbach			9 Feb 1871
Dobson	Arthur Dudley	2793	1874	1875–1888 NZ; Victoria to 1892
Don	Prof. John	4286	5 Dec 1900	Removed 1906
Duigan	James	2894	1 Dec 1875	Journalist, Wanganui
Dutton	Daniel	3246	23 Mar 1881	Primitive Methodist Minister, Wellington
Edwards	Walter Cleeve	3929	10/01/1894	Greymouth
Ekberg	Benjamin Pherson	4116B	balloted 24 Feb 1897; not elected	Ballarat & NZ
Ellam	Albert Spencer	4116A	not elected 1897	Auckland; FRGS
Farquharson	Robert Alexander	4756	17 Nov 1909; Reinst. 8 Nov 1916	Resigned 21 Dec 1938, NZ; W. Aust.; Africa
Ferrar	Hartley T.	4485	7 Dec 1904	d. CM 13, 27 Apr 1932
Finlayson	Alex Moncrieff		*c.*1910	1851 scholar; d. 23 Jul 1917
Fraser	Rev. Charles	2384	3 Apr 1867	Christchurch, NZ
Gordon	Henry A.	3471	2 Dec 1885	Engineer, Wellington
Haast	(Sir) Julius Francis von	2069	18 Feb 1863	Obit. *PGS*, 1888.
Hay	Robert	3804	25 Mar 1891	Civ. Engr., Dunedin
Hector	Sir James	1978	10 Apr 1861	Obit. *QJGS*, **64**, pp. lxi–lxii
Hilgendorf	F.W.	4552	21 Feb 1906	? to 1913
Hill	Henry	3535	9 Feb 1887	Inspector of Schools, Napier
Hood, aka Cockburn-Hood	Thomas	2070	18 Feb 1863	Queensland 1863; NZ 1870–1875
Hosking	George Francis	3841	23 Dec 1891	Bendigo, Otago
Hutchison	Lawrence G.	4596	19 Dec 1906	Auckland
Hutton	Captain Frederick Wollaston	1938	16 May 1860	Sandhurst [Bendigo Victoria]; NZ mainly
Kirkcaldy	Norman Melville	4422	29 Apr 1903	Civil & Mining Eng., Dunedin
Maclaren	J. Malcolm	4337	20 Nov 1901	d. CM, 27 Mar 1935
Mantell	Walter Baldock Durant	1844	1 Dec 1858	Researcher and Public Servant
Marshall	Patrick	3969	5 Dec 1894	Obit. *QJGS*, **106**, 1951, pp. lxviii–lxix
McKay	Alexander	3578	2 Feb 1888	Res. CM 14, 5 Nov 1913
Paulin	Robert	3842	23 Dec 1891	Surveyor/Prospector, Dunedin

(*Continued*)

Appendix 2. *Continued*

Surname	First names	Application no.	Election date	Additional Information
Pinfold	Rev. James Thomas	3944	21 Feb 1894	Dunedin
Rawlin(g)s	Charles Champion (Campion)	3567	7 Dec 1887	Aust. & NZ
Speight	Robert	4647	5 Nov 1907	NZ, including offshore islands
Stephens	Francis Beaumont	4215	26 Apr 1899	
Taylor	Rev. Richard	1279	11 Mar 1840	1843–1873 Bay of Islands
Thomas	A. P. W.	3632	5 Dec 1888	Prof Biol., Auckland
Thompson	Joseph Henry	3210	17 Nov 1880	Auckland & Melbourne
Thomson	James Allen	4640	19 Jun 1907	Obit. *QJGS*, 1929, pp. lxiii–lxiv
Ulrich	Georg(e) Heinrich Friedrich	2404	6 Nov 1867	d. CM 24, 7 Nov 1900 (Otago)
Vincent	William Slade	4075	2 Dec 1896	Rem. *PGS*, 1908
Wood	Charles Sturtivant [?Sturtevant]	2007	8 Jan 1862	d. CM 16, 9 Nov 1864

Abbreviations: d., deceased; Rem., removed; Reinst., reinstated; CM, GSL Council Minute; *PGS*, *Proceedings of the Geological Society, London*; *QJGS*, *Quarterly Journal of the Geological Society, London*.

Archives

ATL: Alexander Turnbull Library, Wellington, New Zealand.

BRANAGAN, D. F. & BRANAGAN, C. D. 1997. Vallance Microsoft Excel Database, prepared from Card Index of T. G. Vallance (1928–1993). Earth Sciences History Group, Geological Society of Australia. The database contains information on more than 3700 geologists, mining engineers and associated Australian and New Zealand workers, from the eighteenth century to the present (including a few living geologists). Copies of the database are held by the National library of Australia and the State Library of New South Wales. The cards are now lodged in the office of the Australian Dictionary of Biography (ADB), Canberra (at Australian Dictionary of Biography, Research School of Social Sciences, The Australian National University, Canberra, ACT 0200, Australia: tel.: (02) 6249 2676; fax (02) 6257 1893. e-mail: adb@coombs.anu.edu.au.

GSL: Geological Society of London.

ML: Mitchell Library, Sydney.

References

ALDERMAN, A. R. 1976. Tate, Ralph (1840–1901). *In*: NAIRN, B. (general ed.) *Australian Dictionary of Biography*, Volume 6, (1851–1890). Melbourne University Press, Melbourne, 243–244.

ANON. 1935. Obituary, James Malcolm Maclaren. *The Mining Journal*, **187**, 195.

AUROUSSEAU, M. 1968. *The Letters of F.W. Ludwig Leichhardt Collected and Newly Translated.* Hahkluyt Society, Second Series, **CXXXV** (3 vols). Cambridge University, Cambridge.

AYRES, P. 1999. *Mawson – A Life*. Miegunyah Press, Melbourne.

BATHER, F. A. 1927. Anniversary Address of the President. *Quarterly Journal of the Geological Society, London*, **83**, lx–lxi.

BEAGLEHOLE, J. (ed.). 1962. *The Endeavour Journal of Joseph Banks 1768–1771.* Public Library of New South Wales and Angus & Robertson, Sydney.

BENSON, W. N. 1950. Obituary: Patrick Marshall. *Transactions of the Royal Society of New Zealand*, **79**, 152–155.

BISHOP, G. 2008. *The Real McKay: The Remarkable Life of Alexander McKay, Geologist.* Otago University Press, Dunedin.

BLAINEY, G. 1969. *The Rush that Never Ended.* Melbourne University Press, Melbourne.

BOLTON, G. C. 1965. *Richard Daintree: A Photographic Memoir.* Jacaranda Press and Australian National University, Brisbane.

BOLTON, G. C. 1988. Rason, Sir Cornthwaite Hector William James (1858–1927). *In*: SEARLE, G. (general ed.) *Australian Dictionary of Biography*, Volume II (1891–1939). Melbourne University, Melbourne, 333–334.

BORDER, R. 1967. Scott, Thomas Hobbes Scott (1783–1860). *In*: PIKE, D. (general ed.) *Australian Dictionary of Biography*, Volume 2 (1788–1850). Melbourne University, Melbourne, 431–433.

BOWDEN, A. J. 2009. Geology at the crossroads: aspects of the geological career of Dr John MacCulloch. *In*: LEWIS, C. L. E. & KNELL, S. J. (eds) *The Making of the Geological Society of London.* Geological Society, London, Special Publications, **317**, 255–278.

BRANAGAN, D. F. 1974. Strzelecki's geological map. *Records of the Australian Academy of Science*, **2**, 68–70.

BRANAGAN, D. F. 1985. Phillip Parker King, Colonial Anchor-Man. *In*: WHEELER, A. & PRICE, J. H. (eds) *From Linnaeus to Darwin: Commentaries on the History of Biology and Geology*. Society for the History of Natural History, London, 179–193.

BRANAGAN, D. F. 1986. Strzelecki's geological map of south eastern Australia: An eclectic synthesis. *Historical Records of Australian Science*, **6**(3), 44–58.

BRANAGAN, D. F. 1990*a*. A history of New South Wales Coal Mining. *In*: BRANAGAN, D. F. & WILLIAMS, K. L. (eds) *Coal in Australia*. The Third Edgeworth David Symposium, University of Sydney, Sydney, 1–15.

BRANAGAN, D. F. 1990*b*. Alfred Selwyn – 19th century trans-Atlantic connections via Australia. *Earth Sciences History*, **9**(2), 143–157.

BRANAGAN, D. F. 1992. Samuel Stutchbury and the Australian Museum. *Records of the Australian Museum Supplement*, **15**, 99–110.

BRANAGAN, D. F. 1993. Samuel Stutchbury: A natural history voyage to the Pacific, 1825–1827 and its consequences. *Archives of Natural History*, **20**(1), 69–91.

BRANAGAN, D. F. 1994. Ludwig Leichhardt: Geologist in Australia. *In*: LAMPING, H. & LINKE, M. (eds) *Australia: Studies on the History of Discovery and Exploration*. Frankfürter Wirstchafts und Sozialgeographische Schriften, Heft, **65**, 105–122.

BRANAGAN, D. F. 1995. The Pole and the Australian Permian. (Keynote paper) Strzelecki International Symposium on the Permian of Eastern Tethys: Biostratigraphy, Palaeogeography and Resources. *Proceedings of the Royal Society of Victoria*, **110**(1/2), 1–30.

BRANAGAN, D. F. 1996. Julian Edmond Tenison Woods: Late nineteenth Century Geology in Asia. *In*: WANG HONGZHEN, ZHAI YUSHENG, SHI BAOHENG, & WANG CANSHENG, (eds) *Development of Geoscience Disciplines in China*. The Council of History of Geology, Geological Society of China. China University of Geosciences Press, 54–62.

BRANAGAN, D. F. (ed.). 2000. *Science in a Sea of Commerce. The Journal of a South Seas Pearling Expedition, 1825–1827*. David Branagan, Northbridge, New South Wales.

BRANAGAN, D. F. 2005*a*. Stutchbury, Samuel (1798–1859). *Oxford Dictionary of Biography*. Oxford University Press, Oxford, **53**, 258–259.

BRANAGAN, D. F. 2005*b*. *T.W. Edgeworth David: A Life*. National Library of Australia, Canberra.

BRANAGAN, D. F. 2008*a*. Captain Eugene de Hautpick – a Russian ghost. *Proceedings of the Royal Society of Victoria*, **120**, 118–136.

BRANAGAN, D. F. 2008*b*. Australia – a Cenozoic history. *In*: GRAPES, R. H., OLDROYD, D. R. & GRIGELIS, A. (eds) *History of Quaternary Geology and Geomorphology*. Geological Society, London, Special Publications, **301**, 189–213.

BRANAGAN, D. F. 2009. Some nineteenth- and twentieth-century Australian geological clerics. *In*: KÖLBL-EBERT, M. (ed.) *Geology and Religion: A History of Harmony and Hostility*. Geological Society, London, Special Publications, **310**, 171–196.

BRANAGAN, D. F. & MOORE, D. 2008. W. H. Fitton's geology of Australia's coasts, 1826. *Historical Records of Australian Science*, **19**(1), 1–51.

BRANAGAN, D. F. & PACKHAM, G. H. 2000. *Field Geology of New South Wales*. New South Wales Department of Minerals & Energy, St Leonards.

BRETT-CROWTHER, M. R. 1979. Buddicom, Robert Arthur, also known as Bedford, Robert (1874–1951) *In*: NAIRN, B. & SEARLE, G. (general eds) *Australian Dictionary of Biography*, Volume 7, (1891–1939). Melbourne University, Melbourne, 475–476.

BRITISH MUSEUM. 1904. *The History of the Collections Contained in the Natural History Departments of the British Museum*, Volume 1. Trustees of the British Museum, London.

BUCKLAND, W. 1821. Notice on the geological structure of a part of the island of Madagascar, founded on a collection transmitted to the Right Honourable, the Earl Bathurst, by Governor Farquhar, in the year 1819; with observations from the interior of New South Wales collected during Mr Oxley's expedition to the River Macquarie, in the year 1818 and transmitted also to Earl Bathurst. *Transactions of the Geological Society, London*, 1st series, **5**, 476–481.

BUREK, C. 2009. The status of women and the first female Fellows. *In*: LEWIS, C. L. E. & KNELL, S. J. (eds) *The Making of the Geological Society of London*. Geological Society, London, Special Publications, **317**, 373–407.

BURNETT, R. I. M. 1936. *The Life and Work of Sir James Hector with Special Reference to the Hector Collection*. M.A. thesis, University of Otago.

BURROWS, C. J. 2005. *Julius Haast in the Southern Alps*. Canterbury University, Christchurch, New Zealand.

CALVERT, J. 1853. *The Gold Rocks of Great Britain and Ireland and a General Outline of the Gold Regions of the World, with a Treatise on the Geology of Gold*. Chapman & Hall, London.

CAMERON, H. C. 1952. *Sir Joseph Banks*. Angus & Robertson, Sydney.

CLARKE, W. B. 1837*a*. On the geological structure and phenomena of Suffolk, and its physical relations with Norfolk and Essex. *Proceedings of the Geological Society*, **II**(50), 528–534. (See also *Transactions of the Geological Society, London*, **V**, 359–384, 1840.)

CLARKE, W. B. 1837*b*. On the geological structure and phenomena of the Northern part of Cotentin and particularly in the immediate vicinity of Cherbourg. *Proceedings of the Geological Society*, **II**(48), 454–455 (1838).

CONINSBY, R. J. 1895. *The Discovery of Gold in Australia*. William Mulligan, London.

COOPER, J. 1993. McKay, Alexander (1841–1917). *The Dictionary of New Zealand Biography*, Volume 2. Bridget Williams Books, Auckland & Department of Internal Affairs, 290–291.

DAVID, T. W. E. 1883. On the evidence of glacial action in South Brecknockshire and East Glamorganshire. *Quarterly Journal of the Geological Society, London*, **39**, 39–54.

DAVID, T. W. E. 1887. On the evidence of glacial action in the Carboniferous and Hawkesbury Series, New South Wales. *Quarterly Journal of the Geological Society, London*, **43**, 190–196.

DAVID, T. W. E. 1895. Evidences of glacial action in Australia and Tasmania. Presidential Address Section

C. *Report of the Australasian Association for the Advancement of Science*, Brisbane, **VI**, 58–95.

DAVID, T. W. E. 1896. Evidence of glacial action in Australia in Permo-Carboniferous time. *Quarterly Journal of the Geological Society, London*, **52**, 289–301.

DAVID, T. W. E. 1906. Glaciation in Lower Cambrian, possibly in Pre-Cambrian time. *Compte Rendu du X Congres International, Mexico*, **1**, 271–274.

DAVID, T. W. E. 1914. Chapter 7. Geology of the Commonwealth. *Handbook of the Commonwealth of Australia*. Government Printer, Melbourne, 241–325.

DAVID, T. W. E. 1926. Salient features in the stratigraphy, tectonic structure and physiography of the Commonwealth of Australia. *Quarterly Journal of the Geological Society, London*, **82**, cv–cvii.

DAVID, T. W. E. & WILSON, J. T. 1914. The Talgai skull. *Scientific Australian*, **10**(1), 4–5.

DAVID, T. W. E. & WOOLNOUGH, W. G. 1926. Cretaceous glaciation in Central Australia. *Quarterly Journal of the Geological Society, London*, **82**, 333–350.

[DAVID, T. W. E. & KING, W. B. R.] 1922. *The Work of the Royal Engineers in the European War. 1914–1919: Geological Work on the Western Front*. Institute of Royal Engineers, Chatham.

DEAN, D. 1999. *Gideon Mantell and the Discovery of the Dinosaurs*. Cambridge University Press, New York.

DE HAUTPICK, E. 1928. *Are There Oil Prospects in Australia?* de Hautpick, Adelaide.

DELL, R. K. 1990. Hector (Sir) James (1834–1907). *The Dictionary of New Zealand Biography*, Volume 1, 1769–1869. Allen & Unwin, Wellington & Department of Internal Affairs, H15, 183–184.

ETHERIDGE, R. JNR & JACK, R. L. 1881. *Catalogue of Works, Papers, Reports, and Maps on the Geology, Palaeontology, Mineralogy, Mining and Metallurgy, etc. of the Australian Continent and Tasmania*. Stanford, London.

FEEKEN, E. H. J., FEEKEN, G. E. E. & SPATE, O. H. K. L. 1970. *The Discovery and Exploration of Australia*. Thomas Nelson, Melbourne.

FITTON, W. H. 1825. An account of some geological specimens collected by P. P. King, in his survey of the coasts of Australia, and by Robert Brown, Esq., on the shores of Carpentaria, during the voyage of Captain Flinders. *Transactions of the Geological Society*, **2**, 11.

FITTON, W. H. 1826. Instructions etc. *In*: KING, P. P. (ed.) *Narrative of a Survey of the Intertropical and Western Coasts of Australia ...*, Volume 2. John Murray, London, 623–629.

FLEMING, C. A. 1986. The contributions of New Zealand Geoscientists to the Development of Scientific Institutions. *Earth Sciences History*, **5**(1), 3–11.

FLEMING, C. A. 1987. *Science, Settlers and Scholars: The Centennial History of the Royal Society of New Zealand*. The Royal Society of New Zealand, Wellington.

FOSTER, W. C. 1985. *Sir Thomas Livingston Mitchell and his World 1792–1855, Surveyor General of New South Wales 1828–1855*. The Institution of Surveyors N.S.W. Incorporated, Sydney.

GRAINGER, E. 1982. *The Remarkable Reverend Clarke*. Oxford University Press, Melbourne.

GREENOUGH, G .B. 1841. Anniversary Address. *Journal of the Royal Geographical Society*, **11**, lxii–lxiii.

GREY, G. 1841. *Journals of Two Expeditions of Discovery in Northwest and Western Australia*. T. & W. Boone, London.

GREY, M., MORTON, A. & EVANS, A. (eds). 2001–2002. Sir Frederick McCoy. *The Victorian Naturalist: McCoy Special Issues* (Parts 1 & 2), **118**(5–6).

HAAST, H. F. VON. 1948. *The Life and Times of Sir Julius von Haast ... Explorer, Geologist, Museum Builder*. H.A. Haast, Wellington, New Zealand.

HAMILTON-SMITH, E. 1996. Father Julian Edmund Tenison Woods – A Pioneer Karst Scientist. *In*: HAMILTON-SMITH, E. (ed.) *Abstracts of Papers: Karst Studies Seminar*. Naracoorte, South Australia.

HARKER, A. 1918. Obituary of Alexander McKay. *Quarterly Journal of the Geological Society, London*, **74**, lx.

HAVARD, W. 1940. Sir Paul Edmund de Strzelecki. *Journal of the Royal Australian Historical Society*, **26**, 20–96.

HERBERT, S. 2005. *Charles Darwin – Geologist*. Cornell University Press, Ithaca.

HERRIES DAVIES, G. L. 2007. *Whatever is Under the Earth: The Geological Society of London 1807 to 2007*. Geological Society, London.

HOCKEN, A. 2002. The resignation of Pat Marshall. *Historical Studies Group Newsletter. Geological Society of New Zealand*, **24**, (March), 31–36.

HOCKEN, A. G. 2003. *Geology at the University of Otago: The First 100 Years*. Geological Society of New Zealand, Miscellaneous Publications, **115**, iv.

HORDERN, M. 1989. *Mariners are Warned! John Lort Stokes and H.M.S. Beagle in Australia 1837–1843*. Miegunyah, Melbourne.

HOWITT, A. W. 1879. Notes on the physical geography and geology of North Gippsland, Victoria. *Quarterly Journal of the Geological Society, London*, **35**, 1–41.

JACK, F. 2008. *Putting Queensland on the Map: The Life of Robert Logan Jack*. University of New South Wales, Kensington, NSW.

JACK, R. L. 1921. *Northmost Australia* (2 vol). Simpkin, London.

JERVIS, J. 1944. Rev. W. B. Clarke, M.A., F.R.S., F.G.S., F.R.G.S.: 'The Father of Australian Geology'. *Journal and Proceedings of the Royal Australian Historical Society*, **30**(6), 345–458.

JOHNSTON, M. 2007. *Mettle & Mines: The Life and Times of Colonial Geologist Edward Heydelbach Davis (1845–1871)*. Nikau, Nelson, New Zealand.

JUKES, J. B. 1847. *Narrative of the Surveying Voyage of H.M.S. Fly ... During the Years 1842–1846* (2 vols). T. & W. Boone, London.

JUKES, J. B. 1850. *A Sketch of the Physical Structure of Australia, So Far as it is at Present Known*. T. & W. Boone, London.

KING, P. P. 1826. *Narrative of a Survey of the Intertropical and Western Coasts of Australia ... 1818 and 1822* (2 vols). John Murray, London.

KNELL, S. J. 2009. The road to Smith: how the Geological Society came to possess English geology. *In*: LEWIS, C. L. E. & KNELL, S. J. (eds) *The*

Making of the Geological Society of London. Geological Society, London, Special Publications, **317**, 1–47.

LAUBE, S. 1990. *Robert Bedford of Kyancutta*. Wednesday, Norwood, South Australia.

MACLEOD, R. 2009. *Imperial Science under the Southern Cross: Archibald Liversidge, FRS and the Culture of Anglo-Australian Science*. University of New South Wales, Kensington, NSW.

MACMILLAN, D. 1967. Nicholson, (Sir) Charles (1808–1903). *In*: PIKE, D. (general ed.) *Australian Dictionary of Biography, Volume 2 (1770–1850)*. Melbourne University, Melbourne, 283–285.

[MAGRA, J.]. 1771. *Journal of a Voyage Around the World in His Majesty's Ship Endeavour in the Years 1768, 1769, 1770, and 1771*, T. Becket & P. A. De Hondt, London, 74–75.

MARKS, E. N. 1988. Skertchly, Sydney Barber Josiah (1850–1926) *In*: SEARLE, G. (ed.) *Australian Dictionary of Biography*, Volume 11 (1891–1939). Melbourne University, Melbourne, 621–622.

MASON, A. 1990. Hector's Little Empire: Geological Survey of New Zealand, Colonial Museum and Laboratory, revisited. *Historical Studies Group Newsletter. Geological Society of New Zealand*, **1**, (September), 7–9.

MASON, A. 1994. Malcolm Maclaren: A forgotten New Zealand geologist. *Historical Studies Group Newsletter. Geological Society of New Zealand*, **8**, (March), 28–35.

MASON, A. 1996, In the Beginning. *Historical Studies Group Newsletter. Geological Society of New Zealand*, **12**, 3–6.

MASON, A. 2003. Malcolm Maclaren: New Zealand's most noteworthy geologist. *Historical Studies Group Newsletter. Geological Society of New Zealand*, **27**, (September), 3–11.

MASON, A. 2007. Reverend Richard Taylor (1805–1873) missionary geologist. (Part 1). *Historical Studies Group Newsletter, Geological Society of New Zealand*, **34**, (September), 6–15.

MASON, A. 2008. Reverend Richard Taylor (1805–1873) missionary geologist (Part 2). *Historical Studies Group Newsletter, Geological Society of New Zealand*, **35**, (March), 9–16.

MASON, A. & WATTERS, W. 1996. The rocks of Cape Colville Peninsula. *Historical Studies Group Newsletter. Geological Society of New Zealand*, **12**, 19–24.

[MASON, A.]. 1995. Alexander Mackay, Government Geologist. *Historical Studies Group Newsletter Geological Society of New Zealand*, **10**, (March), 5–13.

MCCARTHY, G. (ed.). 1991. *Guide to the Archives of Science in Australia: Records of Individuals*. Thorpe/Australian Science Archives Project & National Centre for Australian Studies, Port Melbourne.

MCKENZIE, D. W. (ed.). 1987. *Heinemann New Zealand Atlas*. Heinemann and Department of Survey and Land Information, Auckland.

MITCHELL, T. L. 1831. An account of the limestone caves at Wellington valley, and of the situation near one of them where fossil bones have been found. *Proceedings of the Geological Society, London*, **1**, 321–322.

MITCHELL, T. L. 1838. *Three Expeditions into the Interior of Eastern Australia* (2 vols). Longman, London.

MITCHELL, T. L. 1848. *Journal of an Expedition into the Interior of Tropical Australia in Search of a Route from Sydney to the Gulf of Carpentaria*. Longman, London.

MOORE, D. T., THACKRAY, J. C. & MORGAN, D. L. 1991. A short history of the museum of the Geological Society of London, 1807–1911, with a catalogue of British and Irish accessions, and notes on surviving collections. *Bulletin of the British Museum of Natural History (Historical Series)*, **19**, 51–160.

MOYAL, A. 1993. *A Bright and Savage Land: Scientists in Colonial Australia*. Penguin Books Australia, Ringwood, Victoria.

MOYAL, A. 2003. *The Web of Science: The Scientific Correspondence of the Rev. W. B. Clarke, Australia's Pioneer Geologist* (2 vols). Australian Scholarly Publishing, Melbourne.

OLDROYD, D. 2007. In the footsteps of Thomas Livingstone Mitchell (1792–1855): Soldier, explorer, geologist, and probably the first person to compile geological maps in Australia. *In*: WYSE JACKSON, P. N. (ed.) *Four Centuries of Geological Travel: The Search for Knowledge on Foot, Bicycle, Sledge and Camel*. Geological Society, London, Special Publications, **287**, 343–373.

ORGAN, M. 1994. Bibliography of the Reverend W. B. Clarke (1798–1878). *Journal and Proceedings of the Royal Society of New South Wales*, **127**(3–4), 85–134.

ORGAN, M. 1997. *Reverend W. B. Clarke 1798–1878: Chronology and Calendar of Correspondence*. M. Organ, Wollongong.

OSBORNE, R. A. L. 1991. Red earth and bones: The history of cave sediment studies in New South Wales, Australia. *Earth Sciences History*, **10**, 13–28.

OWEN, J. M. R. 1990. Taylor, Richard (1805–1873). *The Dictionary of New Zealand Biography*, Volume 1 (1769–1869). Allen & Unwin, Wellington & Department of Internal Affairs, 437–438.

PASZKOWSKI, L. 1997. *Sir Paul Edmund de Strzelecki*. Arcadia Australian Scholarly Publishing, Melbourne, **I–xiii**, 370.

PEARL, C. 1978. *Five Men Vanished: The Bermagui Mystery*. Angus & Robertson, Sydney.

PHILLIPS, J. 1858. On the goldfield of Ballaarat. *Quarterly Journal of the Geological Society, London*, **14**, 538–540.

PRIESTLY [SIC], R. & DAVID, T. W. E. 1912. Geological notes of the British Antarctic Expedition, 1907–1909. *Compte Rendu du XI Congres International, Stockholm*, 767–811.

PROUDFOOT, H. 1992. Westmacott, Robert Marsh (*c*. 1801–1870). *In*: KERR, J. (ed.) *The Dictionary of Australian Artists: Painters, Sketchers, Photographers and Engravers to 1870*. Oxford University, Oxford.

RITCHIE, J. (ed.). 1971. *The Evidence to the Bigge Reports: New South Wales Under Governor Macquarie* (2 vols). Heinemann, Melbourne.

RODERICK, C. 1988. *Leichhardt the Dauntless Explorer*. Angus & Robertson, North Ryde, NSW.

ROSALES, H. 1855. On the goldfields of Ballaarat, Eureka and Creswick Creek. *Quarterly Journal of the Geological Society, London*, **11**, 497–503.

SCOTT, T. H. 1824. Sketch of the geology of New South Wales and Van Diemen's Land. *Annals of Philosophy* (new series), **7**, 461–462.

SCOTT, T. H. 1831. Geological remarks on the vicinity of Swan River and Isle Bauche or Garden Island, on the coast of Western Australia. *Proceedings of the Geological Society, London*, **1**(21), 330–331.

SELWYN, A. R. C. 1854. Geology and mineralogy of Mt. Alexander and the country lying between the River Loddon and River Campaspe (Victoria). *Quarterly Journal of the Geological Society, London*, **10**, 299–303.

SELWYN, A. R. C. 1858. The geology of the goldfields of Victoria. *Quarterly Journal of the Geological Society, London*, **14**, 533–538.

SELWYN, A. R. C. 1860. Notes on the geology of Victoria. *Quarterly Journal of the Geological Society, London*, **16**, 145–150.

SHACKLETON, E. H. 1909. *The Heart of the Antarctic: Being the Story of the British Antarctic Expedition, 1907–1909* (2 vols). William Heinemann, London.

SILVER, L. R. 1986. *A Fool's Gold? William Tipple Smith's Challenge to the Hargraves Myth*. Jacaranda, Milton, Queensland.

SOLLAS, W. J. 1905. *Rocks of the Colville Peninsula*. Government Printer, Wellington, New Zealand.

SORRENSON, M. P. K. 1990. Mantell, Walter (1820–1895). *In*: OLIVER, W. H. (ed.) *The Dictionary of New Zealand Biography, Volume 1 (1769–1869)*. Allen & Unwin, Wellington & Department of Internal Affairs, 267–268.

STEPHEN, G. S. 1854. On the gems and gold crystals of the Australian colonies. *Quarterly Journal of the Geological Society, London*, **10**, 303–308.

STRZELECKI, P. E. 1845. *Physical Descriptions of New South Wales and Van Diemen's Land*. Longmans, London, 426.

STUART, J. M. 1864. *Explorations in Australia, The Journals of John McDouall Stuart During the Years 1858, 1859, 1860, 1861 & 1862*, HARDMAN SAUNDERS, W. (ed.). Otley & Co., London.

TEE, G. 1998. Relics of Davy and Faraday in New Zealand. *Notes and Records of the Royal Society of London*, **52**, 93–102.

TEE, G. 2007. Science on the map; places in New Zealand named after scientists. *The Rutherford Journal*, **2**. www.rutherfordjournal.org

VALLANCE, T. G. 1975. Presidential address: Origins of Australian geology. *Proceedings of the Linnean Society of New South Wales*, **100**, 13–43.

VALLANCE, T. G. 1981. The fuss about coal. Troubled relations between palaeontology and geology. *In*: CARR, D. J. & CARR, S. G. M. (eds) *Plants and Man in Australia*. Academic, Sydney, 136–176.

VALLANCE, T. G. 1984. Gideon Mantell (1790–1852): A focus for study in the history of geology at the Turnbull Library. *In*: HOARE, M. E. & BELL, L. G. (eds) *In Search of New Zealand's Scientific Heritage*. Royal Society of New Zeland, Bulletin, **21**, 91–100.

VALLANCE, T. G. 1986. Achievement in isolation: A. W. Howitt, pioneering investigator of metamorphism in Australia. *Earth Sciences History*, **5**, 39–46.

WALLACE, A. 1890. *Jottings Referring to the Early Discovery of Gold in Australia*. G. Murray & Co., Sydney.

WATTERS, W. 1996. Marshall, Patrick (1869–1950). *The Dictionary of New Zealand Biography*, Volume 3. Auckland University with Bridget Williams Books, Auckland, 333–334.

WATTERS, W. 1997. Patrick Marshall 1869–1950. *Historical Studies Group Newsletter. Geological Society of New Zealand*, **15**, (September), 4–16.

WATTS, W. W. (recorder). 1909. The Centenary of the Geological Society of London Celebrated September 26th to October 3rd 1907. Longmans, Green and Co., London.

WEST, F. 1967, Owen Stanley (1811–1850). *In*: PIKE, D. (general ed.) *Australian Dictionary of Biography*, Volume 2 (1788–1850). Melbourne University, Melbourne, 470–471.

WOODS, J. E. T. 1858. Observations on some metamorphism rocks in South Australia. *Transactions of the Philosophical Institute of Victoria*, **2**, 168–176.

WOODS, J. E. T. 1860. On some Tertiary rocks in the colony of South Australia. *Quarterly Journal of the Geological Society, London*, **16**, 253–260 (1859).

WOODS, J. E. T. 1865. On some Tertiary Deposits in the colony of Victoria, Australia. *Quarterly Journal of the Geological Society, London*, **21**, 389–394.

WOODS, Rev. J. E. 1862. *Geological Observations in South Australia: Principally in the District South-East of Adelaide*. Longman, London.

WOODWARD, H. B. 1907. *The History of the Geological Society of London*. Geological Society, London.

ZELLER, S. & BRANAGAN, D. 1994. Australian–Canadian links in a geological chain: Sir William Logan, Dr Alfred Selwyn and Henry Y. L. Brown. *In*: MACLEOD, R. & JARRELL, R. (eds) *Dominions Apart: Reflections on the Culture of Science and Technology in Canada and Australia 1850–1945*. Scientia Canadensis **17**(1–2), 71–102.

The first female Fellows and the status of women in the Geological Society of London

CYNTHIA V. BUREK

Centre for Science Communication, University of Chester, Parkgate Road, Chester CH1 4BJ, UK (e-mail: c.burek@chester.ac.uk)

Abstract: Women were first permitted to become Fellows of the Geological Society of London in 1919. Eight joined in May of that year and then in June a further two were admitted. By February 1922 there were 21 female Fellows. The Geological Society had opened its doors to women and, after an initial rush, there was a slow trickle. However, there were a number of highly regarded female geologists before this time, and several of them received grants, medals and, indeed, submitted papers, although they were not always permitted to read these themselves. Some of the first female Fellows have disappeared without trace, but the contributions of others are significant. As well as being educationalists, they were expert in many different areas of geology, with palaeontology and stratigraphy featuring strongly. Few, however, stayed on in this male-dominated arena once they married and had children. It seems there was no common reason for these women to seek membership of the Geological Society, other than their love of geology, but membership brought them recognition and status. They led the way as role models for future female Fellows and were the first of many women to play a significant role in the Geological Society's history.

The Geological Society started its formal life at 5pm on 13 November 1807 in the Freemasons' Tavern in Great Queen Street, London (Herries Davies 2007; Knell 2009; Lewis 2009). While men were admitted as full Fellows when the Society received its charter in 1825 (Boylan 2009), it took a further 94 years before women were permitted to join and become Fellows. The possible reasons for this time lag are considered, the context set and the motives behind why these women joined are discussed. Each woman is given a short biography, examining the role she played and her area of contribution to the geological sciences.

History of female fellowship – why so long?

The contribution of women throughout the history of geology while not greatly publicized, has, nevertheless, been significant in the development of the science for the variety of roles that they have played (Kölbl-Ebert 2001; Burek & Higgs 2007). In the Society's early years the majority of Fellows did not want women present at their meetings, but, as the Society matured, questions were asked from about the 1860s as to why women were not admitted.[1] Between then and the turn of the twentieth century the question was either put, or contemplated being put, several times by various members. When Leonard Horner (1785–1864) was President (1860–1862),[2] both he and his son-in-law, Charles Lyell (1797–1875), supported the proposal that the 'fair sex' be admitted to attend meetings, but it was considered the Society's meeting room in Somerset House (Boylan 2009) was inadequate for the 'probable influx of a large number of visitors' (Woodward 1907, p. 242). Later supporters of the cause were Thomas Vincent Holmes (1840–1923) who was spurred on by the papers presented, but not read, by Jane Donald (1856–1934) and Catherine Raisin (1855–1945) in 1887; Henry Woodward (1832–1921) in 1895 and 1906; and William Whitaker (1836–1925) in 1900. However, such questions were always defeated until Archibald Geikie (1835–1924) finally introduced two women into an Ordinary Meeting in 1901, and by 1904 women could be guests of members attending Ordinary and Annual General Meetings. One woman – Catherine Raisin, Professor at Bedford College London – availed herself of this opportunity and attended over 50 percent of the meetings

[1] Details of the eventual admittance of women into meetings have been discussed in Herries Davies (2007, pp. 159–163) and will not be dealt with here.

[2] Horner had joined the Society in March 1808, just a few months after its inception, and remained one of its most active members (Lewis 2009).

From: LEWIS, C. L. E. & KNELL, S. J. (eds) *The Making of the Geological Society of London*. The Geological Society, London, Special Publications, **317**, 373–407.
DOI: 10.1144/SP317.21 0305-8719/09/$15.00

between 1909 and 1912. Thus, it was well received by female geologists when the President of the Society, George William Lamplugh (1859–1926), announced in his report that:

> On March 26 1919, a Special General Meeting was held, at which the resolution of the Council affirming the desirability of admitting women as Fellows of the Society was carried by 55 votes to 12. Fourteen have so far availed themselves of the opportunity thus afforded.
>
> (Lamplugh 1920*a*)

This decision, so long in coming, was forced in part by the 1918 Representation of the People Act and the Sex Disqualification (Removal) Act of 1919, which removed legal barriers to the admission of women by bodies governed by charter. Section 1 of the latter Act stated that:

> A person shall not be disqualified by sex or marriage from the exercise of any public function, or from being appointed to or holding any civil or judicial office or post, or from entering or assuming or carrying on any civil profession or vocation, or for admission to any incorporated society (whether incorporated by Royal Charter or otherwise) [. . .].[3]

The Geological Society acted just in time to avoid the stigma of being forced by law to admit women. However, it is interesting to note that, although the first two years of admission saw a slow stream of women being elected Fellows, it was by no means the flood that the male Fellows had feared. After 1919 the numbers remained fairly steady until after World War II. Was the Geological Society unusual in its attitude or was it merely reflecting the views of society at large?

Comparison with other institutions

The Royal Society

The Royal Society is the oldest scientific society in the world. It started life in 1662, but did not allow women to read papers until 1904. However, in 1902, a situation had arisen concerning an application to join the society. It was taken to the legal experts to see if this was permissible as the candidate was a married lady. The answer came back: 'We are of the opinion that married women are not eligible as Fellows of the Royal Society [. . .] A woman, if elected, would become disqualified by marriage'. Further corroboration of this situation can be seen in a quote from a letter from Eleanor Mildred Sidgwick (1845–1936) of Newnham College to Professor McKenny Hughes (1832–1917), known to be a supporter of female emancipation:

> It is clear that no lady ever has been elected a Fellow of the Society and though a bust of the distinguished Mary Somerville stands in the place of honour in the library of the Society, the archives of the Society so far as can be ascertained, contain no evidence that either she or any other woman was ever even nominated for the Fellowship.[4]

Altogether, women had waited 280 years when they were finally allowed into the Royal Society in 1919 under the legislation (Sex disqualification (Removal) Act), it was another 41 years, in 1943, before they could become Fellows of the Royal Society and even then it was another two years before the first female Fellows were actually elected. In 1945 Kathleen Lonsdale (1903–1971) and Marjory Stephenson (1885–1948) were the first women to be so rewarded for their contribution to science.

The Geologists' Association

The Geologists' Association started life in 1857, exactly 50 years after the founding of the Geological Society. It regarded itself as a more field-oriented and amateur-based society than the Geological Society (Burek 2008*a*). Women were permitted to join from the beginning, so that on 8 February 1859 they were not only elected but placed first in the members list, regardless of alphabet. They were elected as 'Miss Falkner, Mrs Wiltshire, Mrs Heslop'. A month later, on 8 March 1859, two further women were admitted, but now the names were mixed in with the men, perhaps recognizing that they should just be accepted as members and not singled out for special treatment. These women were listed as:

- Miss Ann Taylor Slatter
- Mrs Jane Bowles.

By March 1861 there were seven female members and thereafter there was a steady increase at one or two a year. The Geologists' Association always had field trips as a core tenet of their activities (Green 2008) and the first women attended field excursions in 1861 (Sweeting 1958). In addition, they were allowed to serve on Council and to hold positions such as librarian, secretary and field excursion organizers. They are prominently shown on old field-trip photographs from the Association's photographic archive, as illustrated on the front cover of *The Role of Women in the History of Geology* (Burek & Higgs 2007). An analysis of the early

[3] SMA: Royal Society re The admission of women to the Fellowship Case for the opinion of Counsel 1902. MS O113.
[4] SMA: Letter from E. M. Sidgwick to Professor McKenny Hughes, dated 2 February 1902. Uncatalogued.

history of the Association is readily available elsewhere (Sweeting 1958; Burek 2007; Green 2008).

General educational context

Within the broader context of female education, the important turning points in the nineteenth century are 1870 and 1875, which are coincidently repeated in the late twentieth century by 1970 and 1975 (Table 1). The introduction of the 1870 Elementary Education Act initiated universal, public elementary education in place of, or in addition to, the private schools already set up. This encouraged local areas to look at their education provision and to set up education boards. Consequently, a need for teachers arose. Also, for the first time, women had a vote in the election for school boards and, indeed, many were nominated, with four women being elected across the UK. Out of this need for teachers grew female teacher training colleges, including the first non-denominational Edge Hill College in Liverpool set up in 1885 (Montgomery 1997). The importance of female colleges (specifically Bedford College, London and Newnham College, Cambridge) is dealt with elsewhere (Burek 2007, 2008*b*) but summarized in Table 2.

It is worth emphasizing the importance of these institutions within British Victorian Society. Bedford College was the first to be set up in 1849, long before the Elementary Education Act of 1870, but it suffered in its early years from a lack of suitably qualified students. There were few female schools teaching science to the appropriate level and, therefore, few students available to attend. Cheltenham's Women College is one of the few that did fulfil this role (Burek 2007). Bedford College's location in the capital allowed it to serve the middle classes, and also made it possible for some women to attend as day pupils. Girton College and Newnham College, Cambridge, set up in 1869 and 1871, respectively, served a different clientele that was mostly residential, wealthy and of a higher class. These opportunities for women wanting an education were in contrast to that on the Continent. In Germany, for example, a rigid female gender model idealized household duties and motherhood in a climate that was hostile to intellectual women (Kölbl-Ebert 2007). The second date of significance for women's educational emancipation is 1875, with the admittance of women into universities (Table 1), hitherto the preserve of masculinity. For completeness, it is noted that in 1970 the Equal Pay Act was passed and in 1975 the Sex Discrimination Act. Both sought to provide equality of opportunity to both sexes and strengthened the position of women at least legally, if not immediately socially.

Table 1. *Important dates in women's education*

1849	Bedford College opens
1864	Cambridge Open Exam opened to women
1869	Girton College opened
1870	Elementary Education Act
1871	Newnham College opened
1875	University College opened to women
1876	First British degree to a woman
1879	University of London degrees opened to women
1883	Owen's College, Manchester, and Cavendish Labs, Cambridge, opened to women
1900	Bedford College becomes part of London University
1902	Trinity College, Dublin, opened to women
1904	First 'steamboat ladies' sail to Dublin
1919	Sex Disqualification (Removal) Act
1948	Newnham College becomes part of Cambridge University
1970	Equal Pay Act
1975	Sex Discrimination Act

Table 2. *Comparison of significant dates for Bedford College, London, and Newnham College, Cambridge*

Bedford College		Newnham College	
1849	Founded	1871	Founded
1885	Geology taught as a separate subject	1877	Geology first taught
1884	First graduates including geology as a subject	1881	Newnham students allowed into Tripos exams
1900	Bedford College becomes part of University of London	1921	Degrees awarded to women
1901	Morton Sumner Legacy secures geology teaching	1926	Women professors; female students eligible for studentships and prizes
		1948	Becomes part of Cambridge University

Female contributions to the Geological Society prior to Fellowship

For many women the key to gaining knowledge in the nineteenth century, and indeed earlier, came from attending meetings of societies (if they were admitted) and from reading and writing letters both to eminent geologists and to each other. A good example of this was the recently discovered correspondence (1892–1903) between Maria Ogilvie Gordon (1864–1939) and Professor Charles Lapworth (1842–1920), which is now preserved in the Lapworth Archive in Birmingham University.[5] So although women were not allowed to fully participate in the activities of the Geological Society before 1919, this did not stop them carrying out geological work which was then recognized by Society members. Possibly the first to benefit from this was Mary Anning (1799–1847), whose work was noted at a Society meeting after her death (Torrens 1995). Furthermore, the Society erected a stained glass window in her memory in the parish church of Lyme Regis (Woodward 1907, p. 115), where she had lived. The work of Etheldred Benett (1776–1845) was also acknowledged and, indeed, her sections on Chicksgrove Quarry, Wiltshire,[6] were presented to the Society for their records (Burek 2001). Women also contributed to the work of the Society through donations of rocks, fossils, books and money. By examining both the publication record and the awarding of funds and medals, women's contribution to the Society, prior to them becoming Fellows, can be assessed.

The first ever publication by a woman in any of the Society's publications occurred in 1824. Mrs Maria Graham (1785–1842) wrote a letter to Henry Warburton (1784–1858) (then on the Society's Council) about her first-hand experiences of an earthquake in Chile on 19 November 1822, which was subsequently read to members by Warburton. As it was of interest to a wider audience, it was decided to publish it in the Society's *Transactions* (Graham 1823; Kölbl-Ebert 1999). Another 38 years passed before the next publication by a woman, Miss Elizabeth Hodgson (1813–1877), entitled 'On a deposit containing *Diatomaceae*, leaves etc in the Iron ore Mines Dalton-in Furness, near Ulverston' (Hodgson 1863). This was read in 1862 by the then President Sir Andrew Crombie Ramsey (1814–1891) and published in the *Quarterly Journal* of the Society. A further 25 years passed, then, between 1887 and 1907, 13 women published 31 single-authored papers and six joint papers (Table 3). Just six women dominated the publication list of female authors until 1900. They were Jane Donald (later Mrs Longstaff), Gertrude Elles (1872–1960), Ethel Wood (1871–1945) (later Dame Shakespear), Maria Ogilvie Gordon, Margaret Gardiner (1860–1944) and Catherine Raisin. Their names come up time and time again in the geological world of the late nineteenth and early twentieth century, although most have now been forgotten (Burek 2008*b*).

One explanation for such women becoming forgotten or 'lost' was caused by them getting married and changing their names. This is a problem even today with the continuity of many women's publications being difficult to trace on marriage if they change their name. This difficulty was first tackled by the Geological Society in 1899 with Maria Ogilvie Gordon who had married in 1895. The problem was solved by adding her married name in brackets, thus the authorship of the paper reads Maria M. Ogilvie (Mrs Gordon). In 1906 both Ethel Wood (Mrs Shakespear) and Jane Donald (Mrs Longstaff) were now married, having established a publication record under their maiden names, so the precedent set by Maria Ogilvie Gordon was adhered to (Table 3). This early solution by the Geological Society allowed continuity.

The fact that women received financial contributions towards their research from as early as 1893 (Table 4), and that some were honoured with medals of the Society, is significant. What higher honour can the Society give to a colleague than a financial contribution towards their research, or the public award of a medal? Although Catherine Raisin had in some ways already achieved recognition from Bedford College by becoming Head of Department for Geology and then a Vice-Principal in 1898, to be the first woman ever to receive money from the Lyell Fund in 1893, awarded on the basis of noteworthy published research, must have marked an important development in her career. Similarly, awards to the amateur Jane Donald of Carlisle in 1898 and the two recent graduates Helen Drew (1881–1927) and Ida Slater (1881–1969) in 1906 and 1907, respectively, must have raised their confidence significantly.

[5] For a discussion of geological education available to women in the first half of the nineteenth century see Kölbl-Ebert (2002) and Burek (2007, 2008*b*).

[6] GSL: LDGSL3/21. Section of Chicksgrove Quarry.

Table 3. *Early female-authored papers read to and/or published by the Geological Society (only selected examples given after 1907)*

Dates read (and published)	Name	Paper	Communicated by
1887, May 25 (1887)	Jane Donald[1]	Notes upon some Carboniferous species of *Murchisonia*	J. G. Goodchild of Her Majesty's Geological Survey
1887, June 23 (1887)	Catherine A. Raisin	Notes on the metamorphic rocks of South Devon	Professor T. G. Bonney
1888, June 20 (1888)	Margaret I. Gardiner	The Greensand bed at the base of the Thanet Sands	J. J. H. Teall
1889, Feb 20 (1889)	Catherine A. Raisin	On some nodular felstones of the Lleyn	Professor T. G. Bonney
1889, June 19 (1889)	Jane Donald	Descriptions of some new species of Carboniferous Gasteropoda	J. G. Goodchild of Her Majesty's Geological Survey
1890, May 14 (1890)	Margaret Gardiner[2]	Contact-alteration near New Galloway	J. J. H. Teall
1891, February 21 (1891)	Catherine Raisin	On the lower limit of the Cambrian series in N. W. Caernarvonshire	Professor T. G. Bonney
1892	Jane Donald	Notes on some new and little known species of Carboniferous *Murchisonia*	J. G. Goodchild of Her Majesty's Geological Survey
1892, June 22 (1892)	Maria M. Ogilvie	Contributions to the geology of the Wengen & St. Cassian strata in Southern Tyrol	Professor Charles Lapworth
1893, January 11 (1893)	Catherine Raisin	Variolite of the Lleyn, and associated volcanic rocks	Professor T. G. Bonney
1895, February 20 (1895)	Jane Donald	Notes on the Genus *Murchisonia* and its allies: with a revision of the British Carboniferous species, and description of some new forms	J. G. Goodchild of Her Majesty's Geological Survey
1896	Gertrude Elles & Ethel Wood	On the Llandovery and associated rocks of Conway (North Wales)	Dr J. E. Marr, Secretary of the Geological Society
1896	Ethel Skeat & Margaret Crosfield	The geology of the neighbourhood of Carmarthen	Dr J. E. Marr, Secretary of the Geological Society
1897, February 3 (1897)	Gertrude Elles	The sub genera Petalgraptus & Cephalograptus	Dr J. E. Marr
1897, February 24 (1897)	Catherine Raisin	On the nature & origin of the Rauenthal serpentine	Professor T. G. Bonney

(Continued)

Table 3. *Continued*

Dates read (and published)	Name	Paper	Communicated by
1898*a*	Jane Donald[3]	Observations on the Genus *Aclisina*, de Kon., with descriptions of British species and of some other Carboniferous Gasteropoda	J. G. Goodchild of Her Majesty's Geological Survey
1898, May 4 (1898)	Gertrude Elles	The graptolite fauna of the Skiddaw Slates	Dr J. E. Marr
1899	Jane Donald	Remarks on the genera *Extomaria*, Koken and *Hormotoma*, Slater, with descriptions of British Species	J. G. Goodchild of Her Majesty's Geological Survey
1899 (1899)	Bonney, T. G. & Catherine Raisin	On varieties of serpentine and associated rocks in Anglesey	Professor T. G. Bonney
1899, December 20 (1900)	Gertrude Elles	Zonal classification of Wenlock Shales of the Welsh borderland	Dr J. E. Marr
1899, December 21 (1899)	Maria M. Ogilvie (Mrs Gordon)	The Torsian-structure of the Dolomites	Professor W. Watts
1900, March 21 (1901)	Ethel Shakespear	The Lower Ludlow formation and its graptolite fauna	Professor C. Lapworth
1900, March 21	Gertrude Elles	The zonal classification of the Wenlock shale of the Welsh borderland	Dr J. E. Marr
1901, February 6 (1901)	Igerna Sollas[4]	Fossils in the Oxford University MuseumV: on the structure and affinities of the Rhaetic plant *Naiadita*	Professor W. J. Sollas
1901, November 7 (1901)	Catherine Raisin	On certain altered rocks from near Bastagne and their relation to others in the district	Professor T. G. Bonney
1902, March 12 (1902)	Jane Donald	On some of the Proterozoic Gastropoda which have been referred to *Murchisonia* and *Pleurotomaria* with descriptions of new subgenera and species	J. G. Goodchild of Her Majesty's Geological Survey
1903, March 11 (1903)	Catherine Raisin	Petrological notes on rocks from Southern Abyssinia	Professor T. G. Bonney
1903, November 18 (1904)	Maud Healey	Notes on Upper Jurassic Ammonites with special reference to Specimens in University Museum Oxford	Professor W. J. Sollas
1904, January 6 (1904)	Clement Reid & Eleanor Reid	On a probable Palaeolithic floor at Prah Sands (Cornwall)	Clement Reid

1905, March (1905*a*)	Jane Donald	Observations on some of the Loxonematidae, with descriptions of two new species	Professor Theodore Groom
1905, March 8 (1905*b*)	Jane Donald	On some gastropoda from the Silurian rocks of Llangadock (Caemarthenshire)	Professor Theodore Groom
1905	T. G. Bonney & Catherine Raisin	The microscopic structure of minerals forming serpentine and their relationship to its history	Professor T. G. Bonney
1906	Gertrude Elles & Ida Slater[5]	The highest Silurian rocks of the Ludlow district	Professor T. McKenny Hughes
1906	Ethel Wood	On graptolites from Bolivia collected by Dr. J. W. Evans in 1901–1902	Dr. J. W. Evans
1906, February 21 (1906)	Ethel Wood (Mrs Shakespear)[6]	The Tarannon series of Tarannon	Professor C. Lapworth
1906, May 9 (1906)	Jane Donald (Mrs Longstaff)[6]	Notes on the Genera *Omopira, Lophospira* and *Turritoma*: with descriptions of the new Proterozoic species	Professor E. J. Garwood, Secretary of the Geological Society
1907, June 19 (1907)	Marie Stopes[7]	The flora of the Inferior Oolite of Brora (Sutherland)	Professor J. W. Judd
1910	Helen Drew & Ida Slater	Notes on the geology of the district around Llansawel, Carmarthenshire	
1911, January 11 (1912)	G. R. Watney & E. G. Welch[8]	The Salopian rocks of Cautley & Ravenstonedale	Dr J. E. Marr
1912, March 13 (1912)	Jane Longstaff (née Donald)	Some new lower Carboniferous Gasteropoda	Dr G. B. Longstaff
1914, February 25 (1914)	Rachel McRobert (née Workman)	On acid and intermediate intrusions and associated ash-necks in the neighbourhood of Melrose, Roxburghshire	E. B. Bailey
1917, May 2 (1918)	Jane Longstaff (née Donald)	Supplementary notes on Aclisina De Koninck and *Aclisoides* Donald with descriptions of new species	Dr G. B. Longstaff
1918, December 18 (1919)	Eleanor M. Reid	Preliminary description of the plant-remains	Charles Taylor Trechmann
1918	Edmund Johnston Garwood & Edith Goodyear	On the geology of the Old Radnor District, with special reference to an algal development in the Woolhope Limestone	Professor E. J. Garwood

(*Continued*)

Table 3. *Continued*

Dates read (and published)	Name	Paper	Communicated by
1920, 24 March (1920)	Eleanor M. Reid	3. On two preglacial floras from Castle Eden (Co. Durham) 4. A comparative review of Pliocene Floras based on the study of fossil seeds	Herself
1920, November (1921)	Marjorie Elizabeth Jane Chandler	The Arctic Flora of the Cam Valley, at Barnwell, Cambridge	Professor J. E. Marr
1921, January 5 (1922)	Miss Edith Bolton & Miss M C. Tuck	The carboniferous limestone of the Wickwar–Chipping Sodbury area (Gloucestershire)	Principal R. Franklin Sibly
1921, March 9 (1922)	Gertrude Elles	The Bala Country: its structure and rock succession	Herself
1921, May 4 (1922)	Agnes Irene McDonald & A. E. Trueman	The Evolution of certain Liassic Gastropods with special reference to their use in Stratigraphy	Read by Trueman, but Agnes McDonald answered questions
1922, April 12 (1923)	Marjorie Elizabeth Jane Chandler	The geological history of the Genus *Stratiotes*: an account of the evolutionary changes which have occurred within the Genus during the Tertiary & Quaternary Eras	Mrs. E. M. Reid
1924, May 7 (1925)	Ethel Skeat (Mrs Woods)[6] & Margaret Crosfield	The Silurian rocks of the central part of the Clwydian range	Mrs Woods

[1]Mr Goodchild stated that 'Miss Donald had not only studied most carefully the fossils of the northern district in which she resides, but she had also pursued her researches during visits to many museums in this country and also on the continent of Europe' (Goodchild 1887).
[2]Miss Margaret Gardiner is listed as 'Science Mistress, St. Leonard's School, St. Andrews [Scotland]'.
[3]'The Author thanked the speakers for the very kind reception that they had given to her paper' (Donald 1898*a*).
[4]This paper was read by her father.
[5]Professor Hughes expressed his appreciation of the advantage which the Society now enjoyed, of having the results of the excellent work done by women in Geology from themselves and discussing the conclusions arrived at with them (p. 221).
[6]These papers were read and published by women who had established a publication record under their maiden names. The Geological Society therefore decided to carry on in the short term by adding their married name in brackets for continuity.
[7]'Dr. A Strahan congratulated the Authoress upon her discovery of this plant bed at Brora which was not only interesting in itself, but important as affording further evidence for the correlation of the rocks of Brora with those of the Yorks. Coast.' Strahan was in the Chair and Vice-President of the Geological Society at the time. Professor E. J.Garwood 'regretted that he could add nothing to the discussion on the subject of this paper. He would, however, like to take this opportunity of welcoming this first communication from his old pupil, the distinguished Authoress. As many of the Fellows knew, Miss Stopes had interested herself for some time in palaeobotany, and had just received from the Royal Society a substantial grant to enable her to visit the deposits containing coal balls in Japan. He felt sure that the Fellows of the Society would join him in wishing every success to her expedition and would welcome future communications from her after her return' (Garwood 1907; Strahan 1907).
[8]'We also wish to thank Professor Charles Lapworth for the interest which he has shown in our work; Mrs Shakespear for help in identification of fossils and for the drawings of the graptolites figured in this paper and Miss G. L. Elles D.Sc for much help in the past' (Watney & Welch 1912).

Table 4. *Funding and medals awarded to early female geologists*

Year	Name	Award Fund	Medal	Collected by
Before women were admitted				
1893	Catherine Raisin	Lyell £24 16s 3d		Professor Bonney
1898	Jane Donald	Murchison £28 14s 3d		Mr Newton
1900	Gertrude Elles	Lyell £19 6s 0d		Professor McKenny Hughes
1903	Elizabeth Gray	Murchison £22 15s 10d		
1904	Ethel Wood (Shakespear)	Wollaston £34 6s 10d		Professor Marr
1906	Helen Drew	Daniel Pidgeon		Sent
1907	Ida Slater	Daniel Pidgeon		Sent
1908	Ethel Skeat	Murchison £25 8s 4d		Herself
After women were admitted				
1919	Gertrude Elles	Murchison		Herself
1919	Eleanor Mary Reid		Murchison	
1920	Ethel Wood (Shakespear)	Murchison		Herself
1920	Marjorie Chandler		Daniel Pidgeon	
1927	Marjorie Chandler		Wollaston	
1930	Helen Muir-Wood	Lyell		
1932	Maria Ogilvie (Gordon)	Lyell		Herself
1936	Eleanor Mary Reid	Lyell		Herself
1958	Helen Muir-Wood		Lyell	

The place of women at the first Centenary of the Geological Society

The Geological Society began the celebrations of its first Centenary on Thursday 26 September 1907, with a banquet at the Whitehall Rooms of the Hotel Metropole (Fig. 1), and finished with excursions on Saturday 28 September. The cost of a ticket to the banquet was a guinea (21 shillings), which was expensive for the time. It is enlightening to look at the context of women at this banquet and interesting to note that none were placed at the high table. There were 263 guests, of which 34 were women; 20 were the wives or daughters of academics (several of the guests were from overseas and eight were accompanied by women) and only nine were there in their own right, in other words, not accompanying spouses or relatives. These nine were Catherine Raisin (representing Bedford College), Miss Mary K. Andrews (1852–1914) (representing Queen's College, Belfast), Gertrude Elles, Ethel Wood (Mrs Shakespear), Margaret Chorley Crosfield (1859–1952), Dorothea Bate (1878–1951), Ida Slater, Maud Healey (1877, d. after 1952) and Enid Goodyear (1908–1960). Although Igerna Sollas (1877–1965) was accompanying her father, she was also an academic mathematics lecturer who had studied geology at Newnham College, along with Gertrude Elles.

Table 5 shows the status of all women present. There are some notable absences; in particular, Mary Sophia Johnston (1875–1955), Maria Ogilvie Gordon and Ethel Skeat (1865–1939). Ethel Skeat would have been teaching at the Queen's school in Chester and Maria Ogilvie Gordon was perhaps continuing her fieldwork in the Tyrol, which she tried to do every year (Wachtler & Burek 2007). Perhaps the constraints on their time were too great, or the cost of the ticket too high.

The first female Fellows

From having been a very exclusive male 'Club' the situation at the Geological Society eased slightly when, in 1904, women visitors were allowed into meetings if introduced by Fellows. In addition, from 1907, 'any woman who has distinguished herself as a geological investigator or who has

Fig. 1. Attendees at the Centenary dinner in 1907. Mrs Herries is seated just off-centre on the first table at the front of the photograph. On the third table closest to camera and on the right is Mrs Ethel Shakespear, next but one is Miss Maud Healey, next but one to her is Miss M. Andrews, and next but one again is Miss Gertrude Elles. Opposite Miss Andrews is Miss Margaret Crosfield. The other geological ladies present are not clearly visible on the photograph.

Table 5. *Women present at the Centenary celebrations of the Geological Society*

Name	Status
Mrs Adams	Wife of Professor
Miss Mary Andrews	Queen's College Belfast delegate & Belfast Naturalists' field club (geology)
Miss Dorothea Bate	Palaeontologist associated with the British Museum (Natural History)
Frau Bergeat	Wife of Professor
Mrs Clark	Wife of Professor
Mrs J. M. Clarke	Wife of Professor
Miss Margaret Crosfield	Member of the GA & Palaeontological Society
Mrs Cross	Wife of Dr Cross
Madame Dollfas	Wife of Professor
Miss Gertrude Elles	Lecturer Newnham College, Cambridge
Miss Enid Goodyear	University College Geology Museum Assistant to Professor Garwood
Miss Grabham	Relative of Dr Grabham
Mrs Groom	Wife of Dr Groom
Mrs Hague	Wife of Dr Hague
Miss Maud Healey	Lady Margaret Hall, Oxford, first female to read her own paper on ammonites
Mrs Herries	Wife of Mr Herries
Mrs Mary Hughes	Wife of Professor
Madam Lacroix	Wife of Professor
Mrs C Lapworth	Wife of Professor
Mrs A Lapworth	Wife of Dr Lapworth
Madam Lemoine	Wife of Mons. Lemoine
Mrs McKellar	Wife of Mr McKellar
Fröken Nathorst	Wife of Professor
Fräulein Penck	Daughter of Professor
Miss Catherine Raisin	Head of Dept. Bedford College; Morton-Sumner lectureship in geology
Mrs Seeley	Wife of Professor
Mrs Ethel Shakespear	Research Assistant to Professor Lapworth
Miss Ida Slater	Scholar at Newnham College, Cambridge
Miss Igerna Sollas	Lecturer at Newnham College, Cambridge; daughter of Professor
Mrs Spencer	Wife of Mr Spencer
Fru Thoroddsen	Wife of Dr Thoroddsen
Frau Walther	Wife of Professor
Miss Woodward	Daughter of Dr Woodward

GA, Geologists' Association.

shown herself able and willing to communicate to the Society original and important geological information' could be admitted as an Associate on payment of one guinea, although a proposal to *elect* them as Associates was negated at a Special General Meeting held to consider the matter on 15 May 1907 (Woodward 1907, p. 242). The Society was then just a few months short of being 100 years old. This Associateship allowed women to use the facilities of the Geological Society, to attend Ordinary and Annual General Meetings, but not to vote, and it was limited to 40 female members at any one time (Wallace 1969). It was to be another 12 years before women could be elected to Fellowship.

The momentous but forced decision to allow women to become Fellows of the Geological Society was instigated on 26 March 1919. The

President, George William Lamplugh, was in the chair. The significance of the statement makes it worth quoting in full:

> It will be within the recollection of most of the Fellows that the question of the admission of Women to candidature for the Fellowship of the Society has been raised on more than one occasion in the past. It was considered in 1889 and 1901, and again more systematically in 1908–1909, when a poll of the Fellows was taken and three Special General Meetings were held, with inconclusive results.
>
> It is generally recognised that the course of events since these dates has materially changed the situation. Women have been welcomed to our meetings as visitors, and we have had many examples of their qualifications for Fellowship in the excellent papers which they have from time to time contributed to the Society. The value of these papers has been appreciated by all geologists, and has been repeatedly acknowledged by the Council in its Awards. Therefore, in the opinion of the Council, it is no longer reasonable to maintain a sex-bar against qualified candidates for the Fellowship of the Society, and I am empowered by the Council to submit the above mentioned Resolution for your consideration.
>
> A ballot was then taken and the Resolution was declared carried by 55 votes against 12.
>
> The following addition to the Bye-Laws was then proposed – Section XXIII Interpretation – In the interpretation of these Bye-Laws words in the masculine gender only, shall include the feminine gender also.
>
> A ballot having been taken, this addition was carried by 50 votes against 5.
>
> (Lamplugh 1920*b*).

Thus, after 112 years since the Geological Society was founded, women could at last take their rightful place as Fellows. At the meeting on 21 May 1919, Lamplugh read out the list of new Fellows, including the following women:[7]

> Margaret Chorley Crosfield, Undercroft, Reigate (Surrey); [...] Gertrude Lilian Elles, D.Sc., Newnham College, Cambridge; [...] Maria Ogilvie Gordon, D.Sc., 1, Rubislaw Terrace, Aberdeen; Mary Sophia Johnston, 90 Wimbledon Hill, S.W.19; [...] M. Jane Longstaff (née Donald), F.L.S., Highlands, Putney Heath, S.W. 15; [...] Rachael Workman, Lady MacRobert, B.Sc., Colney Park, Radlett (Hertfordshire); Mildred Blanche Robinson, B.Sc., 5 Gledhow Gardens, S.W.5; and Ethel Gertrude Woods (née Skeat), D.Sc., 5 Barton Road, Cambridge, were elected Fellows of the Society.
>
> (Lamplugh 1920*c*)

This was just the first cohort and more joined later the same year. It marked a change in attitudes towards women participating in geological events as now they had the professional recognition that some of them, such as Gordon, felt was long overdue (Creese 1996; Burek 2005). Details of some of the first female Fellows have been reported elsewhere,[8] but a short summary of them all is included here (Table 6), with a more detailed contribution for each of those not documented elsewhere.

Women elected on 21 May 1919

Margaret Chorley Crosfield

Margaret Crosfield (1859–1952) was born on 7 September 1859 in Reigate where she lived all her life. She was educated at The Mount School, York, a Quaker school founded in 1785 and one of the first to send girls to university. Crosfield entered Newnham College, Cambridge, in 1879 at the age of 20 but left a year later. Her studies were interrupted by ill health and when she returned to complete them 10 years later in 1890, with the permission of the authorities, she only took geology not the usual natural tripos. She made some lifelong friends at Cambridge, despite being significantly older than them: Ethel Skeat, with whom she published on the Welsh Borders, was six years younger, and Mary Johnston, with whom Crosfield published on Carmarthen, was 16 years younger. She was also an academic contemporary of Gertrude Elles and Ethel Wood who were, respectively, 13 and 12 years younger than Crosfield. It is interesting that in her will Crosfield mentions both Gertrude Elles and Mary Johnston. All these women were influenced in their studies by Dr (later Professor) John Marr (1857–1933) and Professor Thomas McKenny Hughes.

Crosfield was an active member of the Geologists' Association, which she joined in 1892, 17 years before she was elected a Fellow of the Geological Society. She became a Geologists' Association Council member in 1918, and was its librarian between 1919 and 1923. She was also a member of the Palaeontological Society from 1907 to 1932. Crosfield published three main papers; the first on Carmarthen with Ethel Skeat (Skeat & Crosfield 1896). The quality of this paper was such that it was used by the British Geological Survey in their memoir of the Carmarthen area. In 1914 she published work on the Wenlock limestone of Shropshire with her friend Mary Johnston (Johnston & Crosfield 1914) and then again with Ethel Skeat, now Mrs Woods, on the geology of the Silurian rocks of the Clwydian range (Woods & Crosfield 1925). Crosfield also

[7] The gaps represent the names of men who also joined at this time.

[8] Creese & Creese (1994, 2006); Burek (2001, 2002, 2003*a*, *b*, 2004, 2005, 2007, 2008*b*); Kölbl-Ebert (2001, 2007); Chaloner (2005); Burek & Higgs (2007); Burek & Malpas (2007).

Table 6. *The first 21 women elected to join the Geological Society as Fellows*

Dates	Maiden name	Married name	Education	Age	Place of residence (at time of joining the Society)
		Elected 1919			
1859–1952	Margaret Chorley Crosfield	Single	Incomplete degree	59	Reigate
1872–1960	Gertrude Lilian Elles	Single	DSc	46	Cambridge
1864–1939	Maria Ogilvie	Maria Gordon	DSc, PhD	54	Aberdeen
1875–1955	Mary Sophia Johnston	Single	None	44	Wimbledon
1856–1935	(Mary) Jane Donald	Jane Longstaff	None	63	Putney
1884–1954	Rachel Workman	Lady Rachel MacRobert	BSc	34	Hertfordshire
Unknown	Mildred Blanche Robinson	Single	BSc	–	London
1865–1939	Ethel Gertrude Skeat	Ethel Woods	DSc	54	Cambridge
1855–1945	Catherine Alice Raisin	Single	DSc	64	London
1886–1931	Margaret Flowerdew MacPhee	Margaret Romanes	BSc	34	West Byfleet
1885–1955	Mary Kingdom Heslop	Single	MSc	35	Newcastle upon Tyne
Unknown	Dorothy Margaret Woodhead	Single	BA	–	Dartford
		Elected 1920			
Unknown	Florence Pitts	Single		–	Dorking
1860–1953	Wynne Edwards	Eleanor Mary Reid	BSc	59	Milford on Sea, Hants
1871–1945	Ethel Mary Reader	Dame Ethel Wood	BSc	47	Birmingham
Unknown	Lucy Ormrod	Single	BA	–	Newport, Monmouthshire
		Elected 1922			
1890–1931	Irene Helen Lowe	Single	MSc	32	Kingston Hill, Surrey
1879–1959	Edith Goodyear	Single	BSc	43	London
1896–1968	Helen Marguerite Muir-Wood	Single	MSc	27	London
Unknown	Isabel Ellie Knaggs	–	None	–	London
1880–1958	Marie Carmichael Stopes	1) Gates (annulled) 2) Marie Roe	DSc, PhD	42	Leatherhead, Surrey

had a great interest in education and served on the Reigate Borough Council Education Committee, as well as being a school governor of the Reigate Girls County School (Johnston 1953). In addition, she was a great promoter of women's suffrage. Indeed, some of her field notes are written on the back of suffragette notepaper.[9] She led two field excursions to Reigate for the Geologists' Association in 1899 and 1911, and lectured to the local Holmesdale Natural History Society and other local societies on many subjects. She travelled widely and attended many meetings. Her training at Cambridge ensured that she kept meticulous notes, field notebooks and specimen locations. These can still be seen today at the British Geological Survey in Keyworth[10] and at the Grosvenor Museum in Chester.[11] It was the identification of her labels on some of the specimens that resulted in finding her lost specimens in the museum (Burek & Malpas 2007). She also presented specimens from the Silurian of Wenlock Edge to Liverpool University Geology Department in 1919, and, mostly invertebrates, to the Natural History Museum in London between 1906 and 1930. She presented Palaeozoic invertebrate fossils to them in 1937 at the age of 78. She often exhibited material – for example, in 1919 at the Geologists' Association annual meeting – where she displayed 'Worked flakes from the fields near Savernake Forest and from the neighbouring Wiltshire Down' (Anon. 1919).

The paper she wrote with Ethel Skeat was read in May 1924 to the Geological Society by Ethel Skeat (now Woods) and was published in 1925 (Woods & Crosfield 1925); however, research for this paper had started much earlier. They visited sites referenced in the paper between 1906 and 1909 (Fig. 2) while Ethel was teaching at the Queen's School, Chester, and again in 1911 and 1922 after Ethel had married (Burek & Malpas 2007). There is also evidence that they had help with identification of the graptolites from Gertrude Elles and Ethel Wood.

Crosfield was a spinster of independent means who was lucky enough to be able to finance her geological research from her own pocket, without recourse to family members. To her, Fellowship of the Society meant recognition for her work by a professional body.

Gertrude Lilian Elles

Gertrude Elles (1872–1960) was born in Wimbledon, Surrey, near London, on 8 October 1872, one of six children of Jamison Elles, a Scottish businessman and dealer in Chinese goods, and Mary Chesney. She spent part of her childhood in Scotland and died there in 1960 (Burek 2003*a*). After attending Wimbledon High School, in 1891 she went up to Newnham College, Cambridge, at the age of 19 where she passed the honours degree in geology in the Natural Sciences Tripos four years later. She travelled to Trinity College, Dublin, as one of the 'Steamboat Women' to receive her DSc on 6 July 1905 (Higgs & Wyse Jackson 2007). At the time, Cambridge University did not award degrees to women, but an arrangement was in place whereby graduates of Cambridge, or indeed Oxford, could obtain their degrees from Trinity College, Dublin. Between 1904 and 1907 successful female students therefore took advantage of this by taking the 'steamboat' to collect them, after which the practise was discontinued (Higgs & Wyse Jackson 2007). Working at the Woodwardian Museum in Cambridge, Gertrude Elles became part of the team researching Lower Palaeozoic rocks and fossils, and was awarded a scholarship from Newnham College to pursue this, which involved travelling to Sweden to study the Lower Palaeozoic rocks and fossils.

Working under the editorship of Professor Charles Lapworth, Elles prepared the mammoth monograph on British graptolites with her colleague Ethel Wood, which was finally published in 1918 (Elles & Wood 1901–1918). The names of Elles and Wood will always be linked to the research and study of British Graptolites. She received the Lyell Geological Fund from the Geological Society of London in 1900 (Table 4) for her work on three classic papers on different genera of graptolites (Elles 1897, 1898, 1900) 'adding much to our knowledge of the characters and range of those fossils' through her 'important paper on the graptolite-fauna of the Skiddaw Slates' (Whitaker 1900). The third paper was yet to be published, nevertheless, the President of the Geological Society, William Whitaker (1836–1925), remarked on 'her knowledge of the Wenlock shales of the Welsh borderlands' (Whitaker 1900). She was awarded the fund in absentia because of the Society's rules against women attending meetings at that time, so her Professor, McKenny Hughes, received the award on her behalf, stating that she 'has shown herself to be a clear-sighted stratigraphist and an astute palaeontologist over a much wider field than might appear, from the mention of the work for which this Award has been made'

[9] BGS: Crossfield notes, field notebooks and specimen locations.

[10] BGS: Crossfield notes, field notebooks and specimen locations.

[11] Grosvenor Museum, Chester: Crossfield notes, field notebooks and specimen locations.

Fig. 2. Geologists' Association field trip to Oswestry in 1908. Margaret Crosfield is the second woman from the left, second row from the back; Ethel Skeat is sitting next to her, third woman from the left; Rachel Workman is at the centre of the front with the bow tie.

(McKenny Hughes 1900). It must be remembered that this praise was from the founder of the Sedgwick Museum in Cambridge. Elles was also interested in other issues such as the stratigraphy of Lower Palaeozoic rocks in general (Elles & Slater 1906; Elles 1922) and those of Scotland in particular. She saw the importance of the wider context of fossils and studied them as communities. Her presidential talk to the 1923 British Association for the Advancement of Science, Section C, was, for example, on stratigraphic palaeontology as highlighted by the evolution of both ancient fossil and modern faunal groups. During her lifetime she significantly raised the awareness and identification of the extinct fossil group of graptolites.

Studying graptolites – those small indistinct creatures – was considered too tedious for men, who lacked the dedication to detail necessary to carry out the work. Both Elles and Wood were therefore considered suitable for the painstaking identification and classification work that required patience and fine detail, precisely because they were women: 'One qualification for the task of monograph compilation then, was a capacity for drudgery' (Gould 1998). Furthermore, the reward for production of the monograph was not financial but 'the gratitude of future, if not present generations' (Woodward 1911). In other words, perhaps it was perceived as a suitable occupation for women who did not require financial gain and were prepared to seek knowledge for its own sake, which would not necessarily give immediate academic recognition. It is significant that Elles and Wood took over 18 years to complete their *Monograph on British Graptolites* (Elles & Wood 1901–1918). In her reply to receiving financial aid in the form of the Lyell Fund from the Geological Society in 1900, Elles stated: 'I will strive my very utmost to make the work which I may do in the future worthy of the confidence which such an Award seems to imply' (Elles 1900). She certainly did that, and much more. Her work was recognized by the Geological Society with the first award to a woman of a Society Medal – the Murchison Medal in 1919 – just months before women were admitted to Fellowship: 'The Council desires to acknowledge the important and sustained efficiency of your efforts in advancing Geological Sciences' (Lamplugh 1920*d*).

In 1919 Gertrude Elles was included in the first batch of Fellows of the Geological Society, at the age of 46. She is recognized as a role-model teacher and researcher (Burek 2007, 2008*b*). At the age of 54 she was the first female to be awarded a university readership in Cambridge. She continued lecturing and researching until retiring in 1938 when she was made a Reader Emeritus, and she continued to supervise undergraduates for many years. She also helped research workers with 'her knowledge, her books and sometimes one suspects, her money' (White 1961). For 20 years she continued to occupy her research room in the Sedgwick Museum until her memory started to fail her. She also became very deaf.

At this time it was common for middle-class women in Britain to carry out 'good works' alongside paid employment. For her services to the British Red Cross during World War I Elles was awarded the MBE. She was remembered as an enthusiastic teacher, a lover of music and fishing (W. B. R. K. 1961), and as a superb role model for both men and women who need dedication and determination to study extinct organisms in detail.

Gertrude Elles died on 18 November 1960, four days after arriving in a Helensburgh nursing home in her beloved Scotland at the age of 88 years. In the obituary written for the Newnham College roll letter, she is acknowledged as a moving force in both the college and the department:

> By the kindness of the Provost and Fellows of King's College a memorial service was held in the Chapel on 16th December, a simple service of great beauty which will long be remembered by those who attended. The perfect singing of the choir in that unique setting which she had loved so much, and the gathering of family and friends to do her honour, would surely have delighted her who had so deeply appreciated both beauty and friendship.
>
> (White 1961)

Elles never married; it has been surmised that she was married to her students and geology (White 1961). For her, Fellowship of the Geological Society meant recognition of her work by her peers.

Maria Ogilvie (Mrs Gordon)

Maria or May Ogilvie (1864–1939) was born in Monymusk, Aberdeenshire, Scotland, on 30 April 1864, to Reverend Alexander Ogilvie (1831–1904) LL.D, an eminent educationalist, and his wife, Maria Matilda Nicol (*c.* 1837–1916). Her older brother Francis, later Sir Francis Ogilvie (1858–1930), was also destined to become a geologist. From 1886 to 1900 he was Principal of Herriot Watt College in Edinburgh and this was to have an influence on his sister in later years. Having spent some time from the age of 18 studying piano at the Royal Academy of Music in London after leaving school, Maria Ogilvie returned north. In Edinburgh she pursued the first part of a degree at Herriot Watt College, returning to London to undertake further study in geology, botany and zoology at University College, where she obtained her BSc in 1890. After a visit to the South Tyrol in 1891 with family friends, Baron Ferdinand Freiherr von Richthofen (1833–1905) and his wife, she decided to start mapping the complex geological area. Her research, published in the *Quarterly Journal of the Geological Society* (Ogilvie 1892), formed the basis of her DSc submission to the University of London in 1893. She was the first female to obtain such a distinction. From 1891 to 1895 she moved to the University of Munich, but returned home to London on her marriage to John Gordon (Creese 1996; Burek 2005). However, she continued her research and in 1900 was the first woman to obtain a PhD there, having undertaken all the examinations in German. She is now acknowledged as one of the foremost research geologists of Scotland at the end of the nineteenth century, having published over 19 scientific contributions by 1900. Initially, she corresponded with eminent geologists in this country such as Archibald Geikie and Charles Lapworth who were a great help to her, especially with publications. Letters recently discovered in the Lapworth Archive at Birmingham University show they corresponded for over 10 years between 1892 and 1903, and that Lapworth was of enormous help to her not only in reading manuscripts but also in commenting on and discussing geological ideas with her:[12] 'I think, Sir, I may congratulate you on having made a doctor out of me'.[13] In 1895 she

[12] LA: O39. Letter to Lapworth from Maria Ogilvie Gordon, 21 October 1893, has a note written on the top that states 'This note to be placed on the mantel shelf until answered'.

[13] LA: O37. Letter to Lapworth from Maria Ogilvie Gordon, no date.

Table 7. *Highlights of Maria Ogilvie Gordon's achievements*

Date	Achievements
1893	First female to receive a DSc from University of London
1900	First female to receive a PhD from Munich University
1919	Formed the Council for the representation of women in the League of Nations
1919	Among first group of women to be elected to the Geological Society of London
1932	Received Lyell Medal from Geological Society of London
1920	First JP and chairman of the Marylebone Court of Justice
1935	Dame of the British Empire (DBE)

married Dr John Gordon, a medical doctor, and had three children, but this did not stop her working in the Austrian mountains. When they were older she often took the children with her[14] or left them with servants while they were at the Robert Gordon College in Aberdeen. It took World War I to halt her work. Many of Gordon's publications were in German and as such were not quite so accessible to English-speaking geologists. Thus her recognition by the British was late in coming, and not until after awards and accolades in the form of honorary memberships had been given to her by the Vienna Geological Society and the Natural History Museum of Trento, Italy, both in 1931, as well as honorary doctorates from the Universities of Innsbruck (1928), Edinburgh (1935) and Sydney (1938). Despite her outstanding contribution to geological knowledge, she was not present in 1907 at the Centenary celebrations of the Geological Society (Burek 2005). Gordon had many other calls on her time and was involved in many other activities, especially those concerned with her offices both at the national and international level in women's associations and organizations. She was also about to publish her *Handbook of Employment Specially Prepared for the Use of Boys and Girls on Entering the Trades, Industries and Professions* (Gordon 1908).

Maria Ogilvie Gordon was awarded the Lyell medal from the Geological Society for her work on the Dolomites when she was 68. She died seven years later at the age of 75. Table 7 highlights her outstanding achievements. Further accounts of her activities can be found in Creese (1996), Burek (2005, 2008*b*) and Wachtler & Burek (2007). In 2006 Gordon was further honoured by having a Middle Triassic (Anisian) Fern *Gordonopteris lorigae* from the Dolomites of northern Italy named after her by Johanna van Konijnenburg-van Cittert, Evelyn Kustatscher and Michael Wachtler: 'Derivation of name. After Dr Maria Ogilvie Gordon who was one of the pioneers of Triassic palaeobotany in the Dolomites' (van Konijnenburg-van Cittert *et al.* 2006).

To summarize, Maria Gordon basically had three careers: music, geology and public works, the last being primarily education, franchise and the health of women. She was an extraordinary woman, initially supported by her husband and later, after his death, at least intellectually, by her brother and her mentors Sir Archibald Geikie and Professor Charles Lapworth. She was of independent means and therefore had the ability to carry on her work with the help of servants and family members.

Mary Sophia Johnston

Mary Johnston (1875–1955) was born in Folkestone, Kent, on 29 October 1875. She was the second daughter of Reverend William A Johnston (d. 1898), the rector of Acrise near Folkestone. She was educated privately and developed a love of geology, which she never lost. She never married and lived in Folkestone until 1898, the year her father died, when she and her mother moved to Wimbledon Hill, London. From the rural area of Kent, she was suddenly given the opportunities that the Capital offered her; at the age of 23 she joined the Geologists' Association and became a geology student at University College, London. Her strong and enthusiastic support of the Association, both financial and temporal, is shown in the positions she held and the work she did for them. She became the voluntary Illustrations Secretary (1910–1925) and often adopted the role of 'official' photographer. She was also librarian (1932–1936) and a member of council (1918–1924). She was elected an honorary member in 1939 at the age of 64. In the photographs of the Association she is often seen at the rock face examining the materials with her great friend Margaret Crosfield who is mentioned in Johnston's will of 1950, but whom she sadly outlived by three years. She was a keen supporter of the fieldwork they undertook. She maintained and brought to each annual meeting the albums of fieldtrip photographs, a tradition that is continued today by Mrs Marjorie Carrick, the

[14] Robert Gordon College Archive, Schoolhill, Aberdeen. An article in the Gordonian school magazine, 1911, was written by Gordon's son, John, on a holiday to the Tyrol in Austria when he was 12.

keeper of the Geologists' Association photographic archive. Mary Johnston's support of the Association also continued after her death, since in her will she left them £1000 in a named legacy.[15]

Johnston was also a member of the Palaeontographical Society, to which she also left a legacy of £1000. When she became a Fellow of the Geological Society in 1919 she is listed with no degree (Lamplugh 1920*a*); however, she was both a Fellow of the Zoological Society and the Royal Geographical Society. By that time, she had moved to larger premises in Kew. Here she worked on and studied her geological material and encouraged others to do the same. Her extensive collection of maps, books, guides and photographs were often borrowed to illustrate other geologists' lectures.

Johnston was a great traveller, attending many of the International Geological Union meetings in Spain, France, South Africa and the USA, and the International Geological Congress in Toronto, Canada (A. L. L. 1956). She travelled to New Zealand and Egypt, all the time collecting geological material, specifically different forms of silica. However, most of her publications are of a palaeontological nature. She shared her knowledge freely with others, giving talks on her travels, exhibiting materials at Geologists' Association meetings (Anon. 1919), and writing several articles in the *Geological Magazine* and local society journals (Johnston 1901, 1927). Her most important article was written with her friend Margaret Crosfield in 1914 on the Ballstones of the Wenlock limestone, which was published in the *Proceedings* of her beloved society, the Geologists' Association (Johnston & Crosfield 1914).

Johnston donated many of her invertebrate fauna, including foraminifera and bryozoan specimens, to the British Museum Natural History from 1932 to 1939 and again in 1942. Her executors presented additional fossils to them in 1956 in accordance with her wishes.[15] She also donated material to other national and regional museums and universities. She actively supported the South-Eastern Union of Scientific Societies, acting as their honorary treasurer for several years, and it is interesting to note that in her will of 1950 she left them £1000, but this was revoked by a codicil on 18 February 1953.[16]

Johnston was able to support many non-geological causes, including rowing organizations and schools, and she also had a great interest in both church history and heraldry. She was one of those women of independent means who devoted her life to the betterment of others, but also provided a fantastic archive for future generations in her maintenance of the Geologists' Association's photographic record. Mary Sophia Johnston had a warm and benevolent nature as is shown in both her obituaries (A. L. L. 1956; H. D. T. 1955). She became a Fellow of the Geological Society to better understand the natural world and to associate with like-minded people. She died on 23 January 1955 in London after a short illness.

(Mary) Jane Donald (Mrs Longstaff)

Jane Donald (1856–1935) was born in Carlisle, the eldest of four children, to Matthewman Hodgson Donald. She was educated privately in London and then went to the Carlisle College of Art. From an early age, her passion within natural history was the mollusca phyla. She collected, drew and described modern genera and species, and read her first paper on them at the Cumberland Association for the advancement of Literature and Science in Carlisle in 1881, at the age of 25. She then turned her attention to fossil examples of the same phyla. Encouraged by Mr John George Goodchild (1844–1906) of the Geological Survey, she published a paper on some Carboniferous gastropods in the *Transactions of the Cumberland & Westmoorland Association* in 1885 (Donald 1885). In all, she published about 20 papers on Palaeozoic gastropods, specializing in three families. Between 1887 and 1918, she published seven papers in the *Quarterly Journal of the Geological Society* (Table 3), the last just two years before she died. Although never having any formal training in geology and always acting as an amateur, her knowledge and reputation grew. Her early art training held her in good stead for drawing diagrams for her papers (Fig. 3). She became the leading expert in the systematic palaeontology of the Murchisoniidae, Pleurotomariidae and Loxonematidae families of gastropoda (Table 3). She was meticulous in her work, which was of a very high standard. This was recognized when, in 1898, she received the Murchison Fund from the Society, being only the second woman to receive such financial help. The President, Henry Hicks (1837–1899), addressed her recipient, Mr Newton, as follows: 'On the present occasion a lady who has attained distinction as a palaeontologist has been selected by the Council to receive an award from the Murchison Fund [...]'. He goes on to list the 'five important papers by Miss Donald' (Donald 1887, 1889, 1892, 1895, 1898*a*) and then tells of her travels to study foreign collections. She is 'untiring in her

[15] Will, Probate Office, London.

[16] Will, amended by codicil, Probate Office, London.

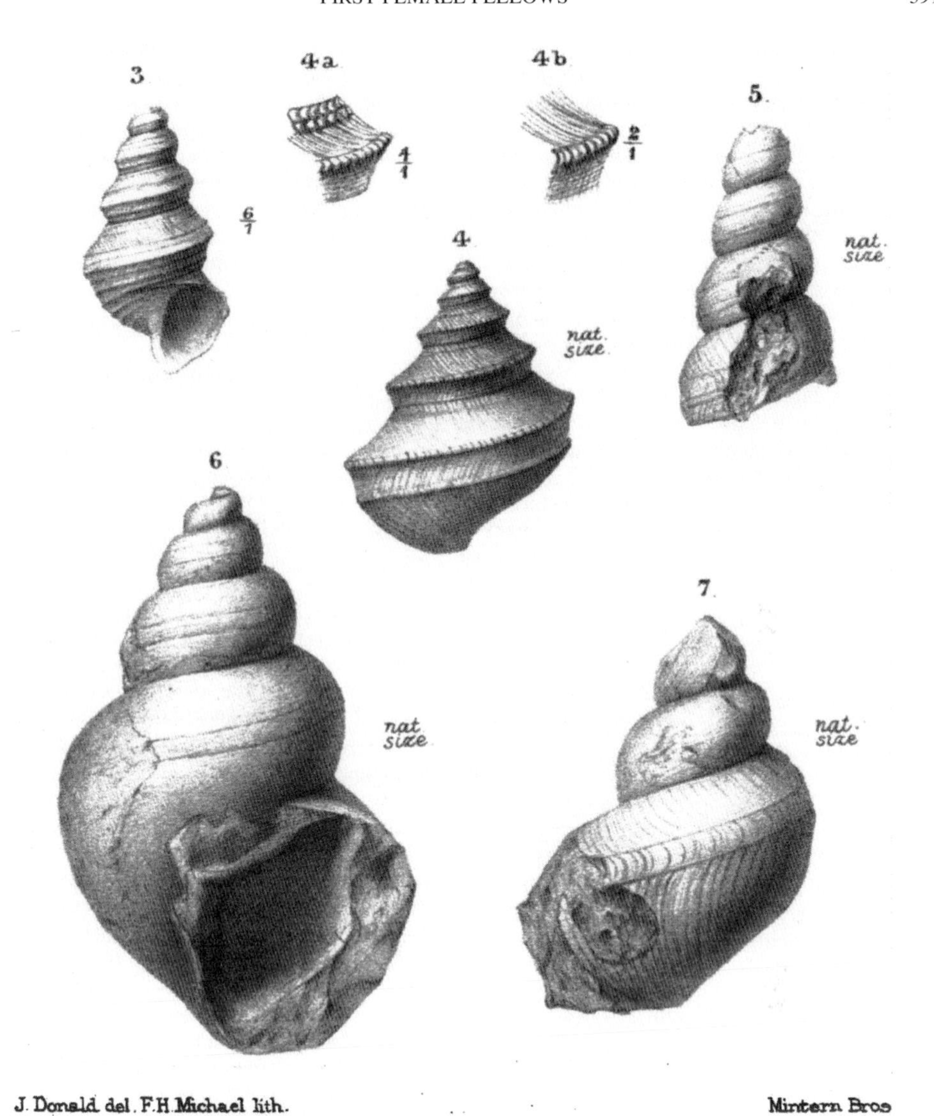

Fig. 3. Carboniferous gastropods drawn by Jane Donald.

zeal in collecting information for future work [. . .] the Council hope it will be accepted not only as a token of appreciation of the excellent work which she has already accomplished, but in the hope that it may be some incentive to her to continue her palaeontological researches among the Palaeozoic rocks' (Hicks 1898).

In her reply in a letter read by Mr Newton, Donald states:

> The news came to me as a great surprise for I had previously deemed it no small honour that my papers should have been considered worthy of publication in the Quarterly Journal of the Society and this higher recognition will certainly prove an encouragement to further research and I hope better work. My studies have been a source of great pleasure to me and I feel that there is still much to be found out.
>
> (Donald 1898*b*)

In 1906, at the age of 50, Donald married Dr George Blundell Longstaff (1849–1921) of Putney Green, London, well known as an eminent entomologist and author of *Butterfly Hunting in Many Lands* (1912) (Cox 1935). They had no children and her husband died in 1921. She lived for a further 14 years, collecting and working on her favourite shells and fossils. In 1919, at the age of 63, she became a Fellow of the Geological Society.

Of a financially sound disposition, through an inheritance, Donald travelled widely throughout her life, visiting museums and collections as far away as South America and South Africa. Everywhere her passion for understanding fossils and their behaviour took hold. She even tried some breeding experiments with the large South African land snail. Her extensive collection of recent shells was eventually given to the British Museum by her nephew in 1936 (L. R. C. 1936). At the time of her death in 1935, she was the longest surviving member of the Geologists' Association (53 years), having joined in 1883 at the age of 27. Her shells are still to be seen today at the Natural History Museum in London, and her understanding and contribution to the taxonomic advancement of the Gastropoda is significant.

Rachel Workman (Lady MacRobert)

Rachel Workman (1884–1954) was born in Worcester, Massachusetts, USA, the only daughter of a wealthy American physician, Dr William Hunter Workman (1847–1937) and his wife, Fanny Bullock Workman (1859–1925), who were both famous Himalayan explorers (W.C.S. 1955). Her mother was exceptional, being honoured for her climbing achievements by the Royal Geographical Society among others (Millington 1972). Workman's grandfather had been Governor of Massachusetts. Rachel had a brother, Siegfried, born in Germany when she was five years old, but who died when she was only nine. Much of her early life was spent travelling with her parents around Europe and she was educated in Dresden, as well as by a governesses. After the death of their son, Workman's parents became very restless and effectively had no fixed abode. However, they were independently wealthy as a result of a fortune inherited through Fanny's mother, Eliza Hazard, who was part of the Connecticut explosives manufacturers family – the 'Hazards of gunpowder' (Mitchell 1979) – thus Workman was sent to finish her education at Cheltenham's Women College (W. C. S. 1955). She disliked the strict regime of boarding school immensely, as she had been brought up by her parents to believe in freedom, equality for women and in women's education (Millington 1972). Later she was to support the suffragette movement. This schooling was followed by Royal Holloway College where Workman obtained a BSc – a 'second class honours in Geology as an internal student' from London University (13 December 1911) – before going on to Edinburgh University where she studied geology and political economy (W. C. S. 1955). She later studied at the Royal College of Science and the Royal School of Mines. The interest she caused when attending the latter is noted in a letter to her fiancée and future husband, Sir Alexander MacRobert of Douneside:

> Our class goes three times a week this year [1910] to the Royal School of Mines for classes under Prof. Cox so of course I went too, as they are all about gold mining, placer deposits etc [...]. I noticed considerable surprise on Cox's face on entering the room, and much whispering among the students. The next day Prof. Watts rushed up to me with an amused grin and said 'so you are attending the mining lectures I hear, well you have broken all records. No woman has ever attended mining lectures before! Are you going on?' So I said yes, and assured him I had no idea they were not admitted but that of course I was going on. If I hear any more I shall require to see the Statutes which exclude women. Of course there are none, and it simply has not occurred before! But I don't mean to be turned out. Especially as they happen to be very good. But I can see I have created a sensation.[17]

Later she went to the Mineralogical Institute of Oslo under Victor Moritz Goldschmidt (1888–1947). She was also a Fellow of the Geological Society of Stockholm. She carried out considerable geological research along the Scottish Borders and in the iron-bearing region of Sweden, as well as in the Indian goldfields (Workman 1911; McRobert 1912, 1914). She attended many International Congresses and her letters on the 1910 International Congress in Stockholm are very interesting. She was one of only five women attending the Congress field trips to Lapland and northern Sweden, and she was often the only woman on the outdoor excursions (Millington 1972). This did not bother her:

> Everyone is most awfully nice, but I had to overcome the usual annoyance men have when women are about on scientific expeditions. Now that they have found that I am not a drag on them or bore them with talk they are very pleasant.
>
> Everyone is most awfully kind to me and I am having a splendid time. It is most inspiring to meet so many eminent geologists and to hear their discussion etc. in the field.[18]

Workman spoke many languages including French, Italian, Spanish, Danish, Norwegian and German, having lived on the Continent, and, indeed, her mother used to write to her while at school in French or German and these languages were a

[17] MRA: MS Section 2 correspondence 9/5/4. 1910. Letter to Sir Alexander of Douneside from Rachael Workman.

[18] MRA: MS Section 2 correspondence 9/5/7. 1910. Letter to Sir Alexander of Douneside from Rachael Workman.

Fig. 4. International Geological Congress 1913. Second row: Catherine Raisin is second from the left and Rachel Workman is fourth from the left.

great asset when travelling: 'I also associate with the French and German circle. My languages stand me in good stead'.[19]

It was at the International Congress of 1910 that her interest in metamorphism and glaciation grew, and where she developed an interest in garden design and alpine plants:

> I am getting a most lovely collection of metamorphic rocks. They are quite new to me. I find it rather difficult to follow, but I am in luck in my companions that I mainly roam with [...]. I am acquiring a good deal of Alpine and glacial botany as well as the rare flowers that are found and brought to me.[20]
>
> The glacial erosion of Lake terraces etc. are quite wonderful and I am leaving this party next week to go to do some special work on the remains of the ice age in the extreme north. [...] I have discovered I have a special interest in this line and I of course know no better opportunity is offered in any country.
> (Millington 1972)

She knew and corresponded with Catherine Raisin and Maria Ogilvie Gordon. Indeed, she is pictured seated next but one to Catherine Raisin at the International Congress in 1913 in Toronto (Fig. 4). And in the Douneside archives there is a letter to Workman from Maria Ogilvie Gordon, dated just before her marriage. It is clear from this letter that, although they did not often meet, they were friends and collaborators in geology.[21]

Even after her Quaker marriage to Alexander MacRobert (1854–1922) on 7 July 1911 and the birth of their children (Fig. 5), Workman continued her geological work (Figs 6 and 7). Her husband, a self-made millionaire, was 30 years older and often away for months at a time in India. As she did not always go with him, there was plenty of time to pursue her own interests. On one of these occasions, in 1913, she had been left with her small son at Douneside, so she decided to return to London to keep her geological contacts up to date: 'there is no doubt one needs to be in touch with the current thought and to attend lectures and hear and discuss things and keep abreast of the times'.[22] At the time, she was writing a follow-up

[19] MRA: MS Section 2 correspondence 9/5/9. 1910. Letter to Sir Alexander of Douneside from Rachael Workman.
[20] MRA: MS Section 2 correspondence 9/5/13. 1910. Letter to Sir Alexander of Douneside from Rachael Workman.
[21] MRA: MS Section 4 correspondence 16/6/2. 1911. Letter to Sir Alexander of Douneside from Rachael Workman.
[22] MRA: MS Section 2 correspondence 9/5/50. 1913. Letter to Sir Alexander of Douneside from Rachael Workman.

Fig. 5. Rachel Workman and children (courtesy of MRA).

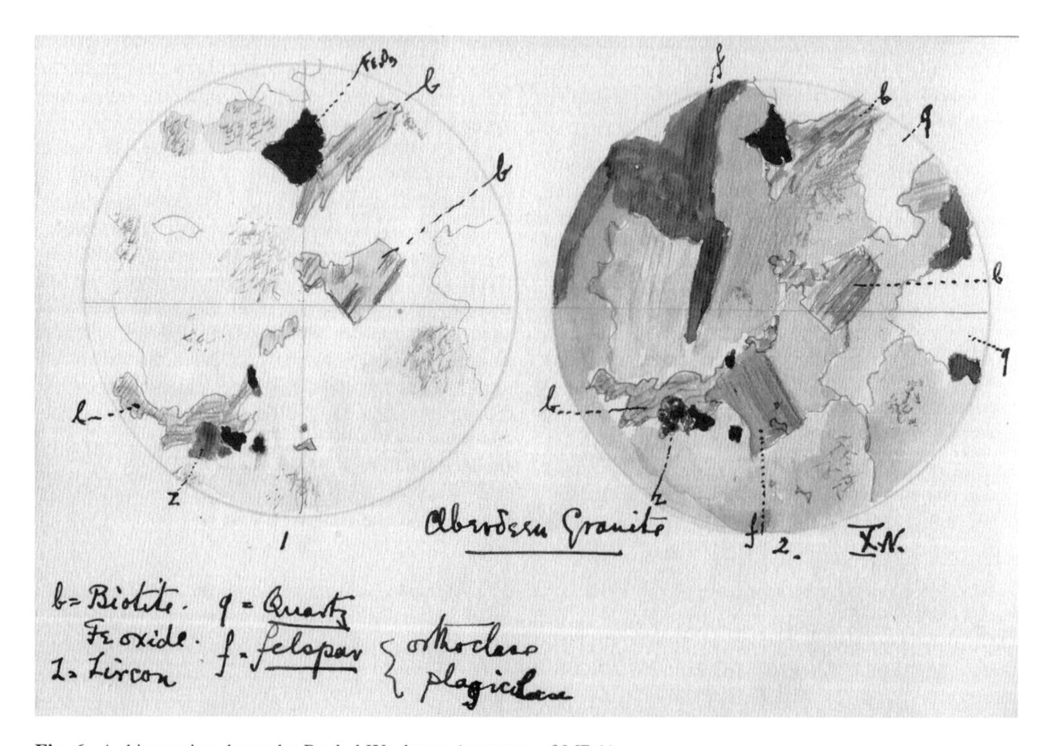

Fig. 6. A thin section drawn by Rachel Workman (courtesy of MRA).

Fig. 7. Rachel Workman's rock box showing samples from the Lake District.

paper to the one previously published in the *Geological Magazine* on a nepheline-syenite boulder dredged from the Atlantic (McRobert 1912). While in London, she attended an Annual General Meeting of the Geological Society. What happened there further illustrates her feisty nature:

> An attempt was made to eject me. The Secretary rushed up and said I was not a Fellow, so I explained this was through no fault of mine but the Society's and waved him aside and marched in. Then another man rushed up and gave Professor Watts as my host and he quite collapsed! They need not try any tricks with me because I am a women, I have always gone to the Annual Meetings and intend to do so if in London!
> (Millington 1972)

Workman joined the Geological Society as a Fellow in 1919 in the first batch of eight women; at 34 she was one of the youngest. Her equality training by her mother often showed its face and her next remarks emphasize how badly she thought women were treated:

> I am very much amused at the first list of 16 women admitted to the F.G.S. There are one or two notabilities the others merely wives. It is obvious why they were admitted at this juncture. They are badly needing additional subscriptions so the female subscriber has a financial value if none other. Poor downtrodden race! [23]

Her husband died of a heart attack in 1922 and she survived him by 32 years. She tragically lost all three of her pilot sons: the eldest, Alisdair, in a flying accident in 1938; and the two younger ones during World War II, within a few months of each other (Dodd 1980). In May 1941 Roderic, a flight lieutenant in the RAF, was killed fighting for Britain over Iraq. Less than a year later, his younger brother, Iain, a pilot officer in the RAF, was reported missing in action over the North Atlantic. Stricken, Lady MacRobert made what she called 'a mother's immediate reply'.[24] Enclosing a cheque for £25 000 to buy a Stirling bomber, she wrote to the Air Ministry:

> It is my wish, as a mother, to reply in a way my sons would applaud – attack with great fire power, head on and hard. They would be happy that their mother would avenge them and help to attack the enemy. I, therefore, feel that an appropriate name for the bomber would be the MacRobert's Reply.[25]

A year later, when Iain was confirmed dead, Lady MacRobert sent another cheque to buy four Hurricane fighters to be named after her three sons and

[23] MRA: MS Section 2 correspondence 9/5/52. 1919. Letter to Sir Alexander of Douneside from Rachael Workman.

[24] *Time* magazine, Monday 13 September 1954.

[25] Imperial College London: a hundred years of living science, p. 12. http://www3.imperial.ac.uk/pls/portallive/docs/1/18187720.PDF.

Fig. 8. Rachel Workman's (Lady MacRobert's) gravestone at Douneside, Aberdeenshire.

the fourth to be called the 'Lady', saying 'I have no more sons to give'.[26] Subsequently, she set up the MacRobert Trust and donated a vast fortune for it to look after former RAF pilots (Dodd 1980).

Her love of nature, and especially geology, grew as a result of her having few friends of her own age when she was young. She was always happiest with older people or alone with Nature. However, her generosity and belief in equality for women, even in times of war, was inspirational. She received a letter in 1942 from 'the medical women of Russia'. This letter highlights her strength in times of adversity and how others looked up to her:

> We women doctors of a Military Hospital from far off Leningrad, greet you, British mother-patriot, who by your bravery, have set an example of courage to the whole world. A mother who has lost three sons in the struggle against the brutal gang of Fascists is not simply a mother of three sons, but is a mother of three fighters who made the supreme sacrifice for us [...] and the right to live freely.[27]

Later in life her overriding hobby was breeding Aberdeen Angus cattle and gardening. On her death on 1 September 1954 she left over £500 000, having inherited her parent's fortune in America and her husband's business interests in India. With the passing of time, her geological studies waned as social work took over her life and her geological activities became part of her younger accomplishments. Lady MacRobert is buried in the garden in her beloved Douneside, Aberdeenshire (Fig. 8).

A note of caution: when Workman's husband became a baronet, just a few months before he died, he changed the spelling of his name from McRobert to MacRobert (Millington 1972). This can cause confusion regarding his wife's name on publications and letters, given first as Lady McRobert and later on as Lady MacRobert.

Mildred Blanche Robinson

Nothing is known of Mildred Robinson, except that she was a graduate with a BSc living in SW London at the time of her election as a Fellow of the Geological Society. Further research is needed.

Ethel Gertrude Skeat (Mrs Woods)

Ethel Gertrude Skeat (1865–1939) was the third daughter born to Professor Walter William Skeat (1835–1912), the Elrington and Bosworth Professor of Anglo-Saxon in Cambridge, and Bertha Jones. Ethel's sisters were Bertha and Margaret, her brothers, Walter and Arthur. She was educated from the age of 12 at Bateman House, a private school in Cambridge, until she left in 1885 at the age of 20. Dr John Marr and Professor McKenny Hughes, colleagues of her father at the University, probably kindled her interest in geology during this time (Burek & Malpas 2007). In 1885 she travelled south to St Leonards on Sea, East Sussex, to a private boarding school for a year until she was 21. At the age of 25 (1891) she went up to Cambridge to Newnham College, and in 1894 became the Arthur Hugh Clough Scholar. She completed the Natural Science Tripos certificate part 1, gaining a Class 1 at the age of 29, but without being awarded a degree. She joined the Geologists' Association in 1893 and while a student collaborated with her long-time friend, Margaret Crosfield, on their first paper on Wales (Carmarthen), which was published in the *Quarterly Journal of the Geological Society* (Skeat & Crosfield 1896). While at Newnham, in 1895–1897, she was elected to a Bathurst research studentship and went to the University of Munich, initially for two years, to work under Professor Karl A. von Zittel (1840–1904); the first female student to do so.[28] She had continuous associations with Newnham College from 1896 until her marriage in 1910, being first a scholar of Newnham College, and later an Associate (Burek & Malpas 2007).

In Munich she investigated the palaeontology of Jurassic Neocomian glacial boulders from Denmark with Dr Victor Madsen (1873–1941), who later had a glacier in Greenland named after him, and this led to an important paper in a Danish Geological Survey publication (Skeat & Madsen 1898). She also travelled to Paris, Geneva and Lausanne as a

[26] Cameron, E. 'Tragic heroine of Tarland' (Aberdeen) *Evening Express* article, 8 September 1998, p. 11.

[27] MRA: MS Section 4/19. Other printed material 19/4, pp. 54–58. Greetings from Russian Medical Women serving in a military hospital in Leningrad.

[28] Pers. comm. Kõlbl-Ebert 2007.

geological research worker. While in Munich she was the first woman to be admitted as a guest to scientific lectures at Munich University after a petition by Professor Zittel to the Senate.[29] Thus she must have paved the way for and may have met – after all, they were roughly the same age and both British – Maria Ogilvie, who received her doctorate in 1900. At this time it was unusual for women to be attending university to study geology and they would have found each other's company welcoming.

By 1898 Skeat (now 35 years old) was a science schoolmistress boarding at 68 Stanwell Road, Penarth, Glamorgan with a Mrs Emma Perry.[30] Here she met her lifelong friend Beatrice Elizabeth Clay, the Head Mistress of Seconder Public School.[31] From South Wales Ogilvie moved north to take the job of Second Mistress, Head of Science and Geography, and was in charge of the sixth form at Queen's School, Chester, where she started in January 1904.[32] Subjects she was responsible for were botany, geology, nature study, geography, chemistry and zoology, as listed on her original records at the school.[33] That she was an excellent teacher is in no doubt and in that respect she must have been responsible for instilling an interest in the natural sciences, and geology in particular, in a generation of girls 'not only by the excellent school reports, but – far better – by the enthusiasm she has evoked from her pupils. She has always entered, heart and soul, into the interests, the work and the play of the School, and she will be sorely missed'.[34] This is sadly missing in the present curriculum of Queen's School, as geology does not figure in the National Curriculum.

Skeat donated books to the sixth form library, such as books of poems by Wordsworth and Keats, and participated in the life of the city by joining The Chester Society of Natural Science, Literature and Arts in 1904–1909. She gave at least three talks to the Society in 1905 on 'Jurassic shorelines; or a fragment of world history', in 1909 on the 'Bernese Oberland' and in 1910 on 'Life's failures', each one illustrated with lantern-slides (Burek 2008*a*). On 7 April 1905 she gave a demonstration to schoolchildren at the Grosvenor Museum on recent exploration in Greenland, using materials lent to her by the Danish Authorities.

Interestingly, her affiliation for her talk on 14 April 1905 under the geological section of the society is listed as Assoc. Newnham Coll., Mem. Geol. Association, but by 31 March 1910 she has become Miss E. G. Skeat, DSc.[35] On 29 April 1905 Ethel Skeat received a doctorate from Trinity College, Dublin[36] in 'recognition of her contribution to geological research'.[37]

Skeat's research continued while she was teaching at Chester, and in recognition of her achievements she was awarded the Murchison Fund in 1908 (Table 4) by the Geological Society of London, an amount of £25 6s 4d. She was the third woman to receive this fund after Jane Donald and Elizabeth Gray.[38] The citation by Archibald Geikie said:

> The Council has this year awarded to you the Balance of the Proceeds of the Murchison Geological Fund as a mark of appreciation of your geological work, especially among the glacial deposits of Denmark and the Lower Palaeozoic rocks of Wales. It is with much gratification that we hail in you another woman who is worthily placed on the roll of those who have gained our awards.
>
> (Geikie 1908)

Skeat was congratulated by her colleagues at the Queen's School, where this unusual honour for a woman is stressed: 'Miss Skeat is to be congratulated on an honour rare in the case of a woman. She has been awarded the Murchison Fund by the Geological Society of London'.[39] In the roll call of Newnham College after her death in 1939, Gertrude Elles in her obituary erroneously states that Skeat was the first recipient of the Fund,

---

[29] Pers. comm. Kõlbl-Ebert 2007.

[30] 1901 Census.

[31] Clay became the executor of Skeat's will and it was to her that Skeat left all her jewellery and wearing apparel.

[32] QSC. 1904. Teacher registration form.

[33] QSC. 1904. Teacher registration form.

[34] QSC. *Have Mynde* (school newsletter).

[35] Chester Society of Natural Science, Literature and Art 39th annual report & proceedings,1909–1910, p. 9.

[36] Pers. comm. Bettie Higgs, 2006.

[37] QSC. *Have Mynde* (school newsletter), 1905.

[38] Elizabeth Gray (née Anderson) (1831–1924) is not discussed here since she was not a member of the Geological Society; however, she made an important contribution to Scottish geology and is regarded as one of the foremost Scottish fossil collectors. She received the Murchison Fund from the Geological Society of London in 1903 at the age of 72 for her lifetime contribution to the study of Ordovician and Silurian stratigraphy through her meticulous collecting, recording and labelling of her fossil collection. She was a great friend of Jane Donald and Charles Lapworth.

[39] QSC. *Have Mynde* (school newsletter), 1908.

furthermore, the Newnham College Register (volume 1) for the year 1891 incorrectly states that it was the Murchison *Medal* she received.

Skeat stayed in Chester until August 1910 when she moved back to Cambridge, the home of her birth, to marry Professor Henry Woods (1868–1952), the famous palaeontologist, whom she had met in Cambridge during her student days (Burek & Malpas 2007). He had entered St John's College in 1887 and become a demonstrator in 1892. She was wished well by all her friends and colleagues at Chester. In 1911 she became a lecturer at the Cambridge Training College for Women, later known as Hughes Hall, and remained there for two years. She returned after World War I as Registrar and Honorary Secretary, a post she held from 1919 until 1937, only a year before she died.

During the war her knowledge of German saw her working in the postal censorship code department. This was also the case for Maria Ogilvie Gordon. After the war, although carrying on with her research in North Wales with Margaret Crosfield, her interests expanded to include physical geography and she published *Principles of Geography: Physical and Human* in 1923, which was further revised in 1926 (Woods 1923). This was regarded as a model textbook at the time. She also published a book on the Baltic region after her experience of working in that area earlier in her life (Woods 1932).

Around the time Skeat moved to Chester she researched with her friend and colleague from Cambridge undergraduate days, Margaret Crosfield. This culminated in their publication of a paper in the *Quarterly Journal of the Geological Society* in 1925, which Skeat read on 7 May 1924 as Mrs Woods, entitled 'The Silurian rocks of the Central part of the Clwydian range' (Woods & Crosfield 1925). On 27 May 1926 Alfred Newstead, Curator and Librarian of the Chester Society, noted in his annual Curator and Librarian's report for 1925–1926 that 'Mrs. Ethel Woods and Miss Margaret Crosfield had donated their collection of graptolites to the Grosvenor Museum in Chester and a copy of their reprint on Silurian rocks'. It is a 'valuable addition' states Newstead[40] (Burek 2008*a*).

When Skeat died Elles wrote her obituary for the *Proceedings of the Geologists' Association* and Newnham College, referring to Crosfield's comment about her which stated: 'I should have liked to have given you some idea of the friendly, companionable, delightful, and equable partner she was' (Elles 1940), thus it is clear that all three knew each other.

Women elected on 25 June 1919

Catherine Alice Raisin

Catherine Alice Raisin (1855–1945) was born on 24 April 1855, the youngest child of Daniel Francis Raisin and Sarah Catherine Woodgate. Catherine had three brothers who were substantially older than herself as her mother was 45 when she was born. Raisin was educated at the North London Collegiate School, one of the oldest private schools for girls in Britain. Her geological education began at the age of 18 when she started classes at University College London, studying first geology and then mineralogy, the subject that was to become her particular research field. By 1878 London University had opened its doors more widely to women and there she studied under Thomas G. Bonney (1833–1923), also attending Thomas Huxley's (1825–1895) zoological lectures at the Royal School of Mines. In 1884 Raisin was awarded a BSc in both geology and zoology. She received a DSc in 1898.

Raisin published over 24 journal articles on metamorphic petrology during her academic life, mostly on the petrology of serpentines. Her first paper, in 1887, was read to the Geological Society of London by Bonney. At this time, she was working as a demonstrator in botany at Bedford College. Her work was followed up by Bonney and a pattern started to emerge where Raisin did the fieldwork and Bonney did the microscopy. Raisin spent her whole academic career at Bedford College, founded in 1849 as a girls' college. In 1890, she took over as Head of the Department of Geology, the following year being appointed Head of the Department of Botany as well. She became one of the most important early female geologists and her full story is told elsewhere (Burek 2003*b*, 2004, 2007).

Raisin was 64 years old when elected a Fellow of the Geological Society, having already served a lifetime of dedication to education and research. She was a brilliant teacher, and became the role model for the geologists Doris Reynolds (1899–1985), Helen Muir Wood (1895–1968) and Irene Lowe (b. 1890), all of whom she taught at Bedford College. In 1898 she became the first female Vice-Principal of a college, a position she held for three years, and in 1902 she was elected a Fellow of University College London. Raisin was the first woman to be awarded financial support from the Geological Society, receiving the Lyell Fund in 1893 for her research on metamorphism (Table 4). She was also one of only two women to be present

[40] Chester Society of Natural Science, Literature & Art. 55th Annual Report & Proceedings: Curator and Librarian's report for 1925–1926, p. 8.

at the Centenary celebrations of the Geological Society in 1907 as representatives of educational institutions, the other being Mary Andrews who represented Queen's College, Belfast. Her place next to the High Table on the seating plan is testimony to the esteem in which she was held. She served as a role model for research, teaching and administration throughout the last decade of the nineteenth century and the first two decades of the twentieth century. She would have been delighted to receive Fellowship of the Geological Society as she passionately believed in the education of women and equality for all. In her will she left £300 to the Bedford College Geography Department for an annual prize to be given to a student for good work who by default would have been female. This is still given today at Royal Holloway College. She also left £500 to be used for the benefit of non-smoking students (Burek 2007). She had been a member of the Geologists' Association for 67 years (Creese 2004), attending their lectures until 1932 (Burek 2003*b*).

As Raisin approached her final years, she increasingly became an invalid and for the last three years of her life was confined to one room. She finally died of cancer at the age of 90 in Ash Prior's Nursing Home, Cheltenham, on 12 July 1945. In her obituary of Raisin in *Nature* (1945) Doris Reynolds summed up her mentor as 'not only a stimulating and enthusiastic teacher, who worked ungrudgingly to promote their interests, but also as a generous, brave and sympathetic woman whom they loved'.

Margaret Flowerdew MacPhee (Mrs Romanes)

Margaret MacPhee (1886–1931) was born in Helensburgh, Scotland, on the edge of the Highlands. She entered Newnham College Cambridge in 1905 to join the School of Geology and was awarded the Women's Harkness Scholarship for proficiency in 1909. She left Cambridge to work in Glasgow University as a demonstrator. Here she published a paper on 'An algal limestone from Angola' in the *Transactions of the Royal Society of Edinburgh* which she read in June 1915 just before she was married. But since it was not published until 1916, she used her married name, Romanes, in publication. She then started work on the distribution of graptolites in the Riccarton Beds of Southern Scotland, but this research was never completed, perhaps due to her change in status. Her husband, James Romanes who had been a contemporary of hers at Cambridge, travelled overseas a great deal and this may have led to her inability to carry on the work as she had to look after the five children – two boys and three girls (A. B. T. 1943). MacPhee was always willing and able to go out into the field to accompany others (Elles 1932). She had a critical eye and her advice was always readily received. She remained a frequent attendant at the Geological Society, which she joined when the family were living in West Byfleet. Margaret was then 34 years. She remained friends with many of her fellow Cambridge acquaintances and, indeed, it was Gertrude Elles who wrote her obituary for the anniversary address of the President of the Geological Society in 1932 (Elles 1932). She died on August 24 1931, only 45 years old and not having reached her full potential.

Women elected on 3 December 1919

Mary Kingdon Heslop

Born and educated in Egypt, Mary Heslop (1885–1955) moved to Newcastle in England to start her academic scientific training. She graduated from Armstrong College, part of the University of Durham, in 1906 with a degree in physics and geology. She was a research fellow there until 1909, receiving an MSc for her work on igneous petrology (J. A. S. 1955). She published two academic papers on this in 1910 and 1912. Heslop was a pioneer in the use of colour photomicrography – the process of photographing images through a microscope. She became a demonstrator at Newcastle after which she moved to Bedford College, London, under Catherine Raisin (Burek 2003*b*, 2007). However, owing to a lack of opportunities for female geologists at this time she changed her direction towards that of geography and undertook a postgraduate diploma in Oxford in 1916. Heslop then taught full-time at Newcastle, the University of Leeds and Kenton Lodge training college, Newcastle, a post she held for 27 years. She was an excellent teacher. At the time of her election as a fellow of the Geological Society, she was teaching geography in Newcastle. She died in 1955 after a lifetime of service to education.

Dorothy Margaret Woodhead

Nothing is known of Dorothy Woodhead, other than she lived in Dartford, Kent at the time of her election as Fellow and was a graduate with a BA.

Women elected in 1920

Florence Pitts

Little is known about Florence Pitts other than she lived in Dorking in Surrey at the time of her election. She does not appear to have been a graduate.

Wynne Edwards (Mrs Eleanor Mary Reid)

Eleanor Mary Reid (1860–1953) was born Wynne Edwards in Denbigh, North Wales. She was educated at Westfield College, London, where she obtained a BSc in 1892, while teaching maths at Cheltenham Women College under a Miss Beale. In 1897 she married a member of the Geological Survey, Clement Reid (1853–1916). Between 1907 and 1915 she published 12 joint papers with her husband on fossil Quaternary and Tertiary plants and seeds, in one case, in 1908, doubling the number of species recognized from the Cromer Forest Bed. This was published by the Linnaean Society (Reid & Reid 1908). In 1919 she was awarded the Murchison Fund for her work in palaeobotany: 'The Council has awarded to Mrs. Eleanor Reid the Murchison Fund, in appreciation of the service which she has rendered by her prolonged and skilful investigation of Tertiary and Pleistocene plant remains' (Lapworth 1920).

After she became a Fellow of the Geological Society, she collaborated closely with Miss Marjorie Chandler, publishing the *Monograph on the London Clay Flora* in 1933. She received the Lyell Medal from the Geological Society in recognition of her work in 1936. She was 76 years old:

> At the time when, in cooperation with your honoured husband, you began to study fossil plants, you realised the importance of an accurate knowledge of fruits and seeds, then very imperfectly understood. Commencing with a series of papers on the Holocene and Pleistocene plants, you passed on to the Pliocene, and especially worthy of mention are the important joint monograph on the Pliocene floras of the Dutch–Prussian Border and your review of Pliocene floras published in our Journal in 1920. [...] Meanwhile, amid much work on the earlier Tertiary plants, you applied your mathematical knowledge to their statistical distribution. Steadily advancing, in the true spirit of science, from the known to the unknown, in 1933 you were responsible with Miss Chandler, for the great memoir on the London Clay flora, one of the most valuable contributions to palaeobotany ever written.
>
> (Green 1936)

This presentation by the President of the Geological Society, John Green, shows the high esteem in which Eleanor Reid was held.

Following her husband's death in 1916, she lived on for a further 40 years undertaking palaeobotanical research (Reid 1919, 1920*a*, *b*; Reid & Chandler 1933). She presented specimens of her Pliocene seeds and fruits to Kew and to the British Museum Natural History in 1937, at the age of 77. She worked well into her 80s and her material was used for the *Lower Tertiary Floras of Southern Britain* from 1960 to 1964. At 88 she gave up cycling and eventually died peacefully in her bed at the age of 93.

Ethel Reader Wood (Dame Shakespear)

Ethel Reader Wood (1871–1945) was born on 17 July 1871 at Biddenham, near Bedford. She attended Bedford High School and went up to Newnham College, Cambridge, in 1891, the same year as Ethel Skeat and Gertrude Elles (Newnham College 1923), where she obtained a First Class degree specializing in geology. However, rather like Maria Ogilvie Gordon, she nearly became a professional pianist, geology only just winning out in the end. Her lifelong friendship with Gertrude Elles began in those hallowed halls of Newnham College. They both undertook a further year of research on the rocks of Conway in North Wales for Dr Marr, subsequently published in the *Geological Magazine* (Elles & Wood 1895). After that, in 1896, Ethel Wood moved to Birmingham University to work as a research assistant to Professor Lapworth. She held this post until her marriage in 1906 to Gilbert Shakespear, who had been a fellow geology student at Cambridge and a fellow lecturer in physics at Birmingham.

Her research led to three outstanding papers on aspects of the ecological context of graptolites and how they could be used to decipher the stratigraphy of North Wales and the Welsh Borders (Wood 1900, 1906; Elles & Wood 1901–1918). Her first paper on the Ludlow Formation and its graptolites showed the value of using the fauna to classify an otherwise difficult sequence of mudstones in the Lower Ludlow. Her second paper on the Tarannon Series demonstrated for the first time that these beds were merely a graptolite facies of the Upper Llandovery. Both these papers were communicated to the Society by Lapworth, and then published in the *Quarterly Journal*; they were of the highest quality and the first resulted in her being awarded the Wollaston Fund in 1904. It was:

> an acknowledgement of the value of her contributions to our knowledge of the Graptolites and of the rocks in which these organisms occur. Her papers furnish an excellent example of the application of zonal stratigraphy to groups of rocks [...]. We had looked forward with pleasure to seeing her among us here today; but she has been unavoidably prevented from coming to London.
>
> (Lapworth 1904)

Wood's last paper was published by the Palaeontological Society as one of its monographs. In 1920, as well as being awarded the DBE and becoming a Dame,[41] she was presented with the Murchison

[41] In 1918, she had previously been awarded the MBE for her services to the country.

Medal, together with the sum of 10 guineas from the Murchison Geological Fund for:

> the quality and promise of your work on the Graptolites and Graptolite-bearing rocks of this country [...]. Your special knowledge has also been most usefully applied to the examination of fossils of this class obtained from distant countries. Fortified with your palaeontological knowledge, your fieldwork on the Silurian rocks of the Welsh borderlands, has taught us much regarding the sequence and correlation of this difficult system.
> (Lamplugh 1920*a*)

That same year Wood was elected a Fellow of the Geological Society at the age of 47.

Like two other first female Fellows, Marie Stopes and Rachel MacRobert, Wood went on to gain national recognition not for her geological work but for her social activities, specifically her efforts during World War I. She was a co-founder of the War Pensions Committee, as she was so moved by seeing the plight of soldiers returning after World War I (Elles 1946*a*). This was recognized by President Lamplugh when he presented her with the Murchison medal in 1920:

> We recognise that the strenuous service which you have been doing for the State during and since the War must take precedence; but may we hope that you will, when possible, resume the studies which have proved so profitable to us? In handing you this Medal and Award on behalf of the Council, I ask you to accept also my personal congratulations on your success in advancing geological science.
> (Lamplugh 1920*a*).

An accolade indeed, only a year after women had been admitted as Fellows. However, this wish was not to be, as in 1922 she became a Justice of the Peace for Birmingham and her social activities took over. Wood was a great advocate of equality for women and became President of the Birmingham branch of both the National Council of Women and of the British Federation of University Women, an organization dedicated to helping women in education. She was also concerned about the welfare of girls and women to whom life had given some hard knocks (Elles 1947); having lost her only child in infancy, she felt considerable empathy with them.

In 1929 Wood moved with her husband to Worcestershire and took up farming, which was extremely strenuous during World War II. The long hours and hard work 'wore her out, and when illness struck her, having given all she had to others, she had little strength left to meet it' (Elles 1946*b*). Wood resigned her membership of the Geological Society in April 1942 and died less than four years later at the age of 74. Her funeral in Birmingham filled the church and graveyard with people from all walks of life – from councillors to magistrates, from the civic to the university authorities, from rich to poor girls. They had come to pay their last tribute to a 'wonderful friend'; this was testimony to their respect for her (Elles 1947).

Lucy Ormrod

Nothing is known of Lucy Ormrod other than she lived in Newport, Monmouthshire, was a spinster at the time of her election to the Geological Society and that she was a graduate with a BA.

There were no female Fellows elected during 1921.

Women elected in 1922

In February of 1922 the first of a new batch of women was elected Fellows. Irene Helen Lowe from London (b. 1890, died in India after 1931), a student of Catherine Raisin's and a demonstrator at Bedford College from 1914 to 1919, gained an MSc in 1920. She was appointed an examiner in geology for the University of London in 1920. In March 1922 Edith Goodyear (1879–1959), who by now was senior assistant in the Geology Department of University College and a graduate, and Helen Marguerite Muir-Wood (1896–1968), who had also studied under Catherine Raisin at Bedford College and obtained a MSc, were elected to fellowship. Isabel Ellie Knaggs was also elected at this time, but she appears not to have had a degree and nothing further is known about her.

In June 1992 Marie Carmichael Stopes (1880–1958) was elected a Fellow. Her controversial work on sex within marriage had already been published, and she was just publishing her last two papers on coal, jointly with Robert Wheeler of the Government Department of Scientific and Industrial Research Laboratory: *The terminology in coal research* and *The spontaneous combustion of coal* (Stopes & Wheeler 1923*a*, *b*; Chaloner 1958, 1995, 2008; Watson 2005).

So why did the first early female Fellows join?

After these first 21 women had been elected, the number of female Fellows rose to, and stayed within, the 30s until after World War II. By 1967 the total number was still only just over 100, out of a total membership of 3000 (Wallace 1969), around three percent. Up until that time the total number of new female Fellows often equalled the number leaving due either to marriage and an inability to carry on with their work, or to their being unable to afford the subscription, especially after the substantial rise of 1957. In 2007, the year of the Society's bicentenary, the total number of

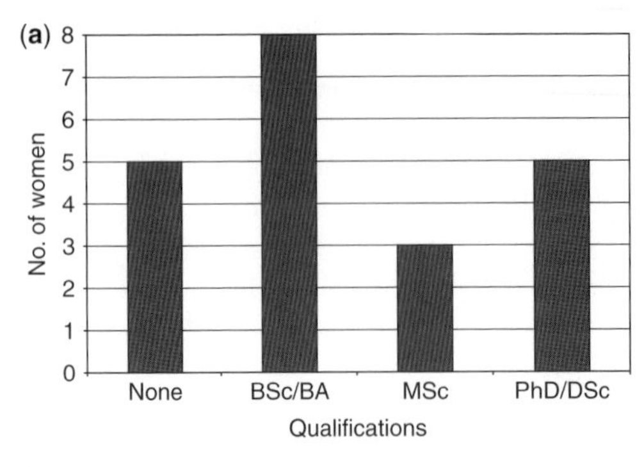

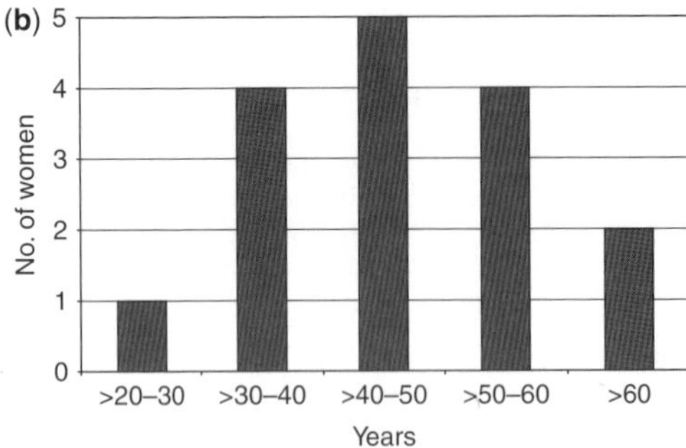

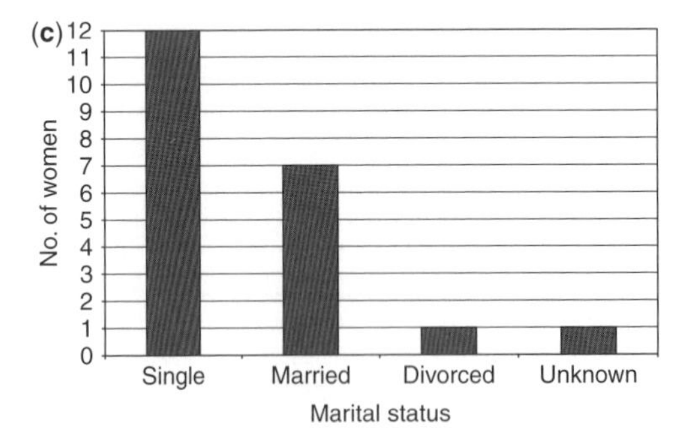

Fig. 9. (**a**) Educational qualifications, (**b**) age and (**c**) marital status of the first 21 female Fellows of the Geological Society of London.

women Fellows stands at 1705 out of a total of 9370, around 18 percent (Geological Society, pers. comm.).

So were there any common denominators between the first 21 female Fellows of the Geological Society at the time of their election? Examination of some of the factors that might have influenced the reasons why these women joined the Geological Society – their educational achievements (Fig. 9a), place of residence at the time of their election (Table 6), their age (Fig. 9b) and their marital status (Fig. 9c) suggests there is no common thread between them, other than their love of geology. It might be assumed that to join a professional organization like the Geological Society required an academic degree, but that does not seem to have been the case. Proximity to London and access to meetings might have been the reason why some joined, but, with several Fellows being located many miles from the capital, this was clearly not the case for them all, although Maria Ogilvie Gordon while listing Aberdeen as her domicile actually did have a flat in London. Interestingly, age was clearly not a factor in determining whether or not a woman joined, for the ages ranged from 27 (Helen Muir-Wood) to Catherine Raisin at 64. Furthermore, the fact that several were married did not prevent them being Fellows, unlike in the Civil Service at that time where women were prohibited from working once they were married, or in the Royal Society, as previously discussed.

Like their male colleagues, fellowship of the Geological Society gave them professional status and initially many of these women may have joined just in order to read their own papers, rather than have others read them on their behalf (probably Rachel Workman). Prior to 1919 they had attended meetings to learn and share their opinions and ideas with others and, undoubtedly, this was also an inducement for them after 1919. However, unlike the Geologists' Association where women had been admitted from its inception and where they held many officer posts (Burek 2008*a*), women did not hold any official positions within the Geological Society until 1982. Indeed, there has only ever been two females to hold such positions: Professor Janet Vida Watson was President from 1982 to 1984, and Professor Lynne Frostick was Secretary between 1988 and 1991. Other women have been on the Council but have not held any of the Officer positions.

While many of the women such as Catherine Raisin, Gertrude Elles, Rachel Workman and Ethel Skeat actively encouraged equality in higher education and some, like Margaret Crosfield, supported the suffragette cause, very few of the first female Fellows stayed professionally within their geological pursuits. Nevertheless, the training they undertook stood them in good stead for endeavours in other areas and many accomplished fame, if not fortune, in different walks of life. Dr Marie Stopes, Dr Maria Ogilvie Gordon and Dame Ethel Shakespear are good examples of this. Geology was, and still is, a very male-dominated field of study and initially they had no female role models to follow.

Conclusions

It took 112 years for the first women to be admitted to the Geological Society, although some debate in the late nineteenth century preceded their election in 1919. They came from all walks of life, from a variety of professions and some had the privilege of being able to be interested in geology just for the love of it. The importance of being one of the first females to join the Society was recognized by geologists and non-geologists alike. Furthermore, the fact that many of their obituarists mention it many years after it happened – for some as much as 40 years later – shows the impact that it made and the lasting impression that it generated:

- Mary Johnston: 'She was one of the first women admitted to Fellowship of the Geological Society of London' (A. L. L. 1956);
- Mary Kingdon Heslop: 'She was elected a Fellow of the Geological Society in December 1919, being among the earliest of the women to receive that honour' (J. A. S. 1955);
- Margaret Crosfield: 'In 1919, Miss Crosfield had the pleasure of being elected the first woman Fellow of the Geological Society' (Johnston 1953);
- Rachel Workman: 'She was one of the first eight women to be elected Fellows of the Geological Society of London on 21 May 1919' (W. C. S. 1955).

Through their publications in *the Quarterly Journal of the Geological Society*, these women contributed substantially to graptolite and gastropoda classification, palaeobotany, geology of the Tyrol, petrology of metamorphic rocks, especially serpentine, and to stratigraphy across a wide range of areas, particularly in Wales.

Many of these women provided an inspiration for the female Fellows who followed them, right up to the present day, when we have only the second female President of the Society, Professor Lynne Frostick. Some of the first female Fellows have disappeared into obscurity, but their presence at those meetings of the Geological Society marked a watershed in attitudes towards professional female geologists. This recognition of equality must continue into the foreseeable future for all our sakes.

First, I would like to acknowledge the time and trouble my two referees, Dr Bettie Higgs and Dr Martina Kölbl-Ebert, and my editor Dr Cherry Lewis, gave to improving this paper, which was presented at the Founding Fathers Bicentenary conference. It developed from an idea following the History of Geology Group's (HOGG) conference at the Geological Society on the Role of Women in the History of Geology in November 2005. I would also like to thank the Sedgwick Museum; Jon Clothworthy of the Lapworth Museum; The British Geological Survey; archivists from Royal Holloway University of London; Penny Hartley, secretary of the Robert Gordon School; Group Captain Mike Sweeney of Douneside House and the MacRobert Trust archivist; secretary of the Queen's school, Chester; and the Library of the Geological Society, especially Wendy Cawthorne. Without their help and access to archives, much of this research would not have materialized and the paper would have been the poorer for it.

The paper is dedicated to all female Fellows of the Geological Society, including my fellow undergraduate colleague Lynne Frostick, to whom I wish every success in her Presidency, but most especially my two daughters, Frances Elizabeth Cubitt FGS and Veronica Louise Caddy (née Cubitt) FGS, who represent the future of female fellowship in the Society.

Archives

SMA: Sedgwick Museum archive, Cambridge University.

GSL: Geological Society of London archive.

BGS: British Geological Survey Palaeontological Archive, Keyworth.

LA: Lapworth Archive, Lapworth Museum, University of Birmingham.

MRA: MacRobert Archive, Tarland, Aberdeenshire, Scotland.

QSC: Queen's School, Chester, archive.

References

A. B. T. 1943. Obituary notices: James Romanes. *Proceedings of the Geological Society*, **xcix**, lxxxviii–lxxxix.

A. L. L. 1956. Annual report of the Council. *Geologists' Association*, **1955**, 197–198.

ANON. 1919. Miss M. S. Johnston: Flint artefacts from Marlborough Downs, The Association's Albums of geological photographs, List of exhibitors, November 7th 1919, report of the session. *Proceedings of the Geologists' Association*, **30**, 15.

BOLTON, E. & TUCK, M. C. 1922. The carboniferous limestone of the Wickwar–Chipping Sodbury area (Gloucestershire). *Quarterly Journal of the Geological Society*, **77**, i–ci.

BONNEY, T. G. & RAISIN, C. A. 1899. On varieties of serpentine and associated rocks in Anglesey. *Quarterly Journal of the Geological Society*, **55**, 276–304.

BONNEY, T. G. & RAISIN, C. A. 1905. The microscopic structure of minerals forming serpentine and their relationship to its history. *Quarterly Journal of the Geological Society*, **61**, 690–715.

BOYLAN, P. J. 2009. The Geological Society and its official recognition, 1824–1828. *In*: LEWIS, C. L. E. & KNELL, S. J. (eds) *The Making of the Geological Society of London*. Geological Society, London, Special Publications, **317**, 319–330.

BUREK, C. V. 2001. The first lady geologist or collector par excellence? Women in the history of geology II. *Geology Today*, **17**(5), 192–194.

BUREK, C. V. 2002. Etheldred Benett (1776–1845). *Journal of the Cork Geological Association*, **5**, 22–23.

BUREK, C. V. 2003*a*. Gertrude Elles. *In*: LERNER, K. L. & LERNER, B. W. (eds) *World of Earth Science*, Gale Cengage, Farmington Hills, MI.

BUREK, C. V. 2003*b*. Catherine Raisin, a role-model professional geologist. *Women in the History of Geology III. Geology Today*, **19**(3), 107–111.

BUREK, C. V. 2004. Raisin, Catherine Alice (1855–1945). *In*: LIGHTMAN, B. (ed.) *Dictionary of Nineteenth*

Century British Scientists (4 vols). Bristol Thoemmes Continuum Press, Bristol.

BUREK, C. V. 2005. Who were they? The lives of geologists 5: Dame Maria Matilda Ogilvie Gordon – a Britisher – and a woman at that (1864–1939). *Teaching Earth Science*, **30**, 42–44.

BUREK, C. V. 2007. The role of women in geological higher education – Bedford College, London (Catherine Raisin) and Newnham College, Cambridge, UK. *In*: BUREK, C. V. & HIGGS, B. (eds) *The Role of Women in the History of Geology*. Geological Society, London, Special Publications, **281**, 9–38.

BUREK, C. V. 2008*a*. The role of the voluntary sector in the evolving geoconservation movement. *In*: BUREK, C. V. & PROSSER, C. D. (eds) *The History of Geoconservation*. Geological Society, London, Special Publications, **300**, 61–89.

BUREK, C. V. 2008*b*. The role women have played in developing the science of geology 1797 to 1918–1919 in Britain. *Open University Geological Society Journal*, **29**, 18–25.

BUREK, C. V. & HIGGS, B. (eds). 2007. *The Role of Women in the History of Geology*. Geological Society, London, Special Publications, **281**.

BUREK, C. V. & MALPAS, J. A. 2007. Rediscovering and conserving the Lower Palaeozoic 'treasures' of Ethel Woods (née Skeat) and Margaret Crosfield in northeast Wales. *In*: BUREK, C. V. & HIGGS, B. (eds) *The Role of Women in the History of Geology*. Geological Society, London, Special Publications, **281**, 203–226.

CHANDLER, M. E. J. 1921. The Arctic Flora of the Cam Valley, at Barnwell, Cambridge. *Quarterly Journal of the Geological Society*, **77**, 4–22.

CHANDLER, M. E. J. 1923. The geological history of the Genus *Stratiotes*: An account of the evolutionary changes which have occurred within the Genus during the Tertiary & Quaternary Eras. *Quarterly Journal of the Geological Society*, **79**, 117–138.

CHALONER, W. G. 1958. Obituary, Dr. Marie Stopes. *Proceedings of the Geological Association*, **70**, 118–120.

CHALONER, W. G. 1995. Marie Stopes (1880–1958): The American connection. *In*: LYONS, P. C., MOREY, E. D. & WAGNER, R. H. (eds) *Historical Perspectives of Early Twentieth Century Carboniferous Paleobotany in North America*. Geological Society of America, Memoirs, **185**, 127–134.

CHALONER, W. G. 2005. The Palaeobotanical work of Marie Stopes. *In*: BOWDEN, A. J., BUREK, C. V. & WILDING, R. (eds) *History of Palaeobotany Selected Essays*, Geological Society, London, Special Publications, **241**, 127–135.

CHALONER, W. G. 2008. Marie Stopes—palaeobotanist. *Open University Geological Society Journal*, **29**, 26–30.

COX, L. R. 1935. Mrs Mary Jane Longstaff, Obituary notices. *Proceedings of the Quarterly Journal of the Geological Society*, **91**, xcvii–xcviii.

CREESE, M. R. S. 1996. Maria Ogilvie Gordon (1864–1939). *Earth Sciences History*, **15**, 68–75.

CREESE, M. R. S. 2004. Raisin, Catherine Alice (1855–1945). *Oxford Dictionary of National Biography Online*. Oxford University Press.

CREESE, M. R. S. & CREESE, T. M. 1994. British women who contributed to research in the geological sciences in the nineteenth century. *British Journal of the History of Science*, **27**, 23–54.

CREESE, M. R. S. & CREESE, T. M. 2006. British women who contributed to research in the geological sciences in the nineteenth century. *Proceedings of the Geologists' Association*, **117**, 53–85.

DODD, F. L. 1980. The MacRobert Story. *Gordonian*, **1980**, 161–163.

DONALD, M. J. 1885. Notes on some Carboniferous Gasteropoda from Penton and elsewhere. *Transactions of the Cumberland Association*, **11**, 150–151.

DONALD, M. J. 1887. Notes upon some Carboniferous species of *Murchisonia*. *Quarterly Journal of the Geological Society*, **43**, 617–631.

DONALD, M. J. 1889. Descriptions of some new species of Carboniferous Gastropoda. *Quarterly Journal of the Geological Society*, **45**, 619–625.

DONALD, M. J. 1892. Notes on some new and little-known species of Carboniferous *Murchisonia*. *Quarterly Journal of the Geological Society*, **48**, 562–582.

DONALD, M. J. 1895. Notes on the Genus *Murchisonia* and its allies: With a revision of the British Carboniferous species, and description of some new forms. *Quarterly Journal of the Geological Society*, **51**, 210–234.

DONALD, M. J. 1898*a*. Observations on the Genus *Aclisina*, de Kon., with descriptions of British species, and of some other Carboniferous Gastropods. *Quarterly Journal of the Geological Society*, **54**, 45–72.

DONALD, M. J. 1898*b*. Award of the Murchison Geological Fund reply. *Proceedings of the Geological Society*, **54**, xliv.

DONALD, M. J. 1899. Remarks on the genera *Ectomaria*, Koken and *Hormotoma*, Slater, with descriptions of British species. *Quarterly Journal of the Geological Society*, **55**, 251–272.

DONALD, M. J. 1902. On some of the Proterozoic Gastropoda which have been referred to *Murchisonia* and *Pleurotomaria* with descriptions of new subgenera and species. *Quarterly Journal of the Geological Society*, **58**, 313–339.

DONALD, M. J. 1905*a*. Observations on some of the Loxonematidae, with descriptions of two new species. *Quarterly Journal of the Geological Society*, **61**, 564–566.

DONALD, M. J. 1905*b*. On some gastropoda from the Silurian rocks of Llangadock (Caemarthenshire). *Quarterly Journal of the Geological Society*, **61**, 567–578.

DONALD, M. J. 1906. Notes on the genera Omospira, Lophespira and Turritoma: With descriptions of the new Proterozoic species. *Quarterly Journal of the Geological Society*, **62**, 552–572.

DREW, H. & SLATER, I. L. 1910. Notes on the geology of the district around Llansawel, Carmarthenshire. *Quarterly Journal of the Geological Society*, **66**, 402–419.

ELLES, G. L. 1897. The sub genera Petalgraptus & Cephalograptus. *Quarterly Journal of the Geological Society*, **53**, 186–212.

ELLES, G. L. 1898. The graptolite fauna of the Skiddaw Slates. *Quarterly Journal of the Geological Society*, **54**, 463–539.

ELLES, G. L. 1900. Zonal classification of Wenlock Shales of the Welsh borderland. *Quarterly Journal of the Geological Society*, **56**, 370–414.

ELLES, G. L. 1922. The Bala Country: Its structure and rock succession. *Quarterly Journal of the Geological Society*, **78**, 132–175.

ELLES, G. L. 1932. Obituary: Margaret Flowerdew Romanes (née MacPhee), Anniversary address. *Proceedings of the Geological Society*, part 3, lxxxiii.

ELLES, G. L. 1940. Obituary: Dr. Ethel Gertrude Woods. *Proceedings of the Quarterly Journal of the Geological Society*, **xcv**, cviii–cix.

ELLES, G. L. 1946*a*. Obituaries: Dame Ethel Shakespear D.B.E. *Nature*, **157**, 256–257.

ELLES, G. L. 1946*b*. Obituary notices: Dame Ethel Mary Reader Shakespear (née Wood). *Proceedings of the Quarterly Journal of the Geological Society*, **cii**, xlvi–xlvii.

ELLES, G. L. 1947. Dame Ethel Shakespear J. P. D.Sc (Birm). *Newnham College Roll Letter*, 47–49.

ELLES, G. L. & SLATER, I. L. 1906. The highest Silurian rocks of the Ludlow district. *Quarterly Journal of the Geological Society*, **62**, 195–222.

ELLES, G. L. & WOOD, E. M. R. 1895. Supplementary notes on Drygill Shales. *Geological Magazine*, **2**, 216–249.

ELLES, G. L. & WOOD, E. M. R. 1896. On the Llandovery and associated rocks of Conway (North Wales). *Quarterly Journal of the Geological Society*, **52**, 273–288.

ELLES, G. L. & WOOD, E. M. R. (Mrs Shakespear) 1901–1918. *Monograph of British Graptolites*. Palaeontological Society, London.

GARDINER, M. I. 1888. The Green sand bed at the base of the Thanet Sand. *Quarterly Journal of the Geological Society*, **44**, 755–760.

GARDINER, M. I. 1890. Contact-alteration near New Galloway. *Quarterly Journal of the Geological Society*, **46**, 569–581.

GARWOOD, E. J. 1907. Discussion in Stopes M. 1907. The Flora of the Inferior Oolite of Brora, Sutherland. *Quarterly Journal of the Geological Society*, **63**, 382.

GARWOOD, E. J. & GOODYEAR, E. 1918. On the geology of the Old Radnor District, with special reference to an algal development in the Woolhope Limestone. *Quarterly Journal of the Geological Society*, **74**, 1–30.

GEIKIE, A. 1908. Award of the Murchison Geological Fund. *Proceedings of the Quarterly Journal of the Geological Society of London*, **64**, xlviii.

GOODCHILD, J. G. 1887. Discussion in Donald M. J. 1887: Notes upon some Carboniferous species of *Murchisonia*. *Quarterly Journal of the Geological Society*, **43**, 631.

GORDON, M. M. & OGILVIE, 1908. *Handbook of Employment Specially Prepared for the Use of Boys and Girls on Entering the Trades, Industries and Professions*. Rosemount Press, Aberdeen.

GOULD, S. J. 1998. *Leonardo's Mountain of Clams and Diet of Worms*. Harmony Books, New York.

GRAHAM, M. 1823. An account of some effects of the late Earthquake in Chile. Extracted from a Letter to Henry Warburton, Esq V.P.G.S. *Transactions of the Geological Society, London*, series II, **1**, 413–415.

GREEN, C. P. 2008. The Geologists' Association and geoconservation – history and achievements, *In*: BUREK, C. V. & PROSSER, C. (eds) *The History of Geoconservation*. Geological Society, London, Special Publications, **300**, 91–102.

GREEN, J. 1936. Award of Lyell Medal. *Proceedings of the Geological Society*, **92**, liv–lv.

H. D. T. 1955. Obituary Notices: Miss Mary Sophia Johnston. *Proceedings of the Geological Society*, **1529**, 141–142.

HEALEY, M. 1904. Notes on Upper Jurassic Ammonites with special reference to Specimens in University Museum Oxford. *Quarterly Journal of the Geological Society*, **60**, 54–64.

HERRIES DAVIES, G. L. 2007. *Whatever is Under the Earth. The Geological Society of London 1807 to 2007*. Geological Society, London.

HICKS, H. 1898. Anniversary Meeting Award of the Murchison Geological Fund. *Proceedings of the Quarterly Journal of the Geological Society*, **54**, xliii.

HIGGS, B. & WYSE JACKSON, P. N. 2007. The role of women in the history of geological studies in Ireland. *In*: BUREK, C. V. & HIGGS, B. (eds) *The Role of Women in the History of Geology*. Geological Society, London, Special Publications, **281**, 137–154.

HODGSON, E. 1863. On a deposit containing *Diatomaceae*, leaves etc in the Iron ore Mines Dalton-in Furness, near Ulverston. *Quarterly Journal of the Geological Society*, **19**, 19–35.

J. A. S. 1955. Obituary Notices: Mary Kingdon Heslop. *Proceedings of the Geological Society*, **1529**, 139–140.

JOHNSTON, M. S. 1901. Some geological notes on Central France. *Geological Magazine*, **8**, 59–65.

JOHNSTON, M. S. 1927. Geological notes on Spain and Majorca. *Proceedings of the Liverpool Geological Society*, **14**, 340–342.

JOHNSTON, M. S. 1953. Margaret Chorley Crosfield. *Annual Report of the Council Geologists' Association*, **1953**, 62–63.

JOHNSTON, M. S. & CROSFIELD, M. C. 1914. A study of Ballstone and the Associated Bed in the Wenlock Limestone of Shropshire. *Proceedings of the Geologists' Association*, **25**, 193–224.

KNELL, S. J. 2009. The road to Smith: how the Geological Society came to possess English geology. *In*: LEWIS, C. L. E. & KNELL, S. J. (eds) *The Making of the Geological Society of London*. Geological Society, London, Special Publications, **317**, 1–47.

KÖLBL-EBERT, M. 1999. Observing Orogeny – Maria Graham's account of the earthquake in Chile in 1822. *Episodes*, **22**(1), 36–40.

KÖLBL-EBERT, M. 2001. On the origin of women geologists by means of social selection: German and British comparisons. *Episodes*, **24**, 182–193.

KÖLBL-EBERT, M. 2002. British geology in the early 19th century: a conglomerate with a female matrix. *Earth Sciences History*, **21**, 3–25.

KÖLBL-EBERT, M. 2007. The role of British and German women in early 19th century geology: A comparative assessment. *In*: BUREK, C. V. & HIGGS, B. (eds)

The Role of Women in the History of Geology. Geological Society, London, Special Publications, **281**, 155–164.

LAMPLUGH, G. W. 1920*a*. Report of the Council for 1919. *Annual Report of the Geological Society*, Part 1, **xi**, xlii–xliii.

LAMPLUGH, G. W. 1920*b*. Minutes of March 26th, 1919. *Proceedings of the Quarterly Journal of the Geological Society*, **lxxv**, (part 1), xcvii–xcviii.

LAMPLUGH, G. W. 1920*c*. Minutes of May 21st, 1919. *Proceedings of the Quarterly Journal of the Geological Society*, **lxxv**, (part 1), ci.

LAMPLUGH, G. W. 1920*d*. Award of the Murchison Medal. *Proceedings of the Quarterly Journal of the Geological Society*, **lxxv**, xliv.

LAPWORTH, C. 1904. Award of the Wollaston Donation-Fund. *Proceedings of the Quarterly Journal of the Geological Society*, **60**, xlv–xlvi.

LAPWORTH, C. 1920. Award from the Murchison Fund. *Proceedings of the Quarterly Journal of the Geological Society*, **75**(part 1), xlix.

LEWIS, C. L. E. 2009. Doctoring geology: the medical origins of the Geological Society. *In*: LEWIS, C. L. E. & KNELL, S. J. (eds) *The Making of the Geological Society of London*. Geological Society, London, Special Publications, **317**, 49–92.

LONGSTAFF, G. B. 1912. *Butterfly hunting in Many Lands, Notes of a Field Naturalist*. Longmans, Green & Co., London.

LONGSTAFF, M. J. 1912. Some new lower Carboniferous Gasteropoda. *Quarterly Journal of the Geological Society*, **68**, 295–309.

LONGSTAFF, M. J. 1918. Supplementary notes on *Aclisina* De Koninck and *Aclisoides* Donald with descriptions of new species. *Quarterly Journal of the Geological Society*, **73**, 59–83.

L. R. C. 1936. Mrs Mary Jane Longstaff (née Donald). *Annual Report of the Council Geologists' Association*, **47**, 97.

MCDONALD, A. I. & TRUEMAN, A. E. 1922. The evolution of certain Liassic gastropods with special reference to their use in stratigraphy. *Quarterly Journal of the Geological Society*, **77**, 297–344.

MCKENNY HUGHES, T. 1900. Award of the Lyell Fund. *Proceedings of the Quarterly Journal of the Geological Society*, **56**, xlviii–xlix.

MCROBERT, R. 1912. A nepheline–syenite boulder dredged from the Atlantic. *Geological Magazine*, **IX**, 1–4.

MCROBERT, R. 1914. Acid and intermediate intrusions and associated ash-necks in the neighbourhood of Melrose (Roxburghshire). *Quarterly Journal of the Geological Society*, **70**, 303–314.

MILLINGTON, R. 1972. *The MacRobert Story, Volume 2*. MacRobert Trust, Tarland, Aberdeenshire, 220–415.

MITCHELL, I. C. 1979. *The Douneside Story*. MacRobert Trust Pamphlet, Tarland, Aberdeenshire.

MONTGOMERY, F. A. 1997. *Edge Hill University College: A History 1885–1997*. Phillimore & Co., Chichester, West Sussex.

NEWNHAM COLLEGE. 1923. *Register 1891, Volume 1, 1871–1923*. Newnham College, Cambridge, 112.

OGILVIE, M. M. 1892. Contributions to the geology of the Wengen & St. Cassian strata in Southern Tyrol. *Quarterly Journal of the Geological Society*, **49**, 1–78.

OGILVIE, M. M. (MRS GORDON). 1899. The Torsian-structure of the Dolomites. *Quarterly Journal of the Geological Society*, **55**, 560–634.

RAISIN, C. A. 1887. Notes on the metamorphic rocks of South Devon. *Quarterly Journal of the Geological Society*, **43**, 715–733.

RAISIN, C. A. 1889. On some nodular felstones of the Lleyn. *Quarterly Journal of the Geological Society*, **45**, 247–269.

RAISIN, C. A. 1891. On the lower limit of the Cambrian series in N. W. Caernarvonshire, *Quarterly Journal of the Geological Society*, **47**, 329–342.

RAISIN, C. A. 1893. Variolite of the Lleyn, and associated volcanic rocks. *Quarterly Journal of the Geological Society*, **49**, 145–165.

RAISIN, C. A. 1897. On the nature and origin of the Rauenthal serpentine. *Quarterly Journal of the Geological Society*, **53**, 246–268.

RAISIN, C. A. 1901. On certain altered rocks from near Bastagne and their relation to others in the district. *Quarterly Journal of the Geological Society*, **57**, 55–72.

RAISIN, C. A. 1903. Petrological notes on rocks from Southern Abyssinia. *Quarterly Journal of the Geological Society*, **59**, 292–306.

REID, C. & REID, E. M. 1904. On a probable Palaeolithic floor at Prah Sands (Cornwall), *Quarterly Journal of the Geological Society*, **60**, 106–112.

REID, C. & REID, E. M. 1908. Preglacial flora of Britain. *Journal of the Linnean Society*, **38**, 206–227.

REID, E. M. 1919. Preliminary description of the plant-remains. *Quarterly Journal of the Geological Society*, **75**, 197–200.

REID, E. M. 1920*a*. 3. On two preglacial floras from Castle Eden (County Durham). *Quarterly Journal of the Geological Society*, **76**, 104–110.

REID, E. M. 1920*b*. 4. A comparative review of Pliocene floras based on the study of fossil seeds. *Quarterly Journal of the Geological Society*, **76**, 145–161.

REID, E. M. & CHANDLER, M. E. J. 1933. *London Clay Flora*. British Museum Natural History, London.

REYNOLDS, D. L. 1945. Dr Catherine Alice Raisin. *Nature*, **156**(1945), 327–328.

ROMANES, M. F. 1916. Note of an Algal limestone from Angola. *Transactions of the Royal Society of Edinburgh*, **LI**(111), 581–584.

SHAKESPEAR, E. M. R. 1901. The Lower Ludlow formation and its graptolite fauna. *Quarterly Journal of the Geological Society*, **56**, 415–492.

SHAKESPEAR, E. M. R. 1906. The Tarannon series of Tarannon. *Quarterly Journal of the Geological Society*, **62**, 644–701.

SKEAT, E. G. & CROSFIELD, M. C. 1896, The geology of the neighbourhood of Carmarthen. *Quarterly Journal of the Geological Society*, **LII**, 523–541.

SKEAT, E. G. & MADSEN, V. 1898. The Jurassic, Neocomian and Gault boulders found in Denmark. *Danmarks Geologiske Undersogelse*, **2**, 1–213.

SOLLAS, I. B. J. 1901. Fossils in the Oxford University Museum, V: On the structure and affinities of the

Rhaetic plant *Naiadita*. *Quarterly Journal of the Geological Society*, **57**, 307–312.

STOPES, M. C. 1907. The flora of the Inferior Oolite of Brora, Sutherland. *Quarterly Journal of the Geological Society*, **63**, 375–382.

STOPES, M. C. & & WHEELER, R. V. 1923*a*. Terminology in coal research. *Fuel Bulletin*, **1**, 5–9.

STOPES, M. C. & WHEELER, R. V. 1923*b*. The spontaneous combustion of coal. *Fuel Bulletin*, **1**, 1–125.

STRAHAN, A. 1907. Discussion in: Stopes, M. C. 1907. The flora of the Inferior Oolite of Brora, Sutherland. *Quarterly Journal of the Geological Society*, **63**, 382.

SWEETING, G. S. (ed.). 1958. *The Geologists' Association 1858–1958*. The Geologists' Association, London.

TORRENS, H. 1995. Mary Anning (1797–1847) of Lyme: The greatest fossilist the world ever knew. *British Journal for the History of Science*, **28**, 257–84.

VAN KONIJNENBURG-VAN CITTERT, J. H. A., KUSTATSCHER, E. & WACHTLER, M. 2006. Middle Triassic (Anisian) ferns from Kühwiesenkopf (Monte Prá della Vacca), Dolomites, Northern Italy. *Palaeontology*, **49**(5), 943–968.

WACHTLER, M. & BUREK, C. V. 2007. Maria Matilda Ogilvie Gordon (1864–1939): Scottish researcher in the Alps. *In*: BUREK, C. V. & HIGGS, B. (eds) *The Role of Women in the History of Geology*. Geological Society, London, Special Publications, **281**, 305–318.

WALLACE, P. 1969. Fifty years of feminine fellowship. *Proceedings of the Geological Society of London*, **1658**, 209–214.

WATNEY, G. R. & WELCH, G. 1912. The Salopian rocks of Cautley & Ravenstonedale. *Quarterly Journal of the Geological Society*, **67**, 238–262.

WATSON, J. 2005. One hundred and fifty years of palaeobotany at Manchester University. *In*: BOWDEN, A. J., BUREK, C. V. & WILDING, R. (eds) *History of Palaeobotany*. Geological Society, London, Special Publications, **241**, 229–257.

W. B. R. K. 1961. Gertrude Lilian Elles Obituary. Annual report of the Council. *Proceedings of the Geologists' Association*, **72**, 168–170.

W. C. S. 1955. Obituary Notices: Lady (Rachel Workman) MacRobert. *Proceedings of the Quarterly Journal of the Geological Society of London*, **1529**, 146.

WHITAKER, W. 1900. Award of the Lyell Geological Fund. *Proceedings of the Quarterly Journal of the Geological Society*, **56**, xlviii.

WHITE, A. B. 1961. Gertrude Lilian Elles. *Newnham College Roll Letter*. Newnham College, Cambridge, 45–49.

WOOD, E. M. R. 1900. The Lower Ludlow formation and its graptolite fauna. *Quarterly Journal of the Geological Society of London*, **56**, 422–485.

WOOD, E. M. R. 1906. On graptolites collected by Dr. J. W. Evans in 1901–1902. *Quarterly Journal of the Geological Society of London*, **62**, 431–432.

WOODS, E. G. 1923. *The Principles of Geography: Physical and Human*. Oxford University Press, Oxford.

WOODS, E. G. 1932. *The Baltic Region: A Study in Physical and Human Geography*. Methuen, London.

WOODS, E. G. & CROSFIELD, M. C. 1925. The Silurian rocks of the central part of the Clwydian Range. *Quarterly Journal of the Geological Society, London*, **81**, 170–194.

WORKMAN, R. 1911. Calcite as a primary constituent of igneous rocks. *Geological Magazine*, **8**, 193–201.

WOODWARD, H. B. 1911. *The History of the Geological Society of London*. Geological Society, London.

A year to remember

TED NIELD

The Geological Society, Burlington House, Piccadilly, London W1J 0BG, UK
(e-mail: ted.nield@geolsoc.org.uk)

Abstract: The Geological Society of London is the oldest geological society in the world. On the 13 November 2007 it commemorated its 200th anniversary. Throughout 2007, even into 2008, lectures, dinners, speeches, field trips and a whole host of other events across the country celebrated the occasion.

Writing in the Society's *Annual Report 2007*, President Richard Fortey, FRS, said of the Bicentenary: 'I think everyone is agreed that our Bicentenary was a great success; not merely because 'nothing went wrong'... but because we achieved and even surpassed all our stated objectives'. In other words, the Society enjoyed itself more than somewhat – but also did one or two things that will live on afterwards.

It all began with the simultaneous release of 4567 balloons from the Courtyard of Burlington House on 10 January 2007 (Fig. 1), marking not only its own 200th birthday, but also the 4567 millionth birthday of planet Earth.

The event was witnessed by children from local schools, the Executive Board of the United Nations International Year of Planet Earth, Fellows of the Society, staff members at Burlington House and members of the public. President Richard Fortey said:

> Two hundred years ago this year, less than a mile from this very spot, the Geological Society of London was founded by 13 men sitting in a pub on Long Acre.[1] It was also a Friday 13th (of November), which I think shows a creditable disregard for cant and superstition – something that has been a hallmark of geologists and their subject ever since.
>
> In its 200 years this Society, the oldest national Society in the world for Earth science and now by far the largest in Europe, has seen its subject achieve maturity in a shorter time than any other science.
>
> The challenges that lie ahead for Earth scientists are greater today than they have ever been. Everything that humanity needs that cannot be grown has to be dug from the Earth – and therefore discovered by a geologist. Nearly all the energy, raw materials and water we depend on, come to us thanks to the geosciences. Our challenge for the next 200 years is to use our understanding of the Earth system so that we can continue to use these riches in a wise and sustainable way – to achieve a healthier, wealthier society for all.
>
> We all inhabit the Earth courtesy of its geology. From finding oil, gas and uranium to disposing of radioactive waste, charting past climate change, forecasting volcanic eruptions and warnings of tsunamis – the work of geologists is often unfashionable because it reminds people of uncomfortable truths they would rather ignore. But that is the nature of all science, which does not exist to please us, or to make us feel good, but to help us understand uncompromising nature. Moreover, because science 'works', it enables us to use that understanding to make people's lives better – and, if we try really hard, without destroying our world in the process.
>
> So when we look at the Thames Barrier and worry about rising sea levels, or to sub-Saharan Africa and worry about increasing drought and water shortages, or when we wonder where the energy we need for our very existence will come from in the future, we are asking geological questions. Earth scientists have a unique capacity to solve these problems because they think on the Earth's timescale, measurable in millions and billions of years, not ours, measured in mere centuries and millennia. Above all, Earth science helps us to see how robust, yet also how changeable our Earth is – and so to understand that when it comes to our climate, constancy is the one thing we have no right to expect. We must adapt.

Half of the all-natural, fully biodegradable balloons bore the Society's logo, while the other half bore that of the United Nations International Year of Planet Earth. The latter was proclaimed by the UN General Assembly for 2008, so that the three years of activity that straddle it could begin with the anniversary of foundation of the world's first national learned society dedicated to furthering Earth science.

[1] This 'pub' was in fact the Freemasons' Tavern on Great Queen Street which, as Lewis (2009) shows, was much more an expensive dining club than a pub. It is frequently confused with the pub of the same name on Long Acre. Also, as many in this volume show, only 11 men sat down to dine that night – the two others were there in spirit only.

From: LEWIS, C. L. E. & KNELL, S. J. (eds) *The Making of the Geological Society of London*.
The Geological Society, London, Special Publications, **317**, 409–415.
DOI: 10.1144/SP317.22 0305-8719/09/$15.00

Fig. 1. Children with just a few of the 4567 balloons released on 10 January 2007 to mark the beginning of the Bicentenary year.

Speaking of the Year, TV presenter Professor Aubrey Manning, University of Edinburgh (Fig. 2), said:

> This United Nations Year will be in 2008, but as its activities span three years it is beginning now, to coincide with the birthday of the world's oldest Geological Society. In fact among the crowd here today we welcome Dr Eduardo de Mulder and his colleagues from the International Year's Board of Directors, drawn from all over the world, who have chosen to have their first meeting here in Burlington House today.

Fig. 2. Professor Aubrey Manning, University of Edinburgh.

He continued:

> The Earth sciences in the twenty-first century are reaching a peak of interest and influence – and I speak as a biologist, whose subject is also said to be entering a golden age just now. I believe this is no coincidence. As we learn more about the planet that – uniquely as far as we know – has given rise to life, the more we realise that old discipline barriers – biology, geology, chemistry and so on – are meaningless. Earth and life cannot be separated. Life and the planet on which it arose are a unit.
>
> We have no other home. The way we use it – for use it we must – is crucially important to the people who come after us. We hold this planet in trust for future generations, for the children we see here today. And the key to this sustainable use, to learning to live within our environmental means without destroying something that has evolved over millions of years, is understanding the Earth system itself.
>
> This is the challenge facing the world today, a challenge for research and education that I hope the United Nations International Year of Planet Earth will go a long way towards meeting.

The balloons were finally released by TV presenter Dr Iain Stewart (University of Plymouth), who revealed the answer to the question 'how old is the

Earth?', which he had posed earlier to the schoolchildren who had spent the morning learning about Earth science in the Society's apartments.

The guiding light of the Society's year-long celebration was Professor Paul Henderson, who chaired the Steering Committee. The Fundraising Committee, chaired by former President Sir Mark Moody Stuart, made available those things without which none of the Steering Committee's plans would have been possible. The generosity of the Fellowship was also surprising. The President's letter of appeal, which was circulated three times, netted about a quarter of a million pounds (which makes it the best-earning piece that I ever wrote, by a very long way).

On 15 February a reception at the House of Lords was held by the Corporate Affiliates Committee to launch its bicentennial campaign to increase the number of corporate Affiliates. Former President Lord Oxburgh hosted the evening, which heard brief addresses from Richard Fortey, Oxburgh himself and another former President, Richard Hardman. Over 100 attended; many representing potential new recruits.

Legacy projects

From early on, perhaps even as early as 1996, the Society had decided that it wanted to celebrate its past at the same time as looking forward to the future both of the science and of itself. Later, it realized it wanted to engage the public as well as the Fellowship all over the country, and internationally. Crucially, it wanted the Bicentenary to leave a *legacy*.

One such legacy project had, of course, been set in train over a decade before – Gordon Herries Davies's spirited and idiosyncratic Bicentenary history *Whatever is Under the Earth*. Despite, or perhaps because of being the only book ever published by the Society with a title beginning with 'Whatever', it was easily the Society's best-seller in 2007 and achieved exactly what its commissioners wanted – it brought the Society's history to life for the non-historians among us. History undoubtedly – even, perhaps, rather alarmingly – also came alive during the History of Geology Group's costume evening (Moody 2009), held on the 12 November, the night before the Society's Anniversary Dinner. This culminated in the unveiling of a plaque on the front of the New Connaught Rooms, Great Queen Street (Fig 3), which now occupies the site of the Freemasons' Tavern where the Society was founded. Those who attended both it and the gala dinner at the Natural History Museum on the following night were left in no doubt that legacies weren't everything – the Society intended to have a ball on its 200th birthday.

True to the spirit of 'outreach' that came to dominate the year, the London Lectures to the general public, under Shell sponsorship, played to packed houses and went on (under extended sponsorship) to form a UK contribution to the UN International Year of Planet Earth in 2008. All these lectures were made available for view on a new Society website, so converting them into a lasting (global) resource. Not only London benefited, Shell also sponsored a set of lectures held all over the country – the Regional University Tour – which began in the autumn of 2007 and overlapped into 2008.

The Society's new website was a major Bicentenary legacy project, providing the portal to the Bicentenary's most exciting and ambitious legacy – the Lyell Collection – a highly functional online archive of 230 000 pages published by the Society since 1845. It was a major undertaking in its own right, with a central role in making the Society's books and journals available to Fellows and the global Earth science community. Funded by Foundation Sponsors, Shell and BP, supported by Schlumberger, this digital library of all the Society's major published material became available to the world – and was free to approved higher education institutions in developing countries. It went live on the same night as the new website that hosted it, which was also the night that the refurbished Lyell Room opened – with bespoke furniture designed by Luke Hughes – within the Society's newly refurbished apartments. Like many another Bicentenary event, the launch of the Lyell Collection in the Lyell Rooms – the Collection's physical embodiment – was partly lubricated by the serving of a specially produced Bicentenary beer, *Baker's Dozen Founders Ale*, made by the Cambridge Brewery.

Publishing

The Lyell Collection was just one of the innovations in information services that the Society achieved during 2007. The website included a complete online catalogue of the Society's book, journal and map library holdings, and an innovative visual tool for browsing the index data associated with map series via a Geographic Information System (GIS), both of which are open to Fellows and others. However, the monumental achievement that was the Lyell Centre did not overshadow conventional sales, with 28 new titles published and all journals produced on schedule in the Bicentenary year. Further good reading could be found in 17 bicentennial reviews published in the *Journal of the*

Fig. 3. Dr Cherry Lewis, Chairman of the History of Geology Group, with Professor Richard Fortey, President of the Geological Society of London, after the unveiling of the plaque to commemorate the founding of the Geological Society. It was erected on the outside of the New Connaught Rooms, which now stands on the site of the Freemasons' Tavern.

Geological Society. These were specially commissioned by the Editorial Board to celebrate topical issues in the Earth sciences.

As Edmund Nickless, the Society's Executive Secretary, wrote in the *2007 Annual Report:*

> It is difficult for geologists, used to reflecting upon the impermanence of things, to consider a span of decades as 'long term'; but we also did things in 2007 that will live on in centuries. Day three of our Bicentennial Conference will live in the minds of some of the hundreds of young people who came to see the challenges of a career in Earth sciences, and as they progress through life, they will encounter this Society, just as we did in our time. Catering for their needs as aspirant and professional geoscientists, providing continuity through their careers, this Society will be for them the portal that gives access to unrivalled archives of geoscientific information.
>
> The Lyell Collection will live, longer than buildings or individuals, to be a service to future generations. It will last because it will keep growing for as long as this Society exists, benefiting our Fellows and, especially, geologists in developing countries.
>
> We can – and perhaps should – allow ourselves a moment or two of pride. The Bicentenary was, as we intended, not just about history, but our future. For that I must thank everyone who helped: our generous sponsors, the Fellows who donated and participated, the volunteers who gave their time and effort, the committees who worked so hard, and the staff in the Publishing House and Burlington House, without whom none of it would have happened.

Local heroes

Although (as in 1907) the burden of organizing Bicentenary events and projects made huge demands upon the staff, not everything in the Bicentenary was organized centrally. The *Local Heroes* initiative, brainchild of Professor Joe Cann FRS (former Honorary Secretary, Foreign and External Affairs), proved hugely popular. The Society provided grants, publicity and logistical support for regional events celebrating geologists whose scientific contributions reached far beyond the happy highways of home. This initiative also continued through 2008 as a contribution to the 150th anniversary of the Geologists' Association. *Local Heroes* set out to celebrate UK geological pioneers and their achievements in those places most closely associated with them. Each was organized by a local group with support from the Society, which also co-ordinated and administered the initiative.

More than 20 *Local Heroes* events were organized by other societies, regional groups, branches of the Geologists' Association, university departments, museums and others. Many of the heroes celebrated were well known (James Hutton, William Smith, Mary Anning), but others were not so prominent in wider geological memory, and it was good to be reminded of the achievements of Martin Te Punga, Bill Ramsbottom and John Cadman. One event even celebrated a fossil, *Charnia*, while others marked the achievements of groups of geologists – ensuring that the headcount of the celebrated well exceeded the number of events. Audiences varied, but several drew attendances of over 100, and the initiative as a whole reached a much wider community than is normal for the Society.

Nor did the Bicentenary neglect fieldwork, and perhaps the Society's most ambitious ever excursion – to South America *In Darwin's Footsteps* – provided the photographic backdrop for the *2007 Annual Report*. As President Fortey wrote: '[the field trip] served to remind us that right from our foundation we have always been a truly international Society'.

Restorations

William Smith's 1815 map is one of the Society's greatest treasures and is certainly the most visited, thanks to Simon Winchester's book *The Map that Changed the World* which brought the map to the attention of the wider public. When the restored map was unveiled on 1 February 2007, in the presence of Nigel Press whose company (NP Satellite Mapping) generously sponsored its restoration, the Society marked the beginning of the final stage of its apartments' long refurbishment by placing this great monument to the power of human imagination alongside Greenough's 1820 map, which represents the Society's first, and perhaps greatest, achievement (Knell 2009) – in a place where the public at large could more easily admire both. Professor Hugh Torrens, world expert on William Smith's life and work, delivered a short lecture about Smith's work. Subsequently, as part of the finishing touches to the apartments' restoration, MacCulloch's map of Scotland, reissued in facsimile by the British Geological Survey, was also hung on the Society's main staircase. The gift formed part of the BGS's support for the Society's Bicentenary.

One of the most remarked-on aspects of the Burlington House refurbishment was a new 1.7 tonne reception desk in the reinstated formal entrance from Piccadilly. Devised by conservation architects Julian Harrap, this monumental addition to the Piccadilly entrance was designed to say 'rocks are us', and has proved hugely popular with visitors (and most, although not all, Fellows!). Seventeen slabs of stone, roughly two metres long, were acquired and stacked to create the desk. All 17 are classic British building stones that contribute to our built heritage and, fortunately, remain available. Eric Robinson, expert on UK building stones, wrote: 'For the Society's Bicentenary, it is of interest that all would have been in production in 1807 when the Society was born. How better could the Society celebrate its Bicentenary?'.

Conference

In September 2007 the Society's Bicentenary Conference welcomed over 1000 individuals to the Queen Elizabeth II Conference Centre in Westminster, bringing together an internationally renowned cast of invited-only speakers, all exploring and reviewing the state of play in forefront issues of Earth science. If Fellows had wanted a three-day refresher course in their science, they could not have asked for one better than the Bicentenary Conference.

During the conference, President Richard Fortey received, on behalf of the Society, the Leopold von Buch Medal of the German Society for Geosciences (DGG) – the highest honour that can be bestowed by the DGG, and never before presented to a Society. Presenting the award, Dr Heinz-Gerd Roehling (Treasurer of the DGG) said it was being given 'in honour of 200 years of outstanding achievements in promoting geosciences in Great Britain and throughout the world'. The American Geological Institute's Explorer Award was also exceptionally awarded to the Society on the occasion of its 200th birthday. It was presented by Larry Woodfork, a former President of AGI and Chair of International Year of Planet Earth Inc., to Richard Fortey who accepted it on behalf of the Society (Fig. 4).

The Bicentenary Conference also included a successful careers track on Day 3. While the main conference continued in plenary sessions, over 200 school students from all over the South East spent their time pestering representatives of Shell, BP, Anglo, BG Group and many more about careers in geoscience. Day 3 itself embodied an ambitious idea – to draw out the results of geoscience research and explore their implications for pressing societal problems of our age, this being done with one eye cocked across the road at the Palace of Westminster. Sadly, few members of that House attended, partly

Fig. 4. President Richard Fortey (left) with the American Geological Institute's Explorer Award presented by Larry Woodfork, a former President of AGI and Chair of International Year of Planet Earth Inc.

at least because of a timing difficulty; the House was in recess at the time.

British Association

Another timing difficulty involved the British Association for the Advancement of Science, whose Annual Festival was running during the same week as the Society's Bicentenary Conference. This had the effect of emptying the capital of its daily and weekly science media, which was unfortunate, although not foreseeable, as our timetable had been set many years before the BA's. However, taking this problem as an opportunity and a challenge, the Society decided to take the Bicentenary Conference to the BA, with the help of the BA's Recorder for Geology, Dr Richard Waller (Keele University).

Thus, as the Bicentenary Conference concluded in London, some hardy souls got on trains to York. They included Professor Peter Styles (Keele University), Dr Gabi Schneider (Director, Geological Survey, Namibia), Dr Jon Gluyas (Fairfield Energy), Dr Chris Carlon (Anglo American), Dr Richard Fortey (President), Dr Cherry Lewis (University of Bristol), Profesor Bill McGuire (University College London Benfield Hazard Research Centre) and Dr John Ludden (Director, BGS). The mix of historical and modern geology provided a fascinating programme for all.

The Bicentenary Dinner

While the Bicentenary was spectacularly launched with balloons slightly after the beginning of the Bicentenary year, the culmination of the celebrations had to come slightly before its end – and could come on no other day than the anniversary, the evening of 13 November 2007.

The gala dinner at the Natural History Museum main hall attracted about 700 people who, in addition to the usual speeches, were also pleased to witness the presentation of prizes in the Student Essay Competition. The essay title was: *How will the geosciences contribute to achieving a sustainable energy supply in the 21st century and beyond?* The judges were Professor Richard Fortey; Dr Mike Naylor, Vice-President Technical, Shell Exploration & Production; and Dr Ted Nield, Editor, *Geoscientist* and Chair, Association of British Science Writers. The final result was announced by guest speaker Professor Aubrey Manning. First prize went to Caroline Burberry (Imperial College London). Pete Rowley (Royal Holloway, University of London) and Helen Jones (Open University) were awarded joint second prize. Before the awards were presented, the Society's President, gave the welcome speech:

> My lords, distinguished guests, ladies and gentlemen;
>
> At an event marking the culmination of a Bicentenary Year that has been many years in the planning, it is

difficult to know where to begin acknowledging the many people here tonight to help us launch the Society's *next* 200 years. First, I must acknowledge our Guest of Honour, Professor Aubrey Manning, who though not a geologist himself, has done as much as anyone in this country to bring Earth science to a wider public through a series of brilliant and engaging television documentaries. Professor Manning has graciously agreed to favour us with an after-dinner speech this evening, to which I am looking forward very much.[2]

We are honoured by the presence this evening of no less than eight past presidents of the Society: Professors Howel Francis, Charles Holland, Bernard Leake, Derek Blundell, Tony Harris, Charles Curtis; Lord Oxburgh and Sir Mark Moody-Stuart. Three Honorary Fellows also number among us, namely: distinguished science documentarist Sir David Attenborough, together with Professors Knut Olav Bjørlykke and Gerald Friedman. From our kindred societies in the United States we welcome: from the American Association of Petroleum Geologists, Will Green, President, and Ray Thomasson, Past President; from the American Institute of Professional Geologists, Kelvin Buchanan; and from the Geological Society of America, Jack Hess, Executive Director.

I must salute, with deepest gratitude, all representatives of our Bicentenary Year Platinum Sponsors, BP and Shell – as well as Schlumberger, in the person of CEO Andrew Gould. From around the courtyard of Burlington House, I welcome representatives from our neighbour learned societies. And last, but in no sense least: the many Fellows and Officers who have spearheaded our various Bicentenary projects; members of the Society's faithful and dedicated staff, without whom none of our Bicentenary ambitions would have been realised; members of Councils past and present, and all Fellows of the Society, I say – to one and all – welcome.

After a short pause, Professor Fortey continued:

Dear Friends,

Exactly 200 years ago this very night, our Society was founded – at a dinner in a pub. A hundred years later, in 1907, the Society celebrated its centenary with a *Conversazione* in this cathedral of Natural History[3] – the high spot of a year that was, for them, every bit as packed with interest and excitement as this year has been for us. And so it is appropriate that we should celebrate those 200 years, and all the generations of geologists that have come and gone, with a grand dinner in this hallowed place, under the gaze not only of our extinct friend here, but also the effigies of those other giants of Earth science, Charles Darwin, Thomas Henry Huxley, and Richard Owen – all proud Fellows (and one proud President).

Our founding document, signed on that night 200 years ago, reminds us that fellowship – with a small 'f' as well as a large one – lies at the heart of our science. True, that document also speaks of dull (though important) matters like adopting standard nomenclatures and so on; but mainly, it speaks of making geologists acquainted with one another; of stimulating zeal; of communication; of ascertaining what is known and what remains to be discovered. It speaks of friendship, collaboration, communication, and the excitement of exploring the history and structure of our world, *together*. It speaks in other words of community, common purpose, common cause.

As anyone will know who has read Gordon Herries-Davies's bicentenary history, our Society's fortunes have waxed and waned many times over the two centuries it has existed. We are now fortunate to be living through a 'waxing' phase. But always, through fat times and lean, the things that have *united* us as Earth scientists have proved stronger than the things that tended to drive us apart. We must now use our great history as the inspiration for our *next* 200 years – two more centuries of making like-minded friends across the globe, and publishing what we discover to stimulate research by our colleagues, and those who come after us.

At the heart of all those activities will forever remain – fellowship: binding us together in the great collaborative enterprise of science. That, above all, is what we must celebrate here tonight.

Ladies and gentlemen, I offer you a toast – to Fellowship.

References

KNELL, S. J. 2009. The road to Smith: how the Geological Society came to possess English geology. *In*: LEWIS, C. L. E. & KNELL, S. J. (eds) *The Making of the Geological Society of London*. Geological Society, London, Special Publications, **317**, 1–47.

LEWIS, C. L. E. 2009. Doctoring geology: the medical origins of the Geological Society. *In*: LEWIS, C. L. E. & KNELL, S. J. (eds) *The Making of the Geological Society of London*. Geological Society, London, Special Publications, **317**, 49–92.

MOODY, R. T. J. 2009. Dining with the Founding Fathers: a personal view. *In*: LEWIS, C. L. E. & KNELL, S. J. (eds) *The Making of the Geological Society of London*. Geological Society, London, Special Publications, **317**, 439–448.

WATTS, W. W. 1909. *The Centenary of the Geological Society of London*. Longmans, Green, and Co., London.

[2] Unfortunately, we were unable to locate a record of this speech.

[3] For further details of this event, see Watts (1909, p. 149).

Walk with the Founding Fathers

MARTIN J. S. RUDWICK

Department of History and Philosophy of Science, University of Cambridge, Free School Lane, Cambridge CB2 3RH, UK

(e-mail: mjsr100@cam.ac.uk)

Abstract: Part of the History of Geology Group's Bicentenary celebrations included a field trip to the Isle of Wight entitled 'Walk with the Founding Fathers'. It followed in the footsteps of Thomas Webster as he attempted to understand the island's complex geology almost 200 years ago. The material used for the field trip – prepared by Martin Rudwick – is reprinted here. The logistics of the weekend were organized by History of Geology Group committee members, John Mather and Dick Moody, and the trip was led by Professors Martin Rudwick and Hugh Torrens.

Field trip on the Isle of Wight: 9–11 November 2007

Introduction

This geological field trip (Fig. 1), unlike most such events, will be consistently *historical* in character. We will try, as far as possible, to study the topography of the island, and its rocks and fossils, *through the eyes* of the early nineteenth-century geologists who first learnt how to make sense of its structure and geohistory. In other words, we will adopt the methods that *historians* of the sciences routinely use, rather than judging past ideas by constant reference to what 'we now know' from modern research. Only by this kind of imaginative time travel can we hope to understand and appreciate the remarkable scientific achievements of our 'Founding Fathers' and their contemporaries.

The field trip will focus on the most important early geological study of the island, in order to understand *in depth* the field practice of (some) geologists two centuries ago, and the kinds of reasoning that they used while learning how to make sense of what they observed. Thomas Webster (1773–1844) – an architect and artist who, in 1812, was appointed the Geological Society's first salaried officer (as librarian, curator and draughtsman) – was commissioned by a wealthy patron, Sir Henry Englefield (*c.* 1752–1822), to visit the Isle of Wight in 1811–1812 to work out its geology and make drawings of its natural features and historic buildings. Webster later found important reasons to return in 1813. He reported his geological conclusions in a paper read to the Society in 1813–1814 and promptly published in its *Transactions* (1814); his letters to Englefield were later printed, together with his superb field drawings, in his patron's handsome coffee-table book about the island (1816). On several important issues Webster's work was one of the first major pieces of British research to confirm and extend what had already been claimed by geologists in Continental Europe: notably the vast scale of movements of the Earth's crust, even at geologically recent periods, and the magnitude of subsequent erosion; and the reality of complex sequences of changing environmental conditions in the distant past.

The field trip will also consider more briefly the work of geologists of the next generation, such as Charles Lyell and Gideon Mantell, in unravelling the stratigraphy of the island, relating it to that of mainland Britain and using it to help reconstruct the history of life on Earth.

Fig. 1. Martin Rudwick leading the field trip on the Isle of Wight.

From: LEWIS, C. L. E. & KNELL, S. J. (eds) *The Making of the Geological Society of London*.
The Geological Society, London, Special Publications, **317**, 417–438.
DOI: 10.1144/SP317.23 0305-8719/09/$15.00

Plan for the field trip

(Subject to modification in case of vile weather or other contingencies.)

- **Friday 9th November**
  Travel to Wellington Hotel, Ventnor.
  Dinner in hotel.
  Introductory talks by field trip leaders.

- **Saturday 10th November**
  09:00 Coach from hotel in Ventnor around the SW coast to Freshwater Bay.

  Walk up on to Tennyson Down and along to The Needles: east–west ridge of white Chalk with bedding planes accentuated by lines of black flints, tilted at high angles; general view of topography to north and south, and (if weather is clear) of whole island in relation to Solent and mainland beyond.

  Walk down to Alum Bay village: en route, views of vertical strata in cliffs of Alum Bay, and of horizontal strata in Headon Hill beyond.

  Walk down to shore of Alum Bay, and south along beach: study vertical clays and sands, abutting Chalk. Walk back up to village (total morning walk *c.* seven kilometres).

  Lunch at Needles Park restaurant at Alum Bay.

  14:00 Coach from Alum Bay village back along coast road to Compton Chine.

  Walk down to beach: study strata in cliffs and follow them back (NW) to junction with Chalk. Double back SE along beach, studying further strata in cliffs, to rocks at Hanover Point. If tidal conditions are favourable, find huge fossil footprints; low tide at *c.* 17:30. Walk up to coast road (total afternoon walk *c.* four kilometres).

  Coach from coast road above Hanover Point to reception at Dinosaur Isle museum in Sandown.

  18:30 Coach from Dinosaur Isle to hotel in Ventnor.

  Dinner at hotel.
  Discussion session with leaders (if requested).

- **Sunday 11th November**
  09:00 Coach from hotel in Ventnor back to Alum Bay village (pack and bring all luggage from hotel).Walk down to shore; check again on vertical strata abutting Chalk of The Needles (seen yesterday); walk back (north) towards Hatherwood Point, to study the relation between vertical and horizontal strata; climb from shore up to top of Headon Hill (rough scrambling), studying sequence of horizontal strata and their fossils; walk down to Alum Bay village (total morning walk *c.* two kilometres).

  Lunch at Needles Park restaurant at Alum Bay.

  14:00 Coach from Alum Bay to Ryde, for ferry to Portsmouth and train to London.

Primary sources

Source 1

Ordnance Survey map of the topography of the western end of the Isle of Wight, first edition (on the scale of one inch to one mile), published in 1810 as part of the new survey organized by the military Board of Ordnance for use in the defence of the south coast of England against the threatened invasion by Napoleon's forces (Fig. 2).

Source 2

Three of Webster's letters to his patron Englefield, written to report on his fieldwork and the inferences he was drawing from it, while still on the spot (but probably tidied up later for publication). Three of his 12 letters are concerned with the areas covered on our field trip.

(a) Letter V. (Englefield 1816, pp. 148–156) describes his traverse along the SW coast, particularly the cliffs to the south of the 'middle range' of Chalk hills ('Brixton' is now Brighstone; 'Brook point' is now Hanover Point).

LETTER V

FROM ST CATHERINE'S TO FRESHWATER

Freshwater, June 7th, 1811

DEAR SIR

The road through Chale, Kingston, and Shorwell, lies over the stratum of ferruginous sand; but the soil, where it is not covered by gravel, varies from red to yellow, according to the particular bed; and the cliffs, on the sea shore, which are not generally above thirty or forty feet in height, are chiefly sections of this stratum. They are much acted upon by the sea, and

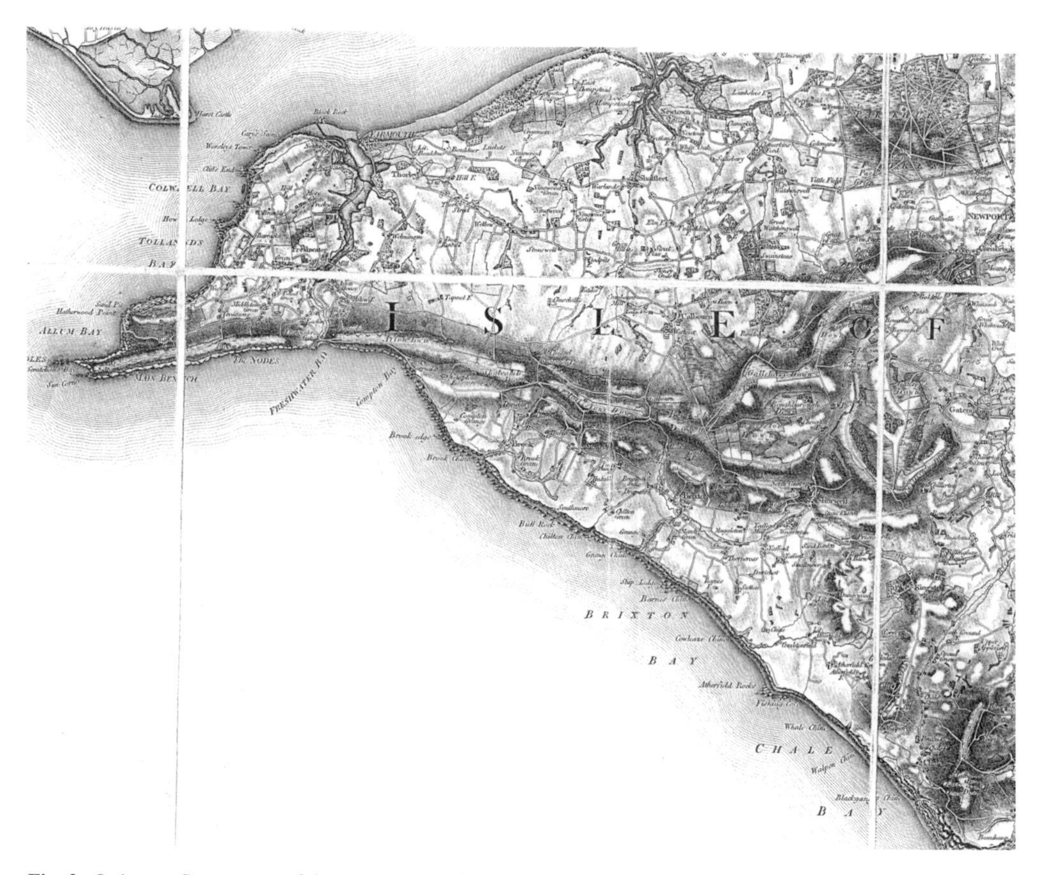

Fig. 2. Ordnance Survey map of the western end of the Isle of Wight, 1810.

the land is wearing away very fast from Blackgang to Freshwater gate.*

When I arrived at Brixton, and approached the middle range of hills, I felt anxious to ascertain whether the green sandstone, which I had observed at the Culver, Undercliff, and St. Catherine's, accompanied the chalk through its whole course; and in this case, I expected to find it in my way to the top of the hills, as I should then pass over the edges of the highly inclined strata. Instead, therefore, of proceeding to Motteston, I struck off to the right, and ascended Brixton down.

I soon fell in with the red and white sand rock, which forms the general substratum of this part of the country; the road being cut into it. It dipped to the north, inclining therefore the same way, though not so much, as the chalk of the middle range.

On reaching the top of this down, I perceived a valley between it and the middle hills, which extended westward to Compton and I concluded that its situation was occupied by the dark marl, although this was totally concealed by the soil and herbage.

At the foot of the middle range, near to a very conspicuous monumental building lately erected, a road leads up to a chalk pit; and a few hundred yards below this pit a wall crosses the road. This wall, I observed, was constructed of stones from the stratum I was in search of, being sandstone and limestone. Hence, I concluded, although I could not perceive any appearance of the rock, that it existed under the soil; these stones being probably such as lay loose upon the surface. On arriving at the pit, I found this was cut in the lower chalk, which was entirely without flints, and dipped rapidly to the north, having the same inclination as all the chalk in the middle hills.

Continuing my course westwards for about half a mile further, I came to another chalk pit, exactly similar to the last, and also having a road leading

* *From some fossil organic remains, which I have since seen in the possession of a stonemason, who got them at Atherfield rocks, I apprehend that the latter consists of a portion of the Green sandstone. The fossils were extremely beautiful, and I was informed, were very abundant. Not having visited these rocks, which can only be seen properly at low Water, I am unable to speak of their position from actual observation.*

down from it; on this I descended, again crossing the edges of the strata at right angles; and in the place where I expected it, I found the sandstone rock. The road was cut several feet deep into it, and the stratum corresponded to its general appearance in other parts of the island.

In search of the same strata I examined several roads which led up to these hills, and never failed to find the sandstone; the lower chalk appearing in the hill above,

The hills immediately above Brixton, Motteston, and Compton, form, therefore, a second and lower range, between which and the chalk hills is the valley above mentioned. All these lower hills consist of the red ferruginous sand stratum, of which Longstone is a portion that is coarse grained and indurated.

The section of this valley I conceived must be interesting; and having therefore proceeded to the cliff on the sea-shore below Compton, the stratum of dark grey marl, which I expected, appeared in the face of it. It had been more acted upon than the other strata, perhaps by the same cause that rounded the hills; and a deep hollow had been formed, that however had been nearly filled up again by a deposition of gravel more than fifty feet in thickness.

I had now arrived at the spot where you wished me to observe the junction of the nearly vertical chalk with the other strata of this end of the island. The section, afforded by the cliff, shewed that all the strata here, inclined in the same direction as those in the hills. To the east of the dark blue marl, was the ferruginous sand, with more than its usual, quantity of iron; to the west of it, was the green sandstone; and then the white marl, and the two beds of chalk.

So far, the order of the strata corresponded with what had been before observed in Sandown bay.

The shore at this place is covered with very large fragments of the green sandstone stratum, which, as well as the others, is constantly falling, the sea in stormy weather dashing furiously against the cliffs. The subordinate beds of limestone are here in smaller quantity; and I did not perceive any chert. Many of the fallen masses contained much green sand; and others consisted of a singular breccia, composed of pieces of sandstone and limestone, cemented by calcareous matter and sand.

Other masses of rock consisted of distinct layers of this breccia separated by beds of sand, or soft sandstone, in which were numerous pieces of red limestone, and ferruginous sandstone, which, on examination, proved to be rounded, and apparently water-worn, casts of the cornua ammonis, and also of various bivalve shells. Some of these were a foot or two in diameter. Besides this, the number and variety of fossil shells in these rocks, lying in strata between the beds, was very considerable.

The awful grandeur of this place (which is, I believe, but little visited) arising from the threatening aspect of many huge masses of rock just ready to fall, with the noise and fury of the waves, formed a scene altogether but little fitted for minute examination. Ideas of another kind crowded upon the imagination: and it was impossible to observe these strata with water worn fossils, and the numerous pieces which had belonged to rocks older than those I was viewing, without reflecting upon the strange revolutions which our planet has undergone, and the many different states it has been made to assume, before the present order of things became established.

In one of the former states of the earth, the spaces between the beds of these rocks seemed to have been the shores of a sea, upon which were still lying vestiges of submarine animals that had belonged to a period of still greater antiquity. Or, perhaps, various processes, had been going on, even during the formation of the present rocks; and systems of destruction and renovation had been acting alternately.

During the short time that I passed there, I collected a considerable number of fossil shells belonging to the green sand; and I am convinced that this spot will afford much interest to the fossilist.[1]

Turning to the east, I kept on my left the Cliff of dark red ferruginous sandstone, which contained here also beds of black clay and yellow sand; a deep and gloomy chasm in it is called Compton chine, at the top of which is a small barrack. ([Englefield 1816] Plate XXII.)

It was near to this place, that I had been informed, fossil fruits had been found in great abundance, and which were vulgarly called in the island, Noah's nuts: I made a point of endeavouring to see them in situ. Near the top of this cliff lie numerous trunks of trees, which, however, were not lodged in the undisturbed strata, but buried eight or ten feet deep under sand and gravel. Many of them were a foot or two in diameter, and ten or twelve feet in length. Their substance was very soft, but their forms and the ligneous fibre were quite distinct: round them were considerable quantities of small nuts, that appeared similar to those of the hazel. None of the wood nor fruits were at all mineralised, but, resembled the state of such substances when they have lain long in bogs, or marshes. Among them was some phosphat of iron.

No hazel whatever now grows upon the island; nor has the subversion of these trees been an event of recent occurrence. Their situation, under such deep beds of earth and gravel, points out its remote antiquity. Yet pieces are sometimes found so fresh, as to bear being worked into furniture; and a farmer in the neighbourhood shewed me a table which he had made of this wood.

The ferruginous sand Cliffs continued some way farther, preserving nearly the same inclination. But the strata succeeding to it, and which dipped with a gradually decreasing angle, until they were at last nearly horizontal near Brook, were very different. They consisted of a succession of beautifully coloured plastic clays alternating with beds of red and yellow sand, sandstone, slate clay with fossil shells, and also limestone containing veins of calcareous spar.

[1] These, with the other specimens collected in the Isle of Wight, were presented, by Sir Henry Englefield, to the Museum of the Geological Society.

At Brook point the cliffs interested me much. They were about thirty feet in height; and were composed chiefly of clay resting upon a bed of soft sandstone, which contained a considerable quantity of sulphur, arising from the decomposition of pyrites.

At this place I observed many masses of a coaly blackness, bearing the exact form and resemblance of trunks of trees that had been charred, lying on the beach, and imbedded in the clay cliffs, and also in the rock.

In some parts, the ligneous fibre was still evident. In other parts, the wood had been converted into a substance much resembling Jet; its blackness being intense, its cross fracture conchoidal, and its lustre very great. Other parts of the trees were entirely penetrated by pyrites and considerable groups of crystals of this substance were frequently attached to the outside.

They were imbedded in clay of various colours, white, grey, yellow and red; and lay in irregular horizontal strata of several inches in thickness, being often pressed flat by the incumbent weight.

Over this stratum of clay, which is about eight or ten feet thick, there is another, of the same depth, of sand and gravel highly ferruginous; and the water which filters from it is strongly impregnated with sulphat of iron.

On lifting up some of the sea weeds which grew upon the shore between high and low water marl, I was surprised to find almost all the rocks below them composed of petrified trees, which still retained their original forms. They were of various sizes, from eight or ten feet long and two feet in diameter, to the size of small branches. The knotty bark, and the ligneous fibre, were very distinct; and they were frequently imbedded in masses of clay now indurated and in the state of an argillaceous rock. Some parts of these trees were converted into iron stone; and other parts consisted of a great variety of substances, being partly calcareous, siliceous, ferruginous, pyritous, bituminous, and ligneous; and the whole exhibited a beautiful example of the astonishing processes of nature in converting vegetables into coal, and in filling their substance with solid rock.

These changes have no doubt taken place at a period too remote for human conjecture; and whilst the trees were yet buried under the strata which still partially cover them; their present situation on the shore arises merely from the sea having made gradual inroads upon the land, and, after having washed away the soil above, exposed them to view.

Having seen, from the top of St. Catherine's, the general nature of the coast eastward of this place, I did not consider it necessary to go farther in this direction; and therefore returned back, and proceeded towards Freshwater gate.

Afton down is almost entirely covered with small flints, among which are an infinity of fragments of silicified Alcyonia, both of the globular and ramified species. The hill is abruptly cut down on the south side towards the sea; and in the face of this cliff the lower chalk appears: but as the sea washes close up to the foot of it, it is impossible to go below round to Freshwater gate. To be seen distinctly, it should therefore be viewed from the water. These cliffs are much worn into hollows by the waves; and on the top of the nearly vertical strata of chalk and flints, appears a horizontal bed of gravel, about twelve feet in thickness, which probably extends all over Afton down, and has furnished the fossil Alcyonia which are found on the top.

Freshwater gate is a place of much interest; and the particular nature of the flinty chalk may be studied here with great convenience. The real dip of the chalk strata is to the north-west, at an angle of about 70°. The small inn stands upon a considerable deposition of clay and gravel, which appears to have partly filled up a deep hollow formed in the chalk by which Afton down and the high down on the other side of the inn are separated from each other. This clay and gravel may be seen in the face of the cliffs on each side of the inn, at the part where they begin to rise from the flat; and it wends gradually with the chalk, frequently filling up the fissures. The spring which rises here, and the marsh, owe their existence, in all probability, to a bed of clay in this place.

From this bay, which is also filled with shingles, the extensive and magnificent line of chalk cliffs, well known by the name of Mainbench, is seen stretching westwards to the Needles; but as I did not view them from the sea, from which only they can be distinctly observed, I am not certain whether they consist entirely of the chalk with flints, or whether there may not be some portions of the lower chalk. I was informed by the fishermen, that a sandstone rock appears above water at ebb tide, at a small distance south of Scratchell's bay. This is not an improbable circumstance; as it is exactly the spot where the stratum of green sand may be expected. The bottom of the sea, at that part, consists of black mud; probably arising from the dark grey marl. The place is much resorted to by the fishermen, from all the south side of the island, for catching skate and thornbacks, which they employ as bait for the lobster fishery.

The road to Freshwater church presented no geological phenomenon of importance; and from this place I propose to proceed to Alum bay and the Needles.

I am, &c.

T. WEBSTER.

(b) Letter VI. (Englefield 1816, pp. 157–163) describes the vertical strata in Alum Bay and (very briefly) the horizontal strata to the north, before recording his eventful boat trip out to The Needles to draw the views looking back at the island.

LETTER VI

ALUM BAY, COLWELL BAY, AND THE NEEDLES

DEAR SIR,

Yarmouth, June 12, 1811.

ALUM bay is so extraordinary a place, that I am unable to express, in adequate terms, the surprise I felt on first viewing it. This scene is indeed of a species unique in this country; and nothing that I had previously seen, bore the least resemblance to it.

The south side ([Englefield 1816] Pl. XVII. No. 3, Pl. XIX. No. 3) is screened by the line of chalk hills which runs out into the sea almost due west, terminated by the insulated masses called the Needles, that lie in an exact line with the north face of the chalk cliff. At the bottom of the bay are the very remarkable vertical strata of clay and sand, which are in close contact with the flinty chalk,

These clay and sand cliffs consist of a great variety of vertical beds, which differ considerably in their natures, and which it would require a very attentive examination to describe with great accuracy of detail. But as my primary object was not to ascertain the particular nature of these vertical strata, I shall not at present attempt to enumerate them with perfect precision. They consist, however, generally, of a vast number of alternations of layers of very pure clay, and pure sand, with ferruginous sand and shale. Of these beds, some are several feet, whilst others are not an eighth of an inch in thickness.

Next to the chalk, is a vertical bed of chalk marl; then one of clay of a deep red colour, or sometimes mottled red and white. This is succeeded by a very thick bed of dark blue clay with green earth, containing nodules of marl or argillaceous limestone with fossil shells. Then follows a vast succession of alternating beds of sand of various colours, white, bright yellow, green, red and grey; plastic clay, white, black, grey, and red; ferruginous sandstone and shale, together with several beds of a species of coal, or lignite, the vegetable origin of which is evident.

The number and variety of these vertical layers is quite endless, and I can compare them to nothing better than the stripes on the leaves of a tulip. On cutting down pieces of the cliffs, it is astonishing to see the extreme brightness of the colours, and the delicacy and thinness of the several layers of white and red sand, shale and white sand, yellow clay and white or red sand, and indeed almost every imaginable combination of these materials.

These cliffs, although so highly coloured that they could scarcely come within the limits of picturesque beauty, were not, however, without their share of harmony. The tints suited each other admirably; and their whole appearance, though almost beyond the reach of art to imitate, was extremely pleasing to the eye.

Their forms, divested of colour, when viewed near, and from the beach, were often of the most sublime class; resembling the weather worn peaks of Alpine heights. This circumstance they derive from the same source as those primitive mountains; for the strata being vertical, the rains and snow water enter between them; and wear deep channels, leaving the more solid parts sharp and pointed.

The north end of these curiously coloured cliffs is terminated by a very thick vertical stratum of blackish clay with much green earth, which contains septaria and a great number and variety of fossil shells. These fossils I recognised to correspond with those in the Hordwell cliff in Hampshire, on the opposite shore of the Solent, which are so well described in the publication of Brander and Dr. Solander. This circumstance is extremely interesting; as it serves to connect this stratum, though vertical, with that of the Hordwell cliffs, which is horizontal. In the middle of the stratum, a deep hollow is worn, through which a stream of water comes down; and a path leads up, and passes through a rabbit warren.

To the north of this stratum a dirty yellow sand joins; but which appears not to he vertical, but rather dips about 45° to the north: this, however, was not very evident: On this last sand leans a very thick, bed of sand ([Englefield 1816] Pl. XXIII) which is extremely white and pure, and which is very nearly horizontal; it may be traced to the north, quite round Totland Bay.

It is this horizontal bed of white sand, which they dig and export to the glass houses, where it is employed for making the best glass. Over it, is a stratum of black clay, upon which rest a series of very singular strata, consisting of marl, shale, and soft calcareous stone, the whole containing numerous fossil shells, chiefly in fragments; above these, and a little removed back, are cliffs of similar horizontal strata of considerable height: and these continue with interruptions round the north-west part of the island.

The lower part of this series of horizontal strata is bent upwards, and suddenly broken off, where it comes up against the vertical cliffs of Alum bay ([Englefield 1816] Pl. XXIII and XXIV) just in the same manner as the calcareous rock in Whitecliff bay; and they correspond also, in the nature of the substance, and the fossils which they contain.

These points of agreement are remarkable; and determine the causes of the curvature of the strata, and the circumstances under which they were formed, to have been the same, in both places.

They also point out, in a satisfactory manner, the continuity of these strata; and shew, that the clay cliffs of Whitecliff bay and Alum bay are the sections of the two ends of a long line of vertical strata, which accompany the chalk and sandstone, from the east to the west end of the island. It is true, that the clay in the first of these bays is not so much striped with various colours, being chiefly a yellowish clay mixed with sand of the same colour; but the bed immediately adjoining the chalk is of the same dark red colour as the corresponding bed at Alum bay.

I have no doubt, that the shelly calcareous rock extends over great part of the north side of the island; since I found the cliffs round the bays of Totland and Colwell composed of it, and I have already mentioned that it occurs also at Gurnet bay, and Whitecliff bay.

The cliffs in Colwell bay ([Englefield 1816] Pl. XVII. No. 2.) are, like those at the bottom of Sandown bay, formed by the section of a valley; the strata on each side of which dip in a contrary direction, rising towards the middle of the valley.

From Colwell bay, I procured a small fishing boat to take me out to the Needles. These insulated masses of chalk, which from the island appear much less striking, proved, on approaching them, to be rocks of great magnitude.

My visit to them happened at an interesting juncture. The Pomone, a frigate of fifty guns, returning home from Persia, after an absence of three years, had, but the day before, struck upon the point of the most western Needle. The chalk rocks having pierced through the bottom of the ship, she remained immovable; and, filling with water, instantly became a complete wreck. The crew and passengers, among whom were some Persian princes, fortunately got safe on shore.

The vessel afforded me a scale, by which to judge of the size of the Needle's; and I was surprised to find that the hull of the frigate did not reach one fourth of their height. Viewing this scene merely as picturesque, and independently of the feelings of regret naturally excited by the loss of so valuable a ship, it was one of the grandest I have ever witnessed. The view of the end of the Isle of Wight, from the Needles, at any time, is one of the most uncommon, and at the same time one of the most magnificent scenes, in Great Britain; but such a wreck, on a spot so extraordinary, formed a combination, which, though not strictly accordant with the rigid laws of picturesque composition, yet was in nature highly sublime.

There being no chance whatever of saving the ship, it was determined to endeavour to take out the guns, and carry off such part of the wreck as could be broken up. The masts were cut down, and lay overboard; and thirty or forty cutters, gun boys, and other vessels, were either lying very near, or were sailing backwards and forwards to the dockyard at Portsmouth; whilst many spectators, in pleasure yachts and other boats, attracted by this extraordinary occurrence, were viewing the scene. The ship's boats, manned by the unfortunate sailors, lay on their oars at a little distance.

The officers were seen on the wreck, giving orders to the carpenters, who were cutting down the rigging, and whatever was about the deck. Others were lowering the guns into vessels which conveyed them to the boys, and the sea around was strewed with the floating fragments of the ship; whilst, at a proper distance, the fishing boats of the island were busily employed in picking up such pieces as would otherwise have drifted to sea.

When the flood tide is in, there is frequently a very great swell at the Needles; which occasioned such an agitation among the various boats and vessels, as made it almost impossible to approach the Pomone without danger, but increased very much the picturesque effect of the groupes; while the foam of the waves, almost constantly dashing over the ship, spouted to a great height, running back again through her gun ports.

Sailing round the Needles, I had a full view into Scratchell's bay. The form of the chalk cliff; over which, when at a sufficient distance, is seen the Lighthouse, is singularly elegant; and the advancing line of these magnificent detached masses, the Needles, formed a whole that is scarcely to he equalled. The lines of flints in the chalk were distinctly to be seen; shewing, that at the north side, the strata are nearly vertical, dipping about eighty degrees; the angle lessening towards the south corner, where they dip about sixty degrees.

I took the opportunity of the beginning of the ebb tide, to search for some place where I could land upon the foot of the Needles, in order to make the drawing which you requested of this end of the island; but could not accomplish it with any degree of safety.

At length, I observed a small rock, a little to the north of the middle Needle, appearing just above water. Upon this I effected a landing with some difficulty; and, though completely wetted through by the spray, I remained about two hours, and completed a sketch. ([Englefield 1816] Pl. XXV.)

In the middle of the view, is the easternmost Needle; and behind that, an arch in the chalk, there being a space between it and the adjoining Needle sufficient to admit a boat. At the bottom of the drawing, is seen the rock upon which I sat, which is about six or seven feet across, and which is the stump of the lofty spire of chalk which chiefly gave the name to these rocks, and, which being worn through by the waves, fell down in 1764. A view of it, drawn when it was still standing, is seen in Sir Richard Worsley's History of the Isle of Wight.

The view of Alum bay ([Englefield 1816] Pl. XIX. No. 3.), shewing the coloured cliffs, is taken from the same spot; and may be joined to the left side of the last, to complete the scene.

The return of the flood occasioned me to take to the boat, which had been lying anchored at a little distance; and I was informed, that on the first gale, the Pomone would be dashed to pieces; which accordingly happened not many days afterwards.

I am, &c.

T. WEBSTER.

(c) Letter XII. (Englefield 1816, pp. 226–234) written from an unspecified place on the Isle of Wight, describes his unanticipated return visit, prompted by seeing the specimens that the Parisian geologist Alexandre Brongniart had sent (at the height of Napoleon's war with Britain!) to the French refugee (and 'founder' member) Comte de Bournon in London. The specimens were to illustrate the already famous memoir on the 'Paris Basin' (1808/1811) by Brongniart and his Parisian colleague Georges Cuvier, and Brongniart's paper on fossil shells of freshwater molluscs (1810). This time Webster's fieldwork was focused on the horizontal strata in Headen (now Headon) Hill above Alum Bay, because he now recognized in them a sequence of strata with such freshwater shells, alternating with other strata containing marine fossils, which roughly paralleled the Parisian sequence and, therefore, greatly extended its significance.

LETTER XII

FRESHWATER FORMATION

DEAR SIR,

Isle of Wight, February 11, 1813

You will, no doubt, be a little surprised at receiving a letter from me, dated from this place, and at this season of the year. But I could not resist the pleasure of communicating to you an account of the discoveries I have been making here, and the occasion which led to them.

I mentioned in my last letter on the subject of the Isle of Wight, the opinion which I entertained, that the north side of the island, consisting of a series of horizontal strata, had been deposited in an immense basin; and that it was not an improbable circumstance, since they were of later formation than the chalk, that they might be found to correspond with some of those in the basin of Paris lately described.

I feel assured that you will participate in my satisfaction, on finding this conjecture confirmed by the opportunity I have since had of viewing a series of specimens of the French strata; a comparison of which with those of the Isle of Wight renders their identity no longer problematical.

These specimens bad been presented by M. Brongniart, in illustration of his Memoir, to the Count de Bournon, who had deposited them in the Museum of the Geological Society. A circumstance promising so much interest demanded a careful examination; and I have undertaken another journey to this place, with the view of studying with some attention the contents of the Isle of Wight basin, for the purpose of instituting a comparison between them and those of the basin of Paris.

The hill called Headen, adjoining to Alum bay, I recollected as the place where this is best examined; the lofty cliffs affording the finest natural sections, and exhibiting a great variety of strata. Thither, therefore, I directed my first inquiries; and the following is a more minute enumeration of the strata which compose this hill.

I formerly described the lowest of them to be a very white sand which is employed in the manufacture of glass. Immediately over this is one of very dark blue clay without shells.

On this rests a series of beds of sand and sandy marl with a good deal of a brown coaly substance. These contain vast quantities of shells, chiefly in fragments, but which prove to have been entirely of freshwater origin. They belong to the Planorbis, and Lymneus, of Larmarck.

Over these beds is a very thick stratum of a greenish marl, which contains an immense quantity of fossil shells, many of which are in perfect preservation. These are entirely marine.*

The stratum over this consists of the calcareous rock which I formerly mentioned to have been employed here in building, together with a friable calcareous marl both of which contain prodigious numbers of fossil freshwater shells. Many of these are quite perfect. They consist of several species of the Lymneus and Planorbis, exactly corresponding to those described by Lamarck in the basin of Paris; and they agree, in general, at least, with the recent freshwater shells of this country. It is with the casts of these shells that the stone of Bembridge and Gurnet are filled.

Over this considerable stratum of calcareous rock and marl, there is a thin bed of blue clay with many fragments of shells: and also one of calcareous sandstone without any shells. This is covered by another bed of calcareous concretions, containing a few freshwater shells; and this last bed, which is the termination of the series, is extremely hard and dense, and some parts have even a porcellanous character.

A very thick bed of flint gravel forms the top of the hill.

All these strata may he distinctly traced along the cliffs of Headen, and quite to Totland bay, which, together with Colwell bay, is covered by the lowest bed of fine sand. The whole series, however, can only be seen at Headen. In the other parts of the western, as well as the northern parts of the island, only portions of the series remain; so that, to understand them, it is necessary first to learn their characters at Headen.

The stratum filled with marine shells, interposed between two strata containing only those of freshwater, is an extremely curious phenomenon; and corresponds to what we are told of the Paris basin.

The strata exhibiting the remains of freshwater animals render it probable that this place was the situation of a lake, in some former state of the earth; and the alternation of marine strata with these appear to shew, that the same place has been a lake, and a part of the sea in succession.

Singular as these conclusions may seem to be, it is impossible to withold our assent to them. To deny them, would be to refuse the recognition of causes from their effects; and the evidence of these formations in fresh water, rests upon the same ground as that of the rest of the strata in the bosom of the ocean, viz. the exuviae of animals which they contain. At the same time that our judgment must be convinced by such proofs, our astonishment must be excited, on considering the changes that must have taken place.

The bottom of our Isle of Wight basin, we have seen, was the clay and sand over the chalk, or the London clay. These strata are of marine origin, as is evident from the fossil organic bodies found in them: and we have also seen that the subversion of the chalk, and such a change of position as gave rise to the basin, took place at a period subsequent to the deposition of the clay stratum.

If it be safe to speculate on a subject so difficult of explanation, it would seem not improbable, that this hollow was at first a gulph, communicating with the sea on the east, where it is still open. From the same or similar causes which at present often produce the same effects, this gulph or arm of the sea, might have been converted into an extensive lake of fresh water; in which a series of strata was

* Since my return to London, I have, by the kind assistance of Mr. James Parkinson, been enabled to describe these shells. They are various species of the Cerithium; viz. plicatum, lapideum, semicornatum, mutabile, turritellatum, and tricarinatum : various Ostreae approaching to Ost. deltoidea, Cyclas deltoidea, Cytherea scutellarea, Ancilla buccinoides, Ancilla sabulata, Ampullaria spirata, &c.

Fig. 3. (**a**) View of the western tip of the Isle of Wight, seen from The Needles. Chalk dipping steeply to the north (left) (Englefield 1816, plate 25). (**b**) Continuation of the same view to the north, with vertical strata in Alum Bay (right) and almost horizontal strata in Headen Hill (left) (Englefield 1816, plate 19). (a) and (b) are reproduced courtesy of the Eyles Collection, University of Bristol Library Special Collections.

formed, enclosing an infinity of Planorbes and Lymnei. This is the first mentioned series seen in the lower part of Headen, and may be called the oldest or lower freshwater formation.

Another alteration afterwards took place in the relative level of the land and sea in this part of the globe. The ocean again resumed its dominion, and the site of the lake was covered with salt water. Numerous shells, many of which were different from those left by the former sea, were produced, and buried in newly formed strata.

Again this spot must have been subjected to other changes; the former state of things returned: another freshwater lake was formed, and was destined to commemorate its existence by the petrifaction of the animated beings contained in it.

When we compare this succession of events with what has been so ably described by MM. Cuvier and Brongniart, we cannot but be struck with their correspondence; nor entertain a doubt, that the gulph and the basin of the Isle of Wight were influenced by a part of the same causes that operated in forming that extraordinary series of strata still existing in the basin of Paris.

Some differences between them are, however, to be noticed. The plastic clay and sand is mentioned as

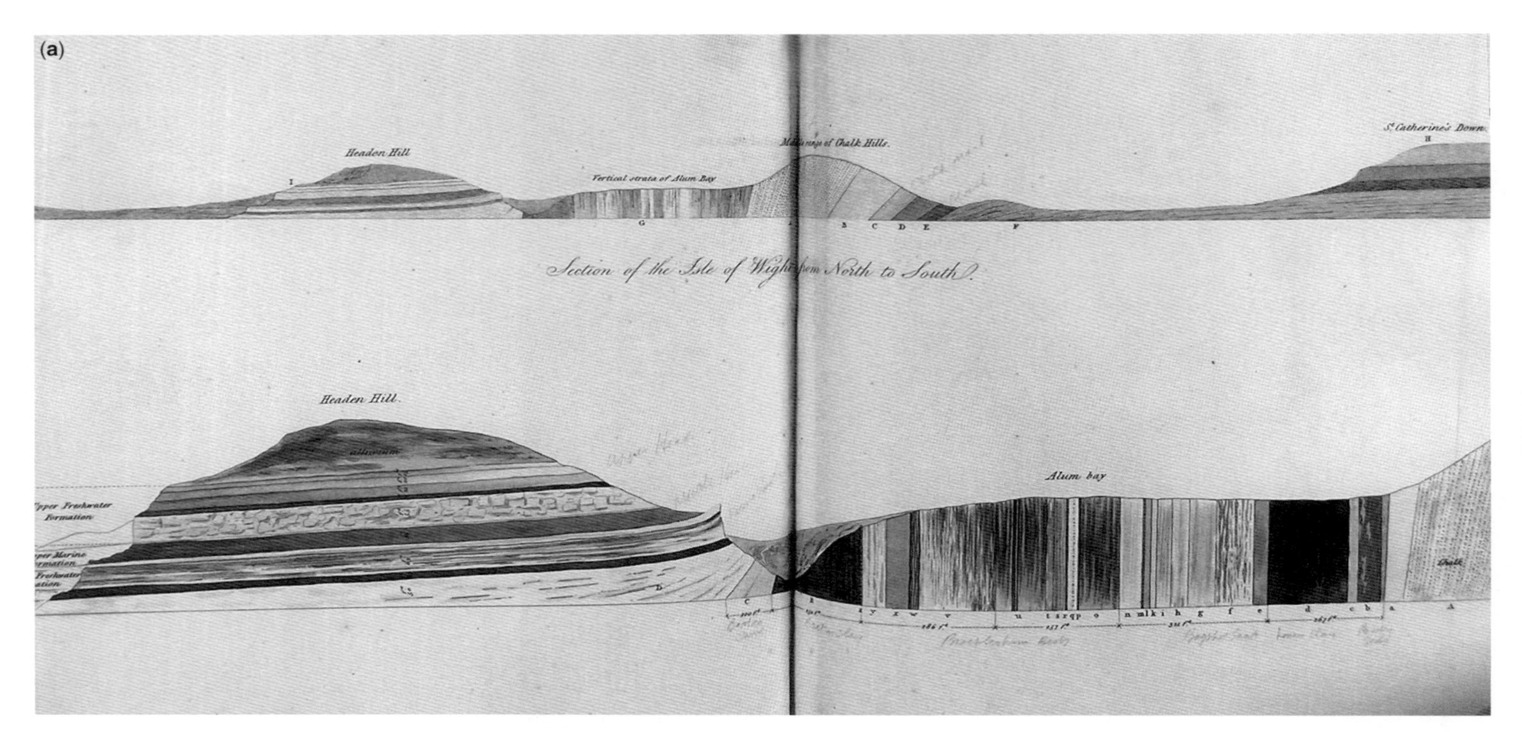

Fig. 4. Two of Webster's interpretative drawings based on his fieldwork. (**a**) Two sections through the west end of the Isle of Wight. Upper section from St Catherine's Down near Ventnor (south, right), through to the 'middle range' of near-vertical Chalk and vertical strata in Alum Bay, to near-horizontal strata in Headen Hill (north, left). Lower section, on a larger scale, showing vertical strata (some with marine fossils) in Alum Bay, and in Headen Hill two 'Freshwater Formations' separated by another 'Marine Formation' (Webster 1814, plate 11). Reproduced courtesy of the Eyles Collection, University of Bristol Library Special Collections.

(b)

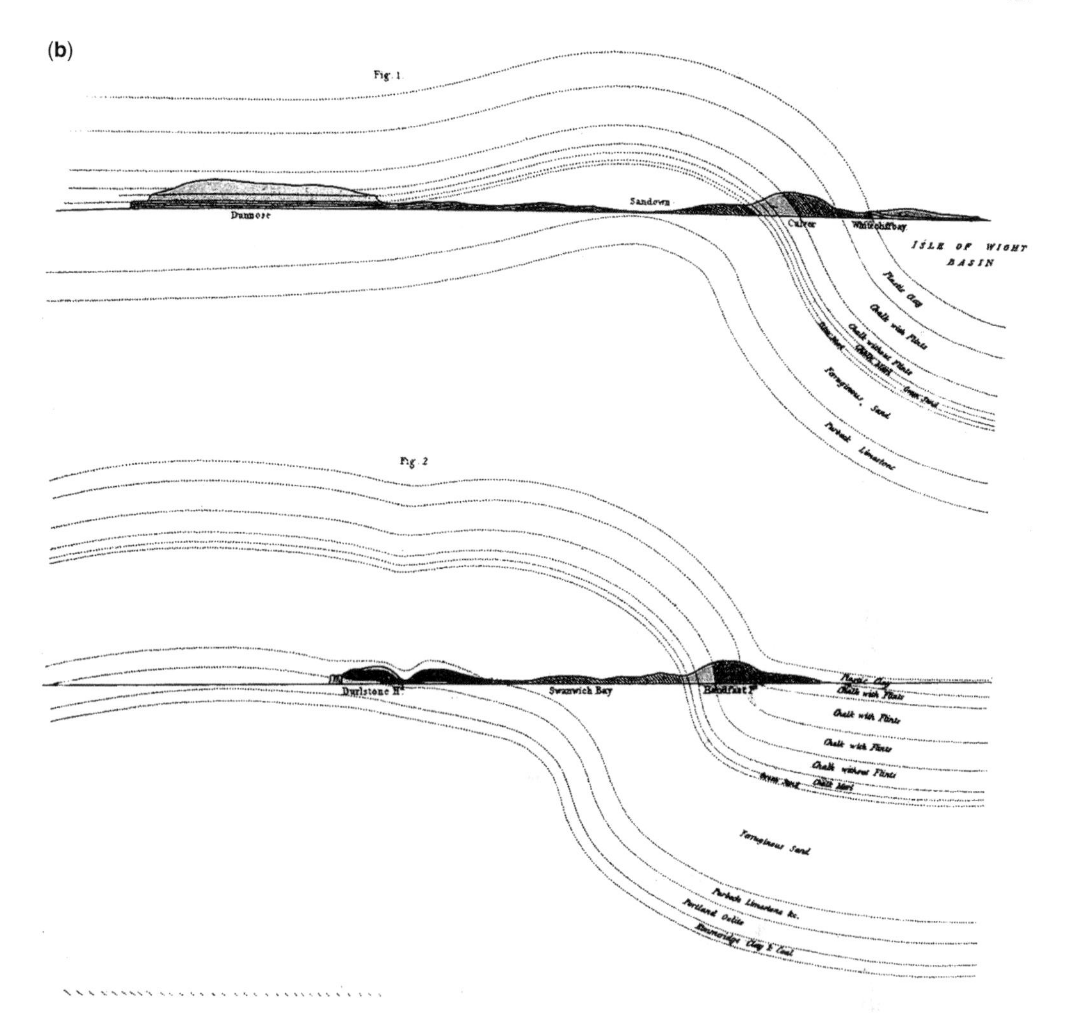

Fig. 4. (**b**) Two 'theoretical sections' through the Isle of Wight (upper section) and 'Isle' of Purbeck further west (lower section), showing Webster's inferred east–west fold structure, implying major crustal movement *after* the formation of the Alum Bay strata but possibly *before* the Headen Hill strata, and thousands of feet of subsequent erosion (Englefield 1816, plate 47). (a) and (b) are reproduced courtesy of the Eyles Collection, University of Bristol Library Special Collections.

the first deposition in the French basin; whereas our basin was not formed till after the deposition of these strata.

In the Parisian series we find also vast beds of gypsum, and nodular concretions and beds of siliceous matter. It was in vain, however, that I looked for these in the Isle of Wight. Instead of beds of gypsum I could only find crystals of selenite; and of siliceous concretions, I met with no trace in any of the strata.

In Colwell bay, portions of the alternations of freshwater and marine strata are to be seen; and in the cliffs of this bay is a remarkable bank of fossil oyster shells belonging to the latter.

The western shore of the island, from this place to Gurnet point, presents less interest; being generally flat, and composed of slopes of marl or clay produced by the mouldering of the banks.

At Gurnet bay, however, we find again a rock containing freshwater shells, which I consider as belonging to the lower freshwater formation; although more calcareous and more indurated than that of Headen. There were several beds of this stone alternating with clays.

Here I found in great abundance the small fluted globular fossil bodies called by Lamarck, Gyrogonites, and considered by him as a minute shell. I have since observed them in the walls of that ancient Saxon ruin in Southampton, which, in your description of that place, you have conjectured was formerly a palace; a circumstance which shews that at a very remote period this stone was extensively employed.

At West and East Cowes there are no vertical cliffs, but the shores are composed chiefly of mouldering slopes. The strata of these appear to belong to the lowest freshwater formation; although occasionally are found alternating beds of marine shells, produced probably by this more ancient lake having been subject to encroachments of the sea. Over this may be distinguished, though in ruins, the upper marine formation, by its numerous fossil shells.

In the grounds of Lord Henry Seymour, these strata assume a new character; consisting of a siliceous limestone, or a limestone containing a great quantity of siliceous matter in the state of sand. This is called Rag, and is an extremely durable stone. His Lordship has employed it as the chief material in the erection of his mansion; and also in a fine wall and terrace which he has constructed to prevent the farther encroachments of the sea. Many parts of this stone were filled with casts of shells resembling Helix vivipara Linn and other freshwater turbinated shells.

Above this lie several beds of stone, composed solely of the hollow moulds of the fragments of shells cemented together by sparry matter. It is singular that the shells themselves seem to have disappeared, leaving cavities in their stead, as if they had been absorbed into the substance of the rock.

The quarries here were extremely interesting, and afforded a good opportunity of observing some of the effects of the last revolutions to which the country has been subjected. For ten or twelve feet below the surface, the stratum consisted of sand containing imbedded in it large detached blocks of the siliceous limestone I have mentioned, which lie in a very irregular manner, and had evidently been rounded by the action of water.

From Cowes to near Ryde, the shore presents little geological interest. It is flat, and consists of blue clay, sometimes containing fossil shells which are usually marine; such as Cerithia, oysters, and Cytherea.

The quarries of Binstead, once so famous, are on the road from Ryde to Newport; but are now scarcely worked. They exhibit, however, proofs of, their ancient importance, in the extensive vestiges of the pits which are filled up. They afforded a good section. The lowest stratum is a white sand, in which are numerous springs. Over this is a bed of about a foot thick, of the same siliceous limestone which I had seen at Lord Henry Seymour's; then a thin bed of sand; another of limestone; and a bed of a few inches, of a very white marl, wholly composed of fragments of shells, too much comminuted to ascertain their species. Next follow several considerable beds of the sparry stone with the hollow moulds of shells, similar to what I have described at Cowes; and on the top, detritus consisting of blue clay intermixed with large blocks of calcareous stone filled with casts of the Lymneus and other freshwater shells, which appeared to be analogous to the rock forming the upper freshwater formation at Headen.

The fragment stone of Binstead, therefore, probably belongs to the lower freshwater formation.

It was this last stone that had been employed for the mouldings in the abbey of Quarr, the ruins of which are near to this place; and, although at first sight, it would seem to be very unfit for such a purpose, yet it has proved extremely durable; the angles of the mouldings being nearly as sharp as on the day in which they were cut. The other stone with the Lymnei, &c. had been used in the walls. The fragment stone is at present burnt for lime.

This rock may be traced along the shore eastward; and at Bembridge, the rock with Lymnei occurs. It is the bent stratum that comes up near to the vertical clay.

Having now traced this freshwater formation round the edges of the island, I endeavoured to ascertain whether it could be found in the interior.

It is evident, from what has been said respecting the construction of the country, that it could only be expected on the north side of the middle range of hills. Considerable quarries of this stone are accordingly worked in the neighbourhood of Calbourne; and the rock may be frequently seen, in the roads between this place and Yarmouth. In these quarries, I found the Planorbes of very great size, being full two inches in diameter. The Lymnei and Ampullariae also were numerous and large. This probably belonged to the upper freshwater formation, which, I have little doubt, is met with in several other places; but it would require much time to ascertain all its localities. When we reflect upon the proofs we have just seen, of the successive changes that have taken place in this basin, from salt to fresh water, it is probable that these were not confined to this spot. The causes, whatever they were, no doubt extended their influences to some distance. Hence it will be highly interesting to trace the vestiges of the same effects in other parts of England. In all the strata over the chalk, particularly in the counties of Hampshire, Sussex, Kent, Essex, Suffolk, Norfolk, Lincolnshire, and Yorkshire, circumstances may perhaps be observed that may extend the analogies I have described.

Permit me to add, that under your auspices I have prosecuted this inquiry: and, if by the hints I have furnished, should any thing be permanently added to the general stock of science, it must chiefly be attributed to the spirit which prompted you to begin the investigation, and the liberal Manner in which you have enabled me to pursue it.

I am, &c.

T. WEBSTER.

P. S. The following table [Table 1] exhibits a general view or abstract of the whole series of observations which I have lately made; and contains the order of, superposition of the strata of the southeast part of Great Britain, of which the above Letters may be considered as furnishing the proofs.

Table 1. *Order of the Upper Strata of the South-East Part of England, deduced from a Series of Observations made for Sir* HENRY ENGLEFIELD *in the Years* 1811, 1812, *and* 1813, *by* THOMAS WEBSTER, M.G.S.

Alluvium.	The ruins or detritus of regular strata, formed either by the present existing causes, or by some extraordinary and unknown agents. It is composed chiefly of water-worn fragments of flints, mixed with sand and clay in various proportions.
Upper Freshwater Formation.	This, in the Isle of Wight, consists of a calcareous rock, in which numerous fossil freshwater shells are imbedded. See Letter 12th. It agrees in character and situation with the corresponding formation in the basin of Paris, and other parts of the continent of Europe. Traces of a freshwater formation are to be observed also in the London basin, between the alluvium and London clay, consisting of marl with freshwater shells, and containing also numerous bones of land animals, as the elephant, hippopotamus, buffalo, elk, ox, &c. These have been found chiefly at Sheppey, Brentford, Essex, Suffolk, and Norfolk. In other places, as at Sheppey, Emsworth in Sussex, &c. vast quantities of the fruits of tropical countries have been found in a corresponding situation.
Upper Marine Formation.	This bed consists of bluish or greenish marl and clay, containing a great number of fossil marine shells, which, in general, are different from those found in the London clay. It is known in this country, with certainty, only in the Isle of Wight.
Lower Freshwater Formation.	This formation is ascertained in the Isle of Wight. It is placed under the last, and consists of clay, marl, and sand, with vegetable matter resembling an imperfect coal, or peat, and contains numerous fragments of freshwater shells. At the bottom is found a mixture of marine with freshwater shells. As the alternation of marine with freshwater strata has not been observed in any other part of this country except the Isle of Wight, the traces of a freshwater formation in the London basin, cannot perhaps be referred to this.
Sand without Shells.	In the Isle of Wight this sand is extremely pure; it is dug at Alum bay, and is used for making the best glass. The Bagshot sand, perhaps belongs to this; and, possibly, the Greyweathers; but the positions of these have not yet been accurately determined.
London Clay.	This is the blue clay of London, Highgate, Brentford, Sheppey, Portsmouth, Stubbington, Hordwell, Southend, Harwich, &c It is distinguished by its septaria, and its beautiful and numerous organic remains. In Alum bay, it is the most northerly of the vertical strata. Bognor rocks are subordinate to this bed. It agrees in its fossils, and geognostic situation, with the lower beds of the *calcaire grossier* of the Paris basin.
Plastic Clay and Sand.	The clay in this formation is often extremely pure, and fit for the potter. It is much employed in the potteries in Staffordshire. It is seen in Alum bay, the trough of Poole, and at the bottom of the blue clay in many parts of the London basin. An imperfect coal, or lignite, also frequently occurs in it. This formation corresponds to the French plastic clay, which lies over their chalk.
Chalk with Flints.	This formation in England extends from Flamborough head in Yorkshire to a little beyond Lyme Regis in Devonshire, and, where it is not covered by the beds above, forms chalk hills or downs. It is distinguished by the regular layers of flint nodules.
Chalk without Flints.	The inferior bed of chalk: in the south-east part of England is always without flints. When the chalk with flints is wanting, it forms the surface. The relations of both may be seen at the Culver and Compton bay, in the Isle of Wight, Handfast point, Beachy head, Guildford, Dorking, &c. It differs from the former only in the absence of flints, in the beds being thicker, and the chalk being sometimes a little harder.
Chalk Marl.	This bed consists of chalk and an intimate mixture of clay; it is always found below the two last strata, It may be readily distinguished from chalk, by its falling to pieces on being wetted and dried again: Some varieties of it, when burnt, form an excellent cement for building. It is also a valuable manure.

(*Continued*)

Table 1. *Continued*

Green Sandstone.	The formation to which I have given this name, consists of silicious sand united by calcareous matter, and contains also mica and green earth. From the variety in the proportion of the latter ingredient, it is by some divided into the green sand and grey sand, a distinction which cannot always be made, since these alternate and pass into each other. It is found in the wealds of Kent and Sussex, at the foot of the chalk downs, and is dug at Rygate and Measham for firestone. It is seen also at Folkstone, Beachy head, the Culver and Compton bay, in the Isle of Wight, Pewsey in Wiltshire, &c. Alternating with it, are often beds of limestone, as at Maidstone in Kent where they are called Kentish rag, also in the Undercliff Isle of Wight. Beds of chert occur in it. It abounds in organic remains.
Blue Marl.	This bed may be seen under the former very distinctly in the Isle of Wight, as at Sandown bay, many parts of the Undercliff, Niton, and Compton. It contains very few fossils.
Ferruginous Sand.	This denomination is given also to an alternating series of silicious sandstone, clay, and lime-stone. The sand-stone contains always more or less oxide of iron, sometimes in such quantity, as in the wealds of Kent and Sussex, that it was formerly employed as an iron ore. The clay tracts of the wealds belong to it. This formation may be also seen at Sandown bay, Blackgang and Compton chines, Swanwich bay, Hastings, Tunbridge Wells, &c. Fossil shells are rarely found in it: but carbonized wood is met with in abundance.
Purbeck Shell Limestone.	This formation consists of numerous beds of shells and fragments of shells, cemented together by calcareous spar, and alternating with shell and marl. The Purbeck, and perhaps the Petworth marbles, form part of this series: and it is farther remarkable for containing numerous freshwater shells, and bones of the turtle: hence it is not improbable, that part of it may have been formed in fresh water.
Clay with Gypsum.	At Swanwich in Dorsetshire, this is dug under the shell limestone. The gypsum does not occur in great quantity, but is employed for plaister.
Portland Oolite.	This includes the stone of Tillywhim and Windspit quarries, called the Purbeck Portland, and that from Portland island. It is entirely calcareous, and is formed of small grains or concretions adhering together. It is the only stone used for tile. Fronts public buildings in London. Some of its beds contain many marine fossils, also fossil wood and chert.
Bituminous Shale containing the **Kimmeridge Coal**.	This may be seen at Kimmeridge Encombe, and the isle of Portland. It is the lowest stratum visible in that part of the country to which the above observations have extended.

Source 3

Two of Webster's field drawings – later engraved and published in Englefield's (1816) book – of the western tip of the Isle of Wight, up to Alum Bay (see Fig. 3a and b).

Source 4

Excerpts from Webster's paper 'On the Fresh-water Formation in the Isle of Wight', read at the Geological Society at five meetings between 18 June 1813 and 7 January 1814 (i.e. after his return from his second visit), and published later in 1814 in the *Transactions* (a photocopy of these excerpts will be available in the hotel[2]). In his 'Introduction' (pp.164–170) and 'Concluding Observations' (pp. 245–254), Webster summarizes the 'geognostic' (i.e. stratigraphical) sequence on the island and nearby mainland, focusing on the strata above the Chalk; and he discusses the tectonic and geohistorical

[2] They are not reproduced in this chapter.

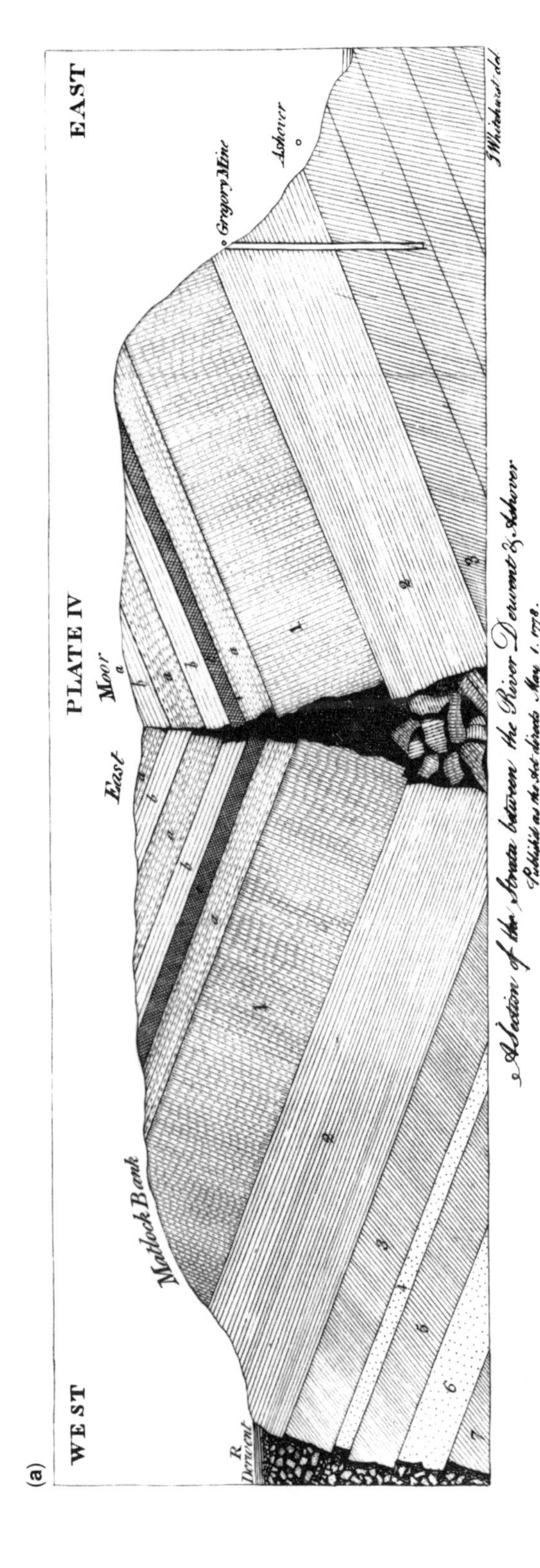

Fig. 5. Relevant 'background' images. (**a**) A section from Whitehurst (1778) through part of Derbyshire, to show how tilted 'Secondary' rocks were commonly depicted with ruled straight lines and often interpreted as broken rather than folded.

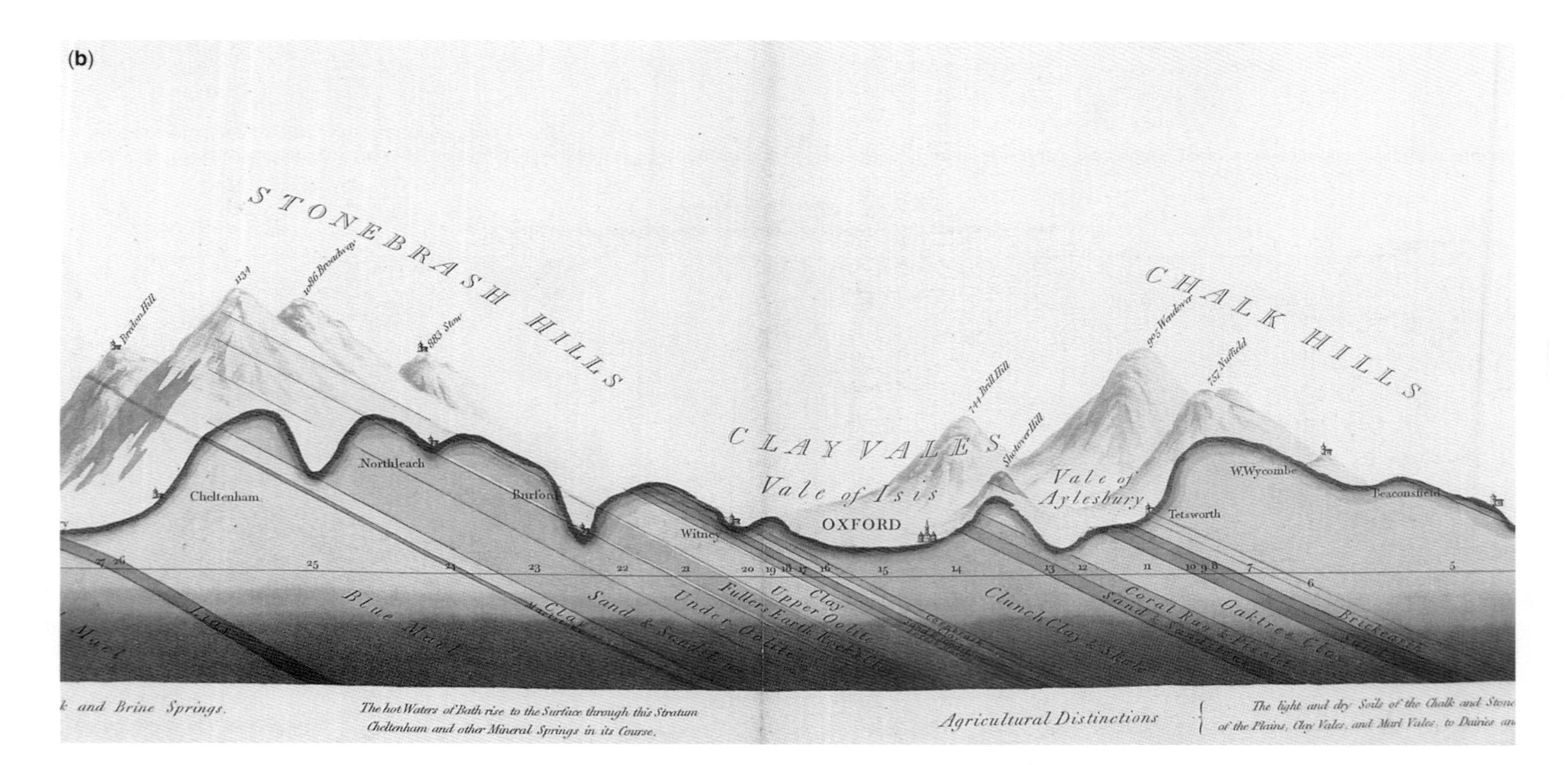

Fig. 5. (**b**) A section from Smith (1817) through the Cotswold (Stonebrash) and Chiltern (Chalk) Hills, with Oxford in between, showing dipping ‘Secondary’ strata ruled in straight lines, extrapolated down to great depths. Such sections, by William Smith (1756–1835) and his colleague John Farey (1766–1826), were already circulating in manuscript, years before this one was published.

Fig. 5. (**c**) A view from Saussure (1779) of the most famous huge fold structure in the Alps, at Nant d'Arpenaz (between Geneva and Chamonix), showing how solid 'Secondary' rock strata had apparently been contorted into a huge S-shaped fold (the upper limb is in the distant cliff).

implications of what he had seen, specifically in the light of Cuvier & Brongniart's Parisian research,

Source 5

Two of Webster's interpretative drawings based on his fieldwork, which are presented here as Figure 4(a) and (b).

Source 6

Relevant 'background' images from earlier geological works with which Webster and his contemporaries would have been familiar:

(i) Comparable fold structures. Earlier work on fold structures that would have been of use to Webster would have included Whitehurst (1778), Saussure (1779) and Smith (1817) (see Fig. 5a–c).

(ii) Significant fossils. Earlier work on fossils that would have been available to Webster would have included Brander (1766) and Brongniart (1810) (see Fig. 5d and e).

(iii) Comparisons with the Paris Basin. Comparisons with the geology of the Paris Basin would have been available from the published work by Cuvier & Brogniart (1811) (see Fig. 5f and g).

Secondary source

A short passage in Rudwick 2005 (pp. 512–521) covers Webster's work on the Isle of Wight in its immediate context of research by Cuvier & Brongniart on the 'Paris Basin' (1808/1811) and by James Parkinson on the 'London Basin' (1811). The passage includes reproductions of four of Webster's illustrations, with detailed explanatory captions. (A copy of this book will be available in the hotel, for background reading.)

(d)

Fig. 5. (**d**) Fossil mollusc shells from Brander (1766) from the coastal cliffs at Hordwell (now Hordle) on the English mainland west of the Isle of Wight, similar to living marine species; the same fossils were found by Webster in the Alum Bay strata.

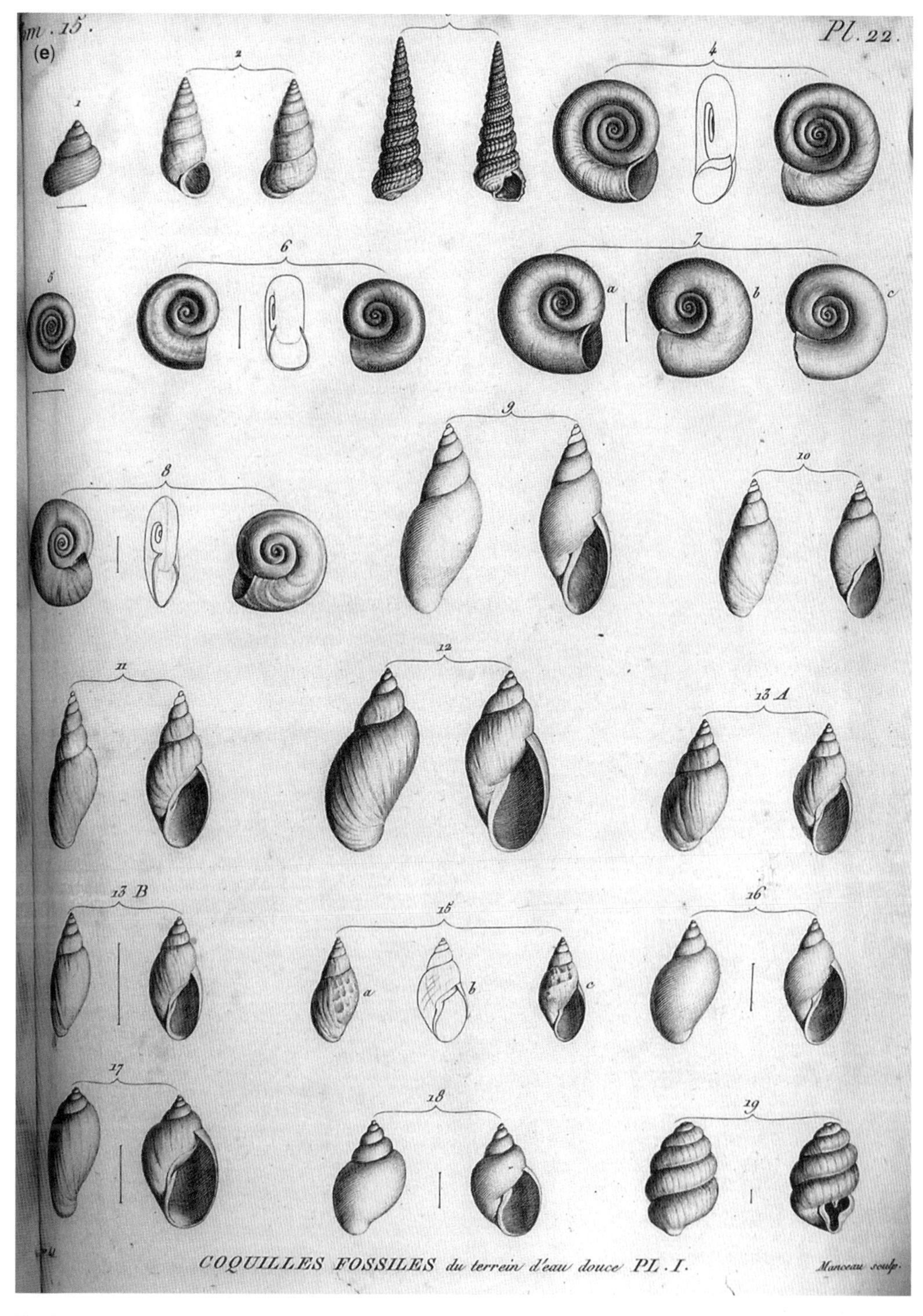

Fig. 5. (**e**) Fossil mollusc shells from Brongniart (1810) from the Paris Basin, similar to living freshwater species, and therefore used by him and Cuvier as markers of freshwater episodes; the same fossils were found by Webster in the Headen Hill strata.

(f)

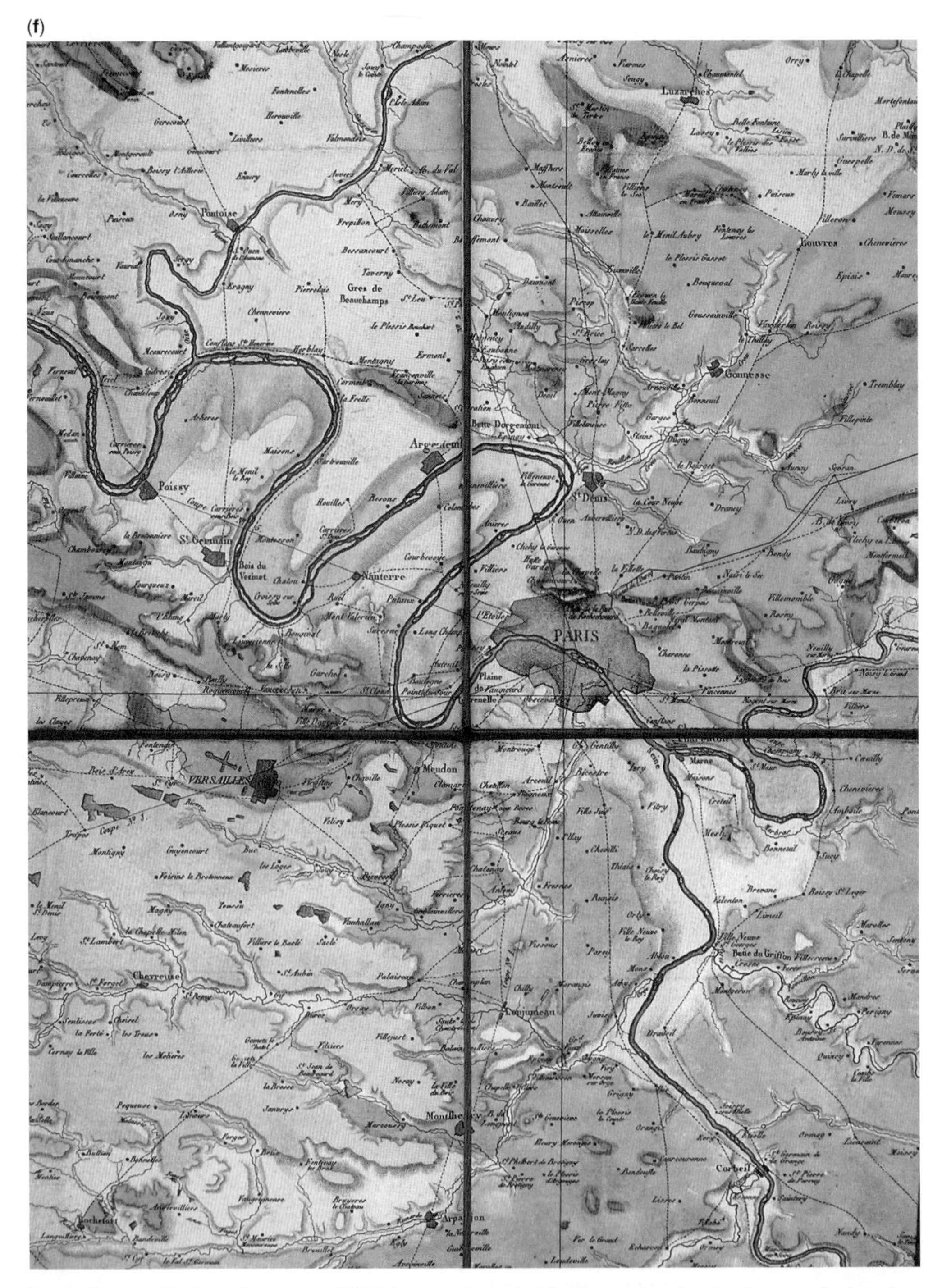

Fig. 5. (**f**) Part of Cuvier & Brongniart's (1811) 'geognostic map' of the Paris region, showing river gravels along the valley of the Seine, and outcrops of horizontal strata on the hills to north and south.

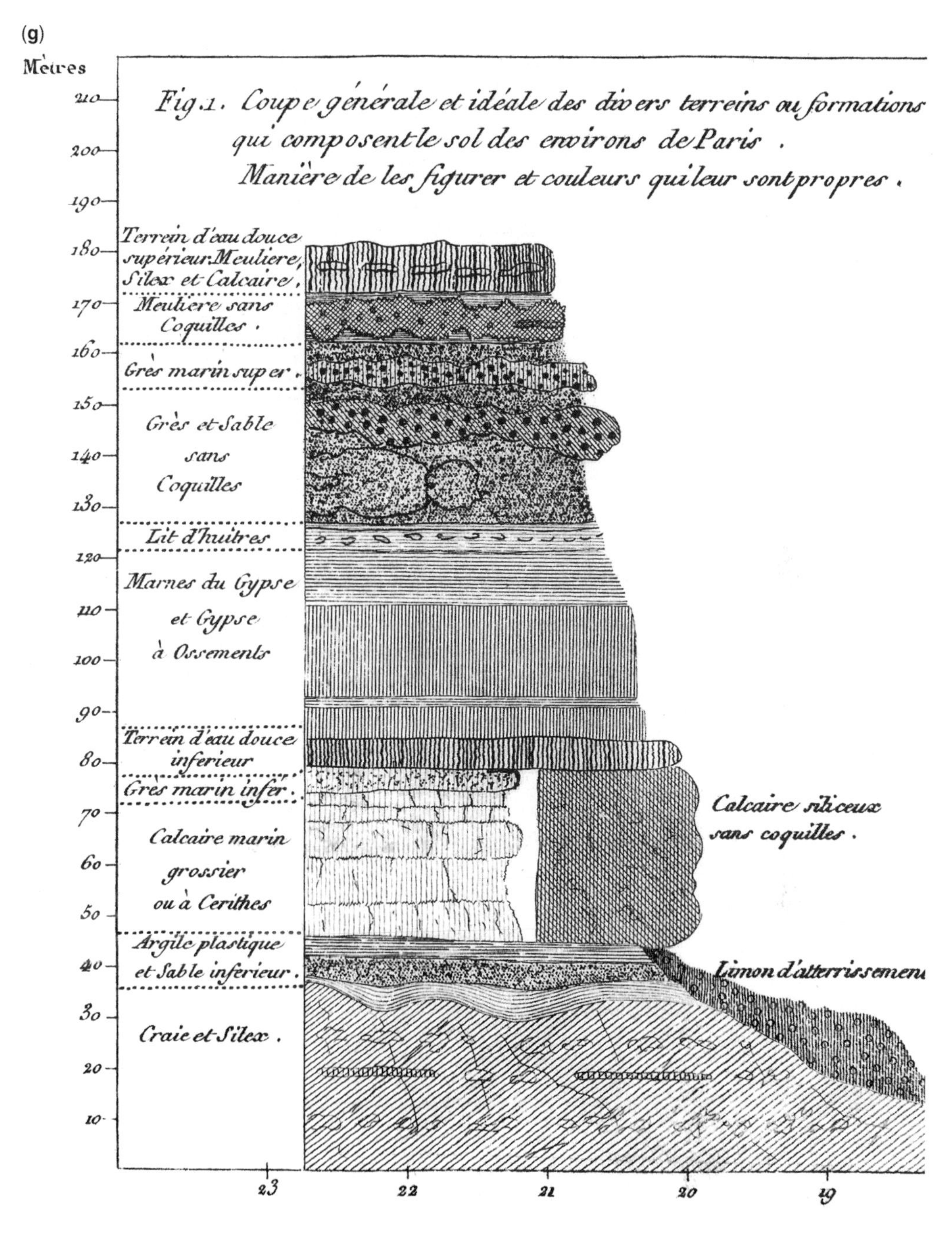

Fig. 5. (**g**) An 'ideal section' of strata in the Paris Basin, summarizing Cuvier & Brongniart's (1811) interpretation of the post-Chalk sequence as an alternation of marine (*marin*) and freshwater (*d'eau douce*) periods in that region. This was used by Webster as a standard against which to interpret the sequence on the Isle of Wight. (a), (b) and (d)–(g) are reproduced courtesy of the Eyles Collection, University of Bristol Library Special Collections.

References

BRANDER, G. 1766. *Fossilia Hantoniensia collecta, et in MusaeoBrittanico deposita.* London.

BRONGNIART, A. 1810. Sur les terrains qui paroissent avoir été formés sous l'eau douce. *Annales des Muséum d'Histoire Naturelle* **15**, 357–405, plates 22–23.

CUVIER, G. & BRONGNIART, A. 1808. Essai sur la géographie minéralogique des environs de Paris [preliminary version]. *Journal des mines*, **23**, 421–458. [English translation in Rudwick 1997, 133–156].

CUVIER, G. & BRONGNIART, A. 1811. [Full version.] *Mémoires de l'Institut Impérial de France, Classe des Sciences Naturelles et Mathématiques*, **1810**(1), 1–278.

ENGLEFIELD, H. C. 1816. *A Description of the principal picturesque beauties, antiquities, and geological phenomena, of the Isle of Wight, with additional observations on the Strata of the island ... by T. Webster.* Payne & Foss, London.

PARKINSON, J. 1811. Observations on some of the Strata in the Neighbourhood of London, and on the Fossil Remains contained in them. *Transactions of the Geological Society*, **1**, 324–354.

RUDWICK, M. J. S. 1997. *Georges Cuvier, Fossil Bones and Geological Catastrophes.* University of Chicago Press, Chicago, IL.

RUDWICK, M. J. S. 2005. *Bursting the Limits of Time: The Reconstruction of Geohistory in the Age of Revolution.* University of Chicago Press, Chicago, IL.

SAUSSURE, H.-B., DE. 1779–1796. *Voyages dans les Alpes, précédés d'un Essai sur l'Histoire naturelle des Environs de Genève* (4 vols). Samuel Fauche, Neuchâtel.

SMITH, W. 1817. *Stratigraphical System of Organized Fossils, With Reference to the Specimens of the Original Geological Collections in the British Museum, Explaining Their State of Preservation and Their use in Identifying the British Strata.* E. Williams, London.

WEBSTER, T. 1814. On the freshwater formations in the Isle of Wight, with some observations on the strata above the Chalk in the south-east part of England. *Transactions of the Geological Society*, **2**, 161–254.

WHITEHURST, J. 1778. *An Inquniry into the Original State and Formation of the Earth: Deduced from Facts and the Laws of Nature: to Which is Added an Appendix, Containing Some General Observations on the Strata in Derbyshire, with Sections of them [...].* Printed for the author and W. Bent, by J. Cooper, London; and sold at G. Robinson's.

Dining with the Founding Fathers: a personal view

RICHARD T. J. MOODY

School of Earth Sciences and Geography, Kingston University, Penrhyn Road, Kingston, Surrey KT1 2EE, UK (e-mail: rtj.moody@virgin.net)

Abstract: The History of Geology Group's (HOGG) Bicentenary celebrations included a dinner on the site of the Freemasons' Tavern where the Geological Society was founded, entitled *Dine with the Founding Fathers*. It was organized by HOGG committee member, Dick Moody, who recounts the history of the Freemasons' Tavern, the difficulties leading up to the dinner and the highlights of a memorable evening.

The organization of the Founding Fathers' dinner at 5pm on 13 November 1807 (Knell 2009; Lewis 2009) must have been a fairly casual affair: a messenger to book a table at the Freemasons' Tavern in Great Queen Street and a dozen or so letters to the interested parties. One assumes the menu, that cost 15 shillings, was focused on fresh meat and vegetables, washed down with ale or wine. The landlord would have been most agreeable and met individual requests with a smile. The majority of the Founding Fathers lived within walking distance, or were a short carriage ride away.

In the early 1800s the City of London was sited north of the River Thames with swathes of green fields south of the Waterloo Road and Greenwich Road highways. Edgware Road divided the City from the countryside to the west, and Regents Park and the Sommer's Town area defined the urban boundary to the north. The Mile End Road and Commercial Road to the east were like narrow corridors, lined intermittently with houses and commercial properties predating the expansion of commercial interests and the growth of the dockland areas. By 1807 the population of London was over a million and the docklands area had grown extensively with the building of the new docks on the Isle of Dogs, which was started in 1799. The City itself was laid out along similar lines as today, with the Strand, St Martin's Lane, Holborn and Chancery Lane surrounding Covent Garden and Lincoln's Inn Fields. Great Queen Street has been described as London's first real street, having changed from a bridleway to a Royal thoroughfare in the 1620s. Its early residents had an 'enviable view of the pastoral charm of North London'.[1] By the time Robert C. Rowe (1775–1843) had published his famous map of London in 1804, Drury Lane and Long Acre had become the major thoroughfares through Covent Garden (Fig. 1).

Great Queen Street has been the headquarters for Freemasons in England since 1717, although at that time they met in taverns and livery halls around the area. It was not until 1775 that the Freemasons established their first permanent lodge in the street. [2] In April 1774 the premier Grand Lodge purchased the freehold of number 61, the property then consisting of a house fronting onto Great Queen Street, behind which lay a large garden and a second house (Riley & Gomme 1914). A competition[3] was held for the design of a Grand Masonic Hall that would join the two houses together; the winning design was by Thomas Sandby (1723–1798). The back house became offices and meeting rooms, and the house fronting on to Great Queen Street was temporarily leased to a John Brooks, a frequent bankrupt! This became the Freemasons' Tavern and Coffee Shop, which opened in 1775 with a Luke Reilly as innkeeper (Clements 2006).

The building housing the Freemasons' Tavern stood on the site until 1788 when it was replaced by a four-storey structure designed by Sandby & William Tyler (1760–1801). It was this later building, beautifully depicted by John Nixon's (1760–1881) watercolour of 1803 (Fig. 2), in which the Geological Society was founded. Clements (2006) notes that the fortunes of the Tavern then fluctuated under different tenants or leaseholders over a 50-year period, although John Cuff was successful enough to have amassed a fortune of £120 000 by the time of his death in November 1848. The 1788 building was replaced in 1864 (Fig. 3)

[1]In and around Covent Garden: www.coventgarden.uk.com.

[2]The history of Freemasons' Hall: www.ugle.org.uk/ugle/the-history-of-freemasons-hall.htm.

[3]Library and Museum of Freemasonry: FMH PAP/172. Advertisement to architects of the Freemasons' Hall competition. Undated, *c*. February 1863.

From: LEWIS, C. L. E. & KNELL, S. J. (eds) *The Making of the Geological Society of London*. The Geological Society, London, Special Publications, **317**, 439–448.
DOI: 10.1144/SP317.24 0305-8719/09/$15.00

(a)

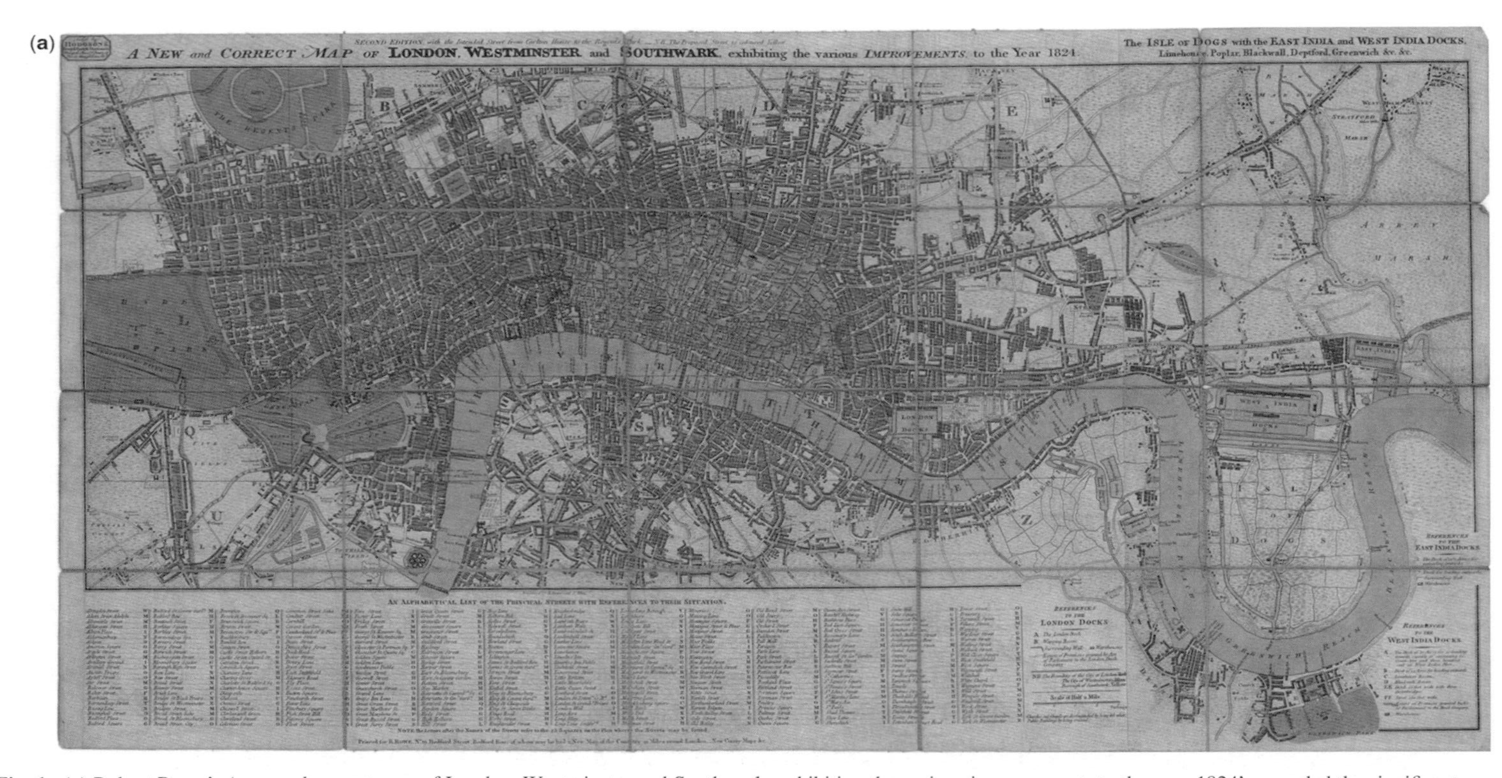

Fig. 1. (**a**) Robert Rowe's 'new and correct map of London, Westminster and Southwark, exhibiting the various improvements to the year 1824', recorded the significant development of the docklands area to the east.

(b)

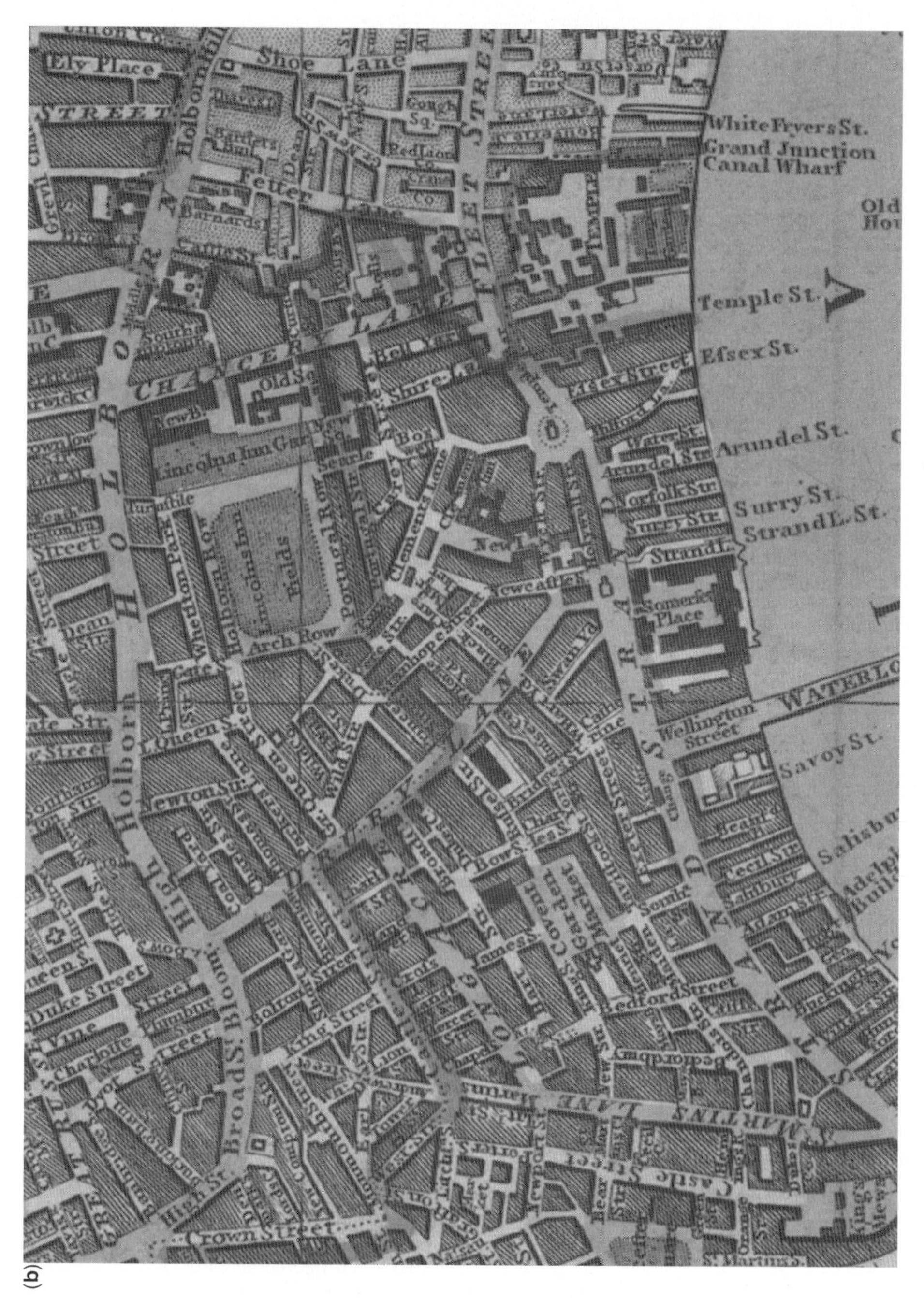

Fig. 1. (b) Detail of the area around Covent Garden. Courtesy: Edwin C. Bolles Collection at Tufts University.

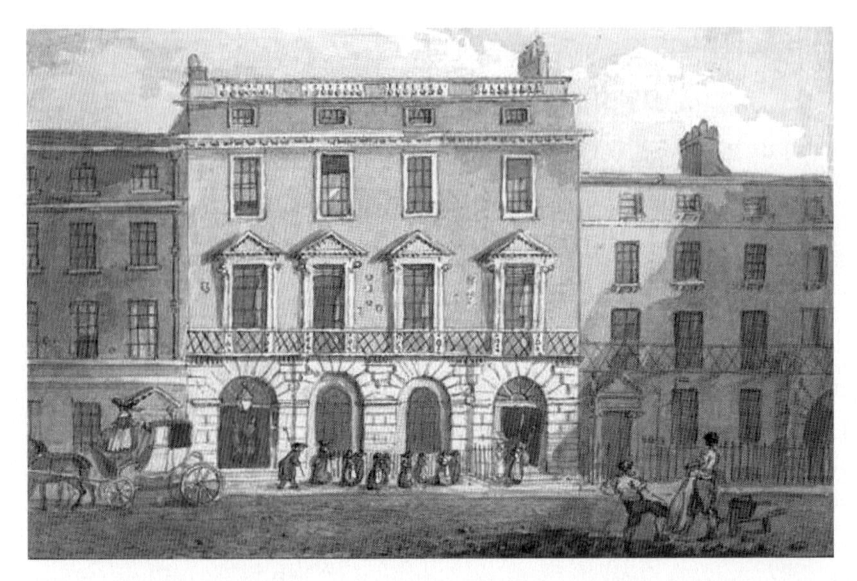

Fig. 2. 1803 watercolour of the four-storied building at 61 Great Queen Street, London, by John Nixon. The fourth storey was added to the original building in 1788–1789 when rebuilding took place under the direction of Thomas Sandy and William Tyler. Courtesy: Freemasons' Lodge.

and a new company, formed on behalf of the Grand Lodge of the Freemasons, was required to raise £65 000, by the sale of 6500 £10 shares, to facilitate the purchase of the lease, stock and goodwill of the Freemasons' Tavern from the current tenants, David Shrewsbury and George Elkington. Clements (2006) records that the company's prospectus drew attention to the fact that for nearly a century the Grand Hall had been used to hold 'great public meetings of a large number of the political, religious, and educational societies', such as the Humane Society (Fig. 4).

Fig. 3. The Victorian frontage of the Freemasons' Hall, rebuilt in 1864. Courtesy: Freemasons' Lodge.

In 1909 the Grand Lodge spent £30 000 on the renovation of the Tavern and changed the name to the Connaught Rooms in honour of the then Grand Master, the Duke of Connaught and Strathearn (Clements 2006). The New Connaught Rooms is a product of many changes – it has recently undergone a million-pound facelift, and there are now 29 meeting rooms that cater for groups of between 14 and 550 people. It was an ideal venue in which to remember the characters that played such a lasting role in the development and promotion of our science. Thus, the History of Geology Group's (HOGG) Bicentennial Dinner on the 12 of November 2007, was held in the Cornwall Room, with the magnificently-domed Crown Room hosting the reception.

In order to do justice to the occasion, the HOGG committee (Fig. 5) agreed to the design and printing of a Bicentennial brochure advertising the dinner, and a series of invitation cards to each of the dinner, conference and field trip.[4] The target for the dinner was 200 diners, with HOGG and the

[4]These were designed by Marcus Landau of Conker Design.

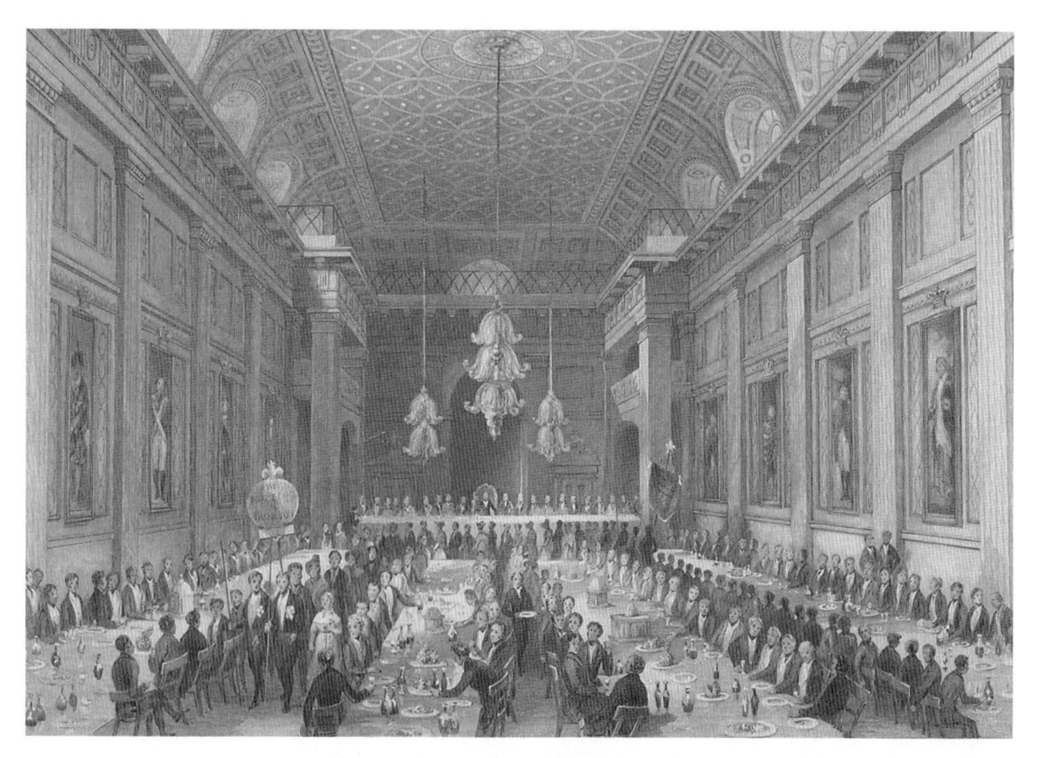

Fig. 4. The Great Hall of the Freemasons' Lodge was used for many functions. This undated print is captioned: 'Dinner of the Royal Humane Society, with a Procession of the Persons saved during the Year from Drowning'. The banner carried by the procession reads: 'We praise God and Thank you'. In 1777, James Parkinson (1755–1824), the only Freemason among the Geological Society's founders (Lewis 2009), was awarded the Royal Humane Society's silver medal for resuscitating Bryan Maxey who had hanged himself. Courtesy: Fry Collection, University of Bristol Library Special Collections.

Geological Society Club jointly sponsoring the evening. In the early stages bookings for the dinner and the conference lagged behind those for the Isle of Wight field trip that was soon fully subscribed, but as the date approached, late-comers were begging to be 'squeezed in'.

Organizing a dinner at an establishment with over 225 years of trading experience should have been a simple affair, but, from the time I became involved with the organization until the moment we sat down to eat, I dealt with five separate event managers, none of whom recorded my name and address on the house computer. As a consequence, at 3.30 pm, on the afternoon of the dinner, nobody setting up the dining room had an up-to-date copy of my seating plan. Fortunately, the staff allocated to the event on the night were amenable to change. With an hour to go, however, they were still working on the final arrangement of the tables and had still failed to produce a lay-out that matched my seating plan.

Unlike the Society's black-tie dinner held the following day, *Dine with the Founding Fathers* was intended to be more informal – a meeting of friends and colleagues, a celebration. Period costume was encouraged and HOGG committee member, Peter Tandy, took on the task of finding enough costume companies to meet the demand at reasonable prices. Many of the costumes were delivered to the dinner venue early on the 12th, and ladies and gentlemen were assigned separate dressing areas in the Oxford and Coronet Rooms. The excitement of dressing up was palpable when a flood of people arrived 40 minutes before the opening of the pre-dinner reception. The costumiers struggled to cope with the rush, and the staff of the Crown Room had to be hurried into opening the bar early. Some fully costumed individuals even ventured out to drink in the public houses of Great Queen Street, standing side by side with bowler-hatted, briefcase-carrying Freemasons who had been visiting the Grand Temple.

By 7.30 pm the guests of honour had arrived, the musicians and photographer were in place, the costumes were paraded and the Crown Room increasingly became a vibrant, noisy venue for the first of

Fig. 5. Members of the HOGG Committee on the night of the 12 November 2007. From left to right: Peter Tandy (Newsletter Editor), Nic Bilham, Patrick Boylan, Cherry Lewis (Chairman), John Mather, Alan Bowden (Treasurer), Anne O'Connor (Secretary), Richard Moody. Missing members: Beris Cox and Anthony Brook. Courtesy: Ricky at Artistic Images, Ai Events (modified by R. T. J. Moody).

the two dinners to celebrate the inauguration of the oldest geological society in the world. At 7.50 pm there was a short interlude for the presentation of three embossed scrolls by Dr Irena Malakova (Russian Academy of Sciences) to the Geological Society of London from three Russian institutes: the Russian Academy of Sciences, signed by academicians Nickolay P. Laverov (Vice-President), Yuri G. Leonov (Secretary to the Geosciences) and Gennady V. Kalabin (Director of the Vernadsky State Geological Museum); the Moscow Society of Naturalists and the Moscow State University, signed by Victor A. Sadovnichy (President of the Moscow Society of Naturalists and the Rector of the University) and Eugene E. Milanosky (Vice-President of the Moscow Society of Naturalists); and the Mineralogical Society of Russia signed by the President of the society, Dmitry V. Rundquist. The President, Professor Richard Fortey, received the scrolls on behalf of the Geological Society of London (Fig. 6).

Professor Fortey (Fig. 7) then unveiled a plaque that had been erected outside the New Connaught Rooms, indicating the site of the Society's inauguration on the 13 of November 1807. Standing on the pavement on a cold and windy November evening, with crowds of people passing by, Professor Fortey marked the unveiling with a short speech:

> Friends and guests,
>
> On this very spot, exactly 200 years ago tomorrow night, 11 of the men listed on the plaque I am about to unveil met to eat, drink, smoke, debate and inaugurate the Geological Society of London.
>
> Of course, history records that there were 13 founders, two of whom – [William Haseldine] Pepys and [William] Phillips – were unable to be present in person. I mention this footnote to history because, of course, this is a History of Geology Group event and I have an eye to the many 'referees' standing before me, ready to pounce on any inaccuracies. (With referees in mind it is perhaps appropriate to remember that the Freemasons' Tavern, in which we were founded and which once stood on this site, also saw the foundation of many other societies, including the Football Association, in 1863.)
>
> I am tickled by the fact that our Society was founded by 13 men on Friday 13th, because it seems to me to be an apt metaphor for a learned society that has been the enemy of cant and superstition since that very night – and it makes me proud. So – with cant and superstition in mind – I should just like to point out the amazingly

Fig. 6. The President, Professor Richard Fortey, receiving scrolls from various Russian Institutions, presented by Dr Irena Malakova. Courtesy: Cherry Lewis.

Fig. 7. Professor Richard Fortey at HOGG's Bicentennial Dinner. Courtesy: Peter Tandy.

> 'intelligent design' that has created, on this spot 200 years later, a great hotel with a dining room capacious enough to allow the successors of those 13 men the opportunity to meet in much greater numbers and celebrate their achievement.
>
> Ladies and gentlemen (and especially the Committee of the History of Geology Group and its members, whose idea this excellent occasion was and who have made it a reality), we salute the generations of bold-thinking Earth scientists that have come and gone since 13 November 1807, and I urge you now to raise your glasses – and in the absence of glasses, your hearts – to the Geological Society of London. We shall use our great and noble history as a springboard for another 200 years, for 'making geologists acquainted with one another, of stimulating their zeal, facilitating the communication of new facts, and ascertaining what is known and what yet remains to be discovered'.[5]
>
> Ladies and gentlemen, the rest of our Society's journey into the yet unknown regions of Earth history begins tonight.

Dinner was served slightly late in the Cornwall Room, but the vibrancy and jovial nature of the reception was carried over to the dining area. As in the reception, where the music of the Alicia Hunt Duo was totally masked by loud conversation, the strings of the Abraxas Ensemble in the Cornwall Room rarely extended beyond the tables immediate to their podium. Actors dressed as Mary Anning and William Smith mingled almost inconspicuously among the diners, half of whom were wearing period costume. Guests dined on poached salmon with a hollandaise sauce, garnished with asparagus, which was followed by venison steak with a puree of potato and pumpkin, red cabbage and a wild cranberry jus. Unfortunately, the choice of venison seemed beyond the capabilities of the chef and large pieces of 'firm' meat saddened the accompanying vegetables. Fortunately, the advent of the National Health Service and good dentistry avoided the obvious problems that would have been rife 200 years earlier. The meal concluded with a triple chocolate mousse served with a raspberry coulis.

The break between the main course and dessert offered an opportune moment for the President of the Society to welcome the guests, and Lord Robin Derwent, guest of honour with his wife Sybille, toasted the Society:

> Among so many distinguished geologists, I feel somewhat like the cuckoo in the nest. We are celebrating this

[5]Quote from the objectives of the society, given in the first minutes. Geological Society of London archive: Ordinary Minute Book 1, 13 November 1807.

evening the Founding Fathers of the Geological Society. My connection with the world of geology is that it was my great-great-grandfather, Sir John Johnstone, who employed William Smith, became his Patron, introduced him to the Scarborough Philosophical Society and subsequently gave the stone to enable the Scarborough Rotunda to be built to Smith's design. As we all know, of course, during this period Smith was less than welcome at the Geological Society of London and indeed Greenhough [sic] (in Yorkshire we say Green-ugh!) plagiarised his map and was perhaps indirectly responsible, therefore, for his bankruptcy and committal to debtors' prison. However, all's well that ends well. In due course, Smith received the first Woolaston Medal from the Geological Society and his contribution to geology was generously acknowledged by it.

It therefore gives me enormous pleasure to have this opportunity to thank the Geological Society for their moral support for the restoration of the Rotunda and to thank the Society even more for its gift of a reproduction of Smith's map of England which will have an honoured place in the Rotunda when it reopens as the William Smith Museum of Geology next May. We are most grateful. I was delighted to learn that the History of Geology Group will be visiting Scarborough next October when we will be able to show you what we have achieved and you will see the map in place.

Finally, on behalf of my wife and fellow guests may I thank our hosts for a delicious and convivial dinner. May I ask you to join me in the toast to the Geological Society of London and its Founding Fathers – coupled with the name of William Smith!

Mr Ray Thomasson, past President of the American Geological Institute and the American Association of Petroleum Geologists, and dressed as President Jefferson's representative to the Court of His Britannic Majesty during November 1807, then spoke on behalf of the Institute (Fig. 8):

Lord Derwent, President Fortey, Chairman Moody, esteemed Fellows and honored guests:

It is with great pleasure that I bring congratulations from President Jefferson on the launching of what I am certain will become the flagship geological society in the world. President Jefferson has had a keen interest in geology and especially palaeontology for many years. As a past President of the American Philosophical Society, Mr Jefferson was particularly interested in a set of bones sent to him as those of a mammoth. He was able to determine they were [in fact] the bones of a clawed quadruped which he described in detail. Dr Wistar and Dr Charles Peale of Philadelphia have identified the bones as those of a giant sloth. He was flattered to have it named *Megalonyx jeffersoni* – the first such mammal found in the new world. Mr Jefferson, even though an amateur, was willing to challenge Comte de Buffon's assertion that human and animal life in America was both degenerative and inferior to that of Europe. To counter Buffon's argument he collected and described the bones of a mastodon. He then had them shipped to Paris as his evidence that animals are actually larger in America.

(a)

(b)

Fig. 8. (**a**) Mr Ray Thomasson, past President of the American Geological Institute and the American Association of Petroleum Geologists, dressed as (**b**) William Pinkney, Minister Plenipotentiary, President Jefferson's representative to the Court of His Britannic Majesty during November 1807. Image (a) courtesy: Ricky at Artistic Images, Ai Events.

May I say I am particularly excited to be Mr Jefferson's representative to this group of ardent geological scientists. Mr John Playfair's recent publication in Scotland has helped us to better understand the thoughts of Dr James Hutton. As amateurs, both Mr Jefferson and I found Dr Hutton's thoughts revolutionary but sometimes hard to grasp. If you visit our new nation, Mr Jefferson would be honored for you to visit his home, Monticello, in Virginia. There his

Fig. 9. The plaque erected outside the New Connaught Rooms, marking the site of the Geological Society's inauguration on the 13 of November 1807. Courtesy: Cherry Lewis.

interest in natural history caused him to commission William Clark of the Lewis and Clark expedition, which explored our recent purchase of Louisiana from the French, to collect a mastodon head which he has displayed in the entry hall at Monticello.

In closing I would like to say that I can only hope that we Americans, who owe so much to Great Britain's leadership in government, the arts and science, will in time develop a sister geologic organization which could affiliate with you. Possibly, it could be called the American Geological Institute.

Thank you.

William Pinkney, Minister Plenipotentiary.

After the toasts, the evening passed quickly, but no one was in a hurry to leave early. The industry-sponsored tables, occupied by B G Group plc, Centrica Plc, Hess Ltd, Premier Oil Plc and SASOL Petroleum International (Pty) Ltd, were buoyant until midnight when the music of the Abraxas String Ensemble finally started to make an impression. As the new day dawned, a feeling of relief came over me and I could relax until the next time I volunteered for a new challenge. The evening of the 13th would transport many off to a new venue at the Natural History Museum. Two dinners, two nights to remember – 2007 had been a year for celebration, a time to salute the Founding Fathers (Fig. 9), and a year to broadcast the nature and importance of our science worldwide.

The author would like to thank Susanne Belovari, Archivist for Reference and Collections, and Jennifer Phillips, Archives and Research Assistant, at the Digital Collections and Archives, Tufts University, Medford, MA, USA, for their help in sourcing a digital copy of the Robert Rowe map of London in 1804/1824. Thanks are also due to Diane Clements, Director of The Library and Museum of Freemasonry at the Freemasons' Hall, London for her help in sourcing the images of the Nixon watercolour of the Freemasons' Tavern and the rebuilt tavern of 1864. Diane was also an inspiration in terms of the factual history of the Tavern over the past 200 years.

References

CLEMENTS, D. 2006. The hall in the garden. *MQ Magazine*, **18**, 250–253. Magazine of the United Grand Lodge of England: www.mqmagazine.co.uk.

KNELL, S. J. 2009. The road to Smith: how the Geological Society came to possess English geology. *In*: LEWIS, C. L. E. & KNELL, S. J. (eds) *The Making of the Geological Society of London*. Geological Society, London, Special Publications, **317**, 1–47.

LEWIS, C. L. E. 2009. Doctoring geology: the medical origins of the Geological Society. *In*: LEWIS, C. L. E. & KNELL, S. J. (eds) *The Making of the Geological Society of London*. Geological Society, London, Special Publications, **317**, 49–92.

RILEY, W. E. & GOMME, Sir E. 1914. Freemasons' Hall. *In*: *Survey of London. Volume 5: St Giles-in-the-Fields*. The London County Council, London, Part 2, 59–83.

Appendix I

Geological Inquiries
[The Geological Society's first publication]

1808

Introduction

GEOLOGY relates to the knowledge of the system of our earth, of the arrangements of its solid, fluid, and aeriform parts, their mutual agencies, and the laws of their changes.

In this point of view, it is necessarily connected with many branches of Natural Science, but it is more particularly dependent upon Mineralogy, which distinguishes the species of inorganic bodies; and upon Chemistry, which investigates the intimate nature of matter and its hidden properties.

Geology in its comprehensive sense is consequently a sublime and difficult science; but fortunately for its progress it is susceptible of division into many different departments, several of which are capable of being extended by mere observation.

The knowledge of the general and grand arrangements of nature must be collected from a number of particular and minute instances, and on this ground the slightest information relating to the structure of the earth is to be regarded as of some importance. —To reduce Geology to a system demands a total devotion of time, and an acquaintance with almost every branch of experimental and general Science, and can be performed only by Philosophers; but the facts necessary to this great end, may be collected without much labour, and by persons attached to various pursuits and occupations; the principal requisites being minute observation and faithful record. —The Miner, the Quarrier, the Surveyor, the Engineer, the Collier, the Iron Master, and even the Traveller in search of general information, have all opportunities of making Geological observations; and whether these relate to the metallic productions, the rocks, the strata, the coal of any district; or to the appearances and forms of mountains, the direction of rivers, and the nature of lakes and waters, they are worthy of being noticed.

It is with a view to facilitate and in some measure to direct general research, that the Members of the Geological Society have collected from different sources and put together the annexed inquiries; and, as insulated remarks and local information can be of no avail, unless preserved and arranged, they venture to offer a repository for any facts that may be communicated to them. One great end of their association is to afford means by which this kind of knowledge may be concentred; and they conceive that by the labours and talents of many individuals thus united and assisted, several important objects may be easily attained; that Mineralogical maps of districts, which are now so much wanting, may be supplied; that the nomenclature of the science may be gradually amended by the selection of the most current and significant terms; that theoretical opinions may be compared with the appearances of Nature, and above all, a fund of practical information obtained applicable to purposes of public improvement and utility. — They address themselves more especially to their countrymen, and they cannot conclude without noticing the extraordinary facility of obtaining Geological information, afforded by the territory of the British Isles, and the peculiar interest which ought to be attached by their inhabitants to such inquiries. In no equal space is so great a surface of the earth laid bare by nature and by art; no country is richer in those mineral productions on which some of the most important of our manufactures, and the most useful of the arts of life depend; and the present time is one in which we are particularly called upon to explore and employ the whole of our native riches and internal resources.

From: LEWIS, C. L. E. & KNELL, S. J. (eds) *The Making of the Geological Society of London*.
The Geological Society, London, Special Publications, **317**, 449–456.
DOI: 10.1144/SP317.25 0305-8719/09/$15.00

Geological Inquiries

§ I. Concerning Mountains and Hills

Are they solitary, or in groups, or do they form a chain?

If Solitary,

The general figure, as conical, pyramidal, &c. —more particularly of the summits?

The height above their base, and above the level of the sea?

The length, breadth, and general form of a horizontal section passing through the base, or the *ground plan*; and the points of the compass between which the long diameter lies?

The degree of declivity on every side with regard to the circumjacent plain?

Do they present on any side abrupt craggy faces, and to what points of the compass are these opposed?

Do these precipices extend to the foot of the mountain, or are there at their bottom sloping banks of loose fragments?

Is the surface smooth or rugged? —dry or marshy?

To what height does vegetation ascend, and what are the prevailing plants in different parts of the ascent?

The springs, streams, lakes, hollows, gullies, caverns?

Whether any loose blocks of stone are found on the surface, different from those of which the mountain is composed?

In addition to the preceding Inquiries,
If in a group,

Are the component mountains of nearly the same height?

Which are highest, the central or external ones?

If in a Chain,

The outline of the chain?

Its highest point?

Its length?

Whether straight or curved, and extended between what points of the compass?

Whether any lateral ridges proceed from the main chain?

§ II. Concerning Vallies.

Their geographical boundaries?

Their length, breadth, depth?

Are they occasionally dilated and contracted, or do their sides preserve an uniform parallelism?

Is the bottom or floor even or rugged? —nearly level or much inclined? If inclined, whether regularly or interruptedly, and in what direction?

Are the slopes that form their sides smooth and gentle, or rugged and precipitous?

Do the opposite sides consist of the same kind of rocks, and do they correspond in the inclination of their beds or strata?

Are there on their sides depositions of water-worn and rounded pebbles, either loose or compacted, and to what height do they reach?

Are the detached fragments, by which the bottom is overspread, angular or rounded? of the same species of rock as composes the sides of the valley, or different?

Of what description is the solid rock or base upon which these rest?

Are they open or closed at one or both extremities?

Do any subordinate lateral vallies open into the main one, and what remarkable circumstances occur at their junction?

Do streams rise in or flow through them, and in what direction?

§ III. Concerning Plains.

Their shape and extent, with the nature, height, and general appearance of the hills or mountains by which they may be bounded?

The degree and direction of their inclination or slope?

The nature and character of the different soils by which they are covered?

Whether dry or abounding in springs and standing waters?

If traversed by streams, in what direction do they flow?

Are the beds of rounded pebbles (if such occur) composed of minerals similar to those which form the surrounding mountains?

Have any opportunities presented themselves in sinking shafts or wells, cutting canals, excavating docks and quarries, and digging foundations, of examining the subjacent strata, and what are the results of such observations?

§ IV. Concerning Rivers.

Their source, their mouth?

The direction and length of their course, and whether these are the same now as formerly?

Their breadth, depth, and rapidity?

What is the rate of their descent or fall? is it uniform or interrupted?

The amount of their periodical increase or decrease?

The colour, temperature, and other properties of the water?

Whether any part of their course is subterranean?

Do they run in the same direction as the strata, or cross them, and at what angle?

The nature of the bed, whether rock, mud, sand, or gravel? Are the pebbles of the same rock as that of the adjacent country?

§ V. Concerning Lakes, Springs, and Wells.

1.—Lakes.

The extent, depth, temperature, and other properties of the water?

The periods and amount of their greatest annual increase and decrease?

Whether supplied by springs or streams, and whether any streams flow out of them?

Of what is the bason composed?

Are there any appearances that indicate the extent to have been formerly different from what it is at present; and does this alteration seem to have been gradual or sudden?

Are there shoals of gravel and low islands in those parts where streams flow in; and do these increase from year to year?

2.—Springs.

The physical and chemical properties of the water—the nature of its deposit?

The quantity discharged in a given time, and the degree to which this is affected by dry or wet seasons?

The kind of rock from which the water issues?

3.—Wells.

Their depth?

The number, thickness, and species of strata pierced through in sinking, and the order of their position?

Whether all the wells of a district derive their water from the same stratum?

Whether when the water first flows, it rises rapidly and accompanied by sand?

Is the water liable to periodical increase or decrease?

§ VI. Concerning Shores or Coasts.

If the shore is flat, to what extent? and whence are the sand and pebbles derived? Are they part of the adjacent cliffs, or brought down by rivers, or deposited by the sea? in what quantity and of what description?

If the coast is precipitous, the form and elevation of the cliffs, with the nature and disposition of the rocks which compose them?

§ VII. Concerning the Sea.

Its depth, tides, currents, inlets, nature of the bottom, &c.

The height to which it rises?

What effects has it produced on the adjacent rocks, &c.?

Are there any indications of its having formerly had a different level?

§ VIII. Concerning Rocks.

Their horizontal outline?

Are they separated from each other by thin bands of clay, or other extraneous substances? or slightly joined to one another? or firmly welded together?

When two rocks of different species come in contact, is any difference in colour, hardness, &c. observable between the adjacent surfaces and other portions of the same rock?

When a rock terminates at the surface of the earth, are any fragments of it to be traced in the form of gravel, &c.? —Does it re-appear after such interruption, and what is the nature of the intervening substance?

The form of their broken ends?

Are any rocks observed to terminate constantly together? and what are they?

If Stratified.

Is the stratification distinct or indistinct?

What is the number and thickness of the strata, and the order of their position?

Do they alternate or recur at regular intervals?

Do they, whether straight or waved, preserve their parallelism throughout, or are they cuneiform, &c.?

When vertical, what points of the compass are opposed to their sides, and what to their edges?

What is the amount of their dip, or the angle which they form with the horizon, and is it the same throughout their whole extent?

To what point of the compass do they decline?

What several strata, of the same species, are incumbent on each other, do they differ in thickness or consolidation?

Where veins, dykes, or fissures occur, are the strata depressed, elevated, contorted, or altered in any other way?

How far does the external form of the mountain correspond with the position of the strata?

If the stratum contains broad and thin distinct particles (such as mica), do these all lie in the same direction?

Note—Care must be taken in examining strata, not to be deceived by distance or perspective, or by mistaking fissures for stratification, and fallen strata for strata in their natural position; and it should be kept in mind, that before the inclination of a stratum can be determined with certainty, it is necessary that it should be seen on two of its adjacent sides.

If Unstratified.

Are they amorphous, columnar, or in globular concretions?

Do they split with the same ease in all directions, or have they what is called a grain?

Do they abound in fissures, and what is the direction and extent of these?

§ IX. Concerning the Materials of Rocks.

Are they composed of one mineral substance, or of more? In the latter case, which has impressed the other?

Are they composed of parts cemented together, or adhering to each other without a cement?

Are they granular, slaty, porphyritic, amygdaloidal, or any compound of these? If Breccia, are the included nodules large or small, entire or broken, &c.?

Do they contain fragments of other rocks, and of what description? Sand? Shells? Corals? Vegetable impressions, or any thing that appears to belong to a different formation?

Are there hollow nodules, and in what manner are they lined?

Is there any character, by which substances found in one stratum can be distinguished from similar substances found in another? or by which, what have been called primary strata may be distinguished from secondary strata, and strata of transition?

What minerals are found to be generally concomitants of others?

How are the several species affected by the combined action of air and moisture? Where large fragments have been torn by torrents from known rocks, what is the progress of their decomposition, and is there any re-aggregation?

What are the characteristic forms of each species of rock —in mountains?

in detached blocks?

How are they affected by peat moss lying on them?

What are the plants, the presence or absence of which indicates the nature of the soil?

By what local denominations are the different rocks distinguished, and to what economical purposes are they applied?

§ X. Concerning Veins.

Are they of the same materials as the rock in which they occur, or of any contiguous rock?

What is their direction with regard to the points of the compass, and the inclination of the adjacent strata?

Are they vertical, horizontal, or inclined, and at what angle?

What are their several dimensions?

Are they nearly the same thickness at different depths? Do they terminate in a wedge and this, at the top or bottom of the vein?

Is their longitudinal course straight or curved?

Is it of uniform breadth, or does it enlarge and diminish?

Do they ramify, and in what direction? Do the branches re-unite?

In what order are the minerals arranged, of which the vein is composed?

Are there any fragments of other rocks, any pebbles, any organic remains among them?

When a vein comes in contact with a different species of rock from that in which it was first observed, is the vein abruptly cut off, raised, depressed, turned aside, or are its materials altered?

If a vein is cut off, or shifted by the interposition of a stratum or mass of rock, does it re-appear or recover its direction on the other side of the interposed body?

Is it shifted or cut off without any apparent cause?

Are neighbouring veins composed of the same materials?

Have veins, consisting of similar materials, the same direction?

What proportion do the several veins bear to the rock in which they are found?

Do they run parallel to each other?

Do they tend to a common centre?

Do they cross each other, and what phenomena occur under these circumstances?

What is the nature of their floor, sides and roof?

Do the veins seem to have produced any change on the adjacent part of the containing rock, as indurating it, disturbing the regularity of its stratification, &c.?

Can they be traced to beds composed of the same materials as themselves?

§ XI. Concerning Organic Remains.

To what class, and species, do they belong?

Do they conform to the direction of the strata in which they occur?

Do particular shells, &c. affect particular strata?

What change have they undergone? Are the vegetables compressed, carbonized, bituminized, silicified, or penetrated with pyrites in whole or in part? Do the shells retain their enamel? The bones their phosphoric acid, &c.?

Do the shells or other organic remains appear perforated or worm eaten?

What is the nature of the rock or bed in which they are found?

Are the bones disposed in entire skeletons? Are those of different animals mingled together?

Are the shells worn, broken, crushed, or thrown out of their natural position? Are the different species confusedly intermixed?

Does this mixture extend not merely to species and tribes, but even to classes? i.e. are the remains of fish and sea shells accompanied by those of land animals and vegetables?

Are any analogous living species now found, or known to have been formerly found in their vicinity or elsewhere?

Among the various organic remains, can any traces be observed of the existence of man?

FINIS

W. Phillips, printer, George yard, Lombard-street.

Appendix II

Preface to:
Complete Treatise of Carbonated Lime and Aragonite
by
Comte de Bournon
1808

Translated from the French original by
Margaret Morgan, Royal Cornwall Museum

ADVERTISEMENT

When the author undertook the writing of these three volumes, he had conceived the project of fulfilling the same task with regard to all mineral substances, which would have formed a treatise of mineralogy as complete as mineralogical knowledge of this period could allow it to be done, and infinitely more than any of those which would have preceded it. He had gathered, on this subject, an immense quantity of material, such that he dares to say no European cabinet could show similar. Mineralogical scientists can judge by the facts that these three volumes contain, with regard to carbonated lime and aragonite; facts of which almost all the objects on which they are based are in his hands. It is the same in almost all the other mineral substances, of which a very large number have offered him such a multiplicity of facts, either not described or to be corrected, that he is not afraid to suggest that several of them must be completely altered. Events, which he could fear but which his reason refused to contemplate, forcing him to end his mineralogical work, he changed the first title from *traité de minéralogie* [*Mineralogical treatise*] which he had first given to these first three volumes, to *traité complet de la chaux carbonatée et de l'arragonite* [*Complete treatise of carbonated lime and aragonite*], a title which perfectly suits this work and which he dares to believe he has fulfilled.

From: LEWIS, C. L. E. & KNELL, S. J. (eds) *The Making of the Geological Society of London*.
The Geological Society, London, Special Publications, **317**, 457–464.
DOI: 10.1144/SP317.26 0305-8719/09/$15.00

TO HIS IMPERIAL MAJESTY

ALEXANDER I

Emperor and Ruler of all the Russias

SIRE,

The generous and enlightened protection that your IMPERIAL MAJESTY accords to the sciences, which he loves and encourages with such success throughout his empire, has entitled him to the profound gratitude of all men who devote themselves enthusiastically to their study. This sentiment, which I feel in all its breadth, has made me ardently desire that your IMPERIAL MAJESTY should allow me to offer him the homage of the fruit of my labours in a science which owes part of its progress to the discoveries which are made daily in Russia. He has had the kindness to receive my request favourably and to authorise that I should inscribe this work, of which some particular circumstances have delayed the publication, with the name of AUGUST PROTECTOR, who is indeed willing to allow it to appear under his auspices. This favour inspires me with confidence, by giving me more hope and right to the indulgence that I need.

I am with the deepest respect,

SIRE,

Of your IMPERIAL MAJESTY

the very humble, very obedient

and very respectful servant

Count de Bournon

LE C[TE] DE BOURNON

PRELIMINARY DISCOURSE

From the first years of the springtime of my life, the study of Nature has had infinite charm for me. From this same period, my sweetest pleasures have been to seek to know the diversity of the always interesting objects that she holds in her bosom, to converse with her and to deserve, by the constancy of my dedication, that she should reward my work by some of the favours that she usually grants to those who cultivate her zealously. After giving myself up for some time to the study of botany, then to that of antomology, the first essay on crystallography of Romé de l'Isle, when I read it, directed my attention for ever towards the study of mineralogy. This first glance, cast on one of the most interesting facts of Nature, whose games of chance it caused to disappear, children of our ignorance and previously formerly so often advanced, caused the strong desire to know minerals to be born in me.

Wishing, consequently, to study mineralogy methodically, and needing to acquire the preliminary knowledge necessary for this study, I went to find Romé de l'Isle in Paris. This excellent man, to whom mineralogy has been so greatly indebted and whose character was openness and kindness itself, was very willing to be my guide. I had the satisfaction of keeping him afterwards as my friend, until death came to take him away from Science, who mourned him and from his friends, whose souls are still in mourning. Throughout his life, I kept up a very active correspondence with him, which I have had the satisfaction of bringing out with me from France and which filled the intervals that I put between the journeys that I made to him. This correspondence in no small way contributed to sustain and even to intensify my zeal for mineralogy. Through it, I communicated all my observations to him as I made them; I also had conveyed to him all the mineralogical objects that I thought ought to interest him, and principally the new crystals that I was able to observe.

At that time, I lived in Grenoble, the capital city of Dauphiné, situated at the foot of the Alps. How precious this position was for the study which occupied and charmed my leisure hours! Chance favoured me sufficiently to mark my first steps in the mineralogical career by some new observations which were then appreciated by the mineralogists of Paris, much above their value.

This appreciation flattered my vanity and increased my zeal. Everything tended to increase my taste for mineralogy; and so it became almost a passion; but also never did passion present so much enjoyment from its beginning to its greatest development. Happy, as much as it was possible to be, in myself, having friends and knowing no other society but theirs, living in a happy and flourishing kingdom, governed by the best, the sweetest and the most virtuous, as well as having been the most unfortunate of kings, what situation could be more fortunate for the study of the sciences, for which the calm of the soul, tranquillity and security form the most important benefits and the most absolute of needs.

My situation at the foot of the Alps, which made them for me the richest and most instructive of cabinets, put me in the position of often making journeys there. Let me linger for a moment on the memory of all the pleasure which accompanied me there. To all the enjoyment offered to me by such an imposing, and at the same time so interesting, a spectacle presented by those colossal deposits of one part of the treasures and the secrets of Nature, was joined that of the most agreeable company. Mr Villars, first class botanist, Mr Schreiber, then director of the Allemond silver mine and of all those of the Dauphinois Alps, who joined to a profound knowledge of the working of mines that of mineralogy. Mr l'Abbé Du Cros, librarian of the Grenoble public library and director of the mineralogical cabinet of that establishment, have usually gone with me into these mountains, frequently all three and always at least some of them. I shall never forget the delicate, preventative care with which they diverted from me all difficulties and obstructions, to allow me to enjoy in its entirety and to enjoy completely the pleasure of being with them: may they at least know that I have preserved its memory and gratitude.

After studying Nature in the great mass of the Alps and carefully examining those of the masses of which the rock belongs either to the primitive stony products, or to the secondary ones; after observing the way in which these products behave with regard to the other great deposits made at their base or on their slopes, or even on their summits; after travelling at different periods through a great part of the mountains, as much shell-bearing as chalky etc. of a part of France such as those of Dauphiny, Lyonnais, Burgundy, Franche-Comté, Champagne, Lorraine etc. it remained for me to observe the great chains of low primitive mountains. I was also very anxious to know the major facts about extinct volcanoes. These facts which today are a subject of discussion between mineralogists whom they have divided into two different groups, whose members, as is normally the case, are extreme in their opinions and equally reject the opinion of the very-wise who wish to be placed between them. These facts, I say, deserve indeed to be studied carefully. Already those presented by the extinct volcanoes of Auvergne have changed the opinion that Dolomieu, one of the scientists who has concerned himself the most with volcanoes, had adopted in the study that he had made of Italian ones. Observing these same extinct volcanoes would, I think, also have a great effect on the opinion of those mineralogists who, having adopted the Neptunist system in all its exclusive rigour, would make up their minds to travel through them and study them before giving an opinion. The mountains of Auvergne, together with those of Forez, of Velay, of Vivarais and of the Cévennes, gave my wishes in this regard ample satisfaction and doubtless even with more benefit than could any of the other European mountain chains. They also offered me another important interest in the coal mines which at St Etienne and at Rive de Giers in Torey are so numerous and so abundant there. I particularly wanted to study the low primitive mountains and principally the granite ones. These mountains, uncovered much later than the high ones, covered again later and, perhaps for some, at a period very near that of their formation, by the product of the disintegration and the destruction of others must, by that very fact, have preserved most perfectly the character of the last products of their formation. They seemed to me to have, consequently, to promise facts which would be peculiar to them and of which the high mountains could not give information, and the outcome perfectly fulfilled my expectations. Consequently, I was to spend several summers in succession in these provinces and in various excursions that I made there, the enjoyment, which had up till then accompanied the study that I was making of mineralogy, followed me there. I received in the places in these provinces where I stayed, evidence of courtesy and kindness of which the memory has not been effaced and has served sometimes since to compensate for contrary feelings when I experienced them.

It was at that period that I gave a short note on some objects connected to the mineralogy of Forez in a small work entitled *Essai sur la Lithologie des environs de St. Etienne etc* [*Essay on the Lithology of the St Etienne area*]. This work was originally only intended to accompany a crate of the mineralogical products of that district that I was sending to Romé de l'Isle and to act as an explanation about them for him and it was only at his request, as that of some other mineralogists, that I decided to have it printed. If I were beginning this little work today, there are many things that would remain in it, my opinion about them having in no way changed since, but there are also many that I no longer see in the same way and which I would rectify. Eh! What would indeed be the use of studying and growing old, if the knowledge one acquires in this way did not perfect the judgment and did not put it in the position to rectify mistakes committed previously by it. I am, therefore, not afraid to admit that I have very frequently varied in regard to the opinions that the study of mineralogy had caused me to embrace at different times and I would not answer to being free of new variations, if ever the Creator of Nature allowed me a certain number of years to observe it. This essay, in which I have inserted some observations on the formation of coal mines in this part of Forez, also contains some details on the action of fire on schists, as on the sandstone which accompanies these mines, in one of them known as Ricamari, which was burning then and of which the conflagration went back to a very early period of which tradition itself gave no trace: these details, that I think I am the first to have given, appear to me still to deserve some interest today.

Several years had already passed since the study of mineralogy occupied extremely pleasantly all my leisure moments. I had already made a very large number of observations and collected a very sizeable mineralogical collection. I then formed the plan of occupying myself with the writing of my observations and decided to settle in a very pleasant estate that I had near Metz (*note not translated).* The proximity of a new chain of low primitive mountains, the Vosges, like the neighbouring mountains of the Electorate of Treves and of Coblenz, of Mayence etc. and those of the mercury mines of the Duchy of Deux Ponts, promised me new crops, new observations and consequently new enjoyment.

It is at this period, when I believed I had fixed firmly around me the most assured means of a lasting and complete felicity, that I saw this edifice of happiness, raised so voluptuously, disappear in an instant, as the happiest dream fades when one awakes. I had surrounded myself with all that could satisfy my desires and my tastes and prolong my enjoyment even into old age. To the agreeable occupations that I had had up till then, I had added agricultural work for which I had always had a strong inclination. Without ambition, asking nothing, wanting only peace with my neighbours and the friendship of some of them, having sometimes the happiness of being useful to men of the working class by whom I was surrounded, who would not have thought that I had indeed found that philosopher's stone for mankind – constant happiness? The revolution that covered France with blood, ashes and tears and has since brought the same evils to the whole continent, began its ravages. I had to abandon country, fortune and friends, become a mere inhabitant of the world, of total indifference to the new beings that I was to meet there and having for them, as my only recommendation, the so discredited one of misfortune. After wandering for a long time in Germany with my family, the precious possession that, at least, I have been able to preserve, and with so many other illustrious companions in the same misfortune, I finally happily landed on the shore of true liberty: it had become in this moment, also that of the refuge and protection granted to misfortune and persecution.

I was then granted new obligations to mineralogy. Up till then, it had charmed my leisure hours; becoming from this moment my foster mother, it was preparing for me enjoyment of which I was still not aware, that of being able to be, by myself and directly through the product of my work, the support of all my family. This enjoyment, unknown to fate, and which rarely it can appreciate, is a powerful compensation for its hardship, so true is it that sources of happiness always remain for man when he knows how to draw on them.

Some time after my arrival in London, the care, the arrangement as well as the increase in the three largest cabinets of that city, Mr Greville's, Sir John St. Aubyn's and Sir Abraham Hume's were placed in my hands at the same time. One of them, Mr Greville's, which was already very sizeable, can today be regarded as one of the most complete in Europe, and even, in a number of respects, such as in the gem section, which at that time were very neglected in London and were quite generally poor throughout, it is quite clearly the leading one. A great number of the other collections have successively been placed in my care for classifying and arrangement; and that at a time when the different objects connected with mineralogy, for which several hawkers had grown up in Europe, could find somewhere to be placed advantageously only in England where they arrived from all directions. This position, as well as the constant work which it demanded on my part, has made me feel how precious, in a study which in particular necessitates the repeated use of sight, is the multiplicity of objects which are presented to it, with a sufficient time lapse for each of them to pass in turn and take the true place that belongs to it. I had seen much mineralogy since, in addition to my own collection of which the number of pieces rose to nearly 14 thousand, I had examined with a great deal of care a large number of the French cabinets, several of the German ones and all those in Paris worthy of being cited. However, this huge new mass of samples of the various mineral substances and their varieties, which passed through my hands, as well as the possibility of comparing them with each other, contributed greatly in increasing the knowledge that I could have acquired up till then, with regard to those same substances. The writing that I was contemplating of the former observations that the study of mineralogy had caused me to make and which had been so cruelly interrupted by the disasters in France, precursors of those in Europe had, therefore, acquired new material. The knowledge that I had of minerals had greatly increased; several objects which were merely doubtful for me had been clarified; many others had been rectified. I finally felt that, in reality, I had the possibility of being useful to the science to which I had owed, up till then, pleasure and help. For that, the cessation of the occupations which were so necessary to me, cessation of which from day to day I glimpsed the approach more distinctly, must allow it. That time has arrived: through it, I have been given back to myself, and to that sweet solitude which was so necessary to me, that I have always loved and which had acquired new charms for me. So, I have been able to give myself up entirely to the long-planned assembling of the notes and observations that I might have made, as well as research and particular work which it was necessary for me to do to make them as useful as possible; the principal goal towards which I aspired.

That work completed, I have for a long time been uncertain about the method that I would adopt to present it to the public. First I wanted to make a simple resumé of the observations which were completely mine, but I was not long in feeling that this resumé would present for each subject only isolated facts and which, deprived in this way of any connection at all with any of the other already known facts in the same substances, could offer only extremely dry details almost totally deprived of interest. Moreover, each of these facts, isolated in this way, would thereby have lost all the merit which it could have and perhaps even a great part of its usefulness. Then I planned to give only the simple crystallographical facts with regard to which my observations are immense; but this work would have had the same dryness, principally for those mineralogists who have not given themselves to the study of crystallography and it would have been impossible for me to lead gradually to some confidence in this science: these same facts would have been, at the least, equally deprived of interest. Moreover, this method would have made me leave aside all those observations which would not directly concern crystallography, to which I could not persuade myself. However, I was feeling a strong disinclination to make another treatise of mineralogy: so many have been appearing for some years! From my point of view, I also felt that, with the method that had to be followed in a treatise of this sort, this work would then become large and would demand, consequently, quite a long time before it was finished and my present position, which leaves me no resources except work, could not allow me to envisage calmly the interval which must of necessity elapse between the beginning of this work and its end.

It is, however, this last that I decided upon as the most appropriate to reach the goal I was contemplating. I, therefore, determined to make in effect a complete treatise of mineralogy, but at the same time to give it in separate sections. By this method it will be possible, I hope, for me to be able to continue it to its conclusion and that in such a way that each part will appear, although incomplete in relation to the general treatise, is, however, perfectly complete with regard to the objects which will be dealt with by adopting the order of classification that I think the most appropriate to be followed in this science.

[Omitted: from start of third paragraph on page ix of the original to end of last two lines at top of page xi.]

In most of the mineralogical works that appear nowadays, their authors consider that they should restrict themselves almost completely to the descriptive part of this science, its *oryctognosy* [i.e. a study based on specimens], and not concern themselves at all with that which considers the situation of each of these mineral substances in relation to the different [rock] masses of which the globe is composed, the role that they play there, the relations either of origin or of formation that they have with one another and thus with the great masses that contain them, and the effects that time and various other specific causes can have had on them: [all this being] the part [of mineralogy] known under the name of *geology*. . . .

Thus we have a great number of materials relevant to geology, and every day adds to their number; but we still do not have enough of them, or at least our observations on them are not at all adequately controlled, and are neither sufficiently complete or circumscribed to enable us to establish its theory. Only the collaboration of mineralogists who devote themselves to its study, their agreement on their observations, on the way in which they should be made, and on how they are written up, can accelerate greatly the moment when we shall be in a position to be able to establish this theory. With what satisfaction, therefore, should we view the establishment of the Geological Society that has just been formed in London; a society for which these considerations are the foundation, and of which the position is extremely favourable and the means for the future are incalculable.*

Our efforts, with regard to geology, ought therefore to be directed at this time towards the collection of the greatest possible number of facts. In this case, is not the mineralogist who, in speaking of each substance, neglects to mention the principal geological facts – the existence of which his observations, or those made reliably by others, have taught him – failing to cooperate in bringing forward the time when the geological study of the globe, brought to perfection, will make treatises on mineralogy complete? They will then be divided naturally into two parts, the one dealing simply with the oryctognostic study of minerals [i.e. in the laboratory], the other with their geological study [i.e. in the field].

The present epoch is well suited for devoting zealous activity to this subject [of geology], and for being committed to uniting its means with those – so powerful and satisfactory – that geology is receiving daily from the famous savants who now direct towards this subject the improvement of the knowledge that is given to us by physics, conchology, comparative anatomy and even botany.

[Omitted: from beginning of last sentence on page xii of original to end of first paragraph on page xxxiv.]

These two volumes being completed, I still had to find the means to have them printed, and in a way that would allow me to be compensated for the time that their writing had taken me, and thereby, at the same time, to gain the possibility of being able to continue my work. Moreover, my position and that of my family, to the support of which my work has been devoted, was making this an obligation and indeed a duty. This was not the easiest matter; and it must be admitted that the situation – as unhappy as it is astonishing, inconceivable and above all humiliating for mankind – in which the whole of Europe has been plunged, a situation so far from favourable for cultivating the sciences, was not such as to give me hope of success.

After several unfruitful attempts I was on the point of giving up and turning my attention towards quite another kind of occupation, when, finding myself one day with Dr Babington, one of the most renowned doctors in London, and Mr Allen, professor of physics at the Royal Institution, and talking with them about having found it impossible to have my work printed in any profitable way, and thus about the necessity with which I would probably have to give it up, these two friends, joined at the same instant by Mr W. and Mr. R Phillips, offered, in a manner that was as noble as it was generous, to put themselves at the head of a subscription list, which they proposed to limit to sixteen persons, the purpose of which would be to underwrite the expenses of my work, while assigning me the whole of the profit. To this generous proposal they added some extremely flattering remarks about their certainty that they would see this subscription list completed very swiftly by the mineralogists in London, and that I should consider myself completely detached from the matter, since they alone would take care of it.

*This society has scarcely been established (its first session dates at most from six months ago), yet already a very considerable number of savants, in all parts of the sciences, count themselves among its members. A zeal driven both by a love of the sciences and by a deep patriotic feeling – which so eminently distinguishes the English character – has brought them together. The same zeal is the guarantee of their work: what should not be expected, then, of the science that is its object?

I accepted with immediate gratitude this flattering evidence of their esteem and their friendship; and it is infinitely valuable for me to be able to record my gratitude here. Their project has been completely successful: the subscription list, at the head of which they have placed themselves, has been completed with little delay. The initial obligation that I had towards them does not, however, leave me any less aware of the extent of what I owe to those who have been willing to respond to their invitation and to join them: I ask them to accept here the expression of my gratitude in this respect, and to allow me to adorn my work with the names of the scientists to whom today I am able publicly to pay homage.

William Allen, Esq. F.R.S.	Chas. Hatchett, Esq. F.R.S.
Sir J. St. Aubyn, Bart. M.P. F.R.S.	Luke Howard, Esq.
Wm. Babington, M.D. F.R.S.	Richard Knight, Esq.
Robert Ferguson, Esq.	Richard [James] Laird, M.D.
Rt. Hon. Charles Greville, F.R.S.*	William Phillips, Esq.
Geo. B. Greenough, M.P. F.R.S.	Richard Phillips, Esq.
Sir Abm. Hume, Bart. M.P. F.R.S.**	John Williams, Junr, Esq.

To these names I should add that of my worthy and respected friend Dr. Crichton, physician to H. M. the Emperor of Russia; also those of Mr. Novossiltzoff, Prince Czartorinsky and Count Strogonoff, who, having learnt from him that in London there was a list of subscriptions for the printing of my work, have had the great honesty to send me theirs, adding that if this was not sufficient to complete what had been opened, they would transmit a second to me. May this work soon be able to reach them in Russia, to express to them how much I have been flattered and honoured by this procedure on their part, as noble as it was tactful.

May I indeed have achieved this goal *(the goal is to contribute to the study of mineralogy and produce a useful aid to this end)* and after losing everything, completely isolated, without country, without fortune, without status, have been able to acquire some right to the esteem and goodwill of the generous, hospitable people to whom I owe the possibility of continuing the study of the science which is the joy of my life and who, by the wisdom of their laws, the benevolence and steadfastness of their government, surround my life with the tranquillity and security so essential to happiness.

---

*I must testify here to Mr Greville the memory that I have retained of the debt I owed him on my arrival in London as on several other occasions; it made a profound impression on me then. Since, I have taken all possible care to show him as much as I could, that I retained a gratitude which I am pleased here to reiterate. Moreover, I owe several new facts, as well as particular observations which are scattered through this work, to his superb mineralogical collection that I have already said is one of the largest in Europe and, in some respects the most important, of which the arrangement and augmentation were entrusted to me for a very long time.

**That is not the only debt that I have owed and still owe daily to Sir Abraham Hume, to whom I am doubly attached by the bonds of gratitude and those of friendship: it is too satisfying for me to express publicly to him these two sentiments not to profit with real joy from the opportunity which is offered to me.

Index

Page numbers in *italic* denote figures. Page numbers in **bold** denote tables.